AF251426

Neuronal Serotonin

Neuronal Serotonin

Edited by

N. N. Osborne
Nuffield Laboratory of Ophthalmology,
University of Oxford

and

M. Hamon
Neurobiologie Cellulaire et Fonctionelle,
INSERM U 288, Paris

A Wiley–Interscience Publication

JOHN WILEY & SONS
Chichester · New York · Brisbane · Toronto · Singapore

Library of Congress Cataloging-in-Publication Data:

Neuronal serotonin.

  'A Wiley–Interscience publication.'
  Includes index.
  1. Serotonin—Physiological effect. 2. Serotonin—
Receptors. I. Osborne, Neville N. II. Hamon, Michel.
QP801.S4N47   1988   599′.0188   87–27933
ISBN 0 471 91154 2

British Library Cataloguing in Publication Data:

Neuronal serotonin.
  1. Neurochemistry 2. Serotonin—
Physiological effect 3. Central nervous
system
  I. Osborne, Neville N. II. Hamon, M.
  612′.822   QP356.3

  ISBN 0 471 91154 2

Phototypeset by Input Typesetting Ltd, London SW19 8DR
Printed and bound in Great Britain by the Bath Press, Avon

Contents

List of Contributors

A. Beaudet *Departments of Neurology & Neurosurgery and Anatomy, Montreal Neurological Institute, McGill University, Montréal (Québec), Canada.*

K. H. Buchheit *Preclinical Research Department, Sandoz Ltd, CH-4002 Basle, Switzerland.*

A. Calas *Laboratoire de Physiologie des Interactions Cellulaires, Université de Bordeaux I, UA CNRS n 339, Avenue des Facultés, 33405 Talence Cédex, France.*

J. M. Cossery *INSERM U 288, Neurobiologie Cellulaire et Fonctionnelle, Faculté de Médecine Pitié-Salpêtrière, 91 Boulevard de l'Hôpital, 75634 Paris Cédex 13, France.*

A. J. Cross *Astra Neuroscience Research Unit, 1 Wakefield Street, London WC1N 1PL, England*

L. Descarries *Université de Montreal, Centre de Recherche en Sciences Neurologiques, CP 6128 Succ A, Montréal (Québec), Canada H3C 3J7.*

M. M. Dietl *Preclinical Research Department, Sandoz Ltd, CH-4002 Basle, Switzerland.*

S. El Mestikawy *INSERM U 288, Neurobiologie Cellulaire et Fonctionnelle, Faculté de Médecine Pitié-Salpêtrière, 91 Boulevard de l'Hôpital, 75634 Paris Cédex 13, France.*

M. B. Emerit *INSERM U 288, Neurobiologie Cellulaire et Fonctionnelle, Faculté de Médecine Pitié-Salpêtrière, 91 Boulevard de l'Hôpital, 75634 Paris Cédex 13, France.*

C. A. Fornal *Program in Neuroscience, Department of Psychology, Princeton University, Princeton, NJ 08544, USA.*

M. Geffard *Laboratoire de Neuroimmunologie, Institut de Biochimie Cellulaire et de Neurochimie IBCN-CNRS, 1 Rue Camille Saint-Saëns, 33077 Bordeaux Cédex, France.*

H. Gozlan *INSERM U 288, Neurobiologie Cellulaire et Fonctionnelle, Faculté de Médecine Pitié-Salpêtrière, 91 Boulevard de l'Hôpital, 75634 Paris Cédex 13, France.*

D. Graham *Department of Biology, Laboratories d'Études et de Recherches Synthélabo (LERS), 58 Rue de la Glacière, 75013 Paris, France.*

M. Hamon *INSERM U 288, Neurobiologie Cellulaire et Fonctionnelle, Faculté de Médecine Pitié-Salpêtrière, 91 Boulevard de l'Hôpital, 75634 Paris Cédex 13, France.*

D. Hoyer *Preclinical Research Department, Sandoz Ltd, CH-4002 Basle, Switzerland.*

B. L. Jacobs *Program in Neuroscience, Dept of Psychology, Princeton University, Princeton, NJ 08544, USA.*

H. K. Kimelberg *Division of Neurosurgery and Departments of Biochemistry and Pharmacology/Toxicology, Albany Medical College, Albany, NY 12208, USA.*

P. J. Knott *Departments of Psychiatry and Pharmacology, The Mount Sinai School of Medicine, One Gustave L Levy Place, New York, NY 10029, USA.*

W. P. Koella *The Friedrich Miescher Institute, Basle, Switzerland.*

S. Z. Langer *Department of Biology, Laboratoires d'Études et de Recherches Synthélabo (LERS), 58 Rue de la Glacière, 75013 Paris, France.*

D. Le Bars *Unité de Recherches de Neurophysiologie Pharmacologique, de l'INSERM U161, 2 Rue d'Alésia, 75014 Paris, France.*

O. Lutz *INSERM U 288, Neurobiologie Cellulaire et Fonctionnelle, Faculté de Médecine Pitié-Salpêtrière, 91 Boulevard de l'Hôpital, 75634 Paris Cédex 13, France.*

M. Montange *Laboratoire de Physiologie des Interactions Cellulaires, Université de Bordeaux I, UA CNRS n 339, Avenue des Facultés, 33405 Talence Cédex, France.*

N. N. **Osborne** *Nuffield Laboratory of Ophthalmology, University of Oxford, Walton St, Oxford OX2 6AW, England.*

J. M. **Palacios** *Preclinical Research Department, Sandoz Ltd, CH-4002 Basle, Switzerland.*

S. **Patel** *Laboratoire de Neuroimmunologie, Institut de Biochimie Cellulaire et de Neurochimie IBCN-CNRS, 1 Rue Camille Saint-Saëns, 33077 Bordeaux Cédex, France.*

A. **Pazos** *Department of Pharmacology and Therapeutics, School of Medicine, University of Cantabria, 39071 Santander, Spain.*

L. **Peuble** *Laboratoire de Neuroimmunologie, Institut de Biochimie Cellulaire et de Neurochimie IBCN-CNRS, 1 Rue Camille Saint-Saëns, 33077 Bordeaux Cédex, France.*

S. J. **Peroutka** *Departments of Neurology and Pharmacology, Stanford University Medical Center, Stanford, California 94305, USA.*

B. P. **Richardson** *Preclinical Research Department, Sandoz Ltd, CH-4002 Basle, Switzerland.*

E. **Sanders-Bush** *Department of Pharmacology and Psychiatry, School of Medicine, Vanderbilt University, Nashville, TN 37232, USA.*

J. J. **Soghomonian** *Université de Montréal, Centre de Recherche en Sciences Neurologiques, CP 6128 Succ A, Montréal, Québec, Canada H3C 3J7.*

P. **Soubrié** *Département de Pharmacologie, Pitié-Salpêtrière, 91 Boulevard de l'Hôpital, 75013 Paris, France.*

Y. **Takeuchi** *Departments of Pediatrics, Kyoto Prefectural University of Medicine, Kawaramachi, Hirokoji, Kamikyo-ku, Kyoto 602, Japan.*

S. **Tuffet** *Laboratoire de Neuroimmunologie, Institut de Biochimie Cellulaire et de Neurochimie IBCN-CNRS, 1 Rue Camille Saint-Saëns, 33077 Bordeaux Cédex, France.*

Preface

About a year and a half ago we approached the publishers, John Wiley, with the suggestion that we should organize and edit a monograph on serotonin to complement the previous book produced in 1982 and entitled *Biology of Serotonergic Transmission*. The idea was to provide a work which would include the new developments in serotonin research as well as the more established aspects of serotonin function, viz. role in sleep, pain and behaviour, all of which are areas not covered by the first book. As a consequence, the range of topics has been intentionally restricted to certain aspects. We are aware that our choice of subjects is biased and, with hindsight, can acknowledge that there are other areas which might have been included.

A great deal of emphasis is placed in the present monograph on serotonin receptors. Since the beginning of the 1980s much progress has been made in this area, with new information being gathered all the time. Specific serotonin receptors linked to the utilization of inositol phosphate(s) as a second messenger have recently been characterized. Important advances have been made in the field of peripheral serotonin receptors and the production of specific antagonists to these receptors promises to be of clinical relevance for the future. There is now evidence that no fewer than five types of serotonin receptor exist in the mammalian nervous system, although there is disagreement about the nature of the precise 'differences' between the various receptors, their localization and the functional significance of the receptor types. This division of opinion is caused partly by the multitude of experimental strategies employed by different authors, and is illustrated in this book.

There is little doubt that the mode of action of neurones which contain serotonin exhibits characteristics highly specific to this particular class of neurones in mammals. For example, there seems to be a striking difference between neurones containing serotonin and those releasing other transmitters, in that serotonergic neurones can apparently impinge on other neurones through non-synaptic contacts, whereas neurones impinging on serotonin fibres form true synapses at their point of contact. Furthermore, doubt has been expressed as to whether serotonin, like other neurotransmitters, is released exclusively from the nerve terminals. Thus evidence is emerging of serotonin release from cell bodies and dendrites, with important functional significance. To derive most benefit from the many contributors to this book, the editors asked each author to present a specific view of the subject rather than giving a comprehensive account. Future trends in research in each area have been identified, showing that the next decade of serotonin research will be as exciting as the last. *In situ* hybridization studies on serotonin neurones will perhaps explain why some of these cells can be viewed immunohisto-chemically without pharmacological treatment, while others cannot. Specific antibodies will not only identify precisely the localization of different receptor types but also isolate and characterize the receptors in molecular terms. We also need to explain why serotonin is taken up by certain non-neuronal cells in a seemingly specific manner and why some receptors appear to exist on some of these glial elements. The co-localization of serotonin with other putative neurotransmitters including neuropeptides and γ-aminobutyric acid is also an intriguing (but not unique) phenomenon which deserves thorough biochemical and functional investigations. These are just a few of the subjects which might provide material for a future book on serotonin research.

We should like to conclude by thanking most sincerely all the authors of the present book for participating in this venture, and by expressing our gratitude to Dr Michael Dixon of the publishers for his support in its preparation.

NEVILLE OSBORNE and MICHEL HAMON

CHAPTER 1

Production of Antisera to Serotonin and their Metabolites and their Use in Immunocytochemistry

MICHEL GEFFARD, SOPHIE TUFFET, LAURENCE PEUBLE and SARAWASTI PATEL
Laboratoire de Neuroimmunologie
Institut de Biochimie Cellulaire et de Neurochimie IBCN-CNRS
1, Rue Camille Saint-Saëns
33077 Bordeaux Cédex
France

INTRODUCTION

Until 15 years ago, serotonin (HT) was thought to be the sole intrinsic neuroactive indoleamine of the mammalian extrapineal central nervous system (CNS). Since then, the application of various techniques has shown the presence of 5-methoxytryptamine (Björklund *et al.*, 1971; Prozialeck *et al.*, 1978; Boisin *et al.*, 1979), tryptamine (Marsden and Curzon, 1974; Juorio and Durden, 1984; Juorio and Greenshaw, 1985) and *N*-acetyl serotonin (Pulido *et al.*, 1983).

A number of methods have been developed to both visualize and localize amines that have been identified as being associated with neurotransmission and/or neuromodulation. Among these techniques, the Falck–Hillarp and glyoxylic acid techniques have been widely employed for the localization of monoamines (Falck *et al.*, 1962; Dahlström and Fuxe, 1964, 1965; Fuxe, 1965; Fuxe and Jonsson, 1967; Ungerstedt, 1971; Bjöklund *et al.*, 1972; Lorèn *et al.*, 1976; Furness *et al.*, 1977). However, the lack of a selectivity in visualizing the reaction products, a difficulty in tracing serotonergic pathways in spite of pharmacological enhancement (Kuhar *et al.*, 1972) and the lack of precision in the localization of axon terminals represented serious limitations of these techniques. Although the relevant enzymes of neurotransmitter metabolism have been detected by immunocytochemistry (Hökfelt *et al.*, 1973, 1975; Joh *et al.*, 1975; Pickel *et al.*, 1976, 1977), chemical neuronal pathways remained poorly defined. In this respect, the specific localization of neurotransmitters or neuromodulators themselves still needs to be achieved. The direct approach to the observation of neurotransmitter-related molecules has been employed in the case of certain neuropeptides (Hökfelt *et al.*, 1980) and serotonin by using antibodies directed specifically against these molecules (Steinbusch *et al.*, 1978; Lidov *et al.*, 1980; Buffa *et al.*, 1980; Consolazione *et al.*, 1981; Steinbusch, 1981; Takeuchi *et al.*, 1982). Serotonin and small neuropeptides, however, must be made immunogenic before the possibility exists to make antibodies to such chemicals. Thus, chemical manipulation of neurotransmitter-like molecules (Störm-Mathisen *et al.*, 1983; Geffard *et al.*, 1985b) including the indoleamines enabled us to produce specific antisera for immunocytochemical studies (Patel *et al.*, 1986). The following principles were found to be necessary:

1. During immunogen synthesis, the original part of the neurotransmitter molecule had to be preserved. As previously described (Geffard *et al.*, 1985b), glutaraldehyde was preferred as a coupling agent, as it is essentially non-polymerized, reacts quickly, preserves the morphological structure of biological material and, in principle, produces very specific antibodies to chemically related molecules. The coupling of hapten to a protein carrier via a double condensation reaction involves the hapten amino-group and the ε-amino group of lysine residues in the proteins (Avrameas *et al.*, 1978).

2. The immune response was directed towards the original part of each conjugated indoleamine by reducing the double bonds resulting from the condensation reaction and by alternating immunization with different protein carrier conjugates having the same hapten. The double bonds

formed in the condensation reaction need to be reduced to produce an aliphatic chain which is less antigenic than the hapten.

3. The specificity of each antiserum was evaluated using enzyme-linked immunosorbent assay (ELISA) with conjugates which mimic the hapten structure of both the immunogens and tissues. This point is particularly relevant when considering the results reported by Milstein *et al.* (1983) that the rat monoclonal antibody against paraformaldehyde-coupled serotonin recognizes free dopamine, but that this is not the case if the dopamine is first treated with paraformaldehyde. This emphasizes the need for establishing the optimal fixation conditions for a tissue sample before the antibody can be used. During the course of the immunization, antibody titres are measured by evaluating antibody binding with coated conjugates. Cross-reactivity ratios are calculated from displacement curves obtained by competition between the hapten concerned and a set of analogues belonging to the same family.

4. The coupling agent used for immunogen synthesis should also be used in the immunological detection of neurotransmitters in tissues.

In the investigation described in this chapter, the specificity of HT antisera raised either via formaldehyde or glutaraldehyde was compared. Then, taking this into account, antisera against each indoleamine having a free amino group were developed after glutaraldehyde coupling (Fig. 1). The immunocytochemical applications reported here demonstrate that HT is not the sole intrinsic neuroactive indoleamine in the rat brain.

METHODS

Synthesis of immunogenic conjugates

In order to become antigenic, each molecule belonging to the indoleamine family was linked to a protein carrier such as bovine serum albumin (BSA) or human serum albumin (HSA) via either formaldehyde (F) or glutaraldehyde (G).

Formaldehyde conjugates

Tryptophan (W), 5-hydroxytryptophan (HW), tryptamine (T), serotonin (HT), 5-methoxytryptamine (MT), 5-methoxytryptophan (MW), *N*-acetyl serotonin (aHT) and melatonin (aMT) were coupled to BSA using an F condensation reaction, following the method of Ranadive and Sehon (1967) as modified by Grota and Brown (1974). Five milligrams of hapten were dissolved in 1 ml of a mixture containing 3 M sodium acetate solution pH 8/ ethanol (vol/vol). They were then mixed with 10 mg of BSA or HSA in

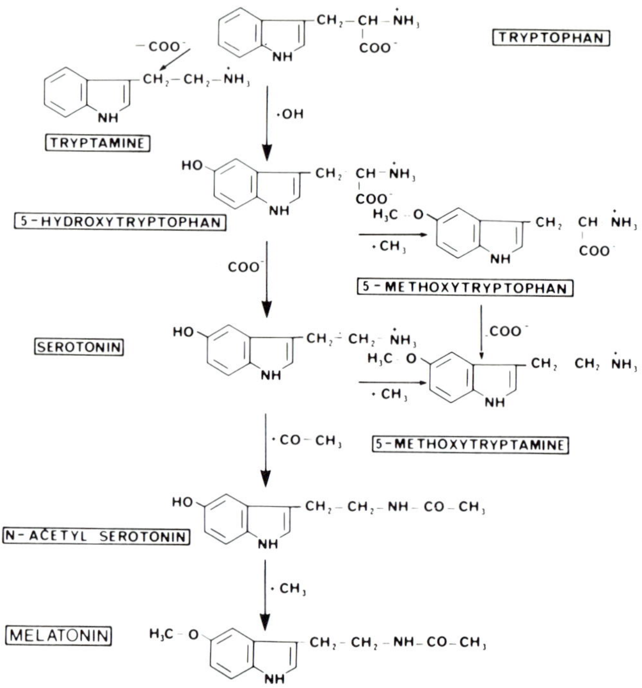

Fig. 1 Structural relationships between indoleaminergic compounds. Each compound differs from the others in one or two chemical groups either in the 5-indole nucleus position or on the side chain. Each molecule, having a free amino group, can be conjugated with formaldehyde or glutaraldehyde to the protein carrier

1 ml of 1.5 M sodium acetate pH 8. Then, 1 ml of 7.5% F was added to the above mixture and stirred for 5 min at room temperature. The yellow solution obtained was dialysed against running tap water for two days at 4°C and clarified by centrifugation.

Glutaraldehyde conjugates

W, HW, MW, HT, MT, and T were also conjugated to BSA or HSA via G under the following experimental conditions. Five milligrams of hapten were dissolved in 1 ml of a mixture containing 3 M sodium acetate solution pH 8/ ethanol (vol/vol) and were mixed with 10 mg of BSA or HSA in 1 ml of 1.5 M sodium acetate solution pH 8. Then, 250 µl of 1% G solution was added to the above mixture. A stable yellow colour indicated the end of coupling reaction. Then, 250 µl of sodium borohydride solution (0.1 M) was added to saturate the double bonds. The solution became translucent and then was dialysed against running acetate buffer (0.1 M) pH 6.2 for 2 days at 4°C. Insoluble material was then removed by centrifugation at 10,000 $\times$ g for 15 min.

Spectral analysis of each conjugate coupled either with F or G was performed to determine the molar coupling ratio by calculating the concentrations of conjugated indoleamine and BSA or HSA at 300 nm and 280 nm respectively and by taking into account the molar extinction coefficients after coupling. The coupling ratios were calculated as conjugated hapten concentration to protein carrier concentration.

Production of antisera against indoleamines

Two rabbits were immunized with each conjugate. They were injected alternatively with 500 µg of immunogen having BSA or HSA as protein carrier, emulsified in the same volume of Freund's complete adjuvant. Every two weeks the rabbits received either an intramuscular or subcutaneous injection of the same antigen. Immunization was continued until a high titre was reached. After the third injection, the rabbits were bled once a week.

Purification of antisera

Two procedures were used. Whatever the method employed, the same results were obtained concerning the elimination of anti-protein carrier treated by G antibodies.

Affinity chromatography method

This method has been used only for G antisera: 4 ml of each crude antiserum was dialysed for 4 days at 4°C against phosphate buffer 10^{-2} M containing NaCl 5×10^{-2} M plus sodium azide 10^{-2} M at pH 7.4. Each dialysed antiserum was then eluted through a DE 52 (Whatman) column (1.5×48 cm). Immunoglobulins G (IgG) were the first protein fraction obtained, well separated from other immunoglobulins and albumin. IgG was concentrated 5 times in an Amicon apparatus before being eluted by affinity chromatography. In a second step, 60 mg of the immunogenic protein carriers (BSA and HSA) used for immunization were coupled to 6 mg of G-actived ultrogel (Act-ultrogel ACA-22, IBF) for one night at 4°C. After coupling, only 3 mg of protein/ml of ultrogel was retained. Sodium borohydride was added to saturate the double bonds. The IgG previously purified on a DE 52 column was then eluted for 4 h through a column (1 cm $\times$ 6 cm, IBF) containing proteins coupled to Act-ultrogel with the previous buffer. The first elution peak corresponded to non-specific IgG and to the ones recognizing the conjugated hapten, such as HT-G. This fraction was then used for the specificity studies tested using the ELISA method described below. Antibodies to the protein carrier were eluted through a glycine buffer 0.2 M, pH 3. They usually represented 8.5% of the total IgG population.

Adsorption method

Antisera were purified by adsorption with the protein carriers previously treated with F or G. A solution containing 5 mg of lyophilized F- or G-treated BSA and HSA was mixed with 1 ml of crude antiserum for 18 h at 4°C. The serum was then centrifuged at 10,000 × *g* for 10 min to discard the pellet containing the immune complexes. The purified supernatant was used to study the antibody specificity and affinity by ELISA tests.

ELISA method

Affinity and specificity tests were carried out by competition experiments using the ELISA method (Engvall and Perlmann, 1972; Butler *et al.*, 1978; Geffard *et al.*, 1985a). Polystyrene well-plates (Nunc) were coated with a solution containing either an indoleamine conjugate or a conjugated protein treated by either F or G. After using several proteins, such as HSA, BSA and haemoglobin, the best coating was obtained with BSA conjugates.

The well-plates were then coated with 200 µl of an indoleamine–G–BSA or G–BSA solution at 1 µg/ml in 0.05 M carbonate buffer, pH 9.6, for 16 hours at 4°C. For the F-conjugates, the well-plates were coated with a solution at 0.1 µg/ml. These concentrations of conjugates adsorbed on the well-plates corresponded to those which gave the optimal absorbance. After this period, in order to saturate the well-plates perfectly and to prevent the non-specific binding of antisera, they were filled with 200 µl of phosphate-buffered saline (PBS) containing 0.05% Tween 20 plus 1% pig serum (PBS-Tween-Ps) and left for 30 min at 37°C.

The well-plates were then rinsed three times with PBS-Tween. Purified antisera, preincubated for 16 h at 4°C with or without indoleamine compounds, diluted in PBS-Tween-PS, were applied to the well-plates and incubated at 37°C for 2 h. The final dilutions of the conjugates varied between 10^{-5} M and 10^{-10} M. Again the well-plates were washed three times with PBS-Tween and filled up with 200 µl of 1:5,000 or 1:10,000 goat anti-rabbit labelled with horseradish peroxidase (HRP) diluted in PBS-Tween-PS. After 1 h incubation at 37°C, the well-plates were rewashed three times with PBS-Tween. Peroxidase was assayed by incubating with a 0.4 mg/ml *o*-phenylenediamine solution in 0.1 M acetate 0.2 M phosphate buffer pH 5, containing 1% hydrogen peroxide (Merck, 30 volumes) for 10 min in darkness. The reaction was stopped by adding 50 µl of a 4 N H_2SO_4 solution per well. The optical density was read at 492 nm with a Multiskan Titertek apparatus. Experimental values were corrected by deducting blank values from well-plates coated with F-BSA or G-BSA.

Immunocytochemical method

Three-month-old male Wistar rats of 200 g body weight were kept under a 12 h darkness/12 h light cycle (the phase commencing at 7.00 a.m.). The rats were perfused (between 9.00 and 10.00 a.m.) through the aorta with 50 ml of a 1:1 mixture of 5% G and 0.5 M allyl alcohol in 0.1 M cacodylate buffer, pH 11 (McRae-Degueurce and Geffard, 1986). The allyl alcohol ensures better penetration of the G into rat brain. The rats were then perfused with 500 ml of 0.5 M G solution to 0.1 M cacodylate buffer (pH 7.5) containing 0.9 % sodium metabisulphite. The brains were then removed and post-fixed for 45 min in 0.5 M G solution containing 0.9% sodium metabisulphite, and then washed thoroughly with 0.05 M Tris buffer (pH 7.5). Vibratome sections of 50 μm thickness were cut through the region of the raphe nuclei and then placed in 0.05 M Tris buffer (pH 7.5) containing 0.9 % sodium metabisulphite. The sections were reduced using 0.1 M sodium borohydride (in the same buffer), washed thoroughly and processed for immunocytochemistry using the peroxidase/anti-peroxidase (PAP) method (Sternberger *et al.*, 1970). The sections were incubated overnight at 4°C in a 1:10,000 or 1:2,500 dilution of either antibody to HT, antibody to MT or antibody to T. The antisera had previously been purified on the protein carriers used during immunization as described above. Following the primary incubation, the sections were washed and incubated for 45 min at 37°C in 1:100 dilution of goat anti-rabbit IgG (Miles-Yeda), washed again, incubated for 45 min at 37°C with 1:1,000 dilution of rabbit PAP (Miles-Yeda), washed and developed in a solution of 0.05% 3-3'-diaminobenzidine tetrahydro-chloride (DAB) containing 0.01% H_2O_2. In order to determine the specificity of staining, HT, T and MT antisera were incubated overnight at 4°C with each of the following conjugates: HT–G–BSA, MT–G–BSA and T–G–BSA. Each mixture was then centrifuged at 10,000 × g for 15 min, and the super-natants were used for the primary incubations of the immunocytochemical procedure.

RESULTS

Spectral analysis of immunogenic conjugates

The original structure of the hapten was preserved during the immunogen synthesis. After purification the molar extinction coefficients of each conjugate were calculated. Identical adsorbances at 280 nm and 300 nm were recorded for either F or G protein plus non-conjugated indoleamine and for the F or G conjugate at the same concentrations. At these wavelengths, the optical density of the conjugated indoleamine was superior to that given by BSA. The ultraviolet spectrum of indoleamine remained unchanged in the conjugates, indicating that the original part of each molecule was preserved.

Results of the ultraviolet spectrophotometric studies of the immunogens HT–F–BSA and HT–G–BSA are shown in Figs 2 and 3. The molar coupling ratios of these conjugates were calculated as described in the previous section. They were 45 and 23, respectively.

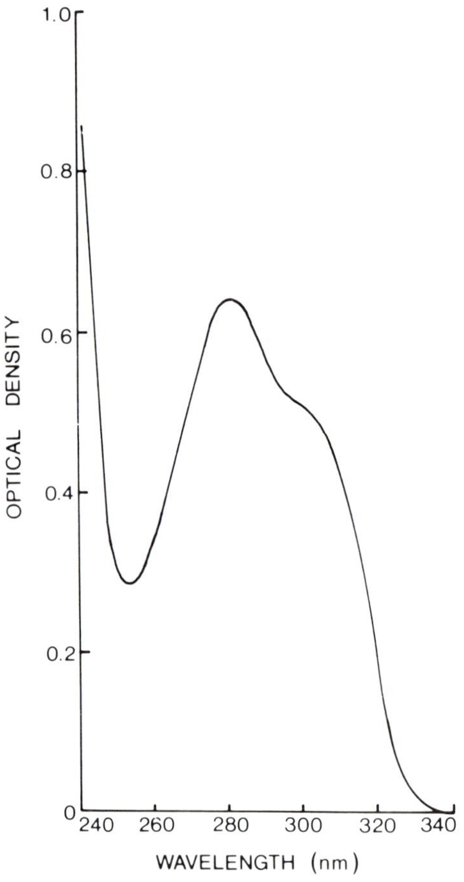

Fig. 2 Ultraviolet spectrum of HT–F–BSA. The coupling molar ratio was determined by calculating the concentration of HT–F and BSA at 300 nm and 280 nm, respectively: HT = 8.6×10^{-5} M; BSA = 1.9×10^{-6} M. For this conjugate, the ratio is 45

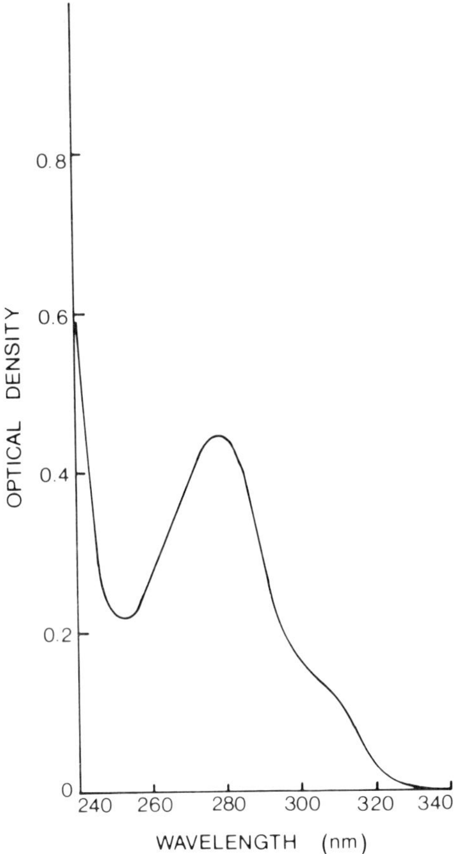

Fig. 3 Ultraviolet spectrum of HT–G–BSA. The coupling molar ratio was determined by calculating the concentration of HT–G and BSA at 300 nm and 280 nm, respectively: HT = 1.1×10^{-4} M; BSA = 4.75×10^{-6} M. For this conjugate, the ratio is 23

Specificity studies of the antisera

The specificity of each antiserum was tested by competition using the ELISA method. The result of a competition between each conjugated indoleamine coated on the well-plates, and conjugated or non-conjugated indoleamine pre-incubated with the antiserum was referred to as B. Under these conditions the optical density decreased. B_0 was the optical density corresponding to the antibody binding without competition. The value of B_0 was adjusted to approximately 1 at 492 nm by varying the dilution of each antiserum and of the anti-rabbit labelled with peroxidase. The ratio B/B_0 allowed us to analyse more precisely the results than with the absorbance alone and to calculate the average of three experiments (Geffard *et al.*, 1985a).

Antisera against F conjugates

Antibodies were obtained after coupling HT via F to BSA or HSA. Purified antiserum was tested against conjugated compounds such as: W–F–BSA, HW–F–BSA, MW–F–BSA, T–F–BSA, HT–F–BSA, aHT–F–BSA, aMT–F–BSA, MT–F–BSA and HT. Displacement curves obtained after competition experiments between coated HT–F–BSA and each indole-amine–F–BSA or free HT preincubated with an antiserum dilution showed that the best displacement was recorded for HT–F–BSA. This displacement occurred between 10^{-9} M and 10^{-5} M. The most immunoreactive compound from HT–F–BSA was MT–F–BSA, with a factor of 9 (Table 1). The other compounds were poorly recognized and no antibody affinity was found for aHT–F–BSA and aMT–F–BSA. Unconjugated HT was poorly recognized, underlining the importance of the F residues. In view of these results and in agreement with the previous ones obtained (Geffard *et al.*, 1984) we can conclude that anti-HT–F antibodies poorly discriminate the residue HT–F and MT–F, indicating their relatively weak specificity. These limitations led us to use G preferentially instead of F for raising other antisera against the closer HT-compounds.

Table 1 Specificity of anti-HT–F antibodies

Compound		Cross-reactivity ratio[a]
5-Hydroxytryptamine–F–BSA	(HT–F–BSA)	1
5-Methoxytryptamine–F–BSA	(MT–F–BSA)	1:9
5-Methoxytryptophan–F–BSA	(MW–F–BSA)	1:344
Tryptamine–F–BSA	(T–F–BSA)	1:2,343
Tryptophan–F–BSA	(W–F–BSA)	1:5,452
5-Hydroxytryptophan–F–BSA	(HW–F–BSA)	
N-Acetyl serotonin–F–BSA	(aHT–F–BSA)	
Melatonin–F–BSA	(aMT–F–BSA)	>1:10,000
5-Hydroxytryptamine	(HT)	

[a] HT–F–BSA concentration/non-conjugated HT or conjugated indoleamine concentration at half-displacement. These ratios are obtained from displacement curves established with ELISA tests. Competition experiments were performed between HT–F–BSA coated at 0.1 µg/ml and unconjugated or conjugated indoleamines: HT–F–BSA; MT–F–BSA; MW–F–BSA; T–F–BSA; W–F–BSA; HW–F–BSA; aHT–F–BSA; aMT–F–BSA; HT previously incubated for 16 h at 4°C with anti-HT–F antibodies.

Antisera against G conjugates

Antibodies were obtained after coupling each indoleamine to BSA or HSA via G. Each purified antiserum was tested against both its corresponding non-reduced conjugate and non-conjugated molecule, then each conjugated indoleamine.

Anti-HT–G antibody specificity The displacement curves obtained by competition between coated HT–G–BSA and each indoleamine–G–BSA showed that the most immunoreactive compound was HT–G–BSA with a self-displacement between 10^{-9} M and 10^{-7} M. HT=G=BSA, the non-reduced conjugate, was only 10 times less recognized than the reduced one (Table 2). Non-conjugated HT exhibited no immunoreactivity. This result is in agreement with the antibody recognition site corresponding to 3 or 4 amino acid residues (Geffard *et al.*, 1984). From HT–G–BSA, the most immunoreactive compounds were MT–G–BSA and HW–G–BSA, at factors of 42 and 50 respectively. MW–G–BSA and T–G–BSA were poorly recognized and W–G–BSA exhibited no immunoreactivity at all.

Table 2 Specificity of anti-HT–G antibodies

Compound		Cross-reactivity ratio[a]
5-Hydroxytryptamine–G–BSA	(HT–G–BSA)	1
5-Hydroxytryptamine=G=BSA	(HT=G=BSA)	1:10
5-Methoxytryptamine–G–BSA	(MT–G–BSA)	1:42
5-Hydroxytryptophan–G–BSA	(HW G BSA)	1:50
5-Methoxytryptophan–G–BSA	(MW–G–BSA)	1:198
Tryptamine–G–BSA	(T–G–BSA)	1:538
Tryptophan–G–BSA	(W–G–BSA)	1:2,500
5-Hydroxytryptamine	(HT)	1:10,000

[a] HT–G–BSA concentration/non-conjugated HT or conjugated indoleamine concentration at half-displacement. These ratios are obtained from displacement curves established with ELISA tests. Competition experiments were performed between HT–G–BSA coated at 1 μg/ml and unconjugated or conjugated indoleamines: HT–G–BSA; HT=G=BSA; MT–G–BSA; HW–G–BSA; MW–G–BSA; T–G–BSA; W–G–BSA; HT previously incubated for 16 h at 4°C with anti-HT antibodies.

W–G antiserum specificity The best displacement was observed with W–G–BSA. This self-displacement occurred between 10^{-9} M and 10^{-7} M. Unconjugated W was poorly recognized. The most immunoreactive compounds from W–G–BSA were HW–G–BSA and MW–G–BSA, at factors of 106 and 251, respectively (Table 3). No antibody affinity was found for T–G–BSA, HT–G–BSA and MT–G–BSA. W=G=BSA, the non-reduced conjugate, was 444 times less immunoreactive than W–G–BSA. From these competition results, it appears that the anti-W affinity is directed towards conjugated W.

Table 3   Specificity of anti-W–G antibodies

Compound		Cross-reactivity ratio[a]
Tryptophan–G–BSA	(W–G–BSA)	1
5-Hydroxytryptophan–G–BSA	(HW–G–BSA)	1:106
5-Methoxytryptophan–G–BSA	(MW–G–BSA)	1:251
Tryptophan=G=BSA	(W=G=BSA)	1:444
Tryptophan	(W)	1:3,230
Tryptamine–G–BSA	(T–G–BSA)	1:11,100
5-Methoxytryptamine–G–BSA	(MT–G–BSA)	<1:50,000
5-Hydroxytryptamine–G–BSA	(HT–G–BSA)	

[a] W–G–BSA concentration/non-conjugated W or conjugated indoleamine concentration at half-displacement. These ratios are obtained from displacement curves established with ELISA tests. Competition experiments were performed between W–G–BSA coated at 1 µg/ml and unconjugated or conjugated indoleamines: W–G–BSA; HW–G–BSA; MW–G–BSA; W=G=BSA; W; T–G–BSA; MT–G–BSA; HT–G–BSA previously incubated for 16 h at 4°C with anti-W antibodies.

Anti-HW–G antibody specificity The most immunoreactive compound, HW–G–BSA, showed a self-displacement between 10^{-9} M and 10^{-7} M. From HW–G–BSA, MW–G–BSA was the most immunoreactive conjugate at a factor of 252 (Table 4). W–G–BSA, HT–G–BSA, MT–G–BSA and T–G–BSA were not recognized at all. Under these conditions, the antibody affinity remained directed towards the original part of conjugated HW.

Table 4   Specificity of anti-HW–G antibodies

Compound		Cross-reactivity ratio[a]
5-Hydroxytryptophan–G–BSA	(HW–G–BSA)	1
5-Hydroxytryptophan=G=BSA	(HW=G=BSA)	1:5
5-Methoxytryptophan–G–BSA	(MW–G–BSA)	1:252
Tryptophan–G–BSA	(W–G–BSA)	1:13,420
5-Hydroxytryptamine–G–BSA	(HT–G–BSA)	
5-Methoxytryptamine–G–BSA	(MT–G–BSA)	<1:50,000
5-Hydroxytryptophan	(HW)	
Tryptamine–G–BSA	(T–G–BSA)	

[a] HW–G–BSA concentration/non-conjugated HW or conjugated indoleamine concentration at half-displacement. These ratios are obtained from displacement curves established with ELISA tests. Competition experiments were performed between HW–G–BSA coated at 1 µg/ml and unconjugated or conjugated indoleamines: MW–G–BSA; HW=G=BSA; W–G–BSA; HT–G–BSA; MT–G–BSA; HW; T–G–BSA previously incubated for 16 h at 4°C with anti-HW antibodies.

MW–G antiserum specificity As for the other W derivatives, MW–G was highly antigenic. The best recognition was obtained with MW–G–BSA. The non-conjugated MW exhibited a slight immunoreactivity, whereas HW–G–BSA, W–G–BSA, MT–G–BSA, HT–G–BSA, and T–G–BSA showed no immunoreactivity at all. The cross-reactivity ratio between MW–G–BSA and each indoleamine conjugate proved to be more than 7,700 (Table 5), demonstrating the high antibody specificity and affinity for conjugated MW.

Table 5 Specificity of anti-MW–G antibodies

Compound		Cross-reactivity ratio[a]
5-Methoxytryptophan–G–BSA	(MW–G–BSA)	1
5-Methoxytryptophan=G=BSA	(MW=G=BSA)	1:13
5-Methoxytryptophan	(MW)	1:608
5-Hydroxytryptophan–G–BSA	(HW–G–BSA)	1:7,762
Tryptophan–G–BSA	(W–G–BSA)	1:42,680
5-Methoxytryptamine–G–BSA	(MT–G–BSA)	
5-Hydroxytryptamine–G–BSA	(HT–G–BSA)	<1:50,000
Tryptamine–G–BSA	(T–G–BSA)	

[a] MW–G–BSA concentration/non-conjugated MW or conjugated indoleamine concentration at half-displacement. These ratios are obtained from displacement curves established with ELISA tests. Competition experiments were performed between MW–G–BSA coated at 1 μg/ml and unconjugated or conjugated indoleamines: MW–G–BSA; MW=G=BSA; MW; HW–G–BSA; W–G–BSA; MT–G–BSA; HT–G–BSA; T–G–BSA previously incubated for 16 h at 4°C with anti-MW antibodies.

Anti-MT–G antibody specificity The antigen MT–G had the capacity to induce an immune response. MT–G–BSA was the most immunoreactive compound with a self-displacement between 10^{-9} M and 10^{-7} M. MT=G=BSA and non-conjugated MT were rather well recognized (Table 6). W–G–BSA and HW–G–BSA were not recognized. The other compounds showed a poor immunoreactivity. The antibody site studied here can bring a new contribution to the molecular detection of MT in tissues.

Anti-T–G antibody specificity T–G–BSA, the most immunoreactive compound, showed a self-displacement between 10^{-9} and 10^{-7} M. In this case, T=G=BSA was poorly recognized, 1,826 times less than T–G–BSA. Non-conjugated T was also poorly recognized (2,000 times less). The cross-reactivity ratios, calculated at half-displacement from T–G–BSA, were for HT–G–BSA and MT–G–BSA 1:56 and 1:60 respectively. MW–G–BSA, W–G–BSA and HW–G–BSA were not recognized at all (Table 7). This antibody site was therefore able to discriminate conjugated-T from the other conjugated indoleamines present in tissues.

Table 6   Specificity of anti-MT–G antibodies

Compound		Cross-reactivity ratio[a]
5-Methoxytryptamine–G–BSA	(MT–G–BSA)	1
5-Methoxytryptamine=G=BSA	(MT=G=BSA)	1:12
5-Methoxytryptamine	(MT)	1:22
5-Hydroxytryptamine–G–BSA	(HT–G–BSA)	1:26
Tryptamine–G–BSA	(T–G–BSA)	1:30
5-Methoxytryptophan–G–BSA	(MW–G–BSA)	1:40
Tryptophan–G–BSA	(W–G–BSA)	1:495
5-Hydroxytryptophan–G–BSA	(HW–G–BSA)	1:1,902

[a] MT–G–BSA concentration/non-conjugated MT or conjugated indoleamine concentration at half-displacement. These ratios are obtained from displacement curves established with ELISA tests. Competition experiments were performed between MT–G–BSA coated at 1 μg/ml and unconjugated or conjugated indoleamines: MT–G–BSA; MT=G=BSA; MT; HT–G–BSA; T–G–BSA; MW–G–BSA; W–G–BSA; HW–G–BSA previously incubated for 16 h at 4°C with anti-MT antibodies.

Table 7   Specificity of anti-T–G antibodies

Compound		Cross-reactivity ratio[a]
Tryptamine–G–BSA	(T–G–BSA)	1
Tryptamine=G=BSA	(T=G=BSA)	1:32
5-Hydroxytryptamine–G–BSA	(HT–G–BSA)	1:50
5-Methoxytryptamine–G–BSA	(MT–G–BSA)	1:60
Tryptamine	(T)	1:2,047
5-Methoxytryptophan–G–BSA	(MW–G–BSA)	1:5,763
Tryptophan–G–BSA	(W–G–BSA)	1:37,200
5-Hydroxytryptophan–G–BSA	(HW–G–BSA)	<1:50,000

[a] Tryptamine–G–BSA concentration/non-conjugated T or conjugated indoleamine concentration at half-displacement. These ratios are obtained from displacement curves established with ELISA tests. Competition experiments were performed between T–G–BSA coated at 1 μg/ml and unconjugated or conjugated indoleamines: T–G–BSA; T=G=BSA; HT–G–BSA; MT–G–BSA; T; MW–G–BSA; W–G–BSA; HW–G–BSA previously incubated for 16 h at 4°C with anti-T antibodies.

Immunocytochemistry

In order to visualize each indoleamine compound, rat brain was G-fixed according to the procedure previously described by Degueurce and Geffard (1986). The immunocytochemical applications of G-antisera previously described by Patel *et al.* (1986) enabled us to investigate the distribution of T- and MT-containing neurons in rat raphe nuclei and to compare it with that of HT-containing neurons. Of the three antisera tested, the antiserum against conjugated HT produced the most intense staining (Fig. 4), which was in agreement with previous results obtained using antisera raised with F

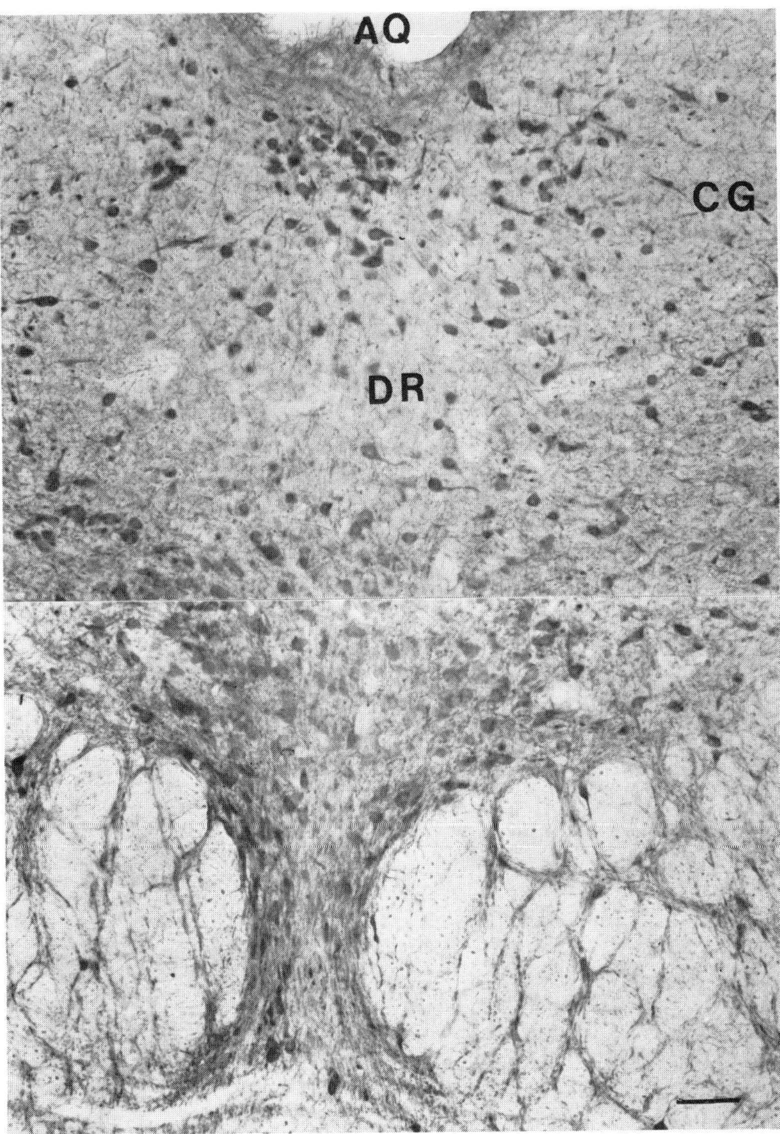

Fig. 4 Transversal 50 μm vibratome sections of treated G-fixed rat brain in the region of the dorsal raphe nuclei. Sections were borohydride reduced and incubated with HT–G antiserum at 1:10,000. This figure shows many HT-containing cell bodies in the dorsal raphe nucleus. Note the presence of very thin fibres also on the ependymal layer. AQ, cerebral aqueduct; CG, central grey; DR, dorsal raphe nucleus. Scale bar: 100 μm

conjugates (Steinbusch *et al.*, 1978; Steinbusch, 1981), HT-like immunoreactivity was found in more cell bodies than were MT-like and T-like immunoreactivities (Figs 5, 6). These results are in accordance with previous investigations (Björklund *et al.*, 1971; Prozialeck *et al.*, 1978; Boisin *et al.*, 1979; Juorio and Durden, 1984; Juorio and Greenshaw, 1985). But the cell bodies containing MT-like, HT-like and T-like immunoreactivity had a similar appearance. Immunoreactive perikarya were most abundant in the upper portion of the dorsal raphe, being less numerous in the interpeduncular nucleus. Immunoreactive soma were less densely distributed in median raphe nuclei than in the dorsal raphe nuclei. The smallest number of positive cell bodies was found in the B9 group. Immunoreactive processes appeared to emanate from the majority of positively labelled cell bodies. The basal portion of the dorsal raphe nucleus contained a dense bundle of positive fibres. The bundle seemed to be intermingled with a bundle emanating from the median raphe nucleus. The dense group of fibres was vertically oriented. As the fibres traversed the dorsal raphe nucleus, they fanned out, eventually assuming a horizontal position. In the B9 group of cells, the processes of the cell bodies also had an almost horizontal orientation. Considering these results it appears that some HT cell bodies in the raphe region contain also MT and/or T.

The specificity of each immunoreactivity was tested by substituting the primary antisera with antisera that had been pre-incubated with either HT, MT and T conjugated to protein carrier with G. The procedure used has been previously described by Patel *et al.* (1986). Whatever antiserum is concerned, the specific staining was abolished only after pre-adsorptions with the appropriate conjugate. For example, the reactions obtained with HT antisera pre-adsorbed with MT and T conjugates were indistinguishable from the immunoreactivity produced by HT antiserum that had not been pre-adsorbed. The same procedures were used for reactions obtained with MT and T antisera, and the identical conclusions were drawn.

From these results and the previous work of Patel *et al.* (1986), it can be concluded that the indoleaminergic compounds HT, MT and T are present in raphe nuclei neurones and they may therefore play a major role in neurotransmission or neuromodulation.

DISCUSSION

The development of antibodies directed against the molecules from the indoleamine family offers prospects of identifying each metabolite in biological samples. The first antisera raised against HT have been obtained by using F as a coupling agent. These anti-HT–F antibodies have been widely utilized in immunocytochemical studies on both peripheral and central nervous systems (Buffa *et al.*, 1980; Lidov *et al.*, 1980; Steinbusch *et al.*, 1981; Lauder *et al.*,

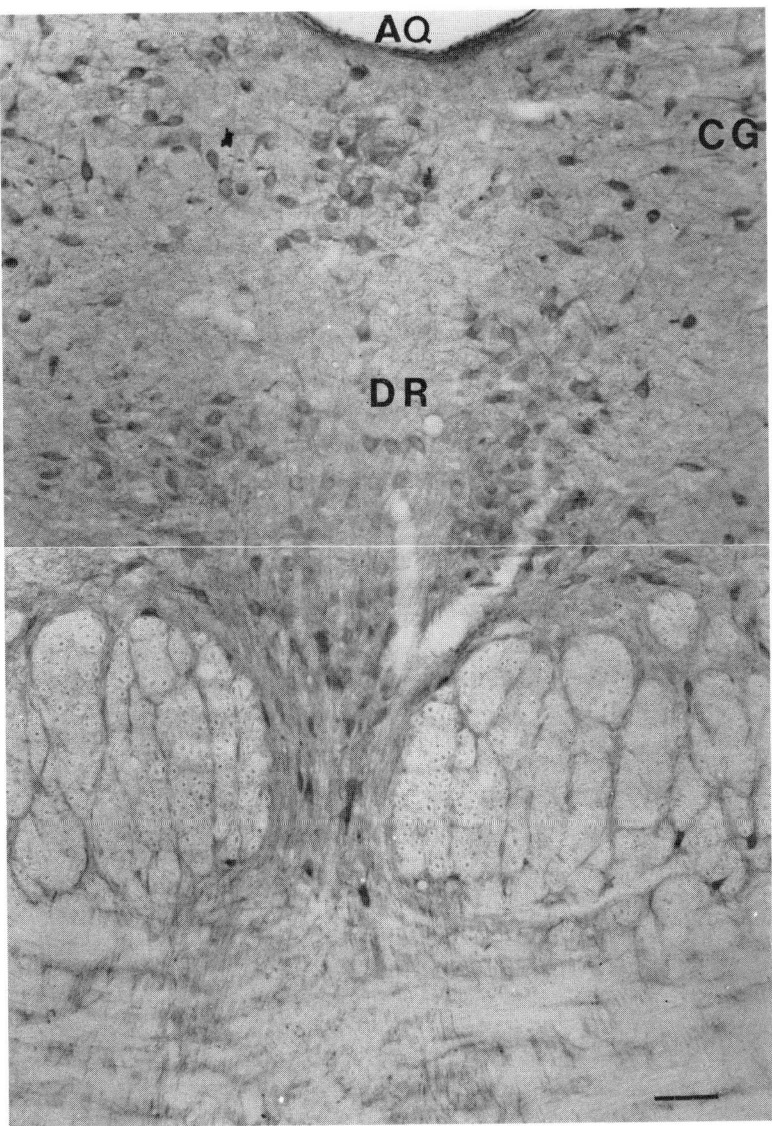

Fig. 5 MT-containing neurons in the region of the dorsal raphe nuclei. Sections were borohydride reduced and incubated with MT–G antiserum at 1:10,000. Although the MT distribution overlaps the HT one, MT immunoreactivity is weaker than HT immunoreactivity. The MT immunoreactive cells appeared to be of the same morphological type as the HT-immunoreactive cells. AQ, cerebral aqueduct; CG, central grey; DR, dorsal raphe nucleus. Scale bar: 100 μm

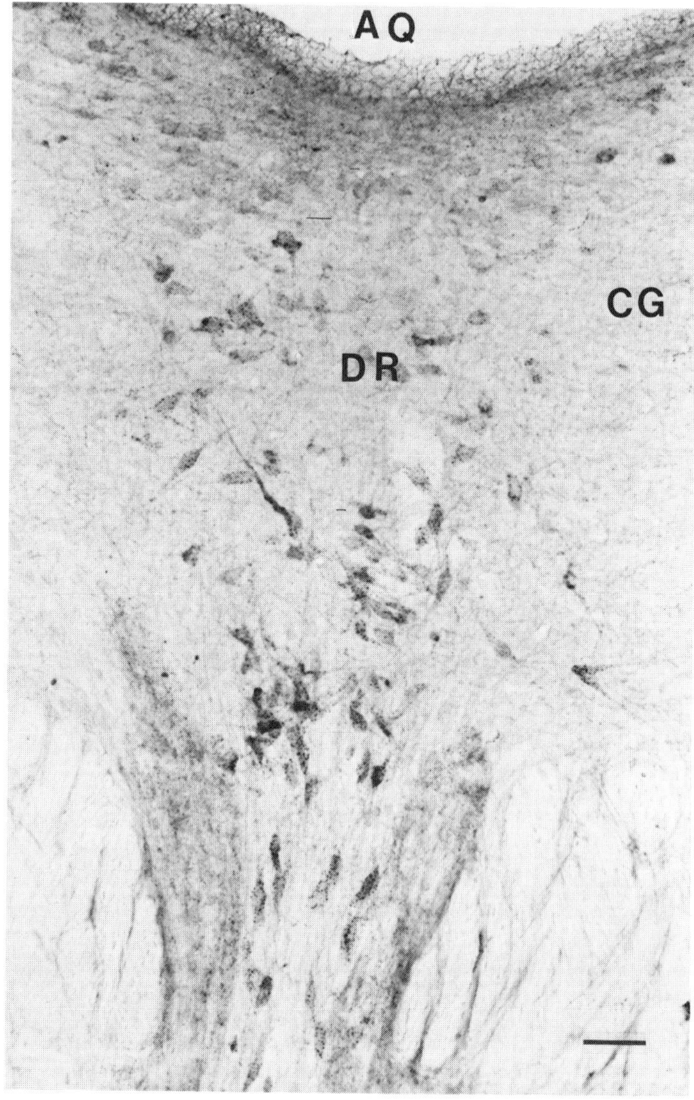

Fig. 6 Immunocytochemical application of anti-T–G antibodies. As for MT, T immunoreactivity was found in rat dorsal raphe nuclei. The distribution of T-cell bodies overlaps this HT one. But as for MT, the T immunoreactivity is weaker than HT immunoreactivity. AQ, cerebral aqueduct; CG, central grey; DR, dorsal raphe nucleus. Scale bar: 100 μm

1982; Takeuchi *et al.*, 1982). But, the antibody recognition sites have not been clearly defined. The recent employment of ligands which are either radioactive or do not mimic the antigenic structures in the immunogen tested by equilibrium dialysis, showed that these anti-HT–F antibodies are not as specific as they should be (Geffard *et al.*, 1984). Moreover, the ELISA results presented here from another F-antiserum confirm that conjugated HT and conjugated MT are poorly discriminated. It appears that the HT–F antisera are not specific enough to distinguish between these conjugated molecules in tissues. So, the localization and interaction of different metabolites of HT are not reliable. We have therefore developed another approach using G as a coupling agent and subsequently obtained more specific antibodies directed against each molecule belonging to the indoleamine family (Geffard *et al.*, 1985a).

The G antisera reported here illustrate the remarkable power of the antibody molecules. Each indoleamine having a free amino group is different from the other because of the presence or absence of one or two chemical groups, placed either in the 5-indole nucleus position or on the side chain. A slight structural difference such as hydroxyl or *o*-methoxy or carboxylic group allows the development of a good immune response, highly specific for each of the six conjugated molecules. These results are in close agreement with the work of Landsteiner (1945) and Goodman (1975) and more recent studies (Störm-Mathisen *et al.*, 1983; Geffard *et al.*, 1985a,b). For each indoleamine conjugate (Tables 2–7), the non-reduced compound was less immunoreactive than the reduced one at a factor around 10. However, for $W=G=BSA$ and $T=G=BSA$ the factors were 444 and 32 respectively. Thus, to obtain the best immunodetection of the conjugated hapten in biological samples, sodium borohydride must be used as a reducing agent.

The cross-reactivity ratios, calculated from the displacement curves (Tables 2–7), gave a good approximation of the specificity of the antisera. From the displacement curves it was observed that the conjugates appeared in a regular decreasing order, depending on their structural differences. It can be concluded that when the difference between the molecular structure of the various conjugates is important, the cross-reactivity ratio is small. In the same way the antisera directed against W, HW and MW recognized their conjugates better than the antisera directed against HT, MT and T, because of the presence of a carboxylic group on their lateral chain. The negative electric charge of the carboxylic group is regarded as a determinant factor in the study of specificity (Goodman, 1975). Obviously, in the case of HT and MT the role of hydroxyl and *o*-methoxy groups appears less determinant when the specificity of the corresponding antisera is studied. Finally, although the antibody site recognizes a structure equivalent to three amino acid residues, the optimal affinity is observed for the residues belonging to the lateral chain and the indole moiety, a fact which may explain the important role played by the carboxylic group in the conjugates and the loss of affinity for the non-reduced conjugates.

For histological studies, it is an advantage to use G as bifunctional coupling agent because it favours the retention and preservation of small sized molecules in the tissues, as previously described for dopamine (Buijs *et al.*, 1984; Onteniente *et al.*, 1984; Kah *et al.*, 1984; Takeda *et al.*, 1986), gamma-aminobutyric acid (GABA) (Störm-Mathisen *et al.*, 1983; Seguela *et al.*, 1984) and other neurotransmitters (Campistron *et al.*, 1986a,b,c). The G antisera obtained represent tools for studies aimed at understanding the anatomical and functional relationships between neurotransmitters in the nervous systems. In the same way, recent studies have contributed to allowing a redefinition of the indoleaminergic network in the CNS (Patel *et al.*, 1986) from animals which were not pharmacologically treated. Thus, HT is not considered as the sole intrinsic neuroactive indoleamine of the extrapineal CNS; MT and T are also present (see Figs 4–6). In this respect, we put forward the hypothesis that some HT-immunoreactive cells also contain MT and some HT-immunoreactive cells also have T. These new anatomical data must be included in further pharmacological, biochemical and physiological studies concerning the indoleaminergic neurotransmission and neuro-modulation.

ACKNOWLEDGEMENTS

This work was supported by grants from CNRS and INSERM.

REFERENCES

Avrameas, S., Ternynck, T., and Guesdon, J. L. (1978) Coupling of enzymes to antibodies and antigens, *Scand. J. Immun.*, **8**, 7–23.

Björklund, A., Falk, B., and Steveni, U. (1971) Microscopectofluorimetric character-ization of monoamines in the central neurons system: evidence for a new neuronal monoamine-like compound, *Prog. Brain. Res.*, **34**, 63–73.

Björklund, A., Lindvall, O., and Svensson, L. A. (1972) Mechanisms of fluorophore formation in the histochemical glyoxylic acid method for monoamines in brain, *Histochemie*, **32**, 113–131.

Boisin, T. R., Jonsson, G., and Beck, O. (1979) On the occurrence of 5-methoxytryp-tamine in brain, *Brain Res.*, **173**, 79–88.

Buffa, R., Crivelli, O., Lavarini, C., Sessa, F., Verme, G., and Solcia, E. (1980) Immunohistochemistry of brain 5-hydroxytryptamine, *Histochemistry*, **68**, 9–15.

Buijs, R. M., Geffard, M., Pool, C. W., and Hoorneman, M. D. (1984) The dopami-nergic innervation of the supraotic and paraventricular nucleus. A light and electron microscopical study, *Brain Res.*, **323**, 65–72.

Butler, J. E., Feldush, P. L., Mc Givern, P. L., and Stewart, N. (1978), The enzyme-linked immunoadsorbent assay: a measure of antibody concentration or affinity, *Immunochemistry*, **15**, 131–136.

Campistron, G., Geffard, M., and Buijs, R. M. (1986a) Immunological approach to the detection of taurine and immunocytochemical results, *J. Neurochem.*, **56**, 862–868.

Campistron, G., Buijs, R. M., and Geffard, M. (1986b) Specific antibodies against aspartate and their immunocytochemical application in the rat brain, *Brain Res.*, **365**, 179–184.

Campistron, G., Buijs, R. M., and Geffard, M. (1986c) Immunological approach to glycine and immunocytochemical results, *Brain Res.*, **376**, 400–405.

Consolazione, A., Milstein, C., Wright, B., and Cuello, A. C. (1981) Immunocytochemical detection of serotonin with monoclonal antibodies, *J. Histochem. Cytochem.*, **29**, 1425–1430.

Dahlström, A., and Fuxe, K. (1964) Evidence for the existence of monoamine-containing neurones in the central nervous system 1. Demonstration of monoamines in the cell bodies of brain stem neurones, *Acta Physiol. Scand.*, **62**, 1–55.

Dahlström, A., and Fuxe, K. (1965) Evidence for the existence of monoamine neurons in the central nervous system 1. Demonstration of monoamines in the cell bodies of brain stem neurones, *Acta Physiol. Scand.*, **64**, 1–36.

Engvall, E., and Perlmann, P. (1972) Quantitation of specific antibodies by enzyme labelled anti-immunoglobin in coated tubes, *J. Immunol.*, **109**, 129–135.

Falck, B., Hillarp, H. A., Thieme, G., and Thorpe, A. (1962) Fluorescence of catecholamines and related compounds condensed with formaldehyde, *J. Histochem. Cytochem.*, **10**, 348–354.

Furness, J. B., Costa, M., and Blessing, W. W. (1977) Simultaneous fixation and production of catecholamine fluorescence in central nervous tissue by perfusion with aldehydes, *Histochem. J.*, **9**, 745–750.

Fuxe, K., (1965) Evidence for the existence of monoamine neurons in the central nervous system, *Acta Physiol. Scand.*, **64** (suppl. 247), 39–85.

Fuxe, K., and Jonsson, G. (1967) A modification of the histochemical fluorescence method for the improved localisation of 5-hydroxytryptamine, *Histochem.*, **11**, 161–166.

Geffard, M., Dulluc, J., and Heinrich-Rock, A. M. (1985a) Antisera against the indolealkylamines: tryptophan, 5-hydroxytryptophan, 5-hydroxytryptamine, 5-methoxytryptophan and 5-methoxytryptamine tested by an ELISA method, *J. Neurochem.*, **44**, 1221–1228.

Geffard, M., Heinrich-Rock, A. M., Dulluc, J., and Seguela, P. (1985b) Antisera against small neurotransmitter-like molecules, *Neurochem. Int.*, **7**, 403–413.

Geffard, M., Seguela, P., and Buijs, R. M. (1984) Immunorecognition of anti-serotonin antibodies by using a radiolabelled ligand, *Neuroscience Letters*, **50**, 217–222.

Goodman, J. W. (1975) Antigenic determinants and antibody combining sites, in *The Antigens* (Ed. M. Sela), pp. 127–187, Academic Press, New York.

Grota, L. J., and Brown, G. M. (1974) Antibodies to indolealkylamines: serotonin and melatonin, *Canad. J. Biochem.*, **52**, 196–202.

Hökfelt, T., Fuxe, K., and Goldstein, M. (1973) Immunohistochemical localization of aromatic-L-amino acid decarboxylase (DOPA-decarboxylase) in central dopamine and 5-hydroxytryptamine nerve cell bodies of the rat brain, *Brain Res.*, **53**, 175–180.

Hökfelt, L., Fuxe, K., and Golstein, M. (1975) Applications of immunocytochemistry to studies on monoamine-cell systems with special reference to nervous tissues, *Ann. N. Y. Acad. Sci.*, **254**, 407–432.

Hökfelt, T., Johansson, O., Ljungdahl, A., Lundberg, J. M., and Schultzberg, M. (1980) Peptidergic neurons, *Nature*, **284**, 515–521.

Joh, T. H., Shikimi, T., Pickel, V. M., and Reis, D. J. (1975) Brain tryptophan hydroxylase: purification and production of antibodies to and cellular and ultra-structural localization of serotoninergic neurons of rat midbrain, *Proc. Natl. Acad. Sci. USA*, **72**, 3575–3579.

Juorio, A. V., and Durden, D. A. (1984) The distribution and turnover of tryptamine in the brain and spinal cord, *Neurochem. Res.*, **9**, 1281–1293.

Juorio, A. V., and Greenshaw, A. J. (1985) Tryptamine concentrations in areas of 5-hydroxytryptamine terminal innervation after electrolytic lesions of midbrain raphe nuclei, *J. Neurochem.*, **45**, 422–426.

Kah, O., Chambolle, P., Thibault, J., and Geffard, M. (1984) Existence of dopaminergic neurons in the preoptic region of the goldfish, *Neurosci. Lett.*, **48**, 293–298.

Kuhar, M. J., Aghajanian, G. K., and Roth, R. M. (1972) Tryptophan hydroxylase activity and synaptosomal uptake of serotonin in discrete brain regions after midbrain raphe lesions: correlation with serotonin levels and histochemical fluorescence, *Brain. Res.*, **44**, 165–176.

Landsteiner, K. (1945) *The Specificity of Serological Reactions*, 2nd edn, Harvard University Press, Cambridge, MA.

Lauder, J. M., Petrusz, P., and Wallace, J. (1982) Combined immunocytochemistry of serotonin and [3H]-thymidine autoradiography, *J. Histochem. Cytochem.*, **30**, 788–794.

Lidov, H. G. W., Grzanna, R., and Millwer, M. E. (1980) The serotonin innervation of the cerebral cortex in the rat. An immunohistochemical analysis, *Neuroscience*, **5**, 207–229.

Lorèn, I., Björklund, A., Falk, B., and Lindvall, O. (1976) An improved histofluorescence procedure for freeze-dried-paraffin-embedded tissue based on combined formaldehyde-glyoxylic acid perfusion with high magnesium content and acid pH, *Histochemistry*, **49**, 177–192.

McRae-Degueurce, A., and Geffard, M. (1986) One perfusion mixture for immunocytochemical detection of noradrenaline, dopamine, serotonin and acetylcholine in the same rat brain, *Brain Res.*, **376**, 217–219.

Marsden, C. A., and Curzon, G. (1974) Effects of lesions and drugs on brain tryptamine, *J. Neurochem.*, **23**, 1171–1176.

Milstein, C., Wright, B., and Cuello, A. C. (1983) The discrepancy between the cross-reactivity of a monoclonal antibody to serotonin and its immunohistochemical specificity, *Molec. Immunol.*, **20**, 113–123.

Onteniente, B., Geffard, M., and Calas, A. (1984) Immunocytochemical electron microscopical study of dopaminergic innervation of the rat lateral septum with anti-dopamine antibodies, *Neuroscience*, **13**, 385–393.

Patel, S., Dulluc, J., and Geffard, M. (1986) Comparison of serotonin and 5-methoxytryptamine immunoreactivity in rat raphe nuclei, *Histochemistry*, **85**, 259–263.

Pickel, V. M., Joh, T. H., and Reis, D. (1976) Monoamine synthesizing enzymes in central dopaminergic, noradrenergic and serotonergic neurons. Immunohistochemical localization by light and electron microscopy, *J. Histochem. Cytochem.*, **24**, 792–806.

Pickel, V. M., Joh, T. H., Reis, D. (1977) A serotoninergic innervation of adrenergic neurons in nucleus of locus coeruleus: demonstration by immunocytochemical localization of the transmitter specific enzymes tyrosine and tryptophan hydroxylase, *Brain Res.*, **131**, 197–214.

Prozialeck, W. C., Boemme, D. H., and Vogel, N. H. (1978) The fluorometric determination of 5-methoxytryptamine in mammalian tissues and fluids, *J. Neurochem.*, **30**, 1471–1477.

Pulido, O., Brown, G., and Grota, L. J. (1983) An immunohistochemical method for the localization of N-acetylserotonin (NAS) in the central nervous system. Description, validation and application of the technique, *J. Histochem. Cytochem.*, **31**, 1343–1350.

Ranadive, N. S., and Sehon, A. H. (1967) Antibodies to serotonin, *Canad. J. Biochem.*, **45**, 1701–1710.

Seguela, P., Geffard, M., Buijs, R. M., and Le Moal, M. (1984) Antibodies against GABA: specificity studies and immunocytochemical results, *Proc. Natl. Acad. Sci. USA*, **81**, 3888–3892.

Steinbusch, H. W. M. (1981) Distribution of serotonin immunoreactivity in the central nervous system of the rat cell bodies and terminals, *Neuroscience*, **6**, 557–618.

Steinbusch, H. W. M., Verhofstad, A. A. J., and Joosten, H. W. F. (1978) Localization of serotonin in the central nervous system by immunocytochemistry: description of a specific and selective technique and some applications, *Neuroscience*, **3**, 811–819.

Sternberger, L. A., Hardy, P. H., Cuculis, J. J., and Meyer, H. G. (1970) The unlabelled antibody enzyme method of immunocytochemistry. Preparation and properties of soluble antigen-antibody complex (Horseradish peroxidase-antihorseradish peroxidase) and its use in identification of spirochetes, *J. Histochem. Cytochem.*, **18**, 315–338.

Störm-Mathisen, J., Leknes, A. K., Bore, A. T., Vaaland, J. L., Edminsson, P., Haug, F. H. S., and Ottersen, O. P. (1983) First visualization of glutamate and GABA in neurons by immunocytochemistry, *Nature*, **301**, 517–520.

Takeuchi, Y., Kimura, H., and Sano, Y. (1982) Immunohistochemical demonstration of the serotonin containing nerve fibers in the cerebellum, *Cell Tiss. Res.*, **226**, 1–12.

Takeda, S., Vieillemaringe, J., Geffard, M., and Remy, C. (1986) Immunohistochemical evidence of dopamine cells in cephalic nervous system of silkworm *Bombyx mori*. Coexistence of dopamine and α-endorphin-like substance in neurosecretory cells of the suboesophageal ganglion, *Cell Tiss. Res.*, **243**, 125–128.

Ungerstedt, U. (1971) Stereotaxic mapping of the monoamine pathways in the rat brain, *Acta Physiol. Scand.* (Suppl. 367), 1–47.

Neuronal Serotonin
Edited by N. N. Osborne and M. Hamon
© 1988 John Wiley & Sons Ltd

CHAPTER 2

Distribution of Serotonin Neurons in the Mammalian Brain

YOSHIHIRO TAKEUCHI
Department of Pediatrics
Kyoto Prefectural University of Medicine
Kawaramachi, Hirokoji
Kamikyo-ku
Kyoto 602
Japan

INTRODUCTION

Current knowledge concerning the distribution of serotonergic neurons and fibers has been obtained using various techniques: (a) formaldehyde-induced fluorescence (FIF) histochemistry; (b) biochemical determination of serotonin; (c) autoradiographic tracing methods; (d) autoradiography following superfusion or infusion of tritiated serotonin; (e) immunohistochemistry using antibodies to biosynthetic enzymes, such as tryptophan hydroxylase; and (f) immunohistochemical methods using antibodies against serotonin coupled with carrier proteins i.e. bovine serum albumin, bovine thyroglobulin and limulus hemocyanin. Among these methods, the last has proved to be most pertinent in neurobiological studies because of the high sensitivity and specificity (Steinbusch *et al.*, 1978; Wallace *et al.*, 1982; Takeuchi *et al.*, 1983a).

Immunofluorescence and immunoperoxidase methods, including peroxidase–antiperoxidase (PAP) and avidin–biotin peroxidase complex (ABC) methods have revealed that serotonergic neuronal elements have a wider and denser distribution than expected from earlier experiments. The distribution of the serotonergic neuron system of the rat was extensively studied by Steinbusch (1981). Concerning the serotonergic pathways, comprehensive reviews were established by Consolazione and Cuello (1982), Steinbusch (1984), and recently by Nieuwenhuys (1985). The present chapter is a review of morphological data on serotonin neurons, obtained using immunohistochemical methods. The interspecific differences among various species are given particular attention.

LOCALIZATION OF SEROTONIN

To determine the distribution of serotonin neurons, it is fundamental that the immunohistochemical staining should be specific for serotonin and that the staining reflects the physiological state of the serotonin neurons. Therefore, the available data were those obtained in the absence of pharmacological pretreatments, such as an administration of a monoamine oxidase inhibitor (MAOI), tryptophan or colchicine.

Under physiological conditions, serotonergic neuronal somata are organized according to certain basic patterns from the level of the pyramidal decussation to the level of the nucleus ruber, in all mammals – rat (Steinbusch, 1981; Lidov and Molliver, 1982; Takeuchi *et al.*, 1982b), mouse (Ishimura, 1985a), rabbit (Howe *et al.*, 1983a; Yamada and Sano, 1985), cat (Takeuchi *et al.*, 1982b; Jacobs *et al.*, 1984; Marson and Loewy, 1985), dog (Kojima *et al.*, 1983), monkey (Takeuchi *et al.*, 1982a; Saavedra *et al.*, 1983), and the human fetus (Takahashi *et al.*, 1984).

Serotonin neurons are distributed mainly within specific cytoarchitectonic entities, namely the nuclei of the raphe, but are by no means congruent within them. Therefore, Dahlström and Fuxe (1964) introduced a new classification of the indoleamine-containing cells in the brain of the rat – nine cell groups numbered B1–B9, on the basis of data obtained by the FIF method. Most of these groups have also been immunohistochemically detected (cf. Figs 1, 2).

Serotonin-immunoreactive cell bodies

B1: Serotonin-immunoreactive cell bodies of this group are localized mainly within the nucleus raphe pallidus in the medulla oblongata. Other cell bodies are present in the reticular formation dorsally to the nucleus olivaris accessorius dorsalis. Caudally, this group extends to the level of the pyramidal decussation. Rostrally, cell bodies belonging to this group are continuous with the caudal part of group B3.

B2: Serotonin-immunoreactive somata of the B2 group are situated at the same level as group B1 and occupy a more dorsal position. Most locate within the confines of the nucleus raphe obscurus and are characterized by two symmetrical paramedian rows of cell bodies.

B3: The majority of serotonin-immunoreactive cell bodies are present within the nucleus raphe magnus while others constitute a laterally extending band along the fiber bundles of the corpus trapezoideum.

These three caudal serotonergic neuron groups (B1–B3) appear to be less conspicuous in the carnivore and primate than in the rodent.

B4: This group is reportedly situated at the level of the medial vestibular nucleus lying just dorsal to the nucleus prepositus hypoglossi, but no special comparison with a raphe nucleus is possible immunohistochemically.

B5: The main portion of this group coincides with the nucleus raphe pontis at the level of the nucleus motorius nervi trigemini. More dorsally, a number of serotonin neurons locate between the bilateral fasciculi longitudinale mediale.

B6: The cells of this group are close to and in the midline beneath the rostral part of the fourth ventricle in the area dorsomedial to the nucleus tegmenti dorsalis (of Gudden). This group seems to be an extension of the nucleus raphe dorsalis, or of the group B7.

B7: This is a large mesencephalic group of serotonin neurons localized mainly within the nucleus raphe dorsalis, situated in and ventral to the central gray matter (periaqueductal gray) from the level of the nucleus tegmenti dorsalis to the caudal pole of the nucleus nervi oculomotorii (Fig. 3). The serotonergic neurons of this group can be classified into three subgroups: dorsomedian, ventromedial and lateral. The cell bodies of the lateral subgroup often extend to the lateral border of the nucleus raphe dorsalis and are characterized by being of a large size, multipolar with long processes and intensely stained – throughout various species.

B8: This group seems to locate in the nucleus centralis superior and the nucleus linearis of the lower mesencephalon. Additional cells locate outside the border of these nuclei and scatter in the reticular formation as rostrally as the beginning of the nucleus interpeduncularis.

B9: The cell bodies belonging to this group locate within and around the lemniscus medialis and extend as a horizontal band, laterally to the lemniscus lateralis. This group does not correspond with any raphe nuclei (Fig. 4).

In addition to these groups (B1–B9), other serotonin neurons groups are detectable immunohistochemically, under physiological conditions.

(a) Serotonergic neurons are seen in the area just ventral to the surface of

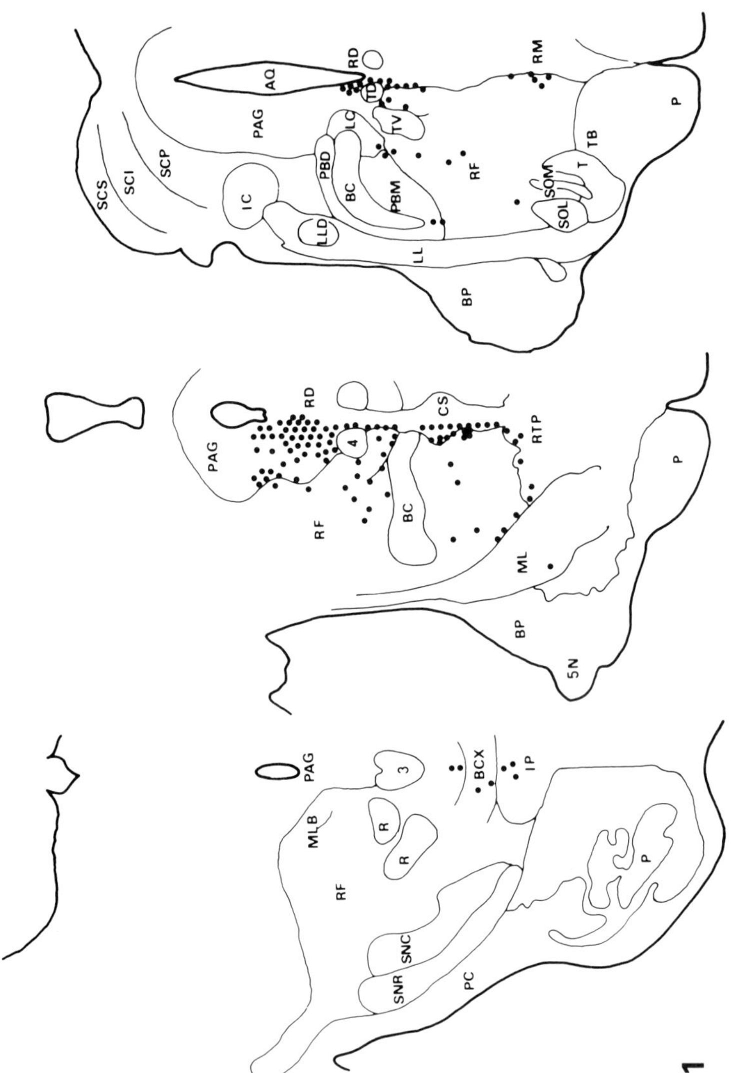

Figs 1 and 2 Schematic drawings of frontal brain sections of the monkey (*Macaca fuscata*) to illustrate the distribution of serotonin-immunoreactive cell bodies. Abbreviations from Takeuchi *et al.*, 1982a. (Reproduced by permission of S. Karger AG, Basle.) (For abbreviations, see facing page)

1

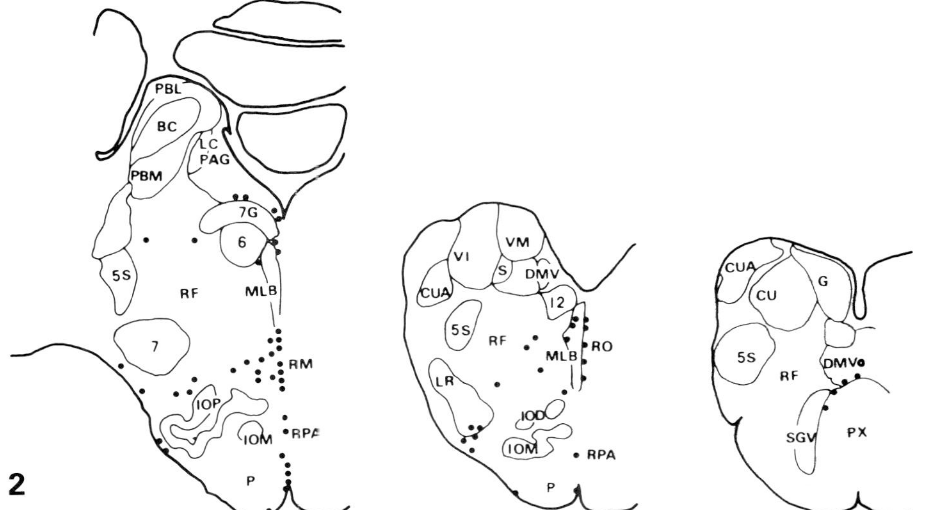

Abbreviations (Figs 1 and 2): AQ aqueductus mesencephali; BC brachium conjunctivum (pedunculus cerebellaris superior); BCX decussatio brachi conjunctivi (decussatio pedunculorum cerebellarium superiorum); BP brachium pontis (pendunculus cerebellaris medius); CS nucleus centralis superior; CU nucleus cuneatus; CUA nucleus cuneatus accessorius; DMV nucleus dorsalis n. vagi; G nucleus gracilis; IOD nucleus olivaris accessorius dorsalis; IOM nucleus olivaris accessorius medialis; IOP nucleus olivaris inferior; IP nucleus interpeduncularis; LC locus ceruleus; LL lemniscus lateralis; LLD nucleus lemnisci lateralis dorsalis; LR nucleus reticularis lateralis; ML lemniscus medialis; MLB fasciculus longitudinalis medialis; P tractus corticospinalis; PAG periaqueductal gray (central gray matter); PBD nucleus parabrachialis dorsalis; PBL nucleus parabrachialis lateralis; PBM nucleus parabrachialis medialis; PC crus cerebri; PX decussatio pyramidum; R nucleus ruber; RD nucleus raphe dorsalis; RF reticular formation (formatio reticularis); RM nucleus raphe magnus; RO nucleus raphe obscurus; RPA nucleus raphe pallidus; RTP nucleus reticularis tegmenti pontis; S tractus solitarius; SCI colliculus superior, stratum intermedium; SCP colliculus superior, stratum profundum; SCS colliculus superior, stratum superficiale; SCG nucleus substantiae griseae ventralis; SNC substantia nigra, zona compacta; SNR substantia nigra, zona reticularis; SOL nucleus olivaris superior lateralis; SOM nucleus olivaris superior medialis; T nucleus corporis trapezoidei; TB corpus trapezoideum; TD nucleus tegmenti dorsalis; TV nucleus tegmenti ventralis; VI nucleus vestibularis inferior; VM nucleus vestibularis medialis; 3 nucleus n. oculomotorii; 4 nucleus n. trochlearis; 5N nervus trigeminus; 5S nucleus tractus spinalis n. trigemini; 6 nucleus n. abducentis; 7 nucleus n. facialis; 7G genu n. facialis; 12 nucleus n. hypoglossi

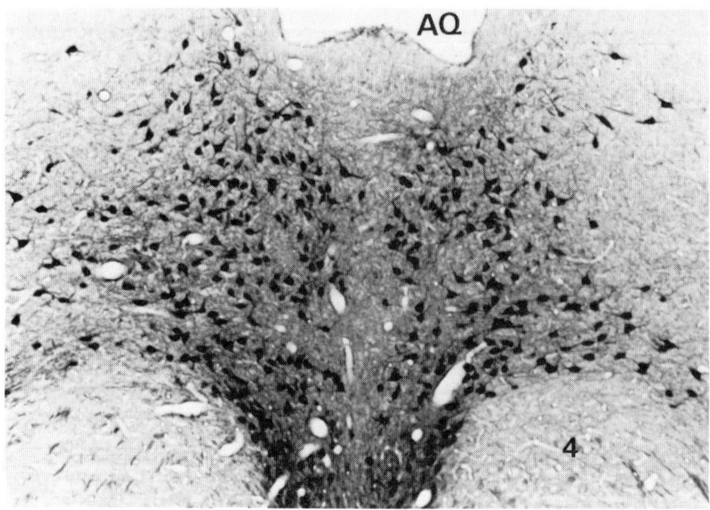

Fig. 3 Nucleus raphe dorsalis forms bilateral clusters dorsal to the nucleus trochlearis. Serotonin-immunoreactive cells are seen in the periaqueductal gray. AQ: aqueductus mesencephali. *Macaca fuscata*, ×42

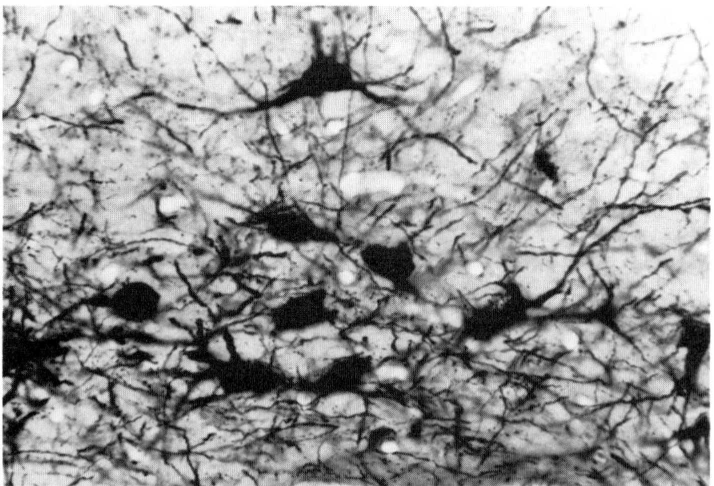

Fig. 4 A number of serotonin-immunoreactive cells with long processes are located dorsal to the lemniscus medialis. *Macaca fuscata*, ×255

the medulla oblongata. In the rat, these cell bodies are concentrated in the area surrounded by the origins of the abducens, hypoglossal, glossopharyngeal, and vagus nerves. The highest number of superficial serotonin-containing cell bodies appears along the basilar sulcus, where they represent the most ventral extension of B1 group.

(b) Some serotonin neurons lie intermingled with noradrenaline cells of the locus ceruleus, the subceruleus area and the nucleus parabrachialis.

(c) Serotonin neurons are scattered in the wing-like structures surrounding the nucleus interpeduncularis and the area ventralis tegmenti.

(d) Serotonin-immunoreactive cell bodies are localized in the dorsomedial portion of the nucleus interpeduncularis. These neurons in the apical subnucleus are considered to project to the hippocampus (Wirtshafter *et al.*, 1986).

The only comprehensively quantitative analysis on serotonin-immunoreactive cell bodies has concerned data on the mouse. In mice not given pharmacological pretreatment, serotonin-immunoreactive cell bodies were calculated to be approximately 26,000 in total number, of which 66.9% are located in the raphe system: that is to say, nucleus raphe pallidus (B1), 6.2%; nucleus raphe obscurus (B2), 2.6%; nucleus raphe magnus (B3), 6.1%; nucleus raphe pontis (B5), 4.5%; nucleus centralis superior (B8), 8.1%; nucleus linealis, 3.8%; and nucleus raphe dorsalis (B7), 35.5%. Other than the raphe system, additional serotonin-immunoreactive neurons are seen in the following regions: reticular formation, 18.9%; within and around the lemniscus medialis (B9), 8.5%; ventral surface of the medulla oblongata, 4.0%; and periaqueductal gray, 0.2% (Ishimura, 1985b).

Masked indoleamine cells

Other serotonin-immunoreactive cells have been detected in the following regions of several mammals after pharmacological manipulations, such as MAOI, tryptophan or colchicine treatment. These cells were designated 'masked indoleamine cells' by Nishida *et al.* (1985).

(a) Relatively small monopolar cells in the area postrema of the rat (colchicine treatment – Steinbusch, 1981; MAOI treatment – Takeuchi and Sano, 1983c; Nishida *et al.*, 1985; Lanca and Kooy, 1985).

(b) Serotonin neurons in the medial nuclei of the solitary tract of the rat (Calza *et al.*, 1985, intracisternal administration of colchicine).

(c) Serotonin-immunoreactive cell bodies in the substantia nigra and the area ventralis tegmenti of the rat (Steinbusch *et al.*, 1982, MAOI and tryptophan treatment).

(d) Serotonin-immunoreactive cell bodies in the nucleus dorsomedialis hypothalami of the rat (Frankfurt *et al.*, 1981, Steinbusch *et al.*, 1982, MAOI and tryptophan treatment) and cat (Ueda *et al.*, 1983c, MAOI treatment). An intraventricular injection of dopamine, noradrenaline and adrenaline as well as isoproterenol also reveal serotonin-immunoreactive cell bodies in this nucleus of the MAOI pretreated rat (Arezki *et al.*, 1985).

(e) Serotonin-immunoreactive neurons in the ventral part of the middle to the caudal lateral hypothalamic area of the cat (Ueda *et al.*, 1983c, MAOI treatment).

Overall, despite some differences in the distribution pattern in the rodent, carnivore and primate, close similarities in the organization of serotonin neurons are evident in these diverse mammalian species.

Serotonin-immunoreactive nerve fibers and terminals

The majority of serotonin-immunoreactive axons are unmyelinated, though myelinated serotonin-immunoreactive fibers have been detected in some areas, i.e. in the medial forebrain bundle of the rat and monkey (Azmitia and Gannon, 1983) and in the nucleus raphe dorsalis of the monkey (Kapadia *et al.*, 1985).

As for type grouping of serotonin-containing nerve fibers, many data have been reported (Steinbusch, 1981; Ruda *et al.*, 1982; Sano *et al.*, 1982a; Sano, 1983).

Unmyelinated serotonin-immunoreactive fibers are, in general, divided into two groups:

(a) Fibers with no or only a few varicosities. These fibers are long and preferentially oriented in a rostrocaudal direction, as found in the cingulum bundle and the ventral serotonergic bundle (Fig. 5). These long and straight fibers are also present in the neocortex and the striatum, especially in the primate. They are considered to correspond to the 'tract fibers' designated as such by Sano *et al.* (1982a).

(b) Thin fibers with varicosities showing no special direction and spread out in almost all areas of the brain. In very densely innervated regions, such as the globus pallidus, the substantia nigra, and some areas of the inferior olivary complex of the primate, intervaricose segments of serotonin fibers are rarely detectable.

Telencephalon

Serotonin-immunoreactive fibers are present throughout all layers in the olfactory bulb of mammals, but there are species-related differences in the pattern of serotonin distribution. In case of the rat and cat, a plexus of serotonin fibers is diffuse in the glomerulus (Fig. 6), while in the monkey,

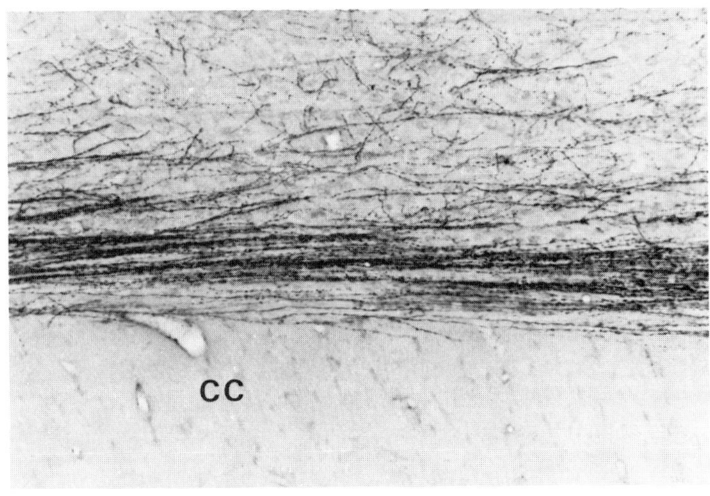

Fig. 5 Cingulum bundle of serotonin fibers in sagittal section. CC: corpus callosum. Rat, ×255

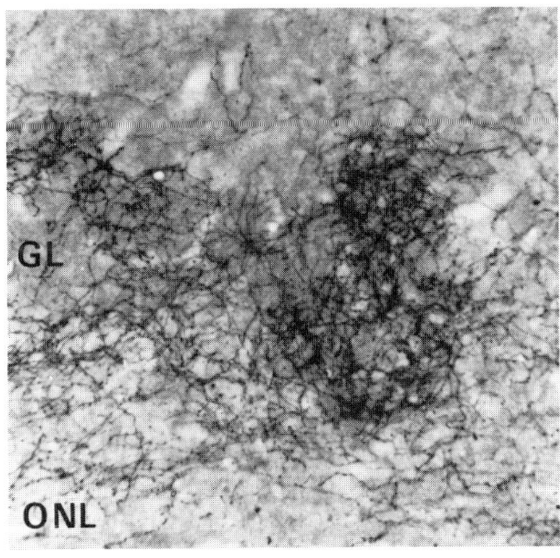

Fig. 6 A diffuse as well as dense plexus of serotonin fibers in the glomerular layer (GL). ONL: olfactory nerve layer. Rat, ×212

the distribution of serotonin fibers is scanty and partial, as is seen in the accessory olfactory bulb of the rat (Takeuchi *et al.*, 1982c).

In the hippocampus, there is a relatively restricted laminar pattern of serotonin innervation. In the alveus, there are few serotonin fibers, whereas in the stratum oriens and radiatum, numerous serotonin fibers spread out, in various orientations. The stratum pyramidale has only a few fibers, and the stratum lacunosum and moleculare have a very high density of serotonin fibers (Takeuchi *et al.*, 1983a). Abundant serotonin fibers with extremely dense varicosities in the area fasciola cinerea of the rat have been evidenced (Zhou and Azmitia, 1986). A widespread and dense distribution of serotonin fibers is also revealed in the retrohippocampal region (Köhler *et al.*, 1981). This innervation is heterogeneous with regard to the morphological characteristics, the density, and the spatial orientation of serotonin fibers.

The serotonin innervation of the septum is heterogeneous with regard to both morphology of the individual processes and the density of distribution in different parts of the septum. Apart from thin convoluted fibers with varicosities and thick relatively straight fibers with few varicosities, pericellular plexuses with large varicosities are found in close association with perikarya in the lateral septum (Köhler *et al.*, 1982; Gall and Moore, 1984). Three areas of the septum of the rat receive a prominent innervation by serotonin fibers: the diagonal band of Broca, the ventral part of the lateral septum, and an area bordering the medial edge of the island of Calleja (Köhler *et al.*, 1982).

In the neostriatum (nucleus caudatus and putamen) of all mammals, the concentration of serotonin fibers is high in the ventral, medial and caudal neostriatum (Figs 7, 8). Especially in the area bound by the globus pallidus, serotonin fibers are abundant and compactly arranged along the nucleus. In the neostriatum of the monkey, a few thick fibers are intermingled and are also observed outside the medial medullary lamina and in the central position of the medial pallidal segment (Pasik and Pasik, 1982; Sano, 1983; Mori *et al.*, 1985b). The paleostriatum (globus pallidus and entopeduncular nucleus) of the rat and cat as well as the medial pallidal segment of the monkey are diffusely innervated with serotonin fibers, while in the lateral pallidal segment of the monkey the distribution of fibers is sparse (Mori *et al.*, 1985b).

On the basis of electron microscopic observation of the neostriatum of the primate, Pasik and Pasik (1982) proposed the attractive hypothesis that serotonin is released as a neurotransmitter from synapsing boutons (serotonin receptor 2), and as a neuromodulator from non-synapsing varicosities (serotonin receptor 1).

Concerning serotonergic innervation of the amygdala complex, the only data available are on the rat (Steinbusch, 1981). In general, the amygdala complex has a high density of serotonergic innervation; particularly the rostral and medial parts of the nucleus amygdaloideus basalis are densely

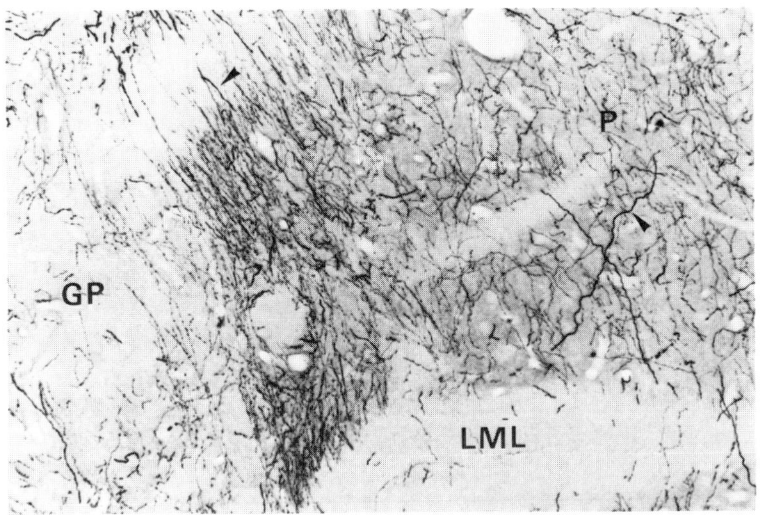

Fig. 7 Serotonin fibers in the neostriatum. Thick fibers (tract fibers, arrowheads) are intermingled in this area of the primate. GP: globus pallidus, LML: lateral medullary lamina, P: putamen. *Macaca fuscata*, ×136

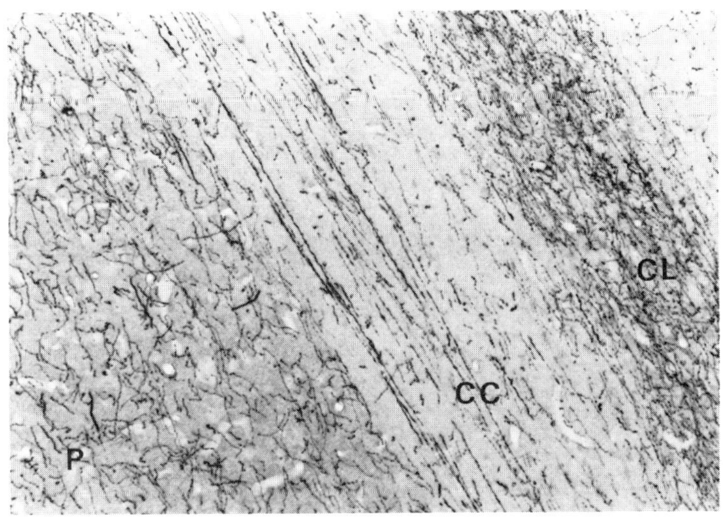

Fig. 8 Thick serotonin fibers are present in the corpus callosum (CC). A dense serotonergic innervation is seen in the claustrum (CL). P: putamen. *Macaca fuscata*, ×136

innervated. In the most caudal part of the amygdala complex, a very high density of serotonergic innervation is confined to the nucleus amygdaloideus medialis posterior and its medial and lateral part.

The entire neocortex receives serotonergic innervation which spreads over all cortical layers, while a laminar pattern of serotonin distribution is evident in the phylogenetically old cortex, i.e. mesocortex, archicortex and paleo-cortex (Lidov *et al.*, 1980; Steinbusch, 1981; Takeuchi *et al.*, 1982c; Takeuchi *et al.*, 1983a). In the rodent, serotonergic innervation in the neocortex is relatively uniform (Lidov *et al.*, 1980), but in the primate, different cytoarchitectonic fields show considerable differences in density and laminar distribution pattern of the serotonergic innervation (Takeuchi and Sano, 1983b). Among the cortical areas of the monkey (Figs 9, 10) – areas 4, 3-1-2, 17, 18, 41 and 42 – the primary visual cortex (striate cortex, area 17) contains the highest density of serotonin fibers, while the primary motor cortex (area 4) possesses the lowest concentration (Fig. 11). In the area 3–1–2, 18, 41 and 42, a fairly uniform density of serotonin fibers is observed across the six cortical layers, apart from a relatively dense plexus of fine serotonin fibers in layers IV-V (Takeuchi and Sano, 1983b). As for the primary visual cortex with the most outstanding findings in serotonergic innervation, further data have been reported. Morrison *et al.* (1982) proposed that serotonergic and noradrenergic projections exhibit a high degree of laminar complementarity

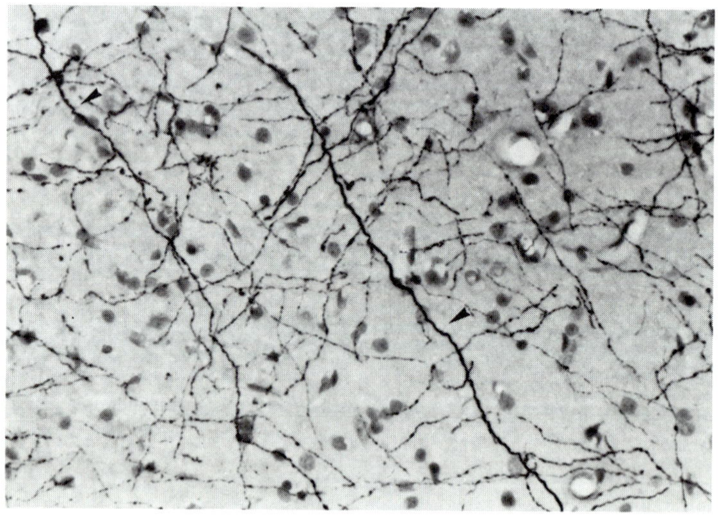

Fig. 9 Serotonin fibers in a tangential section through layer I of primary somatic sensory cortex (area 3-1-2). The arrowheads indicate tract fibers among plexus composed of fine varicose fibers. Counterstained with cresyl violet. *Macaca fuscata*, ×255

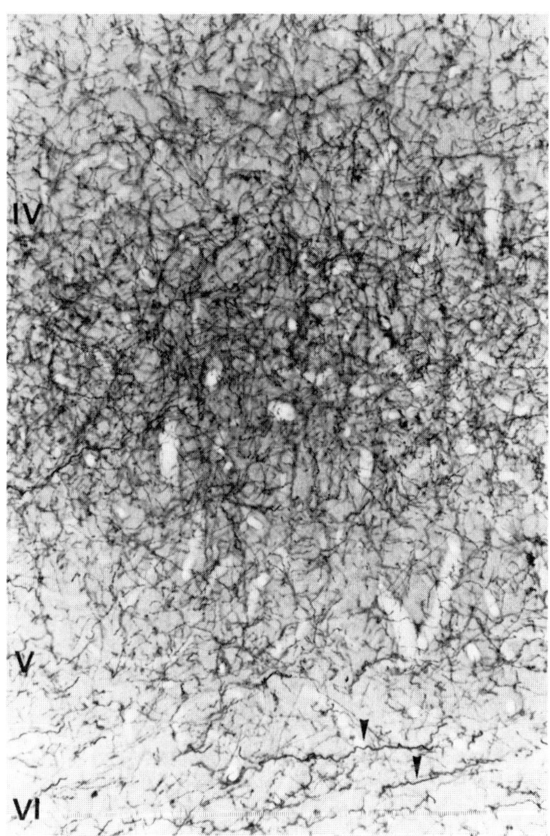

Fig. 10 Serotonin fibers in the primary visual cortex (area 17). The highest density of serotonin fibers occurs in layer IVc. Tract fibers (arrowheads) run in tangential direction, within layers V and VI. *Macaca fuscata*, ×136

– raphe-cortical serotonergic projection preferentially innervates the spiny stellate cells of layers IVa and IVc, whereas the ceruleo-cortical noradrenergic projection innervates pyramidal cells. However, an immunoelectron-microscopical analysis shows that serotonin-immunoreactive varicosities make contact with both stellate and pyramidal cells (Takeuchi and Sano, 1984). Furthermore, Morrison and Foote (1986) refer to serotonergic and noradrenergic innervation of cortical, thalamic and tectal visual structures in Old and New World monkeys and report that serotonergic innervation is more dense than noradrenergic innervation. The serotonergic innervation exhibits distinct, but limited, regional and nuclear specialization which is not so striking as the specialization of noradrenergic innervation.

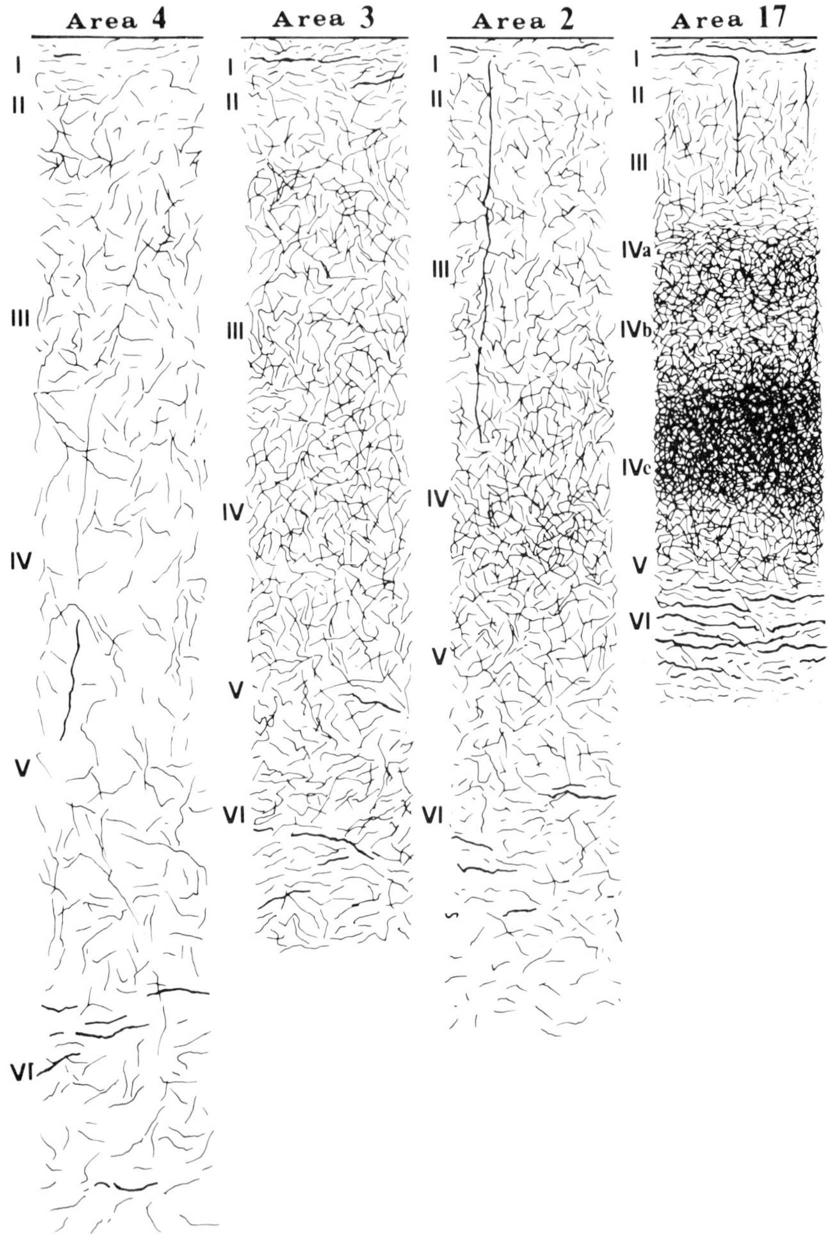

Fig. 11 Characteristic distribution of serotonin fibers in some typical areas of the primate neocortex: the primary motor cortex – area 4; the primary somatic sensory cortex – areas 3 and 2; the primary visual cortex – area 17. (Reproduced by permission of Springer-Verlag, Heidelberg, from Takeuchi and Sano, 1983b)

In a quantitative analysis, the overall numerical density of serotonin-immunoreactive varicosities in the monkey striate cortex is approximately 770,000/mm^3 and the highest concentration of immunoreactive varicosities (*ca.* 1,400,000/mm^3) is observed in the upper portion of layer IVc, the next highest concentration being in layer IVb (*ca.* 1,180,000/mm^3) (Takeuchi and Sano, 1984).

Diencephalon

In the thalamus of the rat, the nucleus periventricularis thalami, nucleus medialis thalami (pars medialis), and the nucleus geniculatus lateralis ventralis receive a very high density of serotonergic innervation (Steinbusch, 1981). In addition, after pretreatment with tryptophan and MAOI, the density of innervation is also greatest in the nucleus rhomboideus and nucleus reuniens (Cropper *et al.*, 1984b).

As for the serotonergic innervation in the nucleus geniculatus lateralis, comparative data are available. In mammals, the pattern of serotonin innervation is denser in the ventral than in the dorsal area. In the rat, serotonin fibers are evenly distributed in the form of a dense network, but serotonin fibers are most numerous in the C complex and the medial interlaminal nucleus of the cat, and in the S layer and interlaminal zone of the monkey (Ueda and Sano, 1986). In mammals, serotonin fibers are especially dense in the target area of W-cells of retinal ganglion cells, thereby suggesting the possibility of an interaction between the geniculate visual system mediated by W-cells and the serotonergic neuron system (Ueda and Sano, 1986). Thus it is noteworthy that serotonergic axonal varicosities are characteristically concentrated within certain relay centers of the visual system, across phylogeny.

In the subthalamic nucleus, fine varicose fibers are present and species differences are also noted. The overall density of serotonin fibers is highest in the monkey, and lowest in the cat. In the rat and cat, serotonin fibers are diffuse in the subthalamic nucleus, whereas in the monkey they are particularly abundant in the ventral and medial portions of the subthalamic nucleus (Mori *et al.*, 1985a).

Immunohistochemical studies on the serotonergic innervation of the hypothalamus of various mammals have been reported (Steinbusch and Nieuwenhuys, 1981; Westlund and Childs 1982; Sano *et al.*, 1982b; Jennes *et al.*, 1982; Ueda *et al.*, 1983a, b; Sawchenko *et al.*, 1983; Leranth *et al.*, 1983; Simerly *et al.*, 1984; Kawata *et al.*, 1984; Warembourg and Poulain, 1985). The highest concentration of serotonin fibers is found in the nucleus ventromedialis hypothalami, and the nucleus mamillaris. In the nucleus suprachiasmaticus, serotonin fibers are dense in the rat and cat (Fig. 12), yet this area in the monkey is practically devoid of such fibers. The nucleus hypothalamic

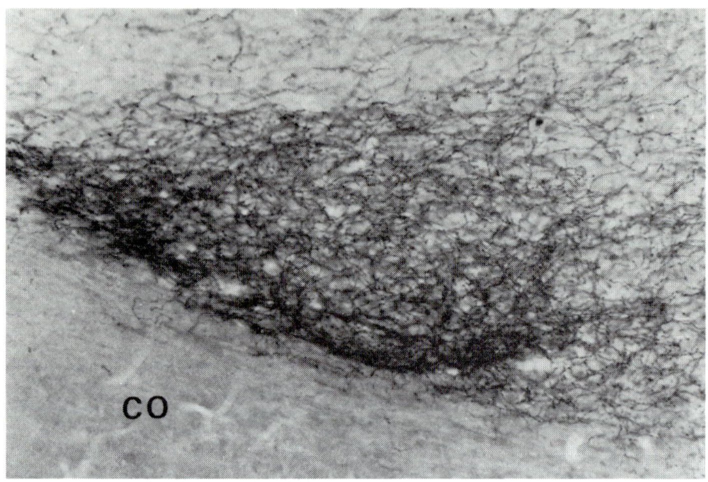

Fig. 12 The ventral portion of the nucleus suprachiasmaticus shows a very high density of serotonin fibers. CO: chiasma opticum. Sagittal section. Rat, ×170

anterior has thin varicose fibers, but the density differs with the species – the highest density is seen in rats, then monkeys, while cats have the lowest density (Ueda *et al.*, 1983c). A moderate supply of serotonin fibers is present in the ventral portion of the nucleus mamillaris medialis of the rat and other mamillary nuclei, such as nuclei premamillaris, supramamillaris and mamillaris lateralis, which contain few varicose fibers. In the monkey, Kawata *et al.* (1984) obtained evidence for an abundant distribution of serotonin fibers throughout the complex of the mamillary nucleus, in particular, the dorsal portion of the nucleus mamillaris medialis. However, in the cat and dog the mamillary nuclei, including the nucleus mamillaris medialis, nucleus mamillaris lateralis and nucleus supramamillaris, display a low density of serotonin fibers (Ueda *et al.*, 1983b).

It is of interest that the density of serotonin fibers in the paraventricular nucleus and the supraoptic nucleus of the rat is low relative to that in the immediately surrounding neuropils, in striking contrast to noradrenergic input into these nuclei (Sawchenko *et al.*, 1983).

The medial preoptic nucleus is a sexually dimorphic complex composed of three architecturally distinct subdivisions, each of which contains a character-istic density of serotonin fibers (Simerly *et al.*, 1984). These observations indicate that serotonin plays an important role in the control of two sexually dimorphic reproductive functions – gonadotrophin release and mating behavior. Apart from the medial preoptic nucleus, sexual dimorphism in the distribution of serotonin fibers has heretofore not been detected in the mammalian brain, whereas it was clarified, in detail, in the anterior column of spinal cord (Kojima *et al.*, 1983).

A dense plexus of serotonin fibers occurs in the ventrolateral portion of the median eminence of the rat (Jennes *et al.*, 1982). Sano *et al.* (1982b) demonstrated the presence of serotonin fibers in the internal and external layers of the cat infundibulum as well as in the intermediate and posterior lobes of the hypophysis. These same findings were noted in the monkey (Kawata *et al.*, 1984). In rats, the neural lobe exhibits a large number of serotonin fibers with fine varicose fibers observed among cells of the pars intermedia (Saland *et al.*, 1986). The presence of serotonin fibers in the rat anterior lobe has also been described (Westlund and Childs, 1982; Leranth *et al.*, 1983).

Mesencephalon

In the tectum of the mesencephalon, serotonin fibers distribute throughout the superior colliculus and these fibers have a characteristic arrangement corresponding to the laminar structures of this region. The stratum griseum superficiale of the rat, hamster, cat, and monkey receive the most extensive serotonergic innervation, though in the monkey the innervation is less conspicuous. In the chipmunk, serotonin fibers are denser in the stratum griseum intermedium than in the stratum griseum superficiale (Ueda *et al.*, 1985).

In the tegmentum of the mesencephalon, the substantia nigra and the dorsal portion of the nucleus interpeduncularis receive a very high density of serotonergic innervation. In these areas intervaricose segments of the serotonin-immunoreactive fibers can be rarely identified.

An analysis of serotonin-immunoreactive fibers among the midbrain periaqueductal gray of the rodent revealed the lowest overall volume fraction in the dorsal region and the highest overall volume fraction in the ventromedial region, which constitutes 2.6% of the neuropil volume (Clements *et al.*, 1985).

Rhombencephalon

Except for the extraocular muscle nuclei, the motor nuclei of the cranial nerves (nucleus motorius n. trigemini, nucleus n. facialis, nucleus ambiguus, nucleus n. accessorii, and nucleus n. hypoglossi) receive rich inputs from serotonin neurons in the mouse, rat, guinea pig, dog, cat, and monkey. The motoneurons of the primate are innervated by varicose serotonin fibers in a manner different from that of other species, i.e. their cell bodies and proximal dendrites are tightly encircled by a large number of serotonin fibers (Fig. 13) (Takeuchi *et al.*, 1983b). Such a species-related difference in serotonergic innervation was also noted in anterior horn motoneurons (Kojima and Sano,

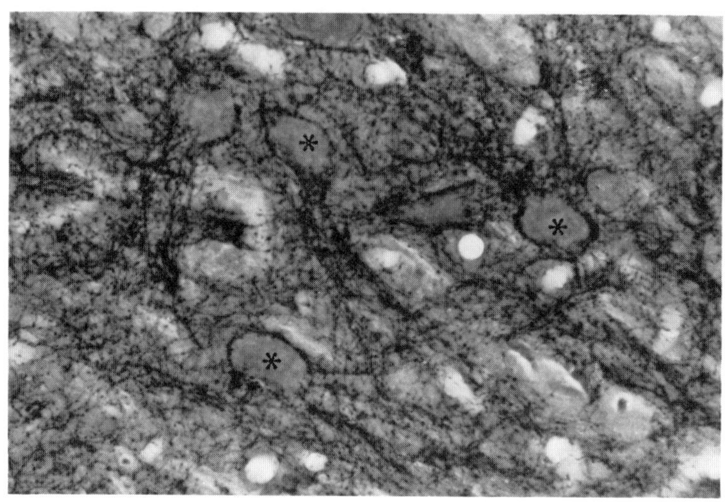

Fig. 13 The majority of serotonin-immunoreactive structures are tightly apposed to the cell bodies (asterisk) and proximal dendrites of the nucleus nervi facialis. *Macaca fuscata*, ×230

1983). Bloom (1981) reported a marked contrast between the intense nor-adrenergic innervation of the somatic efferent cranial nerve nuclei, with sparse to minimal serotonergic innervation, and visceral efferent cranial nerve nuclei, with more intense serotonin fibers and far less marked noradrenergic fibers. Cropper *et al.*, (1984a) noted that the trigeminal motor nucleus of the rat contained as many serotonin fibers as did the subnucleus caudalis, and that these fibers are thicker and varicosities more irregularly spaced than in the subnucleus caudalis, thereby suggesting a difference in serotonergic effects on the motor and sensory nuclei. In addition, in the sensory nuclei of the trigeminal nuclear complex, serotonergic innervation appears to be dense in areas primarily related to nociceptive afferent activity (Cropper *et al.*, 1984a).

In the rat rhombencephalon, a very high density of serotonin fibers was observed in the lateral portion of the reticular formation, such as the dorso-medial part of the nucleus reticularis pontis oralis.

The entire inferior olivary complex of mammals contains serotonin fibers, and species-related differences in the localization of serotonin fibers are significant (Fig. 14). In the rat, the overall density of serotonin fibers is sparser than in other species, cats and monkeys; the highest concentration of varicose serotonin fibers is observed in the lateral portion of the dorsal accessory olive (Takeuchi and Sano, 1983a; Bishop and Ho, 1984). In the opossum, the densest serotonin-immunoreactive elements are present in subnucleus b of the caudal medial accessory nucleus, and the rostral principal

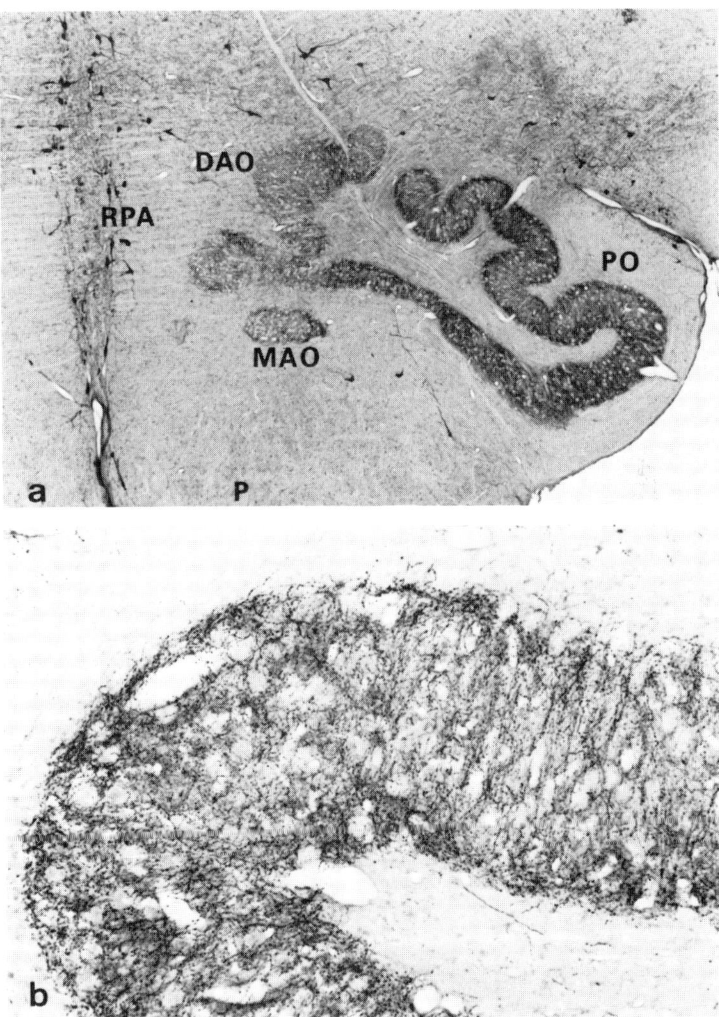

Fig. 14(a) The entire inferior olivary complex is supplied with serotonin fibers establishing a distinct contrast to the adjacent neuronal structure. DAO: dorsal accessory olivary nucleus, MAO: medial accessory olivary nucleus, P: pyramidal tract, PO: principal olivary nucleus, RPA: nucleus raphe pallidus. *Macaca fuscata*, ×21. (b) The principal olivary nucleus displaying high density of serotonin fibers. *Macaca fuscata*, ×136

olive is sparsely populated with serotonin fibers (King *et al.*, 1984). In the cat, the highest concentration of serotonin fibers occurs in the caudal portion of the medial accessory olive, the dorsomedial cell column and the lateral portion of the dorsal accessory olive. In the monkey, the caudal medial accessory olive, the lateral portion of the dorsal accessory olive and the dorsal as well as the lateral lamella of the principal olive show a maximum density of serotonin fibers, thereby suggesting that serotonin plays a more extensive functional role throughout the inferior olivary complex of the primate (Takeuchi and Sano, 1983a).

Biochemical determinations revealed that both the cerebellar cortex and nuclei contain the least amount of serotonin, among various areas in the central nervous system, thus only immunohistochemical procedures will yield a precise evaluation of serotonergic innervation in the cerebellum (Fig. 15). The pattern of innervation of the cerebellum differs among the cortical layers and also among the animal species (Fig. 16). In the rat, the tangential serotonin fibers are predominant in the molecular layer and a few vertical and oblique fibers are present, while in the granular layer, relatively short and oblique fibers are present. No serotonin-immunoreactive mossy fiber rosettes are found. The pattern of serotonergic innervation differs somewhat within the different lobules, despite a uniform structure of the cerebellar cortex (Takeuchi *et al.*, 1982d). These findings in the rat cerebellum were

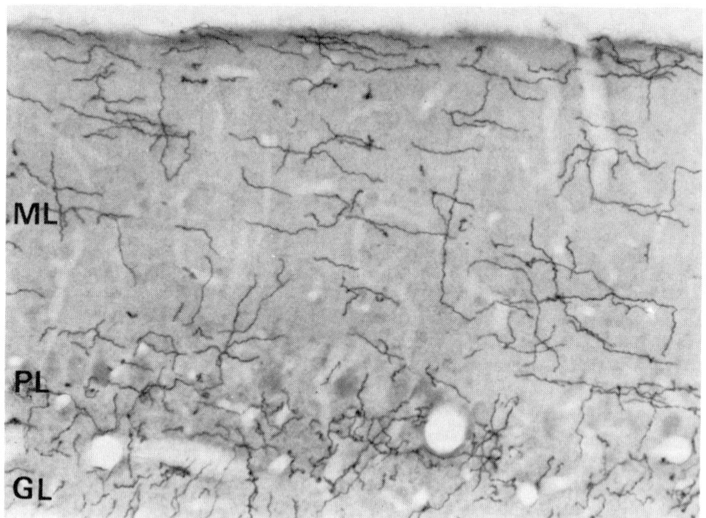

Fig. 15 Vermal zone of the cerebellum. Serotonin fibers are predominant in the molecular layer (ML) and Purkinje cell layer (PL). Tangential serotonin fibers are numerous in the molecular layer. GL: granular layer. Guinea pig, ×170

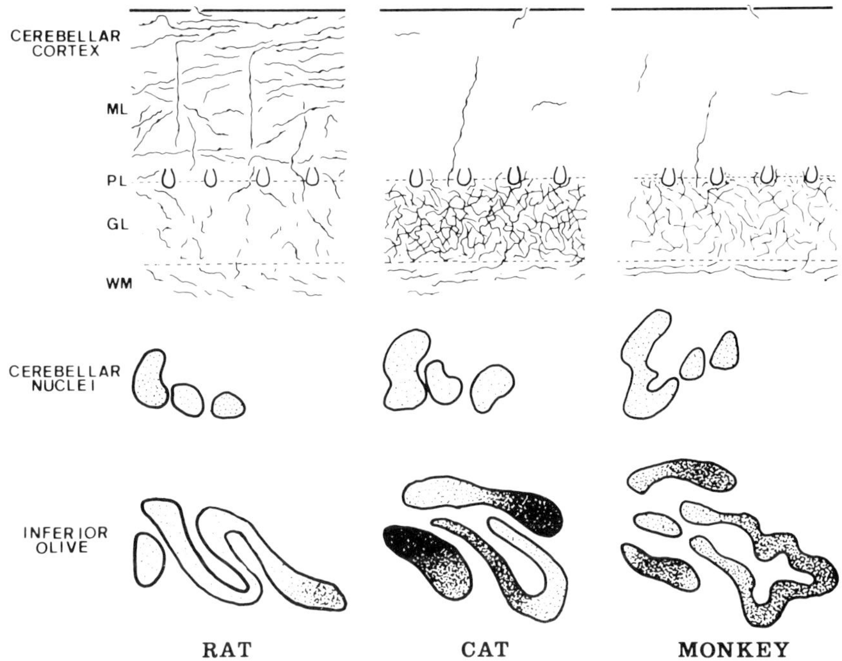

Fig. 16 Schematic drawing showing characteristic distribution of serotonin fibers in the cerebellar cortex, nuclei and inferior olivary complex of three species

confirmed and extended by Bishop and Ho (1985). On the other hand, in the case of the cat, only a few serotonin fibers were found in the molecular layer of the cerebellar cortex, and dense networks of serotonin fibers were present in the granular layer (Takeuchi *et al.*, 1982d). Species differences were also noted in the opossum cerebellum – the densest distribution of serotonin occurs in the Purkinje cell–granular cell border and little serotonin-immunoreactivity is present in the molecular layer (Bishop *et al.*, 1985). These findings suggest that in the cat and opossum, neuronal elements in the Purkinje cell and granular cell layers are likely targets for serotonergic afferents, while in the rat, the potential exists for more generalized effects throughout the cerebellum.

The cerebellar nuclei receive a relatively rich serotonergic innervation and there is a differential distribution of serotonin within the nuclei (Takeuchi *et al.*, 1982d; Bishop and Ho, 1985; Bishop *et al.*, 1985).

Circumventricular organs and supraependymal plexus

Most of the circumventricular organs are considered to participate in neuro-humoral regulations; thus immunohistochemical studies on serotonin distribution in these organs have been numerous (Jennes *et al.*, 1982; Tramu *et al.*, 1983; Takeuchi and Sano, 1983c; Matsuura *et al.*, 1985). Various densities of serotonin fibers are found in the organum vasculosum laminae terminalis, subfornical organ, subcommissural organ and area postrema, but serotonin-containing cell bodies are evident in the area postrema only after MAOI treatment. Serotonergic supraependymal fibers are present on the surface of the subfornical organ, but not on the subcommissural organ and area postrema. The dense serotonergic plexus of the basal portion of the subcommissural organ appears to be continuous with the serotonergic supraependymal plexus (Takeuchi and Sano, 1983c). In the pineal gland, serotonin-immunoreactive parenchymal cells and nerve fibers are present and some fibers are considered to be of central nervous system origin (Matsuura and Sano, 1983; Matsuura *et al.*, 1983). In the monkey, serotonin fibers are clearly restricted to the ventro-proximal portion of the pineal organ (Matsuura and Sano, 1986).

Recent immunohistochemical techniques also confirmed and extended the findings of serotonergic supraependymal plexus obtained by FIF histochemistry and autoradiography (Steinbusch, 1981; Lorez and Richards, 1982; Sano, 1983; Takeuchi and Sano, 1983a; Tramu *et al.*, 1983; Matsuura *et al.*, 1985). The serotonergic supraependymal plexus is present in all regions of the ventricular systems, except on the third ventricle floor and in the choroid plexus (Figs 17, 18). The density of the serotonergic supraependymal plexus is higher in small species, such as mice and guinea pigs (Matsuura *et al.*, 1985). Concerning the pattern of serotonin distribution in the supraependymal plexus, species-specific differences are not so distinct and the highest density of serotonergic supraependymal fibers is observed on the surface of the caudate nucleus in the lateral ventricle in the mouse, rat, guinea pig, rabbit, cat, dog, and monkey (Matsuura *et al.*, 1985). Lorez and Richards (1982) summarized regional differences in the density of the rat serotonergic supraependymal plexus as follows: serotonergic supraependymal plexus is sparse around the floor of the third ventricle, on the area preoptica, and the roof of the foramen interventriculare. In contrast, very high densities occur mainly on the floor of the interventricular foramen, nucleus medialis habenula, recessus pinealis, most regions of the fourth ventricle floor and on all linings of the Luschka's foramen. Matsuura *et al.* (1985) verified these findings and noted a few serotonergic supraependymal plexus in the medial part of the fourth ventricle roof and in the central canal of the medulla oblongata and spinal cord, which had not been hitherto described. Steinbusch (1981, 1984) reported the presence of supraependymal serotonergic neurons in the wall of the fourth ventricle and the cerebral aqueduct of the rat brain, but this finding was challenged by Lorez and Richards (1982), Sano (1983), and Matsuura *et al.* (1985). Most recent combined radioautographic and immunohistochemical studies showed that the majority of the supraependymal fibers are reactive to GABA and serotonin antisera (Harandi *et al.*, 1986).

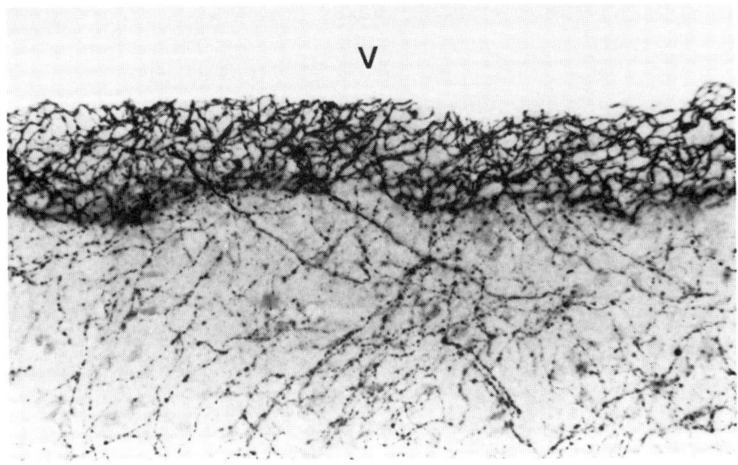

Fig. 17 Serotonergic supraependymal plexus of the lateral ventricle (V). Rat, ×255

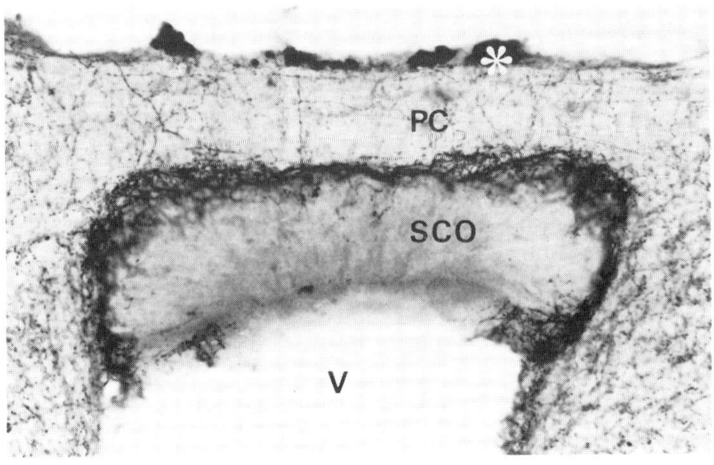

Fig. 18 Markedly dense plexus of serotonin fibers is observed in the basal portion of the subcommissural organ (SCO). Above the posterior commissure (PC), serotonin-immunoreactive cells of the lamina intercalalis are scattered (asterisk). V: third ventricle. Rat, ×187

Apart from the supraependymal plexus, serotonergic innervation of the pia mater was noted in various areas on the brain surface, such as the medulla oblongata (Sano *et al.*, 1982c), cerebellum (Takeuchi *et al.*, 1982d), cerebral cortex (Takeuchi and Sano, 1983b), and human cerebral blood vessels (Griffith and Burnstock, 1983). In the rat, the existence of serotonin neurons and fibers in a superficial position just beneath the external glial limiting membrane of the medulla oblongata was confirmed, using autoradiography and immunohistochemistry (Gorcs *et al.*, 1985).

AXONAL RAMIFICATION

On the basis of immunohistochemical investigations on the serotonin neuron system in the central nervous system of various vertebrates, Sano *et al.* (1982a) proposed a new concept concerning the morphology of serotonin neurons, especially their axonal ramification. Serotonin-immunoreactive fibers frequently branch off and anastomose to form a dense and extensive three-dimensional network or reticulum that is distributed throughout the brain. This reticulum is apparently different from the so-called 'telodendron' which is observed in the terminal portions of the usual neurons. As seen by light microscopy, serotonergic varicose fibers anastomose to form ring-like structures of various sizes. On the surface of the ependymal layer of the mammalian ventricular walls serotonin fibers form a dense network, the so-called supraependymal plexus. These fibers penetrate the ependymal layer at various points of the ventricular walls and connect with the plexus in the ventricular lumen (Matsuura *et al.*, 1985). The existence of a true anastomosis, network or circular meshwork, composed of serotonergic supraependymal fibers is elucidated by scanning electronmicroscopy (Matsuura *et al.*, 1985).

Immunohistochemical techniques were also used to study the serotonergic raphe nucleus of the newborn rat and mouse *in vitro*, and cultures of these tissues demonstrated a dense serotonin-immunoreactive plexus (Figs 19, 20), indicating the formation of an interneuronal syncytium of serotonin neurons (Takeuchi *et al.*, 1983c). These findings strongly suggest that Golgi's reticular theory has to be re-evaluated and a new concept of the morphology of neurons defined.

FUTURE TRENDS

As described above, serotonergic neurons in the brainstem are provided with highly branched fibers which innervate virtually the entire central nervous system, comprising the most expansive central neuronal network yet described. This is a most significant character of serotonin, as compared to findings with other putative neurotransmitters. To clarify the functional

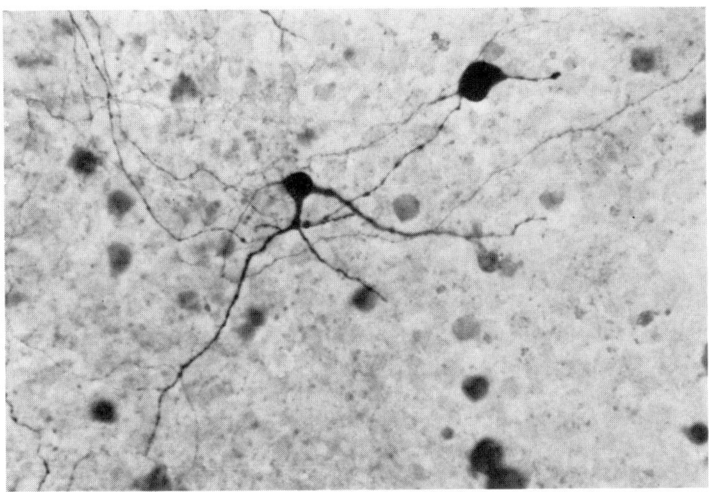

Fig. 19 Serotonin-immunoreactive neurons in seven-day-old culture
of mouse brainstem. ×255

significance of serotonin neurons morphologically, the following points have
to be considered:

(a) The exact distribution of the serotonin neuron system, under physio-
    logical conditions, has to be determined in various species, particularly
    in the primate.

(b) The correlations between serotonin and other putative neurotransmitters
    have to be evidenced at the ultrastructural level, with greater precision
    and a higher specificity.

(c) In order to compare the data obtained by biochemical or pharmacological
    methods, morphological approaches should be used to investigate change
    in the serotonin neuron system under the experimental conditions, for
    instance malnutrition (Howe *et al.*, 1983b; Ishimura, 1985b), intracer-
    ebral injection of quinolinic acid (Aldinio *et al.*, 1985), etc.

(d) To describe serotonin localization more precisely and objectively, and
    to examine morphological changes occurring in serotonin neurons under
    different experimental or pathological states, quantitative studies of sero-
    tonin-immunoreactive elements have to be made. Various investigators
    have described serotonergic neuronal somata or varicosities in a quanti-
    tative fashion (serotonergic neuronal somata: Agnati *et al.*, 1982a,b,
    1984; Daszuta and Portalier, 1985; Marson and Loewy, 1985;
    serotonergic varicose fibers: Takeuchi and Sano, 1984; Ueda *et al.*, 1985;
    Clements *et al.*, 1985).

(e) The dynamics of serotonin neuron system are at present being studied

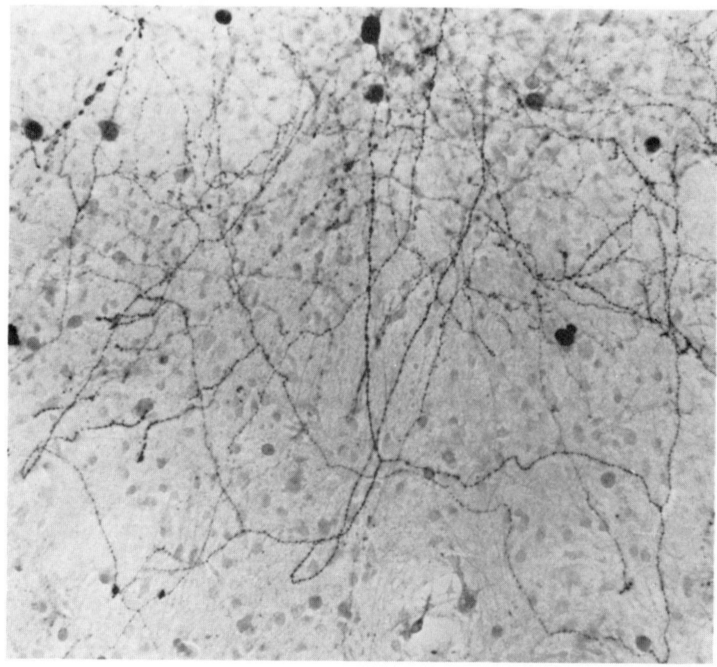

Fig. 20 Anastomoses of serotonergic axons originating from different neurons. Ten-day-old culture of mouse brainstem. ×170. (Reproduced by permission of T. Fujita from Takeuchi *et al.*, 1983c)

using *in situ* hybridization techniques and these findings will add new dimensions to the understanding of serotonin-related neurology.

REFERENCES

Agnati, L. F., Fuxe, K., Hökfelt, T., Benfenati, F., Calza, L., Johansson, O., and De May, J. (1982a) Morphometric characterization of transmitter-identified nerve cell groups: Analysis of mesencephalic 5-HT nerve cell bodies, *Brain Res. Bull.*, **9**, 45–51.

Agnati, L. F., Fuxe, K., Zini, I., Benfenati, F., Hökfelt, T., and De May, J. (1982b) Principles for the morphological characterization of transmitter-identified nerve cell groups, *J. Neurosci. Meth.*, **6**, 157–167.

Agnati, L. F., Fuxe, K., Calza, L., Zini, I., Hökfelt, T., Steinbusch, A., and Verhofstad, A. (1984) A method for rostrocaudal integration of morphometric information from transmitter-identified cell groups. A morphometrical identification and description of 5-HT cell groups in the medulla oblongata of the rat, *J. Neurosci. Meth.*, **10**, 83–101.

Aldinio, C., Mazzari, S., Toffano, G., Köhler, C., and Schwarcz, R. (1985) Effects of intracerebral injections of quinolinic acid on serotonergic neurons in the rat brain, *Brain Res.*, **341**, 57–65.

Arezki, F., Bosler, O., and Steinbusch, H. W. M. (1985) Morphological evidence that serotonin-immunoreactive neurons in the nucleus dorsomedialis hypothalami could be under catecholaminergic influence, *Neurosci. Lett.*, **56**, 161–166.

Azmitia, E., and Gannon, P. (1983) The ultrastructural localization of serotonin immunoreactivity in myelinated and unmyelinated axons within the medical forebrain bundle of rat and monkey, *J. Neurosci.*, **3**, 2083–2090.

Bishop, G. A., and Ho, R. H. (1984) Substance P and serotonin immunoreactivity in the rat inferior olive, *Brain Res. Bull.*, **12**, 105–113.

Bishop, G. A., and Ho, R. H. (1985) The distribution and origin of serotonin immunoreactivity in the rat cerebellum, *Brain Res.*, **331**, 195–207.

Bishop, G. A., Ho, R. H., and King, J. S. (1985) Localization of serotonin immunoreactivity in the opossum cerebellum, *J. Comp. Neurol.*, **235**, 301–321.

Bloom, F. E. (1981) A comparison of serotonergic and noradrenergic neuro-transmission in mammalian CNS, in *Serotonin Neurotransmission and Behaviour* (Eds B. L. Jacobs and A. Gelperia), pp. 403–424, The MIT Press, Cambridge, Mass.

Calza, L., Giardino, L., Grimaldi, R., Rigoli, M., Steinbusch, H. W. M., and Tiengo, M. (1985) Presence of 5-HT-positive neurons in the medial nuclei of the solitary tract, *Brain Res.*, **347**, 135–139.

Clements, J. R., Beitz, A. J., Fletcher, T. F., and Mullett, M. A. (1985) Immunocytochemical localization of serotonin in the rat periaqueductal gray: A quantitative light and electron microscopic study, *J. Comp. Neurol*, **236**, 60–70.

Consolazione, A., and Cuello, A. C. (1982) CNS serotonin pathways, in *Biology of Serotonergic Transmission* (Ed. N. N. Osborne), pp. 29–61, Wiley, London.

Cropper, E. C., Eisenman, J. S., and Azmitia, E. C. (1984a) 5-HT-immunoreactive fibers in the trigeminal nuclear complex of the rat, *Exp. Brain Res.*, **55**, 515–522.

Cropper, E. C., Eisenman, J. S., and Azmitia, E. C. (1984b) An immunocytochemical study of the serotonergic innervation of the thalamus of the rat, *J. Comp. Neurol.*, **224**, 38–50.

Dahlström, A., and Fuxe, K. (1964) Evidence for the existence of monoamine-containing neurons in the central nervous system. I. Demonstration of monoamines in cell bodies of brain neurons, *Acta Physiol. Scand.*, **62** (Suppl. 232), 1–55.

Daszuta, A., and Portalier, P. (1985) Distribution and quantification of 5-HT nerve cell bodies in the nucleus raphe dorsalis area of C57BL and BALBc Mice. Relationship between anatomy and biochemistry, *Brain Res.*, **360**, 58–64.

Frankfurt, M., Lauder, J. M., and Azmitia, E. C. (1981) The immunohistochemical localization of serotonergic neurons in the rat hypothalamus, *Neurosci. Lett.*, **24**, 227–232.

Gall, C., and Moore, R. Y. (1984) Distribution of enkephalin, substance P, tyrosine hydroxylase, and 5-hydroxytryptamine immunoreactivity in the septal region of the rat, *J. Comp. Neurol.*, **225**, 212–227.

Gorcs, T. J., Liposits, Z., Palay, S. L., and Chan-Palay, V. (1985) Serotonin neurons on the ventral brain surface, *Proc. Natl. Acad. Sci.*, **82**, 7449–7452.

Griffith, S. G., and Burnstock, G. (1983) Immunohistochemical demonstration of serotonin in nerves supplying human cerebral and mesenteric blood-vessels – Some speculations about their involvement in vascular disorders, *Lancet*, **8324**, 561–562.

Harandi, M., Didier, M., Aguera, M., Calas, A., and Belin, M. F. (1986) GABA and serotonin (5-HT) pattern in the supraependymal fibers of the rat epithalamus: Combined radioautographic and immunocytochemical studies. Effect of 5-HT content on [³H]GABA accumulation, *Brain Res.*, **370**, 241–249.

Howe, P. R. C., Moon, E., and Dampney, R. A. L. (1983) Distribution of serotonin nerve cells in the rabbit brainstem, *Neuroscience Lett.*, **38**, 125–130.

Howe, P. R. C., Rogers, P. F., King, R. A., and Smith, R. M. (1983b) A biochemical and immunohistochemical study of central serotonin nerves in rats with chronic thiamine deficiency, *Brain Res.*, **270**, 19–28.

Ishimura, K. (1985a) Immunohistochemical study on the serotonin neuron system in the developing brain. I. A. quantitative analysis on the distribution of serotonin-containing cell bodies in the mouse brain, *J. Kyoto Pref. Univ. Med.*, **94**, 1301–1312 (in Japanese).

Ishimura, K. (1985b) Immunohistochemical study of the serotonin neuron system in the developing brain. II. A quantitative analysis on the effect of undernutrition, *J. Kyoto Pref. Univ. Med.*, **94**, 1313–1323 (in Japanese).

Jacobs, B. L., Gannon, P. J., and Azmitia, E. C. (1984) Atlas of serotonergic cell bodies in the cat brainstem: An immunocytochemical analysis, *Brain Res. Bull.*, **13**, 1–31.

Jennes, L., Beckman, W. C., Stumpf, W. E. and Grzanna R. (1982) Anatomical relationships of serotoninergic and noradrenergic projections with the GnRH system in septum and hypothalamus, *Exp. Brain Res.*, **46**, 331–338.

Kapadia, S. E., De Lanerolle, N. C., and La Motte, C. C. (1985) Immunocytochemical and electron microscopic study of serotonin neuronal organization in the dorsal raphe nucleus of the monkey, *Neuroscience*, **15**, 729–746.

Kawata, M., Takeuchi, Y., Ueda, S., Matsuura, T., and Sano, Y. (1984) Immunohistochemical demonstration of serotonin-containing nerve fibers in the hypothalamus of the monkey, *Macaca fuscata*, *Cell Tissue Res.*, **236**, 495–503.

King, J. S., Ho, R. H., and Burry, R. W. (1984) The distribution and synaptic organization of serotoninergic elements in the inferior olivary complex of the opossum, *J. Comp. Neurol.*, **227**, 357–368.

Köhler, C., Chan-Palay, V., and Steinbusch, H. (1981) The distribution and orientation of serotonin fibers in the entorhinal and other retrohippocampal areas. An immunohistochemical study with anti-serotonin antibodies in the rat's brain, *Anat. Embryol.*, **161**, 237–264.

Köhler, C., Chan-Palay, V., and Steinbusch, H. (1982) The distribution and origin of serotonin-containing fibers in the septal area: A combined immunohistochemical and fluorescent retrograde tracing study in the rat, *J. Comp. Neurol.*, **209**, 91–111.

Kojima, M., and Sano, Y. (1983) The organization of serotonin fibers in the anterior column of the mammalian spinal cord – an immunohistochemical study, *Anat. Embryol.*, **167**, 1–11.

Kojima, M., Takeuchi, Y., Goto, M., and Sano, Y. (1983) Immunohistochemical study on the distribution of serotonin-containing cell bodies in the brain stem of the dog, *Acta Anat.*, **115**, 8–22.

Lanca, A. J., and Van Der Kooy, D. (1985) A serotonin-containing pathway from the area postrema to the parabrachial nucleus in the rat, *Neuroscience*, **14**, 1117–1126.

Leranth, C. S., Palkovits, M., and Krieger, D. T. (1983). Serotonin immunoreactive nerve fibers and terminals in the rat pituitary – Light- and electron-microscopic studies, *Neuroscience*, **9**, 289–296.

Lidov, H. G. W., Grzanna, R., and Molliver, M. E. (1980) The serotonin innervation of the cerebral cortex in the rat – An immunohistochemical analysis, *Neuroscience*, **5**, 207–227.

Lidov, H. G. W., and Molliver, M. E. (1982) An immunohistochemical study of serotonin neuron development in the rat: Ascending pathways and terminal fields, *Brain Res. Bull.*, **8**, 389–430.

Lorez, H. P., and Richards, J. G. (1982) Supra-ependymal serotoninergic nerves in mammalian brain: Morphological and functional studies, *Brain Res. Bull.*, **9**, 727–741.

Marson, L., and Loewy, A. D. (1985) Topographic organization of substance P and monoamine cells in the ventral medulla of the cat, *J. Autonomic Nervous System*, **14**, 271–285.

Matsuura, T., and Sano, Y. (1983) Distribution of monoamine nerve fibers in the pineal organ of the dog. Fluorescence and immunohistochemical studies, *Cell Tissue Res.*, **234**, 519–531.

Matsuura, T., and Sano, Y. (1986) Characteristic pattern of monoaminergic nerve fibers in the pineal organ of the monkey, *Macaca fuscata*, *Cell Tissue Res.*, **245**, 453–456.

Matsuura, T., Takeuchi, Y., and Sano, Y. (1983) Immunohistochemical and electron microscopical studies on serotonin-containing nerve fibers in the pineal organ of the rat, *Biomed. Res.*, **4** (3), 261–270.

Matsuura, T., Takeuchi, Y., Kojima, M., Ueda, S., Yamada, H., Nojyo, Y., Ushizima, K., and Sano, Y. (1985) Immunohistochemical studies of the serotonergic supraependymal plexus in the mammalian ventricular system, with special reference to the characteristic reticular ramification, *Acta Anat.*, **123**, 207–219.

Mori, S., Takino, T., Yamada, H., and Sano, Y. (1985a) Immunohistochemical demonstration of serotonin nerve fibers in the subthalamic nucleus of the rat, cat, and monkey, *Neurosci. Lett.*, **62**, 305–309.

Mori, S., Ueda, S., Yamada, H., Takino, T., and Sano, Y. (1985b) Immunohisto-chemical demonstration of serotonin nerve fibers in the corpus striatum of the rat, cat and monkey, *Anat. Embryol.*, **173**, 1–5.

Morrison, J. H., Foote, S. L., Molliver, M. E., Bloom, F. E., and Lidov, H. G. (1982) Noradrenergic and serotonergic fibers innervate complementary layers in monkey primary visual cortex: An immunohistochemical study, *Proc. Nat. Acad. Sci. USA*, **79**, 2401–2405.

Morrison, J. H., and Foote, S. L. (1986) Noradrenergic and serotoninergic inner-vation of cortical, thalamic, and tectal visual structures in old and new world monkeys, *J. Comp. Neurol.*, **243**, 117–138.

Nieuwenhuys, R. (1985) *Chemoarchitecture of Brain*, Springer-Verlag, Berlin.

Nishida, K., Ueda, S., and Sano, Y. (1985) Immunohistochemical studies of masked indoleamine cells in the area postrema of the rat, *Histochemistry*, **82**, 101–106.

Pasik, T. and Pasik, P. (1982) Serotoninergic afferents in the monkey neostriatum, *Acta Biol. Acad. Sci. Hung.*, **33**, 277–288.

Ruda, M. A., Coffield, J. and Steinbusch, H. W. M. (1982) Immunocytochemical analysis of serotonergic axons in laminae I and II of the lumbar spinal cord of the cat, *J. Neurosci.*, **2**, 1660–1671.

Saavedra, J. P., Pasik, T., and Pasik, P. (1983) Immunocytochemistry of seroton-inergic neurons in the central nervous system of monkeys, in *Neural Transmission of Learning and Memory* (Eds. R. Captto and C. Ajmone Marsan), pp. 81–95, Raven Press, New York.

Saland, L. C., Wallace, J. A., and Comunas, F. (1986) Serotonin-immunoreactive nerve fibers of the rat pituitary: Effects of anticatecholamine and antiserotonin drugs on staining patterns, *Brain Res.*, **368**, 310–318.

Sano, Y. (1983) Morphological aspects of the serotonin neuron system in the central nervous system – An immunohistochemical approach, in *Gunma Symposia on*

Endocrinology 20, Progress and Prospects in Endocrinology, pp. 137–152, Center for Acad. Publ., Japan, Tokyo.

Sano, Y., Takeuchi, Y., Kimura, H., Goto, M., Kawata, M., Kojima, T., Matsuura, T., Ueda, S., and Yamada, Y. (1982a) Immunohistochemical studies on the processes of serotonin neurons and their ramification in the central nervous system, with regard to the possibility of the existence of Golgi's rete nervosa diffusa, *Arch. Histol. Jap.*, **45**, 305–316.

Sano, Y., Takeuchi, Y., Matsuura, T., Kawata, M., and Yamada, Y. (1982b) Immunohistochemical demonstration of serotonin nerve fibers in the cat neurohypophysis, *Histochemistry*, **75**, 293–299.

Sano, Y., Takeuchi, Y., Yamada, H., Ueda, S., and Goto, M. (1982c) Immunohistochemical studies on the serotonergic innervation of the pia mater, *Histochemistry*, **76**, 277–280.

Sawchenko, P. E., Swanson, L. W., Steinbusch, H. W. M., and Verhofstad, A. A. J. (1983) The distribution of cells of origin of serotonergic inputs to the paraventricular and supraoptic nuclei of the rat, *Brain Res.*, **277**, 355–360.

Simerly, R. B., Swanson, L. W., and Gorski, R. A. (1984) Demonstration of a sexual dimorphism in the distribution of serotonin-immunoreactive fibers in the medial preoptic nucleus of the rat, *J. Comp. Neurol.*, **225**, 151–166.

Steinbusch, H. W. M. (1981) Distribution of serotonin-immunoreactive in the central nervous system of the rat-cell bodies and terminals, *Neuroscience*, **6**, 557–618.

Steinbusch, H. W. M. (1984) Serotonin-immunoreactivity neurons and their projections in the CNS, in *Handbook of Chemical Neuroanatomy*, Vol. 3, *Classical Transmitters and Transmitter Receptors in the CNS*, Part II (Eds A. Björklund, T. Hökfelt and M. J. Kuhar), pp. 68–125, Elsevier, Amsterdam.

Steinbusch, H. W. M. and Nieuwenhuys, R. (1981) Localization of serotonin-like immunoreactivity in the central nervous system and pituitary of the rat, with special reference to the innervation of the hypothalamus, in *Serotonin; Current Aspects of Neurochemistry and Function. Adv. Exp. Med. Biol.*, Vol. 133 (Eds B. Haber, S. Gabay, M. R. Issidorides and S. G. A. Alivisatos), pp. 7–36, Plenum, New York.

Steinbusch, H. W. M., Verhofstad, A. A. J., and Joosten, H. W. J. (1978) Localization of serotonin in the central nervous system by immunohistochemistry: description of a specific and sensitive technique and some applications, *Neuroscience*, **3**, 811–819.

Steinbusch, H. W. M., Verhofstad, A. A. J., Joosten, H. W. J., and Goldstein, M. (1982) Serotonin-immunoreactive cell bodies in the nucleus dorsalis hypothalami, in the substantia nigra, and in the area tegmentalis ventralis of Tsai: observations after pharmacological manipulations in the rat, in *Cytochemical Methods in Neuroanatomy* (Eds S. Palay and V. Chan-Palay), pp. 407–421, Alan Liss, New York.

Takahashi, H., Nakashima, S., Takeda, S., Ohama, E., and Ikuta, F. (1984) Topography of the serotonin neurons in the brain stem of human fetus: An immunohistochemical study, *NoShinkei*, **36**, 697–708 (in Japanese).

Takeuchi, Y., and Sano, Y. (1983a) Immunohistochemical demonstration of serotonin-containing nerve fibers in the inferior olivary complex of the rat, cat and monkey, *Cell Tissue Res.*, **231**, 17–28.

Takeuchi, Y., and Sano, Y. (1983b) Immunohistochemical demonstration of serotonin nerve fibers in the neocortex of the monkey (*Macaca fuscata*), *Anat. Embryol.*, **166**, 155–168.

Takeuchi, Y., and Sano, Y. (1983c) Serotonin distribution in the circumventricular organs of the rat – an immunohistochemical study, *Anat. Embryol.*, **167**, 311–319.

Takeuchi, Y., and Sano, Y. (1983d) Immunohistochemical detection of serotonin in the central nervous system, in *Structure and Function of Peptidergic and Aminergic Neurons* (Eds Y. Sano, E. A. Zimmerman, and Y. Ibata), pp. 335–340, JSSP, Tokyo and VNU Sci. Press, Utrecht.

Takeuchi, Y., and Sano, Y. (1984) Serotonin nerve fibers in the primary visual cortex of the monkey – Quantitative and immunoelectron microscopical analysis, *Anat, Embryol.*, **169**, 1–8.

Takeuchi, Y., Kimura, H., Matsuura, T. and Sano, Y. (1982a) Immunohistochemical demonstration of the organization of serotonin neurons in the brain of the monkey (*Macaca fuscata*), *Acta Anat.*, **114**, 106–124.

Takeuchi, Y., Kimura, H. and Sano, Y. (1982b) Immunohistochemical demonstration of the distribution of serotonin neurons in the brainstem of the rat and cat, *Cell Tissue Res.*, **224**, 247–267.

Takeuchi, Y., Kimura, H. and Sano, Y. (1982c) Immunohistochemical demonstration of serotonin nerve fibers in the olfactory bulb of the rat, cat and monkey, *Histochemistry*, **75**, 461–471.

Takeuchi, Y., Kimura, H. and Sano, Y. (1982d) Immunohistochemical demonstration of serotonin-containing nerve fibers in the cerebellum, *Cell Tissue Res.*, **226**, 1–12.

Takeuchi, Y., Kimura, H., Matsuura, T., Yonezawa, T. and Sano, Y. (1983a) Distribution of serotonergic neurons in the central nervous system. A peroxidase-antiperoxidase study with anti-serotonin antibodies, *J. Histochem. Cytochem.*, **31**, 181–185.

Takeuchi, Y., Kojima, M., Matsuura, T., and Sano, Y. (1983b) Serotonergic innervation on the motoneurons in the mammalian brainstem. Light and electron microscopic immunohistochemistry, *Anat. Embryol.*, **167**, 321–333.

Takeuchi, Y., Matsuura, T., Yonezawa, T., and Sano, Y. (1983c) Immunohistochemical observation of serotonin neurons in newborn mouse and rat brainstem cultures, *Arch. Histol. Jap.*, **46**, 79–86.

Tramu, G., Pillez, A. and Leonardelli, J. (1983) Serotonin axons of the ependyma and circumventricular organs in the forebrain of the guinea pig. An immunohistochemical study, *Cell Tissue Res.*, **228**, 297–311.

Ueda, S., and Sano, Y. (1986) Distributional pattern of serotonin-immunoreactive nerve fibers in the lateral geniculate nucleus of the rat, cat and monkey (*Macaca fuscata*), *Cell Tissue Res.*, **243**, 249–253.

Ueda, S., Ihara, N., and Sano, Y. (1985) The organization of serotonin fibers in the mammalian superior colliculus. An immunohistochemical study, *Anat. Embryol.*, **173**, 13–21.

Ueda, S., Kawata, M. and Sano, Y. (1983a) Identification of serotonin and vasopressin immunoreactivities in the suprachiasmatic nucleus of four mammalian species, *Cell Tissue Res.*, **234**, 237–248.

Ueda, S., Kawata, M., Takeuchi, Y., and Sano, Y. (1983b) Immunohistochemical demonstration of serotonin nerve fibers in the hypothalamus of the cat, *Anat. Embryol.*, **168**, 315–330.

Ueda, S., Nishida, K., Nojyo, Y., Takeuchi, Y., and Sano, Y. (1983c) Serotonin containing neurons in the rat and cat brain, especially in the hypothalamus, following monoamine oxidase inhibitor pretreatment – an immunohistochemical study using anti-serotonin antiserum, *Arch. Histol. Jap.*, **47**, 405–410.

Wallace, J. A., Petrusz, P., and Lauder, J. M. (1982) Serotonin immunocytochemistry in the adult and developing rat brain: methodological and pharmacological considerations, *Brain Res. Bull.*, **9**, 117–129.

Warembourg, M., and Poulain, P. (1985) Localization of serotonin in the hypothalamus and the mesencephalon of the guinea-pig. An immunohistochemical study using monoclonal antibodies, *Cell Tissue Res.*, **240**, 711–721.

Westlund, K. N., and Childs, G. V. (1982) Localization of serotonin fibers in the rat hypothalamus, *Endocrinology*, **111**, 1761–1763.

Wirtshafter, D., Asin, K. E., and Lorens, S. A. (1986) Serotonin-immunoreactive projections to the hippocampus from the interpeduncular nucleus in the rat, *Neurosci. Lett.*, **64**, 259–262.

Yamada, H., and Sano, Y. (1985) Distribution of serotonin nerve cells in the rabbit brain – Immunohistochemistry by the two-step ABC technique using biotin-labeled rabbit serotonin-antibody, *Arch. Histol. Jap.*, **48**, 343–354.

Zhou, F. C., and Azmitia, E. C. (1986) Induced homotypic sprouting of serotonergic fibers in hippocampus. II. An immunocytochemistry study, *Brain Res.*, **373**, 337–348.

CHAPTER 3

Ultrastructural Relationships of Central Serotonin Neurons

JEAN-JACQUES SOGHOMONIAN[1], ALAIN BEAUDET[2] and LAURENT DESCARRIES[1]
Centre de Recherche en Sciences Neurologiques
Département de Physiologie
Faculté de médecine
Université de Montréal[1]
and
Departments of Neurology & Neurosurgery and of Anatomy
Montreal Neurological Institute
McGill University[2]
Montréal
Québec
Canada

INTRODUCTION

The appellation serotonin (5-hydroxytryptamine, 5-HT) neurons refers to those nerve cells that synthesize, take up and presumably utilize 5-HT as their (or one of their) neurotransmitter(s). The cell bodies of these neurons are mostly concentrated within the raphe nuclei of the brainstem and give rise to an extensive network of both ascending and descending projections, which arborize profusely but heterogeneously within the different regions of the neuraxis (Fuxe, 1965; Azmitia, 1978; Parent *et al.*, 1981; Steinbusch, 1981; for reviews, see Beaudet and Descarries, 1981; Consolazione and Cuello, 1982; Beaudet and Descarries, 1987).

Techniques such as fluorescence histochemistry, radioautography after uptake of [³H]5-HT and immunohistochemistry using antibodies against endogenous 5-HT (references in Beaudet and Descarries, 1987) have yielded precise information on the light microscopic distribution of central 5-HT nerve cell bodies and axon terminals, notably in the rat, cat and monkey (e.g. Parent *et al.*, 1981; Wiklund *et al.*, 1981b; Consolazione and Cuello, 1982; Takeuchi *et al.*, 1982a,b; Azmitia and Gannon, 1986; see also Takeuchi, Chapter 2 of this volume). The application of some of these techniques at the electron microscopic level has permitted the characterization of the fine structure and cellular relationships of 5-HT neurons within some of their nuclei of origin and territories of projection (for review, see Beaudet and Descarries, 1987). More recently, information pertaining to the afferent and/ or efferent connections of 5-HT neurons has been added as a result of the development of combined techniques for transmitter identification and/or tracing of neuronal pathways. Combined techniques have also allowed for the simultaneous cellular visualization of various coexistent chemical messengers in certain groups of 5-HT-containing nerve cell bodies or axon terminals.

The present chapter considers the cytological organization of central 5-HT neurons in various CNS regions, with emphasis on synaptology and relational aspects. For further information on the regional distribution and intrinsic ultrastructural features of these neurons, the reader is referred to earlier reviews by Steinbusch (1984) and Beaudet and Descarries (1987).

CELL BODIES AND DENDRITES

Distribution and transmitter identity

In the mammalian CNS, most 5-HT-containing nerve cell bodies are located in the brainstem. The distribution of these neurons was first demonstrated in the rat by Dahlström and Fuxe (1964) using the formaldehyde-induced fluorescence technique. A similar distribution pattern was later observed in the rat as well as in other species, including cat and monkey, by means of histofluorescence, radioautographic and immunohistochemical techniques

(Hubbard and Dicarlo, 1974; Jacobowitz and McLean, 1978; Poitras and Parent, 1978; Lackner, 1980; Parent *et al.*, 1981; Steinbusch, 1981; Wiklund *et al.*, 1981b; Takeuchi *et al.*, 1982a). In these later studies, 5-HT nerve cell bodies were shown to spread further from the midline, reaching into the locus coeruleus and subcoeruleus, for example. 5-HT neurons were also identified in the area postrema and the interpeduncular complex. A group of 5-HT-containing cell bodies was described in the pars ventralis of the nucleus dorsomedialis hypothalami, first by radioautography after [^{3}H]5-HT uptake (Beaudet and Descarries, 1979) and then with 5-HT immunohisto-chemistry after L-tryptophan loading and inhibition of monoamine oxidase activity (Frankfurt *et al.*, 1981; Steinbusch and Nieuwenhuys, 1983). Whether these hypothalamic neurons have the ability to synthesize 5-HT remains unclear, however, since a recent immunohistochemical investigation has failed to detect tryptophan-hydroxylase immunoreactive cell bodies within that part of the hypothalamus (Weissmann *et al.*, 1987). In contrast, trypt-ophan-hydroxylase immunoreactivity was found in a small group of nerve cell bodies occupying the tubero-mamillary region of the basolateral hypo-thalamus. Lastly, a small number of 5-HT-immunoreactive neurons intrinsic to the spinal cord were identified in rat and monkey, in animals pretreated or not with L-tryptophan and a monoamine oxidase inhibitor (Lamotte *et al.*, 1982; Bowker, 1986a; Newton *et al.*, 1986).

In all these areas including the raphe nuclei, 5-HT nerve cell bodies are found among numerous non-5-HT neurons. For instance, in rat nucleus raphe dorsalis (NRD), 5-HT cells account for approximately one-third of a total population of 36,000 neurons (Descarries *et al.*, 1982). Furthermore, there is an increasing body of evidence indicating the presence of other putative transmitters within at least some 5-HT neurons. In rat or cat, substance P, thyrotropin-releasing hormone (TRH), enkephalin or gamma-aminobutyric acid (GABA) have been shown to coexist with 5-HT in various raphe nuclei (Chan-Palay *et al.*, 1978; Hökfelt *et al.*, 1978; Glazer *et al.*, 1981; Johansson *et al.*, 1981; Cuello *et al.*, 1982; Hunt and Lovick, 1982; Belin *et al.*, 1983; Gamrani *et al.*, 1984; Léger *et al.*, 1986; Mâgoul *et al.*, 1986). There is even conclusive evidence for a triple co-localization of substance P, TRH and 5-HT in medullary neurons projecting to the spinal cord in the rat (Johansson *et al.*, 1981). The proportion of 5-HT neurons showing either one of these combinations of transmitters appears to vary between the various raphe nuclei. Approximately half of 5-HT immunoreactive neurons in nuclei raphe pallidus and obscurus of the cat were shown to exhibit enkephalin-like immu-noreactivity, whereas only few such double labeled cells were detected in the NRD (Léger *et al.*, 1986). Coexistence with GABA has been particularly well established in rat NRD which contains the greatest number of neurons exhibiting immunoreactivity against both 5-HT and glutamic acid decarboxy-lase (GAD) (Belin *et al.*, 1983). About one-third of the 5-HT-containing neurons in this nucleus appear to be GAD positive. The exact role of these different co-localizations in relation with 5-HT transmission is still largely unknown.

Axonal inputs

Most available data concerning the identity and cytological organization of terminals impinging upon 5-HT neurons are derived from ultrastructural studies in the rat or the cat. In both species, 5-HT perikarya and dendrites in NRD are studded with afferent axon terminals (Descarries *et al.*, 1982; Pecci-Saavedra *et al.*, 1982; Clements *et al.*, 1985; Kapadia *et al.*, 1985; Chazal and Ralston, 1987). In the rat, the mean number of incoming synaptic boutons on 5-HT neurons has been evaluated at 7 per 100 μm of somatic membrane and 11 per 100 μm of dendritic membrane (Descarries *et al.*, 1982). Using X-ray analytical electron microscopic detection of chromium-tagged, glutaraldehyde-condensed monoamines, Felten and Sladek (1983) showed that, in the monkey, only 35% of the 5-HT nerve cell bodies in NRD receive numerous axon terminals. Most 5-HT neurons have less than 5% of their membrane abutted by axon terminals and are mostly covered with astrocytic processes and oligodendroglial satellites. In the NRD of *Macaca fascicularis*, some 5-HT immunoreactive dendritic spines were found to receive characteristic crest synapses from pairs of unlabeled terminals (Kapadia *et al.*, 1985).

The ultrastructural heterogeneity of afferent axon terminals terminating on 5-HT nerve cell bodies and their dendrites has often been taken as an indication of the diversity of their afferent inputs, both in terms of anatomical origin and of transmitter identity (Descarries *et al.*, 1982; Clements *et al.*, 1985; Kapadia *et al.*, 1985; Chazal and Ralston, 1987). Afferents to NRD have been traced back to the nucleus of the solitary tract, the superior vestibular nucleus, the locus coeruleus complex, the parabrachial nuclei, the nucleus laterodorsalis tegmenti pontis, the griseum centralis ponti, the lateral habenula, the substantia nigra, several hypothalamic and preoptic areas, the prefrontal cortex and the hippocampus (Nauta, 1958; Aghajanian and Wang, 1977; Mosko *et al.*, 1977; Pasquier *et al.*, 1977; Sakaï *et al.*, 1977; 1978; Kawasaki and Sato, 1981; Kalén *et al.*, 1985). The NRD also receives projections from other raphe nuclei, namely from nuclei raphe medianus and magnus (Bobillier *et al.*, 1976). Some of these afferents could be 5-HT in nature, although 5-HT-containing terminals inside the NRD may also originate from recurrent collaterals issued by intrinsic 5-HT neurons (Mosko *et al.*, 1977; Kalén *et al.*, 1985). Irrespective of their origin, 5-HT axon terminals are scarce in NRD (Descarries *et al.*, 1982; Chazal and Ralston, 1987) and only rarely show synaptic specializations. In a detailed radioautographic study of rat NRD, only 2 out of 100 [³H]5-HT-labeled axonal profiles visualized in single sections were found to establish a synaptic junction (Descarries *et al.*, 1982). On the other hand, some synaptic contacts between [³H]5-HT-labeled axon terminals and tyrosine hydroxylase immunolabeled dendrites have been recently demonstrated in rat NRD (Fig. 1), indicating that dopamine neurons in this nucleus (Descarries *et al.*, 1986) may be directly influenced by 5-HT (Beaudet *et al.*, in preparation).

Only few of the known afferents to the NRD have been demonstrated to contact 5-HT neurons. 5-HT nerve cell bodies and dendrites have been both shown to receive a noradrenaline (NA) input by combining [³H]NA

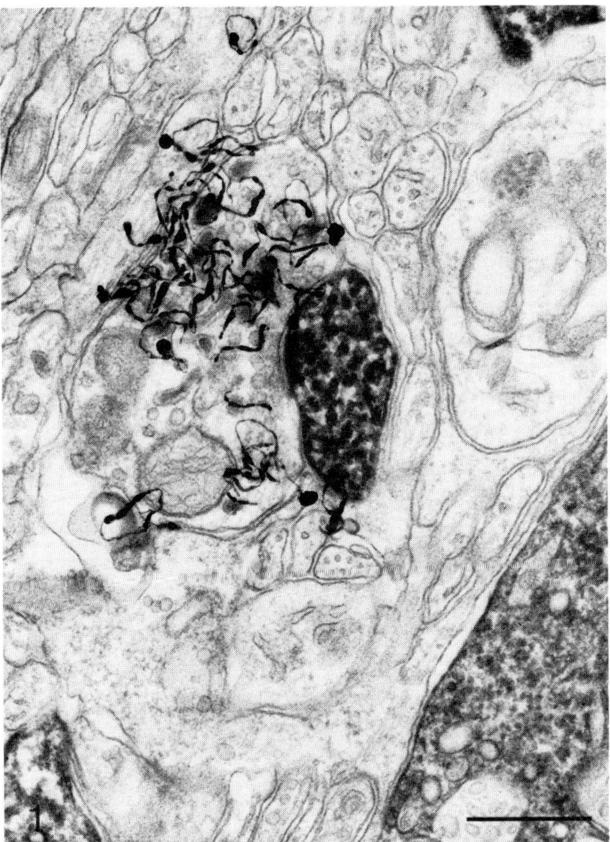

Fig. 1 5-HT axon terminal (varicosity) in synaptic contact with a dopamine dendrite in the nucleus raphe dorsalis of an adult rat. The 5-HT ending was radioautographically labeled with [³H]5-HT and the dopamine process immunostained with tyrosine hydroxylase antibody. Small, clear synaptic vesicles, a few larger dense-cored vesicles and two mitochondrial profiles fill the 5-HT terminal. The active zone is characterized by clustering of the clear synaptic vesicles and an asymmetrical junctional complex. Scale bar: 0.5 μm

radioautography with the ultrastructural identification of 5,7-dihydroxytrypt-amine lesioned 5-HT nerve cell bodies or dendrites (Baraban and Aghajanian, 1981). This NA input appears to be preferential in that more than half of the [^{3}H]NA-labeled endings were found to contact degenerating 5-HT elements. 5-HT terminals have not been observed on 5-HT nerve cell bodies or dendrites in rat NRD, but some would be present in the cat (G. Chazal, personal communication). Finally, substance P immunoreactive fibers have been visualized in the vicinity of 5-HT-containing somata by light-microscopic immunohistochemistry (Ljungdahl *et al.*, 1978). Biochemical evidence suggests that these substance P afferents originate from the habenula (Neckers *et al.*, 1979). Ultrastructural examination will be needed to determine if they directly contact the 5-HT neurons.

Dendro-dendritic and dendro-somatic relationships

In rat, cat and monkey, 5-HT-containing perikarya and dendrites are often directly apposed to other neuronal somata or dendrites. In most instances, these contacts are solely characterized by an interruption of the interneuronal glial sheath and juxtaposition of the plasma membranes (Descarries *et al.*, 1982; Kapadia *et al.*, 1985; Pecci-Saavedra *et al.*, 1986; Chazal and Ralston, 1987). In cat NRD, however, Chazal and Ralston (1987) have observed vesicular clusters near the apposed plasma membrane of 5-HT immunoreac-tive dendrites (Fig. 2). Even though synaptic membrane specializations are not always found at these points, the presence of aggregated vesicles suggests that 5-HT might be released to act upon adjacent dendrites. This mechanism might account for the fairly large amounts of 5-HT released *in vivo* within cat NRD (Héry *et al.*, 1982).

In the rat, most of the soma/dendrites that are directly apposed to the perikarya or dendrites of 5-HT neurons belong to non-5-HT neurons. In the rostral pole of rat NRD, some of these apposed cells have been identified as dopamine neurons, by combined radioautography and immunocytochemistry (Beaudet *et al.*, in preparation). These 5-HT/DA dendro-dendritic or dendro-somatic appositions were never seen with membrane differentiations suggestive of a synaptic junction; a few 5-HT dendrites did, however, exhibit small contingents of clustered synaptic vesicles, usually located at some distance from the zone of contact (Beaudet *et al.*, in preparation).

In the NRD of cat and monkey, differentiated contacts have been observed between adjacent 5-HT dendrites (Chazal and Ralston, 1987; Kapadia *et al.*, 1985). Dendritic bundling involving direct juxtaposition of 5-HT dendrites has also been reported in the rat, but these appositions never showed junc-tional specializations (Descarries *et al.*, 1982). The presence of synaptic contacts between 5-HT dendrites in the NRD of cat and monkey is particu-larly interesting in view of the scarcity of 5-HT axonal input onto these 5-HT cells (see above). These dendro-dendritic junctions could indeed serve

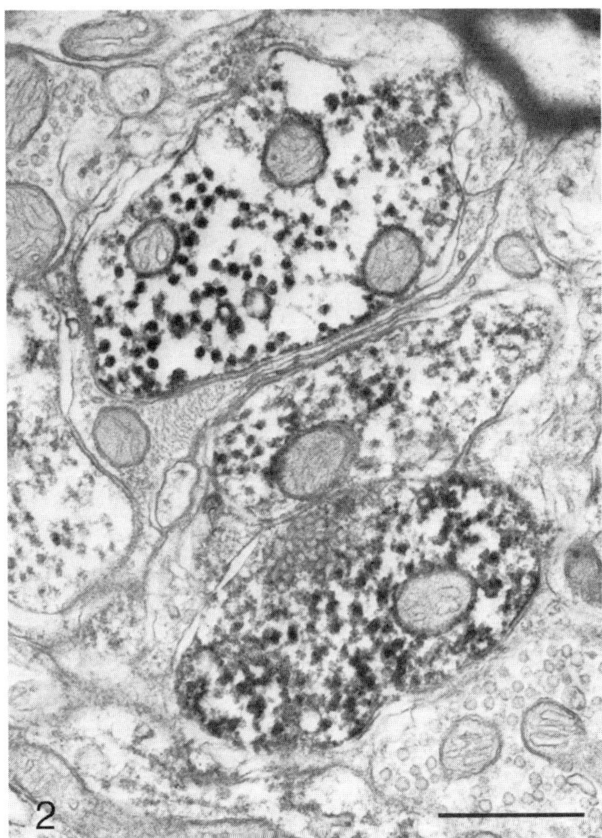

Fig. 2 Three adjacent 5-HT immunoreactive dendrites in the nucleus raphe dorsalis of a cat. The lower profile is juxtaposed to the middle one and exhibits a cluster of synaptic vesicles in the zone of apposition. Serial sections across this dendro-dendritic contact failed to show any membranous differentiation. A glial leaflet separates the upper dendrite from the middle one. Scale bar: 0.5 μm. Courtesy of Geneviève Chazal (unpublished)

as a morphological substrate for the powerful auto- and mutual inhibitory mechanisms already known to control the firing rate of 5-HT neurons in rat NRD (Aghajanian *et al.*, 1972; Wang and Aghajanian, 1978, 1982).

TERMINAL AXONS

Spinal cord

5-HT axon terminals are present in all laminae of the gray matter of the spinal cord (Steinbusch, 1981). These axons mainly originate from the brainstem, namely from the medullary raphe complex and from sources located in the central gray and reticular formation of the midbrain (Bowker *et al.*, 1981, 1987). As shown in cat spinal cord (Hylden *et al.*, 1985), descending 5-HT axons travel within the dorsolateral and ventral funiculi to innervate the dorsal and ventral horn, respectively. A small proportion of spinal 5-HT varicosities appears to originate from intrinsic 5-HT neurons (Lamotte *et al.*, 1982, Bowker, 1986; Newton *et al.*, 1986). In the monkey, spinal 5-HT cell bodies are mostly located ventral to the central canal in laminae X and send axonal processes to the intermediate gray and medial portion of the ventral horn (Lamotte and De Lanerolle, 1983). Some of these processes are adjacent to large blood vessels or to the ependymal lining of the central canal, suggesting that 5-HT might be involved in the control of spinal blood flow or cerebrospinal fluid (CSF) composition. At the level of the sacral cord, a few 5-HT cells have also been found in the marginal layer and in the substantia gelatinosa of the dorsal horn. In the rat, 5-HT cell bodies have been detected dorsal to the central canal, in laminae VII and X (Newton *et al.*, 1986). Spinal 5-HT neurons in the rat show less predilection for the vicinity of blood vessels or the central canal, and it has been suggested that they may rather be acting as interneurons in autonomic spinal circuitry (Newton *et al.*, 1986).

The *superficial layers* of the dorsal horn (Rexed's laminae I and II) contain the densest network of 5-HT axon terminals in the spinal cord (Steinbusch, 1981; Lamotte and DeLanerolle, 1983). The synaptology of these axon terminals has been investigated in a number of mammalian species using radioautographic and immunocytochemical ultrastructural approaches. In cat and monkey, most 5-HT varicosities were thus found to establish symmetrical synaptic contacts with the shaft of either proximal or distal dendrites (Ruda and Gobel, 1980; Ruda *et al.*, 1982; Lamotte and DeLanerolle, 1983; Light *et al.*, 1983). Fewer synapses were observed on neuronal perikarya, although examples of perikarya contacted by multiple 5-HT endings were reported in cat lamina I (Ruda *et al.*, 1982). Axo-spinous synapses appear to be the least frequent type of contact established by 5-HT axon terminals. Axo-axonic synapses involving 5-HT terminals have never been convincingly demonstrated in the superficial layers. In contrast, 5-HT axon terminals have often been observed in apposition with other nerve endings in both cat and monkey (Ruda *et al.*, 1982; Lamotte and DeLanerolle, 1983) as well as in the rat (Maxwell *et al.*, 1983). Such axo-axonal appositions might provide an anatomical substrate for the reported presynaptic effects of 5-HT on the excitability of primary afferents (Carstens *et al.*, 1981).

In the rat, as opposed to cat and monkey, very few synaptic differentiations have been observed between 5-HT endings and either dendrites or nerve cell bodies (Maxwell *et al.*, 1983). Whether this difference in synaptic incidence reflects a bias in sampling or a true interspecies difference remains to be determined.

The morphological heterogeneity of dendrites synaptically contacted by 5-HT endings in the cat has led to the suggestion that several types of intrinsic neurons receive a 5-HT input (Ruda *et al.*, 1982). The presence, within layers I and II of the cat caudal medulla, of membranous appositions and synaptic contacts between [³H]5-HT labeled axon terminals and enkephalin immunoreactive dendrites (Glazer and Basbaum, 1984) indicates that enkephalin-containing neurons constitute one such recipient type. Since most enkephalin-containing neurons in the spinal cord are local circuit interneurons, it is likely that dorsal horn intrinsic neurons represent a synaptic target for 5-HT axon terminals. In addition, light-microscopical observations indicate that 5-HT axon terminals in the cat terminate on physiologically identified nociceptive-specific and wide dynamic range neurons projecting to the mesencephalon and the thalamus (Hoffert *et al.*, 1983; Miletic *et al.*, 1984; Hylden *et al.*, 1986).

Laminae III and IV also contain a relatively dense network of 5-HT axons (Steinbusch, 1981). A combined retrograde tracing and immunohistochemical study in these laminae has indicated that nearly all retrogradely labeled neurons which project to the dorsal column nuclei in the monkey and the cat have 5-HT axon terminals apposed to their somata or proximal dendrites (Nishikawa *et al.*, 1983). In the electron microscope, some of these 5-HT terminals were seen to establish symmetrical synaptic junctions with the perikarya and dendrites of the retrogradely labeled cells. Since dorsal column post-synaptic neurons correspond to either one of two physiological types, wide dynamic range and low-threshold mecanoreceptive, it may be assumed that both types receive 5-HT afferents (Nishikawa *et al.*, 1983).

In the rat *lateral horn*, retrograde labeling of spinal neurons has been combined with light-microscopic 5-HT immunohistochemistry to demonstrate 'intimate contacts' between 5-HT axon terminals and sympatho-adrenal neurons (Appel *et al.*, 1986). Some 5-HT axons in this location have been demonstrated to exhibit substance P and/or TRH immunoreactivity in the rat (Appel *et al.*, 1986, 1987), or enkephalin and/or somatostatin immunoreactivity in the guinea pig (Chiba and Masuko, 1987), suggesting that 5-HT and these peptides might jointly regulate the sympathetic outflow. Currently available data on the ultrastructural relationships of 5-HT axon terminals in the lateral horn indicate that some of these endings make appositional or synaptic contacts on dendrites or nerve cell bodies (Misukawa *et al.*, 1986; Chiba and Masuko, 1987). Whether the latter elements belong to sympathetic preganglionic neurons remains to be demonstrated.

The existence of 5-HT terminal arrangements around large neurons in the *ventral horn* of the spinal cord was initially reported in the rat using fluorescence histochemistry (Fuxe, 1965). Immunocytochemical electron-microscopic investigations subsequently confirmed that 5-HT axon terminals synaptically contact the somata or dendrites of presumptive motoneurons (Pecci-Saavedra *et al.*, 1983; Mizukawa *et al.*, 1986). Such direct relationships could account for the excitatory effects of 5-HT on ventral horn neurons as observed in microiontophoretic experiments (White and Neuman, 1980). In the monkey ventral horn, it has been shown by light microscopy that a sub-population of 5-HT endings was also immunoreactive for substance P or TRH (Bowker, 1986b; Harkness and Brownfield, 1986). This suggests that these molecules could interact for regulating the activity of ventral horn neurons.

Brainstem

Cranial nerve nuclei

The existence of a rich 5-HT input to most cranial nerve nuclei was established in early histofluorescence studies (Fuxe, 1965) and later confirmed by micromeasurements of endogenous 5-HT (Palkovits *et al.*, 1974) and 5-HT immunohistochemistry (Steinbusch, 1981). The respective contributions of the different raphe nuclei to these innervations are still largely unknown.

Branchiomeric and somatomotor nuclei The highest density of 5-HT innervation is that of branchiomeric and somatomotor cranial nerve nuclei (Palkovits *et al.*, 1974). In the rat trigeminal motor nucleus, for instance, the number of 5-HT varicosities radioautographically labeled after intracerebral microinstillation of [^{3}H]5-HT has been estimated at 1.8×10^6 varicosities per mm^3 (Schaffar *et al.*, 1984). Within the rat oculomotor nucleus the number of 5-HT terminals is slightly lower, with a reported figure of 1.3×10^6 per mm^3 of tissue (Soghomonian *et al.*, 1986).

At the electron-microscopic level, 5-HT axon terminals directly apposed to or making synaptic contact with neuronal perikarya were described in the facial and oculomotor nuclei of the rat (Aghajanian and McCall, 1980; Soghomonian *et al.*, 1986), as well as in the trigeminal motor, ambiguus, accessory, and hypoglossal nuclei of the monkey (Takeuchi *et al.*, 1983). Axo-somatic and axo-dendritic synapses formed by 5-HT varicosities in the facial nucleus have been viewed as the morphological substrate for the facilitatory effect of 5-HT on the activity of these motoneurons (McCall and Aghajanian, 1979). However, not all of the dendrites contacted by 5-HT boutons in motor nuclei are necessarily those of motoneurons, and 5-HT may therefore exert indirect as well as direct effects on motoneurons within these nuclei (Figs. 3a, b). In the facial (Aghajanian and McCall, 1980), trigeminal motor (Schaffar *et al.*, 1984) and oculomotor (Soghomonian *et al.*, 1986) nuclei of the rat, the overall synaptic incidence of 5-HT varicosities has been estimated at 30%, 4.9% and 15%, respectively. In the rat oculomotor nucleus, [^{3}H]5-HT-labeled axon terminals were often directly apposed to unlabeled endings themselves in synaptic contact or in apposition with the somata or proximal dendrites of motoneurons (Fig. 4). Such a feature suggests an alternate substratum for a possible presynaptic regulation by 5-HT of afferents impinging upon motoneurons.

Sensory nuclei The caudal pole of the spinal trigeminal nucleus (which corresponds to the dorsal horn of the cervical spinal cord), and the nucleus of the solitary tract (NTS) (which receives its afferent input from the VIIth, IXth and X[th] cranial nerves), have both been shown to contain a dense 5-HT innervation (Maley and Elde, 1982; Ruda *et al.*, 1982; Pickel *et al.*, 1984). In the rat, the 5-HT innervation of the NTS arises mainly from medullary raphe nuclei (Rea *et al.*, 1982; Thor and Helke, 1987), although transmitter-specific retrograde transport studies suggest that, in the cat, a contingent could also arise from 5-HT ganglion cells found in the nodose ganglia (Gaudin-Chazal *et al.*, 1982). Within the cat NTS, 5-HT immunoreactive axon terminals observed in serial thin sections were found to rarely establish synaptic contacts (Maley and Elde, 1982). In contrast, within the rat medial NTS, up to 30% of [³H]5-HT labeled axonal profiles were found to exhibit synaptic differentiations in single sections (Pickel *et al.*, 1984). In both species these 5-HT synapses were shown to be mainly of the asymmetric variety and were usually found on large or small dendritic branches (Maley and Elde, 1982; Pickel *et al.*, 1984). A few axo-spinous synapses were also reported. By combining [³H]5-HT radioautography with tyrosine hydroxylase immunohistochemistry, it was demonstrated that approximately 6% of 5-HT synaptic targets within the NTS are immunoreactive for the enzyme (Pickel *et al.*, 1984). This latter finding is consistent with a direct modulation of catecholamine neurons by 5-HT within the rat NTS.

Inferior olive

All subdivisions of the inferior olive have been shown to receive a 5-HT input with some degree of interspecies variability (Wiklund *et al.*, 1977; Takeuchi and Sano, 1983b; King *et al.*, 1984). In the rat, the source of this 5-HT input has been identified as the nucleus reticularis paragigantocellularis (Bishop and Ho, 1986). Indirect immunohistochemical evidence suggests the coexistence of 5-HT and substance P in at least some of these olivary afferents (Paré *et al.* 1987). In the dorsal accessory olive of the rat (Wiklund *et al.*, 1981a), and throughout the inferior olivary complex of the opossum (King *et al.*, 1984), 5-HT axon terminals showed a very low incidence of synaptic contacts. In the rat, less than 5% of 750 [³H]5-HT labeled profiles examined in single thin sections exhibited a synaptic differentiation (Wiklund *et al.*, 1981a). In the opossum, a similarly low incidence (2%) was observed using either [³H]5-HT radioautography or 5-HT immunocytochemistry. In both species, labeled synapses were mostly asymmetric and established on dendrites. The non-synaptic 5-HT axon terminals were found in close

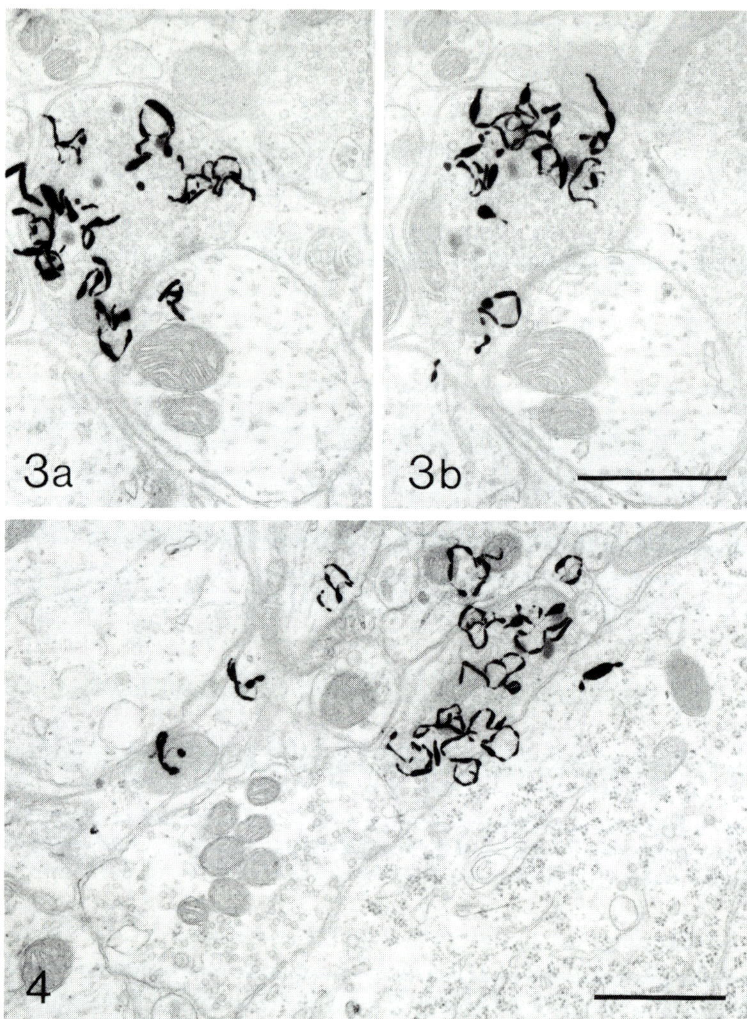

Fig. 3 5-HT axon terminals in the rat oculomotor nucleus. These terminals were labeled for radioautography by local microinstillation of [³H]5-HT. Figs. a and b (adjacent sections) show a well differentiated asymmetrical synaptic junction between the 5-HT ending and a dendritic branch. Scale bar: 1 μm

Fig. 4 5-HT axon terminals in the rat oculomotor nucleus, labeled as in Fig. 3. The labeled 5-HT varicosity is in appositional contact with a large perikaryon exhibiting a well-developed granular endoplasmic reticulum (presumed motoneuron). This 5-HT varicosity is juxtaposed to another terminal, unlabeled, which contains uniformly round, clear synaptic vesicles and shares a comparable relationship with the nerve cell body. Scale bar: 1 μm

apposition with dendritic shafts, appendages or spines (Wiklund *et al.*, 1981a; King *et al.*, 1984). In the rat, non-synaptic and synaptic profiles alike often exhibited clustered microcanaliculi and/or tubular elements against the portion of the membrane juxtaposed to dendritic profiles, suggesting that 5-HT might be released from undifferentiated portions of the varicosities (Wiklund *et al.*, 1981a). In the caudal medial accessory nucleus of the opossum, a substantial number of 5-HT varicosities showed membranous appositions with nerve cell bodies (12%) or axon terminals (8%) (King *et al.*, 1984). These 5-HT varicosities on cell bodies were often surrounded by astrocytic processes.

The synaptic and appositional characteristics of 5-HT axon terminals in the olivary complex of both rat and opossum suggest that 5-HT mainly exerts a distal action on olivary neurons except in the caudal medial accessory nucleus of the opossum, where a proximal action on cell bodies is also possible. Physiological data indicate that 5-HT has a hyperpolarizing action in the olivary complex (Wiklund *et al.*, 1981c). Whether this action regulates the electrotonic coupling of olivary dendrites and explains the critical role of 5-HT in the induction of tremor elicited after harmaline administration (Sjolund *et al.*, 1977) remains to be determined.

Locus coeruleus

The locus coeruleus is the major group of noradrenaline neurons in brain. It receives a very dense 5-HT innervation which was initially demonstrated in the rat using radioautography following *in vivo* uptake of [³H]5-HT (Léger and Descarries, 1978). The origin of these 5-HT afferent appears to be shared between nuclei raphe dorsalis, centralis superior and pontis (Léger *et al.*, 1980). These 5-HT axon terminals have also been identified ultrastructurally using tryptophan hydroxylase immunohistochemistry (Pickel *et al.*, 1977). 5-HT varicosities in the locus coeruleus are often non-synaptic but some do establish typical asymmetric synapses with dendritic branches and spines presumably belonging to noradrenaline neurons (Pickel *et al.*, 1977; Léger and Descarries, 1978). It has been proposed that 5-HT axons may directly influence central noradrenaline function through these contacts.

Midbrain tegmentum

5-HT axons have been ultrastructurally characterized in three regions of the midbrain tegmentum: the substantia nigra, the ventral tegmental area and the red nucleus.

In monkey substantia nigra, pars reticulata, 5-HT immunoreactive varicosities were shown to establish synaptic contacts with large dendritic shafts (Pasik *et al.*, 1984). Such contacts, however, appear to represent only a small proportion of those found on any given dendrite. Most are asymmetrical and a few exhibit sub-synaptic dense bodies. In the rat, 5-HT immunoreactive terminals in either pars reticulata or compacta have been described as in close apposition to peridendritic axon terminals or free in the neuropil with occasional symmetrical synapses 'en passant' made on dendrites and somata (Mori *et al.*, 1987).

In the rat ventral tegmental area (A 10), a relatively large fraction (more than 18%) of 5-HT axon terminals labeled by radioautography following intraventricular infusion of [³H]5-HT was found to establish *bona fide* synapses, most of them on dendritic shafts (Hervé *et al.*, 1987). By dual radioautographic and immunocytochemical labeling, it could be shown that some of the contacted dendrites contained tyrosine hydroxylase, indicating that ascending 5-HT axons could directly regulate mesolimbic and mesocortical dopamine neurons (Hervé *et al.*, 1987).

In the cat red nucleus, 5-HT axon terminals have been identified by electron microscopic radioautography following *in vitro* labeling with [³H]5-HT (Bosler *et al.*, 1983). Some of these terminals were found to establish synaptic contact with small dendritic processes as well as with perikarya and proximal dendrites of magnocellular neurons (Bosler *et al.*, 1983).

Cerebellum

A 5-HT innervation of the cerebellar cortex was first demonstrated by fluorescence histochemistry in the rat (Hökfelt and Fuxe, 1969). Subsequent immunohistochemical studies showed this innervation to be sparse and heterogeneously distributed (Takeuchi *et al.*, 1982b). A small contingent of cerebellar 5-HT afferents corresponds to mossy fibers (Chan-Palay, 1975; Beaudet and Sotelo, 1981), but most belong to a diffuse projection system which arborizes in all three cortical layers. By electron microscope radioautography, almost all mossy-like 5-HT varicosities were found to establish Gray's type I synaptic contacts on the dendritic digits of granule cells (Chan-Palay, 1975; Beaudet and Sotelo, 1981). A few also formed synapses 'en marron' with the dendrites of Golgi cells (Chan-Palay, 1975). In contrast, a very small proportion of the varicosities belonging to the diffuse system were found in synaptic junction (less than 9% according to Beaudet and Sotelo, 1981). In single thin sections from the granular layer, 2 out of 75 nonmossy terminals examined were seen in synaptic contact, both of them with dendrites of Golgi cells. In the molecular layer, 4% of the [³H]5-HT-labeled varicosities were seen in synaptic contact, either on Purkinje cell branchlet spines, or on the dendrites of cortical interneurons, e.g. Golgi, basket or stellate cells (Beaudet and Sotelo, 1981).

In the deep cerebellar nuclei, 5-HT axons are scarce and reportedly contact both the soma of interneurons and the dendrites of large and small projection neurons (Chan-Palay, 1975).

Hypothalamus

The hypothalamus receives a 5-HT innervation of variable density, which mainly originates from mesencephalic nuclei, but also, in the case of the dorsomedial nucleus of the hypothalamus, from intrinsic nerve cell bodies (Beaudet and Descarries, 1979; Frankfurt and Azmitia, 1983). The functional role of 5-HT in the hypothalamus is still largely unknown, but various physiological and pharmacological data indicate that the indoleamine is involved in regulating the release of several other informative molecules such as growth hormone releasing-hormone (GRF), corticotrophin releasing hormone (CRF), gonadotrophin releasing-hormone (LHRH), prolactin, vasoactive intestinal peptide (VIP), GABA and TRH.

Rostrally, the *medial preoptic area* was shown to contain a moderate number of 5-HT terminals (Descarries and Beaudet, 1978; Steinbusch, 1984). In an electron microscope radioautographic examination of this region after intraventricular infusion of [³H]5-HT, approximately one-third of axonal profiles labeled in single thin sections were reported to exhibit a synaptic specialization (Kiss and Halász, 1985). Most of these synapses were asymmetrical and established on dendritic shafts or spines. In double labeling studies, [³H]5-HT-labeled axon terminals were sometimes seen in direct apposition or, occasionally, in synaptic contact with dendrites immunoreactive for LHRH (Kiss and Halász, 1985). Approximately 5% of axon terminals found in contact, synaptic or not, with LHRH-immunostained dendrites contained 5-HT. These data indicate that within the medial preoptic area, 5-HT may act directly on LHRH neurons to regulate pituitary gonadotrophin secretion.

The *suprachiasmatic nucleus* (SCN) receives one of the densest 5-HT inputs in the brain (Descarries and Beaudet, 1978; Steinbusch, 1981). This innervation is believed to arise primarily from the median raphe nucleus (Van de Kar and Lorens, 1979). Electron microscope radioautography has revealed that the 5-HT innervation of the rat SCN is partly synaptic. Indeed, some 10% of axonal profiles radiolabeled in the nucleus following intraventricular injection of [³H]5-HT showed synaptic specializations in single thin sections (Bosler and Beaudet, 1985a). From this figure, it was estimated that approximately 40% of these 5-HT varicosities might establish *bona fide* synapses. By combining [³H]5-HT radioautography and VIP immunocytochemistry, it was shown that a large proportion of 5-HT axon terminals in rat SCN were either directly apposed to, or in synaptic contact with VIP-immunoreactive cells (Kiss *et al.*, 1984; Bosler and Beaudet, 1985a). These contacts were mostly found on dendrites but a few were also observed on neuronal somata (Fig. 5). The fact that more than 60% of all synaptic contacts established by 5-HT axons in the rat SCN were found on VIP-immunoreactive neurons led Bosler and Beaudet (1985a) to suggest that VIP neurons might constitute preferential synaptic targets for 5-HT afferents in this region of the brain.

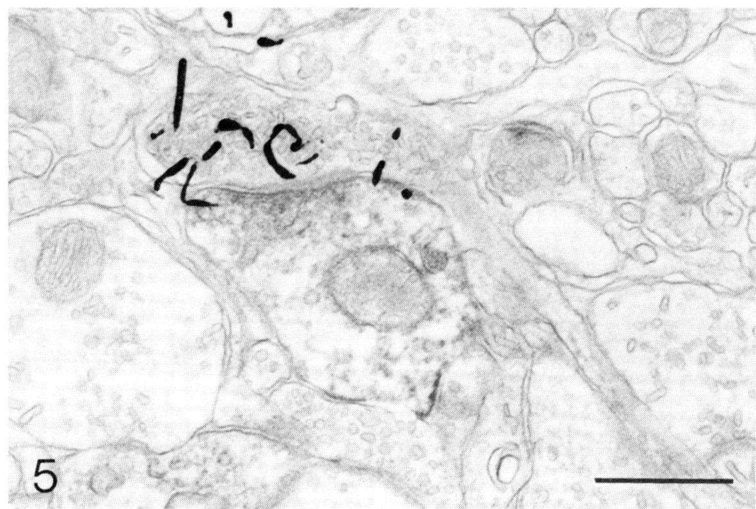

Fig. 5 Combined [³H]5-HT radioautographic and VIP immunocytochemical labeling in the rat suprachiasmatic nucleus. The 5-HT varicosity
is seen emerging from a thin, unmyelinated, axonal segment. It forms
an asymmetrical synaptic contact with the VIP-immunoreactive dendritic
profile in which a large granular vesicle is visible. In this instance, the
synaptic junction spans across most of the contact zone between the labeled
elements. Scale bar: 1 μm. Reproduced from Bosler and Beaudet (1985a)
by permission of Chapman and Hall Ltd

This innervation is likely to be involved in the circadian regulation of gonadotrophin and prolactin secretion (Bosler and Beaudet, 1985a).

5-HT axons within the SCN were also shown by combined radioautography
and immunocytochemistry to be directly apposed to glutamic acid decarboxylase (GAD) immunoreactive axon terminals (Bosler et al., 1984a). Such axo-
axonal contacts, although deprived of any obvious membrane specialization,
might provide a cytological basis for 5-HT/GABA interactions in the SCN
(François-Bellan et al., 1985).

As shown in a recent study by Liposits et al. (1987), the *paraventricular
nucleus* is moderately innervated by 5-HT fibers in its parvocellular subdivision. These 5-HT axons form terminal boutons and synapses 'en passant'
with dendrites and somata of parvocellular neurons. Moreover, using
combined immunocytochemical detection of 5-HT and CRF, it is reported
that some of these terminals make axo-dendritic and axo-somatic synapses
with CRF immunoreactive neurons.

The *arcuate nucleus* receives a 5-HT innervation which mainly originates
from the dorsal and median raphe nuclei (Van de Kar and Lorens, 1979;
Willoughby and Blessing, 1987). In the rat, 5-HT axon terminals radioautographically labeled after intraventricular injection of [³H]5-HT represent no
more than 3% of all nerve endings visible in the arcuate nucleus (Kiss and

Halász, 1986). At the ultrastructural level, very few of these varicosities were found to exhibit junctional specializations. Most of the synapses observed were established on dendritic shafts (Kiss and Halász, 1986). Combined radioautographic and immunocytochemical examination of this region has demonstrated the existence of direct appositions between [³H]5-HT-labeled terminals and the nerve cell bodies and/or dendrites of neurons belonging to the dopamine cell groups A12 and A13 (Bosler *et al.*, 1984b; Kiss and Halász, 1986). In a few cases, asymmetrical membrane specializations were visible at such sites of contact (Kiss and Halász, 1986). These data indicate that 5-HT might be implicated in the proximal regulation of dopamine neurons within the arcuate nucleus, and thereby potentially affects the dopamine regulation of prolactin, gonadotrophin and somatotrophin secretion (see Kordon *et al.*, 1980).

[³H]5-HT-labeled axon terminals have also been detected next to ACTH-immunoreactive perikarya and dendrites within the rat arcuate nucleus (Kiss *et al.*, 1984; Bosler and Beaudet, 1985b). Most of these 5-HT varicosities were merely juxtaposed to the ACTH-immunoreactive elements, although a few were seen to establish asymmetrical junctional complexes on dendritic shafts (Bosler and Beaudet, 1985b).

Circumventricular organs and supra-ependymal fibers

The circumventricular organs are a series of midline structures opened to neuro-hemal exchanges, located around the ventricular system. They comprise the median eminence (ME), the organum vasculosum laminae terminalis (OVLT), the subfornical organ (SFO), the subcommissural organ (SCO), the pineal gland and the area postrema (see Bouchaud and Bosler, 1986).

The 5-HT innervation of the *median eminence* predominates in its outer layer, the inner layer showing only sparsely distributed terminals (Calas, 1977; Jennes *et al.*, 1982). In duck, rat and monkey, the 5-HT terminals in the outer layer were shown by electron microscopic radioautography to contact the basement membrane of portal capillaries (Calas *et al.*, 1974; Calas, 1977; Nakai *et al.*, 1983). Other labeled 5-HT varicosities were found to directly abut tanycytic processes or were closely apposed to other types of axonal varicosities (Calas *et al.*, 1974; Calas, 1977). However, no specialized junctions were ever reported between 5-HT and other terminal types in the ME. Some of the juxtaposed axonal varicosities have been identified in combined immunocytochemical and radioautographic studies as TRH-(Nakai *et al.*, 1983) or tyrosine-hydroxylase containing (Bosler and Beaudet, 1985b). A light-microscopic study of Jennes *et al.*, (1982) has demonstrated a considerable overlap between 5-HT- and LHRH-immunoreactive axon terminals in the rat ME. As yet, however, there is no ultrastructural evidence for axo-axonal contacts between 5-HT and LHRH-positive elements. Taken together, available electron microscopic data thus far suggest that 5-HT within the ME might be acting: (1) by being directly released into the portal system; (2) through modification of the permeability of the capillary basement membrane to various neurosecretory factors; (3) by an action on tanycytic function; and (4) by regulating the release of hypophysiotropic hormones or of other transmitters through axo-axonal appositions (see Calas, 1977).

The relatively dense 5-HT innervation of the organum vasculosum laminae terminalis has been visualized by fluorescence histochemistry, radioautography and immunohistochemistry (Moore, 1977; Bosler, 1978; Parent *et al.*, 1981; Jennes *et al.*, 1982; Takeuchi and Sano, 1983a). After *in vivo* labeling with [³H]5-HT, labeled axon terminals in rat OVLT were found to represent approximately 3% of all axonal profiles (Bosler, 1978). In the juxtavascular zone of the organ, these terminals were never seen in synaptic junction (Bosler and Descarries, in preparation). They often lay in the vicinity of blood vessels, but never abutted directly the parenchymal basement membrane. However, they were often apposed to tanycytic or astrocytic processes with which they occasionally made synaptoid contacts. In the juxta-ventricular zone, both synaptic and non-synaptic 5-HT terminals were visible. Synaptic contacts were symmetrical and asymmetrical and established on dendritic processes (Fig. 6). As in the juxtavascular zone, synaptoid contacts between 5-HT terminals and astrocytic processes were occasionally observed. The light-microscopic distribution of 5-HT immunoreactive fibers within the OVLT shows good overlap with that of LHRH-immunoreactive axons (Jennes *et al.*, 1982). Although there is still no evidence for close contacts between 5-HT- and LHRH-containing elements at the electron microscopic level, this suggests possible presynaptic effects of 5-HT on the release of LHRH into the vascular space (Jennes *et al.*, 1982).

Both radioautographic and immunohistochemical studies indicate that the 5-HT innervation of the *subfornical organ* is relatively sparse (Calas *et al.*, 1978; Takeuchi and Sano, 1983a; Tramu *et al.*, 1983). In the rat, [³H]5-HT-labeled boutons were mainly found beneath the ependyma and within the vascular region. In the latter zone, [³H]5-HT-labeled axon terminals were seen to approximate fenestrated capillaries (Calas *et al.*, 1978). Some of these perivascular fibers even exhibited a discontinuous thickening of their membrane in front of the basal membrane surrounding the pericapillary space. Intraparenchymal 5-HT fibers occasionally showed synaptic-like differentiations (Calas *et al.*, 1978).

The rat *subcommissural organ* also receives a dense 5-HT innervation arising mainly from the midbrain nuclei raphe dorsalis and medianus, with some contribution from the nucleus raphe pontis (Léger *et al.*, 1983). Inside the SCO, 5-HT axon terminals were shown to establish asymmetrical synaptic contacts with both epen- and hypendymocytes (Bouchaud and Arluison, 1977; Bouchaud, 1979; Møllgård and Wiklund, 1979). In fact, more than 75% of axon terminals synapsing upon SCO ependymocytes were reported

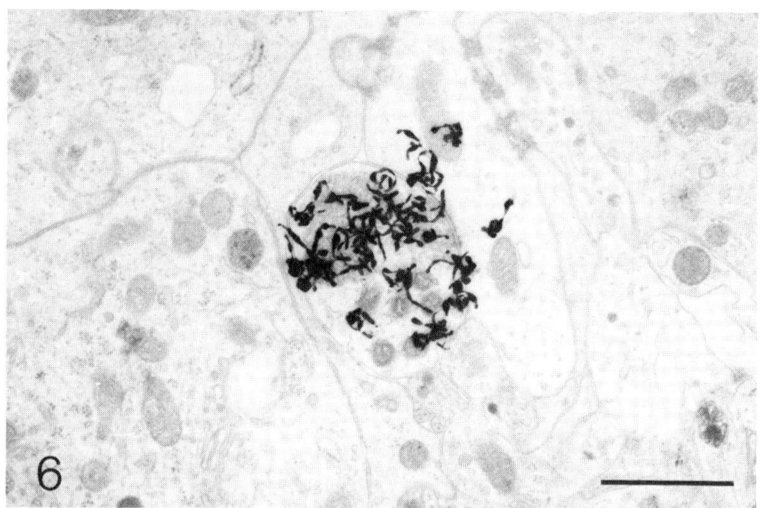

Fig. 6 Electron microscope radioautograph from the rat organum vasculosum laminae terminalis (OVLT) after intraventricular administration of [³H]5-HT. This labeled axon terminal was located in the juxtaventricular zone of the organ, where many were seen to establish symmetrical synaptic contacts with dendritic shafts. This 5-HT ending contains numerous small, round synaptic vesicles, some of which are crowded in an active zone. The labeling appears to be preferentially associated with the vesicles. Calibration bar: 1 μm. Kindly contributed by Olivier Bosler (unpublished)

to contain 5-HT, suggesting that 5-HT plays a major role in the regulation of the secretory activity of these cells. According to Léger *et al.* (1983), the 5-HT afferents from NRD would inhibit the secretory activity of SCO ependymocytes, while those from nucleus raphe medianus would exert opposite effects. It has been shown by Gamrani *et al.* (1981) that some 5-HT axon terminals in the SCO also contain GABA, but the functional significance of this co-localization remains to be determined.

An extensive network of 5-HT axons forms a *supra-ependymal* plexus along the walls of the cerebral ventricles (Richards *et al.*, 1973; Lorez and Richards, 1975; Chan-Palay, 1976). 5-HT supra-ependymal fibers pervade most of the ventricular cavities, including the surface of the OVLT and of the SFO (Takeuchi and Sano, 1983a). They are absent beneath the SCO, but the ones seen along the lateral edge of the organ appear to be continuous with the dense 5-HT plexus of the hypendymal layer (Møllgård and Wiklund, 1979; Takeuchi and Sano, 1983a). Electron microscopic examination of supra-ependymal 5-HT varicosities in rat and monkey has revealed that these are secured to the ependyma by small desmosome-like attachment plaques but are otherwise devoid of synaptic specialization (Chan-Palay, 1976). In the monkey, where supra-ependymal boutons are larger, the desmosomes are also more conspicuous (Chan-Palay, 1976). At the light microscopic level, individual 5-HT axons are seen to criss-cross over one another, forming a lacy network (Chan-Palay, 1976). At nodal points, six or seven axons may join together. In the electron microscope, these nodal points appear as clusters of axonal profiles filled with synaptic vesicles. Neither synapses nor desmosomes have yet been observed between 5-HT axons within such nodal groups. As in the SCO, a majority of 5-HT supra-ependymal fibers are immunoreactive for GABA (Harandi *et al.*, 1986).

Thalamus

Ultrastructural data pertaining to the 5-HT innervation of the thalamus have been mainly obtained in the lateral geniculate (LGN) of the monkey and cat using 5-HT immunocytochemistry (Wilson and Hendrickson, 1987; DeLima Singer, 1987). This 5-HT innervation is likely to arise from a restricted area of the NRD (Moore *et al.*, 1978; Pasquier and Villar, 1982). Previous light-microscopic investigations have indicated that 5-HT terminals predominate within the ventral subdivision of the LGN but some are also present in its dorsal part (Mantyh and Kemp, 1983; Cropper *et al.*, 1984; Ueda and Sano, 1986). In their ultrastructural examination of monkey LGN, Wilson and Hendrickson (1987) found an exceedingly low incidence of junctional 5-HT varicosities, synapses being formed by two out of 100–150 immunoreactive profiles. This low synaptic incidence was confirmed by the survey of a few varicosities in serial sections. A similar conclusion was reached in the dorsal LGN and the perigeniculate nucleus of the cat, where rare synaptic 5-HT terminals were observed to contact small dendritic profiles (DeLima and Singer, 1987). The consequent suggestion that in LGN, as in other brain structures, 5-HT might be non-synaptically released, would be in keeping with the notion that 5-HT is able to depress the spontaneous or synaptically evoked activity of virtually every cell in the LGN (Rogawski and Aghajanian, 1980) in spite of the relatively low density of 5-HT innervation in this part of the brain (Wilson and Hendrikson, 1987).

Basal ganglia

All components of the basal ganglia contain relatively high endogenous levels of 5-HT (Palkovits *et al.*, 1974). 5-HT innervation is particularly dense within the globus pallidus, where a density of 4.8×10^6 axon terminals per mm^3 of tissue has been measured in the rat (Soghomonian *et al.*, 1987). In rat neostriatum, the 5-HT innervation is somewhat weaker (2.6×10^6 axon terminals per mm^3), and 3 to 4 times less dense in dorsal as compared to ventral regions (Soghomonian *et al.*, 1987).

Little is known of the fine structural features of 5-HT axons within the *globus pallidus*. In the monkey, 5-HT-immunolabeled boutons from this brain region have been reported to form asymmetric synapses with dendritic shafts (Pasik *et al.*, 1984). Crest synapses have also been observed, involving an unlabeled terminal opposite the 5-HT immunoreactive one on the same dendritic spine (Pasik *et al.*, 1984).

5-HT axon terminals of the *neostriatum* have been ultrastructurally characterized in rat, cat and monkey (Calas *et al.*, 1976; Arluison and Delamanche, 1980; Pasik and Pasik, 1982). In all species, these 5-HT varicosities showed a relatively low incidence of synaptic specializations (Fig. 8). For instance, in the paraventricular neostriatum of the rat, only 49 of 484 [³H]5-HT-labeled profiles examined in single sections showed a synapse (Soghomonian and Descarries, in preparation). Synaptic junctions were found on dendritic shafts (45%) and dendritic spines (55%) and were mostly of the asymmetrical type (90%) (Figs. 7a, b). In the monkey, synaptic contacts were mainly observed on dendritic spines (Pasik and Pasik, 1982). In rat neostriatum, appositions between 5-HT and unlabeled axon terminals were frequently observed (Soghomonian and Descarries, in preparation). Such axo-axonal contacts might provide support for a presynaptic regulation of transmitter release by 5-HT, as recently demonstrated for acetylcholine (Gillet *et al.*, 1986).

Olfactory bulb

In spite of the considerable attention devoted in recent years to the cytochemical organization of the olfactory bulb (for review, see Macrides and Davis, 1983), there has been only one ultrastructural study on its 5-HT innervation. This study was carried out in the rat, after *in vivo* labeling by local injection of [³H]5-HT (Halász *et al.*, 1978). In the glomerular layer, [³H]5-HT-labeled boutons were shown to form axo-dendritic synaptic contacts with periglomerular cells, whereas in the external plexiform and internal granular layers, they were found to synapse upon the dendritic shafts or spines of granule cells. Synapses on mitral cells were not demonstrated. Such data suggest that 5-HT action in the olfactory bulb is exerted via interneurons. It is not yet known whether the periglomerular cells contacted by 5-HT afferents belong to the distinct subpopulation known to synthesize dopamine (Halász *et al.*, 1977; Ljungdahl *et al.*, 1977).

Cerebral cortex and hippocampus

The 5-HT innervation of the cerebral cortex originates from the midbrain raphe nuclei dorsalis and medianus and spreads to every allo- and iso-cortical area (references in Descarries *et al.*, 1987). The laminar and regional distributions of this cortical input are rather uniform in the rat (Beaudet and Descarries, 1976; Lidov *et al.*, 1980). Both radioautography after [³H]5-HT uptake (Descarries *et al.*, 1975) and 5-HT immunocytochemistry (Molliver *et al.*, 1982) have been used for electron microscopic detection of the 5-HT axon terminals. Only a small fraction of sectional profiles labeled by

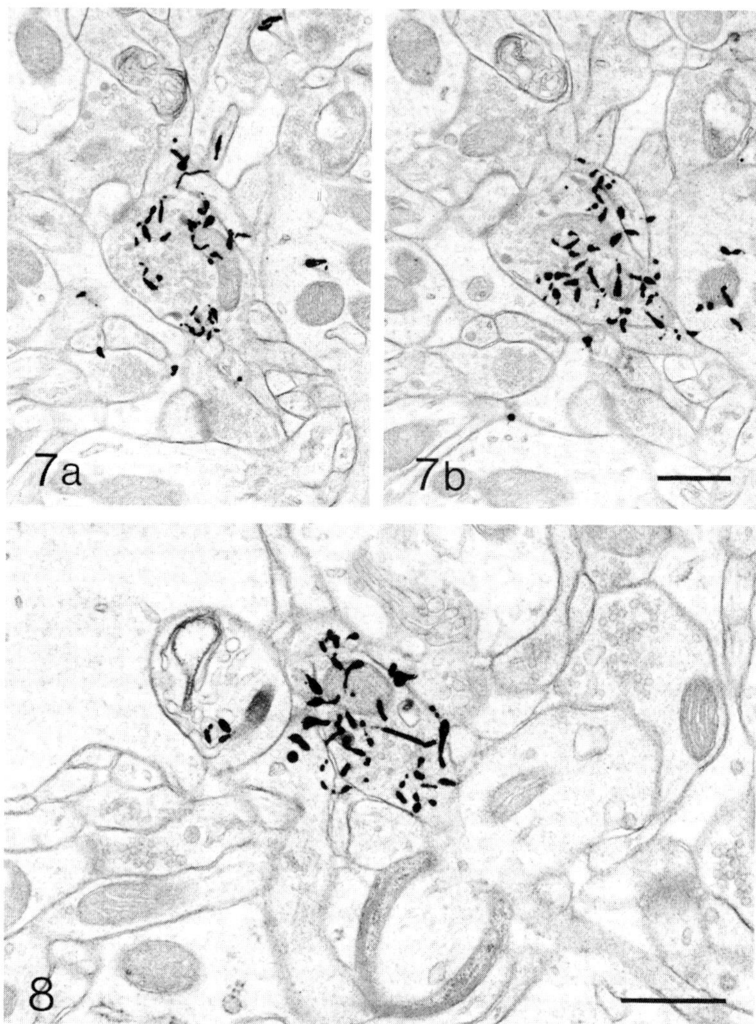

Fig. 7 Electron microscope radioautographs from the rat neostriatum after intraventricular administration of [³H]5-HT. In two adjacent sections, this 5-HT terminal exhibits an asymmetrical synaptic contact with a small dendritic spine. The dendritic spine also receives an unlabeled axon terminal containing slightly larger synaptic vesicles than those in the 5-HT ending. The rest of the 5-HT varicosity is enwrapped by glia. Scale bar: 1 μm

Fig. 8 Radioautographs from the rat neostriatum, as in Fig. 7. This 5-HT axon terminal is also surrounded by neuronal and glial elements but does not show any junctional complex of synaptic differentiation. It is otherwise indistinguishable from that in Figs 7 a, b. Scale bar: 1 μm

radioautography after topical application of [³H]5-HT *in vivo* were found to exhibit synaptic differentiations in the dorsal, fronto-parietal neocortex of adult rat (Descarries *et al.*, 1975). According to probability estimates of seeing these junctions in single thin sections, it was later calculated that only some 15% of these 5-HT varicosities were actually engaged in synaptic relationships (Beaudet and Descarries, 1984). The rare junctional 5-HT varicosities observed at the time (*n* = 13/303) showed either symmetrical or asymmetrical specializations. These synapses were exclusively established on dendritic branches and spines (Descarries *et al.*, 1975).

Three studies have since challenged these original observations. From preliminary results of Lidov using 5-HT-immunocytochemistry in the somato-sensory cortex of the rat, Molliver *et al.* (1982) reported that at least 50% of cortical 5-HT varicosities identified in single thin sections formed conspicuous synaptic contacts with dense postsynaptic membrane specializations. Parna-velas and McDonald (1983) reported similar findings in rat striate cortex after examination in single thin sections of what they interpreted as labeled axon terminals after topical application of [³H]5-HT. These authors wrote that: 'contrary to the reports of Descarries and colleagues, serotonin is present in large vesicle-containing terminals which form type I synapses with spines and dendritic shafts throughout the cortex' (see also Parnavelas *et al.*, 1985). Recently, Papadopoulos *et al.* (1987) have reiterated this conclusion after serial section analysis of 5-HT immunoreactive terminals in rat occipital cortex.

This issue is currently being re-examined in our laboratory using a 5-HT antiserum raised against a serotonin–glutaraldehyde–protein conjugate (Geffard *et al.*, 1985; Patel *et al.*, 1986). Immunostaining with this antiserum can be carried out in tissue adequately fixed by perfusion of 3.5% glutaral-dehyde, and reaches sufficiently deep in Vibratome sections to permit identi-fication of whole 5-HT varicosities in uninterrupted series of thin sections for electron microscopy (Fig. 9). Of the 54 such 5-HT axon terminals serially examined in the molecular layer of the frontal neocortex (Fr1–2 areas, according to Zilles, 1985), only 14 were found to exhibit a junctional complex suggestive of synaptic differentiation (Séguéla and Descarries, in prep-aration). The observed junctions were relatively small, both of the symmetrical and asymmetrical variety, and all were made with dendritic branchlets or spines. The linear relationship between junctional frequency and the number of adjacent thin sections available for examination allowed to estimate the actual synaptic incidence at 35–40% of total. As recently demonstrated in the case of cortical dopamine (DA) endings (Séguéla *et al.*, 1987), it is probable that this proportion varies between layers as well as between different regions of the cortex.

A comparable immunocytochemical analysis in serial thin sections has been carried out on the motor cortex of the monkey using a formaldehyde-albumin-conjugated 5-HT antiserum (DeFelipe and Jones, 1987). These

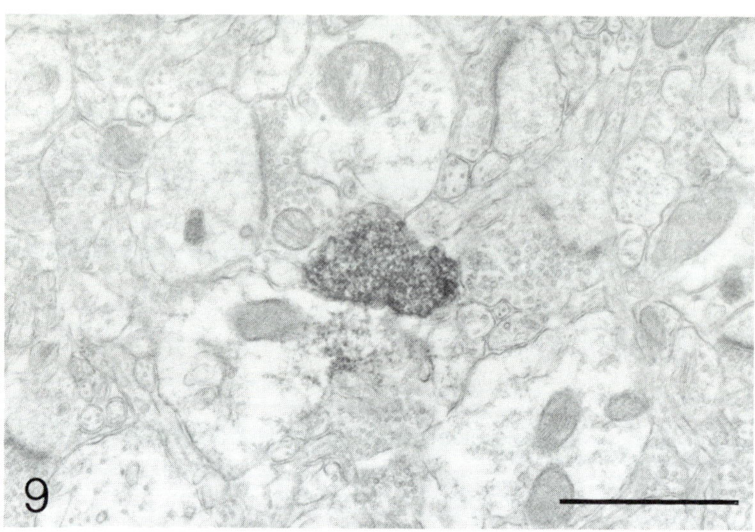

Fig. 9 Immunoreactive 5-HT axonal varicosities from the cerebral cortex of rat. In both rat and monkey (Fig. 10) many such terminals have now been demonstrated by serial sectioning to lack morphologically defined synaptic junctions (see page 82). This micrograph is from the molecular layer of rat frontal cortex (Fr1-2). This 5-HT varicosity exhibits evenly distributed small, round and clear synaptic vesicles. It is directly apposed to adjacent unlabeled axon terminals and dendrites. Scale bar: 1 μm. Contributed by Philippe Séguéla (unpublished)

authors have found a minimal fraction of cortical 5-HT axon terminals engaged in morphologically defined synaptic contacts (DeFelipe and Jones, personal communication; Figs 10a, b). It is thus most likely that in the cerebral cortex of both rat and monkey many 5-HT axon terminals are deprived of synaptic membrane differentiations, as initially proposed.

Even though the hippocampus has been shown by both radioautographic and immunohistochemical methods to receive a relatively dense 5-HT innervation (Azmitia and Marovitz, 1980; Lidov *et al.*, 1980; Köhler, 1984), there is still limited information on the fine structure and cellular relationships of these 5-HT terminals. In the only study published to date on this topic, immunocytochemically identified 5-HT terminals in adult rat dentate gyrus were described as being adjacent to a number of neuropil elements including dendrites, axons and glia (Anderson *et al.*, 1986). Two categories of immuno-reactive profiles were observed: large ones that never appeared to form synaptic junctions, and smaller ones which occasionally showed asymmetrical synapses upon dendritic spines or small dendritic branches (15% of all varicosity profiles in single thin sections). A poor correlation between the density of 5-HT innervation and that of 5-HT binding sites has been reported in rat hippocampus (Köhler, 1984). This apparent mismatch would be

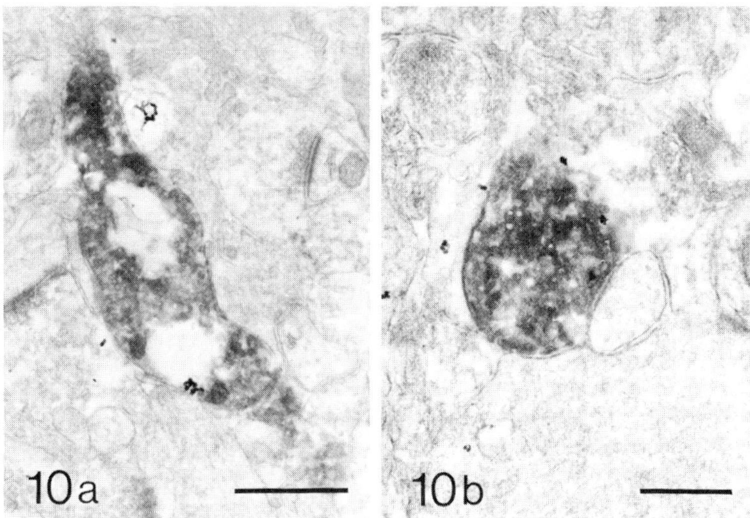

Fig. 10 Immunoreactive 5-HT axonal varicosities from the motor cortex of *Macaca fascicularis* (Cynomolgus). Fig. a illustrates the usual appearance of these varicosities in single thin sections. Their lack of synaptic contact has now been confirmed in twenty instances where such varicosities were completely examined in long series of thin sections. Fig. b shows the only 5-HT terminal which made a possible synaptic contact in this material (symmetrical junction with dendritic spine). Scale bar: 0.5 μm. Personal communication and micrographs by J. DeFelipe and E. G. Jones (unpublished)

consistent with the hypothesis of a non-junctional release of 5-HT within this region. In the absence of visible axo-axonic specialized contacts (Anderson *et al.*, 1986) non-junctional release of 5-HT could also be invoked to explain the presynaptic inhibition of acetylcholine release documented in the dentate gyrus (Maura and Raiteri, 1986).

CONCLUSIONS AND FUTURE TRENDS

A first conclusion to be drawn from the foregoing review is that original observations regarding the paucity of synaptic contacts established by 5-HT varicosities in cerebral cortex (Descarries *et al.*, 1975) have now been extended to several other brain regions. It is also clear, however, that the incidence of synaptic contacts established by 5-HT axons varies between brain regions and may even reach 100% under particular experimental conditions (Beaudet and Sotelo, 1981). Diversity in methodological approaches can no longer be invoked to account for such variations in synaptic incidence, since they have now been observed using techniques as different as [3H]5-HT uptake radioautography and 5-HT immunohistochemistry, sometimes within the same areas of the brain. It thus appears that there exist true regional differences in the synaptic modeling of 5-HT axons, and therefore presumably in the mode of action of 5-HT in the CNS. Why two differently structured modes of 5-HT transmission should exist and whether synaptic varicosities represent more permanent sites of information transfer than non-synaptic ones remains to be determined.

The presence of a membrane differentiation on those 5-HT axon terminals that establish a synaptic junction pinpoints some cellular targets of the released transmitter. These targets can in turn be identified with regard to their origin, projection or neurochemical build-up as exemplified in the present chapter. Double labeling studies have considerably furthered our knowledge of the connectivity of 5-HT neurons in mammalian CNS and are beginning to provide solid morphological grounds for better understanding of the function and the consequences of pharmacological manipulations of the 5-HT systems in brain.

If, as originally proposed, 5-HT can also be released from non-junctional varicosities in the extracellular space, another key to the identification of 5-HT targets should lie with the localization at cellular and sub-cellular levels, of the corresponding 5-HT receptors. It will then be of particular interest to determine whether elements that are directly apposed but not synaptically linked to 5-HT varicosities constitute preferential targets for this biogenic amine.

ACKNOWLEDGEMENTS

J.J.S. is recipient of a Fellowship and A.B. holds a Scientist award from the Medical Research Council of Canada (MRC). Personal studies referred to in the text were also supported in large part by the MRC (grants MT-3544 to L.D. and MA-7366 to A.B.). The authors are indebted to Kathleen Leonard, Sylvia Garcia and Kenneth C. Watkins for excellent technical assistance. They also thank Daniel Cyr for his photographic work and Lucie Perrault and Hélène Collerette for typing the manuscript.

REFERENCES

Aghajanian, G. K., and McCall, R. B. (1980) Serotonergic synaptic input to facial motoneurons: localization by electron-microscopic autoradiography, *Neuroscience*, **5**, 2155–2161.
Aghajanian, G. K., and Wang, R. Y. (1977) Habenular and other midbrain raphe afferents demonstrated by a retrograde tracing technique, *Brain Res.*, **122**, 229–249.
Aghajanian, G. K., Haigler, H. J., and Bloom, F. E. (1972) Lysergic acid diethylamide and serotonin: direct action on serotonin containing neurons, *Life Sci.*, **11**, 615–622.
Anderson, K. J., Holets, V. R., Mazur, P. C., Lasher, R. S. and Cotman, C. W. (1986) Immunocytochemical localization of serotonin in the rat dentate gyrus following raphe transplants, *Brain Res.*, **369**, 21–28.
Appel, N. M., Wessendorf, M. W., and Elde, R. P. (1986) Coexistence of serotonin

and substance P-like immunoreactivity in nerve fibers apposing identified sympatho-adrenal preganglionic neurons in rat intermedio-lateral cell column, *Neurosci. Lett.*, **65**, 241–246.

Appel, N. M., Wessendorf, M. W., and Elde, R. P. (1987) Thyrotropin-releasing hormone in spinal cord: coexistence with serotonin and with substance P in fibers and terminals apposing identified preganglionic sympathetic neurons, *Brain Res.*, **415**, 137–143.

Arluison, M., and De La Manche, I. S. (1980) High-resolution radioautographic study of the serotonin innervation of the rat corpus striatum after intraventricular administration of [³H]5-hydroxytryptamine, *Neuroscience*, **5**, 229–240.

Azmitia, E. C. (1978) The serotonin-producing neurons of the midbrain median and dorsal raphe nuclei, in *Handbook of Psychopharmacology*, Vol. 9, *Chemical Pathways in the Brain* (Eds L. L. Iversen, S. D. Iversen and S. H. Snyder), pp. 233–314, Plenum Press, New York.

Azmitia, E. C., and Gannon, P. J. (1986) The primate serotonergic system: a review of human and animal studies and report on *Macaca fascicularis*, in *Advances in Neurology*, Vol. 43, *Myoclonus* (Eds Fahn *et al.*) pp. 407–463, Raven Press, New York.

Azmitia, E. F., and Marovitz, W. F. (1980) In vitro hippocampal uptake of tritiated serotonin (³H-5-HT): a morphological, biochemical and pharmacological approach to specificity, *J. Histochem. Cytochem.*, **28**, 636–644.

Baraban, J. M., and Aghajanian G. K. (1981) Noradrenergic innervation of serotonergic neurons in the dorsal raphe: demonstration by electron microscopic autoradiography, *Brain Res.*, **204**, 1–11.

Beaudet, A., and Descarries, L. (1976) Quantitative data on serotonin nerve terminals in adult rat neocortex, *Brain Res.*, **111**, 301–309.

Beaudet, A., and Descarries, L. (1979) Radioautographic characterization of a sero-tonin-accumulating nerve cell group in adult rat hypothalamus, *Brain Res.*, **160**, 231–243.

Beaudet, A., and Descarries, L. (1981) The fine structure of central serotonin neurons, *J. Physiol.* (Paris), **77**, 193–309.

Beaudet, A., and Descarries, L. (1984) Fine structure of monoamine terminals in cerebral cortex, in *Monoamine Innervation of Cerebral Cortex* (Eds L. Descarries, T. Reader and H. H. Jasper), pp. 77–93, Alan R. Liss, New York.

Beaudet, A., and Descarries, L. (1987) Ultrastructural identification of serotonin neurons, in *Methods in the Neuroscience, Monoaminergic Neurons: Light Microscopy and Ultrastructure* (Ed. H. W. M. Steinbusch), pp. 265–313, John Wiley & Sons, Chichester.

Beaudet, A., and Sotelo, C. (1981) Synaptic remodeling of serotonin axon terminals in rat agranular cerebellum, *Brain Res.*, **206**, 305–329.

Belin, M. F., Nanopoulos, D., Didier, M., Aguera, M., Steinbusch, H., Verhofstad, A., Maître, M., and Pujol, J. F. (1983) Immunohistochemical evidence for the presence of γ-aminobutyric acid and serotonin in one nerve cell. A study on the raphe nuclei of the rat using antibodies to glutamate decarboxylase and serotonin, *Brain Res.*, **275**, 329–339.

Bishop, G. A., and Ho, R. H. (1986) Cell bodies of origin of serotonin-immunoreac-tive afferents to the inferior olivary complex of the rat, *Brain Res.*, **399**, 369–373.

Bobillier, P., Séguin, S., Petitjean, F., Salvert, D., Touret, M., and Jouvet, M. (1976) The raphe nuclei of the cat brainstem: a topographical atlas of their efferent projections as revealed by autoradiography, *Brain Res.*, **113**, 449–486.

Bosler, O. (1978) Radioautographic identification of serotonin axon terminals in the rat organum vasculosum laminae terminalis, *Brain Res.*, **150**, 177–181.

Bosler, O., and Beaudet, A. (1985a) VIP neurons as prime synaptic targets for serotonin afferents in rat suprachiasmatic nucleus: a combined radioautographic and immunocytochemical study, *J. Neurocytol.*, **14**, 749–763.

Bosler, O., and Beaudet, A. (1985b) Relations ultrastructurales entre systèmes monoaminergiques et peptidergiques dans l'hypothalamus, *Annales d'Endocrinologie* (Paris), **46**, 19–26.

Bosler, O., Beaudet, A., and Tappaz, M. (1984a) Morphological evidence for cellular interactions between serotoninergic terminals and VIP and GABAergic elements in rat suprachiasmatic nucleus, *Excerpta Medica 7th Int. Congr. Endocrinology*, **652**, 486.

Bosler, O. Joh, T. H., and Beaudet, A. (1984b) Ultrastructural relationships between serotonin and dopamine neurons in the rat arcuate nucleus and medial zona incerta: a combined radioautographic and immunocytochemical study, *Neurosci. Lett.*, **48**, 279–285.

Bosler, O., Nieoullon, A., Onteniente, B., and Dusticier, N. (1983) In vitro radioautographic study of the monoaminergic innervation of cat red nucleus. Identification of serotoninergic terminals, *Brain Res.*, **259**, 288–292.

Bouchaud, C. (1979) Evidence for a multiple innervation of subcommissural ependymocytes in the rat, *Neurosci. Lett.*, **12**, 253–258.

Bouchaud, C., and Arluison, M. (1977) Serotoninergic innervation of ependymal cells in the rat subcommissural organ. A fluorescence electron microscopic and radioautographic study, *Biol. Cell.*, **30**, 65–72.

Bouchaud, C., and Bosler, O. (1986) The circumventricular organs of the mammalian brain with special reference to monoaminergic innervation, *Intern. Rev. Cytology*, **105**, 283–327.

Bowker, R. M., (1986a) Intrinsic 5-HT-immunoreactive neurons in the spinal cord of the fetal non-human primate, *Develop. Brain Res.*, **28**, 137–143.

Bowker, R. M. (1986b) Serotonergic and peptidergic inputs to the primate ventral spinal cord as visualized with multiple chromagens on the same tissue section, *Brain Res.*, **375**, 345–350.

Bowker, R. M., Reddy, V. K., Fung, S. J., Chan, J. Y. H., and Barnes, C. D. (1987) Serotonergic and non-serotonergic raphe neurons projecting to the feline lumbar and cervical spinal cord: a quantitative horseradish peroxidase-immunocytochemical study, *Neurosci. Lett.*, **75**, 31–37.

Bowker, R. M., Westlund, K. N., and Coulter, J. D. (1981) Origins of serotonergic projections to the spinal cord in rat: an immunocytochemical-retrograde transport study, *Brain Res.*, **226**, 187–199.

Calas, A. (1977) Radioautographic studies of aminergic neurons terminating in the median eminence, in *Advances in Biochemical Psychopharmacology* (Eds E. Costa and G. L. Gessa), pp. 79–88, Raven Press, New York.

Calas, A., Alonso, G., Arnauld, E., and Vincent, J. D. (1974) Demonstration of indolaminergic fibres in the median eminence of the duck, rat and monkey, *Nature*, **250**, 241–243.

Calas, A., Besson, M. J., Gauchy, C., Alonso, G., Glowinski, J., and Cheramy, A. (1976) Radioautography study of in vivo incorporation of [^{3}H]-monoamines in the cat caudate nucleus: identification of serotoninergic fibers, *Brain Res.*, **118**, 1–13.

Calas, A., Bosler, O., Arluison, M., and Bouchaud, C. (1978) Serotonin as a neurohormone in circumventricular organs and supraependymal fibers, in *Brain-Endocrine Interaction III. Neural Hormones and Reproduction*, 3rd Int. Symp. Würzburg (Eds D. E. Scott and N. Y. Rochester), pp. 238–250, S. Karger, A. G. Basel.

Carstens, E., Klumpp, D., Randic, M., and Zimmermann, M. (1981) Effect of iontophoretically applied 5-hydroxytryptamine on the excitability of single primary afferent C- and A-fibers in the cat spinal cord, *Brain Res.*, **220**, 151–158.

Chan-Palay, V. (1975) Fine structure of labelled axons in the cerebellar cortex and nuclei of rodents and primates after intraventricular infusions with tritiated serotonin, *Anat. Embryol.*, **148**, 235–265.

Chan-Palay, V. (1976) Serotonin axons in the supra- and sub-ependymal plexuses and in the leptomeninges; their roles in local alterations of cerebrospinal fluid and basomotor activity, *Brain Res.*, **102**, 103–130.

Chan-Palay, V., Jonsson, G., and Palay, S. L. (1978). Serotonin and substance P coexist in neurons of the rat's central nervous system, *Proc. Nat. Acad. Sci. USA*, **75**, 1582–1586.

Chazal, G., and Ralston III, H. J. (1987) Serotonin containing structures in the nucleus raphe dorsalis of the cat: an ultrastructural analysis of dendrites, presynaptic dendrites and axon terminals, *Brain Res.* (in press).

Chiba, T., and Masuko, S. (1987) Synaptic structure of the monoamine and peptide nerve terminals in the intermediolateral nucleus of the guinea pig thoracic spinal cord, *J. Comp. Neurol.*, **262**, 242–255.

Clements, J. R., Beitz, A. J., Fletcher, T. F., and Mullett, M. A. (1985) Immunocytochemical localization of serotonin in the rat periaqueductal gray: a quantitative light and electron microscopic study, *J. Comp. Neurol.*, **236**, 60–70.

Consolazione, A., and Cuello, A. C. (1982) CNS serotonin pathways, in *Biology of Serotonergic Transmission* (Ed. N. N. Osborne), pp. 29–61, John Wiley & Sons, London.

Cropper, E. C., Eisenman, J. S., and Azmitia, E. C. (1984) An immunocytochemical study of the serotonergic innervation of the thalamus of the rat, *J. Comp. Neurol.*, **224**, 38–50.

Cuello, A. C., Priestley, J. V., and Milstein, C. (1982) Immunocytochemistry with internally labeled monoclonal antibodies, *Proc. Nat. Acad. Sci. USA*, **79**, 665–669.

Dahlström, A., and Fuxe, K. (1964) Evidence for the existence of monoamine-containing neurons in the central nervous system. I. Demonstration of monoamines in cell bodies of brain neurons, *Acta Physiol. Scand.*, **62**, Suppl. 232, 1–55.

DeFelipe, J., and Jones, E. G. (1987) A light and electron microscopic study of serotonin-immunoreactive fibers and terminals in the monkey sensory-motor cortex, *Exp. Brain Res.* (in press).

DeLima, D. A., and Singer, W. (1987) The serotonergic fibers in the dorsal lateral geniculate nucleus of the cat: distribution and synaptic connections demonstrated with immunocytochemistry, *J. Comp. Neurol.*, **258**, 339–351.

Descarries, L., and Beaudet, A. (1978) The serotonin innervation of adult rat hypothalamus, in *Biologie Cellulaire des Processus Neurosécrétoires Hypothalamiques*, Colloques Internationaux du CNRS (Eds J. D. Vincent and C. Kordon), pp. 135–153, Editions du CNRS, Paris.

Descarries, L., Beaudet, A., and Watkins, K. C. (1975) Serotonin nerve terminals in adult rat neocortex, *Brain Res.*, **100**, 563–588.

Descarries, L. D., Berthelet, F., Garcia, S., and Beaudet, A. (1986) Dopaminergic projection from nucleus raphe dorsalis to neostriatum in the rat, *J. Comp. Neurol.*, **249**, 511–520.

Descarries, L., Doucet, G., Lemay, B., Séguéla, P., and Watkins, K. C. (1987) Structural basis of cortical monoamine function, in *Neurotransmitters and Brain Function* (Eds M. Avoli, T. A. Reader, R. W. Dykes and P. Gloor), Plenum Publ. Corp., New York (in press).

Descarries, L., Watkins, K. C., Garcia, S., and Beaudet, A. (1982) The serotonin neurons in nucleus raphe dorsalis of adult rat: a light and electron microscope radioautographic study, *J. Comp. Neurol.*, **207**, 239–254.

Felten, D. L., and Sladek, J. R. (1983) Monoamine distribution in primate brain V. Monoaminergic nuclei: anatomy, pathways and local organization, *Brain Res. Bull.*, **10**, 171–284.

François-Bellan, A. M., Héry, M., Bosler, O., Barrit, M. C., and Hery, F. (1985) Effects of baclofen on 5-HT metabolism in the suprachiasmatic nucleus of the rat, *Neurosci. Lett.*, Suppl. 22, 5558.

Frankfurt, M., and Azmitia, E. (1983) The effect of intracerebral injection of 5,7-dihydroxytryptamine and 6-hydroxydopamine on the serotonin-immunoreactive cell bodies and fibers in the adult rat hypothalamus, *Brain Res.*, **261**, 91–99.

Frankfurt, M., Lauder, J. M., and Azmitia, E. C. (1981) The immunocytochemical localization of serotonergic neurons in the rat hypothalamus, *Neurosci. Lett.*, **24**, 227–232.

Fuxe, K. (1965) Evidence for the existence of monoamine neurons in central nervous system IV. Distribution of monoamine nerve terminals in central nervous system, *Acta Physiol. Scand.*, **65**, Suppl. 247, 37–85.

Gamrani, H., Belin, M. F., Aguera, M., Calas, A., and Pujol J. F. (1981) Radioautographic evidence for an innervation of the subcommissural organ by GABA-containing nerve fibers, *J. Neurocytol.*, **10**, 411–424.

Gamrani, H., Harandi, M., Belin, M. F., Dubois, M. P., and Calas, A. (1984) Direct electron microscopic evidence for the coexistence of GABA uptake and endogenous serotonin in the same rat central neurons by coupled radioautographic and immunocytochemical procedures, *Neurosci. Lett.*, **48**, 25–30.

Gaudin-Chazal, G., Seyfritz, N., Araneda, S., Vigier, D., and Puizillout, J. J. (1982) Selective retrograde transport of ^{3}H-serotonin in vagal afferents, *Brain Res. Bull.*, **8**, 503–509.

Geffard, M., Dulluc, J., and Rock, A. M. (1985) Antisera against the indolealkylamines: tryptophan, 5-hydroxytryptophan, 5-hydroxytryptamine, 5-methoxytryptophan, and 5-methoxytryptamine tested by an enzyme-linked immunosorbent assay method, *J. Neurochem.*, **44**, 1221–1228.

Gillet, G., Ammor, S., and Fillion, G. (1986) Serotonin inhibits acetylcholine release from rat striatum slices. Evidence for a presynaptic receptor-mediated effect, *J. Neurochem.*, **45**, 1687–1691.

Glazer, E., and Basbaum, A. I. (1984) Axons which take up [^{3}H] serotonin are presynaptic to enkephalin immunoreactive neurons in cat dorsal horn, *Brain Res.*, **298**, 386–391.

Glazer, E. J., Steinbusch, H., Verhofstad, A., and Basbaum, A. I. (1981) Serotonin neurons in nucleus raphe dorsalis and paragigantocellularis of the cat contain enkephalin, *J. Physiol.* (Paris), **77**, 241–245.

Halász, N., Ljungdahl, A., and Hökfelt, T. (1978) Transmitter histochemistry of the rat olfactory bulb. II. Fluorescence histochemical, autoradiographic and electron microscopic localization of monoamines, *Brain Res.*, **154**, 253–271.

Halász, N., Ljungdahl, A., Hökfelt, T., Johansson, O., Goldstein, M., Park, D., and Biberfeld, P. (1977) Transmitter histochemistry of the olfactory bulb. I. Immunohistochemical localization of monoamine synthesizing enzymes. Support for intrabulbar, periglomerular dopamine neurons, *Brain Res.*, **126**, 455–474.

Harandi, M., Didier, M., Aguera, M., Calas, A., and Belin, M. F. (1986) GABA and serotonin (5-HT) pattern in the supraependymal fibers of the rat epithalamus: Combined radioautographic and immunocytochemical studies. Effect of 5-HT content on [^{3}H]GABA accumulation, *Brain Res.*, **370**, 241–249.

Harkness, D. H., and Brownfield, M. S. (1986) A thyrotropin-releasing hormone-containing system in the rat dorsal horn separate from serotonin, *Brain Res.*, **384**, 323–333.

Hervé, D. Pickel, V. M., Joh, T. H., and Beaudet, A. (1987) Serotonin axon terminals in the ventral tegmental area of the rat: Fine structure and synaptic input to dopaminergic neurons, *Brain Res.* (in press).

Héry, F., Faudon, M., and Ternaux, J. P. (1982) In vivo release of serotonin in two raphe nuclei (raphe dorsalis and magnus) of the cat, *Brain Res. Bull.*, **8**, 123–129.

Hoffert, M. J., Miletic, V., Ruda, M. A., and Dubner, R. (1983) Immunocytochemical identification of serotonin axonal contacts on characterized neurons in laminae I and II of the cat dorsal horn. *Brain Res.*, **267**, 361–364.

Hökfelt, T., and Fuxe, K. (1969) Cerebellar monoamine nerve terminals, a new type of afferent fibers on the cortex cerebelli, *Exp. Brain Res.*, **9**, 63–72.

Hökfelt, T., Ljungdahl, A., Steinbusch, H., Verhofstad, A., Nilsson, G., Brodin, E., Pernow, B., and Goldstein, M. (1978) Immunohistochemical evidence of substance P-like immunoreactivity in some 5-hydroxytryptamine-containing neurons in the rat central nervous system, *Neuroscience*, **3**, 517–538.

Hubbard, J. E., and DiCarlo, V. (1974) Fluorescence histochemistry of monoamine-containing cell bodies in the brain stem of the squirrel monkey (*Saimiri sciureus*) III. Serotonin-containing groups, *J. Comp. Neurol.*, **153**, 385–398.

Hunt, S. P., and Lovick, T. A. (1982) The distribution of serotonin, met-enkephalin and β-lipotropin-like immunoreactivity in neuronal perikarya of the cat brainstem, *Neurosci. Lett.*, **30**, 139–145.

Hylden, J. L. K., Hayashi, H., Ruda, M. A., and Dubner, R. (1986) Serotonin innervation of physiologically identified lamina I projection neurons, *Brain Res.*, **370**, 401–404.

Hylden, J. L. K., Ruda, M. A., Hayashi, H., and Dubner, R. (1985) Descending serotonergic fibers in the dorsolateral and ventral funiculi of cat spinal cord, *Neurosci. Lett.*, **62**, 299–304.

Jacobowitz, D. M., and MacLean, P. D. (1978) A brainstem atlas of catecholaminergic neurons and serotonergic perikarya in a pigmy primate (*Cebuella pygmaea*), *J. Comp. Neurol.*, **177**, 397–416.

Jennes, L., Beckman, W. C., Stumpf, W. E., and Grzanna, R. (1982) Anatomical relationships of serotoninergic and noradrenergic projections with the GnRH system in septum and hypothalamus, *Exp. Brain Res.*, **46**, 331–338,

Johansson, O., Hökfelt, T., Pernow, B., Jeffcoate, S. L., White, N., Steinbusch, H. W. M., Verhofstad, A. A. I., Emson, P. C., and Spindel, E. (1981) Immunohistochemical support for three putative transmitters in one neuron: coexistence of 5-hydroxytryptamine, substance P and thyrotropin releasing hormone-like immunoreactivity in medullary neurons projecting to the spinal cord, *Neuroscience*, **6**, 1881–1981.

Kalén, P., Karlson, M., and Wiklund, L. (1985) Possible excitatory amino-acid afferents to nucleus raphe dorsalis of the rat investigated with retrograde wheat germ agglutinin and D-[³H] aspartate tracing, *Brain Res.*, **360**, 285–297.

Kapadia, S. E., DeLanerolle, N. C., and Lamotte, C. C. (1985) Immunocytochemical and electron microscopic study of serotonin neuronal organization in the dorsal raphe nucleus of the monkey, *Neuroscience*, **15**, 729–746.

Kawasaki, T., and Sato, Y. (1981) Afferent projections to the caudal part of the dorsal nucleus of the raphe in cats, *Brain Res.*, **211**, 439–44.

King, J. S., Ho, R. H., and Burry, R. W. (1984) The distribution and synaptic organization of serotoninergic elements in the inferior olivary complex of the opossum, *J. Comp. Neurol.*, **227**, 357–368.

Kiss, J., and Halász, B. (1985) Demonstration of serotoninergic axons terminating on luteinizing hormone-releasing hormone neurons in the preoptic area of the rat using a combination of immunocytochemistry and high resolution autoradiography, *Neuroscience*, **14**, 69–78.

Kiss, J., and Halász, B. (1986) Synaptic connections between serotoninergic axon terminals and tyrosine hydroxylase-immunoreactive neurons in the arcuate nucleus of the rat hypothalamus. A combination of electron microscopic autoradiography and immunocytochemistry, *Brain Res.*, **364**, 284–294.

Kiss, J., Léranth, C. and Halász, B. (1984) Serotoninergic endings on VIP neurons in the suprachiasmatic nucleus and on ACTH-neurons in the arcuate nucleus of the rat hypothalamus. A combination of high resolution autoradiography and electron microscopic immunocytochemistry, *Neurosci. Lett.*, **44**, 119–124.

Köhler, C. (1984) The distribution of serotonin binding sites in the hippocampal region of the rat brain. An autoradiographic study, *Neuroscience*, **13**, 667–680.

Kordon, C., Enjalbert, A., Héry, M., Joseph-Bravo, P. I., Rotsztejn, W., and Ruberg, M. (1980) Role of neurotransmitters in the control of adenohypophyseal secretion, in *Handbook of the Hypothalamus*, Vol. 2 (Eds P. J Morgane and A. J. Panksepp), pp. 253–306, Dekker, New York.

Lackner, K. J. (1980) Mapping of monoamine neurons and fibers in the cat lower brainstem and spinal cord, *Anat. Embryol.*, **161**, 169–195.

Lamotte, C. C., and DeLanerolle, N. C. (1983) Ultrastructure of chemically defined neuron systems in the dorsal horn of the monkey. III. Serotonin immunoreactivity, *Brain Res.*, **274**, 65–77.

Lamotte, C. C., Johns, D. R., and DeLanerolle, N. C. (1982) Immunohistochemical evidence of indolamine neurons in monkey spinal cord, *J. Comp. Neurol.*, **206**, 359–370.

Léger, L., Charnay, Y., Dubois, P. M., and Jouvet, M. (1986) Distribution of enkephalin-immunoreactive cell bodies in relation to serotonin containing neurons in the raphe nuclei of the cat: immunohistochemical evidence for the coexistence of enkephalin and serotonin in certain cells, *Brain Res.*, **362**, 63–73.

Léger, L., and Descarries, L. (1978) Serotonin nerve terminals in the locus coeruleus of adult rat: a radioautographic study, *Brain Res.*, **151**, 1–13.

Léger, L., Degueurce, A., Lundberg, J. J., Pujol, J. F., and Møllgård, K. (1983) Origin and influence of the serotoninergic innervation of the subcommissural organ in the rat, *Neuroscience*, **10**, 411–423.

Léger, L., McRae Degueurce, A., and Pujol, J. F. (1980) Origine de l'innervation sérotoninergique du locus coeruleus chez le rat, *C. R. Acad. Sc. Paris, Série D*, **290**, 807–810.

Lidov, H. G. W., Grzanna, R., and Molliver, M. E. (1980) The serotonin innervation of the cerebral cortex in the rat: an immunohistochemical analysis, *Neuroscience*, **5**, 207–227.

Light, A. R., Kavoodjian, A. M., and Petrusz, P. (1983) The ultrastructure and synaptic connections of serotonin-immunoreactive terminals in spinal laminae I and II, *Somatosensory Res.*, **1**, 33–50.

Ljungdahl, A., Hökfelt, T., Halász, N., Johansson, O., and Goldstein, M. (1977) Olfactory bulb dopamine neurons – the A15 catecholamine cell group, *Acta Physiol. Scand.*, **452**, 31–35.

Ljungdahl, A., Hökfelt, T., and Nilsson, G. (1987) Distribution of substance P-like immunoreactivity in the central nervous system of the rat. I. Cell bodies and terminals, *Neuroscience*, **3**, 861–943.

Lorez, H. P., and Richards, J. G. (1975) 5-HT nerve terminals in the fourth ventricle of the rat brain: their identification and distribution studied by fluorescence histo-chemistry and electron microscopy, *Cell Tissue Res.*, **165**, 37–48.

McCall, R., and Aghajanian, G. K. (1979) Serotonergic facilitation of facial moto-neuron excitation, *Brain Res.*, **169**, 11–27.

Macrides, F., and Davis, B. J. (1983) The olfactory bulb, in *Chemical Neuroanatomy* (Ed. P. C. Emson), pp. 391–426, Raven Press, New York.

Mâgoul, R., Onteniente, B., Oblin, A., and Calas, A. (1986) Inter- and intracellular relationships of substance P-containing neurons with serotonin and GABA in the dorsal raphe nucleus: combination of autoradiographic and immunocytochemical techniques, *J. Histochem. Cytochem.*, **34**, 735–742.

Maley, B., and Elde, R. (1982) The ultrastructural localization of serotonin immuno-reactivity within the nucleus of the solitary tract of the cat, *J. Neurosci.*, **2**, 1499–1506.

Mantyh, P. W., and Kemp, J. A. (1983) The distribution of putative neurotransmitters in the lateral geniculate nucleus of the rat, *Brain Res.*, **288**, 344–348.

Maura, G., and Raiteri, M. (1986) Cholinergic terminals in the rat hippocampus possess 5-HT 1B receptors mediating inhibition of acetylcholine release, *Eur. J. Pharmacol.*, **129**, 333–337.

Maxwell, D. J., Leranth, C., and Verhofstad, A. A. J. (1983) Fine structure of serotonin-containing axons in the marginal zone of the rat spinal cord, *Brain Res.*, **266**, 253–259.

Miletic, V., Hoffert, M. J., Ruda, M. A., Dubner, R., and Shigenaga, Y. (1984) Serotoninergic axonal contacts on identified cat spinal dorsal horn neurons and their correlation with nucleus raphe magnus stimulation, *J. Comp. Neurol.*, **228**, 129–141.

Mizukawa, K., Otsuka, N., and Hattori, T. (1986) Serotonin-containing nerve fibers in the rat spinal cord: electron microscopic immunohistochemistry, *Acta Med. Okayama*, **40**, 1–10.

Møllgård, K., and Wiklund, L. (1979) Serotoninergic synapses on ependymal and hypendymal cells of the rat subcommissural organ, *J. Neurocytol.*, **8**, 445–467.

Molliver, M. E., Grzanna, R., Lidov, H. G. W., Morrison, J. H., and Olschowka, J. A. (1982) Monoamine systems in the cerebral cortex, in *Cytochemical Methods in Neuroanatomy* (Eds V Chan-Palay and S. L. Palay), pp. 255–278, Alan R. Liss, New York.

Moore, R. Y. (1977) Organum vasculosum laminae terminalis: innervation by sero-tonin neurons of the midbrain raphe, *Neurosci. Lett.*, **5**, 297–302.

Moore, R. Y., Halaris, A. E., and Jones, B. E. (1978) Serotonin neurons of the midbrain raphe: Ascending projections, *J. Comp. Neurol.*, **180**, 417–438.

Mori, S., Matsuura, T., Takino, T., and Sano, Y. (1987) Light and electron micro-scopic immunohistochemical studies of serotonin nerve fibers in the substantia nigra of the rat, cat and monkey, *Anat. Embyol.*, **176**, 13–18.

Mosko, S. S., Haubrich, D., and Jacobs, B. L. (1977) Serotonergic afferents to the dorsal raphe nucleus: evidence from HRP and synaptosomal uptake studies, *Brain Res.*, **119**, 269–290.

Nauta, W. J. H. (1958) Hippocampal projections and related neural pathways to the midbrain in the cat, *Brain*, **81**, 319–340.

Nakai, Y., Shioda, S., Ochiai, H. J., Kudo, J., and Hashimoto, A. (1983) Ultrastruc-tural relationship between monoamine- and TRH-containing axons in the rat median eminence as revealed by combined autoradiography and immunocytochem-istry in the same tissue section, *Cell Tissue Res.*, **230**, 1–14.

Neckers, L. M., Schwartz, J. P., Wyatt, R. J., and Speciale, S. G. (1979) Substance P afferents from the habenula innervate the dorsal raphe nucleus, *Exp. Brain Res.*, **37**, 619–623.

Newton, B. W., Maley, B. E., and Hamill, R. W. (1986) Immunohistochemical demonstration of serotonin neurons in autonomic regions of the rat spinal cord, *Brain Res.*, **376**, 155–163.

Nishikawa, N., Bennett, G. J., Ruda, M. A., Lu, G. W., and Dubner, R. (1983) Immunocytochemical evidence for a serotoninergic innervation of dorsal column postsynaptic neurons in cat and monkey: light and electronmicroscopic observations, *Neuroscience*, **10**, 1333–1340.

Palkovits, M., Brownstein, M., and Saavedra, J. M. (1974) Serotonin content of the brain stem nuclei in the rat, *Brain Res.*, **80**, 237–249.

Papadopoulos, G. C., Parnavelas, J. G., and Buijs, R. (1987) Monoaminergic fibers form conventional synapses in the cerebral cortex, *Neurosci. Lett.*, **76**, 275–279.

Paré, M., Descarries, L., and Wiklund, L. (1987) Innervation and reinnervation of rat inferior olive by neurons containing serotonin and substance P: an immunohistochemical study after 5,6-dihydroxytryptamine lesioning, *J. Neurocytol.*, **16**, 155–167.

Parent, A., Descarries, L., and Beaudet, A. (1981) Organization of ascending serotonin systems in the adult rat brain. A radioautographic study after intraventricular administration of [³H]5-hydroxytryptamine, *Neuroscience*, **6**, 115–138.

Parnavelas, J. G., and McDonald, J. K. (1983) The cerebral cortex, in *Chemical Neuroanatomy* (Ed. P. C. Emson), pp. 505–549, Raven Press, New York.

Parnavelas, J. G., Moises, H. C., and Speciale, S. G. (1985) The monoaminergic innervation of the rat visual cortex, *Proc. R. Soc. Lond.*, B., **223**, 319–329.

Pasik, T., and Pasik, P. (1982) Serotoninergic afferents in the monkey neostriatum, *Acta Biol. Acad. Sci. Hung.*, **33**, 277–288.

Pasik, P., Pasik, T., Holstein, G. R., and Pecci-Saavedra, J. (1984) Serotoninergic innervation of the monkey basal ganglia: an immunocytochemical, light and electron microscopy study, in *The Basal Ganglia: Structure and Function* (Eds J. S. McKenzie, R. E. Kenn and L. N. Wilcock), pp. 115–129, Plenum Press, New York.

Pasquier, D. A., Kemper, T. L., Forbes, W. B., and Morgane, P. J. (1977) Dorsal raphe, substantia nigra and locus coeruleus: interconnection with each other and the neostriatum, *Brain Res. Bull.*, **2**, 323–339.

Pasquier, D. A., and Villar, M. J. (1982) Specific serotonergic projections to the lateral geniculate body from the lateral cell groups of the dorsal raphe nucleus, *Brain Res.*, **249**, 142–146.

Patel, S., Dulluc, J., and Geffard, M. (1986) Comparison of serotonin and 5-methoxytryptamine immunoreactivity in rat raphe nuclei, *Histochemistry*, **85**, 259–263.

Pecci-Saavedra, J. P., Brusco, A., and Peressini, S. (1982) Immunocytochemical study of synaptic connectivities of the raphe neurons: the possible presynaptic role of dendrites, *Communicaciones Biológicas*, **1**, 189–197.

Pecci-Saavedra, J. P., Brusco, A., Peressini, S., and Oliva, D. (1986) A new case for a presynaptic role of dendrites: An immunocytochemical study of the n. raphe dorsalis, *Neurochem. Res.*, **11**, 997–1009.

Pecci-Saavedra, J. P., Pasik, T., and Pasik, P. (1983) Immunocytochemistry of serotoninergic neurons in the central nervous system of monkeys, in *Neural Transmission, Learning and Memory* (Eds R. Caputto and C. Ajmone Marsan), pp. 81–96, Raven Press, New York.

Pickel, V. M., Joh, T. H., Chan, J., and Beaudet, A. (1984) Serotoninergic terminals: ultrastructure and synaptic interaction with catecholamine-containing neurons in the medial nuclei of the solitary tract, *J. Comp. Neurol.*, **225**, 291–301.

Pickel, V. M., Joh, T. H., and Reis, D. (1977) A serotonergic innervation of noradrenergic neurons in nucleus locus coeruleus: demonstration by immunocytochemical localization of the transmitter specific enzymes tyrosine and tryptophan hydroxylase, *Brain Res.*, **131**, 197–214.

Poitras, D., and Parent, A. (1978) Atlas of the distribution of monoamine-containing nerve cell bodies in the brain stem of the cat, *J. Comp. Neurol.*, **179**, 699–718.

Rea, M. A., Aprison, M. H., and Felton, D. L. (1982) Catecholamines and serotonin in the caudal medulla of the rat: combined neurochemical-histofluorescence study, *Brain Res. Bull.*, **9**, 227–236.

Richards, J. G., Lorez, H. P., and Tanzer, J. P. (1973) Indolealkylamine nerve terminals in cerebral ventricles: identification by electron microscopy and fluorescence histochemistry, *Brain Res.*, **57**, 277–288.

Rogawski, M. A., and Aghajanian, G. K. (1980) Norepinephrine and serotonin: opposite effects on the activity of lateral geniculate neurons evoked by optic pathway stimulation, *Exp. Neurol.*, **69**, 678–694.

Ruda, M. A., and Gobel, S. (1980) Ultrastructural characterization of axonal endings in the substantia gelatinosa which take up [^{3}H] serotonin, *Brain Res.*, **184**, 57–83.

Ruda, M. A., Coffield, J., and Steinbusch, H. W. M. (1982) Immunocytochemical analysis of serotonergic axons in laminae I and II of the lumbar spinal cord of the cat, *J. Neurosci.*, **2**, 1660–1671.

Sakai, K., Salvert, D., Touret, M., and Jouvet, M. (1977) Afferent connection of the nucleus raphe dorsalis in the cat as visualized by the horseradish peroxidase technique, *Brain Res.*, **137**, 11–35.

Sakai, K., Touret, M., Salvert, D., and Jouvet, M. (1978) Afferents to the cat locus coeruleus and rostral raphe nuclei as visualized by the horseradish peroxidase technique, in *Interactions Between Putative Neurotransmitters in the Brain* (Eds S. Garattini, J. F. Pujol and R. Samanin), pp. 319–342, Raven Press, New York.

Schaffar, N., Jean, A., and Calas, A. (1984) Radioautographic study of serotoninergic axon terminals in the rat trigeminal motor nucleus, *Neurosci. Lett.*, **44**, 31–36.

Séguéla, P., Watkins, K., and Descarries, L. (1987) Ultrastructural features of dopamine axon terminals in the anteromedial and the suprarhinal cortex of adult rat, *Brain Res.* (in press).

Sjolund, B., Björklund, A., and Wiklund, L. (1977) The indolaminergic innervation of the inferior olive. 2. Relation to harmaline induced tremor, *Brain Res.*, **131**, 23–37.

Soghomonian, J. J., Descarries, L., and Lanoir, J. (1986) Monoamine innervation of the oculomotor nucleus in the rat. A radioautographic study, *Neuroscience*, **17**, 1147–1157.

Soghomonian, J. J., Doucet, G., and Descarries, L. (1987) Serotonin innervation in adult rat neostriatum 1. Quantified regional distribution, *Brain Res.*, **425**, 85–100.

Steinbusch, H. W. M. (1981) Distribution of serotonin immunoreactivity in the central nervous system of the rat. Cell bodies and terminals, *Neuroscience*, **6**, 557–618.

Steinbusch, H. W. M. (1984) Serotonin-immunoreactive neurons and their projections in the CNS, in *Handbook of Chemical Neuroanatomy*, Vol. 3, *Classical Transmitters and Transmitter Receptors in the CNS*, Part II (Eds A. Björklund, T. Hökfelt and M. J. Kuhar), pp. 68–125, Elsevier, Amsterdam.

Steinbusch, H. W. M., and Nieuwenhuys, R. (1983) The raphe nuclei of the brain stem: A cytoarchitectonic and immunohistochemical study, in *Chemical Neuroanatomy* (Ed. P. C. Emson), pp. 131–207, Raven Press, New York.

Takeuchi, Y., and Sano, Y. (1983a) Serotonin distribution in the circumventricular organs of the rat, *Anat. Embryol.*, **167**, 311–319.

Takeuchi, Y., and Sano, Y. (1983b) Immunohistochemical demonstration of serotonin-containing nerve fibers in the inferior olivary complex of the rat, cat and monkey, *Cell Tissue Res.*, **231**, 17–28.

Takeuchi, Y., Kimura, H., and Sano, Y. (1982a) Immunohistochemical demonstration of the distribution of serotonin neurons in the brainstem of the rat and cat, *Cell Tissue Res.*, **224**, 247–267.

Takeuchi, Y., Kimura, H., and Sano, Y. (1982b) Immunohistochemical demonstration of serotonin-containing nerve fibers in the cerebellum, *Cell Tissue Res.*, **226**, 1–12.

Takeuchi, Y., Kojima, M., Matsuura, T., and Sano, Y. (1983) Serotonergic innervation on the motoneurons in the mammalian Brainstem, *Anat. Embryol.*, **167**, 321–333.

Thor, K. B., and Helke, C. J. (1987) Serotonin- and substance P-containing projections to the nucleus tractus solitarii of the rat, *J. Comp. Neurol.*, **265**, 275–293.

Tramu, G., Pillez, A., and Leonardelli, J. (1983) Serotonin axons of the ependyma and circumventricular organs in the forebrain of the guinea pig, *Cell Tissue Res.*, **228**, 297–311.

Ueda, S., and Sano, Y. (1986) Distributional pattern of serotonin immunoreactive nerve fibers in the lateral geniculate nucleus of the rat, cat and monkey (*Macaca fuxata*), *Cell Tissue Res.*, **243**, 249–253.

Van de Kar, L. D., and Lorens, S. A. (1979) Differential serotonergic innervation of individual hypothalamic nuclei and other forebrain regions by the dorsal and median midbrain raphe nuclei, *Brain Res.*, **162**, 45–54.

Wang, R. Y., and Aghajanian, G. K. (1978) Collateral inhibition of serotonergic neurons in the rat dorsal raphe nucleus. Pharmacological evidence, *Neuropharmacology*, **17**, 819–825.

Wang, R. Y., and Aghajanian, G. K. (1982) Correlative firing patterns of serotonergic neurons in rat dorsal raphe nucleus, *J. Neurosci.*, **2**, 11–16.

Weissmann, D., Belin, M. F., Aguera, M., Meunier, C., Maître, M., Cash, C. D., Ehret, M., Mandel, P., and Pujol, J. F. (1987) Immunohistochemistry of tryptophan hydroxylase in the rat brain, *Neuroscience* (in press).

White, S. R., and Neuman, R. S. (1980) Facilitation of spinal motoneurone excitability by 5-hydroxytryptamine and noradrenaline, *Brain Res.*, **188**, 119–127.

Wiklund, L., Björklund, A., and Sjölund, B. (1977) The indolaminergic innervation of the inferior olive. I. Convergence with the direct spinal afferents in the areas projecting to the cerebellar anterior lobe, *Brain Res.*, **131**, 1–21.

Wiklund, L., Descarries, L., and Møllgård, K. (1981a) Serotoninergic axon terminals in the rat dorsal accessory olive; normal ultrastructure and light microscopic demonstration of regeneration after 5,6-dihydroxytryptamine lesioning, *J. Neurocytol.*, **10**, 1009–1027.

Wiklund, L., Léger, L., and Persson, M. (1981b) Monoamine cell distribution in the cat brain stem. A fluorescence histochemical study with quantification of indoleaminergic and locus coeruleus cell groups, *J. Comp. Neurol.*, **203**, 613–647.

Wiklund, L., Sjolund, B., and Björklund, A. (1981c) A morphological and functional study on the serotoninergic innervation of the inferior olive, *J. Physiol.* (Paris), **77**, 183–186.

Willoughby, J. O., and Blessing, W. W. (1987) Origin of serotonin innervation of the arcuate and ventromedial hypothalamic region, *Brain Res.* (in press).

Wilson, J. R., and Hendrickson, A. E. (1987) Serotonergic axons in the monkey's lateral geniculate nucleus, *Brain Res.* (in press),

Zilles, K. (1985) *The Cortex of the Rat*, Springer-Verlag, Berlin, pp. 1–21.

CHAPTER 4

Modern Methods for Studying the Release of Serotonin

PETER J. KNOTT
Departments of Psychiatry and Pharmacology
The Mount Sinai School of Medicine
One Gustave L. Levy Place
New York
NY 10029
USA

INTRODUCTION

Progress in neuroscience is to a large extent regulated by the limitations of existing methods and it is particularly striking how the introduction of new techniques so frequently heralds exciting new discoveries. The acceptance of serotonin as a neurotransmitter and the exponential growth of knowledge of its role and pharmacology within the central nervous system appear to validate this observation.

Over thirty years ago, serotonin (5-HT) was first found to be present in brain tissue (Twarog and Page, 1953; Amin, Crawford and Gaddum, 1954). Since that discovery, a wealth of both direct and circumstantial evidence has accumulated to support serotonin as a neurotransmitter in the central nervous system. Evidence of release was essential to establish it as a neurotransmitter candidate, as with any substance found in brain tissue and suspected to have neurohumoral function. A number of methods were developed for investigating release of putative neurotransmitters *in vivo*, all of which depended on the perfusion of different brain structures. Substances released from neurons into the extracellular fluid (ECF) were recovered in the perfusing fluid for subsequent assay by any method with adequate sensitivity (Myers, 1970, 1972; Vogt, 1969). Serotonin release has also been studied *in vitro* using minced, chopped or sliced neuronal tissue, synaptosomes or cultured cells which were incubated, perfused or superfused with physiological solutions. However, because of the extraordinary progress that has been made in recent years on the measurement of serotonin release and metabolism from brain tissue *in vivo*, even in conscious, freely moving animals in parallel with behavior, present discussion will be limited to methods for serotonin release *in vivo*.

METHODS FOR MEASURING NEUROTRANSMITTER RELEASE: GENERAL CONSIDERATIONS

Invasive techniques

The process of perfusion, sample collection and subsequent assay formed the basis of all investigations of serotonin release in the brain until the electrochemical technique of voltammetry was introduced. Voltammetry enables the direct detection and quantification of neurotransmitters in neural tissue at implanted electrodes without removal of samples for subsequent assay. Nevertheless, both perfusion and voltammetry are invasive techniques, requiring insertion of cannulae or electrodes. It is important that these probes produce minimal trauma to brain tissue and so miniaturization and judicious choice of physiologically compatible materials for their construction have been major considerations. With perfusion techniques also, the salt composition and osmolarity of the physiological solution used (or artificial CSF) must approximate closely that of the extracellular fluid. Perfusion of solutions incompatible with cells may either impair neuronal function or provoke abnormal basal firing. Ideal artificial CSF compositions to avoid these problems were therefore developed (e.g. Myers, 1972). Probes should be implanted using standard stereotaxic techniques and sterile conditions for surgery to avoid infection and, at the end of every study, the location of electrodes or cannulae must be verified and any tissue damage evaluated histologically.

Assay procedures for serotonin and 5-HIAA

Development of new and sensitive assays for serotonin (5-HT) and its deaminated acid metabolite 5-hydroxyindoleacetic acid (5-HIAA) has greatly contributed to the ability to monitor release and metabolism of neuronal serotonin. Bioassay (Vane, 1957), frequently used in early perfusion studies, often failed to measure basal serotonin release unless monoamine oxidase inhibitors were added to the perfusion fluid (Vogt, 1969). Another difficulty was that drugs, investigated for effects on serotonin release, might also influence the bioassay preparation and give spurious results. Spectrofluorometric assays adequately sensitive for brain tissue serotonin measurement (e.g. Maickel, Cox, Saillant and Miller, 1968; Jonsson and Lewander, 1970) detected 5-HIAA but not serotonin in cerebral perfusates (Guldberg and Yates, 1968). Other advances, such as gas chromatography–mass spectrometry (GC–MS) assays (Cattabeni, Koslow and Costa, 1972) provide high sensitivity and incomparable selectivity but are arduous and cost-prohibitive for most laboratories. Radioenzymatic assays with sensitivity to serotonin in the picogram range (Saavedra, Brownstein and Axelrod, 1973) were adapted for brain perfusates (Boireau, Ternaux, Bourgoin, Héry, Glowinski and Hamon, 1976) but are laborious, requiring column isolation and double enzymatic conversion of serotonin to labeled melatonin. Other procedures labeled the endogenous pool with isotopic serotonin (Tilson and Sparber, 1972) or with radioactive tryptophan (Glowinski, 1981) and provided valuable pharmacological and physiological information on serotonin release. However, absolute amounts of released serotonin are not measured because the extent to which the endogenous pool is labeled is indeterminable.

Unequivocally, the greatest analytical tool to complement perfusion technology is high-performance liquid chromatography with electrochemical detection (HPLC-EC or LC-EC) due to the innovation of Adams, Kissinger and others (Adams and Marsden, 1982; Mefford, 1981). Major improvements in HPLC columns have culminated in the use of microbore reversed phase columns with the capacity to separate essentially all biogenic amines and their metabolites within a single chromatographic run in less than 15 minutes from injected sample volumes of only a few microliters. Reduction in the detector cell volume and improvement to electrochemical controller circuitry have maximized sensitivity, improved baseline separation of closely eluting substances and reduced noise. Numerous HPLC-EC assays have been developed for serotonin and 5-HIAA in sample volumes as little as 5μl and for femtomole quantities. In most assays reported, detection limits for these 5-hydroxyindoles are in the order of 5 pg injected (e.g. Lackovic, Parenti and Neff, 1981; Mefford and Barchas, 1980; Wagner, Vitali, Palfreyman, Zraika and Huot, 1982). Fluorometric detection coupled with HPLC (HPLC-FD) is generally less sensitive than HPLC-EC, with injected detection limits in the 5–15 ng range for indoles (Graffeo and Karger, 1976). However, because catecholamines and their metabolites have much weaker native fluorescence than the indoles, HPLC with dual electrochemical and fluorimetric monitoring can be a useful tool for peak identification (Cross and Joseph, 1981).

Direct versus indirect methods

The difficulty that early assays frequently failed to detect the low concentrations of serotonin present in brain perfusates, although they were adequately sensitive for tissue measurements of 5-hydroxyindoles, led to alternate approaches for evaluating serotonin release. Numerous indirect methods were developed for studies of brain serotonin metabolism or turnover (Curzon, 1981) and these measures were frequently equated with, or discussed in terms of, serotonin release. Measurements of brain serotonin concentration from extraction and assay of whole tissue will include both the stored intraneuronal amine plus a much smaller extraneuronal (extracellular) pool of indoleamine which has been released. Consequently, these measurements will not differentiate between stored and released serotonin. However, some indication of serotonin turnover or metabolism is provided by concurrent tissue measurement of both serotonin and 5-HIAA. Increases of serotonin release would be expected to concur with increased 5-HIAA production whereas reduced 5-HIAA levels would result from reduced serotonin release. Therefore, evidence of altered serotonin release in response to drugs and environmental change were provided by brain tissue measurement of serotonin and 5-HIAA when animals were killed in groups at set intervals during various treatments (Curzon and Knott, 1976) However, 5-HIAA tissue changes cannot always be equated with altered release. For example, administration of tryptophan, or any treatment which increases brain concentrations of the precursor, although increasing serotonin formation (Curzon and Knott, 1976) may also increase 5-HIAA production by intraneuronal deamination of serotonin that is not sequestered in the storage pool. In other words, it is possible for 5-HIAA to be formed from serotonin *without* release occurring. Another disadvantage of such studies is that *groups* of animals must be killed at intervals should data on time courses of changes be required. However, it is now possible to monitor serotonin release, or at least metabolism, in selected brain regions of conscious, relatively unrestrained, individual animals either acutely or over prolonged periods of study. This has been accomplished by technological improvements in traditional perfusion methods and also by the application of electrochemical techniques *in vivo* to the study of brain neurochemistry. Advances in assay methods for the biogenic amines together with automation by microprocessors have had a major impact both in perfusion and electrochemical technology.

The merits and disadvantages of any technique, to an extent intuitively predictable, are only fully revealed by use. Appraisal of novel techniques for monitoring serotonin release is made by comparison with existing methodology and by the extent to which it provides predictable information in test situations. For example, the technique should reveal increased serotonin release in response to electrical or chemical stimulation of the appropriate raphe nucleus, and conversely a decrease after raphe lesions or pharmacological inhibition of serotonin synthesis. Fulfilment of these criteria will then permit applications to other biological problems. Consequently, discussion of these techniques and their utility also warrants some reference to applications in pharmacological, physiological and behavioral investigations.

PERFUSION AND COLLECTION METHODS

Collecting cup

The collecting cup consists of a small cylinder of plexiglass, plastic or stainless steel placed through a burr hole in the skull so that it rests on the brain surface. Substances diffuse from brain into a physiological solution (artificial CSF), filling the cup which is removed at intervals (or by superfusion and continuous sampling) for analysis, and the cup refilled with fresh artificial CSF. The technique, developed originally for studies of acetylcholine release from the cortical surface (MacIntosh and Oborin, 1953), was also used to measure serotonin release from the sigmoid gyrus of the anesthetized cat, albeit in the presence of a monoamine oxidase inhibitor (Eccleston, Randić, Roberts and Straughan, 1969). Another version of the collecting cup (Besson, Cheramy, Gauchy and Glowinski, 1973) was used to measure dopamine release in the unanesthetized cat 'encéphale isolé' preparation which essentially shows all sleep stages. A sensitive radioenzymatic assay enabled serotonin measurement from the cat caudate nucleus collected by superfusion from its ependymal surface (Fig. 1). This required removal of the cortex overlying the caudate so that the cup could be positioned (Ternaux *et al.*, 1976, 1977) and a flow of oxygenated artificial CSF maintained over the brain tissue was continuously removed for assay. The efflux of serotonin was markedly increased by potassium-provoked depolarization, L-tryptophan or 5-hydroxytryptophan (5-HTP) but was reduced by lysergic acid diethylamide (LSD) which reduces serotonergic neuronal firing. The superfusion cup tech-

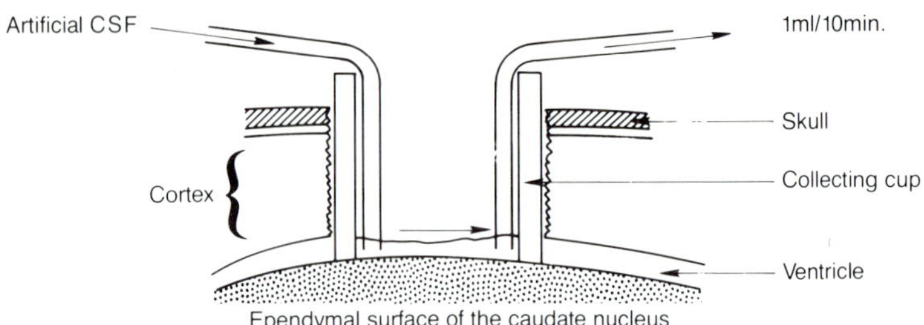

Fig. 1 Superfusion of the cat caudate nucleus. A plastic collecting cup was placed on the ependymal surface of the cat caudate nucleus following surgical removal of overlying cortex. Artificial CSF (pH 7.4), maintained at 37°C and aerated with O_2–CO_2 (95%–5%) mixture pumped into the cup to superfuse the brain tissue was collected every 10 min. Samples collected at 0°C were analyzed for serotonin radio-enzymatically. For details see Ternaux, Héry, Hamon, Bourgoin and Glowinski (1977)

nique was also used to measure serotonin release from the cortical surface which showed similar increases with potassium depolarization and also with the monoamine oxidase (MAO) inhibitor clorgyline (Daszuta, Gaudin-Chazal, Barrit, Faudon, Puizillout and Ternaux, 1979) but diminished during sleep (Puizillout, Gaudin-Chazal, Daszuta, Seyfritz and Ternaux, 1979). A disadvantage is that the technique is limited to areas of the cortical surface or to those structures accessible by removal of overlying tissue. The possibility that the surgery may influence brain function or that inadequate control of hemorrhage may contaminate samples with serotonin released from blood platelets are both potential problems (Héry, Faudon and Fueri, 1986).

Cerebroventricular perfusion and CSF sampling techniques

Feldberg and his colleagues opened up the field by developing several procedures for collection of neurotransmitters and their metabolites released from brain tissue into the cerebroventricular system. They introduced artificial CSF through cannulae implanted into either the third, fourth or lateral ventricles and collected perfusate from either the ventricles, the cisterna or the cerebral aqueduct (Bhattacharya and Feldberg, 1958; Carmichael, Feldberg and Fleischauer, 1964; Feldberg and Myers, 1966; Vogt, 1969). Feldberg and Myers (1966) first detected serotonin by bioassay in effluent ventricular perfusion samples of the anesthetized cat, and this was subsequently confirmed by Portig and Vogt (1969). Later, increased serotonin concentrations in ventricular perfusates were found following raphe stimulation (Ashkenazi, Holman and Vogt, 1972; Holman and Vogt, 1972) or monoamine oxidase inhibition (Goodrich, 1969). In other studies, spectro-fluorometric assays enabled measurement of 5-HIAA in CSF (Guldberg,

Ashcroft and Crawford, 1966) or ventricular perfusates (Guldberg and Yates, 1968). The lack of chemical assays with adequate sensitivity to consistently measure basal serotonin release into perfusates led to the introduction of various radiotracer methods. For example, tritiated serotonin introduced by perfusion to label the endogenous serotonin-releasable pool enabled measurement of minute amounts of the amine (Tilson and Sparber, 1970) but may also result in uptake of the isotope into non-serotonergic neurons if the tracer load is not limited (Gallager, Sanders-Bush, Aghajanian and Sulser, 1975). However, labeling of the neuronal serotonin pool can be selectively achieved by prior infusion or adminstration of the radioactive *precursor* since labeled serotonin is only formed endogenously in cells containing tryptophan hydroxylase (i.e. serotonin neurons). This procedure coupled with cerebroventricular perfusion demonstrated increased serotonin release in the cat upon raphe stimulation (Chiueh and Moore, 1976), whereas reduction of raphe firing in rats following chlorimipramine or LSD was accompanied by decreased serotonin release (Gallager and Aghajanian, 1975).

In other studies, Moore and colleagues used a concentric cannula based on a push-pull perfusion cannula developed by Tilson and Sparber (1970, 1972), and small enough for ventricular sampling in the rat (Nielsen and Moore, 1982a, 1982b; Moore and Nielsen, 1985). This cannula (Fig. 2a) was stereotaxically implanted wtih its tip in the lateral ventricle and, upon recovery from surgery, a four-channel peristaltic pump was used to simul-taneously deliver artificial CSF (flow rate, 20 µl/min) through the inner (push tube) and withdraw CSF through the outer (pull) tube. Samples for HPLC-EC analysis were automatically collected. Readily measurable in these CSF samples, 5-HIAA was found to be fairly constant in a group of rats over a four-week period (ranging from 15.1 ± 5.7 to 17.1 ± 3.6 ng/ml) although differences were noted between animals. Serotonin, however, was undetect-able, possibly due to its rapid metabolism to 5-HIAA (Moore and Nielsen, 1985). As might be anticipated, probenecid, which competes with active transport of acid metabolites at the choroid plexi, sharply increased 5-HIAA concentrations, whereas the monoamine oxidase inhibitor pargyline markedly reduced 5-HIAA (Nielsen and Moore, 1982). Prior treatment with the selec-tive serotonergic neurotoxin, 5,7-dihydroxytryptamine, lowered the basal level of 5-HIAA (but acid metabolites of dopamine were unchanged) and also blocked the increase of 5-HIAA found in response to tryptophan admin-istration (Nielsen and Moore, 1983).

Elghozi and co-workers (Elghozi, Mignot and LeQuan-Bui, 1983; Mignot, Laude and Elghozi, 1984; Danguir, LeQuan-Bui, Elghozi, Devynck and Nicolaidis, 1982; Mignot, Serrano, Laude, Elghozi, Dedek and Scatton, 1985) have also used a concentric cannula arrangement for ventricular

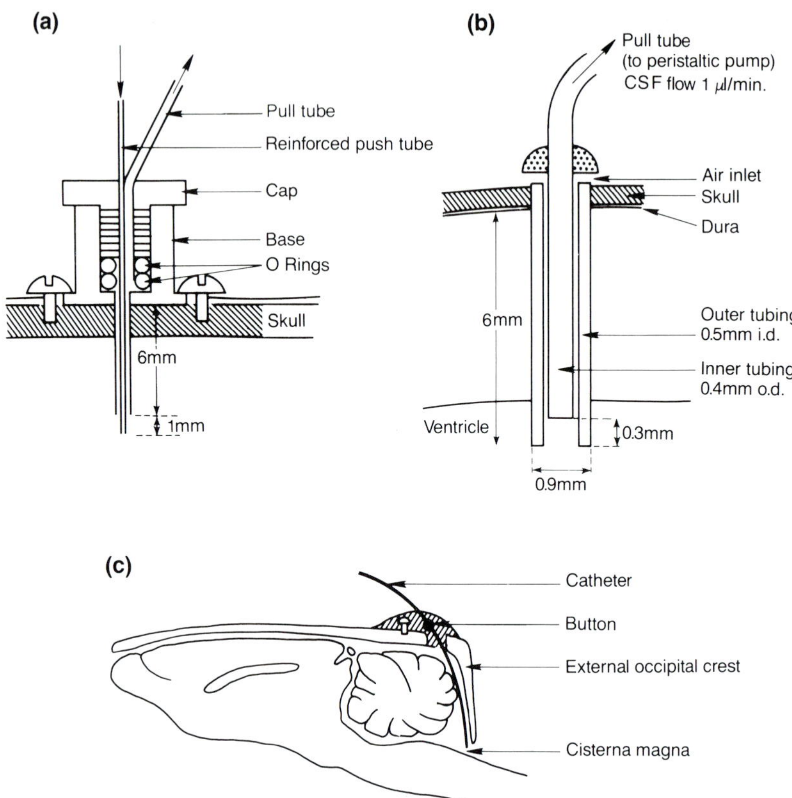

Fig. 2 Techniques for CSF sampling in the rat. (a) Ventricular CSF sampling from the conscious rat. A steel guide cannula (25G) attached to a stainless steel base was chronically implanted with its tip in the rat lateral ventricle. The stainless steel perfusion cap was constructed so that when screwed down into the stainless steel base the central push tube passed into the bore of the guide cannula and projected up to 1 mm beyond its tip into the ventricle. O-rings formed a seal between base and perfusion cap to prevent leaks during infusion. The pull tube also mounted in the perfusion cap opened into the small dead-space between cap and base. (After Moore and Nielsen, 1985.)
(b) Ventricular CSF sampling from the anesthetized rat. A stainless steel guide cannula was placed with its tip in the anterior third ventricle (6 mm below the dura). For sampling the mandrel (present during stereotaxic placement and not shown here) was replaced by a sampling needle shorter than the guide cannula. Serial samples, collected every 15 min by connecting the inner needle to a peristaltic pump (flow rate, 1 µl/min) were analyzed by HPLC-EC. (After Mignot, Laude and Elghozi, 1984.)
(c) Cisternal sampling from the conscious rat. A polypropylene catheter (PP10) runs in a midline slot (made in the skull anterior to the occipital crest) caudally to the cisterna. During implantation, catheters contain enameled wire

sampling from the anesthetized rat (Fig. 2b) but their system uses an open pulling system, unlike that of Moore *et al.* The sampling needle is placed inside a slightly longer guide cannula and ventricular CSF is withdrawn at a rate of 1 µl/min using a peristaltic pump. During sampling, air may enter the system through the open end of the guide cannula to safeguard against excessive negative pressure in the ventricular system during withdrawal (hence 'open pulling system'). Ventricular CSF 5-HIAA concentrations, measured by HPLC-EC, were dramatically increased above their basal concentrations (2.56×10^{-6}M) following probenecid injection. The accumulation rate was therefore also a measure of 5-HIAA elimination and of serotonin turnover (Elghozi, Mignot, and LeQuan-Bui, 1983). Following monoamine oxidase inhibition with tranylcypromine, ventricular CSF 5-HIAA declined exponentially allowing calculation of the serotonin turnover rate (Mignot, Laude, Elghozi, 1984). An exponential decline of 5-HIAA also followed the decarboxylase inhibitor (NSD1015) or the tryptophan hydroxylase inhibitor α-propyldopacetamide which would impair serotonin synthesis and release resulting in reduced amine turnover and 5-HIAA production (Mignot, Serrano, Laude, Elghozi, Dedek and Scatton, 1985).

Another technique for monitoring CSF changes of 5-HIAA based on either repeated sampling or continuous cisternal CSF collection, was developed by Curzon and co-workers for investigations in unanesthetized freely moving rats (Hutson and Curzon, 1987; Sarna *et al.*, 1983, 1984). A polypropylene catheter was surgically implanted in a slot made in the skull and passed caudally to the cisterna (Fig. 2c). Connection of the catheter through a servo-assisted liquid swivel allowed free movement of the rat. CSF was drawn (1 µl/min) by a peristaltic pump to a refrigerated fraction collector and the samples were subsequently assayed for 5-HIAA by HPLC-EC. The measurement of serotonin turnover in individual, freely moving rats was calculated from the linear rise of 5-HIAA following probenecid injection and demonstrated a pronounced reduction following tryptophan hydroxylase inhibition with *p*-chlorophenylalanine (pCPA) (Sarna, Hutson and Curzon, 1983; Hutson, Sarna, Kantamaneni and Curzon, 1984; Hutson, Sarna and Curzon, 1984), which is a situation in which turnover probably parallels serotonin release. These workers also found that serotonin turnover was greater in the dark than the light period so that ultradian and circadian studies are feasible (Hutson, Sarna and Curzon, 1984). Furthermore, investigations of the pharmacokinetics of tryptophan administration in which CSF tryptophan and serotonin metabolism changes were monitored following tryptophan adminis-

---

to facilitate puncture of the dura. The heat-preformed button in the catheter serves both as a depth stop and secures the catheter when embedded in cranioplastic cement keyed to the skull with steel screws. After recovery, CSF sampling (1 µl/min) is achieved either manually or with a peristaltic pump and fraction collector. Details from Sarna, Hutson and Curzon (1984)

tration (Hutson, Sarna and Curzon, 1986; Hutson, Sarna, Kantamaneni and Curzon, 1985) and the correlation of serotonin turnover with social interaction in rats were possible with the technique (Sahakian, Sarna, Kantamaneni, Jackson, Hutson and Curzon, 1987; Hutson and Curzon, 1987), which endorse its considerable utility.

Push-pull perfusion

The collecting cup technique enabled measurement of serotonin release from the cortical surface and larger areas of the cat caudate but is generally unsuitable for study of deeper lying structures or discrete loci. Similarly, ventricular perfusion or CSF sampling allows monitoring of serotonin release, turnover or metabolism in those structures bordering the ventricular system. This may reflect serotonergic changes in larger structures in proximity to the ventricles such as the caudate nucleus but not in discrete nuclei except during those treatments causing a general lowering or elevation of serotonergic activity throughout the CNS.

Efforts to localize the site of neurotransmitter release in the brain led to the development of the push-pull cannula (Gaddum, 1961, 1962). These are generally constructed from two concentric steel tubes so that artificial CSF may be pumped through the inner (push) tube to the desired perfusion site and be removed from that location by withdrawal through the outer (pull) tube. Fluid delivery and removal was usually made with precision syringe pumps for fine control of flow rate. Various modifications to this basic design ranging from use of smaller cannulae, to bevel tips and varying the protrusion of the central push tube, were largely aimed at reducing tissue damage while maximizing contact of perfusion fluid with brain tissue (Myers, 1972; Redgrave, 1985). Serotonin release, measured by bioassay, into perfusates of the superior colliculus of the restrained rhesus monkey (Myers, Kawa and Beleslin, 1969; Beleslin and Myers, 1970a, b) and from various hypothalamic sites in the cat (Veale, Myers and Beleslin, 1973) was accomplished using similar push-pull cannulae. With the closed push-pull perfusion system used in these studies it was imperative to balance the inflow of artificial CSF with its removal from the cannula tip to avoid tissue trauma (e.g. Redgrave, 1985). Thus, expansion lesions, hemorrhage or even hydrocephalus may ensue from occlusion of the outflow cannula, whereas aspiration of brain tissue may occur if withdrawal rate exceeds inflow rate. This problem could be reduced by using carefully matched syringes placed back to back in the syringe drive assembly (Myers, 1972).

Another approach to reduce the danger of tissue damage is by using an open pulling system (Fig. 3; Niéoullon, Chéramy and Glowinski, 1977; Glowinski, 1981). Artificial CSF is pumped into the inner push cannula and withdrawn through the outer cannula by mild suction with a peristaltic pump.

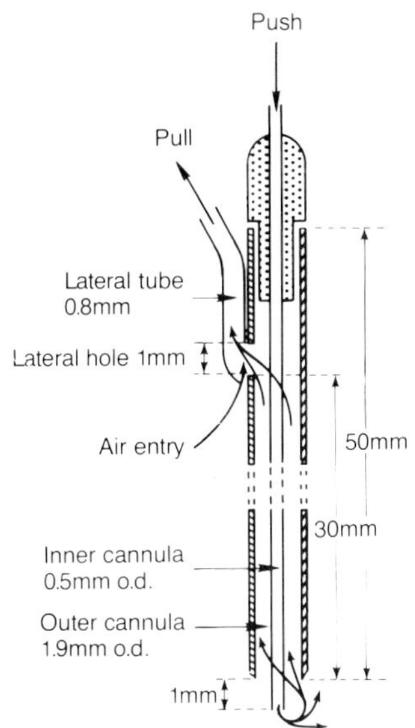

Fig. 3 Push–pull cannula for tissue serotonin release. Superfusion of discrete nuclei and brain sites in the anesthetized cat was made with a push–pull cannula of the type shown here. Artificial CSF, containing [³H]tryptophan, was pumped into the inner (push) cannula (33 µl/min) and withdrawn through the outer (pull) cannula with a peristaltic pump. An air vent in the outer cannula safeguards against excessive negative pressure at the tip. Samples were collected and counted for released [³H]serotonin after chromatographic isolation from the perfusate. Details from Niéoullon, Chéramy, and Glowinski (1977) and Héry, Simonnet, Bourgoin, Soubrié, Artaud, Hamon and Glowinski (1979)

However, the extreme negative pressure at the tip, occasionally problematic with closed systems, is eliminated due to a small hole in the outer cannula which allows air to enter. This cannula has been used by Glowinski's group in their classical studies of dopamine and serotonin release measured concurrently from the cat basal ganglia and substantia nigra. Serotonin release was monitored by the efflux of [³H]serotonin endogenously formed from [³H]tryptophan added to the artificial CSF used for perfusion. This technique clearly measured newly synthesized, apparently preferentially released serotonin since the labeled transmitter appears within 20 min of commencing infusion with the isotopic precursor. The technique offers great sensitivity, but it is important to ensure the isotopic precursor is free of contaminating traces of [³H]serotonin and to completely separate the labeled serotonin formed from the tritiated precursor present in the effluent. Both of these precautionary measures are achieved by chromatography. Serotonin release in the feline caudate nucleus was markedly increased in the presence of the selective serotonin uptake inhibitor, fluoxetine, following depolarization with either potassium or batrachotoxin or by stimulation of the dorsal raphe nucleus with glutamate (Héry, Simonnet, Bourgoin, Soubrié, Artaud, Hamon and Glowinski, 1979). Other studies showed that dopamine applied to the substantia nigra reduced caudate [³H]serotonin release ipsilaterally whereas dopamine synthesis inhibition produced the opposite effect (Bourgoin et al., 1981; Héry et al., 1979, 1980). Sampling from push-pull cannulae at multiple sites enabled investigations of the very complex relationships between nigral and striatal serotonin release during sensory stimulation, which suggested that serotonergic neurons have an important role in integrating sensory information. Visual and auditory stimuli both increased nigral (but not striatal) serotonin release in awake encéphale isolé cats, whereas in the anesthetized preparation, unilateral forepaw stimulation reduced serotonin release in both caudate nuclei and the ipsilateral substantia nigra (Reisine *et al.*, 1982; Soubrié *et al.*, 1981). These techniques are therefore ideally suited to investigations of serotonin release at precise brain loci in relation to physiological manipulations.

The advent of sensitive HPLC-EC assays for biogenic amines and metabolites eliminated the need for radioactive labeling and allowed the measurement of a number of neurotransmitters in the small sample volumes obtained by push-pull perfusions. Indeed, the union of perfusion technology and HPLC has proven to be a perfect marriage for studying brain neurotransmitter release *in vivo*. Loullis and colleagues were among the first to exploit this approach for measurement of 5-hydroxyindoles in perfusates. Perfusion of the lateral hypothalamus in freely moving rats showed that during the behavioral depression following 5-hydroxytryptophan (5-HTP) administration, the perfusate concentrations of 5-HTP, 5-HIAA and serotonin were elevated (Loullis, Hintgen, Shea and Aprison, 1980; Loullis and Hellhammer, 1987). Although conventional push-pull perfusion techniques are still quite widely used, the inherent problems associated with their use are a deterrent for many. This situation has however been reversed by the introduction of a variant of the perfusion approach, intracerebral dialysis.

Intracerebral dialysis

Certain disadvantages of push-pull perfusion, such as pressure lesions, aspiration of erosion of tissue at the perfusion site with damaged cells or tissue fragments being flushed out under the mechanical effects of continuous fluid movement (Myers, 1972; Redgrave, 1985) would be eliminated if the perfusing artificial CSF were separated from the extracellular fluid by a semipermeable or dialysis membrane. Attempts to accomplish this had been made in the early 1970s with the 'dialytrode' (Delgado, DeFeudis, Roth, Ryugo and Mitruka, 1972) which consisted of a small implanted dialysis bag or a hollow dialysis fiber through which the perfusion fluid was pumped (Ungerstedt and Pycock, 1974). These methods, variously termed intracerebral dialysis, intracranial dialysis or microdialysis, showed promise, but, as with push-pull methods, were limited by the sensitivity of available assay procedures. However, the explosion of sensitive HPLC analytical techniques for amines and their metabolites occurring in the late 1970s rekindled interest in microdialysis methods.

Several forms of Ungerstedt's dialysis fiber and Delgado's 'dialytrode' have emerged (Fig. 4). These probes, constructed from commercially available cellulose dialysis tubing (200–350 μm outside diameter) or any other biocompatible tubular, semipermeable membrane, connected to steel tubing allow sample collection from discrete brain regions. Thus appropriately coating the transcerebral 'Vita-Fibre' (Fig. 4a) with epoxy resin limits dialysis to a selected region (Ungerstedt, 1984). The dialysis loop (Fig. 4b) has an exposed diffusion length of 4–12 mm depending on the brain structure to be studied (Ungerstedt, 1984; Clemens and Phebus, 1986; Maidment, Routledge, Martin, Brazell and Marsden, 1986), but the parallel arrangement of the inflow–outflow tubes results in a larger probe than 'dialytrode' types (Fig. 4c; Hutson, Sarna, Kantamaneni and Curzon, 1985, or Fig. 4d; Hernandez, Stanley, and Hoebel, 1986) or recent commercially available concentric dialysis probes (Carnegie Medicin, Solna, Sweden). It is possible with currently available fluid delivery systems to control artificial CSF flow rates through these probes very accurately over the practically useful range of 0.8 μl–10 μl/min. In theory, any molecule small enough to pass through the membrane is recoverable, depending on the membrane used. Most investigators use membranes with a molecular weight (M.W.) limit of 5,000 although others are available up to 50,000 M.W. cut-off. Molecules such as serotonin and 5-HIAA therefore readily diffuse through these membranes into the flowing stream and, furthermore, the rate of diffusion from extracellular fluid into the perfusate is proportional to the concentration gradient. Ungerstedt (1984) has likened the implantation of microdialysis probes to introducing an exogenous 'blood vessel' into the tissue from which samples may be collected. It is normal practice to test dialysis probes *in vitro* by immersing them in standard solutions of indoleamines (and other substances of interest) at concentrations from 10^{-5}M to 10^{-7}M to estimate 'recovery' and select flow conditions that will be employed *in vivo*. The faster the flow rate, the greater will be the concentration gradient from the exterior to the lumen of the probe. Consequently, absolute recovery (e.g. total serotonin or 5-HIAA recovered per unit time) will increase until pressure in the lumen either causes outflow of fluid (and a fall in recovery) or ruptures the membrane. The relative percentage recovery (concentration of effluent/ external concentration $\times$ 100) will however be greatest at the lowest flow rate which allows equilibrium to occur. Relative recoveries of indoleamines also will increase with the membrane surface area or fiber length and values ranging from 5% to 25% have been reported for flow rates between 1–2 μl/min (Zetterstrom, Sharp, Marsden, and Ungerstedt, 1983; Hernandez, Stanley and Hocbel, 1986; Hutson, Sarna, Kantamaneni and Curzon, 1985). If required, brain extracellular fluid concentrations of serotonin and 5-HIAA may then be calculated from a knowledge of relative recovery and the basal brain dialysate values found for a given flow rate.

Microdialysis offers several advantages over conventional push-pull

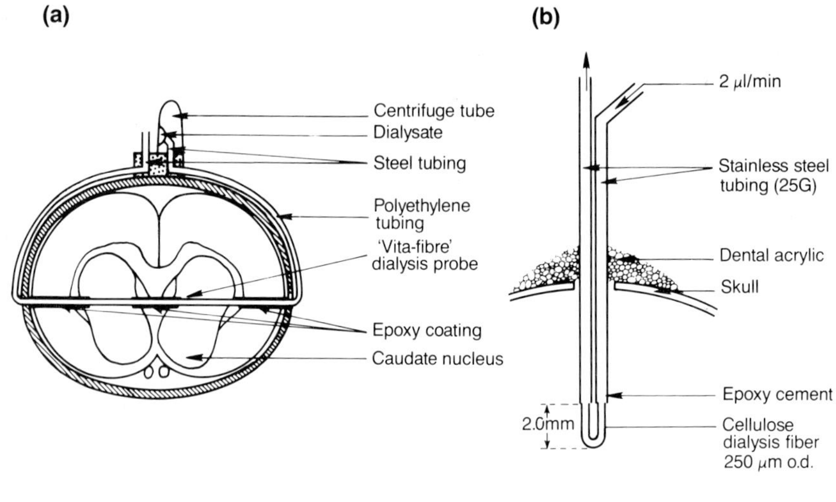

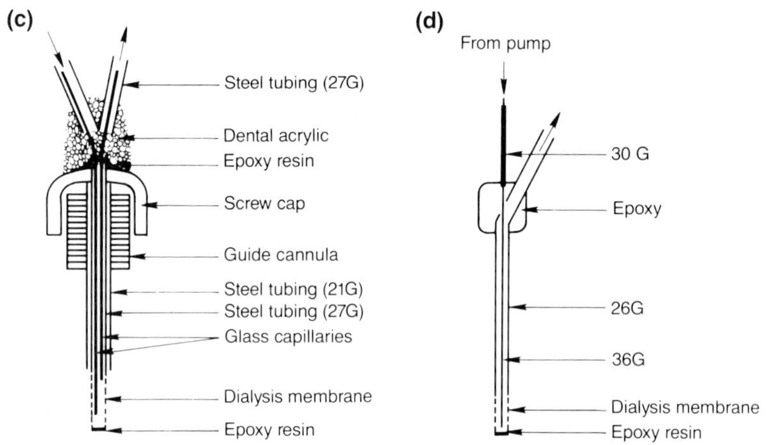

Fig. 4 Intracerebral dialysis probes. (a) The 'Vita-Fibre' for trans-striatal dialysis in the rat has an outside diameter of 340 μm and molecular weight cut-off of 50,000. Coating the probe appropriately with epoxy resin restricts diffusion to the caudate nuclei. Collection of dialysate in a microcentrifuge tube mounted on a cap fixed to the skull may be made by running the effluent subcutaneously in polypropylene (or polyethylene) tubing. (After Ungerstedt and Pycock, 1974.) (b) The dialysis loop, cemented into the ends of two parallel steel tubes, has an overall length of 4–12 mm × 250μm diameter, M.W. cut-off = 5,000. (After Ungerstedt, 1984; Clemens and Phebus, 1986.) (c) Dialytrode type with twin glass capillaries for delivery and withdrawal of artificial CSF from a superfused cylindrical membrane secured to the outer steel tubing with epoxy resin. Dialysis tip = 3.0 mm long × 300 μm diameter, M.W. cut-off = 5,000. (After Hutson, Sarna, Kantamaneni and Curzon, 1985.) (d) Concentric microdialysis probe. Dialysis tip = 2–4 mm long × 200 μm diameter, M.W. cut-off = 6,000. (After Hernandez, Stanley and Hoebel, 1986)

cannula methods. Balancing the inflow with outflow rate, of paramount importance with push-pull perfusion, is not an issue with dialysis probes since artificial CSF pumped through the probe is collected freely at the distal open-ended outflow tube. Suction is unnecessary for collection of the dialysate. Tissue erosion, occasionally a problem with chronic push-pull cannula studies, again is eliminated since the cells surrounding the probe are protected from mechanical damage due to the flowing stream by the dialysis membrane. Probes may be sterilized and the dialysis membrane is an additional safeguard against introducing infection at the dialysis site. Another advantage of the dialysis probe over conventional push-pull technology is that enzymes which may be present, or gain access to the extracellular space from damaged cells and might catabolize recovered substances, are excluded from the dialysate by the membrane. Reducing cell erosion and eliminating recovery of tissue debris provides a remarkably clean sample which contains no protein (at least above M.W. 5,000) and is therefore suitable for direct injection into the liquid chromatograph.

Microdialysis studies in freely moving rats may be made with a liquid swivel in the inflow line and collecting samples from the open-ended outflow tubing in microcentrifuge tubes attached to the skull assembly (Ungerstedt and Pycock, 1974), a small shoulder harness (Sharp, Zetterstom, Herrera-Marschitz, Ljungberg and Ungerstedt, 1986), the spiral spring cover protecting the inflow line (Hernandez, Stanley and Hoebel, 1986) or even running the effluent to a microprocessor-controlled injector for direct on-line HPLC analysis (Johnson and Justice, 1983; Justice and Neill, 1986). An advantage of using conscious animals is that interaction of the anesthetic effect with other drugs will be avoided. This is probable since microdialysis in the rat frontal cortex demonstrated that basal extracellular concentrations of serotonin (10.2×10^{-8}M) and 5-HIAA (6.1×10^{-7}M) were higher in freely moving than in chloral hydrate anesthetized rats (6.2×10^{-8}M and 4.3×10^{-7}M respectively) (Maidment, Routledge, Martin, Brazell and Marsden, 1986).

Other studies in freely moving animals also underscore the utility of this technique in investigations of the effects of a tryptophan load (Hutson, Sarna, Kantamaneni and Curzon, 1985). Thus, the time course of 5-HIAA and tryptophan concentration changes in rat striatal dialysates showed remarkable concordance with both changes in their concentrations in simultaneously collected CSF samples and with striatal tissue concentrations in groups of rats killed at intervals after tryptophan administration. Theoretically, with microdialysis these studies could be repeated over a period from days to weeks in the same animal but for the sacrificed rats that is impossible! Microdialysis therefore has the advantage that substantially fewer animals are required for turnover studies and they remain alive at the conclusion of the experiment and are available for additional studies. Hypothalamic serotonin metabolism has also been investigated in rats on a restricted food intake (Hutson and Curzon, 1987) by implanting guide cannulae for placement of dialysis probes into the hypothalamus. Immediately before food was made available and throughout the feeding period hypothalamic 5-HIAA increased but fell sharply after food removal. This suggested that hypothalamic serotonin metabolism and possibly also release increased prior to and during feeding. Other striking associations of serotonin release and behaviour were found with dialysis probes located in the frontal cortex. A temporal relationship was found between extraneuronal serotonin and the change of serotonin-dependent stereotypy induced by *p*-chloroamphetamine (pCA) which releases serotonin (Hutson and Curzon, 1987). Serotonin release was quadrupled, indicated by an increase of the dialysate concentration from 0.5 to 2.0 pmol/ml in response to pCA and was paralleled by serotonin-mediated behaviors of head weaving and forepaw padding. The miniaturization of these dialysis probes allows multiple implantation, and even neurochemical monitoring concurrently from more than one site in the hypothalamus is possible. Recently, Hoebel and colleagues (Hernandez, Stanley and Hoebel, 1986) using microdialysis measured the extracellular concentration of 5-HIAA concurrently in four brain regions of the conscious rat (i.e. nucleus accumbens, striatum, lateral and paraventricular hypothalamus), providing insight into the range of neurochemical-behavioral studies that may be possible in the future.

VOLTAMMETRY METHODS

General principles

Electrochemistry or voltammetry *in vivo*, an alternative approach to the measurement of neurotransmitters and their metabolites in brain extracellular fluid (ECF), has received considerable interest since its early use by the pioneering groups of Adams and Lane (see Adams and Marsden, 1982; Conti *et al.*, 1978; Hutson and Curzon, 1983; Knott *et al.*, 1985; Marsden *et al.*, 1981; Stamford, 1985, 1986). Based on standard electroanalytical procedures, substances located extracellularly are oxidized at an electrode implanted in the brain by applying a potential difference between it and the surrounding ECF (Fig. 5a). The electro-oxidizable neurotransmitters, dopamine, serotonin, norepinephrine and their metabolites, dihydroxyphenylacetic acid (DOPAC), homovanillic acid (HVA) and 5-HIAA are theoretically measurable, whereas non-oxidizable substances (acetylcholine, γ-aminobutyric acid, glutamic acid, etc.) are not. Molecules oxidize optimally above their characteristic redox potentials and the current generated by passage of electrons into the recording electrode (conventionally termed 'working' electrode) is proportional to the concentration of substances oxidizing at the electrode surface. Oxidizing potentials are applied relative to a standard 'reference'

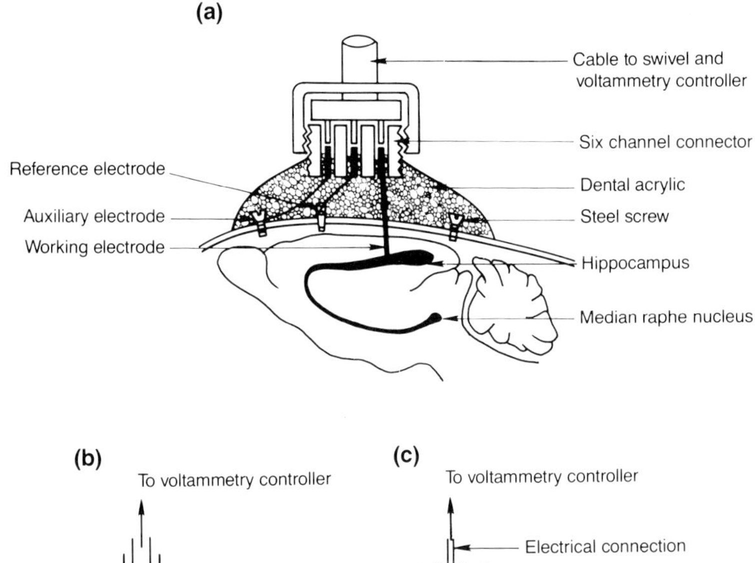

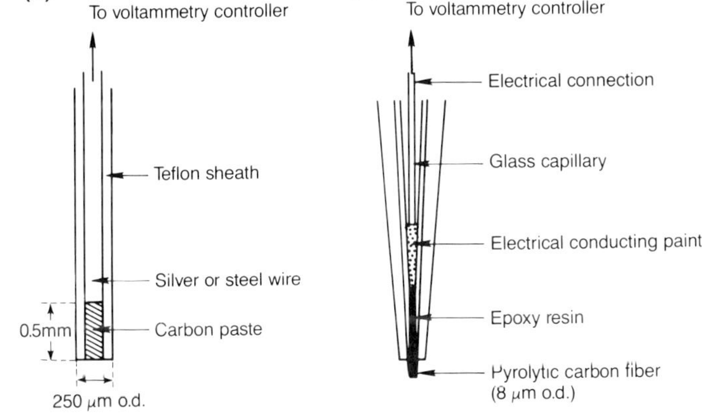

Fig. 5 Voltammetry *in vivo*. (a) Conventional arrangement of three-electrode system for voltammetric recording from the hippocampus. Placement of the steel screw serving as the auxiliary electrode, the silver–silver chloride reference electrode and the working electrode (either carbon paste or pyrolytic carbon fibers – see text).

(b) Carbon paste working electrode. The teflon sheath of insulated silver or steel wire (~250 μm o.d.) is pulled over the end of the core to form a well which is packed with a paste of graphite and oil (silicone or Nujol, i.e. paraffin oil).

(c) Pyrolytic carbon fiber working electrode. Single or multiple 8 μm diameter carbon fibers are cemented with epoxy into the tips of glass micropipet. Electrical continuity between the carbon fiber and the wire for connecting the probe to the voltammetry controller is made either with conducting paint or a mixture of epoxy resin and carbon powder. Electrodes are electrically pretreated to enhance sensitivity and selectivity for ascorbate, DOPAC and 5-HIAA. For details see Ponchon, Cespuglio, Gonon, Jouvet and Pujol (1979) or Sharp, Maidment, Brazell, Zetterstrom, Ungerstedt, Bennett and Marsden (1984)

electrode (e.g. silver–silver chloride) in contact with the brain, and a third, 'auxiliary' electrode ensures rigid control of the applied potential by feedback circuity. It is possible to monitor changes in conscious freely moving animals by cementing this standard three-electrode configuration to the skull and effecting electrical connections from the voltammetry controller and to the recording equipment with a slip-ring (Knott, Andrews and Mueller, 1985). Voltammetry *in vivo* may be likened to inserting the detector electrode of an HPLC-EC system directly into the brain region of interest and measuring all substances concurrently oxidized by the applied potential without chromatographic separation. This would be a most unsatisfactory analytical procedure, except that differences in electrode construction and the manner in which the potential is applied are both tailored to increase selectivity. Electrodes most used for intracerebral measurements are of two basic types, carbon paste (Fig. 5b) and the pyrolytic carbon fiber electrode (Fig. 5c), although many procedures exist to further modify their surface characteristics with resultant changes in sensitivity and selectivity. Typically, carbon paste electrodes are constructed by forming a small well at the end of Teflon-coated silver or steel wire by sliding the insulation over the end of the metal core (tip diameter 250–300 µm) and packing it with a mixture of graphite and oil (silicone or paraffin). Carbon paste can also be mixed with epoxy resin and squeezed into pulled glass capillaries and allowed to cure (tip diameter 50–100 µm).

The carbon fiber electrode is much smaller and more intricately constructed. First described by Ponchon and colleagues (Ponchon, Cespuglio, Gonon, Jouvet and Pujol, 1979) for measurement of catecholamines, it has been very successfully applied to indoleamine measurement. One or more carbon fibers (each of 8 µm diameter) are sealed into a glass capillary (drawn by an electrode puller into a micropipet) with epoxy resin (Fig. 5c). Electrical continuity between the carbon fiber and the wire connecting the electrode to the voltammetry controller is made either with a mixture of resin and graphite powder or conducting paint. The sensitivity and selectivity of these electrodes to catecholamines and indoleamines can be considerably enhanced by a process of electrical pretreatment which modifies the surface of the carbon fiber tip (Gonon, Fombarlet, Buda and Pujol, 1981; Cespuglio, Faradji, Ponchon, Buda, Riou, Gonon, Pujol and Jouvet, 1981; Brazell and Marsden, 1982; Ideda, Miyazaki, Mugitani, and Matsushita, 1984; Sharp, Maidment, Brazell, Zetterstrom, Ungerstedt, Bennett and Marsden, 1984). The advantages of these fiber electrodes over the larger paste electrodes include less tissue damage, greater suitability for implantation into smaller nuclei, and a much smaller charging current (see later), but the smaller tip area also means that generated oxidation currents will be smaller and require superior instrumentation for noise-free recording. However, all voltammetry electrodes so far employed have tip diameters greater than 8 µm, which is too large to penetrate cells (e.g. 0.5–2.0 µm microelectrodes are used for intracellular recording) and therefore will be detecting substances in the ECF. This will include electroactive transmitters and metabolites which have been released from neurons.

Linear sweep voltammetry and semi-differentiated linear sweep voltammetry

Linear sweep voltammetry (LSV) is the technique of *linearly* increasing the potential at the working electrode with a slow voltage ramp to oxidize substances in increasing order of their characteristic oxidation potentials. By displaying the resulting current output from the working electrode as a function of the applied voltage, characteristic 'voltammograms' are produced. As the applied potential reaches the oxidation potential for a particular substance, the peak is produced by the increased flow of current and then, because the unoxidized species in the immediate electrode vicinity is depleted, oxidation current declines, producing the downward slope of the peak. Voltammograms will consist of distinct peaks attributable to individual substances, superimposed on an increasing background current (providing their redox potentials are well separated) and the concentrations of these substances are proportional to their peak heights. Using unmodified carbon-paste electrodes, ascorbate and dopamine co-oxidize between approximately 75 and 200 mV but may be separated from the indoleamines, which oxidize between 300 and 400 mV (Lane, Hubbard and Blaha, 1979; O'Neill, Grün-wald, Fillenz and Albery, 1982; O'Neill *et al.*, 1983; Knott *et al.*, 1985) providing the scan rate is not too fast (e.g. below 20 mV/sec). The precise peak oxidation positions are also influenced by both scan limits and rate. For example, scanning more slowly and/or commencing the oxidizing scan from a more negative point (e.g. below 0 mV) will both tend to shift the peak to a lower potential.

Improved resolution of peaks has been achieved electronically by semi-differentiation of the generated oxidation current using simple additional circuitry for processing the output signal (Lane, Hubbard and Blaha, 1979). A typical semi-differentiated linear sweep voltammogram (scan conditions −200 mV to +500 mV; scan rate 10 mV/sec; Knott, Andrews and Mueller, 1985) from the striatum of the conscious rat shows several peaks which are compared with the location of peaks found by similarly scanning authentic substances *in vitro* (Fig. 6). Peak 1, occurring at ~100 mV *in vivo*, is increased by local microinfusions of ascorbate, dopamine or DOPAC at the electrode. Similarly, peak 2, although not occurring at an identical potential *in vivo* as the indoleamine oxidation *in vitro*, is increased by microinfusions of both serotonin and 5-HIAA close to striatal working electrodes (Mueller, Palmour, Andrews and Knott, 1985; Knott, Mueller and Young, 1986). Differences between peak positions *in vivo* and *in vitro* are explicable by

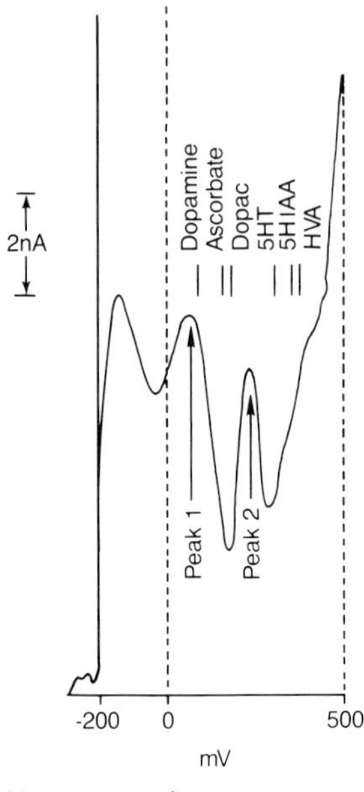

Fig. 6 Semi-differentiated linear sweep voltammogram obtained from rat caudate nucleus. Two distinct peaks due to electrochemical oxidation are observed when scanning the striatum with carbon paste electrodes from −200 mV to +500 mV at 10 mV/sec. The 'peak' below 0 mV as scanning commences depends on charging current due to the capacitance of the electrode. The peak positions for authentic substances from scanning *in vitro* are also indicated. When using standard untreated carbon paste electrodes, striatal peak 1 is due to ascorbate, DOPAC (and possibly dopamine). Peak 2 is where the 5-hydroxyindoles oxidize when infused close to the working electrode *in vivo*. However, uric acid is a major component of peak 2 in striatum with carbon paste electrodes (see Knott, Andrews and Mueller, 1985; Knott, Mueller and Young, 1986)

changes in the electrode surface characteristics produced by exposure to brain tissue (O'Neill, Grünwald, Fillenz and Albery, 1982). The 'peak' occurring at the outset of scanning, in this instance below 0 mV, results from the charging current due to electrode capacitance.

An advantage of LSV and semi-differentiated LSV is that at slow scan rates a degree of resolution of electroactive substances is obtainable. However, slow scanning causes build-up of oxidized material at the electrode surface and depletes substances being measured in the ECF-surrounding electrode. This precludes frequent measurements being obtained (e.g. less than 5 minute intervals) to avoid electrode poisoning and loss of peaks. Lane, Hubbard and Blaha (1979), using semi-differentiated LSV in rats with carbon paste electrodes, reported that the striatal serotonin peak was increased by 5-hydroxytryptophan injection. Others have found that the 'indoleamine' peak (attributed to ECF serotonin and 5-HIAA combined) in the serotonergic-rich rat hippocampus was increased by the selective serotonin uptake inhibitor, fluoxetine, by probenecid (apparently due to the 5-HIAA component) and with *p*-chloramphetamine. The simultaneous measurement of both serotonin plus 5-HIAA was invoked to explain the lack of net change following monoamine oxidase inhibition which would be expected to increase serotonin but reduce 5-HIAA concentrations (Joseph and Kennet, 1981, 1986; Kennet and Joseph, 1982). The hippocampal signal was strikingly increased by immobilization which apparently depended on the availability of tryptophan because these changes were attenuated by *p*-chlorophenylalanine and amino acids which competitively reduce cerebral tryptophan content (Joseph and Kennet, 1983). Others (Echizen and Freed, 1984; Freed, Echizen and Bhaskaran, 1986) reported that 5-HIAA in the rat dorsal raphe nucleus was increased during phenylephrine-induced hypertension and provide evidence that the indoleamine electrochemical signal is changed by agents in unison with changes obtained from tissue measurements using HPLC-EC.

From these studies, the evidence appears virtually overwhelming that using LSV, the peak at ~350 mV reflects either extraneuronal (and therefore released) concentrations of serotonin, or 5-HIAA, or both. However, uric acid in striatal dialysates from microperfusion studies is about 20 times higher than 5-HIAA (Ungerstedt, 1984) and oxidizes at a similar potential to indoleamines with paste electrodes *in vivo* (Knott, Mueller and Young, 1986). Furthermore, treatments aimed at increasing or decreasing uric acid around working electrodes in the striatum produced changes consistent with peak 2 reflecting endogenous uric acid. Thus, microinfusions close to the electrode of uric acid or xanthine oxidase increased, whereas allopurinol treatment and uricase infusion decreased the height of the 'indoleamine' peak (O'Neill, Fillenz, Grünwald, Blomfield, Albery, Jamieson, Williams and Gray, 1984; Mueller, Palmour, Andrews and Knott, 1985; Knott, Mueller and Young, 1986). These findings suggest that in rat striatum 5-HIAA contribution to peak 2 when measured by LSV with paste electrodes is minimal. The relative contributions of 5-HIAA and urate to peak 2 in the hippocampus and raphe nuclei, areas rich in serotonergic terminals and cell bodies respectively, remain to be evaluated.

Another variation of LSV is cyclic voltammetry in which, for example, the voltage ramp is applied from 0 V, first in an anodal direction (when oxidation occurs), and then at a selected potential, and reversed to an equal and opposite cathodal potential (at the same scan rate) before returning to 0 V. During the cathodal-directed sweep, reduction of reversibly oxidized substances will occur. High-speed cyclic voltammetry at carbon fiber micro-electrodes was used to quantify serotonin, ejected iontophoretically into the somatosensory cortex of the anesthetized rat (Kruk, Armstrong-James and Millar, 1980). The short duration of voltage application reduces surface poisoning and means that repeated measurements may be made at short intervals, but rapid scanning provides poor resolution.

Chronoamperometry

In another electrochemical procedure, a constant potential of short duration (e.g. 1 second) is applied at constant intervals (e.g. every 20 seconds or longer) and the current generated is measured and recorded. The term 'chronoamperometry' is used to describe this type of measurement because readings are obtained repeatedly over extended periods. To avoid measurement of the initial charging current due to electrode capacitance, only the current at the end of the applied potential is sampled (e.g. in the last 1/10 second). Chronoamperometry has the advantages that measurements may be made more frequently than with LSV (at 20 sec intervals or greater) and, due to the brevity of each pulse, surface poisoning due to concomitant build-up of oxidized material is minimized. The major disadvantage is that *all* substances at the electrode surface will be measured if their oxidation potentials are less than the applied potential. For example, chronoamperometry at 0.6 V will oxidize ascorbic acid, dopamine, DOPAC, norepinephrine, serotonin and 5-HIAA if present. Nevertheless, with chronoamperometric monitoring in the serotonergic terminal-rich hippocampus, the signal was increased, apparently due to enhanced serotonin release in response to medial raphe stimulation but not in the striatum, which lacks serotonergic projections from this nucleus (Marsden, Conti, Strope, Curzon and Adams, 1979). Increased hippocampal serotonin release measured by chronoamperometry also paralleled serotonin-mediated behavioral changes in response to *p*-chloroamphetamine. Both behavioral and voltammetric signal changes were attenuated by serotonin inhibition with *p*-chlorophenylalanine pretreatment (Marsden, 1979).

Differential pulse voltammetry

The great promise shown by LSV and chronoamperometric monitoring of the indoleamine signal led to exploration of other possible electrochemical techniques. Differential pulse voltammetry (DPV) at electrically pretreated pyrolytic electrodes offers greater selectivity for cerebral 5-HIAA monitoring than other electrochemical techniques developed so far. By superimposing short-duration voltage pulses (e.g. 20–50 ms) of constant amplitude and fixed interval on a linearly increasing voltage ramp at a suitably prepared carbon fiber electrode, it has been possible to differentiate between ascorbic acid, DOPAC and 5-HIAA both *in vitro* and *in vivo* (Fig. 7). Suitable preparation of these electrodes is by electrical pretreatment which involves passing current through the electrode in response to a series of triangular waveform and d.c. potentials while immersed in either phosphate buffered saline (Gonon, Fombarlet, Buda and Pujol, 1981; Cespuglio, Faradji, Ponchon, Buda, Riou, Gonon, Pujol and Jouvet, 1981; Sharp, Maidment, Brazell, Zetterstrom, Ungerstedt, Bennett and Marsden, 1984) or dilute sulphuric acid (Ikeda, Miyazaki, Mugitani and Matsushita, 1984). Electrochemical pretreatment alters the electrode surface so that peak resolution and sensitivity are dramatically enhanced. Furthermore, electrode pretreatment and scanning conditions can be selected for preferential measurement of either ascorbate (peak 1) and DOPAC (peak 2) or else for 5-HIAA (peak 3). When optimizing for 5-HIAA detection, ascorbate and DOPAC are not usually resolved although concurrent measurement of all three substances in the brain may be possible (Ikeda, Miyazaki, Mugitani, and Matsushita, 1984). A differential pulse voltammogram from the striatum (Fig. 7) shows the oxidation peaks due to ascorbate (peak 1, −50 mV), DOPAC (peak 2, +100 mV) and 5-HIAA (peak 3, +300 mV) (Cespuglio, Riou, Buda, Faradji, Gonon, Jouvet and Pujol, 1980; Cespuglio, Faradji, Ponchon, Buda, Riou, Gonon, Pujol and Jouvet, 1981).

Serotonin and 5-HTP also oxidize *in vitro* to produce a peak at 300 mV (the same location as the 5-HIAA peak) but pharmacological evidence strongly suggests that under control conditions *in vivo*, peak 3 reflects principally the

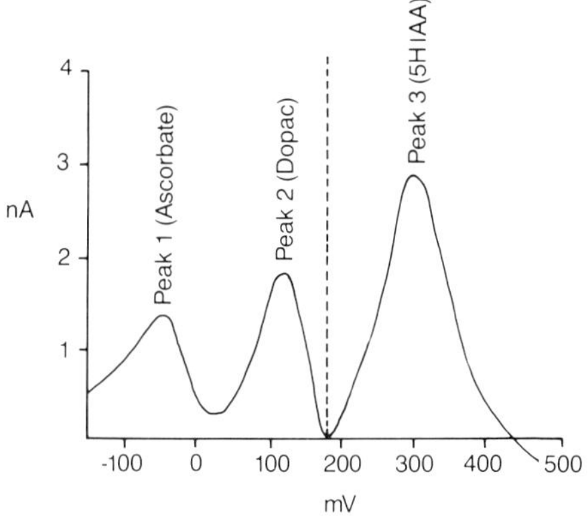

Fig. 7 Differential pulse voltammogram from rat caudate nucleus, obtained with electrically pretreated carbon fiber electrodes using DPV. Selectivity for catechol or indole compounds is determined by electrode pretreatment and electrochemical scanning parameters. Peak 3 is usually recorded under separate conditions from peaks 1 and 2. For example, peak 1 (ascorbate) and peak 2 (DOPAC) may be obtained by scanning from −0.2 to +0.15 V, whereas peak 3 (5-HIAA) may be obtained by scanning from −0.2 to +0.45 V, scan rate 5 mV/s, amplitude 50 mA, frequency 2.5/s (peaks 1 and 2) or 2/s (peak 3) (see Cespuglio, Faradji, Ponchon, Buda, Riou, Gonon, Pujol and Jouvet, 1981; Sharp, Maidment, Brazell, Zetterstrom, Ungerstedt, Bennett and Marsden, 1984)

ECF concentration of 5-HIAA rather than serotonin. For instance, in the rat, unilateral chemical (5,7-dihydroxytryptamine) and electrolytic lesions of the median forebrain bundle produced an ipsilateral striatal loss of peak 3 and in additional studies this peak was increased by tryptophan or reserpine but reduced by the monoamine oxidase inhibitor, clorgyline, or by the decarboxylase inhibitor, NSD 1015 (Cespuglio, Riou, Buda, Faradji, Gonon, Jouvet and Pujol, 1980). Peak 3 therefore responds to these treatments as a deaminated metabolite rather than an indoleamine. DPV with carbon fiber electrodes has been used extensively at numerous sites within the CNS in many other studies which support its capacity to monitor ECF 5-HIAA changes as measured by peak 3.

The effects of raphe stimulation on DPV signals from electrodes located in serotonergic projection areas is one source of evidence that the 300 mV peak is at least a marker for serotonin release. In anesthetized rats, electrical stimulation of the nucleus raphe magnus increased the 300 mV DPV peak in the dorsal horn of the spinal cord. Both the peak and the increase to raphe stimulation were abolished by pCPA pretreatment, indicating that serotonin metabolism was monitored (Rivot, Chiang and Besson, 1982). Other experiments in which the raphe were stimulated have demonstrated similar increases of the 5-HIAA signal. In the rat somatosensory cortex the DPV peak at 300 mV was increased by stimulation of the dorsal raphe nucleus or lateral hypothalamus and reduced by pCPA pretreatment, which also abolished the cortical signal response to electrical stimulation of the ascending serotonergic pathways (Rivot, Lamour, Ory-Lavollee and Pointis, 1983). Moreover, a laminar distribution of the 5-HIAA peak was shown in the rat somatosensory cortex where the hydroxyindole signal decreased as the electrodes were inserted deeper (Lamour, Rivot, Pointis and Ory-Lavollee, 1983).

Release from serotonergic nerve cell bodies has also been explored with DPV. Crespi and Jouvet (1983, 1984) and Cespuglio *et al.* (1983), using DPV with electrically pretreated carbon fiber electrodes, chronically implanted in four raphe nuclei (raphe dorsalis, centralis, pontis, and magnus) of the rat, were able to monitor the 5-HIAA peak for up to eight weeks throughout the sleep–waking cycle. By concurrent polygraphic monitoring they found that the 5-HIAA signal in all four raphe nuclei was maximal during waking, lower during slow-wave sleep, and at a minimum during paradoxical sleep. It was likely that under basal conditions peak 3 in these nuclei was due to 5-HIAA, but the ability to measure serotonin also was indicated, at least under certain conditions. Thus, a biphasic signal change (i.e. of peak 3) following monoamine oxidase inhibitors was seen as an initial fall presumably due to a decrease of 5-HIAA, followed by a large increase presumably due to serotonin (Crespi and Jouvet, 1983). These authors also found that handling the animals increased DPV peak 3 in all four raphe nuclei but, interestingly, a noxious stimulus (tail pinch) increased 5-HIAA in the nucleus raphe magnus only, a structure involved in pain perception which has serotonergic projections to the dorsal horn of the spinal cord (Rivot, Chiang and Besson, 1982).

5-HIAA in rat suprachiasmatic nucleus (SCN), measured by DPV with fiber electrodes in freely moving rats, revealed similar circadian fluctuations to those seen in the raphe. It was also found that drug-induced changes of peak 3 (reserpine or clorgyline) showed good agreement with tissue measurements of 5-HIAA (but not serotonin) assayed by HPLC-FD (Faradji, Cespuglio and Jouvet, 1983). Others (Marsden and Martin, 1986; Maidment, Routledge, Martin, Brazell and Marsden, 1986) have investigated the effects of serotonergic autoreceptor stimulation and blockade on the 5-HIAA peak in the rat SCN. It was found that 8-hydroxy-2-(di-*n*-propylamino)tetralin (8-OH-DPAT), which is an agonist at the serotonin autoreceptor and thereby reduces serotonergic firing, also decreased peak 3 in the SCN. Pretreatment with the putative antagonist ipsapirone (TVX Q 7821) was found to attenuate the usual decrease of peak 3 produced by 8-OH-DPAT. Again this suggested that peak 3 reflects serotonin release by measuring 5-HIAA. The good agreement between the DPV 5-HIAA measurements and tissue levels, however, is not a universal finding. Baumann and Waldmeier (1984) compared the effects of a large number of drugs on DPV signal in conscious rats with 5-HIAA tissue levels measured spectrofluorimetrically under comparable conditions. They found that whereas a good qualitative agreement existed in many cases (e.g. with methiothepin, clonidine and LSD) and also with the effects of these drugs on electrically stimulated serotonin released from cortical slices *in vitro*, for other drugs tested (e.g. cinanserin, metergoline and quipazine) there was either no, or only partial, correlation. It was concluded that DPV preferentially measures extraneuronal changes rather than the overall changes of 5-HIAA derived from tissue biochemistry. Others have directly addressed this problem, by monitoring the DPV peak 3 signal concurrently with microdialysis in striata of the anesthetized rat. The monoamine oxidase inhibitor, tranylcypromine, produced an almost parallel decrease in ECF 5-HIAA (microdialysis) with the change of DPV 5-HIAA signal, but following 5-hydroxytryptophan, although both measurements were increased, the percentage increase of 5-HIAA measured from dialysis was much greater than from voltammetry (Sharp, Maidment, Brazell, Zetterstrom, Ungerstedt, Bennett and Marsden, 1984). One possibility is that another endogenous substance from brain confounds the indole peak. The most likely candidate, uric acid, was found to account for only ~30% of the DPV 300 mV peak in the rat striatum based on its response to uricase infusion (Crespi, Sharp, Maidment and Marsden, 1983) which is considerably less than reported for carbon paste electrodes. It is possible that uric acid results from the trauma of implantation (e.g. damaged cells or hemorrhage) because its contribution to the DPV signal with carbon fiber electrodes is considerably reduced after 7 days implantation (Cespuglio, Gharib, Sarda, Chastrette, Faradji, Delobel and Jouvet, 1985). This contamination from uric acid obviously should be considered in acute experiments but apparently will be less serious in chronic studies. It would still seem that of the electrochemical methods available for measurement of indoleamines in the ECF *in vivo*, DPV with pyrolytic carbon fiber electrodes is the method of choice.

CONCLUSIONS AND FUTURE TRENDS

The past decade has seen significant progress in several key areas of technology in the measurement of neurotransmitter release from neuronal tissue. The introduction of dialysis probes coupled with the expanding appreciation of the merits of rapidly improving HPLC analytical techniques has made studies *in vivo* of serotonin release or metabolism a realistic possibility for most laboratories.

The improvements made in HPLC analytical columns (particularly reversed phase) has contributed greatly, and with the microbore columns more recently introduced it can be predicted that the performance of neurochemical assays will become even more impressive. Microbore columns provide approximately a twenty-fold increase in sensitivity over the traditional larger columns for a given detector (providing the detector dead space is minimized) because the on-column dilution is smaller. With shorter microbore columns resulting in shorter analysis time and considerably enhanced sensitivity (and the additional probability that other improvements will be made to improve detectors and reduce noise), microdialysis measurements of both serotonin and 5-HIAA at much shorter time intervals will be possible. With direct on-line loading of collected samples into the chromatograph it may eventually be possible to obtain data from minute to minute. Although very small dialysis probes are now being constructed, some tissue damage is unavoidable. With chronic implantation, scarring of tissue or gliosis around the probe may result and although there may be some contribution from these cells, presumably detected 5-hydroxyindoles still reflect changes in serotonergic neurons. However, gliosis may also impair and ultimately block diffusion. Conceivably, the development of dialysis membranes with improved biocompatibility, or the use of removable probes which may be cleaned, sterilized and reused will circumvent some of these potential problems.

If progress in voltammetric monitoring of serotonin release continues at its present rate then it also has a healthy future. The difficulties that have been encountered with carbon paste electrodes regarding contamination from uric acid may reflect at least two possibilities. One is that uric acid is the consequence of tissue damage (which might explain the apparently greater contribution with these than with carbon fiber electrodes) and it will be necessary to either limit their size, or eliminate the uric acid contribution. This might be performed with a surface-modified electrode, sensitive to indoleamines but not urate, somewhat analogous to electrode preparations that discriminate ascorbate from dopamine (Knott, Andrews and Mueller, 1985; Gerhardt, Oke, Nagy, Moghaddam and Adams, 1984). Another possibility is that uric acid is derived endogenously from neuronal events. If a significant contribution of hippocampal or striatal voltammetric signal with LSV at paste electrodes is due to urate, then it suggests that uric acid production may parallel ECF 5-hydroxyindoles and is a marker for serotonin release. It would seem that further investigations of this enigma are justified.

Will it ever be possible to develop electrodes able to discriminate between serotonin and 5-HIAA, yet with the sensitivity to monitor serotonin in the extracellular compartment? At present this seems unlikely. However, if asked twenty years ago 'will it ever be possible to develop a technique for on-line monitoring of 5-HIAA in the suprachiasmatic nucleus of the conscious freely moving rat?', how would we have answered?

REFERENCES AND BIBLIOGRAPHY

Adams, R. N., and Marsden, C. A. (1982) Electrochemical detection methods for monoamine measurements in vitro and in vivo, in *New Techniques in Psychopharmacology* (Eds L. L. Iversen, S. D. Iversen and S. H. Snyder) pp. 1–74, Plenum Press, New York.

Amin, A. H., Crawford, T. B. B., and Gaddum, J. H. (1954) The distribution of substance P and 5-hydroxytryptamine in the central nervous system of the dog, *J. Physiol. (London)*, **126**, 596–618.

Ashkenazi, R. J., Holman, R. B., and Vogt, M. (1972) Release of transmitters on stimulation of the nucleus linearis raphe in the cat, *J. Physiol. (London)*, **223**, 255–259.

Baumann, P. A., and Waldmeier, P. C. (1984) Negative feedback control of serotonin release *in vivo*: Comparison of 5-hydroxy-indoleacetic acid levels measured by voltammetry in conscious rats and by biochemical techniques, *Neuroscience*, **11**, 195–204.

Beleslin, D. B., and Myers, R. D. (1970a) A technique for repeated perfusion of withdrawal of fluid from the exposed cerebral cortex of a conscious animal, *Physiol. Behav.*, **5**, 1173–1175.

Beleslin, D. B., and Myers, R. D. (1970b) The release of acetylcholine and 5-hydroxytryptamine from the mesencephalon of the unanesthetized rhesus monkey, *Brain Res.*, **23**, 437–442.

Besson, M. J., Cheramy, A., Gauchy, C., and Glowinski, J. (1973) In vivo continuous estimation of ^{3}H dopamine release and synthesis in the caudate nucleus: effects of α-methyl-*p*-tyrosine and transection of the nigro-striatal pathway, *Arch. Pharmacol.*, **278**, 101–105.

Bhattacharya, B. K., and Feldberg, W. (1958) Perfusion of cerebral ventricles: effects of drugs on outflow from the cisterna and aqueduct, *Br. J. Pharmacol.*, **13**, 156–162.

Boireau, A., Ternaux, J. P., Bourgoin, S., Héry, F., Glowinski, J., and Hamon,

M. (1976) The determination of picogram level of 5-HT in biological fluids, *J. Neurochem.*, **26**, 201–204.

Bourgoin, S., Soubrie, P., Artaud, F., Reisine, T. D., and Glowinski, J. (1981) Control of 5-HT release in the caudate nucleus and the substantia nigra of the cat, *J. Physiol. (Paris)*, **77**, 303–307.

Brazell, M. P., and Marsden, C. A. (1982) Differential pulse voltammetry in the anesthetized rat; identification of ascorbic acid, catechol and indoleamine oxidation peaks in the striatum and frontal cortex, *Br. J. Pharmacol.*, **75**, 539–547.

Carmichael, E. A., Feldberg, W., and Fleischauer, K. (1964) Methods for perfusing different parts of the cat's cerebral ventricles with drugs, *J. Physiol. (London)*, **173**, 354–367.

Cattabeni, F., Koslow, S. H., and Costa, E. (1972) Gas chromatographic-mass spectrometric assay of four indolealkylamines of rat pineal, *Science*, **178**, 166–168.

Cespuglio, R., Faradji, H., and Jouvet, M. (1983) Voltammetric detection of 5-hydroxyindole compounds present at the extracellular levels of the cell bodies and the terminals of the serotonergic system: fluctuations during the sleep-waking cycle in chronically implanted rats, *C.R. Acad. Sci. Paris.*, **296**, 611–615.

Cespuglio, R., Faradji, H., Ponchon, J. -L., Buda, M., Riou, F., Gonon, F., Pujol, J. -F., and Jouvet, M. (1981) Differential pulse voltammetry in brain tissue. I. Detection of 5-hydroxyindoles in the rat striatum, *Brain Res.*, **223**, 287–298.

Cespuglio, R., Faradji, H., Riou, F., Buda, M., Gonon, F., Fujol, J. -F., and Jouvet, M. (1981) Differential pulse voltammetry in brain tissue. II. Detection of 5-hydroxyindoles in the rat striatum, *Brain Res.*, **223**, 299–311.

Cespuglio, R., Gharib, A., Sarda, N., Chastrette, N., Faradji, H., Delobel, B., and Jouvet, M. (1985) Voltamétrie impulsionelle différentielle: mise au point concernant la mesure des composés 5-hydroxyindols et de l'acide urique au niveau du cerveau, *C.R. Acad. Sci. Paris*, **301**, 817–822.

Cespuglio, R., Riou, F., Buda, M., Faradji, H., Gonon, F., Jouvet M., and Pujol, J. -F. (1980) Mesure in vivo par voltamétrie impulsionelle differentielle du 5-HIAA dans le striatum du rat, *C.R. Acad. Sci. Paris*, **298**, 901–906.

Chiueh, C. C., and Moore, K. E. (1976) Effects of dopaminergic agonists and electrical stimulation of the midbrain raphe on the release of 5-hydroxytryptamine from the cat brain *in vivo*, *J. Neurochem.*, **26**, 319–324.

Clemens, J. A., and Phebus, L. A. (1984) Brain dialysis in conscious rats confirms in vivo electrochemical evidence that dopaminergic stimulation releases ascorbate, *Life. Sci.*, **35**, 671–677.

Clemens, J. A., and Phebus, L. A. (1986) In vitro voltammetry versus dialysis: Neurochemical results, *Ann. N.Y. Acad Sci.*, **473**, 556–558.

Conti, J. C., Strope, E., Adams, R. N., and Marsden, C. A. (1978) Voltammetry in brain tissue: Chronic recording of stimulated dopamine and 5-hydroxytryptamine release, *Life Sci.*, **26**, 2705–2716.

Crespi, F., and Jouvet, M. (1983) Differential pulse voltammetry: parallel peak 3 changes with vigilance states in raphe dorsalis and raphe magnus of chronic freely moving rats and evidence for a 5-HT contribution to these peaks after monoamine oxidase inhibitors, *Brain Res.*, **272**, 263–268.

Crespi, F., and Jouvet, M. (1984) Differential pulse voltammetric determination of 5-hydroxyindoles in four raphe nuclei of chronic freely moving rates simultaneously recorded by polarographic technique: Physiological changes with vigilance states, *Brain Res.*, **299**, 113–119.

Crespi, F., Sharp, T., Maidment, N., and Marsden, C. A. (1983). Differential pulse voltammetry in vivo. Evidence that uric acid contributes to the indole oxidation peak, *Neurosci. Lett.*, **43**, 203–207.

Cross, A. J., and Joseph, M. H. (1981) The concurrent estimation of the major monoamine metabolites in human and non-human primate CSF by HPLC with electrochemical and fluorescence detection, *Life Sci.*, **28**, 499–505.

Curzon, G. (1981) The turnover of 5-hydroxytryptamine, in *Central Neurotransmitter Turnover* (Eds C. J. Pycock and P. V. Taberner), pp. 59–81, Croom Helm Ltd, London.

Curzon, G., Hutson, P. H., Kantamaneni, B. D., Sahakian, B. K., and Sarna, G. S. (1985) 3,4-Dihydroxyphenylethylamine and 5-hydroxytryptamine metabolism in the rat: acidic metabolites in cisternal cerebrospinal fluid before and after giving probenecid, *J. Neurochem.*, **45**, 508–513.

Curzon, G., and Knott, P. J. (1976) Environmental, toxicological and related aspects of tryptophan metabolism with particular reference to the central nervous system, *Critical Rev. Toxicol.*, **5**, 145–187.

Danguir, J., LeQuan-Bui, K. H., Elghozi, J. L., Devynck, M. A., and Nicolaidis, S. (1982) LCEC monitoring of 5-hydroxyindolic compounds in the cerebrospinal fluid of the rat related to sleep and feeding, *Brain. Res. Bull.*, **8**, 293–297.

Daszuta, A., Gaudin-Chazal, G., Barrit, M. C., Faudon, M., Puizillout, J. J., and Ternaux, J. P. (1979) Release of endogenous serotonin from 'encéphale isolé' cats. 1. Technical aspects, *J. Physiol. (Paris)*, **75**, 525–530.

Delgado, J. M. R., DeFeudis, F. V., Roth, R. H., Ryugo, D. K., and Mitruka, B. M. (1972) Dialytrode for long term intracerebral perfusion in awake monkeys, *Arch. Int. Pharmacodyn. Ther.*, **198**, 9–21.

Eccleston, D., Randic, M., Roberts, M. H. T., and Straughan, D. W. (1969) Release of amines and amine metabolites from brain by neural stimulation, in *Metabolism of Amines in the Brain* (Ed. G. Hooper), pp. 29–33, Macmillan, London.

Echizen, H., and Freed, C. R. (1984) Altered serotonin and norepinephrine metabolism in rat dorsal raphe nucleus after drug-induced hypertension, *Life Sci.*, **34**, 1581–1589.

Elghozi, J. -L., Mignot, E., and LeQuan-Bui, K. H. (1983) Probenecid sensitive pathway of elimination of dopamine and serotonin metabolites in CSF of the rat, *J. Neural. Trans.*, **57**, 85–94.

Faradji, H., Cespuglio, R., and Jouvet, M. (1983) Voltammetric measurements of 5-hydroxyindole compounds in the suprachiasmatic nuclei in circadian fluctuations, *Brain Res.*, **279**, 111–119.

Feldberg, W., and Myers, R. D. (1966) Appearance of 5-hydroxytryptamine and an unidentified pharmacologically active lipid acid in effluent from perfused cerebral ventricles, *J. Physiol. (London)*, **184**, 837–855.

Freed, C. R., Echizen, H., and Bashkaran, D. (1986) Serotonin metabolism and blood pressure regulation: Insights from brain tissue assays and *in vivo* electrochemical recording, in *Monitoring Neurotransmitter Release During Behaviour* (Eds M. H. Joseph, M. Fillenz, I. A. MacDonald and C. A. Marsden), pp. 133–144, Ellis Horwood Ltd., Chichester.

Gaddum, J. H. (1961) Push-pull cannulae, *J. Physiol. (London)*, **155**, 1–2P.

Gaddum, J. H. (1962) Substances released in nervous activity, in *Pharmacological Analysis of Central Nervous Action* (Eds W. D. M. Paton and P. Londgren), pp. 1–6, Pergamon, Oxford.

Gallager, D. W., and Aghajanian, G. K. (1975) Effect of chlorimipramine and lysergic acid diethylamide on efflux of precursor-formed ^{3}H-serotonin: correlations with serotonergic impulse flow, *J. Pharmacol. Exp. Ther.*, **193**, 785–795.

Gallager, D. W., Sanders-Bush, E., Aghajanian, G. K., and Sulser, F. (1975) An evaluation of the use of intraventricularly administered ³H-5-hydroxytryptamine as a marker for endogenous brain 5-hydroxytryptamine, *Brain Res.*, **93**, 111–122.

Gerhardt, G. A., Oke, A. F., Nagy, G., Moghaddam, B., and Adams, R. N. (1984) Nafion-coated electrodes with high selectivity for CNS electrochemistry, *Brain Res.*, **290**, 390–395.

Glowinski, J. (1981) In vivo release of transmitters in the cat basal ganglia, *Fed. Proc.*, **40**, 135–144.

Gonon, F., Fombarlet, C., Buda, M., and Pujol, J. -F. (1981) Electrochemical treatment of pyrolytic carbon fiber electrodes. *Anal. Chem.*, **53**, 1386–1389.

Goodrich, C. A. (1969) Effect of monoamineoxidase inhibitors on 5-hydroxytryptamine output from perfused cerebral ventricles of anaesthetized cats, *Br. J. Pharmacol.*, **37**, 87–93.

Graffeo, A. P., and Karger, B. L. (1976) Analysis for indole compounds in urine by high-performance liquid chromatography with fluorometric detection, *Clin. Chem.*, **22**, 184–187.

Guldberg, H. C., and Yates, C. M. (1968) Some studies of the effects of chlorpromazine, reserpine and dihydroxyphenylalanine on the concentrations of homovanillic acid, 3,4-dihydroxyphenylacetic acid and 5-hydroxyindole-3-acetic acid in ventricular cerebrospinal fluid of the dog using the technique of serial sampling of the cerebrospinal fluid, *Br. J. Pharmacol.*, **33**, 457–471.

Guldberg, H. C., Ashcroft, G. W., and Crawford, T. B. B. (1966) Concentrations of 5-hydroxyindoleacetic acid and homovanillic acid in the cerebrospinal fluid of the dog before and during treatment with probenecid, *Life. Sci.*, **5**, 1571–1575.

Hahn, Z., Cespuglio, R., Faradji, H., and Jouvet, M. (1983) Temperature dependent variations of 5-hydroxyindoles in ventricular cerebrospinal fluid – an *in vivo* voltammetric study, *Brain Res.*, **289**, 215–222.

Hernandez, L., Stanley, B. G., and Hoebel, B. (1986) A small removeable microdialysis probe, *Life Sci.*, **39**, 2629–2637.

Héry, F., Bourgoin, S., Soubrié, P., Montastruc, J. L., Artaud, F., and Glowinski, J. (1979) In vivo release of newly synthesized ³H-serotonin in the caudate nucleus and the substantia nigra of the rat: possible control by dopaminergic mechanisms, *Neurosci. Lett. Suppl.*, **3**, S238.

Héry, F., Faudon, M., and Fueri, C. (1986) Release of serotonin in structures containing serotonergic nerve cell bodies: Dorsalis raphe nucleus and nodose ganglia of the cat, *Ann. N.Y. Acad. Sci.*, **473**, 239–255.

Héry, F., Simonnet, G., Bourgoin, S., Soubrié, P., Artaud, F., Hamon, M., and Glowinski, J. (1979) Effect of nerve activity on the in vivo release of [³H] serotonin continuously formed from L-[³H]tryptophan in the caudate nucleus of the cat, *Brain Res.*, **169**, 317–334.

Héry, F., Soubrié, P., Bourgoin, S., Montastruc, J. L., Artaud, F., and Glowinski, J. (1980) Dopamine released from dendrites in the substantia nigra controls the nigral and striatal release of serotonin, *Brain Res.*, **193**, 143–151.

Holman, R. B., and Vogt, M. (1972) Release of 5-hydroxytryptamine from caudate nucleus and septum, *J. Physiol. (London)*, **223**, 243–254.

Hutson, P. H., and Curzon, G. (1983) Monitoring in vivo of transmitter metabolism by electrochemical methods, *Biochem. J.*, **211**, 1–12.

Hutson, P. H., and Curzon, G. (1987) Cisternal cerebrospinal fluid sampling and intracerebral dialysis: methods for monitoring amine transmitter metabolism in the conscious rat, *Kurume Med. J.* (in press).

Hutson, P. H., Sarna, G. S., and Curzon, G. (1984) Determination of daily variations of brain 5-hydroxytryptamine and dopamine turnovers and of the clearance of their acidic metabolites in conscious rats by repeated sampling of cerebrospinal fluid, *J. Neurochem.*, **43**, 291–293.

Hutson, P. H., Sarna, G. S., and Curzon, G. (1986) Neuropharmacokinetic applications of *in vivo* monitoring techniques, *Ann. N.Y. Acad. Sci.*, **473**, 549–552.

Hutson, P. H., Sarna, G. S., Kantamaneni, B. D., and Curzon, G. (1984) Concurrent determinations of brain dopamine and 5-hydroxytryptamine turnovers in individual freely moving rats using a repeated sampling of cerebrospinal fluid, *J. Neurochem.*, **43**, 151–159.

Hutson, P. H., Sarna, G. S., Kantamaneni, B. D., and Curzon, G. (1985) Monitoring the effect of a tryptophan load on brain indole metabolism in freely moving rats by simultaneous cerebrospinal fluid sampling and brain dialysis, *J. Neurochem.*, **44**, 1266–1273.

Hutson, P. H., Sarna, G. S., Sahakian, B. J., Dourish, C. T., and Curzon, G. (1986) Monitoring 5-HT metabolism in the brain of the freely moving rat, *Ann. N.Y. Acad. Sci.*, **473**, 321–335.

Ikeda, M., Miyazaki, H., Magitani, N., and Matsushita, A. (1984) Simultaneous monitoring of 3,4-dihydroxyphenylacetic acid (DOPAC) and 5-hydroxyindoleacetic acid (5-HIAA) levels in the brains of freely moving rats by differential pulse voltammetry technique, *Neurosci. Res.*, **1**, 171–184.

Johnson, R. D., and Justice, J. B. (1983) Model studies for brain dialysis, *Brain Res. Bull.*, **10**, 567–571.

Jonsson, J., and Lewander, T. (1970) A method for the simultaneous determination of 5-hydroxy-3-indole-acetic acid (5-HIAA) and 5-hydroxytryptamine (5-HT) in brain tissue and cerebrospinal fluid, *Acta Physiol. Scand.*, **78**, 43–51.

Joseph, M. H., and Kennet, G. A. (1981) In vivo voltammetry in the rat hippocampus as an index of drug effects on extraneuronal 5-HT, *Neuropharmacol.*, **20**, 1361–1364.

Joseph, M. H., and Kennet, G. A. (1983) Stress-induced release of 5-HT in the hippocampus and its dependence on increased tryptophan availability: An in vivo electrochemical study, *Brain Res.*, **270**, 251–257.

Joseph, M. H., and Kennet, G. A. (1986) Serotonin release in rat hippocampus examined by in vivo voltammetry; serotonergic function and tryptophan availability, *Ann. N.Y. Acad. Sci.*, **473**, 256–265.

Justice, J. B., and Neill, D. B. (1986) Interpretations of voltammetry *in vivo* using dialysed perfusion, *Ann. N.Y. Acad. Sci.*, **473**, 170–186.

Kennet, G. A., and Joseph, M. H. (1982) Does in vivo voltammetry in the hippocampus measure 5-HT release? *Brain Res.*, **236**, 305–316.

Knott, P. J., Andrews, C. D., and Mueller, K. J. (1985) Voltammetry measurement in vivo of neurotransmitter release in the freely moving rat, in *In vivo perfusion and release of neuroactive substances* (Eds A. Bayon and R. Drucker-Colin), pp. 141–158, Academic Press, New York.

Knott, P. J., Mueller, K. J., and Young, J. G. (1986) Electrochemical detection in vivo: Indoleamines or uric acid? *Ann. N.Y. Acad. Sci.*, **473**, 215–221.

Kruk, Z. L., Armstrong-James, M., and Millar, J. (1980) Measurement of the concentration of 5-hydroxytryptamine ejected during iontophoresis using multibarrel carbon fibre microelectrodes, *Life Sci.*, **27**, 2093–2098.

Lackovic, Z., Parenti, M., and Neff, N. H. (1981) Simultaneous determination of femtomole quantities of 5-hydroxytryptophan, serotonin and 5-hydroxyindoleacetic acid in brain using HPLC with electrochemical detection, *Europ. J. Pharmacol.*, **69**, 347–352.

Lamour, Y., Rivot, J. P., Pointis, D., and Ory-Lavollee, L. (1983) Laminar distribution of serotonergic innervation in rat somatosensory cortex as determined by in vivo electrochemical detection, *Brain Res.*, **259**, 163–166.

Lane, R. F., Hubbard, A. T., and Blaha, C. (1979) Application of semidifferential electroanalysis to studies of neurotransmitters in the central nervous system, *Electroanal. Chem.*, **95**, 117–122.

Loullis, C. C., Hintgen, J. N., Shea, P. A., and Aprison, M. H. (1980) In vivo determination of endogenous biogenic amines in rat brain using HPLC and push-pull cannula, *Pharmac. Biochem. Behav.*, **12**, 959–963.

Loullis, C. A., and Hellhammer, D. H. (1987) Neurochemical changes in brain during ongoing behavior, *Ann. N.Y. Acad. Sci.*, **473**, 349–365.

Marsden, C. A. (1979) Functional aspects of 5-hydroxytryptamine neurones, *Trends. Neurosci.*, **2**, 230–234.

MacIntosh, F. C., and Oborin, P. E. (1953) Release of acetylcholine from intact cerebral cortex, *Abstr. Int. Physiol. Congr.*, **19**, 580–581.

Maickel, R. P., Cox, R. H., Saillant, J., and Miller, F. P. (1968) A method for the determination of serotonin and norepinephrine in discrete areas of the rat brain, *Int. J. Neuropharmacol.*, **7**, 275–281.

Maidment, N. T., Routledge, C., Martin, K. F., Brazell, M. P., and Marsden, C. A. (1986) Identification of neurotransmitter autoreceptors using measurement of release *in vivo*, in *Monitoring Neurotransmitter Release During Behaviour* (Eds M. H. Joseph, M. Fillenz, I. A. MacDonald and C. A. Marsden), pp. 74–93, Ellis Horwood Ltd, Chichester.

Marsden, C. A. (1979) Functional aspects of 5-hydroxytryptamine neurons. Application of electrochemical monitoring *in vivo*, *Trends Neurosci.*, **2**, 230–234.

Marsden, C. A., and Martin, K. P. (1986) Involvement of 5-HT$_{1A}$- and -α_2 receptors in the decreased 5-hydroxytryptamine release and metabolism in rat suprachiasmatic nucleus after intravenous 8-hydroxy-2-(n-dipropylamino)tetralin, *Br. J. Pharmacol.*, **89**, 277–286.

Marsden, C. A., Bennett, G. W., Brazell, M., Sharp, T., and Stolz, J. F. (1981) Electrochemical monitoring of 5-hydroxytryptamine release in vitro and related in vivo measurements of indoleamines, *J. Physiol. (Paris)*, **77**, 333–337.

Marsden, C. A., Conti, J., Strope, E., Curzon, G., and Adams, R. N. (1979) Monitoring 5-hydroxytryptamine release in the brain of the rat using in vivo voltammetry, *Brain Res.*, **171**, 85–99.

Mefford, I. N. (1981) Application of high performance liquid chromatography with electrochemical detection to neurochemical analysis: Measurement of catecholamines, serotonin and metabolites in rat brain, *J. Neurosci. Methods*, **3**, 207–224.

Mefford, I. N., and Barchas, J. D. (1980) Determination of tryptophan and metabolites in rat brain and pineal tissue by reversed phase high-performance liquid chromatography with electrochemical detection, *J. Chromatog.*, **181**, 187–193.

Mignot, E., Laude, D., and Elghozi, J. -L. (1984) Kinetics of drug-induced changes in dopamine and serotonin metabolite concentrations in the CSF of the rat, *J. Neurochem.*, **42**, 819–825.

Mignot, E., Serrano, A., Laude, D., Elghozi, J. -L., Dedek, J., and Scatton, B. (1985) Measurement of 5-HIAA levels in ventricular CSF (by LCEC) and in striatum (by in vivo voltammetry) during pharmacological modifications of serotonin metabolism in the rat, *J. Neural. Transm.*, **62**, 117–124.

Moore, K. E., and Nielsen, J. A. (1985) Pharmacologically induced changes in the efflux of metabolites of dopamine and 5-hydroxytryptamine from the brains of freely moving rats, in *In Vivo Perfusion and Release of Neuroactive Substances* (Eds A. Bayon and R. Drucker-Colin), pp. 177–200, Academic Press, New York.

Mueller, K. J., Palmour, R., Andrews, C. D., and Knott, P. J. (1985) In vivo voltammetric evidence of production of uric acid by rat caudate, *Brain Res.*, **335**, 231–235.

Myers, R. D. (1970) An improved push-pull cannula system for perfusing an isolated region of the brain, *Physiol. Behav.*, **5**, 243–246.

Myers, R. D. (1972) Methods for perfusing structures of the brain, *Methods. Psychobiol.*, **2**, 169–211.

Myers, R. D., Kawa, A., and Beleslin, D. B. (1969) Evoked release of 5-HT and NEFA from the hypothalamus of the conscious monkey during thermoregulation, *Experientia*, **25**, 705–706.

Nielsen, J. A., and Moore, K. E. (1982a) Measurement of metabolites of dopamine and 5-hydroxytryptamine in cerebroventricular perfusions of anesthetized, freely moving rats, *Pharmacol. Biochem. Behav.*, **16**, 131–137.

Nielsen, J. A., and Moore, K. E. (1982b) Effects of chloral hydrate, γ-butyrolactone and equithesin on the efflux of dopamine and 5-hydroxytryptamine metabolites into cerebroventricular perfusates of rats, *J. Neurochem.*, **39**, 235–238.

Nielsen, J. A., and Moore, K. E. (1983) 6-hydroxydopamine and 5,7-dihydroxytryptamine selectively reduce dopamine and 5-hydroxytryptamine metabolites in cerebroventricular perfusion of rats, *Pharmacol. Biochem. Behav.*, **19**, 905–907.

Niéoullon, A., Chéramy, A., and Glowinski, J. (1977) An adaptation of the push-pull cannula method to study the in vivo release of 3H-dopamine synthesized from 3H-tyrosine in the cat caudate nucleus: effects of various physiological and pharmacological treatments, *J. Neurochem.*, **28**, 819–828.

O'Neill, R. D., Fillenz, M., and Albery, W. J. (1983) The development of linear sweep voltammetry with carbon paste electrodes in vivo, *J. Neurosci. Methods.*, **8**, 263–273.

O'Neill, R. D., Fillenz, M., Grünwald, R. A., Bloomfield, M. R., Albery, W. J., Jamieson, C. M., Williams, J. H., and Gray, J. A. (1984) Voltammetric carbon paste electrodes monitor uric acid and not brain 5 HIAA at the 5-hydroxyindole potential in the rat brain, *Neurosci. Lett.*, **45**, 39–46.

O'Neill, R. D., Grünwald, R. A., Fillenz, M., and Albery, W. J. (1982) Linear sweep voltammetry with carbon paste electrodes in the rat striatum, *Neuroscience*, **7**, 1945–1954.

Ponchon, J. L., Cespuglio, R., Gonon, F., Jouvet, M., and Pujol, J. F. (1979) Normal pulse polarography with carbon fiber electrodes for *in vitro* and *in vivo* determination of catecholamines, *Analyt. Chem.*, **51**, 1483–1486.

Portig, P. J., and Vogt, M. (1969) Release into the cerebral ventricles of substances with possible transmitter function in the caudate nucleus, *J. Physiol. (London)*, **204**, 687–715.

Puizillout, J. J., Gaudin-Chazal, G., Daszuta, A., Seyfritz, N., and Ternaux, J. P. (1979) Release of endogenous serotonin from 'encéphale isolé' cats. II. Correlations with raphe neuronal activity and sleep and wakefulness, *J. Physiol. (Paris)*, **75**, 531–537.

Redgrave, P. (1985) Technical issues associated with push-pull perfusion in anesthetized and unrestrained animals, in *In Vivo Perfusion and Release of Neuroactive Substances* (Eds A. Bayon and R. Drucker-Colin), pp. 11–22, Academic Press, London.

Reisine, T., Soubrié, P., Artaud, F., and Glowinski, J. (1982) Sensory stimuli differentially affect in vivo nigral and striatal [³H] serotonin release in the cat, *Brain Res.*, *232*, 77–87.

Rivot, J. P., Chiang, C. Y., and Besson, M. J. (1982) Increase of serotonin metabolism within the dorsal horn of the spinal cord during nucleus raphe magnus stimulation as revealed by in vivo electrochemical detection, *Brain Res.*, **238**, 117–126.

Rivot, J. P., Lamour, Y., Ory-Lavollee, L., and Pointis, D. (1983) In vivo electrochemical detection of 5-hydroxyindoles in rat somatosensory cortex: effect of the stimulation of the serotonergic pathways in normal and pCPA pretreated animals, *Brain Res.*, **275**, 164–168.

Saavedra, J. M., Brownstein, M., and Axelrod, J. (1973) A specific and sensitive enzymatic isotopic microassay for serotonin in tissues, *J. Pharmacol. Exp. Ther.*, **186**, 508–515.

Sahakian, B. J., Sarna, G. S., Kantamaneni, B. D., Jackson, A. J., Hutson, P. H., and Curzon, G. (1987) CSF tryptophan and transmitter amine turnover may predict social behaviour in the normal rat, *Brain Res.* (in press).

Sarna, G. S., Hutson, P. H., and Cruzon, G. (1983) Determination of brain 5-hydroxytryptamine turnover in freely moving rats using repeated sampling of cerebrospinal fluid, *J. Neurochem.*, **40**, 383–388.

Sarna, G. S., Hutson, P. H., and Curzon, G. (1984) A technique for repeated sampling of cerebrospinal fluid in freely moving rats and its uses, *J. Physiol. Paris.*, **79**, 536–537.

Scatton, B., Serrano, A., and Degueurce, A. (1986) The use of in vivo voltammetry to investigate functional recovery with transplants and neurotransmitter interactions in the rat brain, *Ann. N.Y. Acad. Sci.*, **473**, 284–300.

Sharp, T., Maidment, N. T., Brazell, M. P., Zetterstrom, T., Ungerstedt, V., Bennett, G. W., and Marsden, C. A. (1984) Changes in monoamine metabolites measured by simultaneous in vivo differential pulse voltammetry and intracerebral dialysis, *Neuroscience*, **12**, 1213–1221.

Sharp, T., Zetterstrom, T., Herrera-Marschitz, M., Ljungberg, T., and Ungerstedt, U. (1986) Intracerebral dialysis – a technique for studying dopamine release in the rat brain in relation to behavior, in *Monitoring Neurotransmitter Release During Behaviour* (Eds M. H. Joseph, M. Fillenz, I. A. MacDonald and C. A. Marsden), pp. 94–105, Ellis Horwood, Ltd., Chichester.

Soubrié, P., Montastruc, J. L., Bourgoin, S., Reisine, T., Artaud, F., and Glowinski, J. (1981) In vivo evidence for GABAergic control of serotonin release in the cat substantia nigra, *Europ. J. Pharmacol.*, **69**, 483–488.

Stamford, J. A. (1985) In vivo voltammetry: Promise and perspectives, *Brain Res. Rev.*, **10**, 119–135.

Stamford, J. A. (1986) In vivo voltammetry: Some methodological considerations, *J. Neurosci. Methods.*, **17**, 1–29.

Ternaux, J. P., Boireau, A., Bourgoin, S., Hamon, M., Héry, F., and Glowinski, J. (1976) In vivo release of 5-HT in the lateral ventricle of the rat: effects of 5-hydroxytryptophan and tryptophan, *Brain Res.*, **101**, 533–548.

Ternaux, J. P., Héry, F., Hamon, M., Bourgoin, S., and Glowinski, J. (1977) 5-HT release from ependymal surface of the caudate nucleus in 'encéphale isolé' cats, *Brain Res.*, **132**, 575–579.

Tilson, H. A,, and Sparber, S. B. (1970) On the use of push-pull cannula as a means of measuring biochemical changes during ongoing behavior, *Behav. Res. Meth. Instrum.*, **2**, 131–134.

Tilson, H. A., and Sparber, S. B. (1972) Studies on the concurrent behavioral and neurochemical effects of psychoactive drugs using the push-pull cannula, *J. Pharmacol. Exp. Ther.*, **181**, 387–398.

Twarog, B. M., and Page, I. J. (1953) Serotonin content of some mammalian tissues and urine and a method for its determination, *Am. J. Physiol.*, **175**, 157–161.

Ungerstedt, U. (1984) Measurement of neurotransmitter release by intracranial dialysis, in *Measurement of Neurotransmitter Release in vivo* (Ed. C. A. Marsden), pp. 81–105, John Wiley and Sons Ltd, Chichester.

Ungerstedt, U., and Pycock, C. (1974) Functional correlates of dopamine neuro-transmission, *Bull. Schweiz, Akad. Med. Wiss.*, **30**, 44–55.

Vane, J. R. (1957) A sensitive method for the assay of 5-hydroxytryptamine, *Br. J. Pharmacol.*, **12**, 344–349.

Veale, W. L., Myers, R. D., and Beleslin, D. B. (1973) Effects of calcium on the release of serotonin from the isolated sites within the diencephalon of the cat, *Pharmacol. Biochem. Behav.*, **1**, 259–264.

Vogt, M. (1969) Release from brain tissue of compounds with possible transmitter function: interaction of drugs with these substances, *Br. J. Pharmacol.*, **37**, 325–337.

Wagner, J., Vitali, P., Palfreyman, M. G., Zraika, M., and Huot, S. (1982) Simul-taneous determination of 3,4-dihydroxyphenylalanine, 5-hydroxytryptophan, dopa-mine, 4-hydroxy-3-methoxyphenylalanine, norepinephrine, 3,4-dihydroxyphenyl-acetic acid, homovanillic acid, serotonin and 5-hydroxyindoleacetic acid in rat cerebrospinal fluid and brain by high performance liquid chromatography with electrochemical detection, *J. Neurochem.*, **38**, 1241–1254.

Zetterstrom, T., Sharp, T., Marsden, C. A., and Ungerstedt, U. (1983) In vivo measurement of dopamine and its metabolites by intracerebral dialysis: changes after d-amphetamine, *J. Neurochem.*, **41**, 1769–1773.

CHAPTER 5

Retinal Serotonin and the Co-occurrence of Serotonin with other Neurotransmitters

NEVILLE N. OSBORNE
Nuffield Laboratory of Ophthalmology
University of Oxford
Walton Street
Oxford OX2 6AW
England

INTRODUCTION

During the past decade there has been considerable debate as to whether serotonin (5-hydroxytryptamine) plays a neurotransmitter role in the vertebrate retina. There is now strong support for the idea that serotonin is a transmitter substance in non-mammalian species (i.e. serotonin is present in significant amounts in non-mammalian retina and can be localized in specific populations of amacrine cells by the Falck and Hillarp method or with immunohistochemical techniques (Ehinger, 1982; Osborne *et al.*, 1982;

Osborne, 1984a). However, it still remains to be proved that this amine has a neurotransmitter role in mammalian retina.

The main argument against this hypothesis is that all attempts to localize any serotonin-containing neurones in this tissue have failed (see Ehinger and Florén, 1980; Florén and Hansson, 1980). Ehinger and his colleagues take the view that serotonin is not the neurotransmitter employed by serotonin-accumulating neurones in the mammalian retina, but that the true transmitter may be a substance closely related to this indoleamine. This proposal was based partly on the finding that in the bovine retina specific neurones accumulate ^{3}H-serotonin, yet these neurones are not visible under fluorescence microscopy after incubation in exogenous serotonin and treatment by the Falck and Hillarp method (Ehinger *et al.*, 1981). On the basis of these results, it was suggested that serotonin-accumulating neurones in the bovine retina can metabolize serotonin into some related indoleamine (i.e. the true neuro-transmitter) and are therefore unable to form the serotonin fluorophore with formaldehyde treatment in the Falck and Hillarp process. However, by employing a monoclonal antibody specific for the localization of serotonin, Osborne and Patel (1984) have since demonstrated that the bovine retina has a population of (amacrine) neurones which not only take up, but store, serotonin.

Osborne *et al.* (1982) have argued that serotonin itself is likely to play a neurotransmitter role in the mammalian retina, as they have demonstrated by high-performance liquid chromatography (HPLC) and electrochemical detection (which can discriminate between serotonin and other likely indole derivatives such as 5-hydroxytryptophan, 5,6-dihydroxytryptamine, 5-hydroxyindoleacetic acid and 3-methoxytryptamine) that serotonin is the major indole present in the retina of rabbit, cow and rat. The amount of serotonin is indeed low in comparison with non-mammalian retina (i.e. 9–18 ng/g fresh weight versus 100–500 ng/g), and this may explain the failure to localize serotonergic neurones in mammalian retina by the methods available at present.

Like most other mammalian retinas, the rabbit retina will accumulate exogenous serotonin when exposed to low amounts of this amine (Osborne, 1984a; Osborne and Patel, 1984). Thus, after pretreatment with serotonin, retinal cells which accumulate this amine can be identified by the Falck and Hillarp method, by autoradiography, or by immunocytochemical techniques. These types of study give identical results and have collectively shown that serotonin-accumulating cells are present in the amacrine cell layer of the rabbit retina (Osborne, 1984a; Osborne *et al.*, 1984; Osborne and Patel, 1984). Osborne and colleagues have also shown that serotonin can be synthesized from tryptophan in the rabbit retina, and have thus demonstrated the presence of synthetic enzymes for this amine (Osborne *et al.*, 1984). In addition, Osborne and Patel (1984) report that the uptake of serotonin, thought to be important in the inactivation of this neurotransmitter, is highly specific, and that serotonin accumulated by the serotonergic neurones can be released by potassium depolarization in a calcium-dependent manner. Mitchell and Redburn (1985) have localized binding sites for serotonin in crude membrane fractions of the rabbit retina and have suggested that both 5-HT$_1$ and 5-HT$_2$ sites exist in this tissue. Thus, recent evidence strongly suggests that the rabbit retina has all the molecular components required for a functional serotonergic transmitter system.

The purpose of this chapter is to focus on recent studies on the rabbit retina aimed at obtaining further support for a transmitter role for serotonin. The observation that some GABA-ergic amacrine cells may also utilize serotonin is of interest. This finding will be considered in relation to the broader issue of why some neurones contain more than one neurotransmitter.

DEVELOPMENTAL ASPECTS OF SEROTONIN-ACCUMULATING NEURONES

The capacity of specific neurones in the rabbit retina to take up exogenous serotonin is not restricted to the adult retina (Osborne, 1985a). These neurones appear immediately after birth at a stage when the eyes are unopened and the synaptic connections of the individual neurones have not yet been established. However, the serotonin-accumulating neurones in young rabbits (1–30 days) are not restricted to a population of typical amacrine neurones, as in the adult, but are also associated with spasmodic cells which appear morphologically to resemble interplexiform, bipolar and horizontal cells (Fig. 1). The neurones other than amacrine neurones which accumulate serotonin are more pronounced in the 1–15-day-old retinas, while only the merest signs of their existence are apparent in retinas up to 30 days old.

The finding that processes in the outer plexiform layer and possibly horizontal cells of the vertebrate retina accumulate serotonin is surprising but not unique. Florén and Hendrickson (1984) have shown that a population of horizontal cells in the squirrel retina takes up radioactive serotonin. A system of neuronal processes in the outer plexiform layer of Long–Evans hooded rats (Redburn, 1984) and certain species of elasmobranchs (Brunken *et al.*, 1986) has also been shown to take up exogenous serotonin. The results for the rabbit retina are unusual in that the uptake of serotonin in cells other than amacrine occurs only in young animals. This finding suggests that these cells might play a specific part during the development of the rabbit retina. Could it be that a specific class of serotonergic neurones has a role to play in synaptogenesis? Once they have fulfilled their role, might the neurones differentiate into other types of neurones with a different biochemistry? It is known that cultured sympathetic neurones, for example, may become

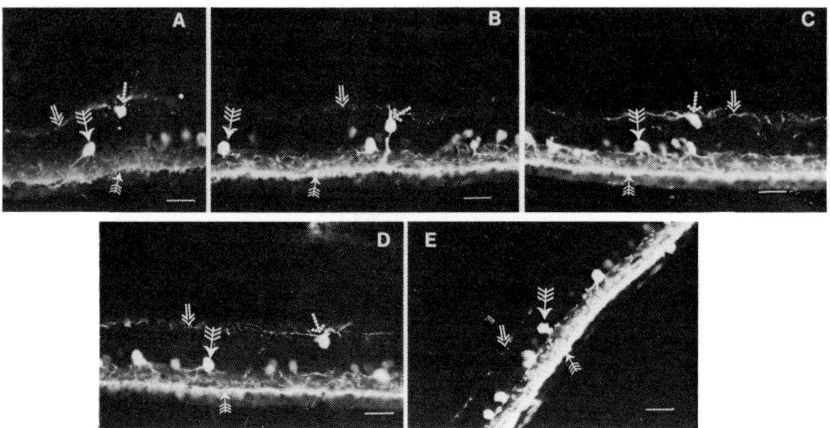

Fig. 1 Immunohistochemical localization of serotonin in 2-day (A), 8-day (B), 10-day (C), 12-day (D) and 15-day (E) rabbit retinas following incubation with exogenous serotonin. These sections show various examples where the indoleamine is associated not only with amacrine perikarya (large arrows) and terminals (small arrows) but also with other cell types (dotted arrows) and the outer plexiform layer. Scale bar = 50 μm. (From Osborne, 1985a)

coded to produce the appropriate neurotransmitter at a critical stage in their development and simultaneously lose their ability to respond to previous instructions (Chun and Patterson, 1977; Patterson *et al.*, 1976). An alternative explanation is that these neurones are misplaced amacrine neurones with no specific function, and as development proceeds they degenerate. This would explain their absence from the process of synaptogenesis (Hume *et al.*, 1983), but this does not apparently occur in the fully developed retina.

STUDIES ON CULTURES OF RETINAL NEURONES

The discovery of the occurrence of serotonin-accumulating neurones in the retina immediately after birth meant that it was possible to study them in a culture system (Osborne and Beaton, 1986a). As anticipated from studies on intact retinas, no positive staining was observed in rabbit retinal cultures subjected to immunohistochemical studies for the localisation of serotonin. However, cultures first incubated in serotonin (1 μM) and then processed by immunofluorescence or the peroxidase conjugate method for serotonin showed that the amine is taken up by certain cells (Fig. 2). As demonstrated in Fig. 3, the serotonin-accumulating cells in culture, like those in the intact retina, also react positively to antiserum PGP 9.5 which is neurone-specific (Thompson *et al.*, 1983). Furthermore, the serotonin-accumulating neurones in both intact retina (Osborne *et al.*, 1984) and cultures (Osborne and Beaton, 1985) do not take up exogenous dopamine. These and other data strongly

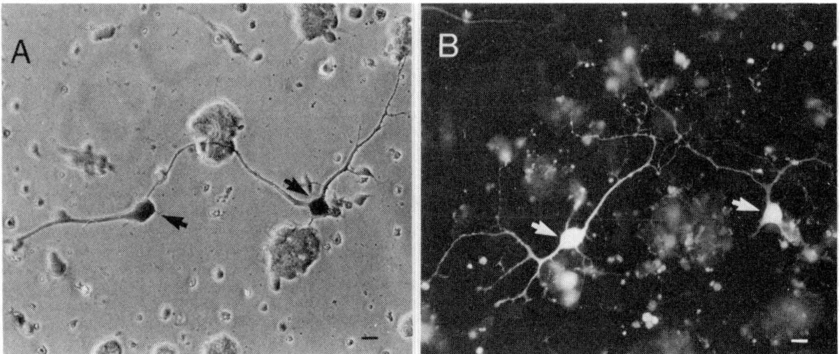

Fig. 2 Serotonin-accumulating cells prepared from 2-day-old postnatal rabbit retinas and maintained *in vitro* for 6 days. Only a few cells (arrows) take up exogeneous serotonin as shown by the peroxidase/anti-peroxidase (A) or immunofluorescent (B) procedures. Scale bar = 20 μm

support the idea that the same cells (i.e. neurones) in intact retina and culture take up exogenous serotonin.

Dissociated retinal cells, usually from 2–3-day-old animals, become attached to the poly-L-lysine coated coverslips in culture conditions within 2–4 hours. Even at this time the serotonin-accumulating cells can be observed (Fig. 4a). At this stage the cells lack processes, showing that transport sites for serotonin are not restricted to the nerve endings. After 3–6 days in culture the neurones which take up serotonin are more mature, with long processes (Fig. 4b). As the neurones grow in culture (Fig. 4c) the processes increase in length. After approximately 12 days in culture, the serotonin-accumulating neurones begin to degenerate (while the glial cells proliferate), with some being present after about 20 days in culture.

ISOLATION AND ANALYSIS OF SEROTONIN-ACCUMULATING CELLS

Since the culture procedure revealed that dissociated retinal cells from 2–3-day-old animals survive and some have the capacity to take up exogenous serotonin, a density centrifugation procedure was used to separate the different cell types within the dissociates (Osborne and Beaton, 1986a). Most methods used for cell sorting of nervous tissue dissociates take advantage of the differences in biophysical properties such as cell size or density to effect separation by gravitational fields. The choice of separation media such as percoll or metrizamide markedly improves the analytical power of the centrifugation methods. The separation of specific classes of cells from a retinal dissociate of 2–3-day-old animals proved difficult simply because most cells have similar physical characteristics in terms of size and shape. The best

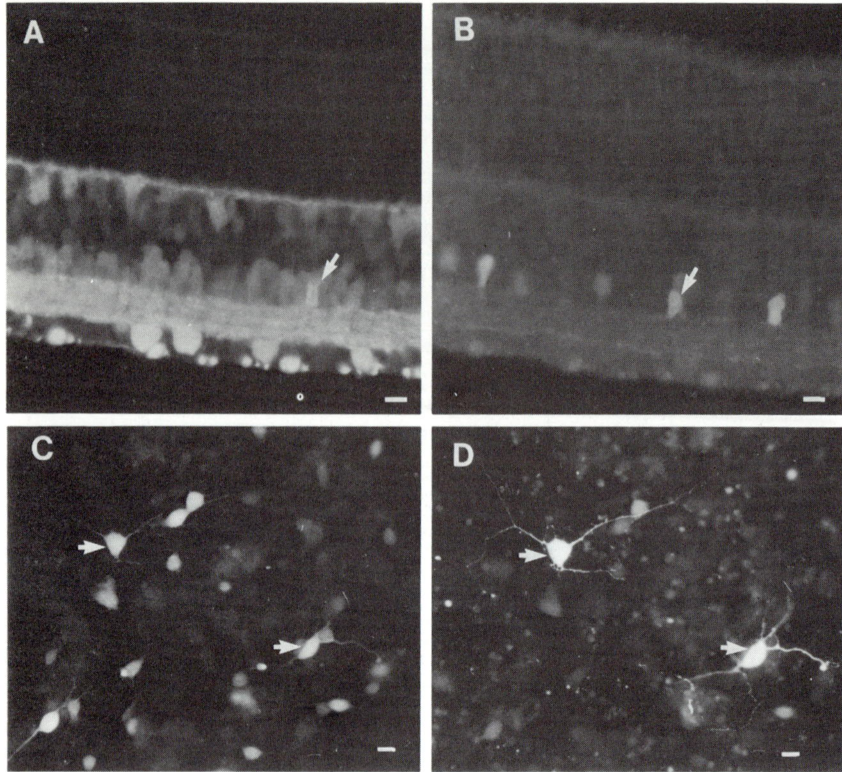

Fig. 3 Double staining of a section of retina (A and B) and cultures of retinal cells (C and D) for the localization of serotonin-accumulating cells (B and D) and PGP 9.5 positive cells (A and C). It can be seen that the serotonin-accumulating cells (arrows) express the PGP 9.5 antigen. Scale bar = 20 μm

results were produced using a metrizamide step gradient where a fourfold increase in serotonin-accumulating cells was found to be associated with the 20–30% interface (Table 1). Cells at the different interfaces of the metriz-amide gradient were then exposed to exogenous serotonin and processed for the immunohistochemical localization of serotonin. The number of cells which accumulated exogenous serotonin compared with the total number of cells in each fraction was then determined. Once it was established that 20–30% interface contained a fourfold enrichment of serotonin-accumulating cells (compared with the original suspension), HPLC and electrochemical analyses for the endogenous serotonin content in each fraction were under-taken. As shown in Table 1, almost all the serotonin detected is associated with the same fraction (20–30% interface) which contains an enriched popu-lation of serotonin-accumulating cells. These data are interpreted as showing

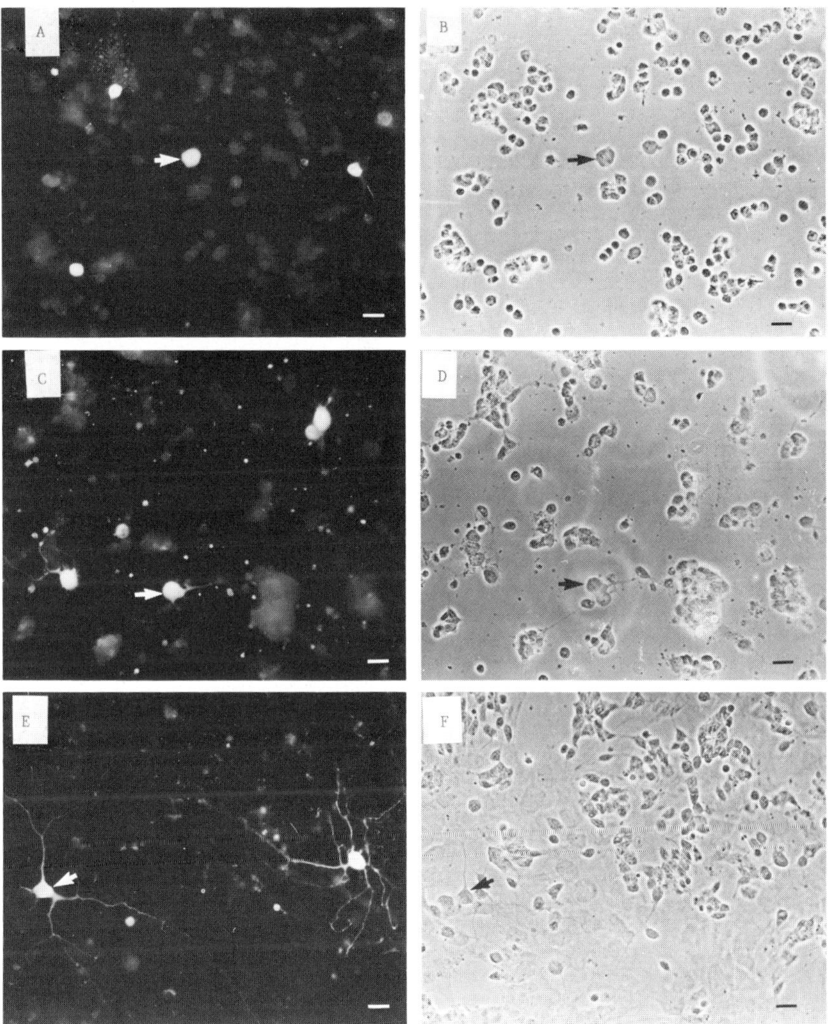

Fig. 4 Serotonin-accumulating cells prepared from 2-day-old postnatal rabbit retinas and maintained *in vitro* for 4 hours (A and B), 4 days (C and D) and 12 days (E and F). The serotonin-accumulating cells can in some cases be identified (arrows) in the accompanying phase-contrast pictures. All the examples are from a single culture. Scale bar = 30 μm

that serotonin is in the cells which take up exogenous amine, thus supporting the tenet that true serotonin-containing neurones exist in the rabbit (and other mammals) as are known to occur in non-mammalian retinas (see Osborne, 1984a).

Table 1 Percentage of serotonin-accumulating cells in dissociates of retina (from 5-day-old animals), related to the endogenous distribution of serotonin. Data given as means ±SEM, where $n = 4$

	Percentage of cells taking up serotonin	Endogenous serotonin (ng/g wet wt.)
Original suspension	0.38 ± 0.07	4 ± 0.6
10–20% metrizamide interface	0.14 ± 0.05	0
20–30% metrizamide interface	1.27 ± 0.25	5 ± 0.9
30–40% metrizamide interface	0.32 ± 0.02	1 ± 0.5

(From Osborne and Beaton, 1986a)

SEROTONERGIC RECEPTORS LINKED TO PHOSPHOINOSITIDES IN THE RETINA

Binding sites for serotonin have been localized in crude membrane fractions of rabbit (Mitchell and Redburn, 1985) and bovine (Osborne, 1981a) retina. From the data of Mitchell and Redburn (1985) it would appear that two types of serotonin-binding sites exist in the rabbit retina. These seem to be of physiological significance. Firstly, electrophysiological data demonstrate that both serotonin and its antagonist methysergide can affect both the spontaneous and light-evoked activity of rabbit retinal ganglion cells (Ames and Pollen, 1969). Secondly, serotonin-sensitive adenylate cyclase has been demonstrated in rabbit retina (Blazynski *et al.*, 1985). Adenylate cyclase is limited to a variety of pre- and post-synaptic events and does not appear to be linked to 5-HT_2 receptor types (Barbaccia *et al.*, 1983). However, 5-HT_2 antagonists reduce the ON responses of ON-centre cells as well as the surround (ON) responses of OFF-centre cells, and enhance the centre (OFF) responses of the latter cells in rabbit retina (Brunken and Daw, 1986). It would therefore seem that both 5-HT_1 and 5-HT_2 receptor types exist in the rabbit retina.

Within the last few years a whole host of neurotransmitters (acetylcholine, noradrenaline, substance P., serotonin, histamine) have been shown to elicit effects through an 'inositol phosphate receptor mechanism' (Berridge and Irvine, 1984; Downes, 1986; Berridge, 1986; Nahorski, 1985). In the case of serotonin it is the 5-HT_2 receptors that utilize inositol triphosphate as a second messenger (Conn and Sanders-Bush, 1984, 1985; Kendall and Nahorski, 1985). This approach therefore provides a means of corroborating the idea that 5-HT_2-like receptors are present in the mammalian retina. Recent studies by Cutcliffe and Osborne (1987) on rabbit retinal slices do indeed confirm the presence of 5-HT_2 receptors. A typical concentration response curve for serotonin and carbachol to stimulate ^3H-inositol phosphate accumulation in rabbit retinal tissues is shown in Fig. 5. Both substances elicited their effects maximally at 1 mM, but it is quite clear from the

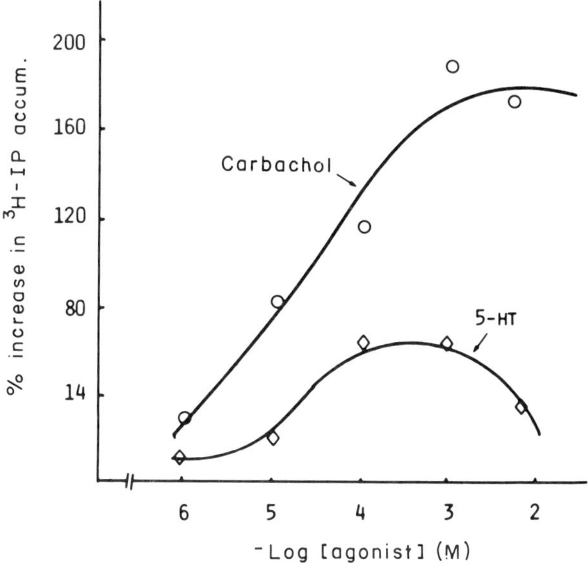

Fig. 5 Dose–response curves for the effect of serotonin (5-HT) and carbachol on the stimulation of inositol phosphate accumulation in rabbit retina. The methodology used is described elsewhere (Cutcliffe and Osborne, 1987)

curves that carbachol is more potent than serotonin. The effect of various antagonists on the serotonin-mediated stimulation of ^{3}H-inositol phosphate is shown in Fig. 6. Antagonists with significant inhibitory effects (by Student's t-test, $p < 0.05$) were cyproheptadine, ketanserin and ritanserin (each at 1 μM), reducing the serotonin-stimulated response to roughly 50% of that observed for 1 mM 5-HT in the absence of any antagonist. While cyproheptadine is known to display a much greater affinity for 5-HT$_2$ than 5-HT$_1$ sites, ketanserin and ritanserin are both highly selective for 5-HT$_2$ sites. Methysergide, a classical serotonin antagonist, did not significantly block the 5-HT-stimulated response. Atropine (1 μM) did not significantly inhibit the 5-HT-mediated response, although the reduction was greater than antici-pated. The antagonists haloperidol and phentolamine (specific for dopamin-ergic and adrenergic receptor sites respectively) at the same concentration had minimal effects. Thus, while the 5-HT-stimulated response in rabbit retina is not inhibited by antagonists selective for cholinergic, muscarinic, dopaminergic or adrenergic sites, results presented show that the response is counteracted by the presence of 5-HT$_2$ antagonists. Since the 5-HT$_1$ agonist 8-OH-DPAT did not stimulate phosphoinositide hydrolysis, even at 1 mM (results not shown), it would appear that the 5-HT response is mediated, at

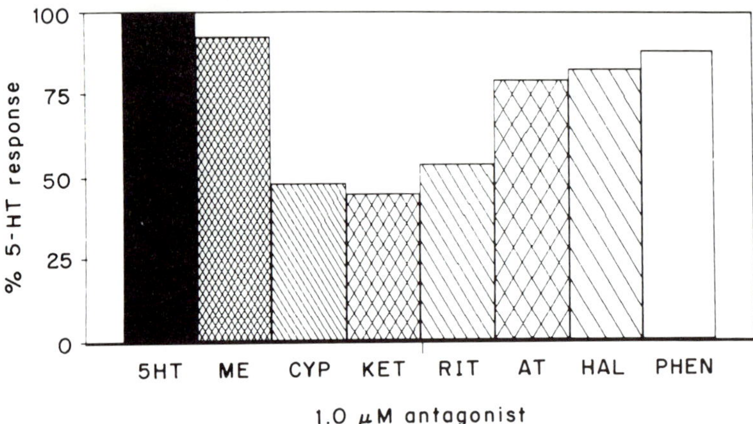

Fig. 6 Effect of various substances on the serotonin-induced stimulation of inositol phosphate accumulation in rabbit retinal slices (see Cutcliffe and Osborne, 1987, for details). Results are expressed as the percentage of the response for 5-HT alone (i.e. percentage of stimulation in ^{3}H-inositol phosphate accumulation when the response elicited by 1 mM 5-HT is standardized as 100%). Results are expressed as means ±SEMs where $n = 3$. Only ketanserin, ritanserin and cyproheptadine significantly inhibited the 5-HT-induced stimulation of inositol phosphate accumulation, i.e. $p < 0.05$ by Student's t-test. ME = methysergide; CYP = cyproheptidine; KET = ketanserin; RIT = Ritanserin; AT = Atropine; HAL = haloperidol; PHEN = phentolamine

least partly, through 5-HT$_2$ receptors. One can not exclude the possibility, furthermore, that some of the 5-HT-induced stimulation of inositol phosphate accumulation is through some non-5-HT$_2$ serotonergic receptor sites. When the stimulatory effects of a series of indoleamines were examined, serotonin itself was shown to evoke the greatest accumulation of inositol phosphate (Fig. 7).

Of particular importance was the finding that the dose–response curve for serotonin-induced phosphoinositide hydrolysis in the rabbit retina was similar to that obtained for the chick retina (results not shown), which, in contrast to mammalian retinal tissue, has been shown to contain significant amounts of the endogenous amine. These results argue in favour of an efficient serotonergic system in the mammalian retina which appears to function (at least partly, via the breakdown of inositol phospholipids) despite low endogenous levels of serotonin.

SEROTONIN-ACCUMULATING CELLS AND GABA

Since cells which take up ^{3}H-GABA and ^{3}H-serotonin in rabbit retinal cultures have similar morphologies, a variety of methods were used to

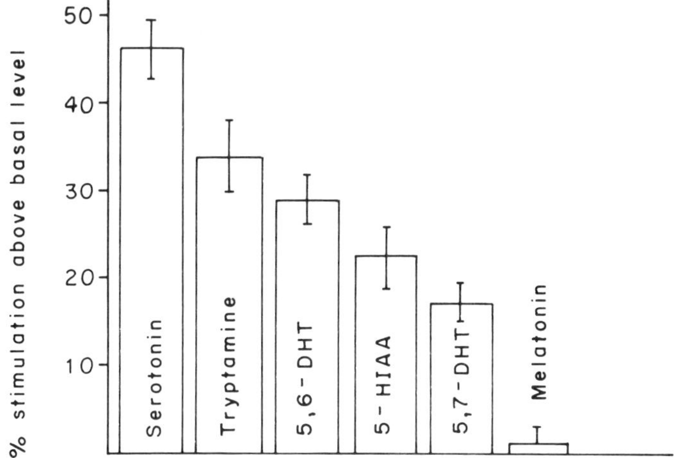

Fig. 7 Comparative effect of serotonin and various analogues of the amine on the stimulation of inositol phosphate accumulation in rabbit retinal slices. The conditions of the experiments were as in Figs. 5 and 6. Results are mean values from three separate experiments each done in triplicate. The SEMs varied from 10–18%. It can be seen that serotonin was the most effective of all the substances tested in stimulating inositol phosphate accumulation

examine this relationship more precisely. One method was to count the number of cells in culture derived from the same dissociate which take up ^{3}H-GABA or ^{3}H-serotonin or a mixture of the two (Osborne and Beaton, 1985). Since the number of cells containing radioactive grains in cultures exposed to a mixture of tritiated compound (3.1%) was not additive, and because many more cells took up ^{3}H-GABA (2.9%) than ^{3}H-serotonin (0.7%), the results were interpreted as suggesting that some cells have the capacity to accumulate both substances (Osborne and Beaton, 1985, 1986a). While the co-localization of neurotransmitter substances has been reported in non-mammalian retinas (Lam *et al.*, 1985), the co-localization of serotonin with another putative transmitter has not been reported in any retina. The findings from rabbit retinal cultures therefore served as an impetus to examine the possibility that certain GABAergic cells in the rabbit retina take up exogenous serotonin.

Intact rabbit retinas or retinal cell cultures were exposed to a mixture of ^{3}H-GABA and unlabelled serotonin. Thereafter, sections were produced from the retina and these, together with the cultures, were processed for the dual localization of the uptake of ^{3}H-GABA by autoradiography and for the immunohistochemical localization of the uptake of serotonin. It can be seen from Figs 8 and 9 that certain cells take up ^{3}H-GABA alone, while other cells take up both substances. This is particularly well documented in the culture studies where the cells can be located precisely.

Evidence of the uptake of serotonin by a subpopulation of GABA cells comes from another series of experiments where it was not necessary to use autoradiography. Retinas, or cultures, were exposed to unlabeled 5,7-dihydroxytryptamine (5,7-DHT) and then processed immunocytochemically for the localization of GABA. The 5,7-DHT is taken up specifically by the serotonin-accumulating cells and can be viewed directly with an excitation filter of 365 nm (barrier filter 420 nm). Since this wavelength is different from the fluorescein fluorophore for visualizing GABA immunoreactivity, both 5,7-DHT uptake and GABA can be viewed in a fluorescence microscope just by changing filters. It was shown by this method (see Osborne and Beaton, 1986b) that some cells in rabbit retinal cultures and sections of retina take up 5,7-DHT (i.e. serotonin-accumulating cells) and also react positively

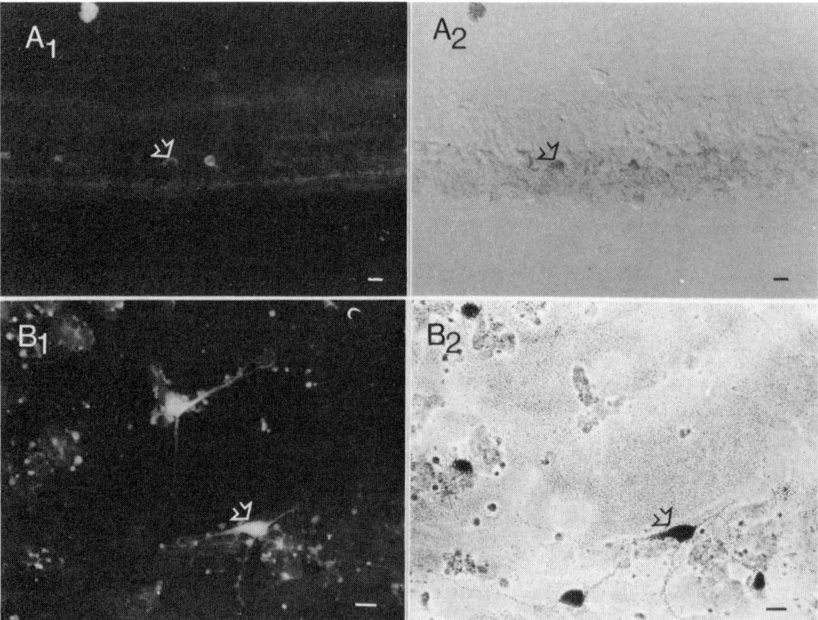

Fig. 8 A_1 and A_2 show the same sections, and B_1 and B_2 are the same retinal cultures after exposure to ^{3}H-GABA and unlabelled serotonin followed by immunostaining for the localization of serotonin. The serotonin taken up by retinal cells could then be viewed by fluorescence microscopy (A_1 and B_1) while the ^{3}H-GABA accumulation could be followed by autoradiography (A_2 and B_2). It can be seen that certain cells take up ^{3}H-GABA. Other cells take up serotonin while some cells take up both ^{3}H-GABA and serotonin (open arrows). Scale bar = 20 μm

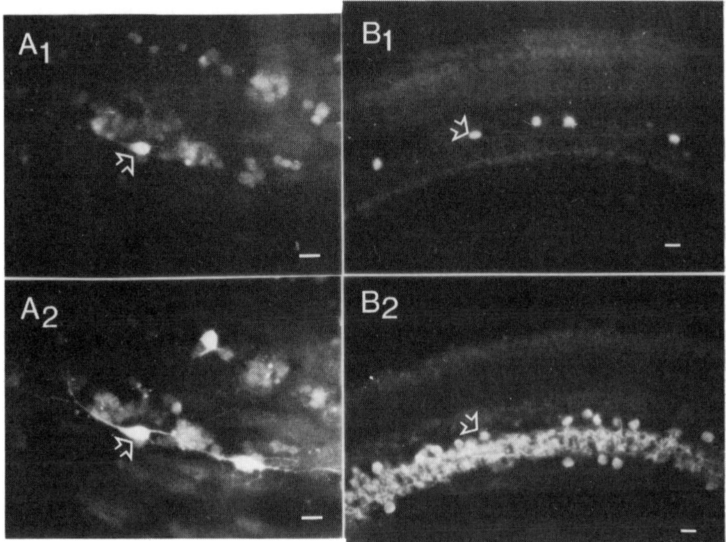

Fig. 9 A_1 and A_2 are the same retinal culture and B_1 and B_2 the same sections after exposure to exogenous 5,7-DHT followed by immuno-staining for the localization of GABA. A_1 and B_1 show specific cells taking up 5,7-DHT. Some of these cells (open arrows) stain positively for the localization of GABA (A_2 and B_2) while other GABA immuno-reactive cells clearly do not take up exogenous 5,7-DHT. Some cells just take up 5,7-DHT. Scale bar = 20 μm

for the localization of GABA. Thus it is concluded that a subpopulation of GABAergic cells has the capacity to take up exogenous serotonin.

Cell counts (of randomly chosen fields under microscopic vision; at least 30 fields were analysed) were made of cultures exposed to 5,7-DHT and stained for the localization of GABA. It was found that 0.85% of all cells were 5,7-DHT-positive and 1.34% stained for the localization of GABA. Also, 80% of all 5,7-DHT-positive cells also contained GABA immunoreactivity, while only 20% of all GABA-positive cells also took up 5,7-DHT.

The results discussed above clearly show that one type of GABAergic cell in the rabbit retina takes up exogenous serotonin, suggesting that a precise interaction between GABA and serotonin occurs at the synaptic level. This idea is only acceptable, however, if the neurones which take up exogenous serotonin are truly serotonergic cells. As discussed above, a great deal of evidence is available to support such a hypothesis. Further support for this comes from studies on the pigeon retina where double labelling immunohisto-chemical experiments show that certain GABAergic amacrine cells contain serotonin (Fig. 10).

Studies on rabbit retinal receptors linked to inositol phosphate metabolism

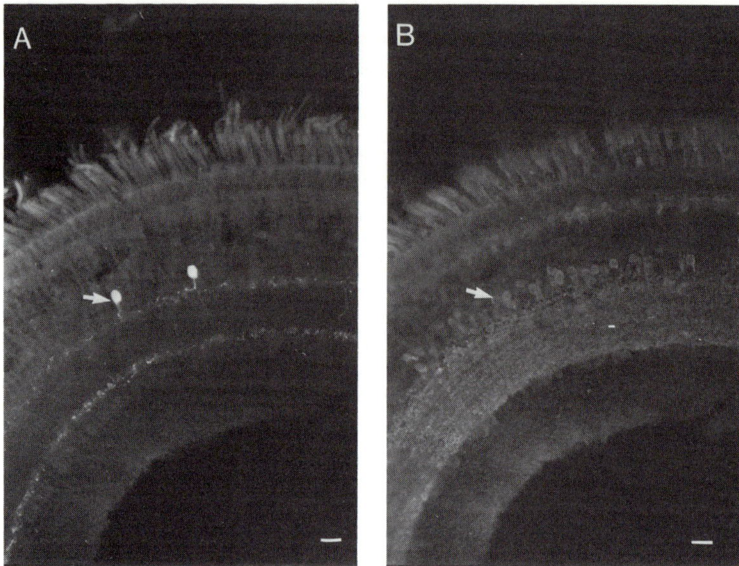

Fig. 10 The same section of a pigeon retina double stained for the localization of serotonin (A) and GABA (B) immunoreactivity. The serotonin
(monoclonal) and GABA (polyclonal) antisera were from rat and rabbit
respectively. The secondary antibodies to recognize each substance were
labelled with rhodamine (A) and fluorescein (B) respectively. A careful
examination of the sections shows that some cells contain both serotonin
and GABA (arrows) immunoreactivities. Scale bar = 20 μm

suggest an interaction between GABA and serotonin (Fig. 10). GABA by
itself does not evoke a significant stimulation of inositol phosphate accumulation. However, GABA significantly potentiates the known 5-HT-induced
stimulation of inositol phosphate accumulation (Fig. 11). These results, therefore, suggest that both GABA and 5-HT may be released from specific
amacrine cells and that they can either elicit effects by themselves or interact
at the synaptic level.

THE SIGNIFICANCE OF THE CO-OCCURRENCE OF SEROTONIN WITH OTHER NEUROTRANSMITTERS

Dale's law, a traditional concept in neurobiology, is thought to state that
each neurone contains only one neurotransmitter. Sir Henry Dale, to whom
Dale's law is attributed, in fact never suggested that neurones contain only
one neurotransmitter. He hypothesized that each neurone releases the same
neurotransmitter at all of its synapses (Dale, 1935; Osborne, 1985b). Whether
Dale's hypothesis is valid is still not proven, although the finding that various

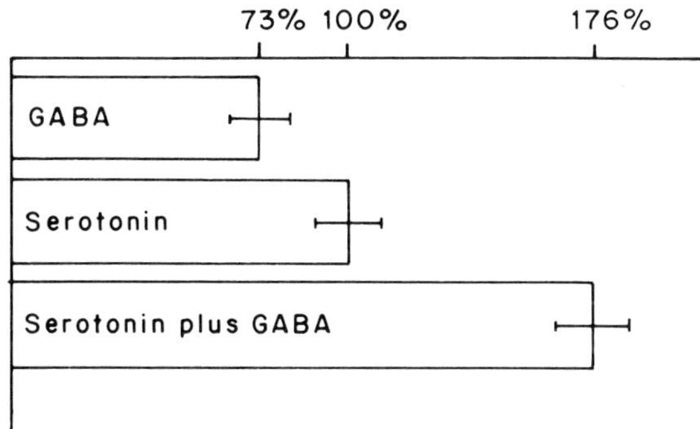

Fig. 11 Relative effect of GABA, serotonin and serotonin plus GABA on the stimulation of inositol phosphate accumulation in retinal slice. Conditions of experiments were as in Figs 5 and 6. GABA (1 mM) only stimulated inositol phosphate accumulation by 28 ± 6% above basal level; this is statistically insignificant by the Student's t-test ($p > 0.05$). In contrast, 1 mM serotonin significantly stimulates the accumulation of inositol phosphate. This effect was further significantly increased by the addition of 1 mM GABA. The results presented are the means of eight separate experiments each carried out in triplicate

neurones contain more than one neurotransmitter does raise a number of points in contradiction to his belief (see Osborne, 1983, 1984b; Cuello, 1982; Chan-Palay and Palay, 1984).

O'Donohue *et al.* (1985) have classified four types of co-transmitter neurones. In Type 1 co-transmitter neurones, multiple neurotransmitters are derived from a common gene coding for a pre-hormone. One example of this is neurones containing the pro-opiomelanocortin derived peptides, α-MSH and β-endorphin. Type 2 co-transmitter neurones contain multiple neuropeptides derived from different genes. Numerous examples of this type of co-transmitter neurone exist in the CNS of both vertebrates and invertebrates. Type 3 co-transmitter neurones contain both neuropeptides and non-peptide transmitters, while Type 4 co-transmitter neurones contain multiple classical non-peptide neurotransmitters. Whether this classification provides any functional insight into the significance of the co-occurrence of neurotransmitters still remains to be determined. However, it does provide a useful means for discussing specific substances. Serotonin, for example, co-occurs with other substances in Type 3 and Type 4 neurones as defined by O'Donohue *et al.* (1985). A list of some of the examples where serotonin

has been shown to co-exist with other neurotransmitter substances is shown
in Table 2.

Table 2 Co-existence of serotonin with either peptide or non-peptide transmitters

	Peptide or non-peptide	Location
Serotonin plus peptide	substance P	brainstem–spinal cord, snail neurone
	substance P and TRH	brainstem–spinal cord
	TRH	brainstem snail neurones
	CCK	
	enkephalin	pons, area prostrema, medulla oblongata
Serotonin plus non-peptide	GABA	raphe neurones, retina, enteric neurones
	acetylcholine	snail neurones
	acetylcholine plus octopamine	snail neurones
	adrenaline	adrenal medullary cells
	dopamine	snail neurones, lobster neurones
	noradrenaline	sympathetic neurones in culture

TRH: thyrotropin-releasing hormone
CCK: cholecystokinin
GABA: gamma-aminobutyric acid

The release of serotonin with either a peptide or non-peptide neurotransmitter from the same neurone has not been demonstrated. Should two
neurotransmitters be released from an ending it is possible to envisage how
they could act on the post-synaptic cells. First, both co-transmitters could
bind to the same active sites on a receptor. Another possibility is that the
co-transmitters may bind to two different receptors on the same target cell.
A third possibility is that the co-transmitters may bind to different transmitters on different target cells. A fourth possibility is that one co-transmitter
modulates the action of a second co-transmitter either at the level of the
synaptic cleft, the transmitter–receptor interaction mechanism, or at some
level on receptors mediated by second messengers (e.g. G-protein site) or
directly on an ion channel linked to the receptor mechanism. A fifth possibility is that one co-transmitter may regulate the release of the other
co-transmitter by acting on pre-synaptic receptors or uptake sites of the first
co-transmitter. Finally, each co-transmitter may be released at different rates,
thus imposing a permutation of different post-synaptic events.

Unequivocal proof of the co-release of two transmitters from the same
neurones is lacking, however. This is due to the difficulty of studying defined

neurones repeatedly and being able to show that what is released from a neurone is a true 'transmitter' (see Osborne, 1984b, for discussion). A physiological stimulation of nigral neurones produces a release of acetyl-cholinesterase, yet the enzyme is not a neurotransmitter (see Greenfield, 1985). The best direct evidence that a neurone utilizes two transmitters (serotonin and acetylcholine) comes from work on a specific serotonergic snail neurone; this has been reviewed elsewhere (Osborne, 1981b, 1984b). As far as vertebrate studies are concerned, indirect evidence suggests that certain brainstem–spinal cord neurones, for example, do employ serotonin, substance P and TRH (thyrotropin-releasing factor hormone) as neur-otransmitters. Some of the first reports of the co-existence of serotonin and substance P in neurones of the medulla oblongata were by Chan-Palay *et al.* (1978) and Hökfelt *et al.* (1978). Subsequent studies in this region showed that serotonergic neurones contain TRH (Hökfelt *et al.*, 1980; Johansson *et al.*, 1981). Experiments with the neurotoxins 5,7-dihydroxytryptamine and 5,6-dihydroxytryptamine revealed that not only serotonergic nerve endings but also substance P- and TRH-positive nerve endings in the ventral horn of the spinal cord were affected by these drugs (Hökfelt *et al.*, 1978; Johansson *et al.*, 1981), and this could also be demonstrated by radioimmunological determinations of amines and peptides (Gilbert *et al.*, 1982; Björklund *et al.*, 1979). These findings strongly suggested that brain serotonergic neurones containing peptides project into the spinal cord. This was substantiated by retrograde tracing experiments combined with immunohistochemistry (see Skirboll *et al.*, 1984). Finally, electron microscopical immunohistochemistry has confirmed that serotonin and substance P are localized in the same nerve endings in the ventral horn (Pelletier *et al.*, 1981).

With proof that some descending fibres in the spinal cord contain sero-tonin, TRH and substance P, a variety of indirect evidence is available to shed light on the functional significance. Barbeau and Bedard (1981) observed a marked activation of the stretch reflex by TRH in chronically spinalized rats or rats pretreated with neurotoxin to serotonin. These results were similar to effects seen after administration of 5-hydroxytryptophan (see Andén *et al.*, 1964). The effect of TRH was blocked by treatment with a 5-HT antagonist, suggesting that the peptide acts at a site closely related to the 5-HT receptor. The finding that substance P, but not TRH, facilitates the release of ^{3}H-serotonin from spinal cord slices in the presence of unlabelled serotonin suggested that substance P acts on pre-synaptic receptors and that its action enhances 5-HT transmission (Mitchell and Fleetwood-Walker, 1981). The combined findings so far discussed indicate, therefore, that certain nerve endings in the ventral horn of the spinal cord release serotonin, TRH and substance P. The released serotonin and substance P act pre-synaptically while some of the released serotonin and TRH act post-synaptically.

Behavioural studies support the findings discussed on the interaction of serotonin, TRH and substance P at certain ventral horn nerve endings. Hansen *et al.* (1983) applied 5-HT, substance P and TRH, alone or in combination, via an intrathecal catheter at the lumbar level and showed that of the various sexual parameters analysed, only intromission and mount latencies were affected. Thus, whereas serotonin and TRH administered separately only prolonged these latencies to a small extent, the combined administration of the two compounds resulted in a dramatic increase. Serotonin and a neuropeptide, therefore, enhance transmission at serotonergic synapses.

CONCLUDING REMARKS

Recent evidence clearly indicates that the small amounts of serotonin present in the mammalian retina are associated with specific neurones. These neurones are of the amacrine type in the adult retina, while in the developing retina they are additionally of the bipolar, interplexiform and horizontal types. Studies designed to analyse the effects of neurotransmitters on inositol phosphate turnover in rabbit retinal slices show clearly that serotonin stimulates inositol phosphate accumulation. All other indoleamines tested were not as effective as serotonin in this process. Since ketanserin specifically, but only partially, counteracted the serotonin effect on inositol phosphate turnover, it is concluded that 5-HT_2 receptors (linked to inositol phosphate) exist in the rabbit retina. There is also some electrophysiological evidence to support the presence of 5-HT_2 receptors. Thus, the idea that serotonin is a transmitter in the mammalian retina is strengthened.

Studies on the serotonin-accumulating neurones of primary cultures of rabbit retinal cells provided the first indication that some of these cells also contained GABA. This was subsequently confirmed in the intact retina where it was clearly shown that some GABA-ergic cells took up exogenous serotonin while others did not. It is of interest to note that examination of the pigeon retina where endogenous serotonin can be localized revealed that some amacrine cells contain both GABA and serotonin. It therefore seems likely that GABA and serotonin co-exist in the rabbit retina, although this could not be proven, of course, given the inability to localize histochemically the small amounts of endogenous serotonin.

Examples of the co-occurrence of serotonin with a non-peptide transmitter such as GABA, or with a peptide transmitter in nervous tissues in general, increase constantly. The significance of the co-occurrence of neurotransmitter-like substances in the same neurone may be complex. To date, however, the evidence that a neurone may utilize two or more transmitters is incomplete; the best example is still the giant snail neurone where acetylcholine also appears to be used as a neurotransmitter. Indirect studies on the neurones in the brainstem–spinal cord containing substance P, TRH and serotonin suggest that these substances are released from such neurone endings with substance P modulating the release of serotonin through a pre-synaptic action, while serotonin and TRH have post-synaptic actions. Studies on rabbit retinal slices also suggest a post-synaptic interaction between two transmitters released from the same neurone. Here it was shown that while GABA does not stimulate inositol phosphate metabolism by itself, it does potentiate, significantly, the serotonin effect, so that it is possible that some neurones release both GABA and serotonin in the retina, and that their post-synaptic interaction gives an alternative means of communication to their individual action upon specific receptors.

FUTURE TRENDS

Although recent findings have considerably strengthened evidence that serotonin is a neurotransmitter in the mammalian retina, it has not yet been unequivocally demonstrated that the amine exists in retinal neurones. There is an obvious need for a procedure more sensitive than those in present use. Radioimmunocytochemistry should theoretically resolve this, but so far a produced radioactive monoclonal antibody to serotonin has given negative results, due, in part, to radioactive impurities (see Osborne *et al.*, 1982). An alternative means of showing the existence of serotonergic cells within the mammalian retina would be to inject, *in vivo*, the precursor tryptophan into the vitreous humour of the eye over a period of time, and then to examine the retina. In such experiments it would be necessary to consider the needs of the tryptophan-hydroxylase enzyme (e.g. co-factors and pH) in order to achieve success.

What is the functional significance of the fact that neurones utilize serotonin within the retina? It is known that the serotonin-accumulating neurones in both cat and rabbit are extensively connected to rod pathways (Holmgren-Taylor, 1982; Ehinger and Holmgren, 1979), which suggests a possible role in scotopic and mesopic rather than photopic vision. Since $5\text{-}HT_2$ type receptors appear to exist in the retina, it would be of interest to compare the electro-physiological effects of serotonergic drugs in light versus dark adapted states of the retina, as the rod pathways will be saturated during photopic conditions. One would predict that serotonergic drugs could have no effect in the photopic state.

While it is impossible to see how unequivocal data could ever be amassed to describe the precise role of any neurotransmitter substance in the retina, the tissue itself offers several advantages for basic research. The clear laminated structure of the retina may well be exploited by using autoradiographical methods to determine the precise localization of receptors employing inositol phosphates as second messengers. Anderson and colleagues (Anderson *et al.*, 1983; Anderson and Hollyfield, 1984) have already used such an approach to examine the effects of light stimulation on the incorporation of ^{3}H-inositol into phosphoinositides in the retina of *Xenopus laevis*. After exposure to light, an enhanced incorporation of ^{3}H-inositol is localized specifically in *Xenopus* horizontal cells. With the anticipated development of an even higher specific activity of ^{3}H-inositol than is at present available, it may well be possible to localize precisely, by electron microscopy, the serotonergic synapses linked to inositol phosphate utilization in the retina. Such methodology can be combined with immunocytochemistry for the localization of different neurone types and/or localization of adenylate cyclase associate synapses. It is therefore possible that in the future the connections made by the serotonin-accumulating cells with other neurones can be precisely determined and the nature of the post-synaptic sites characterized.

The question considered over the last thirty years is whether a neurone can actually employ two transmitters. This question remains unanswered. Undoubtedly, more examples of the co-occurrence of neurotransmitter-like substances in the same cell will emerge in the future. To date, nobody has shown that a neurone contains more than three putative transmitters. This may well change. I anticipate the development of even more types of fluorescent or electron-dense associated secondary antibodies than are available at present. This, combined with autoradiographical methodology, may make it possible to process the same sample for the localization of four or more antigens. It may then be revealed that certain neurones contain four or more putative transmitters. Whether these putative transmitters function as true transmitters will remain moot, of course. While a number of experiments on vertebrate nervous tissue have shown that one substance may modulate the effect of another, as demonstrated for GABA and serotonin in the rabbit retina, such experiments only suggest what may happen at a synapse where the pre-synaptic fibre contains both substances. It seems to me that valuable progress towards the discovery of whether a neurone can use two or more transmitters will come from electrophysiological studies where characterized pre- and post-synaptic elements can be repeatedly studied. In this respect the studies on the giant serotonin cells in the snail still provide the best example (see Pentreath *et al.*, 1982). It is such preparations which offer hope for future studies along the lines discussed.

ACKNOWLEDGEMENTS

The author is extremely grateful to the Wellcome Trust for supporting work in his laboratory for many years.

REFERENCES

Ames, A., and Pollen, D. A. (1969) Neurotransmission in central nervous tissue: a study of isolated rabbit retina, *J. Neurophysiol.*, **32**, 424–442.

Andén, N-E., Jukes, M., and Lundberg, A. (1964) Spinal reflexes and monoamine liberation, *Nature*, 202, 1222–1223.

Anderson, R.E., and Hollyfield, J. G. (1984) Inositol incorporation into phosphoinositides in retinal horizontal cells of *Xenopus laevis*: enhancement by acetylcholine, inhibition by glycine, *J. Cell. Biol.*, **99**, 686–691.

Anderson, R. E., Maude, M. B., Kelleher, P. A., Rayborne, M. E., and Hollyfield, J. G. (1983) Phosphoinoside metabolism in the retina: Localization to horizontal cells and regulation by light and divalent cations, *J. Neurochem.*, **41**, 764–771.

Barbaccia, M. L., Brunello, N., Chuang, D. M., and Costa, E. (1983) Serotonin-elicited amplification of adenylate cyclase in hippocampal membranes from adult rat, *J. Neurochem.*, **40**, 1671–1679.

Barbeau, H., and Bedard, P. (1981) Similar motor effects of 5-HT and TRH in rats following chronical spinal transection and 5,7-dihydroxytryptamine injection, *Neuropharmacology*, **20**, 477–481.

Berridge, M. J. (1986) Cell signalling through phospholipid metabolism. *J. Cell. Sci. Suppl.*, **4** 137–153.

Berridge, M. J., and Irvine, R. F. (1984) Inositol triphosphate, a novel second messenger in cellular signal transduction, *Nature*, **312**, 315–321.

Björklund, A., Emson, P. C., Gilbert, R. T. F., and Skagerberg, G. (1979) Further evidence for the possible coexistence of 5-hydroxytryptamine and substance P in medullary raphe neurons of rat brain, *Br. J. Pharmacol.*, **66**, 112–113.

Blazynski, C., Ferrendelli, J. A., and Cohen, A. R. (1985) Indoleamine-sensitive adenylate-cyclase in rabbit retina – characterization and distribution, *J. Neurochem.*, **45**, 440–447.

Brunken, W. J., and Daw, N. W. (1986) 5-HT$_2$ antagonists reduce ON responses in the rabbit retina, *Brain Res.*, **384**, 161–165.

Brunken, W. J., Witkovsky, P., and Karten, H. J. (1986) Retinal neurochemistry of three elasmobranch species: an immunohistochemical approach, *J. Comp. Neurol.*, **243**, 1–12.

Chan-Palay, V., and Palay, S. L. (Eds) (1984) *Coexistence of Neuroactive Substances in Neurones*, John Wiley & Sons, New York.

Chan-Palay, V., Jonsson, G., and Palay, S. L. (1978) Serotonin and substance P coexist in neurons of the rat's central nervous system, *Proc. Nat. Acad. Sci. USA*, **75**, 1582–1586.

Chun, L. L. Y., and Patterson, P. H. (1977) Role of nerve growth factor in the development of rat sympathetic neurones *in vitro*. I. Survival growth and differentiation of catecholamine production, *J. Cell. Biol.*, **75**, 694–704.

Conn, P. J., and Sanders-Bush, E. (1984) Selective 5-HT$_2$ antagonists inhibit serotonin stimulated phosphatidyl inositol metabolism in cerebral cortex, *Neuropharmacology*, **23**, 993–996.

Conn, P. J., and Sanders-Bush, E. (1985) Serotonin-stimulated phosphoinositide turnover: mediation by S$_2$ binding site in rat cerebral cortex but not in subcortical regions, *J. Pharmacol. Exp. Therap.*, **234**, 195–203.

Cuello, A. C. (Ed.) (1982) *Co-transmission*, Macmillan Press, London.

Cutcliffe, N., and Osborne, N. N. (1987) Serotonergic and cholinergic stimulation of inositol phosphate formation in the rabbit retina. Evidence for the presence of 5-HT$_2$ and muscarinic receptors. *Brain Res.*, **421**, 95–104.

Dale, H. H. (1935) Pharmacology of nerve endings, *Proc. Roy. Soc. Med. Therap. Sect.*, **28**, 319–332.

Downes, C. P. (1986) Agonist-stimulated phosphatidyl inositol 4,5-bisphosphate metabolism in the nervous system, *Neurochem. Int.*, **9**, 211–230.

Ehinger, B. (1982) Neurotransmitter systems in the retina, *Retina*, **2**, 305–321.

Ehinger, B., and Florén, I. (1980) Retinal indoleamine accumulating neurones. *Neurochem. Intern.*, **1**, 209–229.

Ehinger, B., and Holmgren, I. (1979) Electron microscopy of the indoleamine-accumulating neurones, in the retina of the rabbit, *Cell Tissue Res.*, **197**, 175–194.

Ehinger, B., Hansson, Ch., and Tornqvist, K. (1981) 5-Hydroxytryptamine in the retina of some mammals, *Exp. Eye Res.*, **33**, 663–672.

Florén, J., and Hansson, H. C. (1980) Investigations into whether 5-hydroxytryptamine is a neurotransmitter in the retina of rabbit and chicken, *Invest. Ophthal.*, **19**, 117–125.

Florén, I., and Hendrickson, A. (1984) Indoleamine-accumulating horizontal cells in the squirrel monkey retina, *Invest. Ophthal.*, **25**, 992–1006.

Gilbert, R. F. T., Emson, P. C., Hunt, S. P., Bennett, G. W., Marsden, C. A., Sandberg, B. E. B., Steinbusch, H., and Verhofstad, A. A. J. (1982) The effects of monoamine neurotoxins on peptides in the rat spinal cord, *Neuroscience*, **7**, 69–88.

Greenfield, S. A. (1985) The significance of dendritic release of transmitter protein in the substantia nigra, *Neurochem. Int.*, **7**, 887–902.

Hansen, S., Svensson, L., Hökfelt, T., and Everitt, B. J. (1983) 5-hydroxytryptamine-thyrotropin releasing hormone interactions in the spinal cord. Effects on parameters of sexual behaviour in the male rat, *Neurosci. Lett.*, **42**, 299–304.

Hökfelt, T., Ljungdahl, A., Steinbusch, H., Verhofstad, A., Nilsson, G., Brodin, E., Pernow, B., and Goldstein, M. (1978) Immunohistochemical evidence of substance P-like immunoreactivity in some 5-hydroxytryptamine-containing neurons in the rat central nervous system, *Neuroscience*, **3**, 517–538.

Hökfelt, T., Lundberg, J. M., Schultzberg, M., Johansson, O., Ljungdahl, A., and Rehfeld, J. (1980) Coexistence of peptides and putative transmitters in neurons, in *Neural Peptides and Neuronal Communication* (Eds E. Costa and M. Trabucchi), pp. 1–23, Raven Press, New York.

Holmgren-Taylor, I. (1982) Electron microscopic observations on the indoleamine-accumulating neurones and their synaptic connections in the retina of the cat, *J. Comp. Neurol.*, **208**, 144–156.

Hume, D. A., Perry, H., and Gordon, S. (1983) Immunohistochemical localization of a macrophage-specific antigen in developing mouse retina: phagocytosis of dying neurones and differentiation of macroglial cells to form a regular array in the plexiform layers, *J. Cell. Biol.*, **97**, 253–257.

Johansson, O., Hökfelt, T., Pernow, B., Jeffcoates, S. L., White, N., Steinbusch, H. W. M., Verhofstad, A. A. J., Emson, P. C., and Spindel, E. (1981) Immuno-histochemical support for three putative transmitters in one neuron. Coexistence of 5-hydroxytryptamine, substance P, and thyrotropin releasing hormone-like immunoreactivity in medullary neurons projecting to the spinal cord, *Neuroscience*, **6**, 1857–1881.

Kendall, D. A., and Nahorski, S. R. (1985) 5-hydroxytryptamine-stimulated inositol phospholipid hydrolysis in rat cerebral cortex slices: pharmacological characterization and effects of antidepressants, *J. Pharmacol. Exp. Therap.*, **233**, 473–479.

Lam, D. M. K., Li, H. B., Su, Y. Y., and Watt, C. B. (1985) The signature hypothesis: colocalisation of neuroactive substances as anatomical probes for circuitry analysis, *Vison Res.*, **25**, 1353–1364.

Mitchell, R., and Fleetwood-Walker, S. (1981) Substance P, but not TRH, modulates the 5-HT autoreceptor in ventral lumbar spinal cord, *Eur. J. Pharmacol.*, **76**, 119–120.

Mitchell, C. K., and Redburn, D. A. (1985) Analysis of pre- and post-synaptic factors of the serotonin system in rabbit retina, *J. Cell. Biol.*, **100**, 64–73.

Nahorski, S. R. (1985) Inositol phospholipid hydrolysis as a primary response to receptors not linked to adenylate cyclase, *Arzneim. Forsch/Drug. Res.*, **35**, 1886–1909.

O'Donohue, T. L., Millington, W. R., Handelmann, G. E., Contreras, P. C., and Chronwall, B. M. (1985) On the 50th anniversary of Dale's law: multiple neuro-transmitter neurons, *Trends in Pharmacological Sciences*, **6**, 305–308.

Osborne, N. N. (1981a). Binding of ³H-serotonin to bovine membranes of the retina, *Exp. Eye Res.*, **33**, 371–380.

Osborne, N. N. (1981b) Communication between neurones, *Neurochem. Intern.*, **3**, 3–16.

Osborne, N. N. (Ed.) (1983) *Dale's Principle and Communication Between Neurones.* Pergamon Press, Oxford.

Osborne, N. N. (1984a) Indoleamines in the eye with special reference to the serotonergic neurones of the retina, in *Progress in Retinal Research*, Vol. 3. (Eds N. N. Osborne and G. H. Chader), pp. 61–104, Pergamon Press, Oxford.

Osborne, N. N. (1984b). Transmitter specificity in neurones, in *Handbook of Neuro-chemistry*, Vol. 6. (Ed. A. Lajtha), pp. 511–526, Plenum Press, New York.

Osborne, N. N. (1985a). Interplexiform, horizontal and bipolar-like cells of the rabbit retina take up exogenous serotonin during early developmental stages, *Int. J. Devel. Neuroscience*, **3**, 643–646.

Osborne, N. N. (1985b). Communication between neurones: current concepts, in *Selected Topics from Neurochemistry* (Ed. N. N. Osborne), pp. 43–62, Pergamon Press, Oxford.

Osborne, N. N., and Beaton, D. W. (1985) Selecting of uptake of tritiated glycine, GABA, aspartate and serotonin by rabbit retinal cells in cultures, *Biogenic Amines*, **3**, 169–179.

Osborne, N. N., and Beaton, D. W. (1986a) Studies on serotonin-accumulating cells in rabbit retinal cultures, *Brain Res.*, **362**, 287–298.

Osborne, N. N., and Beaton, D. W (1986b) Direct histochemical localization of 5,7-dihydroxytryptamine and the uptake of serotonin by a subpopulation of GABA neurones in the rabbit retina, *Brain Res.*, **382**, 158–162.

Osborne, N. N., and Patel, S. (1984) Postnatal development of serotonin accumu-lating neurones in the rabbit retina and on immunohistochemical analysis of the uptake and release of serotonin, *Exp. Eye Res.*, **38**, 611–620.

Osborne, N. N., Nesselhut, T., Nicholas, D. A., Patel, S., and Cuello, A. C. (1982) Serotonin-containing neurones in vertebrate retina, *J. Neurochem.*, **39**, 1519–1528.

Osborne, N. N., Patel, S., and Beaton, D. W. (1984) Serotonin, a major metabolite of tryptophan in the rabbit retina, in *Progress in Tryptophan and Serotonin Research* (Ed. H. G. Schlessberger, W. Kochen, B. Linzen and H. Steinhart), pp. 275–284, Walter de Gruyter, Berlin.

Patterson, P. H., Reichardt, L. F., and Chun, L. L. Y. (1976) Biochemical studies on the development of primary sympathetic neurones in cell culture, *Cold Spring Harb. Symp. Quant. Biol.*, **40**, 389–393.

Pellcticr, G., Steinbusch, H. W., and Verhofstad, A. (1981) Immunoreactive sub-stance P and serotonin present in the same dense core vesicles, *Nature*, **293**, 71–72.

Pentreath, V. W., Berry, M. S., and Osborne, N. N. (1982) The serotonergic cerebral cells in gastropods, in *Biology of Serotonergic Transmission* (Ed. N. N. Osborne), pp. 457–513, John Wiley & Sons, Chichester.

Redburn, D. A. (1984) Serotonin systems in the inner and outer plexiform layers of the vertebrate retina, *Fed. Proc.*, **43**, 2699–2703.

Skirboll, L., Hökfelt, T., Norell, G., Phillipson, O., Kuypers, H. G. J. M., Bentivoglio, M., Catsman-Berrevoets, C. E., Visser, T. J., Steinbusch, H., Verhofstad, A., Cuello, A. C., Goldstein, M., and Brownstein, M. (1984) A method for specific transmitter identification of retrogradely labeled neurons: immunofluorescence combined with fluorescence tracing, *Brain Res. Rev.*, **8**, 99–127.

Thompson, R. J., Doran, J. F., Jackson, P., Dhillon, A. P., and Rode, J. (1983) PGP 9.5 – a new marker for vertebrate neurones and neuroendocrine cells, *Brain Res.*, **278**, 224–225.

Neuronal Serotonin
Edited by N. N. Osborne and M. Hamon
© 1988 John Wiley & Sons Ltd

CHAPTER 6

Serotonin and Sleep

WERNER P. KOELLA*
*The Friedrich Miescher Institute, Basel and the University of Bern
Switzerland*

INTRODUCTION AND GENERAL CONCEPTS

There is a considerable, if not an overwhelming, amount of evidence that curtailment of (central nervous) serotonergic transmission activity, through, e.g. destruction of serotonergic neurons, inhibition of the amine synthesis or release, or through blockage of (postsynaptic) serotonergic receptors, is followed by insomnia of various degrees. In turn, an increase in serotonergic activity, i.e. enhanced activation of (postsynaptic) 5-HT receptors, achieved through e.g. administration of the amine precursor amino acids, inhibition of its catabolism, or through application of 5-HT to a variety of 'strategic' loci in the brain, is followed by 'dewaking', increased occurrence of various sleep signs (in particular, slow waves and spindles in the EEG) and, in freely behaving animals and man, by an increase in sleep time (in particular, its NREM variety) and/or shortening of sleep latency, in addition to a block of REM sleep.

* Present address: Buchenstrasse 1, CH-4104 Oberwil (BL), Switzerland.

Such evidence initially led to the belief that central serotonergic activity is intimately involved with the making of sleep. In fact the author did not hesitate, close to 20 years ago, to nickname this indoleamine as 'somnotonin' (Koella, 1970).

However, a number of additional experiments made it quite clear that this could not be 'the whole truth'. Recording experiments, aiming at measuring serotonergic activity during sleep and wake, as well as detailed behavioral investigations in waking animals after interruption of serotonergic transmission, clearly indicated that intact serotonergic activity is also of paramount importance for proper performance of a large variety of behavioral activities of waking. And work with a variety of neurotransmitter (NT) systems revealed that other biogenic amines, as well as a number of amino acids and polypeptides, are as important for proper sleeping behavior as is serotonin (Koella, 1981, 1984). In other words, 5-HT systems are a 'source' of influences on sleep and wake and the 'sleep-system' is a 'sink' for influences arising in the 5-HT as well as in other NT systems.

Being confronted with the obvious and undebatably safe fact that serotonergic mechanisms – besides their involvement in such functions as temperature regulation, intrinsic analgesia, and regulation of endocrine activity – play an active role in the organization and regulation of sleeping *and* of waking behavior, it became necessary to find ways and means to explain this double function on the basis of a single and common mechanism of action. And after some intensive probing a 'general theory of vigilance' (Koella, 1982) was established that yielded the conceptual foundation for such a common mechanism. This theory proposed that obviously the main effect of enhanced serotonergic activity is the reduction of vigilance – the level of readiness to 'perform' – in particular in systems of higher and lower function behaviors. It was further postulated that this curtailing influence of serotonin on vigilance is brought about by the (curtailing) influence of this amine on the reactivity of those neuronal networks that subserve the making of such behaviors. And the 'puzzle of the double function' could be resolved by assuming that the organism makes use of this capability of serotonin in both sleep and wake. During the former, in particular during its NREM stages, serotonergic activity produces that particular low level of vigilance in practically all the higher and lower function behavioral systems that indeed characterize these stages of sleep (and this in contrast to the relatively high level of vigilance in the motor systems). During the latter (i.e. waking), serotonergic activity again, yet in a slightly more selective manner, curtails vigilance in all those higher and lower function systems that are, at a given moment, not involved in the making of a particular behavior. Thus 5-HT activity during waking hinders the emergence of uncontrolled and general hyperarousal, and, at the same time, counteracts 'generalization' and saves 'vigilance energy'.

In the following sections the experimental evidence will be described and discussed, i.e. findings in stimulation, lesion, and recording experiments, as well as findings from single-cell work, that led to the interpretation just outlined. The chapter ends with a description of a model machinery – a 'general-purpose' vigilance-controlling apparatus – where serotonergic mechanisms (besides a few others) assume the important role as selective vigilance-reducing instruments.

This same topic – serotonin and sleep – has been dealt with in a number of earlier publications (Koella, 1974, 1979, 1981, 1982, 1984, 1985a, b). The author welcomes this opportunity to write on this problem again, as it provides a chance to elaborate on some new developments in both this field in general and our endeavor towards a still better understanding of this complicated matter.

CHANGES IN SLEEP RELATED TO INDUCED OR SPONTANEOUS CHANGES IN SEROTONERGIC ACTIVITY

Stimulation experiments

'Stimulation experiments' refers here to all those experimental approaches where one enhances, through biophysical or pharmacological means, in a more or less selective manner, serotonergic transmission activity and thus increases the time/space sum of activated (postsynaptic) 5-HT receptors. The latter can also be achieved by direct (local) application of proper receptor agonists.

The systemic injection of L-tryptophan (L-TP) in rats enhances total sleep time, but only marginally increases, if at all, paradoxical sleep (PS), i.e. REM sleep (Hartmann, 1968; Wojcik *et al.*, 1980). The administration of the (direct) serotonin precursor, 5-hydroxytryptophan (5-HTP; 30–50 mg/kg i.p.) to cats and rabbits markedly prolongs slow-wave sleep (SWS = NREM sleep in animals) and suppresses paradoxical sleep (Delorme *et al.*, 1966; Jouvet, 1967, 1972; Koella *et al.*, 1968; Tabushi and Himwich, 1970). Injection of 5-HTP into the solitary tract nucleus of immobilized cats induces a 'sleep-like state' with slow waves and spindles in the cortical EEG (Ledebur and Tissot, 1966). Curtailing the metabolism of serotonin by the injection of monoaminoxidase inhibitors such as nialamide markedly enhances in cats the proportion of slow-wave sleep while virtually eliminating paradoxical sleep for as long as 100 h (Delorme *et al.*, 1966; Jouvet, 1972).

In normal man, L-tryptophan, in doses between 1 and 10 g, is liable to shorten sleep latency and to increase sleeping time, in particular that of stages 3 and 4 (see, e.g. Hartmann, 1968, 1974; Williams *et al.*, 1969; Wyatt *et al.*, 1970) and to produce sleepiness during daytime (Yuwiler *et al.*, 1981).

Injection of very small amounts of serotonin (30–50 ng) into the fourth ventricle of freely moving cats (via a chronically implanted cannula) induces behavioral as well as electrographic signs of sleep. Shortly after injection, a first slow-wave sleep episode of about 1/2 h duration appears which then is followed, after a short interval of obvious wakefulness, by a second, rather prolonged period of almost 100% sleep. For the first 3 to 4 h there is virtually only slow-wave sleep, after which period paradoxical sleep begins, at a rate of up to 20% (Koella, 1970, 1974). The placement of small pieces of filter paper, soaked with serotonin (10^{-6}g/ml) on the area postrema (at the posterior end of the fourth ventricle) induces an abundance of slow waves and spindles in the cortical EEG of cerebellectomized, immobilized (flaxedil) cats (Koella and Czicman, 1966). Injection of 5-HT into the artery that supplies the area postrema induces hypersynchrony (slow waves) in the EEG of immobilized cats (Roth *et al.*, 1970). Intravertebral injection of 5-HT in microgram doses is followed immediately by the appearance of high-amplitude slow waves and spindles in the electrocorticogram of immobilized cats (Koella and Czicman, 1963). Injection of serotonin into the nucleus of the solitary tract (situated just underneath the area postrema) produces sleep in cats (Key and Mehta, 1977). Intracarotid administration of serotonin (again in microgram doses) induces in freely moving cats behavioral signs of sleep (Rothballer, 1957). The same procedure produces, after a short episode of arousal, electrographic signs of sleep in immobilized cats (Koella and Czicman, 1963). Injection of 5-HT into the preoptic area of freely moving cats induces sleep (Yamaguchi *et al.*, 1964).

Finally, it has been shown that low-rate electrical stimulation of the midbrain raphe nuclei of rats is followed by 'sedation' and sleep (Kostowski *et al.*, 1969). It was also shown that electrical stimulation of the anterior raphe nuclei of rats induces a higher yield of 5-hydroxyindole acetic acid and reduced 5-HT concentrations in the forebrain – clear signs of enhanced serotonin release from the 'excited' serotonergic fibers (Aghajanian *et al.*, 1967).

Lesion experiments

In 'lesion experiments' one curtails, by biophysical or pharmacological means, serotonergic transmission activity, or one blocks, with proper antagonists, 5-HT receptors. With either approach one reduces the number of activated 5-HT receptors.

Slowing down the synthesis of serotonin by the application of the tryptophan hydroxylase inhibitor *p*-chlorophenylalanine (PCPA) in the cat is followed by a (dose-dependent) drop in brain serotonin (Koe and Weissman, 1966) and by a dose-dependent reduction of sleep (Delorme *et al.*, 1966; Koella *et al.*, 1968). With a single dose of PCPA the insomnia starts with a delay of about 24 h and reaches its peak on about the third day. With larger doses of PCPA (100 to 200 mg/kg i.p.) it takes as long as 3 to 4 weeks for sleep to return to normal levels. With smaller doses, normosomnia returns somewhat faster and often with a clear intermittent rebound. Also, during the first two days after injection of the enzyme inhibitor there often is a clear transient increase in relative PS times (Koella *et al.*, 1968). Similar observations were made in rats (Torda, 1967) and in monkeys (Weitzman *et al.*, 1968). Repeated administration of PCPA in cats over a period of many days initially leads to the characteristic drop in sleep. Yet, in spite of the continued inhibition of the hydroxylating enzyme, there is a slow return of sleep in the direction of normosomnia which, however, is never completely reached (Dement *et al.*, 1972).

Injection of 5,6-dihydroxytryptamine (5,6-DHT, 1 mg in 200 μl of solvent) into the lateral ventricle of cats leads to a drastic reduction of the 5-HT and 5-HIAA levels in the forebrain, and, after an initial phase of sedation, to a marked and continuous desynchronization of the EEG, as well as to a drastic reduction of SWS and PS. Even 10 days after the treatment, SWS does not return to above 50% of control levels (Froment *et al.*, 1974). Bilateral injection of 5,7-DHT (4 μg/4 μl) into the mesencephalic tegmentum of rats leads to an increase in waking time and a concomitant reduction of SWS, together with a drastic drop in forebrain 5-HT levels (Kiianmaa and Fuxe, 1977). Larger amounts of 5,7-DHT in larger volumes of solvent (200 μg/20 μl) injected into the ventricular system of rats still lead to some insomnia. However, the loss of sleep does not exceed the already considerable insomnia produced by the injection of similar volumes of solvent only (Ross *et al.*, 1976).

Blocking of central 5-HT receptors with methysergide or one of the more selectively acting 5-HT antagonists, methiothepin or metergoline, leads to pronounced reduction of sleep in the rabbit and the cat (Tabushi and Himwich, 1971; Jouvet, 1983).

Electrolytic lesions, aimed at the destruction of the anterior raphe nuclei of cats, not only lead to a drop in forebrain serotonin, but also to a pronounced and long-lasting insomnia. SWS and PS are about equally affected (Renault, 1967; Jouvet, 1969, 1972). In the course of the weeks following surgery, sleep begins to increase again without, however, reaching control levels. There is a marked correlation between the size of the intact raphe tissue, the level of forebrain 5-HT, and the total amount of sleep (Jouvet, 1972). Similar results on sleep of raphectomy were reported for the rat (Kostowski *et al.*, 1968).

Yet, manipulations that curtail (central) serotonergic transmission and its postsynaptic effectiveness affect not only sleep but a variety of waking activities as well. Reduction of central serotonin content or interruption of serotonergic signal transmission to the respective receptors is followed by hyperactivity, hyperaggressivity, hypersexuality, as well as by signs of improvement of memory function (MacKenzie *et al.*, 1978; Yamamoto and

Ueki, 1978; Meyerson and Malmnas, 1978; Essman, 1974). Asin and Fibiger (1983) have shown that rats with electrolytically lesioned median raphe nuclei are about as hyperactive in the 'open field' as are amphetamine-treated animals and significantly more active than control rats. And the surgically lesioned animals are significantly more active than rats in which the serotonergic neurons of the median raphe nuclei are destroyed through local injection of 5,7-DHT. This latter difference is due, in those authors' opinion, to the fact that raphe nuclei contain less than 50% truly serotonergic neurons. Consequently, the activating effect of the electrolytic lesion has to be due to the elimination of elements of non-serotonergic nature in addition to that of elements of serotonergic neurons.

Recording experiments

The serotonin concentration of the brain of rats varies synchronously, yet with a phase-shift of roughly 180°, with the rate of locomotor activity of these night-active animals (Quay, 1965; Scheving et al., 1968). Larger variations in serotonin content, yet with a similar phase relation, are found in the pineal organ of rats (Snyder et al., 1967). The 5-HT content of the serum of rats also varies in a circadian (or rather circasemidian) fashion; a first 'peak' is located towards the end of the nightly activity period; after a short drop at noon a second 'peak' appears which lasts until the beginning of the next activity period (Scheving et al., 1972). The 5-HIAA concentration in the cisternal fluid of freely moving cats increases during SWS (Buckingham and Radulovacki, 1975; Radulovacki et al., 1977). Also in cats, it was found that with the onset of SWS the 5-HIAA content increased in the hippocampus, but not in the thalamus or the striatum (Kovacevic and Radulovacki, 1976). In more recent work with better methodology, it was observed that the yield of 5-HIAA in the hippocampus and (less markedly) in the cortex of rats increased with the transition from waking to slow-wave sleep and dropped below waking levels with the onset of paradoxical sleep (Ogasahara et al., 1980).

However, these results were not entirely confirmed when biophysical techniques, rather than biochemical measures, were used to monitor the activity of serotonergic neurons. With the transition from quiet waking to SWS both the activity of serotonergic units and that of noradrenergic units (recorded from the dorsal raphe nucleus and the locus coeruleus, respectively) were found to drop slightly. And with the onset of paradoxical sleep both activities dropped steeply and remarkably parallel to very low levels (McGinty and Harper, 1976). Also not in line with the results of traditional biochemical analysis are the findings from Jouvet's laboratory (Crespi and Jouvet, 1983), obtained with the recently developed differential pulse voltammetry method. Those authors noted that 'peak 3', assumed to indicate local 5-HT levels, in the records obtained from the nuclei raphe dorsalis and magnus of rats, was maximally high during waking, somewhat lower during SWS, and dropped to minimal amplitude during paradoxical sleep.

It is not easy to explain these discrepancies between the results obtained with the traditional methods – suggesting a high 5-HT activity as well as content, at the onset of, and during, SWS on the one hand – and those obtained with the microelectrode and the voltammetric recordings – suggesting high 5-HT activity during waking, on the other. Experimental artifacts, e.g. contamination with alien chemical activities when using voltammetry, or contamination with activity from non-serotonergic cells when using microelectrode recordings, may be responsible for these conflicting results. Also there is good reason to assume that there is a rather weak correlation between the electrically recorded discharge rate of aminergic somata, on the one hand, and release of transmitter, on the other; and there certainly is a weak correlation between the amount of released transmitter and the magnitude of the postsynaptic effect.

So, although it is the eventual postsynaptic effect of serotonergic transmission that really counts, we have little information about changes of its magnitude with the transition from waking to (slow-wave) sleep and vice versa. At this time we are inclined to assume that there is some substantial serotonergic activity during slow-wave sleep as well as during waking.

THE EFFECT OF SEROTONIN ON NEURONS AND NETWORKS

Bloom *et al.* (1972) have reviewed studies on the effects of serotonin on single neurons, done mainly with the microiontophoretic technique, up to 1971. They state that 'most telencephalic neurons are depressed by serotonin, while many of the spinal or ponto-medullary neurons are excited'. In the review, the authors also pointed out a number of sources of complications that make a straightforward interpretation somewhat problematic. Aside from location (i.e. the particular network that the target cell belongs to) animal species and the type of preparation (anesthesia, immobilization, decerebration) have a decisive influence upon quality (i.e. direction) and quantity of the response of cells to 5-HT. Also the *'Ausgangslage'* – initially active vs. inactive cells, spontaneously active vs. synaptically or glutamate-driven cells – plays an important role for the making of a particular response.

A few examples, in part taken from Bloom and co-workers' review, and in part from more recent findings, should illustrate the situation as far as the influence of serotonin on cells and networks is concerned.

Unidentified, antidromically or aminoacid-driven cells in the pericruciate cortex of the barbiturate- or chloralose-anesthetized cat react to iontophoretically applied 5-HT either with depression or 'no response' (Krnjevic and Phillis, 1963). Randomly selected, spontaneously active cells in the piriform cortex of similar cat preparations react to iontophoretically applied serotonin with depression (Legge *et al.*, 1966). Pyramidal cells in the motor cortex of decerebrate cats, when antidromically driven, react to 5-HT exclusively with a reduction of firing rate (Salmoiraghi and Stefanis, 1967). Also the cells of the striatum of immobilized or anesthetized rabbits and cats react to 5-HT with a slowing of the discharge rate (Herz and Zieglgänsberger, 1968; York, 1970). Identified cells in the hippocampus of rabbits and cats have been found to respond to 5-HT mainly with depression, regardless of whether the units were spontaneously active or were synaptically or antidromically driven (Salmoiraghi and Stefanis, 1967). It was further noted that randomly selected cells of the cat's hypothalamus react to serotonin with acceleration or deceleration in barbiturate-, halothane-, or ether-anesthetized preparations (Bloom *et al.*, 1963). Cells in the visual and auditive relay nuclei of the thalamus (lateral and medial geniculate bodies, respectively) react to iontophoretically applied serotonin with depression only (Curtis and Davis, 1962; Phillis *et al.*, 1967). This suggests that conductance in these relay synapses is reduced under the influence of enhanced serotonergic output. In line with this is the finding that 'modulation' of the lateral geniculate neurons can be induced also by electrical stimulation of the cat's raphe nuclei (Foote *et al.*, 1974).

Electrical stimulation of the dorsal raphe nucleus of immobilized, unanesthetized cats leads to prolonged inhibition of the orthodromic spike generation in the spinal trigeminal nucleus without affecting the antidromically elicited spikes (Sasa *et al.*, 1975). It was assumed that the influence of the raphe-fugal, trigemino-petal pathways is due to a presynaptic inhibition-type mechanism. It was further demonstrated that 92% of the cat's spontaneously active hippocampal pyramidal cells respond with reduced firing to microiontophoretically applied 5-HT, whereas only 48% of those cells respond in this manner to electrical stimulation of the dorsal and median raphe nuclei (Segal, 1975). Pretreatment with PCPA eliminates, and additional application of 5-HTP restores again, the effect of raphe stimulation. Local application of such serotonergic receptor blockers as methysergide and cyproheptadine curtails the effect of raphe stimulation (Segal, 1975).

Electrical stimulation of B_7, the dorsal raphe nucleus, but not of B_8, the median nucleus, inhibits the spontaneous firing rate of cells in the caudate-putamen of rats. This inhibitory effect is well counteracted by the iontophoretic application of methysergide (Olpe and Koella, 1977). After intraventricular injection of 5,7-DHT, cortical neurons within three days become supersensitive (in the sense of enhanced inhibition) to microiontophoretically applied serotonin (Olpe *et al.*, 1981). There is also evidence that the locus coeruleus cells receive a serotonergic input (Pickel *et al.*, 1977) and that this input is inhibitory in nature (Kostowski, 1975).

Quite recently it has been shown by means of intracellular recordings that (microiontophoretically applied) serotonin hyperpolarizes the membranes of cells in the lateral septal nucleus; this effect obviously is due to an increase in the potassium conductance (Joëls *et al.*, 1986). Similar results were reported for cells in the hippocampus (Segal, 1980).

Finally, it was demonstrated that the 'down-regulation' of (central) β-adrenergic receptors in response to desimipramine treatment does not occur after the (central) serotonergic pathways have been lesioned with 5,7-DHT (Janowsky *et al.*, 1982). This clearly indicates that this 'down-regulation' depends on an intact serotonergic input.

AN INTERPRETATION

The main, if not exclusive effect of locally applied serotonin on single neurons in a variety of forebrain areas – the motor, limbic, visual, and auditory cortices, and the hippocampus, the septum, and the striatum – is inhibition of cell discharge. Also, (local) release of 5-HT, as induced by activation of serotonergic fibers projecting to these areas, suppresses neuronal activity. Experimental results obtained through intracellular recordings from cells in the septal area and the hippocampus indicate that this inhibition is effected through the 'classical mechanisms' of membrane hyperpolarization based on an increase of potassium conductance. Obviously, this local inhibition is the very basic mechanism through which serotonin is capable of curtailing, in a widespread fashion, reactivity of the various networks in the structures mentioned above. As these networks, by and large, subserve the actual making and detailed organization of mainly higher and lower function behaviors (cognition, learning, perception, remembering, decision making, and drive, mood and instinctive behavior), as well as complex psychomotor behavior, it can be concluded that (enhanced) serotonergic transmission activity is liable to curtail (local) vigilance in any one of these 'internal' and psychomotor behavioral systems. That is to say, the organism, by increasing serotonergic output in a rather selective fashion, is capable of reducing the readiness – '*Bereitschaft*' in W. R. Hess's terminology – of any one, or all, of these systems to react with a properly organized behavioral act to a series of external and internal behavior-inducing signals (Koella, 1982). But serotonin probably has little such vigilance-suppressing influence on more primitive motor functions, organized mainly through lower brainstem, cerebellar, and spinal mechanisms.

This – as we think rather novel – interpretation of the (or one of the) function(s) of serotonergic 'innervation' furnishes the very basis for a novel view on both the role played by this indoleamine in the organization and regulation of sleep and its role in the restraint of excessive activity during waking.

Before going into any detail about this double function of serotonin, mention must be made of a number of mechanisms involved as vigilance-enhancing instruments, or, in a sense, as true, but more selective, opponents of serotonergic signals. There is a wealth of information that, for example, cholinergic mechanisms are involved as reactivity and thus, as vigilance-enhancing instruments mainly in the realm of such higher function systems as learning and memory read-out (for details and references, see Koella, 1982, 1984). Furthermore, there is good evidence that noradrenergic mechanisms, working mainly through the β-adrenergic receptor function, act as vigilance-enhancing instruments in such higher function behavioral systems as perception, sensory attention, and 'attentive filtering' (see Koella, 1982, 1984, for references). And there is evidence that dopaminergic mechanisms are involved as vigilance-enhancing instruments in motor systems, as well as in systems subserving 'immobile attention' and, possibly, in a variety of other lower function behavioral systems, e.g. reward. This role of dopamine (DA) may be supplemented by some – at this time rather ill-defined – trace-aminergic mechanisms and, possibly, histaminergic mechanisms. There is also some evidence that a number of polypeptides, e.g. substance-P, TRH, vasopressin, enkephalins, somatostatin, are involved as 'adjuvant' instruments, as they seem to be capable of modulating, mainly in a facilitatory direction, the vigilance-enhancing influence of such aminergic mechanisms (Koella, 1982, 1984). And, finally, there is ample evidence – in the reverse direction – that GABAergic mechanisms are involved, not so much as 'diverging' reactivity-reducing instruments projecting to, but rather as 'point-to-point', highly localized inhibitory instruments within, the behavioral systems networks (Koella, 1982, 1984).

With all this information, we are now in the position to sketch the output aspects of the model of a 'general-purpose' vigilance-controlling apparatus (VCA). This apparatus is assumed to act as the sleep organizing and regulating apparatus (SORA) and, at alternate time periods, also as a vigilance-setting machinery in control of the various and manifold behavioral activities of wake. Figure 1 depicts this reflex-type structure of the VCA including, in a somewhat simplified manner, some of the 'logics' of the center and three of the more important inputs.

On the input side of this model one recognizes feedback lines, carrying signals from the behavioral systems networks. These (quite probably) humoral signals – one or several of the mainly polypeptidic 'sleep factors' – inform the center about the 'present state' and the 'recent history' of these networks. In addition, there is input arising in a circadian clock. There is excellent evidence that this '*Zeitgeber*-input' interacts, within a device that may be referred to as integrator, with the humoral feedbacks to produce, in the center, activity that either induces and maintains sleep, or induces and maintains waking (see, e.g. Daan *et al.*, 1984). And, finally, there is input via collaterals from sensory afferent systems, which, in dependence of the quality, quantity and temporal pattern of the afferent signals, either induces and maintains waking, or facilitates the onset and maintenance of sleep. An analyzer is thought to act as the 'decision-making' instrument.

Within the center, besides the analyzer and integrator, there are a wake- (W-) and a sleep- (S-) program; the latter subdivided into a NREM- and a REM-subprogram. They subserve the important role of directing the activity, arising in the integrator and the analyzer, to the nuclei of origin of the various – mainly aminergic – output lines.

The output functions are, in the present connection, of major concern. One recognizes, on the one hand, a cholinergic and a noradrenergic line, projecting, as vigilance-enhancing instruments to the higher-function networks, and dopaminergic lines, projecting, again as vigilance-enhancing instruments, to the motor- and lower-function networks. On the other hand, there are serotonergic lines, projecting, as vigilance-suppressing instruments, to the higher, lower, and (the psychomotor part of the) motor-function networks.

With this arrangement, the organism, through the various switching and integrating functions of the center of the VCA, is capable of producing the proper vigilance profiles, characteristic of the various 'states' (here reduced to wake, NREM sleep, and REM sleep).

During waking, with intermediate to high levels of vigilance in practically all systems, one can expect high levels of activity in the cholinergic, noradrenergic, and dopaminergic systems. Yet, at the same time, the serotonergic output is kept at a reasonably high level to prevent undue hyperarousal in any one of the various behavioral systems, and, probably even more important, to spare all systems and subsystems not 'used' for a given behavioral act from excessively high vigilance, and, thus (as already mentioned in the introduction) to counteract generalization, and, at the same time, to 'save vigilance energy'.

During NREM sleep, with its characteristic low level of vigilance in mainly higher and lower function systems, but a reasonably high level of vigilance in the motor systems, one can expect a rather drastic drop in cholinergic and noradrenergic output, a relatively high dopaminergic output, and a rather high serotonergic output. The latter can be considered, in the sense of a reciprocal-antagonistic innervation pattern, to supplement and enforce in an active fashion the 'permissive' vigilance-reducing effect of the curtailed noradrenergic and cholinergic output to the 'internal' behavior systems.

Finally, during REM sleep, with its characteristic high level of vigilance in the higher and lower function systems (one experiences and learns dreams) and extremely low levels of vigilance in the greater part of the motor systems, one can expect a high noradrenergic and cholinergic output, while the dopaminergic and serotonergic outputs are reduced down to, probably, an 'all-time' low. Low 5-HT activity is bound to support, again in a permissive fashion, the vigilance-enhancing influence of both high cholinergic and high

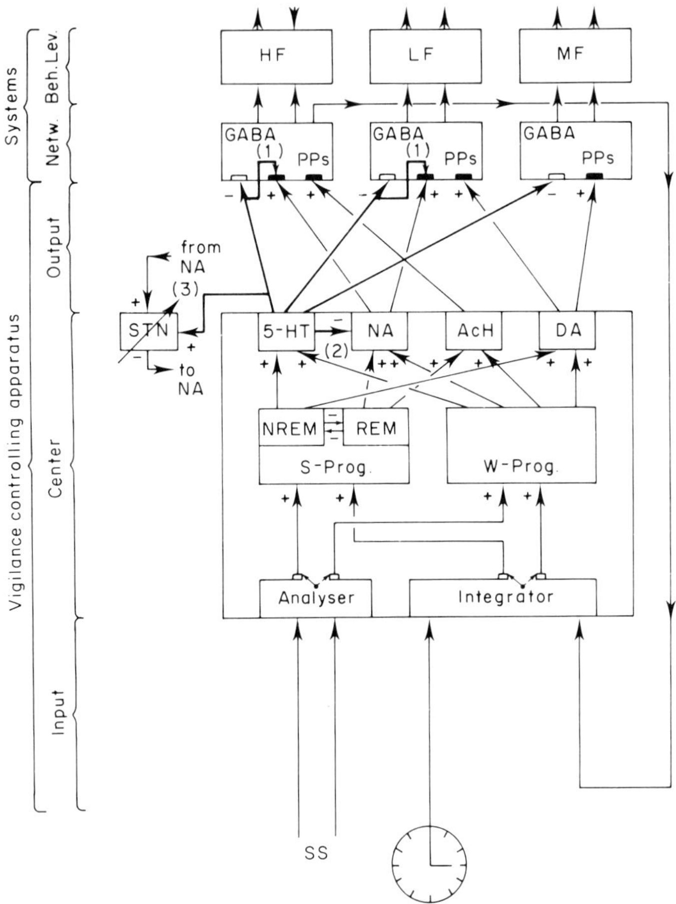

Fig. 1 A simplified model of the vigilance-controlling apparatus (VCA) with its (aminergic) output projections to the various behavioral systems (here 'reduced' to higher, lower, and motor functions; HF, LF, and MF, respectively). Note the three components of VCA: input, center, and output. The *input* includes the (humoral) feedback from the effector networks, the circadian pacemaker, and a variety of sensory afferents (SS). The *center* contains the analyzer (a discriminator for switching sensory signal activity to the W- or the S-program), the integrator (for handling interaction of (humoral) feedback and clock-input), the W- and the S-programs, and the nuclei of origin of the NA, DA, Ach, and 5-HT *output* lines.

Note that the *serotonergic output* – in view of the present topic – is drawn in heavier lines. Included are, in addition to the 'direct', three 'indirect' serotonergic vigilance-suppressing mechanisms:

noradrenergic activity. This also explains that high 5-HT output is bound to counteract the emergence of REM-sleep. One also may speculate that a completely unchecked elevation of vigilance in the systems of higher and lower internal behavior systems is the (or one of the) reason(s) for the uncontrolled – dereistic-type – pattern of thinking during REM sleep.

Thus, there is good reason to assume that with this novel interpretation of serotonergic mechanisms as vigilance-suppressing instruments, the involvement of this indoleamine in the organization and regulation of sleep and in the control of the waking activities, are reasonably well explained.

Yet one has to be aware that serotonin may be involved in the making of low vigilance also through mechanisms outside their direct influence on the reactivity of the effector networks. There is evidence, indeed, that 5-HT, now acting rather as a hormone, after release into the brain tissue and then into the ventricular space, interacts with (extra-blood–brain-barrier) 5-HT receptors in, or close to, the area postrema. Via this point of action it enhances 'amplification' in the nucleus tractus solitarius and, consequently, increases inhibitory feedback in a reticulo–solitario–reticular feedback loop (Koella, 1974). Obviously, intraventricular 5-HT acts as an antiarousal factor and thus supports its (direct) vigilance-suppressing effect.

Furthermore, a serotonergic influence seems to be actively involved in the 'down-regulation' of β-adrenergic receptors (Janowsky et al., 1982). A high serotonergic output is instrumental in the lowering of sensitivity (affinity and number) of these adrenoceptors and, thus, in a reduction of the β-adrenergically mediated vigilance-enhancing effect of noradrenergic activity.

In addition, enhanced 5-HT activity is liable to curtail vigilance – at least in systems receiving a noradrenergic vigilance-enhancing projection – through the well-established inhibitory serotonergic influence on the (noradrenergic) cells of the locus coeruleus (Pickel et al., 1977).

Finally, there is some possibility that serotonin may have an influence on sleep via indole-dependent, but basically non-indolic SWS- and PS-factors (Sallanon et al., 1983; Jouvet, 1983). Those authors have noted that in cats, PCPA, given immediately at the end of a two-day sleep-deprivation (SD)

---

(1) the 'down-regulating' impact on β-adrenergic receptors in the effector networks; (2) the inhibitory influence on the nor-adrenergic cells of the locus coeruleus; and (3) the amplification-enhancing impact on the nucleus of the solitary tract. The latter is assumed to receive a noradrenergic output (representing the output from the ponto-mesencephalic reticular activating system) and to dispatch inhibitory signals to the noradrenergic nuclei of the brainstem (representing again the reticular activating system). Note in the effector networks the (supposedly) modulating poly-peptidergic mechanisms (PPs) and – representing a variety of short-axoned local aminoacidergic neurons – GABAergic mech-anisms. (Redrawn and modified from Koella, 1984)

period, does not impede the appearance of an SWS- and PS-rebound. But if PCPA is administered at the beginning of the SD-period, there is no rebound; in fact, both SWS and PS return only gradually towards control levels. From such evidence, supplemented by information obtained with 5-HT receptor antagonists, the authors deduced that the organism produces, in response to the SD and under the influence of serotonin activity, the two above-mentioned (chemically still unidentified) factors. At first glance this theory looks not unattractive. However, closer scrutiny seems to reveal that the authors' conclusions are not entirely warranted. One has to be aware that in the cat the 5-HT- and the sleep-reducing effects of PCPA (even if given in such large amounts) appear with some delay: 12 to 24 h up to their start and 40 to 60 h until their maxima. Thus, it is not unlikely that during the first two days after the SD-period, i.e. within the first 30 to 48 h after the PCPA application, there is enough serotonin available to allow for an SWS-rebound, and there is no need to postulate the generation of a 'non-indolic' SWS-factor. Yet, if the time of the expected rebound coincides with the maximum of the PCPA effect, after the administration of the enzyme-inhibitor at the beginning of the SD-period, the available 5-HT reserves are not sufficient to bring about a normal sleep pattern, let alone an SWS-rebound. So, at this time, we see little reason to include this SWS-factor in the pool of possible serotonergic mechanisms involved in the organization of sleep. The situation may be somewhat different with respect to the proposed PS-factor. Yet, there too, the available evidence is not sufficient to warrant inclusion of this partial mechanism in a model of SORA or VCA.

CONCLUSIONS

With our admittedly still hypothetical model construct we have been able to assign a number of roles to serotonin within the whole of a machinery that is supposed to organize sleep and control the vigilance levels in the various behavioral systems during waking. Still, we are certain that in many details our 'theory' needs expanding, amending, and altering, both with respect to the whole model and with respect to the exact role of serotonin (and of some of the other transmitter systems). If we can induce some of the readers of this paper to provide new experimental evidence that would suggest such alterations and expansions, or even functional changes in the general concept expounded here, then the elaboration of the model will have already fulfilled one of its main purposes.

REFERENCES

Aghajanian, G., Rosekrans, J. A., and Sheard, M. H. (1967) Serotonin: release in the forebrain by stimulation of midbrain raphe, *Science*, **156**, 402–403.

Asin, K. E., and Fibiger, H. C. (1983) An analysis of neuronal elements within the median nucleus of the raphe that mediate lesion-induced increases in locomotor activity, *Brain Res.*, **268**, 211–223.

Bloom, F. E., Hoffer, B. J., Siggins, G. R., Barker, J. L., and Nicoll, R. A. (1972) Effects of serotonin on central neurons: microiontophoretic administration, *Fed. Proc.*, **31**, 97–106.

Bloom, F. E., Oliver, A. P., and Salmoiraghi, G. C. (1963) The responsiveness of individual hypothalamic neurons to microelectrophoretically administered endogenous amines, *Intern. J. Neuropharmacol.*, **2**, 181–183.

Buckingham, R. L., and Radulovacki, M. (1975) 5-Hydroxyindoleacetic acid in cerebrospinal fluid, an indicator of slow wave sleep, *Brain Res.*, **99**, 440–443.

Crespi, F., and Jouvet, M. (1983) Differential pulse voltammetry: parallel peak 3 changes with vigilance states in raphe dorsalis and raphe magnus of chronic freely moving rats and evidence for a 5-HT contribution to these peaks after monoamine oxidase inhibitors, *Brain Res.*, **272**, 263–268.

Curtis, D. R., and Davis, R. (1962) Pharmacological studies upon neurones of the lateral geniculate nucleus of the cat, *Brit. J. Pharmacol.*, **18**, 217–246.

Daan, S., Beersma, D. G. H., and Borbély, A. A. (1984) Timing of human sleep: recovery process gated by a circadian pacemaker, *Am. J. Physiol.*, **246**; (*Regulatory Integrated Comp. Physiol.*, **15**), R161–R178.

Delorme, F., Froment, J. L., and Jouvet, M. (1966) Suppression du sommeil par la p-chloromethamphétamine et la p-chlorophénylalanine, *C. R. Soc. Biol.*, **160**, 2347–2351.

Dement, W., Mitler, M., and Henriksen, S. (1972) Sleep changes during chronic administration of parachlorophenylalanine, *Rev. Can. Biol.* (Suppl.), **31**, 239–246.

Essman, B. (1974) Brain 5-hydroxytryptamine and memory consolidation, *Adv. Biochem. Psychopharmacol.*, **11**, 265–274.

Foote, W. E., Macrievicz, R. J., and Mordes, J. P. (1974) Effect of midbrain raphe and lateral mesencephalic stimulation on spontaneous and evoked activity in the lateral geniculate of the cat, *Exp. Brain Res.*, **19**, 124–130.

Froment, J. L., Petitjean, F., Bertrand, N., Cointy, C., and Jouvet, M. (1974) Effets de l'injection intracérébrale de 5,6-hydroxytryptamine sur les monoamines cérébrales et les états de sommeil du chat, *Brain Res.*, **67**, 405–417.

Hartmann, E. (1968) On the pharmacology of dreaming sleep (the D-state), *J. Nerv. Mental Dis.*, **146**, 165–173.

Hartmann, E. (1974) Hypnotic effects of L-tryptophan, *Arch. Gen. Psychiatry*, **31**, 394–397.

Herz, A., and Zieglgänsberger, W. (1968) The influence of microelectrophoretically applied biogenic amines, cholinomimetics and procaine on synaptic excitation in the corpus striatum, *Intern. J. Neuropharmacol.*, **7**, 221–230.

Janowsky, A., Okada, F., Manier, D. H., Applegate, C. D., Suiser, F., and Steranka, L. R. (1982) Role of serotonergic input in the regulation of the β-adrenergic receptor-coupled adenylate cyclase system, *Science*, **218**, 900–901.

Joëls, M., Twery, M. J., Shinnick-Gallagher, P., and Gallagher, J. P. (1986) Multiple actions of serotonin on lateral septal neurons of rat brain, *Eur. J. Pharmacol.*, **129**, 203–204.

Jouvet, M. (1967) Mechanisms of the states of sleep: a neuropharmacological approach, in *Sleep and Altered States of Consciousness* (Eds S. S. Kety, E. V. Evarts and H. L. Williams), pp. 86–126, Williams and Wilkins, Baltimore.

Jouvet, M. (1969) Biogenic amines and the states of sleep, *Science*, **163**, 32–41.

Jouvet, M. (1972) The role of monoamines and acetylcholine-containing neurons in the regulation of the sleep-waking cycle, *Erg. Physiol.*, **64**, 166–307.

Jouvet, M. (1983) Hypnogenic indolamine-dependent factors and paradoxical sleep rebound, in *Sleep 1982, Proc. Eur. Congress of Sleep Research*, (Ed. W. P. Koella), pp. 2–18, Karger, Basel.

Key, B. J., and Mehta, V. H. (1977) Changes in electrocortical activity induced by the perfusion of 5-hydroxytryptamine into the nucleus of the solitary tract, *Neuropharmacology*, **16**, 99–106.

Kiianmaa, K., and Fuxe, K. (1977) The effects of 5,7-dihydroxytryptamine-induced lesions of the ascending 5-hydroxytryptamine pathways on the sleep-wakefulness cycle, *Brain Res.*, **131**, 287–301.

Koe, B. K., and Weissman, A. (1966) p-Chlorophenylalanine: a specific depletor of brain serotonin, *J. Pharmacol. Exp. Ther.*, **154**, 499–516.

Koella, W. P. (1970) Serotonin oder Somnotonin? *Schweiz, Med. Wschr.*, **100**, 357–364 and 424–430.

Koella, W. P. (1974) Serotonin – a hypnogenic transmitter and an anti-waking agent, *Adv. Biochem. Psychopharmacol.*, **11**, 181–186.

Koella, W. P. (1979) Vigilance – a concept and its neurophysiological and biochemical implications, in *Pharmacology of the States of Alertness* (Eds P. Passouant and I. Oswald), pp. 171–178, Pergamon Press, Oxford and New York.

Koella, W. P. (1981) Neurotransmitters and sleep, in *Psychopharmacology of Sleep* (Ed. D. Wheatley), pp. 19–52, Raven Press, New York.

Koella, W. P. (1982) A modern neurobiological concept of vigilance, *Experientia*, **38**, 1426–1437.

Koella, W. P. (1984) The organization and regulation of sleep; a review of the experimental evidence and a novel integrated model of the organizing and regulating apparatus, *Experientia*, **40**, 309–338.

Koella, W. P. (1985a) Serotonin and sleep, in *Sleep, Neurotransmitters, and Neuro-modulators* (Eds A. Wauquier, J.-M. Gaillard, J. M. Monti, and M. Radulovacki), pp. 185–196, Raven Press, New York.

Koella, W. P. (1985b) Serotonin and sleep, in *Sleep '84, Proc. Europ. Congress of Sleep Research* (Eds W. P. Koella, E. Rüther, and H. Schulz), pp. 6–10, Gustav Fischer Verlag, Stuttgart.

Koella, W. P., and Czicman, J. S. (1963) Influence of serotonin upon optic evoked potentials, EEG, and blood pressure of cat, *Am. J. Physiol.*, **204**, 873–880.

Koella, W. P., and Czicman, J. S. (1966) Mechanism of EEG-synchronizing action of serotonin, *Am. J. Physiol.*, **211**, 926–934.

Koella, W. P., Feldstein, A., and Czicman, J. S. (1968) The effect of parachlorophenylalanine on the sleep of cats, *Electroenceph. Clin. Neurophysiol.*, **25**, 481–490.

Kostowski, W. (1975) Interactions between serotonergic and catecholaminergic systems in the brain, *Pol. J. Pharmacol. Pharm.*, **27**, 15–24.

Kostowski, W., Giacalone, E., Garattini, S., and Valzelli, L. (1968) Studies on behavioral and biochemical changes in rats after lesion of the midbrain raphe, *Eur. J. Pharmacol.*, **41**, 371–376.

Kostowski, W., Giacalone, E., Garattini, S., and Valzelli, L. (1969) Electrical stimulation of midbrain raphe: biochemical, behavioral, and bioelectrical effects, *Eur. J. Pharmac.*, **7**, 170–175.

Kovacevic, R., and Radulovacki, M. (1976) Monoamine changes in the brain of cats during slow-wave sleep, *Science*, **193**, 1025–1027.

Krnjevic, K., and Phillis (1963) Actions of certain amines on cerebral cortical neurons, *Brit. J. Pharmacol. Chemotherap.*, **20**, 471–490.

Ledebur, I. X., and Tissot, R. (1966) Modification de l'activité électrique cérébrale du lapin sous l'effet de micro-injections de précurseurs des monoamines dans les structures somnogènes bulbaires et pontiques, *Electroenceph. Clin. Neurophysiol.*, **20**, 370–381.

Legge, K., Randić, M., and Straughan, D. W. (1966) The pharmacology of neurones in the pyriform cortex, *Brit. J. Pharmacol.*, **26**, 87–107.

MacKenzie, R. G., Hoebel, B. G., Norelli, C., and Trulson, M. E. (1978) Increased tilt cage activity after serotonin depletion by 5,7-dihydroxytryptamine, *Neuropharmacology*, **17**, 957–963.

McGinty, D. J., and Harper, R. M. (1976) Dorsal raphe neurons: depression of firing during sleep in cats, *Brain Res.*, **101**, 569–575.

Meyerson, B. J., and Malmnas, C. O. (1978) Brain monoamines and sexual behavior, in *Biological Determinants of Sexual Behavior* (Ed. J. B. Hutchison), pp. 521–554, Wiley, New York.

Ogasahara, S., Taguchi, Y., and Wada, H. (1980) Changes in serotonin in rat brain during slow-wave sleep and paradoxical sleep: application of the microwave fixation method to sleep research, *Brain Res.*, **189**, 570–575.

Olpe, H.-R., and Koella, W. P. (1977) The response of striatal cells upon stimulation of the dorsal and median raphe nuclei, *Brain Res.*, **122**, 357–360.

Olpe, H.-R., Ortmann, R., Fehr, B., and Waldmeier, P. C. (1981) Experimentally induced supersensitivity of neocortical neurons to microiontophoretically administered serotonin, *Brain Res.*, **224**, 367–374.

Phillis, J. W., Tebecis, A. K., and York, D. H. (1967) The inhibitory action of monoamines on lateral geniculate neurones, *J. Physiol. (Lond.)*, **190**, 563–581.

Pickel, V., Joh, T. H., and Reis, D. J. (1977) A serotonergic innervation of noradrenergic neurons in nucleus locus coeruleus: demonstration by immunocytochemical localization of the transmitter specific enzymes tyrosine and tryptophan hydroxylase, *Brain Res.*, **131**, 197–214.

Quay, W. B. (1965) Regional and circadian differences in cerebral cortical serotonin concentrations, *Life Sci.*, **4**, 379–384.

Radulovacki, M., Buckingham, R. L., Chen, E. H., and Kovacevic, R. (1977) Similar effects of tryptophan and sleep on cisternal cerebrospinal fluid 5-hydroxyindoleacetic and homovanillic acids in cats, *Brain Res.*, **129**, 371–374.

Renault, J. (1967) Monoamines et sommeil: rôle du système de raphe et de la sérotonine cérébrale dans l'endormissement, Thèse, Université de Lyon.

Ross, C. A., Trulson, M. E., and Jacobs, B. L. (1976) Depletion of brain serotonin following intraventricular 5,7-dihydroxytryptamine fails to disrupt sleep in the rat, *Brain Res.*, **114**, 517–523.

Roth, G. I., Walton, P. L., and Yamamoto, W. S. (1970) Area postrema: abrupt EEG synchronization following close intraarterial perfusion with serotonin, *Brain Res.*, **23**, 223–233.

Rothballer, A. B. (1957) The effect of phenylephrine, methamphetamine, cocaine, and serotonin upon the adrenaline sensitive component of the reticular activating system, *Electroenceph. Clin. Neurophysiol.*, **9**, 409–417.

Sallanon, M., Janin, M., Buda, C., and Jouvet, M. (1983) Serotonergic mechanisms and sleep rebound, *Brain Res.*, **268**, 95–104.

Salmoiraghi, G. C., and Stefanis, C. N. (1967) A critique of iontophoretic studies of central nervous system neurons, *Intern. Rev. Neurobiol.*, **10**, 1–30.

Sasa, M., Munekiyo, K., and Takaori, S. (1975) Dorsal raphe stimulation produces inhibitory effect on trigeminal nucleus neurons, *Brain Res.*, **101**, 199–207.

Scheving, L. E., Dunn, J. D., Pauly, J. E., and Harrison, W. H. (1972) Circadian variation in rat serum 5-hydroxytryptamine and effects of stimuli on the rhythm, *Am. J. Physiol.*, **222**, 252–255.

Scheving, L. E., Harrison, W. H., Gordon, P., and Pauly, J. E. (1968) Daily fluctuation (circadian and ultradian) in biogenic amines of the rat brain, *Am. J. Physiol.*, **214**, 166–173.

Segal, M. (1975) Physiological and pharmacological evidence for a serotonergic projection to the hippocampus, *Brain Res.*, **94**, 115–131.

Segal, M. (1980) The action of serotonin in the rat hippocampal slice preparation, *J. Physiol. (Lond.)*, **303**, 423.

Snyder, S. H., Axelrod, J., and Tweig, M. (1967) Circadian rhythm in the serotonin content of the rat pineal gland: regulating factors, *J. Pharmacol. Exp. Ther.*, **158**, 206–213.

Tabushi, K., and Himwich, H. E. (1970) 5-Hydroxytryptophan and the sleep-wakefulness cycle in rabbits, *Biol. Psychiat.*, **2**, 183–188.

Tabushi, K., and Himwich, H. E. (1971) Electroencephalographic study on the effects of methysergide on sleep in the rabbit, *Electroenceph. Clin. Neurophysiol*, **31**, 491–497.

Torda, C. (1967) Effect of brain serotonin depiction on sleep in rats, *Brain Res.*, **6**, 375–377.

Weitzman, E., Rapport, M. M., McGregor, P., and Jacoby, J. (1968) Sleep patterns of the monkey and brain serotonin concentration: effect of p-chlorophenylalanine, *Science*, **160**, 1361–1363.

Williams, H. L., Lester, B., and Coulter, J. (1969) Monoamines and the EEG stages of sleep, *Activ. Nerv. Sup.*, **11**, 188–192.

Wojcik, W. J., Fornal, C., and Radulovacki, M. (1980) Effect of tryptophan on sleep in the rat, *Neuropharmacology*, **19**, 163–167.

Wyatt, R. J., Engelman, K., Kupfer, D. J., Fram, D. H., Sjoerdsma, A., and Snyder, F. (1970) Effects of L-tryptophan (a natural sedative) on human sleep, *Lancet*, **7678**, 842–846.

Yamaguchi, N., Ling, G. M., and Marczynski, T. J. (1964) The effects of chemical stimulation of the preoptic region, nucleus centralis medialis, or brain stem reticular formation with regard to sleep and wakefulness, *Rec. Adv. Biol. Psychiat.*, **6**, 9–20.

Yamamoto, T., and Ueki, S. (1978) Effects of drugs on hyperactivity and aggression induced by raphe lesions in rats, *Pharmacol., Biochem., Behav.*, **9**, 821–826.

York, D. H. (1970) Possible dopaminergic pathway from substantia nigra to putamen, *Brain Res.*, **20**, 233–249.

Yuwiler, A., Brammer, G. L., Morley, J. E., Raleigh, M. J., Flannery, J. W., and Geller, E. (1981) Short-term and repetitive administration of oral tryptophan in normal men, *Arch. Gen. Psychiat.*, **38**, 610–626.

CHAPTER 7

Serotonin and Pain

DANIEL LE BARS
*Unité de Recherches de Neurophysiologie Pharmacologique de l'INSERM
U161*
2 rue d'Alésia
75014 Paris
France

INTRODUCTION

Pain is a sensation which is difficult to define: by his own experience, man
intuitively understands what the word means but it is not so easy for him to

describe what he feels. Regarding pain in animals, we have no means of knowing what sensations are perceived. The only possible alternative is to observe and describe responses to stimuli which would be likely to cause pain in man. These include motor responses (limb withdrawal, contracture, start. . . .), vocalization and modifications of behaviour (flight, avoidance, aggressiveness. . .). The International Association for the Study of Pain (IASP) defines pain as 'an unpleasant sensory and emotional experience associated with actual or potential tissue damage, or described in terms of such damage' (Merskey *et al.*, 1979). This definition implies verbal communication between humans. Zimmermann (1986) proposed the following definition for animals: 'Pain in animals is an aversive sensory experience caused by actual or potential injury that elicits protective motor and vegetative reactions, results in learned avoidance behaviour, and may modify species specific behaviour, including social behaviour.'

There are many methods for assessing pain in animals, mainly in rodents. The most commonly used of these measure the latency of a motor response to a stimulus which is assumed to be painful: examples of this include the tail-flick, hot-plate, flinch-jump, paw-compression, and tail-pinch tests. The threshold for vocalization following electrical stimulation of the tail is also used. In fact, very few tests involve the observation of behavioural patterns following noxious stimuli; in those cases where this is done, the noxious stimuli generally involve the administration of a chemical algogenic substance into the peritoneum or into a paw (the writhing and formalin tests respectively). With the knowledge that clinical doses of morphine have little effect on the threshold for experimental pain in man (Beecher, 1957), one can see the potentially large gap between these so-called 'pain tests' in animals and clinical pain in man. The word 'nociception' is certainly preferable to 'pain' as far as experimental studies in animals are concerned.

If we restrict our considerations to the most commonly used tests in animals, numerous methodological problems arise. As an example, we will comment briefly on the popular tail-flick test where a light beam is focused on the tail of rats or mice and the latency for withdrawal measured. Most people argue that it is a simple spinal reflex. However, what is recorded is not a reflex in the neurophysiological sense but is in fact a more complicated reaction, for the following reasons:

1. The afferents and efferents involved include S3 to Co3 and L4 to Co2 segments respectively (Grossman *et al.*, 1982).
2. Although tail-flick responses occur both in rats with no spinal cord lesion and in those with transections at the thoracic level when high intensity stimuli are applied (giving latencies in the 2-sec range), it is only present in the intact rat when lower-intensity stimuli are applied (giving latencies

in the 4-sec range), which suggests that supraspinal structures are involved in the latter case (Jensen and Yaksh, 1986a).

3. The test is much more sensitive to systemic morphine when the distal tail section is stimulated rather than the more proximal sections (Yoburn *et al.*, 1984).

4. As will be discussed below, the effects of morphine appear to be increased in mildly stressed animals (Kelly and Franklin, 1984a).

5. As pointed out by Duggan *et al.* (1978), substances or procedures which alter blood flow through the skin may appear to be analgesic merely by reducing the rate of increase of skin temperature produced by the heat stimulus.

6. Similar effects can be attributed to substances or procedures which alter the central temperature of animals as morphine does.

This is not the correct place to comment on all pain tests used in animals but, as with the tail-flick test, most of them have to be considered with care when interpreting the effects of manipulation of serotonergic mechanisms. For instance, when an animal is placed on a hot-plate, several paws are stimulated simultaneously and as we will discuss below, the heterotopic application of several noxious stimuli will result in mutual interactions affecting the different nociceptive responses such that the recorded output will undoubtedly be the result of such interactions. Since such interactions are sustained, at least partly, by serotonergic mechanisms, one can see the difficulties of interpreting such data.

Furthermore, another basic problem regarding the functional implication of serotonergic neurones is their ubiquitous action, suggesting that serotonin is involved in many fundamental functions of the brain. It follows that two questions, one basic and one methodological, arise: (1) the interrelationship between systems involving serotonin, e.g. the sleep–waking cycle, temperature regulation, hormonal equilibrium versus pain and/or analgesia; (2) the fact that pain is a psychological notion which cannot be measured in animals in which a response to a nociceptive stimulus is taken as an index of the supposed sensory experience; this output is always a motor response, which itself is under the influence of 5-HT mechanisms. Thus any signs induced by a general manipulation of 5-HT systems need to be interpreted with caution. For instance, the most used manoeuvre for inactivation of serotonergic neurones employs the systemic administration of *p*-chlorophenylalanine (pCPA) which acts by inhibiting tryptophan hydroxylase. The main advantage of this technique lies in the possibility of antagonizing its effect – a depletion of 5-HT in the brain and spinal cord – by injecting the precursor of serotonin, 5-hydroxytryptophan (5-HTP), albeit with the limitation that 5-HTP can be converted to 5-HT at sites other than serotonergic neurones. Among other symptoms, pCPA-pretreated animals are overactive, insomniac, hyperglycaemic and react excessively to environmental stimuli; we have no way of knowing whether such factors do interact, and if so, to what extent, with the animal's response to a noxious stimulus and the potency of analgesic drugs to change these responses. There exist examples of such interactions in other systems: for example, intravenous infusion of insulin was reported to restore sleep in pCPA-pretreated insomniac rats (Danguir and Nicolaidis, 1985).

When the effects of morphine are studied, the degree of complexity increases as there is a classical notion that opiates not only induce analgesia, but also alter numerous other functions which involve 5-HT mechanisms and could interfere with pain mechanisms and pain-related tests.

In addition to its ubiquitous actions in terms of general functions, the effects of 5-HT are probably also complex at the cellular level: mollusc neurones have been found to show six different types of postsynaptic responses (three excitatory and three inhibitory) to the electrophoretic application of 5-HT (Gerschenfeld and Paupardin-Tritsch, 1974a, b). It is hard to believe that 5-HT mechanisms would be any simpler in the vertebrate (see Chapters 14, 15 and 17 in this volume). Furthermore, in view of the self-regulating capacities of serotonergic neurones as evidenced by Aghajanian's group (see references in Aghajanian, 1981), the question of pre- and post-synaptic modifications induced by any pharmacological treatment is often difficult to assess. In this respect the available pharmacological tools such as 5-HT receptor agonists and antagonists are obviously not sufficiently specific for definite conclusions to be drawn. The interrelationship of 5-HT with other transmitters is another problem which cannot be disregarded and includes both co-localizations and more classical relationships notably involving the other monoamines, substance P, enkephalin, etc. It follows that the data reported herein must be regarded in part as starting points for further studies.

In this review I will deal successively with behavioural, neurochemical and electrophysiological data which provide some evidence for the involvement of both ascending and descending serotonergic pathways in the central regulation of nociception; in each case the effects of morphine will be considered. Indeed, it has been known for years that reserpine antagonizes morphine analgesia in the classical behavioural tests of nociception, and Sparkes and Spencer (1971) suggested that antagonism of morphine analgesia by reserpine may be due to depletion of 5-HT since they observed that intracerebroventricular (ICV) 5-HT restored the analgesic effect of morphine after it had been abolished by reserpine (Fig. 1).

I do not pretend to offer an exhaustive review on the subject of central serotonin and pain and, in particular, I have deliberately omitted headache from this topic because of its very peculiar features. In addition, it is worth pointing out that serotonin exhibits peripheral actions, one of which is an

(a)

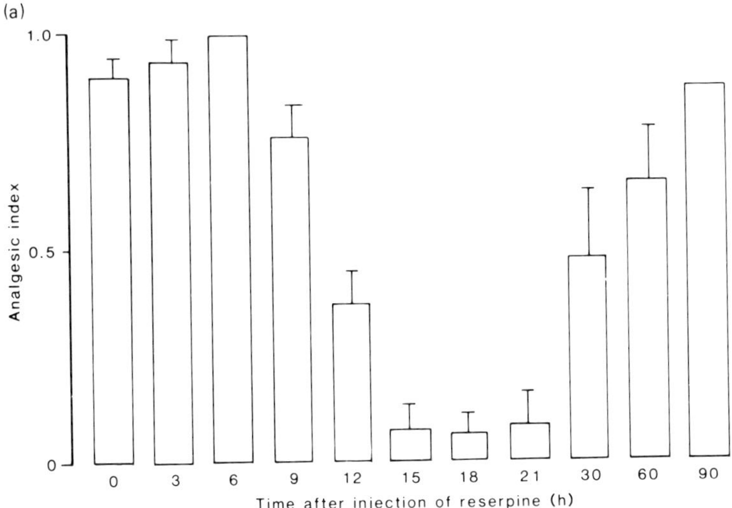

(b)

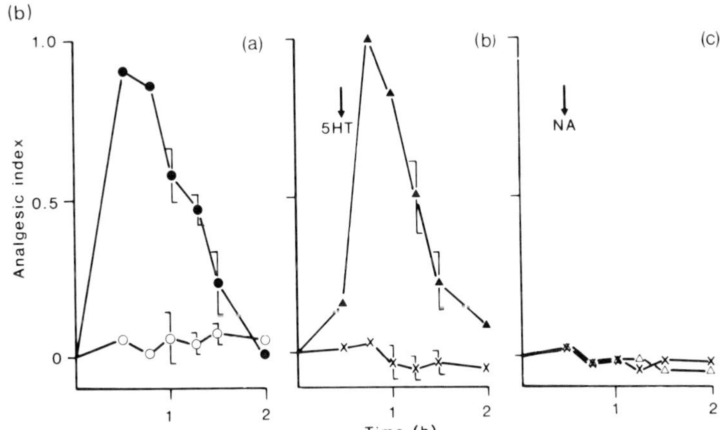

Fig. 1 a – Antinociceptive effect of morphine (8 mg/kg; s.c.) assessed by the paw-compression test, at various times after administration of reserpine (5 mg/kg; i.p.) in the rat. One hour before the time indicated, the nociceptive threshold was determined; the rats were then given morphine and the threshold determined again one hour later.

b – Effects of ICV injections of 5-HT or noradrenaline on the antinociceptive activity of morphine in reserpinized animals (at −16 hours; 5 mg/kg; i.p.).

(a) Effect of morphine (8 mg/kg; s.c.; given at 0 h to rats injected 16 h previously with reserpine (open circles) or saline (solid circles).

(b) All rats were injected with reserpine; 30 min after morphine (solid triangles) or saline (crosses), all animals were given 5-HT (5 μg, ICV).

(c) All rats were injected with reserpine; 30 min after morphine (open triangles) or saline (crosses), all animals were given noradrenaline (20 μg; ICV). (From Sparkes and Spencer, 1971)

algesic action by means of excitation or sensitization of primary afferent neurones (Fjällbrant and Iggo, 1961; Keele and Armstrong, 1964; Sicuteri, 1968; Beck and Handwerker, 1974; Mense and Schmidt, 1974; Fock and Mense, 1976). Such a mechanism could be important in all the conditions where large quantities of 5-HT are liberated by the blood platelets.

The reader will see that the participation of serotonin in pain and/or analgesia is a controversial subject and that the role of 5-HT is probably more complex than might be thought at first sight.

NOCICEPTIVE REACTIONS AND EFFECTS OF MORPHINE THEREON

Various nociceptive reactions have been investigated following depletion of serotonergic neurones, blockade of serotonergic receptors, lesions of raphe nuclei where most 5-HT cells are located, specific destruction of serotonergic neurones, increased availability of serotonin or direct stimulation of serotonin receptors. Since pain in man is specifically alleviated by morphine, with other sensations being unaltered, the effects of the above mentioned pharmacological manipulations were also tested upon the effects of morphine on reactions to noxious stimuli.

Depletion of serotonin

Using the flinch-jump test, Tenen (1967) was first to report an increased sensitivity to high-intensity electrical stimuli in pCPA-pretreated rats; the brain 5-HT of these animals was lowered by 88%. This result was confirmed by several authors (Harvey *et al.*, 1974; Hole and Lorens, 1975; Bodnar *et al.*, 1981) but was disputed by others (Tenen, 1968; Harvey and Lints, 1971; Tilson and Rech, 1974; Berge *et al.*, 1983c). To date there are more than 30 references in which various tests for nociceptive reactions have been used in pCPA-pretreated animals (Table 1Aa) and with negative results being predominantly reported, although Görlitz and Frey (1972) observed an increased sensitivity to a noxious mechanical stimulus and Fibiger and Mertz (1972) reported that pCPA did not change sensitivity to foot shock but increased the reactivity to intense aversive stimulation. Conversely, Bodnar *et al.* (1981) reported an increase of the tail-flick latency following such pretreatment; similarly, pCPA pretreatment was found to increase the threshold for vocalization induced by electrical stimulation of the tail (Crowley *et al.*, 1977; Dickenson and Le Bars, unpublished observations).

Tenen (1968) was first to report that pCPA decreased morphine's antinociceptive effects as assessed by the flinch-jump test. The subsequent literature is again controversial in that the initial result of Tenen was more or less confirmed by several groups using various behavioural tests in rats and mice

Table 1 References related to the effects of serotonin depletors (A: pCPA; B: PCA) on the sensitivity to nociceptive stimuli (a) and the antinociceptive effect of systemic morphine (b)

A. pCPA

(a) Sensitivity to nociceptive stimuli	↗	Tenen, 1967; Fibiger and Mertz, 1972; Görlitz and Frey, 1972; Harvey *et al.*, 1974; Hole and Lorens, 1975; Bodnar *et al.*, 1981.
	○	Tenen, 1968; Saarnivaara, 1969; Fenessy and Lee, 1970; Cheney and Goldstein, 1971; Harvey and Lints, 1971; Major and Pleuvry, 1971; Maruyama *et al.*, 1971; Akil and Mayer, 1972; Fibiger and Mertz, 1972; Ho *et al.*, 1972; Buxbaum *et al.*, 1973; Tilson and Rech, 1974; Vogt, 1974; Akil and Liebeskind, 1975; Dennis and Melzack, 1980; Lin *et al.*, 1980a; Botting and Morinan, 1982; Berge *et al.*, 1983c; Reigle and Barker, 1983; Long *et al.*, 1984; Ogren and Berge, 1984; Archer *et al.*, 1985a; Berge *et al.*, 1985b; Ogren and Johansson, 1985.
	↘	Crowley *et al.*, 1977; Bodnar *et al.*, 1981.
(b) Antinociceptive effect of systemic morphine	↘	Tenen, 1968; Fenessy and Lee, 1970; Major and Pleuvry, 1971; Görlitz and Frey, 1972; Contreras *et al.*, 1973; Vogt, 1974; Takemori *et al.*, 1975; Tulunay *et al.*, 1976; Woolf *et al.*, 1980; Bodnar *et al.*, 1981; Berge, 1982; Botting and Morinan, 1982; Reigle and Barker, 1983; Vonvoigtlander *et al.*, 1984.
	○	Fenessy and Lee, 1970; Cheney and Goldstein, 1971; Maruyama *et al.*, 1971; Buxbaum *et al.*, 1973; Harvey *et al.*, 1974; Tilson and Rech, 1974; Ho *et al.*, 1975; Sugrue, 1979a; Dennis and Melzack, 1980; Long *et al.*, 1984.
	↗	Saarnivaara, 1969; Dennis and Melzack, 1980.

B. PCA

(a) Sensitivity to nociceptive stimuli	↗	Harvey and Lints, 1971.
	○	Sugrue, 1979a; Dennis and Melzack, 1980; Berge *et al.*, 1983c; Archer *et al.*, 1985a; Ogren and Johansson, 1985; Hunskaar *et al.*, 1986.
	↘	–
(b) Antinociceptive effect of systemic morphine	↘	Takemori *et al.*, 1975; Tulunay *et al.*, 1976; Sugrue 1979a; Dennis and Melzack, 1980; Berge *et al.*, 1983c.
	○	Dennis and Melzack, 1980; Berge *et al.*, 1983c.
	↗	Dennis and Melzack, 1980.

but was disputed by others (Table 1Ab). In this context, it is important to note that inverse effects, i.e. facilitation of morphine antinociception, were even reported (Saarnivaara, 1969; Dennis and Melzack, 1980). Differences in species and strains could be the source of the contradictory data in the studies cited above (Tilson and Rech, 1974). Of special interest is the work of Vonvoigtlander *et al.* (1984) in the mouse, which showed that pCPA slightly reduced morphine hypoalgesia but completely blocked the hypoalgesia induced by U50,488H, a selective agonist of kappa opioid receptors, suggesting that the involvement of serotonergic systems in opiate analgesia could be dependent on the subtype of opioid receptor involved.

p-Chloroamphetamine (PCA) has complex actions including an acute effect following a single injection which produces a release of serotonin (see below) and a chronic effect leading to long-lasting depletion of serotonin levels. Chronic depletion of serotonin by PCA was found not to increase the nociceptive threshold assessed by various tests, although hyperalgesic effects were reported with the flinch-jump test (Table 1Ba).

Again, contradictory results were found regarding morphine analgesia following PCA with reports of decreased effects, no effect and facilitatory effects (Table 1Bb).

As an alternative to pharmacologically depleting serotonergic neurones, rats were fed with a tryptophan-free diet and a resultant hypersensitivity to noxious stimuli associated with a reduced effect of morphine was observed (Lytle *et al.*, 1975; Messing *et al.*, 1976; Schlosberg and Harvey, 1978).

Blockade of serotonergic receptors

In an experimental model in the rat, described by Carroll and Lim (1960), electrical stimulation of the tail was used to investigate variations in the thresholds of spinal reflexes, for vocalization during stimulation and for vocalization outlasting the period of stimulation. Herold and Cahn (1968) reported that methysergide (7 or 13 mg/kg, s.c.) did not change the first two thresholds but increased the last, suggesting that the perception of pain, especially its affective component, is sustained by serotonergic mechanisms. Unfortunately such experiments have not been replicated and numerous authors, using less integrated tests – tail-flick, hot-plate and formalin tests – have failed to observe any change in nociceptive thresholds following systemic methysergide (Table 2Aa).

By contrast, Berge and coworkers repeatedly found that, in the rat, systemic metergoline decreased the nociceptive threshold in the tail-flick and hot-plate tests; however, negative results were again reported and, curiously, Fasmer *et al.* (1984) observed in the mouse that metergoline increased and decreased nociceptive reaction in the tail-flick and hot-plate tests respectively (Table 2Ba).

Table 2 References related to the effects of blockers of serotonergic receptors (A: methysergide; B: metergoline; C: cyproheptadine) on the sensitivity to nociceptive stimuli (a) and the antinociceptive effect of systemic morphine (b).

A. Methysergide

(a)	↗	Herold and Cahn, 1968
Sensitivity to nociceptive stimuli	○	Herold and Cahn, 1968: Yaksh *et al.*, 1976; Yaksh and Wilson, 1979; Dennis and Melzack, 1980; Malec and Langwinski, 1980a; Larson, 1983; Kelly and Franklin, 1984a; Romandini and Samanin, 1984.
	↘	–
(b)	↘	Romandini and Samanin, 1984.
Antinociceptive effect of systemic morphine	○	Fennessy and Lee, 1970; Dennis and Melzack, 1980; Malec and Langwinski, 1980a.
	↗	Dennis and Melzack, 1980.

B. Metergoline

(a)	↗	Berge, 1982; Berge *et al.*, 1983a, b; Fasmer *et al.*, 1984; Ogren and Berge, 1984.
Sensitivity to nociceptive stimuli	○	Samanin *et al.*, 1976; Malec and Langwinski, 1980a; Rochat *et al.*, 1982b; Romandini and Samanin, 1984; Ogren and Johansson, 1985.
	↘	Fasmer *et al.*, 1984.
(b)	↘	Malec and Langwinski, 1980a; Rochat *et al.*, 1982b; Romandini and Samanin, 1984.
Antinociceptive effect of systemic morphine	○	Berge *et al.*, 1983b.
	↗	–

C. Cyproheptadine

(a)	↗	Görlitz and Frey, 1972.
Sensitivity to nociceptive stimuli	○	Lee *et al.*, 1979; Yaksh and Wilson, 1979; Malec and Langwinski, 1980a.
	↘	Saarnivaara, 1969.
(b)	↘	–
Antinociceptive effect of systemic morphine	○	Malec and Langwinski, 1980a.
	↗	Saarnivaara, 1969.

Cyproheptadine was found not to alter nociceptive thresholds in the tail-flick and hot-plate tests while it increased nociceptive reactions in the Randall–Selitto test in the rat but decreased such reactions to electrical stimulation of teeth in the rabbit (Table 2Ca).

Finally, Berge *et al.* found an increased reactivity in the tail-flick and hot-plate tests following intraperitoneal administration of mianserin (Berge, 1982; Berge *et al.*, 1983b), but this finding was at variance with the negative results of Malec and Langwinski (1980a).

The antinociceptive effects of systemic morphine have not been influenced by methysergide in most cases, although facilitation and disruption of hypoalgesia have also been reported (Table 2Ab). Interestingly, Yaksh *et al.* (1976) reported that systemic methysergide – and cinanserin – antagonized the antinociceptive action of morphine administered into the periaqueductal grey matter (PAG). Similarly, it was found that systemic cinanserin blocked the antinociceptive effect of morphine microinjected into the nucleus raphe magnus (NRMag) (Dickenson *et al.*, 1979; Azami *et al.*, 1982).

Metergoline has been reported to have no influence on morphine analgesia or to decrease it (Table 2Bb). Cyproheptadine has been reported to have no effect or to facilitate the antinociceptive effects of systemic morphine (Table 2Cb) and to block the analgesia induced by ICV administration of the drug (Lee *et al.*, 1979). Finally it has been reported that neither mianserin nor pizotifen modify morphine analgesia (Malec and Langwinski, 1980a).

An interesting finding is that various blockers of serotonergic receptors (cyproheptadine, ketanserin, pirenperone) do not affect morphine analgesia in the mouse although they block the analgesia induced by U-50,488H, a selective kappa opioid receptor agonist; metergoline and mianserin have been found to be equally inactive in this respect (Vonvoigtlander *et al.*, 1984).

An alternative to the systemic route of administration is to apply the serotonin-receptor blockers directly within the brain intracerebroventricularly (ICV) or within the spinal cord intrathecally (IT). Little attention had been paid to the ICV route; however, in the mouse ICV metergoline was found to have no effect on the tail-flick test and to decrease nociceptive reactions in the hot-plate test (Fasmer *et al.*, 1984) while in the monkey ICV cyproheptadine increased the threshold of the jaw opening reflex (Shyu *et al.*, 1984).

Numerous authors have administered 5-HT receptor blockers intrathecally in the rat using the tail-flick and hot-plate tests. With the exception of Proudfit and Hammond (1981), who reported increased reactions to nociceptive stimuli following IT methysergide, the most common observation has been an absence of effects (Hammond *et al.*, 1980; Larson, 1983; Schmauss *et al.*, 1983; Hammond and Yaksh, 1984; Jensen and Yaksh, 1984, 1986b). IT metergoline has resulted in either no effect (Schmauss *et al.*, 1983) or an increased sensitivity to noxious stimuli (Berge *et al.*, 1983a; Fasmer *et al.*, 1984).

Regarding morphine analgesia, IT methysergide was reported to have no effect following systemic morphine (Proudfit and Hammond, 1981). On the other hand, hypoalgesia resulting from microinjections of morphine within the PAG, NRMag or adjacent regions was found to be blocked by intrathecal methysergide in the tail-flick test but not in the hot-plate test (Yaksh, 1979; Jensen and Yaksh, 1986b).

Lesions of the raphe nuclei

Harvey and Lints (1965) were first to pay attention to the relationship between nociception and telencephalic contents of serotonin. These authors showed that a bilateral lesion of the medial forebrain bundle (MFB) was associated with hypoalgesia as assessed in the flinch-jump test. This finding was consistently replicated by this group and others and found to be reversed by 5-HTP (Lints and Harvey, 1969a, b; Harvey and Lints, 1971; Dennis, 1972; Harvey and Yunger, 1973; Yunger and Harvey, 1973; Harvey *et al.*, 1974). Although such lesions involved several neurotransmitters, the role of serotonin was suggested by the fact that microinjections of 5,7-dihydroxytrypt-amine (5,7-DHT) but not of 6-hydroxydopamine (6-OHDA), mimicked the effect of the lesions (Simansky and Harvey, 1981). However, such effects of 5,7-DHT were not obtained when the hot-plate test was used as the behavioural index of nociception (Hole *et al.*, 1976). Interestingly, MFB lesions were found to facilitate morphine hypoalgesia (Harvey and Yunger, 1973).

One of the features of the raphe nuclei is the high content of neurones containing serotonin, although not all raphe neurones contain 5-HT and conversely not all serotonin-containing neurones are located in the raphe nuclei. However, since most serotonergic projections in the central nervous system originate from these nuclei, electrolytic lesions involving these regions have also been investigated. Using various behavioural tests it was consistently found that electrolytic lesions of nucleus raphe dorsalis (NRD), nucleus raphe medianus (NRMed) or both did not change nociceptive reactions in rats (Table 3Aa and 3Ba).

This lack of effect was also observed following microinjections of 5,7-DHT within NRD or NRMed (Hole and Lorens, 1975; Deakin and Dostrovsky, 1978). However, hyperalgesia following NRMed lesions were reported with the tail-compression and tail-flick tests. On the other hand, Harvey *et al.* (1974) reported that lesions of the two nuclei did not change reaction times in the hot-plate test during the day but increased it during the night, while an important (300%) increase in the threshold for vocalization following stimulation of the ophthalmic division of the trigeminal nerve was also reported following such lesions (York and Maynert, 1978).

Table 3 References related to the effects of electrolytic lesions of raphe nuclei (A: NRMed; B: NRMed + NRD; C: NRMag) on the sensitivity to nociceptive stimuli (a) and the antinociceptive effect of systemic morphine (b).

A. NRMed.

(a) Sensitivity to nociceptive stimuli	↗	Samanin *et al.*, 1970; Samanin and Bernasconi, 1972; Abbott and Melzack, 1982.
	○	Samanin *et al.*, 1970; Samanin and Bernasconi, 1972; Lorens and Yunger, 1974; Pepeu *et al.*, 1974; Adler *et al.*, 1975; Hole and Lorens, 1975; Chance *et al.*, 1978; Miranda *et al.*, 1978; Abbot *et al.*, 1982; Romandini *et al.*, 1986a.
	↘	–
(b) Antinociceptive effect of systemic morphine	↘	Samanin *et al.*, 1970; Samanin and Bernasconi, 1972; Pepeu *et al.*, 1974; Adler *et al.*, 1975; Chance *et al.*, 1978; Miranda *et al.*, 1978; Romandini *et al.*, 1986a.
	○	Lorens and Yunger, 1974.
	↗	Abbott and Melzack, 1982.

B. NRMed + NRD

(a) Sensitivity to nociceptive stimuli	↗	–
	○	Bläsig *et al.*, 1973; Buxbaum *et al.*, 1973; Lorens and Yunger, 1974; Hole and Lorens, 1975; Garau *et al.*, 1975; Minesantone, 1976; Yaksh *et al.*, 1977; York and Maynert, 1978; Abbott *et al.*, 1982.
	↘	Harvey *et al.*, 1974; York and Maynert, 1978.
(b) Antinociceptive effect of systemic morphine	↘	Garau *et al.*, 1975; Yaksh *et al.*, 1977; Abbott *et al.*, 1982.
	○	Bläsig *et al.*, 1973; Buxbaum *et al.*, 1973; Harvey *et al.*, 1974; Lorens and Yunger, 1974; York and Maynert, 1978; Abbott *et al.*, 1982.
	↗	Abbott *et al.*, 1982.

C. NRMag

(a) Sensitivity to nociceptive stimuli	↗	Proudfit and Anderson, 1975; Proudfit, 1981.
	○	Yaksh *et al.*, 1977; Chance *et al.*, 1978; Abbott and Melzack, 1982; Abbott *et al.*, 1982.
	↘	–
(b) Antinociceptive effect of systemic morphine	↘	Proudfit and Anderson, 1975; Yaksh *et al.*, 1977; Chance *et al.*, 1978; Proudfit, 1981; Abbott and Melzack, 1982; Abbott *et al.*, 1982.
	○	Abbott and Melzack, 1982; Abbott *et al.*, 1982.
	↗	–

Lesions of NRMed (Table 3Ab) or NRD and NRMed (Table 3Bb) were found to reduce morphine antinociception, although here again, negative results have also been reported. Furthermore, Abbott and Melzack (1982) reported that NRMed lesions potentiate morphine analgesia assessed by the formalin test. Interestingly, Romandini *et al.* (1986a) confirmed that electrolytic lesions of NRMed blocked morphine hypoalgesia, but lesions with neurotoxins which destroy either all neurones (ibotenic acid) or only 5-HT neurones (5,7-DHT), did not produce such results; these authors concluded that morphine could act on axons passing within or in the vicinity of NRMed (see also Deakin and Dostrovsky, 1978).

The cell bodies of origin of 5-HT axons in the spinal cord are located in the caudal medulla and are clustered into three midline groups, nucleus raphe pallidus, nucleus raphe obscurus and nucleus raphe magnus, which are approximately equivalent to groups B_1, B_2 and B_3 of Dahlström and Fuxe (1964). The first two groups innervate the ventral (motor) horn while the nucleus raphe magnus (NRMag) neurones project to the dorsal (sensory) horn, and have therefore been the subject of attention in studies related to nociception. While hyperalgesia has been reported following electrolytic lesions of or microinjection of tetracaine within NRMag (Proudfit, 1980), negative results have also been reported (Table 3Ca). Such lesions decreased the antinociceptive effects of morphine assessed by the tail-flick or hot-plate tests but were without effect when the formalin test was used (Table 3Cb). Microinjections of 5,7-DHT within the NRMag also reduced morphine hypoalgesia in the hot-plate test but not in the tail-flick test (Mohrland and Gebhart, 1980).

Destruction of serotonergic neurones by ICV or IT neurotoxins

ICV administration of 5,6-DHT or 5,7-DHT was reported to result in hyper-algesia phenomena during the first few days post-injection, with recovery within 2 weeks although 5-HT levels in the brain and spinal cord were still reduced at that time (Bläsig *et al.*, 1973; Lin *et al.*, 1980a; Shyu *et al.*, 1984; Romandini *et al.*, 1986b). During the first few days, the antinociceptive effect of morphine was reported to be reduced (Genovese *et al.*, 1973; Vogt, 1974; Romandini *et al.*, 1986b) or unaltered (Bläsig *et al.*, 1973; Sugrue, 1979a).

IT administration of 5,6-DHT or 5,7-DHT resulted in comparable effects, with hyperalgesia followed by recovery (Fig. 2), Berge *et al.*, 1983a, 1984; Sagen *et al.*, 1983; Fasmer *et al.*, 1983b, 1985; Ogren *et al.*, 1985); however, negative results were also reported (Deakin and Dostrovsky, 1978; Vasco *et al.*, 1984; Ogren *et al.*, 1985). Curiously, Fasmer *et al.* (1985) and Kuraishi *et al.* (1983) even reported hypoalgesia using the formalin test in the mouse and the tail-flick test in the rat after IT administration of 5,6-DHT.

Such procedures were found to reduce morphine-induced hypoalgesia

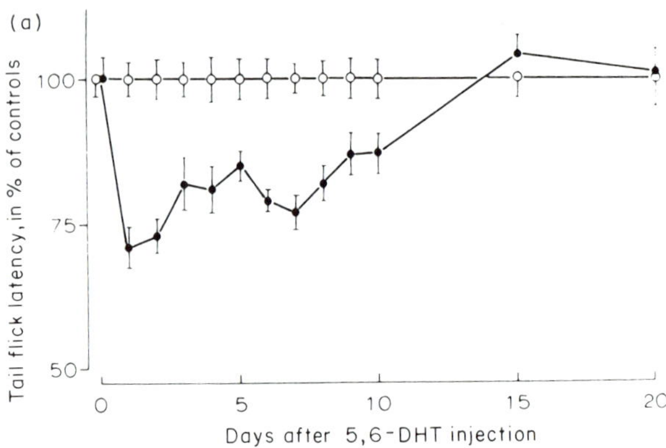

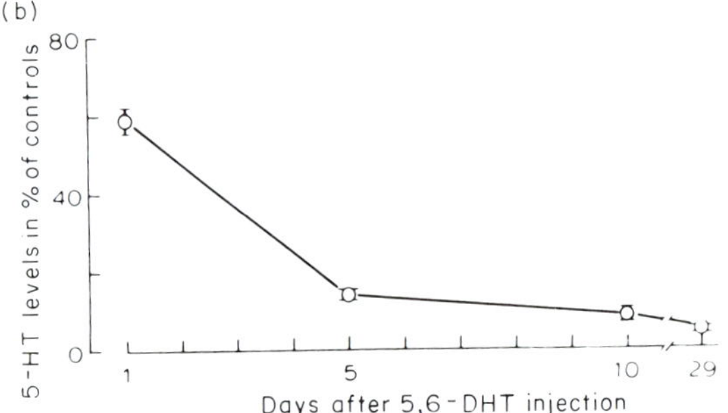

Fig. 2 a – Effect on nociception assessed by the tail-flick
test of 5,6-DHT (20 µg, solid circles) or vehicle (open
circles) given intrathecally in the rat.
b – Spinal levels of 5-HT in 5,6-DHT treated rats
(control level determined in vehicle injected rats was
0.69 ± 0.02 µg/g). (From Berge *et al.*, 1983a)

when the drug was administered either systemically (Deakin and Dostrovsky,
1978; Kuraishi *et al.*, 1983) or within the NRMag (Vasko *et al.*, 1984);
however, negative results were obtained using the tail-pinch test in the rat
(Kuraishi *et al.*, 1983). Negative results were also obtained with the tail-flick
test when morphine was microinjected within the PAG (Johannessen *et al.*,
1982).

Increased availability of serotonin

Precursor loading

Administration of the immediate precursor of serotonin, 5-hydroxy-L-trypto-
phan (5-HTP) increases the amount of serotonin formed and released in the

CNS. However, 5-HTP can be converted to serotonin in cells other than serotonergic neurones. 5-HTP has been studied against various nociceptive reactions. In the 15–250 mg/kg (i.p.) range, 5-HTP was found to increase thresholds in the vocalization, tail-flick, hot-plate and shock-titration tests while negative results were found in the 37.5–105 mg/kg range using the tail-flick, hot-plate and flinch-jump tests (Table 4a). Interestingly, the anti-nociceptive action of 5-HTP was found to be blocked by noradrenaline depletion (Post *et al.*, 1986).

Table 4   References related to the effects of 5-HTP on the sensitivity to nociceptive stimuli (a) and the antinociceptive effect of morphine (b).

(a) Sensitivity to nociceptive stimuli	↗ –	
	○	Dewey *et al.*, 1970; Harvey and Lints, 1971; Major and Pleuvry, 1971; Samanin and Bernasconi, 1972; Pleuvry, 1975.
	↘	Contreras and Tamayo, 1967; Contreras *et al.*, 1970; Lin *et al.*, 1980b; Berge *et al.*, 1981; Post *et al.*, 1986.
(b) Antinociceptive effect of systemic morphine	↘ –	
	○	Takagi *et al.*, 1964; Contreras and Tamayo, 1967; Contreras *et al.*, 1970; Samanin and Bernasconi, 1972.
	↗	Saarnivaara, 1969; Dewey *et al.*, 1970; Major and Pleuvry, 1971; Contreras *et al.*, 1973; Pleuvry, 1975; Takemori *et al.*, 1975; Tulunay *et al.*, 1976.

L-Tryptophan (TRP) is only converted to serotonin in serotonergic neurones, which contain tryptophan hydroxylase. TRP administration was found to decrease the nociceptive reactions in the tail-flick and formalin tests but not in the hot-plate and flinch-jump tests (Dennis and Melzack, 1980; Hole and Marsden, 1975) with equivalent doses (200 mg/kg; i.p.).

The predominant reported action of pretreatment with 5-HTP on morphine antinociception is a potentiation, although negative results have also been reported (Table 4b). Curiously, TRP was found to have no effect (Dennis and Melzack, 1980) or to decrease morphine hypoalgesia (Ho *et al.*, 1975).

Inhibition of serotonin uptake

The inhibition of 5-HT uptake is considered to be a good procedure for increasing the monoamine availability at the receptor. With a few exceptions (Messing *et al.*, 1976; Ogren and Holm, 1980; Lin *et al.*, 1980b; Uzan *et al.*, 1980), the various re-uptake inhibitors including alaproclate, citalopram, femoxetine, fluoxetine, org 6582 and zimelidine when administered system-ically were found to have no effect on the nociceptive thresholds in various tests – mainly tail-flick and hot-plate – (Larson and Takemori, 1977b; Lee *et al.*, 1979; Sugrue, 1979b; Yaksh and Wilson, 1979; Gebhart and Lorens, 1980; Malec and Langwinski, 1980b; Ogren and Holm, 1980; Larsen and Christensen, 1982; Larsen and Arnt, 1984; Ogren and Berge, 1984; Berge *et al.*, 1985 ; Hynes *et al.*, 1985; Larsen and Hyttel, 1985; Ogren and Johansson, 1985). When applied directly within the NRMag by microinjection techniques, zimelidine was found to produce hypoalgesia in the tail-flick test (Llewelyn *et al.*, 1984).

By contrast, these agents have repeatedly been found to potentiate morphine antinociception using the same tests (Larson and Takemori, 1977b; Sugrue, 1979b; Gebhart and Lorens, 1980; Malec and Langwinski, 1980b; Ogren and Holm, 1980; Uzan *et al.*, 1980; Larsen and Christensen, 1982; Hynes *et al.*, 1985; Larsen and Hyttel, 1985). Interestingly, IT citalopram was reported to potentiate the effects of systemic and ICV morphine but not those of IT morphine (Larsen and Christensen, 1982; Larsen and Arnt, 1984); in the same way, IT sertralin potentiated the effects of systemic morphine (Taiwo *et al.*, 1985). Subanalgesic doses of morphine and zimelidine microinjected within the NRMag have been reported to act in synergy and to increase the tail-flick latency (Llewelyn *et al.*, 1984).

Release of serotonin

The acute administration of *p*-chloroamphetamine (PCA, range 1–5 mg/ kg; i.p.) resulted predominantly in an hypoalgesic effect as determined by numerous tests (Kaergaard Nielsen *et al.*, 1967; Görlitz and Frey, 1972; Ogren and Holm, 1980; Tricklebank *et al.*, 1982; Berge *et al.*, 1984; Ogren and Berge, 1984; Ogren and Johansson, 1985; Ogren *et al.*, 1985; Post *et al.*, 1986; Hunskaar *et al.*, 1986); such an effect can be blocked by noradrenaline depletion (Post *et al.*, 1986). By contrast, when the tail-flick test was used as a nociceptive index, PCA appeared to be virtually ineffective (Kaergaard Nielsen *et al.*, 1967; Görlitz and Frey, 1972; Ogren and Holm, 1980; Berge *et al.*, 1984; Ogren and Berge, 1984; Ogren *et al.*, 1985; Ogren and Johansson, 1985; Hunskaar *et al.*, 1986). Similarly, fenfluramine, which is both a 5-HT releaser and an uptake inhibitor, increased the nociceptive threshold in the hot-plate test and decreased it in the tail-flick test; it potentiated morphine antinociception in the former test but had no action in the latter (Rochat *et al.*, 1982b).

Stimulation of serotonin receptors

Since serotonin does not cross the blood–brain barrier, other serotonin agonists were used systemically in the study of nociception.

5-Methoxy-*N*,*N*-dimethyltryptamine (5-MeODMT) induced the serotonergic syndrome in intact rats and decreased nociceptive reactions in the tail-flick, hot-plate and shock-titration tests with doses ranging between 0.5 and 2 mg/kg (s.c.) (Berge, 1982; Berge *et al.*, 1983b; Archer *et al.*, 1985a; Post *et al.*, 1986). Interestingly, such an action could be blocked by the depletion of the noradrenaline content of the spinal cord and by 5-HT receptor antagonists (mianserin, metergoline) in the tail-flick test but not in the other two tests (Archer *et al.*, 1985a; Post *et al.*, 1986). As shown in Fig. 3, 5-MeODMT administered within the ventricles induced dose-dependent hyper- and hypoalgesia as assessed by the tail-flick test (Berge *et al.*, 1980).

Quipazine showed similar effects in the 20–30 mg/kg (i.p.) range (Samanin *et al.*, 1976; Malec and Langwinski, 1980b) while lower doses appeared to be ineffective (Malec and Langwinski, 1980b; Kulkarni and Robert, 1982; see however Post *et al.*, 1986).

Zemlan *et al.* (1980, 1983) recorded nociceptive reflexes in spinal rats; they found that both 5-MeODMT and quipazine decreased, in a metergoline-

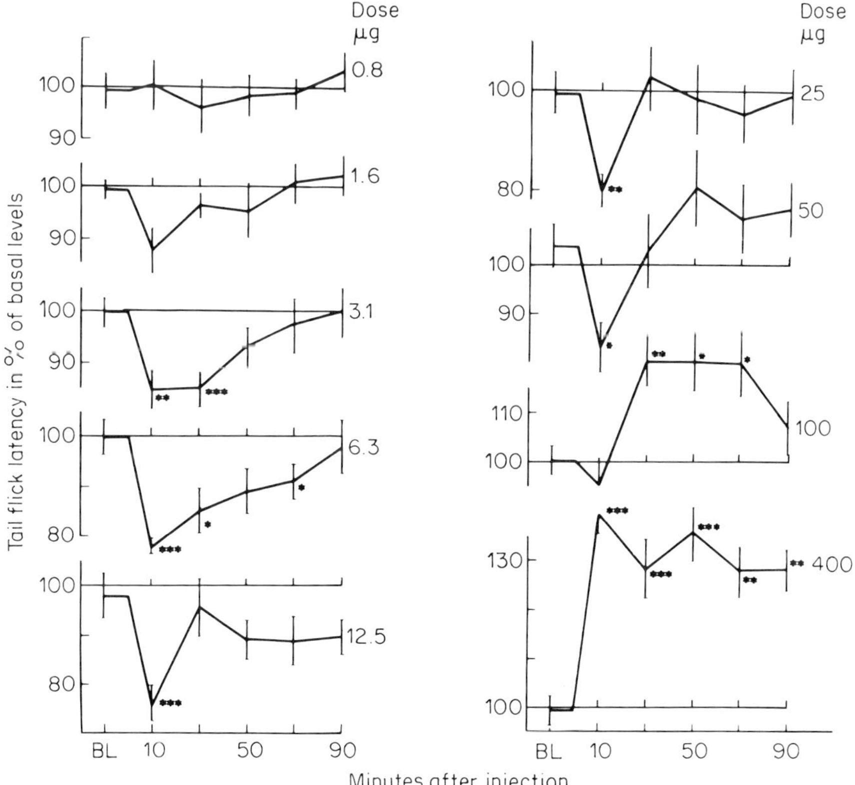

Fig. 3 The dose-dependent hyper- and hypoalgesia induced by ICV administration of 5-MeODMT as assessed by the tail-flick test in the rat. (From Berge *et al.*, 1980)

reversible fashion, the reflex threshold and expanded its receptive field; taken together with the data from intact animals, these findings suggest that 5-HT agonist administration has the opposite effect on a nociceptive flexor reflex in spinal as opposed to intact preparations.

m-Chlorophenylpiperazine was reported to increase and decrease nociceptive reactions in the tail-flick and hot-plate tests respectively (Rochat *et al.*, 1982a), while 8-hydroxy-2-(di-*n*-propylamino)tetralin (8-OH-DPAT) showed no effects in the tail-flick test in the mouse within the 0.06–1 mg/kg range (Berge *et al.*, 1985a; Fasmer *et al.*, 1986). However, 8-OH-DPAT appears to exhibit a complex action on more integrated tests: when the latencies for paw licking or jumping were measured in the hot-plate test, 8-OH-DPAT produced hypo- or hyperalgesia respectively and it also produced, in a dose-dependent fashion, hypo- and hyperalgesia in the formalin test (Fasmer *et al.*, 1986).

Regarding morphine analgesia, quipazine and 8-OH-DPAT were reported to potentiate (Malec and Langwinski, 1980b) and to block it (Berge *et al.*, 1985a) respectively.

The fact that 5-HT does not cross the blood–brain barrier led several workers to apply the amine directly into the brain by the intracerebroventricular route or into the spinal cord by the intrathecal route.

ICV serotonin has been reported to exhibit no effect on (Lee *et al.*, 1979; Sparkes and Spencer, 1971; Sewell and Spencer, 1975; Lee *et al.*, 1979), to decrease (Lin *et al.*, 1980b; Shyu *et al.*, 1984) or to increase (Sparkes and Spencer, 1971) nociceptive reactions. By contrast, Spencer's group has repeatedly reported a potentiation by ICV serotonin of both systemic and ICV morphine antinociception (Sparkes and Spencer, 1971; Sewell and Spencer, 1975; Lee *et al.*, 1979). Microinjections of 5-HT directly within the NRMag produced antinociception in the tail-flick test (Llewelyn *et al.*, 1983).

Although its potency seemed to depend on the test used (Kuraishi *et al.*, 1985), IT serotonin has repeatedly been found to produce antinociception dose-dependently in the rat; this effect was blocked by methysergide or cyproheptadine and potentiated by fluoxetine in the tail-flick test but not in the hot-plate test (Wang, 1977; Yaksh and Wilson, 1979; Schmauss *et al.*, 1983; Archer *et al.*, 1985b; Minor *et al.*, 1985). Interestingly, the 5-HT action could be blocked by a depletion of the noradrenaline content of the spinal cord (Archer *et al.*, 1985b; Minor *et al.*, 1985). IT 5-HTP or quipazine mimicked the effects of 5-HT (Yaksh and Wilson, 1979) while IT 5-MeODMT exhibited an opposite action (Larson, 1982).

In the mouse, IT serotonin produced antinociception in the tail-flick and hot-plate tests (Hylden and Wilcox, 1983; Archer *et al.*, 1985b), while concurrently eliciting a behavioural syndrome consisting of caudally directed biting, licking and scratching, which was potentiated and blocked by IT fluoxetine and serotonergic receptor blockers respectively (Hylden and Wilcox, 1983; Fasmer *et al.*, 1983a; Larson and Wilcox, 1984).

'STRESS-INDUCED ANALGESIA' (SIA)

This term refers to the elevation of thresholds of nociceptive reactions by various stressors in rodents (see references in Tricklebank and Curzon, 1984; Amit and Galina, 1986; Kelly, 1986). SIA has not been found to be unambiguously influenced by pCPA pretreatment since no effects (Bodnar *et al.*, 1981) as well as increased or decreased effects of the 5-HT synthesis inhibitor have been reported. For instance, pCPA enhanced the antinociceptive effects of a brief continuous shock, while leaving the response to prolonged intermittent shocks unaltered (Coderre and Rollman, 1984).

Hypoalgesia induced by 30 seconds of inescapable footshocks was enhanced by pretreatment with pCPA or by depletion of spinal serotonin following injection of 5,7-DHT into the spinal cord or the NRMag; depletion of brain serotonin by injection of 5,7-DHT into the NRMed was without effect (Hutson *et al.*, 1982). Such hypoalgesia was found to be attenuated by 5-HT releasing-drugs (PCA, fenfluramine), by a 5-HT uptake inhibitor (fluoxetine) and by a 5-HT agonist (5-MeODMT) while other agonists (quipazine, trifluoromethylphenylpiperazine) or antagonists (metergoline, methysergide, cyproheptadine, mianserin, methiothepin) were without effect (Hutson *et al.*, 1982, 1983; Tricklebank *et al.*, 1982). By contrast, hypoalgesia induced by a long period (30 min) of inescapable footshocks was attenuated by injection of pCPA (Tricklebank *et al.*, 1984).

A recent very important finding is that, as assessed by the tail-flick test in the rat, analgesia induced by systemic (Kelly and Franklin, 1984a, 1985; Applebaum and Holzman, 1986) or ICV morphine (Applebaum and Holzman, 1986) was potentiated by the stress due to restraint or to a novel environment (Fig. 4a). Since the effects of stress on morphine analgesia were blocked both by methysergide and by valine – which competitively inhibits tryptophan uptake – (Fig. 4b), it was suggested that they were mediated by an increase in serotonin availability, which itself is dependent on sympathetic activity (Kelly and Franklin, 1984a, 1985; Franklin and Kelly, 1986).

Furthermore, NRMag lesion had no effect on unrestrained rats but reduced morphine analgesia in restrained rats (Fig. 4c), suggesting that NRMag is involved in analgesia resulting from an interaction of morphine on stress but not in analgesia induced by morphine alone (Kelly and Franklin, 1984b).

An earlier study (Bhattacharya *et al.*, 1978) reported that pCPA inhibits the increase in tail-flick latency which can be induced by 1–4 hours immobilization. Interestingly, in a more integrated test, the formalin test, the stress of a novel environment was found to reduce the nociceptive behavioural response, and this effect was again reversed by methysergide or valine

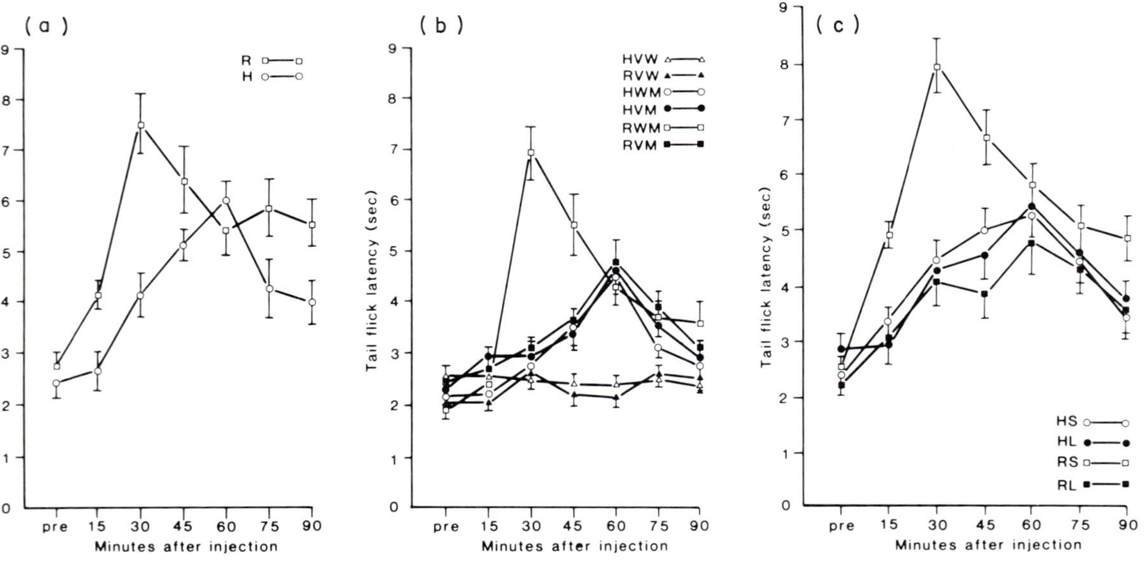

Fig. 4 Hypoalgesia induced by morphine (8 mg/kg; s.c.) in restrained and hand-held rats assessed by measuring the latency with which a rat removed its tail from a hot (55°C) water bath.

(a) Time course of morphine analgesia in rats which were either restrained (R) or hand-held (H) during testing.

(b) Experiment showing that L-valine (200 mg/kg; i.p.; 15–20 min before morphine) prevented the potentiation of morphine analgesia by restraint stress (V and W represent groups receiving valine and its vehicle respectively; M and W represent groups receiving morphine and its vehicle respectively).

(c) Experiments showing that NRMag lesions prevented the potentiation of morphine analgesia by restraint stress (L and S represent the lesioned and sham-operated groups respectively). (From Kelly and Franklin, 1984a, b)

(Abbott *et al.*, 1986). Since many common manipulations of animals used in behavioural experiments related to nociception are mild stressors (e.g. the tail-flick test is most often performed on restrained animals), misleading observations could result and be the source of many of the discrepancies reported above.

HYPOALGESIA PRODUCED BY SOMATIC STIMULATION

Although SIA has often been triggered by somatic stimuli, other experiments have been performed with such conditioning stimulation with the aim of reproducing in animals procedures used for relieving pain in man.

It has been reported that hypoalgesia induced in mice by electrostimulation of the tail and assessed by the hot-plate test, could be enhanced by 5-HTP and reduced by pretreatment with pCPA or administration of methysergide; however, serotonin reuptake inhibitors (fluoxetine, zimelidine) were without effect (Woolf *et al.*, 1980; Bucket, 1981; Shimizu *et al.*, 1981). Similarly, an increase in tail-flick latency produced by electrostimulation of the forepaws was blocked by pretreatment with intrathecal 5,7-DHT (Watkins *et al.*, 1984).

The neuronal basis of acupuncture analgesia has been reviewed by Han and Terenius (1982) who presented evidence for an important, if not prevalent, role for central serotonergic mechanisms in mediating acupuncture analgesia. Briefly, electrolytic lesions of raphe nuclei (NRD, NRD + NRMed, NRMag), section of the dorsolateral funiculus (DLF) of the spinal cord, chemical lesioning with 5,6-DHT of descending or ascending 5-HT fibres, pCPA, cyproheptadine and cinanserin, all to some extent block acupuncture hypoalgesia in various species; conversely, 5-HTP and serotonin uptake inhibitors such as clomipramine were reported to potentiate acupuncture analgesia.

In order to perform a detailed investigation of counter-irritation phenomena, in which peripheral nociceptive stimuli are used to overcome pain originating elsewhere (see below), a behavioural model was developed in the rat; this model involved the concomitant application of two noxious stimuli used in classical pharmacological tests. It was found that the threshold for vocalization induced by electric shocks to the tail was increased by the intraperitoneal injection of the algogenic agent, phenylbenzoquinone (PBQ). This is in keeping with other behavioural experiments which have shown that hypertonic saline injected intraperitoneally, sustained pinch applied to the paws or tail, or electrical stimulation of the tail at currents sufficient to produce vocalization, also induce analgesic effects when the test for analgesia is applied to other areas of the body; in the cat, stimulation of tooth pulp at an intensity which is obviously painful, was reported to greatly increase the threshold for escape behaviour induced by foot shock (see references in Le Bars *et al.*, 1984).

The participation of 5-HT mechanisms in counter-irritation phenomena has been investigated using the model described above. It was found that pretreatment with 5-HTP strongly potentiated, in a dose-dependent fashion over the 5.5–50 mg/kg range (i.p.), the PBQ-induced rise in threshold (Kraus *et al.*, 1982): whereas intraperitoneal injection of PBQ in non-pretreated rats resulted in a mean threshold increase of 30–40% (with an upper limit of 100% rarely, if ever, being exceeded), following pretreatment with 50 mg/ kg 5-HTP, there was a mean increase of 130% (Fig. 5a) with extremely dramatic results (threshold increase: 400–500%) in several animals. The specificity of this effect was confirmed by its suppression by the serotonin receptor blocker, cinanserin (Fig. 5b).

The antinociceptive effect secondary to vaginal probing also seems to be dependent, at least in part, on serotonergic mechanisms since probing increases the efflux of 5-HT into spinal cord superfusates; in addition, the antinociceptive effect of probing is antagonized by intrathecal administration of methysergide (Steinman *et al.*, 1983). However, negative results were obtained following pretreatment with pCPA (Crowley *et al.*, 1977).

'STIMULATION-PRODUCED ANALGESIA' (SPA)

Since the initial observation by Reynolds (1969) that electrical stimulation of the periaqueductal grey matter (PAG) in the rat could induce antinociceptive effects strong enough to allow a laparotomy to be performed, this type of analgesia – which has been termed stimulation-produced analgesia (SPA) – has been extensively studied and reviewed (see references in Basbaum and Fields, 1984; Willis, 1984; Besson and Chaouch, 1987).

There is general agreement that antinociceptive effects can be induced by stimulation of the PAG and the NRMag. However, both in animals and in man, stimulations of the dorsal and lateral parts of the PAG and of nearby regions of the dorsal midbrain tegmentum produce other behavioural effects, notably aversive reactions (see references in Fardin *et al.*, 1984b), whereas diffuse analgesia over the whole body and devoid of side-effects can be elicited from the ventral PAG, including the NRD and, especially from the NRMag and surrounding regions, corresponding to the B_3 group of 5-HT cells (Oliveras *et al.*, 1979; Fardin *et al.*, 1984a, b).

These results point to the raphe nuclei as being the prime candidates for inducing analgesia by brain stimulation and suggest that the reported effects from the whole PAG could have been a consequence of aversive reactions triggered from dorsal regions, thus possibly related to stress-induced analgesia (see above). Indeed, the frequently used tail-flick test is generally carried out on restrained animals, a situation which could mask the side-effects induced by PAG stimulation. In addition, in the context of the neuro-physiology of pain, the function of the dorsal PAG has classically been

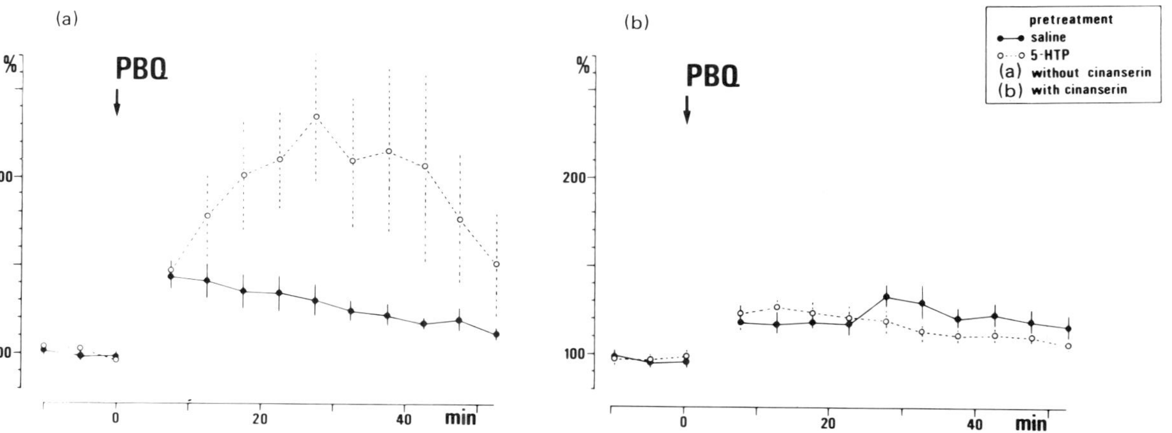

Fig. 5 Involvement of serotonergic mechanisms in counter-irritation phenomena.
(a) The threshold for vocalization induced by electric shocks to the tail of rats was increased by the i.p. injection of the algogenic agent, phenylbenzoquinone (solid line); such an effect was potentiated by pretreatment with 5-HTP (50 mg/kg; i.p.; 75 min before testing; dotted line).
(b) These effects were blocked by cinanserin (10 mg/kg; i.p.; 75 min before testing). (From Kraus et al., 1982)

regarded as being part of the central pain pathways; thus a 'central pain' triggered by the stimulation of these regions could have blocked the pain from peripheral origin (see below). In fact, antinociception associated with aversive responses can be elicited from many sites in the rat brain (Prado and Roberts, 1985). Interestingly, bar pressing to escape from aversive dorsal PAG stimulation is increased by pCPA pretreatment (Kiser *et al.*, 1978).

In any case, it is generally believed that the analgesic effects induced by stimulation of the ventral PAG and the NRMag result from activation of descending fibres from NRMag origin and travelling in the dorsolateral funiculus (DLF) of the spinal cord (Basbaum *et al.*, 1976, 1977; Prieto *et al.*, 1983). A great number of the descending fibres from NRMag origin, travelling through the DLF and innervating the dorsal horn, are serotonergic (Skagerberg and Björklund, 1985).

Analgesia triggered by stimulation of the ventral PAG was reported to be markedly or partially reduced following administration of pCPA (Akil and Mayer, 1972; Akil and Liebeskind, 1975) or D-lysergic acid diethylamide (LSD) (Hayes *et al.*, 1977) respectively. In the cat, repetitive stimulation of the NRMag caused a marked reduction in the analgesic action of such stimulation and this 'tolerance effect' could be reversed by systemic administration of 5-HTP (Oliveras *et al.*, 1978). A similar phenomenon of tolerance has been reported for PAG stimulation in humans and this could be reversed by tryptophan (Hosobuchi, 1980).

SPA induced by stimulation of the NRMag or the adjacent nucleus reticularis paragigantocellularis (NRPG) was found to be attenuated by intrathecal administration of methysergide in the tail-flick and pinch tests (Jensen and Yaksh, 1984; Hammond and Yaksh, 1984). However, these results were not confirmed using the hot-plate test (Jensen and Yaksh, 1984) or, in the case of NRMag stimulation, by using the pinch test and, in the case of NRPG stimulation by using the tail-flick test or the pinch test (Barbaro *et al.*, 1985). Also, SPA induced by stimulation of the PAG was not altered by prior intrathecal administration of 5,7-DHT (Johannessen *et al.*, 1982).

NEUROCHEMICAL EXPERIMENTS

There have been few neurochemical studies regarding the effects of nociceptive stimulation on serotonergic functions. The prolonged application of intense nociceptive electrical stimuli to the tail of anaesthetized rats induced a rise in 5-HT synthesis which was dependent on the part of the CNS being considered: the dorsal horn of the cord was the most sensitive (Fig. 6), the ventral cord and the brainstem were affected to a lesser extent and the forebrain was not significantly affected; by contrast, the prolonged application of innocuous electrical stimuli to the tail was not followed by any detectable changes in 5-HT synthesis (Weil-Fugazza *et al.*, 1984). Interest-

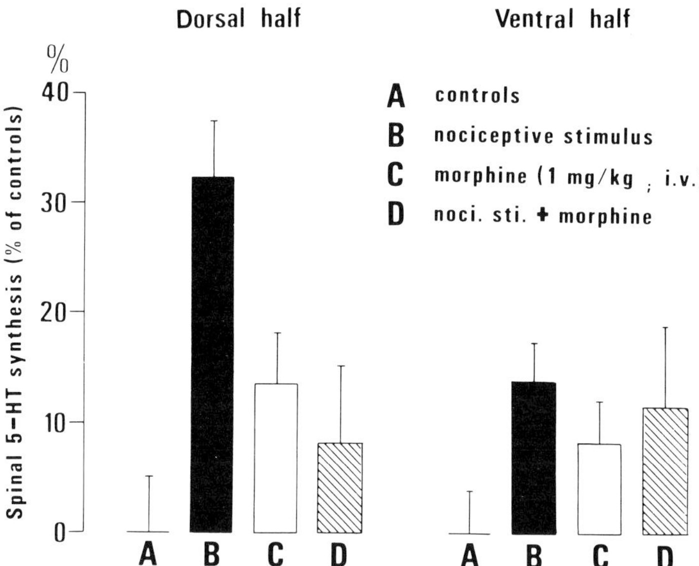

Fig. 6 Interaction between the effects of a prolonged nociceptive stimulus applied to the tail of rats and the effects of morphine (1 mg/kg; i.v.). Saline was administered 35 min before sacrifice in a group of non-stimulated rats (A: controls) and a group of stimulated rats (B); morphine was administered 35 min before sacrifice in a group of non-stimulated rats (C) and a group of stimulated rats (D). In the dorsal cord (left) it was found that morphine counteracted very significantly the increase in 5-HT synthesis induced by the nociceptive stimulus. In the ventral cord (right) 5-HT synthesis was affected to a lesser extent by the nociceptive stimulus and morphine did not interact with such an effect. Note that neither the endogenous TRP levels nor the specific activity of TRP were significantly modified by the two factors of variation (not shown). (From Weil-Fugazza *et al.*, 1981b; see also Weil-Fugazza *et al.*, 1984)

ingly, electrical stimulation of the NRMag also produced an increase in the synthesis of 5-HT within the cord (Bourgoin *et al.*, 1980).

The idea of descending serotonergic pathways being activated by noxious stimuli is supported by other lines of evidence. It has been shown that bilateral high-intensity – but not low-intensity – stimulation of the sciatic nerve induces a release of 5-HT in superfusates from the spinal cord of the cat; thoracic cold block prevented this effect (Tyce and Yaksh, 1981). Since a rise in 5-HT release was also observed following stimulation of the infra-orbital branch of the trigeminal nerve, one can suggest that there is a diffuse release of the monoamine within the cord whatever part of the body is stimulated; interestingly, 5-HT release could be induced by electrical stimulation of the DLF (Tyce and Yaksh, 1981), the NRMag or the NRPG

(Rivot *et al.*, 1982; Hammond *et al.*, 1985; Pilowsky *et al.*, 1986). Unilateral stimulation of the sciatic nerve in the rat at an intensity which activates Aδ- but not C-fibres was reported to have no effect on tryptophan hydroxylase activity in the spinal cord (Nissbrandt *et al.*, 1982).

In an animal model of chronic pain, the polyarthritic rat, a marked increase in the free tryptophan level is observed in serum which leads to a general increase in 5-HT synthesis throughout the central nervous system during the acute phase of the disease (Weil-Fugazza *et al.*, 1979, 1981a; Godefroy *et al.*, 1987). Thereafter, 5-HT synthesis in the brain returns to normal values but is still increased in the spinal cord, especially in its dorsal part. These observations suggest that the descending serotonergic system projecting to the dorsal horn is preferentially activated during chronic pain, in agreement with clinical data showing an increase of 5-hydroxyindoleacetic acid (5-HIAA) levels in the cerebrospinal fluid of chronic suffering patients (Almay *et al.*, 1980; Ghia *et al.*, 1981; Costa *et al.*, 1984; see however Hyppä *et al.*, 1985).

A variety of stressful factors (e.g. restraint, foot-shock, fear, housing in unnatural living quarters, diethyl ether inhalation, etc.) have been shown to induce changes in 5-HT synthesis. In fact, most of the data to which we are referring are related to the effects of large or repetitive exposures to stress; in most cases an increase in 5-HT synthesis was observed in the whole brain or in more restricted areas such as the forebrain, brainstem, mesencephalon, hippocampus, hypothalamus; nevertheless, these responses appeared to be more complex, probably depending on the duration and type of stress, since stress-induced decreases in 5-HT synthesis were also reported (see references in Weil-Fugazza *et al.*, 1984). To the best of our knowledge, there are no available data on the effects of stress on 5-HT metabolism in the spinal cord. In any case, it can be postulated that an increase in serotonergic tone is probably responsible, at least partly, for SIA (see above).

Many authors have demonstrated that acute systemic administration of morphine increases brain 5-HT synthesis and turnover; such effects were generally observed for high doses of morphine but were reported to be dose-related and naloxone-reversible (Yarbrough *et al.*, 1971, 1972, 1973; Görlitz and Frey, 1972; Haubrich and Blake, 1973; Goodlet and Sugrue, 1974; Papeschi *et al.*, 1975; Pérez-Cruet *et al.*, 1975; Bensemana and Gascon, 1978; Garcia-Sevilla *et al.*, 1978; Larson and Takemori, 1978; Fennessy and Laska, 1979; Ahtee and Carlsson, 1979; Weil-Fugazza *et al.*, 1979; Spampinato *et al.*, 1985; see also Lee and Fennessy, 1970; Fennessy and Lee, 1972). The basis of this increase is more probably a rise in brain tryptophan (Larson and Takemori, 1977a, 1978; Messing *et al.*, 1978a, b) than a change in the activity of tryptophan hydroxylase (Yarbrough *et al.*, 1973; Knapp and Mandell, 1972). However, stimulation by morphine of newly synthesized serotonin has been explained by an increase in tryptophan hydroxylase activity (Pérez-Cruet *et al.*, 1975). Since free tryptophan in blood and transport of the amino acid from blood to brain are unchanged following morphine (Goodlet and Sugrue, 1974; Larson and Takemori, 1977a) a decrease in protein synthesis or an increase in protein breakdown have been suggested for the rise in brain tryptophan (Larson and Takemori, 1978; Messing *et al.*, 1978a), but these hypotheses need to be substantiated, especially in view of the clear increase of free tryptophan seen in serum following systemic administration of morphine in the arthritic rat (Weil-Fugazza *et al.*, 1981c). Finally, morphine has been shown to have no clear direct effects on the uptake or release of 5-HT in synaptosomal preparations (Mennini *et al.*, 1978).

Since morphine activates serotonergic projections in many parts of the brain which are not commonly thought to be associated with pain and/or analgesia (Messing *et al.*, 1978b; Snelgar and Vogt, 1980; Johnston and Moore, 1983), these effects are not always necessarily related to analgesia.

Interestingly, microinjection of morphine into the NRD raised 5-HIAA levels in the diencephalon, striatum, nucleus accumbens and cortex with no effect in the hippocampus, while microinjection in the NRMed raised 5-HIAA levels only in the nucleus accumbens (Spampinato *et al.*, 1985). A stimulatory effect of morphine on the *in vitro* release of 5-HT has also been reported from PAG slices (Pycock *et al.*, 1981); in the same preparation, the drug had no clear effects on serotonin uptake (Donzanti and Warwick, 1979).

An increase in 5-HT synthesis has also been demonstrated in the spinal cord using small doses of morphine. In the rat, a naloxone-reversible increase in the accumulation of 5-HIAA has been observed in the whole spinal cord 90 min after the administration of 7.5 mg/kg (s.c.) of morphine (Snelgar and Vogt, 1980); such an effect is absent in morphine-tolerant rats (Vasko and Vogt, 1982). Shiomi *et al.* (1978) observed that 10 mg/kg of morphine (s.c.) induced a progressive increase in 5-HIAA levels in the dorsal half of the cord within 60 min following the injection, with a slow decrease thereafter. In the ventral half of the cord, 5-HIAA levels also showed a slow increase, but there was a plateau and after 150 min 5-HIAA levels were still significantly elevated. These effects appeared to be dose-dependent (range 5–20 mg/kg) and naloxone-reversible. After transection of the spinal cord morphine did not elevate spinal 5-HIAA levels, which suggests that the morphine effect on 5-HT-containing neurones does not occur at the spinal level. Commissiong (1983) reported that this action concerned the dorsal horn, the zona intermedia and the ventral horn to similar extents, suggesting a non-specific action on the three raphe-spinal systems. In addition, in spite of their clear analgesic properties, other mu, delta and kappa opioid receptor agonists did not mimic the effect of morphine (Wood, 1985). Nevertheless, the stimulatory effect of morphine on 5-HT synthesis in the spinal cord is a well established phenomenon; this effect can be blocked by the administration of naloxone and does not occur in morphine-tolerant rats (Godefroy *et al.*, 1980, 1981). The stimulatory effect of morphine occurs with a short latency and is much more marked in the dorsal than in the ventral half of the spinal cord. In the dorsal part, the effect is dose-dependent (range 2.5–10 mg/kg; s.c.) and reaches a maximum in 40 min with total recovery being observed 160 min after the injection (Weil-Fugazza *et al.*, 1981a). Finally, the levels of 5-HT and 5-HIAA have been found to increase in lumbar CSF of conscious dogs following 2.5 mg/kg intravenous morphine (Bardon and Ruckebush, 1984).

There are further data which suggest that the target at which morphine induces these effects might be supraspinal; Yaksh and Tyce (1979) reported that microinjection of morphine (5 μg) into the PAG evoked a release of 5-HT in spinal cerebrospinal fluid, and Vasko and Vogt (1982) found that microinjections of morphine (10 μg) in the vicinity of the NRMag caused a naloxone-reversible increase in 5-HT turnover in the spinal cord. Interestingly, no change in 5-HT turnover in the spinal cord was observed following intrathecal morphine (1–50 μg) which confirmed the lack of effect of morphine on 5-HT terminals in the spinal cord (Vasko *et al.*, 1984).

It has been observed (Bineau-Thurotte *et al.*, 1984) that the release of [^{3}H]5-HT from slices of rat dorsal cord following depolarization by K$^+$ ions was enhanced in a dose-dependent manner (range 2.5–10 mg/kg; s.c.) in slices from animals treated with morphine; this effect did not occur in slices from rats receiving naloxone before morphine administration or from morphine-tolerant rats. Interestingly, the release of [^{3}H]5-HT was not modified when morphine was directly applied onto the slices, further suggesting a supraspinal site of action of the drug. However, Lakoski *et al.* (1980) reported that 7.5 mg/kg (i.p.) morphine did not change the 5-HT turnover in the NRMag or NRPG but increased the substance P content in these structures.

All the studies reported above were performed using rats not exposed to noxious stimulation, i.e. conditions which did not involve activation of pain mechanisms. These studies therefore demonstrate that, in the absence of pain, relatively large doses of morphine activate 5-HT systems and particularly the bulbo-spinal component of these systems. Such observations do not necessarily support the view that similar mechanisms occur during analgesia which, by definition, can be assessed by the use of potentially painful stimuli.

However, as discussed above, nociceptive inputs also increase the 5-HT tone in bulbo-spinal pathways; the interaction between these two variable factors would therefore appear to be a critical point to consider. In fact, morphine very significantly counteracts (Fig. 6), in a naloxone-reversible fashion, the stimulus-induced increase in 5-HT synthesis observed in the dorsal cord and the brainstem (Weil-Fugazza *et al.*, 1984). It is important to note the absence of such phenomena within the ventral cord, which indicates a pharmacological specificity for these effects of morphine in terms of anatomical substrate.

ELECTROPHYSIOLOGICAL EXPERIMENTS

Most electrophysiological experiments regarding 5-HT and pain have been aimed at investigating the role of raphe-spinal systems in controlling the transmission of nociceptive information at the spinal level. Among the relatively few studies dealing with other serotonergic neurones, one can mention the non-specific effect of electrophoretically applied morphine on NRD and NRMed cells (Haigler, 1978).

Effects of 5-HT on dorsal horn neurones

Electrophoretic application of 5-HT inhibited the responses of dorsal horn convergent neurones to nociceptive stimulation, whether the monoamine was administered into the substantia gelatinosa (Headley *et al.*, 1978; Griersmith and Duggan, 1980) or close to the soma of the cell under study (Randic and Yu, 1976; Belcher *et al.*, 1978; Jordan *et al.*, 1978, 1979; Willcockson *et al.*, 1984; Jeftinija *et al.*, 1986). However, excitatory effects on spontaneous activity have also been reported for convergent neurones (Belcher *et al.*, 1978) as well as for superficial neurones (Todd and Millar, 1983) and spino-thalamic neurones located more deeply within the spinal cord (Jordan *et al.*, 1979). Firing evoked by excitatory amino-acids was found to be either facilitated (Belcher *et al.*, 1978) or decreased (Jordan *et al.*, 1978; Willcockson *et al.*, 1984). Interestingly, electrophoretically applied 5-HT was found to reduce excitatory responses of dorsal horn neurones elicited by electrophoresis of substance P (Davies and Roberts, 1981), suggesting an interaction of 5-HT with the substance P receptor. Recently Jeftinija *et al.* (1986) reported that in pCPA-pretreated cats there was an increased proportion of dorsal horn neurones which could be excited electrophoretically by substance P and 5-HT.

Although a presynaptic action of 5-HT has been suggested (Proudfit and Anderson, 1974), a postsynaptic action is more probable since autoradiographical and immunocytochemical studies have shown contacts between 5-HT terminals and dendritic shafts of dorsal horn neurones (see Chapter 3 of this volume).

Raphe-spinal influences on dorsal horn neurones

The fact that brainstem stimulation exerts powerful effects on spinal and trigeminal reflexes evoked by noxious stimulation, suggests that SPA can be mediated, at least partly, by the activation of descending inhibitory mechanisms acting on spinal dorsal horn and trigeminal neurones which are thought to participate in the central transmission of nociceptive information. This point has been repeatedly reported in electrophysiological experiments (see references in Basbaum and Fields, 1984; Willis, 1984; Besson and Chaouch, 1987). However, it is worth pointing out that discrepancies exist between the behavioural studies which show that pure analgesic effects are obtained only from well demarcated structures, whereas inhibition of dorsal horn neurones can be evoked from more extensive regions of the brainstem.

The involvement of serotonergic mechanisms in these inhibitions is still open to question. The inhibition due to the stimulation of the medial diencephalic periventricular grey or the PAG, unlike that of the mesencephalic lateral reticular formation, is reduced by LSD, methysergide or metergoline (Guilbaud *et al.*, 1973; Carstens *et al.*, 1981, 1983; Yezierski *et al.*, 1982; Foong *et al.*, 1985) whereas various 5-HT antagonists do not so easily reduce the inhibitory effect elicited from the NRMag (Belcher *et al.*, 1978; Yezierski *et al.*, 1982).

In animals pretreated by systemic pCPA or intrathecal 5,7-DHT, the effects of reticular formation stimulation are unchanged whereas inhibitions induced by NRMag stimulation are reduced but still detected (Rivot *et al.*, 1980; Chaouch and Besson, 1985). Therefore, the involvement of nonserotonergic raphe-spinal fibres also accounts for these inhibitory effects; in this respect it is important to note that not all, but 40–80% of raphe-spinal neurones are serotonergic (Skagerberg and Björklund, 1985).

The participation of both serotonergic and non-serotonergic neurones is also consistent with the wide range of conduction velocities as measured by antidromic activation techniques (Anderson *et al.*, 1977; West and Wolstencroft, 1977; Fields and Anderson, 1978; Lovick *et al.*, 1978; Wessendorf *et al.*, 1981; Wessendorf and Anderson, 1983; Vanegas *et al.*, 1984a; Willis *et al.*, 1984; Chiang and Pan, 1985); the slow (< 6 m/s) conducting axons are probably serotonergic while the faster ones are not (Wessendorf *et al.*, 1981).

The idea that the NRMag has a prime role in the carrying of information from the PAG to induce inhibition of nociceptive dorsal horn cells has been challenged: the inhibition of spinal nociceptive transmission induced from the PAG is strongly reduced only when both NRMag and the lateral reticular formation are simultaneously blocked by lidocaine microinjection (Gebhart *et al.*, 1983; Sandkühler and Gebhart, 1984) or when the caudal lateral reticular nuclei are lesioned (Morton *et al.*, 1984). It is possible that inhibitions triggered from the dorsal PAG are transmitted via reticular neurones and those triggered from the ventral PAG are sustained by NRMag, since SPA from the dorsal PAG was found to be unaffected by NRMag lesions whereas SPA from the ventral PAG was blocked by identical lesions (Prieto *et al.*, 1983).

Diffuse noxious inhibitory controls

The question arises as to how 5-HT terminals in the spinal cord can be physiologically activated. Serotonin does not seem to be involved in the tonic descending inhibition of neurones involved in the spinal transmission of nociceptive information: tonic inhibition has been found to be unaltered by NRMag lesions, systemic fluoxetine, pCPA pretreatment or electrophoresis of methysergide (Hall *et al.*, 1981; Soja and Sinclair, 1981; Griersmith *et al.*, 1981). However, it may be noted that systemic cinanserin or methysergide have both been reported to increase the responsiveness of dorsal horn convergent neurones to C-fibre activation (Rivot *et al.*, 1987).

By contrast, serotonergic systems seem to be triggered by the activation of nociceptive afferents. Indeed, the activity of certain dorsal horn neurones can be strongly inhibited by noxious inputs via serotonergic pathways. Since such effects do not appear to be somatotopically organized but instead involve the entire body, they have been called diffuse noxious inhibitory controls (DNIC) (Le Bars *et al.*, 1979a). These controls have been observed in the anaesthetized rat: any type of activity in almost all convergent neurones can be inhibited by various noxious stimuli when these are applied to parts of the body distant from the excitatory receptive fields; the strength of these inhibitory effects, which generally outlast the period of conditioning ('post-effects'), is directly related to the strength of the conditioning stimulus (Le Bars *et al.*, 1981a; Villanueva and Le Bars, 1985).

Since DNIC disappear in anaesthetized animals with a cervical section of the spinal cord – 'spinal animals' – (Le Bars *et al.*, 1979b), one can conclude that the underlying mechanisms are not confined to the spinal cord and must ascend to, and redescend from supraspinal levels, implying that a complex loop is activated by painful stimuli. It has thus been postulated (Le Bars *et al.*, 1979b) that such controls involve structures, the electrical stimulation of which might (or in some cases has been shown to) induce profound analgesia. This is supported by several lines of evidence and particularly the probable involvement of 5-HT bulbo-spinal pathways in DNIC. This question has been investigated using several approaches (Le Bars *et al.*, 1982).

Pretreatment with pCPA (300 mg/kg, i.p., daily for 3 days) resulted in a strong reduction of DNIC (Fig. 7a) with the post-effects also being reduced both in terms of magnitude and duration (Dickenson *et al.*, 1981). It is worth pointing out that, although the pretreatment procedure induced a total 5-HT depletion at the spinal level, the blockade of DNIC was not complete; this observation suggests that additional non-serotonergic mechanisms are involved in DNIC. This is in keeping with a study devoted to the effects of electrical stimulation of the NRMag in pCPA-pretreated rats (Rivot *et al.*, 1980). Two 5-HT receptor blockers – cinanserin and metergoline – were also used: both strongly depressed DNIC (Fig. 7b); conversely, the systemic

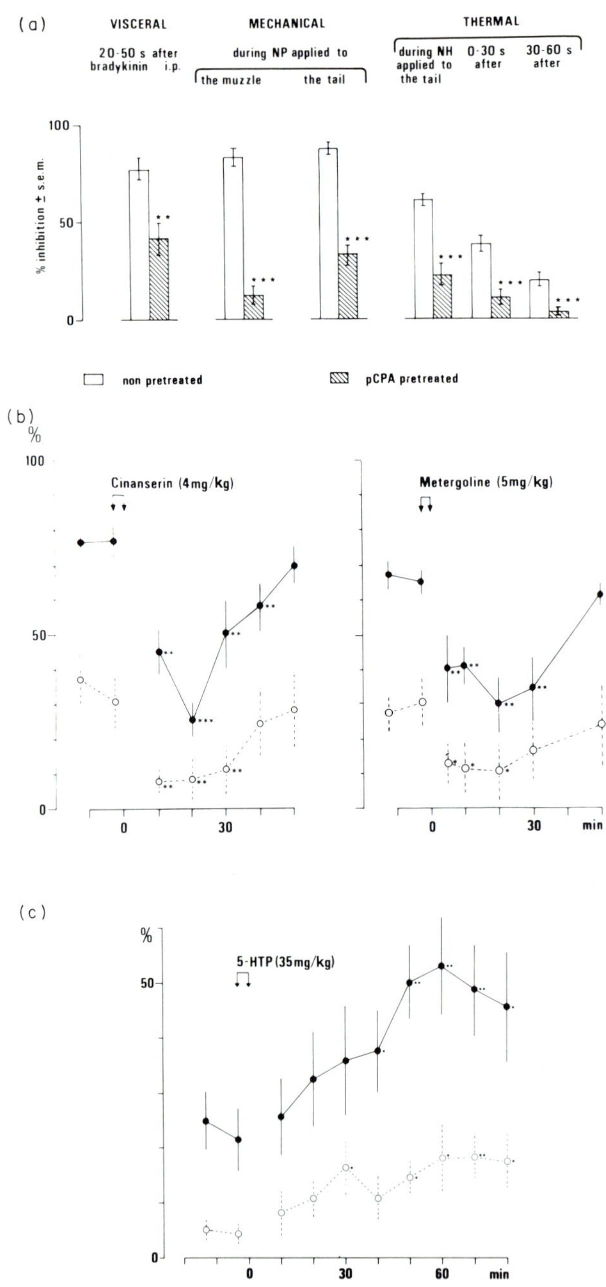
(a)
VISCERAL
MECHANICAL
THERMAL
20-50 s after
bradykinin i.p.
during NP applied to
the muzzle
the tail
during NH
applied to
the tail
0-30 s
after
30-60 s
after
% inhibition ± s.e.m.
100
50
0
**
***
***
***
***
***
non pretreated
pCPA pretreated
(b)
%
100
50
0
Cinanserin (4mg/kg)
**
**
**
***
**
**
**
Metergoline (5mg/kg)
**
**
**
**
*
*
0
30
0
30
min
(c)
%
50
0
5-HTP (35mg/kg)
**
**
**
**
*
**
0
30
60
min

administration of 5-HTP was found to potentiate DNIC (Fig. 7c) (Chitour *et al.*, 1982).

Since there are striking similarities between the inhibitions produced by DNIC and those elicited by stimulation of the NRMag, the involvement of this structure in DNIC has been directly investigated by comparing the inhibitory effects of DNIC on trigeminal nucleus caudalis and dorsal horn convergent neurones before and after electrolytic lesioning of NRMag (Dickenson *et al.*, 1980b). In most neurones recorded in a group of rats in which the cumulative lesions destroyed areas of the brainstem including the NRMag, the ventral part of the NRPG and the midline reticular formation immediately dorsal to the NRMag, a strong reduction of DNIC was apparent immediately after the lesion and also in later tests. Furthermore, it has been recently demonstrated (Villanueva *et al.*, 1986) that the descending projections involved in the triggering of DNIC are confined to the DLF ipsilateral to the neurone under study.

These data, which show that all types of responses of convergent neurones can be inhibited by noxious conditioning stimuli applied to other parts of the body, allow one to speculate further on the role of convergent neurones in pain and nociception.

Clinical correlates

DNIC affect the responses of convergent neurones to noxious stimuli. In other words, when simultaneous noxious stimuli are applied to two distant parts of the body, the pool of convergent dorsal horn units related to the weaker stimulus is inhibited In this respect, we have proposed (Le Bars *et al.*, 1979b, 1984) that DNIC may well form the neural basis of the pain-

---

Fig. 7 Pharmacological evidence for the involvement of serotonergic mechanisms in diffuse noxious inhibitory controls (DNIC).

(a) Inhibitory effects induced by visceral (bradykinin; 8 μg; i.p.), mechanical (NP: noxious pinch) and thermal (NH: noxious heat) heterotopic nociceptive stimuli on the C-fibre evoked response of a lumbar convergent neurone driven from the extremity of a hindpaw: note the strong reduction of inhibitions in pCPA pretreated rats (300 mg/kg; i.p.; 3 days).

(b) Inhibitory effects induced by thermal (immersion of the tail in a 52°C water bath, left) and mechanical (noxious pinch of the muzzle, right) heterotopic stimuli on the C- (solid lines) and A- (broken lines) evoked responses of a lumbar convergent neurone driven from the extremity of a hindpaw: note the reduction of the inhibitions by the intravenous administration of blockers of the serotonergic receptor.

(c) Inhibitory effects induced by a thermal heterotopic stimulus (immersion of the tail in a 46–48°C water bath) on the C- (solid line) and A- (broken line) evoked responses of a limbar convergent neurone driven from the extremity of a hindpaw: note that the intravenous administration of the precursor of serotonin synthesis induced a progressive increase in the inhibitory effects. (From Dickenson *et al.*, 1981; Chitour *et al.*, 1982)

relieving effects of counter-irritation where a peripheral nociceptive stimulus is used against pain originating elsewhere; and it was shown in an earlier section that such hypoalgesic phenomena are sustained by serotonergic mechanisms.

Whether certain types of acupuncture-induced analgesia are equivalent to the counter-irritation phenomenon is not clear, but it seems likely that at least in some cases, the two phenomena share common characteristics (see references in Le Bars *et al.*, 1979b, 1984). From an electrophysiological standpoint, a series of experiments in animals (Chiang *et al.*, 1973; Shen *et al.*, 1975; Du and Chao, 1976) using electroacupuncture versus viscero-somatic reflex discharges strongly supports the idea of the ascending–descending nature of mechanisms subserving this kind of analgesia: inhibitions disappeared in spinal preparations, but remained after decerebration, suggesting that the brainstem is the site of a major link in the mechanism underlying these phenomena. In more precise studies, lesions of the median region of the medulla, and particularly of the NRMag, produced a strong reduction of the inhibitory effects which in addition, according to sectioning experiments, required the ventrolateral and the dorsolateral funiculi as the ascending and descending pathways respectively. These results strongly suggest that both ascending pain pathways and descending serotonergic inhibitory pathways are involved. The analogy between some forms of electroacupuncture and DNIC is further supported by the fact that in the spinal trigeminal nuclei, convergent units are inhibited by electroacupuncture whereas noxious only units are unaffected (Department of Physiology, Kirin Medical College, Changchum, 1977).

In summary, a number of pain-relieving stimuli share common characteristics: the painful or unpleasant nature of the stimulus, widespread analgesic effects, association with long-lasting post-effects, and involvement of raphe-spinal pathways. DNIC may well fit at least some of the requirements of such a system.

Are DNIC involved in the detection of nociceptive information?

On the basis of current knowledge, convergent neurones seem to play a major role in the transmission of painful messages. However, these neurones can exhibit similar or greater levels of activity in response to innocuous peripheral stimulation than in response to noxious stimulation (Le Bars and Chitour, 1983). Therefore, in the intact animal a high rate of impulses can reach supraspinal structures via convergent units when repetitive innocuous peripheral stimuli are applied. On the basis of these observations, and if the spinal mechanisms of nociception are interpreted only in terms of a gain system modulated by spinal and supraspinal inhibitory processes, it is difficult to imagine how the excitatory responses of dorsal horn convergent neurones can be involved in a specific pain-signalling message. We have proposed (Le Bars *et al.*, 1979b, 1983, 1986) that DNIC might be interpreted as an additional filter system allowing the extraction of nociceptive messages emanating from convergent neurones. Indeed, in the freely moving animal, one can envisage that the activity of the whole population of spinal and trigeminal convergent units due to the continual random activation of non-noxious receptors (e.g. hair and touch receptors) would not be negligible. This basic somaesthetic activity transmitted to higher centres would hardly aid the recognition of a meaningful painful signal, but DNIC might provide the means by which a specific pain signal could emerge from such a non-specific system. According to this hypothesis, the pain-signalling message sent to the brain by convergent neurones might result from the contrast between the excitation of a segmental pool of neurones and the DNIC-mediated silencing of the remaining population of convergent neurones. Since serotonergic neurones are a major element in the loop subserving DNIC, these considerations suggest that such neurones are in fact involved in the detection by the brain of nociceptive events (Le Bars *et al.*, 1981). In fact, neurochemical data (see above) are consistent with such an hypothesis.

The effects of morphine on descending controls

The hypothesis that morphine analgesia and SPA share common mechanisms is supported by several findings: (a) some brainstem sites were found to be particularly sensitive to both focal electrical stimulation and the local application of morphine; (b) morphine analgesia and SPA can be blocked by bilateral sections of the DLF; (c) both types of analgesia are believed to be mediated, at least partly, by serotonergic mechanisms (however, see above); (d) cross-tolerance between the two types of analgesia has been reported. A key piece of evidence for the possible existence of common mechanisms is the reported reversal of SPA by the opiate antagonist naloxone both in humans and in animals; however, this latter point is still controversial (see references in Le Bars *et al.*, 1983).

In any case, general agreement has emerged in this field and several functional schemata have been proposed, the basis of which are that descending inhibitory controls emanating from the posterior raphe (mainly the NRMag) and acting on the spinal transmission of nociceptive messages can be electrically or pharmacologically (by morphine) activated, directly or indirectly from the PAG.

However, most of the evidence summarized above supporting the idea of a strengthening of descending inhibitory controls by morphine is, in fact, indirect and in need of further investigation. Behavioural studies have usually been based upon measurement of reaction thresholds which are also largely a function of motor phenomena; the interpretation of such experiments is generally contingent upon two implicit assumptions: (1) that any variations in reaction thresholds in animals are relevant to corresponding pain variations in man, and (2) that the controls of both spinal flexor reflexes and neurones transmitting nociceptive information to the brain are identical.

The hypothesis of an increase in descending inhibitory controls by morphine has been tested electrophysiologically in two ways: (1) by studying the effects of microinjection of this drug into PAG or NRMag, on the responses of spinal nociceptive neurones, and (2) by comparing the effects of systemic morphine on such responses in the presence ('intact' or 'decerebrate' preparation) or absence ('spinal' preparation) of intact descending inhibitory pathways.

Some electrophysiological studies in rats and cats seem to be in accord with such a hypothesis. Bennett and Mayer (1979) reported that microinjections of morphine (4–16 µg) within the rat PAG depressed the responses to noxious stimuli of some dorsal horn neurones. The other studies were performed in the cat, and in one of these, Gebhart *et al.* (1984) showed that microinjection of morphine (10–20 µg) within the PAG clearly reduced the heat-evoked responses of dorsal horn neurones; however, the microinjections were all in the dorsal or the lateral part of the PAG and never in its medioventral aspect, which has repeatedly been reported to be the most effective part for morphine to produce behavioural antinociception; in fact, higher doses of morphine are required in the ventrolateral PAG of this species to elicit antinociceptive reactions (Ossipov *et al.*, 1984). In the study of Clark *et al.* (1983), the precise locations of the microinjection sites were unknown, but high doses of morphine (around 100 µg) into the PAG or the NRMag resulted respectively in a weak inhibition and a more pronounced effect. Finally, Du *et al.* (1984) observed a depression of the nociceptive-evoked responses of dorsal horn neurones following microinjections (10–20 µg) of morphine into the NRMag. However, there are no behavioural data in the cat regarding the effects of morphine microinjections into the NRMag.

By contrast, several other electrophysiological studies do not support the initial hypothesis. For example, in intact and spinal rats the depressive effects of intravenous morphine on C-fibre evoked responses were found to be remarkably similar in the 1–10 mg/kg dose range (Le Bars *et al.*, 1980b). The results of three studies using reversible cold block to temporarily abolish descending inhibition (Le Bars *et al.*, 1976; Duggan *et al.*, 1980; Soja and Sinclair, 1983) are consistent with the hypothesis that morphine decreases tonic descending inhibitions. Four studies using microinjection techniques in the rat, the first within the NRMag (5 µg) (Le Bars *et al.*, 1980a; Llewelyn *et al.*, 1987) and the other two within the PAG (5 and 20 µg) (Dickenson and Le Bars, 1987a, b), have shown that morphine can increase the responses of dorsal horn neurones evoked by noxious stimulation of their receptive fields. Furthermore, we recently found that ICV morphine facilitated the

C-fibre evoked responses of dorsal horn neurones in a dose-related (range 0.6–40 µg) and naloxone-reversible fashion (Bouhassira *et al.*, 1987). This result is consistent with a note by Sinclair (1986) indicating that perfusion of the cat ventricular system with morphine also results in a facilitation of dorsal horn neuronal activities evoked by noxious stimuli. Finally, bilateral lesions of the DLF were found not to affect the depressive effects of systemic morphine on dorsal horn convergent neuronal activities related to pain in the rat (Chitour *et al.*, 1986).

It therefore appears that the evidence for an increase in descending inhibition by morphine is weak, whereas the evidence for the converse is well documented. This assertion is further supported by the reduction in DNIC produced by low doses (0.1–1 mg/kg) of systemic morphine (Le Bars *et al.*, 1981b), low doses (0.6–2.5 µg) of ICV morphine (Bouhassira *et al.*, 1987) and microinjections (5 µg) of this drug within the medio-ventral PAG including the NRD (Dickenson and Le Bars, 1987a). These results show that supraspinal morphine blocks descending inhibitions triggered by noxious stimuli.

What then could be the basis for the antinociception produced by morphine, if indeed these descending inhibitions are reduced?

As explained above, the possibility exists that inhibitory mechanisms such as DNIC have a role in the signalling of pain by convergent neurones. A direct implication of the model proposed would be that a decrease of DNIC (i.e. a rise in basic somaesthetic activity) would lead to hypoalgesia, as morphine does. Thus the blockade of the raphe-spinal 5-HT system could lead to analgesia, and neurochemical experiments do support such a hypothesis (see above).

Recordings of neurones in the nucleus raphe magnus

Attention has focused on the responses of neurones in the NRMag to peripheral stimuli and to systemic administration of opiates (Moolenaar *et al.*, 1976; Anderson *et al.*, 1977; Deakin *et al.*, 1978; Behbehani and Pomeroy, 1978; Fields and Anderson, 1978; Guilbaud *et al.*, 1980; Azami *et al.*, 1981; Tsubokawa *et al.*, 1981; Springfield and Moolenaar, 1983; Fields *et al.*, 1983a, b; Vanegas *et al.*, 1984b; Willis *et al.*, 1984; Barbaro *et al.*, 1986; Liu *et al.*, 1986).

In general, these neurones responded to peripheral noxious stimuli with a complex pattern of inhibition, excitation, a combination of both or no effect. Field's group emphasized the fact that the inhibition induced by noxious heating of the tail in some cells (called 'off-cells') was blocked by morphine (5 mg/kg; i.v.) and suggested that disinhibition of off-cells is a primary process contributing to opiate inhibition of nociceptor-induced reflexes (Fields *et al.*, 1983b). However, the excitation induced by noxious heating of the tail in another group of cells (called 'on-cells') was blocked by a lower dose of systemic morphine (Barbaro *et al.*, 1986). All these effects of systemic morphine were mimicked by microinjections of morphine (5 μg) into the ventral PAG (Cheng *et al.*, 1986). While the former result could provide an argument for the hypothesis that morphine increases descending inhibitory controls, the latter provides a key neuronal link for the inhibition of DNIC by low systemic doses of morphine.

The NRMag receives serotonergic projections from the NRD (see Chapter 3 of this volume) thus allowing 5-HT to act at supraspinal sites to modulate the spinal transmission of nociceptive information, and it has been mentioned earlier that microinjection of 5-HT into the NRMag induces hypoalgesia. However, electrophoretic 5-HT produced either excitation (Llewelyn *et al.*, 1984) or inhibition (Willcockson *et al.*, 1983) of NRMag cells. Interestingly, pretreatment with pCPA led to a marked reduction in the number of cells responding to peripheral stimuli whether by an excitation or by an inhibition; furthermore, the injection of metergoline in normal animals caused a reduction in the magnitude of these neuronal responses to noxious stimuli (Dickenson and Goldsmith, 1986). This suggests that both excitation and inhibition of NRMag cells by noxious stimuli involve serotonergic mechanisms.

However, in all the above mentioned experiments we did not know whether the recorded neurones were serotonergic. In fact, few studies have dealt with identified raphe-spinal serotonergic units. Wessendorf and Anderson (1983) showed in the rat that such units were predominantly excited by noxious pinch applied to the tail, although a subgroup could be inhibited; curiously, noxious heat applied to the tail generally produced no effect. In almost all cases, these neurones were depressed by electrophoretically applied 5-HT. Jacobs's group recorded from serotonin-containing neurones in the NRMag of freely moving cats (Auerbach *et al.*, 1985; Fornal *et al.*, 1985) and found that these neurones could be activated by noxious and stressful stimuli; however, such activation was observed during any period of behavioural arousal, whether or not arousal was due to aversive treatment of the animal. In addition, the activities of these neurones were found not to be increased by an analgesic dose of morphine, a finding which was confirmed in the rat (Chiang and Pan, 1985).

CONCLUSIONS AND FUTURE TRENDS

The review of the literature devoted to serotonin and nociception is somewhat disappointing because the enormous number of references devoted to this problem does not allow one to reach conclusions as to the eventual role(s) of 5-HT in nociceptive processes. The reason for this is to be found in the difficulties of studying pain on the one hand, and integrated 5-HT mechanisms on the other hand (see introduction): nociceptive reactions in animals cannot be interpreted merely as a final product of a single-channelled system responsible for the sensation of pain and, as previously emphasized, 5-HT is involved in several systems potentially able to modulate, directly or indirectly, the transmission and processing of nociceptive information. It is also possible that 5-HT has opposite functions depending on the part of the pain pathway being considered; 5-HT may well participate in the detection of nociceptive events and also in the converse, leading to hypoalgesia. Some of our techniques consider the central nervous system as a black box (e.g. behavioural experiments) where such contradictory events could occur; others consider a single neurone (e.g. electrophysiological experiments) and focus on a single event. All investigators try to get a general view of the system and, according to the discrepancies emphasized in the present review, it is clear that nobody gets the whole truth.

Curiously, by comparison with the overwhelming literature on the subject in animals, there are few data which have been obtained in man. Although there is evidence that both tricyclic antidepressants and monoamine oxidase inhibitors can relieve chronic pain (see Feinmann, 1985), their ubiquitous actions do not reasonably authorize the hypothesis that 5-HT alone is involved in their effects; in addition, both their analgesic action (Bromm *et al.*, 1986) and their capacity to potentiate morphine analgesia (Levine *et al.*, 1986) have been questioned on the basis of experiments related to animal models of chronic pain (Godefroy *et al.*, 1986).

However, relatively selective 5-HT uptake inhibitors such as zimelidine and indalpine were found to produce pain relief or a depression of nociceptive reflexes in man (Johansson and Von Knorring, 1979; Willer *et al.*, 1982). Tryptophan was reported to relieve deafferentation pain (King, 1980) whereas 5-HTP did not alter pain of musculoskeletal origin (Sternbach *et al.*, 1976) and curiously, ritanserin induced an increase in a serotonergic flexion reflex and in a subjective experimental pain threshold (Sandrini *et al.*, 1986). We are thus again confronted with a confused and poorly documented situation.

Perhaps the availability of new pharmacological tools acting on the various types and subtypes of 5-HT receptors could clarify the situation. However, pharmacological approaches should be taken with care because, at least at the present time, 'pure' 5-HT antagonists are not available and most of those which are available exhibit actions against other putative neurotransmitters. In addition, neither the direct application of 5-HT agonists nor the indirect increasing of serotonin availability clearly mimic the physiological function of this substance which occurs in a restricted range of concentrations; thus the unsolved critical question is what constitutes the borderline between pharmacological and physiological actions of endogenous ligands. Finally, the design in animals of new pain tests more closely related to clinical pain in man would be likely to greatly improve our understanding of the role of 5-HT – and other neurotransmitters – in pain.

ACKNOWLEDGEMENTS

The author is grateful to Drs S. W. Cadden and L. Villanueva for reviewing the manuscript and Miss S. Jackson for secretarial help.

REFERENCES AND BIBLIOGRAPHY

Abbott, F. V., and Melzack, R. (1982) Brainstem lesions dissociate neural mechanisms of morphine analgesia in different kinds of pain, *Brain Res.*, **251**, 149–155.

Abbott, F. V., Melzack, R., and Samuel, C. (1982) Morphine analgesia in the tail-flick and formalin pain tests is mediated by different neural systems, *Exp. Neurol.*, **75**, 644–651.

Abbott, F. V., Franklin, K. B. J., and Connell, B. (1986) The stress of a novel environment reduces formalin pain: possible role of serotonin, *Eur. J. Pharmacol.*, **126**, 141–144.

Aghajanian, G. K. (1981) The modulatory role of serotonin at multiple receptors in brain, in *Serotonin, Neurotransmission and Behavior* (Eds B. L. Jacobs and A. Gelperin), pp. 156–185, The MIT Press, Cambridge, Mass.

Ahtee, L., and Carlsson, A. (1979) Dual action of methadone on 5-HT synthesis and metabolism, *Naunyn-Schmiedeberg's Arch. Pharmacol.*, **307**, 51–56.

Adler, M., Kostowski, W., Recchia, M., and Samanin, R. (1975) Anatomical specificity as the critical determinant of the interrelationship between raphe lesions and morphine analgesia, *Eur. J. Pharmacol.*, **32**, 39–44.

Akil, H., and Liebeskind, J. C. (1975) Monoaminergic mechanisms of stimulation-produced analgesia, *Brain Res.*, **94**, 279–296.

Akil, H., and Mayer, D. J. (1972) Antagonism of stimulation-produced analgesia by pCPA, a serotonin synthesis inhibitor, *Brain Res.*, **44**, 692–697.

Almay, B. G. L., Johansson, F., von Knorring, L., Sedvall, G., and Terenius, L. (1980) Relationships between CSF levels of endorphins and monoamine metabolites in chronic pain patients, *Psychopharmacology.*, **67**, 139–142.

Amit, Z., and Galina, H. (1986) Stress-induced analgesia: adaptive pain suppression. *Physiological Reviews*, **66**, 1091–1120.

Anderson, S. D., Basbaum, A. I., and Fields, H. L. (1977) Response of medullary raphe neurons to peripheral stimulation and to systemic opiates, *Brain Res.*, **123**, 363–368.

Appelbaum, B. D., and Holtzman, S. G. (1985) Stress-induced changes in the analgesic and thermic effects of morphine administered centrally, *Brain Res.*, **358**, 303–308.

Appelbaum, B. D., and Holtzman, S. G. (1986) Stress-induced changes in the analgesic and thermic effects of opioid peptides in the rat, *Brain Res.*, **377**, 330–336.

Archer, T., Minor, B. G., and Post, C. (1985a) Blockade and reversal of 5-methoxy-*N,N*-dimethyltryptamine-induced analgesia following noradrenaline depletion, *Brain Res.*, **333**, 55–61.

Archer, T., Arweström, E., Jonsson, G., Minor, B. G., and Post, C. (1985b) Complete blockade and attenuation of 5-hydroxytryptamine-induced analgesia following NA depletion in rats and mice, *Acta Pharmacol. Toxicol.*, **57**, 255–261.

Auerbach, S., Fornal, C., and Jacobs, B. L. (1985) Response of serotonin-containing neurons in nucleus raphe magnus to morphine, noxious stimuli, and periaqueductal gray stimulation in freely moving cats, *Exp. Neurol.*, **88**, 609–628.

Azami, J., Wright, D. M., and Roberts, M. H. T. (1981) Effects of morphine and naloxone on the responses to noxious stimulation of neurones in the nucleus reticularis paragigantocellularis, *Neuropharmacology*, **20**, 869–876.

Azami, J., Llewelyn, M. B., and Roberts, M. H. T. (1982) The contribution of nucleus reticularis paragigantocellularis and nucleus raphe magnus to the analgesia produced by systematically administered morphine investigated with the microinjection technique, *Pain*, **12**, 229–246.

Barbaro, N. M., Hammond, D. L., and Fields, H. L. (1985) Effects of intrathecally administered methysergide and yohimbine on microstimulation-produced antinociception in the rat, *Brain Res.*, **343**, 223–229.

Barbaro, N. M., Heinricher, M. M., and Fields, H. L. (1986) Putative pain modulating neurons in the rostral ventral medulla: reflex-related activity predicts effects of morphine, *Brain Res.*, **366**, 203–210.

Bardon, T., and Ruckebusch, M. (1984) Changes in 5-HIAA and 5-HT levels in lumbar CSF following morphine administration to conscious dogs, *Neurosci. Lett.*, **49**, 147–151.

Basbaum, A. I., and Fields, H. L. (1984) Endogenous pain control systems: brainstem spinal pathways and endorphin circuitry, *Annu. Rev. Neurosci.*, **7**, 309–338.

Basbaum. A. I., Clanton, C. H., and Fields, H. L. (1976) Opiate and stimulus produced analgesia: functional anatomy of a medullo-spinal pathway. *Proc. Natl. Acad. Sci. USA*, **73**, 4685–4688.

Basbaum, A. I., Marley, N. J. E., O'Keefe, J., and Clanton, C. H. (1977) Reversal of morphine and stimulus-produced analgesia by subtotal spinal cord lesions, *Pain*, **3**, 43–56.

Beck, P. W., and Handwerker, H. O. (1974) Bradykinin and serotonin effects on various types of cutaneous nerve fibres, *Pfluegers Arch.*, **347**, 209–222.

Beecher, H. K. (1957) The measurement of pain, *Pharmacol. Rev.*, **9**, 59–209.

Behbehani, M. M., and Pomeroy, S. L. (1978) Effects of morphine injected in periaqueductal gray on the activity of single units in nucleus raphe magnus of the rat, *Brain Res.*, **149**, 266–269.

Belcher, G., Ryall, R. W., and Schaffner, R. (1978) The differential effects of 5-hydroxytryptamine, noradrenaline and raphe stimulation on nociceptive and non-nociceptive dorsal horn interneurones in the cat, *Brain Res.*, **151**, 307–321.

Bennett, G. J., and Mayer, D. J. (1979) Inhibition of spinal cord interneurons by narcotic microinjection and focal electrical stimulation in the periaqueductal central gray matter, *Brain Res.*, **172**, 243–257.

Bensemana, D., and Gascon, A. L. (1978) Relationship between analgesia and turnover of brain biogenic amines, *Can. J. Physiol. Pharmacol.*, **56**, 721–730.

Berge, O. G. (1982) Effects of 5-HT receptor agonists and antagonists on a reflex response to radiant heat in normal and spinally transected rats, *Pain*, **13**, 253–266.

Berge, O. G., and Hole, K. (1981) Tolerance to the antinociceptive effect of morphine in the spinal rat, *Neuropharmacology.*, **20**, 653–657.

Berge, O. G., and Ogren, S. O. (1984) Selective lesions of the bulbospinal serotonergic pathways reduce the analgesia induced by p-chloro-amphetamine in the hot-plate test, *Neurosci. Lett.*, **44**, 25–29.

Berge, O. G., Hole, K., and Dahle, H. (1980) Nociception is enhanced after low doses and reduced after high doses of the serotonin receptor agonist 5-methoxy-N, N-dimethyltryptamine, *Neurosci. Lett.*, **19**, 219–223.

Berge, O. G., Fasmer, O. B., Flatmark, T., and Hole, K. (1983a). Time course of changes in nociception after 5,6-dihydroxytryptamine lesions of descending 5-HT pathways, *Pharmacol. Biochem. Behav.*, **18**, 637–643.

Berge, O. G., Fasmer, O. B., and Hole, K. (1983b) Serotonin receptor antagonists induce hyperalgesia without preventing morphine antinociception. *Pharmacol. Biochem. Behav.*, **19**, 873–878.

Berge, O. G., Hole, K. and Ogren, S. O. (1983c) Attenuation of morphine-induced analgesia by p-chlorophenylalanine and p-chloroamphetamine : test-dependent effects and evidence for brainstem 5-hydroxytryptamine involvement, *Brain Res.*, **271**, 51–64.

Berge, O. G., Fasmer, O. B., Ogren, S. O., and Hole, K. (1985a) The putative receptor against 8-hydroxy-2.(di-n-propylamino) tetralin antagonizes the anti-nociceptive effect of morphine, *Neurosci. Lett.*, **54**, 71–75.

Berge, O. G., Fasmer, O. B., Jorgensen, H. A., and Hole, K. (1985b), Test-dependent antinociceptive effect of spinal serotonin release induced by intrathecal p-chloroamphetamine in mice, *Acta Physiol. Scand.*, **123**, 35–41.

Besson, J. M., and Chaouch, A. (1987) Peripheral and spinal mechanisms of nociception, *Physiol. Reviews*, **67**, 67–186.

Bhattacharya, S. K., Keshary, P. R., and Sanyal, A. K. (1978) Immobilisation stress-induced antinociception in rats: possible role of serotonin and prostaglandins, *Eur. J. Pharmacol.*, **50**, 83–85.

Bineau-Thurotte, M., Godefroy, F., Weil-Fugazza, J., and Besson, J. M. (1984) The effect of morphine on the potassium evoked release of tritiated 5-hydroxytrypta-mine from spinal cord slices in the rat, *Brain Res.*, **291**, 293–299.

Bläsig, J., Reinhold, K., and Herz, A. (1973) Effect of 6-hydroxydopamine, 5,6-dihydroxytryptamine and raphe lesions on the antinociceptive actions of morphine in rats, *Psychopharmacologia (Berl.)*, **31**, 111–119.

Bodnar, R. J., Kordower, J. H., Wallace, M. M., and Tamir, H. (1981) Stress and morphine analgesia: alterations following p-chlorophenylalanine, *Pharmacol. Biochem. Behav.*, **14**, 645–651.

Bodnar, R. J., Kordower, J. H., Reches, A., Wallace, M. M., and Fahn, S. (1984) Reductions in pain thresholds and morphine analgesia following intracerebroventri-cular parachlorophenylalanine, *Pharmacol. Biochem. Behav.*, **21**, 79–84.

Botting, R., and Morinan, A. (1982) Involvement of 5-hydroxytryptamine in the analgesic action of pethidine and morphine in mice, *Br. J. Pharmacol.*, **75**, 579–585.

Bouhassira, D., Villanueva, L., and Le Bars, D. (1988) Intracerebroventricular morphine decreases descending inhibitions acting on lumbar dorsal horn neuronal activities related to pain in the rat (submitted).

Bourgoin, S., Oliveras, J. L., Bruxelle, J., Hamon, M., and Besson, J. M. (1980) Electrical stimulation of the nucleus raphe magnus in the rat: effects on 5-HT metabolism in the spinal cord, *Brain Res.*, **194**, 377–389.

Bromm, B., Meier, W., and Scharein, E. (1986) Imipramine reduces experimental pain, *Pain*, **25**, 245–257.

Bryant, R. M., Olley, J. E., and Tyers, M. B. (1982) Involvement of the median raphe nucleus in antinociception induced by morphine, buprenorphine and tilidine in the rat, *Br. J. Pharmacol.*, **77**, 615–624.

Buckett, W. R. (1981) Pharmacological studies on stimulation-produced analgesia in mice, *Eur. J. Pharmacol.*, **69**, 281–290.

Buxbaum, D. M., Yarbrough, G. G., and Carter, M. E. (1973) Biogenic amines and narcotic effects. I. Modification of morphine-induced analgesia and motor activity after alteration of cerebral amine levels, *J. Pharmacol. Exp. Ther.*, **185**, 317–327.

Carstens, E., Fraunhoffer, M., and Zimmermann, M. (1981) Serotonergic mediation of descending inhibition from midbrain periaqueductal gray, but not reticular formation, of spinal nociceptive transmission in the cat, *Pain*, **10**, 149–167.

Carstens, E., MacKinnon, J. D., and Guinan, M. J. (1983) Serotonin involvement in descending inhibition of spinal nociceptive transmission produced by stimulation of medial diencephalon and basal forebrain, *J. Neurosci.*, **3**, 2112–2120.

Chance, W. T., Krynock, G. M., and Rosecrans, J. A. (1978) Effects of medial raphe and raphe magnus lesions on the analgesic activity of morphine and methadone, *Psychopharmacology*, **56**, 133–137.

Chaouch, A., and Besson, J. M. (1985) Nucleus raphe magnus effect on dorsal horn neurones in the rat after intrathecal 5-7 DHT, *Neurosci. Lett. Suppl.*, **22**, S475.

Cheney, D. L., and Goldstein, A. (1971) The effect of *p*-chlorophenylalanine on opiate-induced running, analgesia, tolerance and physical dependence in mice, *J. Pharmacol. Exp. Ther.*, **177**, 309–315.

Cheng, Z. F., Fields, H. L., and Heinricher, M. M. (1986) Morphine microinjected into the periaqueductal gray has differential effects on 3 classes of medullary neurons, *Brain Res.*, **375**, 57–65.

Chiang, C. Y., and Pan, Z. Z. (1985) Differential responses of serotonergic and non-serotonergic neurons in nucleus raphe magnus to systemic morphine in rats, *Brain Res.*, **337**, 146–150.

Chiang, C. Y., Chang, C. T., Chu, H. L., and Yang, L. F. (1973) Peripheral afferent pathway for acupuncture analgesia, *Scient. Sin.*, **16**, 210–217.

Chitour, D., Dickenson, H., and Le Bars, D. (1982) Pharmacological evidence for the involvement of serotonergic mechanisms in diffuse noxious inhibitory controls (DNIC), *Brain Res.*, **236**, 329–337.

Chitour, D., Villanueva, L., and Le Bars, D. (1986) Lesions of dorsolateral funiculi (DLF) do not affect the depressive effects of systemic morphine upon dorsal horn convergent neuronal activities related to pain in the rat, *Brain Res.*, **377**, 397–402.

Clark, S. L., Edeson, R. O., and Ryall, R. W. (1983) The relative significance of spinal and supraspinal actions in the antinociceptive effect of morphine in the dorsal horn: an evaluation of the microinjection technique, *Br. J. Pharmacol.*, **79**, 807–818.

Coderre, T. J., and Rollman, G. B. (1984) Stress analgesia: Effects of pCPA, yohimbine and naloxone, *Pharmacol. Biochem. Behav.*, **21**, 681–686.

Commissiong, J. W. (1983) Mass fragmentographic analysis of monoamine metabolites in the spinal cord of rat after the administration of morphine, *J. Neurochem.*, **41**, 1313–1318.

Contreras, E., and Tamayo, L. (1967) Influence of changes on brain 5-hydroxytryptamine on morphine analgesia, *Arch. Biol. Med. Exper.*, **4**, 69–71.

Contreras, E., Tamayo, L., and Weitzman, P. (1970) Reduction of the antinociceptive effect of 5-hydroxytryptophan in morphine tolerant rats, *Psychopharmacologia (Berl.)*, **17**, 314–319.

Contreras, E., Quijada, L., and Tamayo, L. (1973) A comparative study of the effects of reserpine and p-chlorophenylalanine on morphine analgesia in mice, *Psychopharmacologia (Berl.)*, **28**, 319–324.

Costa, C., Ceccherelli, F., Bettero, A., Marin, G., Mancusi, L., and Allegri, G. (1984) Tryptophan, serotonin and 5-hydroxyindoleacetic acid levels in human CSF in relation to pain, in *Progress in Tryptophan and Serotonin Research* (Eds H. G. Schlossberger, W. Kochen, B. Linzen and H. Steinhart), pp. 413–416, Walter de Gruyter, Berlin.

Crowley, W. R., Rodriguez-Sierra, J. F., and Komisaruk. B. R. (1977) Monoaminergic mediation of the antinociceptive effect of vaginal stimulation in rats, *Brain Res.*, **137**, 67–84.

Dahlström, A., and Fuxe, K. (1965) Evidence for the existence of monoamine neurons in the central nervous system. II. Experimentally induced changes in the intraneuronal amine levels of bulbospinal neuron systems. *Acta Physiol. Scand. Suppl.*, **247**, 1–36.

Danguir, J., and Nicolaidis, S. (1985) Feeding, metabolism, and sleep: peripheral and central mechanisms of their interaction, in *Brain Mechanisms of Sleep* (Eds D. J. McGinty, R. Drucker-Cloin, A. Morrison and P. L. Parmeggiani), pp. 321–340, Raven Press, New York.

Davies, J. E., and Roberts, M. H. T. (1981) 5-Hydroxytryptamine reduces substance P responses on dorsal horn interneurones: a possible interaction of neurotransmitters, *Brain Res.*, **217**, 399–404.

Deakin, J. F. W., and Dostrovsky, J. O. (1978) Involvement of the periaqueductal grey matter and spinal 5-hydroxytryptaminergic pathways in morphine analgesia: effects of lesions and 5-hydroxytryptamine depletion, *Br. J. Pharmacol.*, **63**, 159–165.

Dennis, M. (1972) Sex-dependent and sex-independent neural control of reactivity to electric footshock in the rat, *Exp. Neurol.*, **37**, 256–268.

Dennis, S. G., and Melzack, R. (1980) Pain modulation by 5-hydroxytryptaminergic agents and morphine as measured by three pain tests, *Exp. Neurol.*, **69**, 260–270.

Department of Physiology, Kirin Medical College, Changchum (1977) The inhibition effect and the mode of action of electroacupuncture upon discharges from the pain-sensitive cells in spinal trigeminal nucleus, *Scient. Sin.*, **20**, 485–501.

Dewey, W. L., Harris, L. S., Howes, J. F., and Nuite, J. A. (1970) The effect of various neurohumoral modulators on the activity of morphine and the narcotic antagonists in the tail-flick and phenylquinone tests, *J. Pharmacol. Exp. Ther.*, **175**, 435–442.

Dickenson, A. H., and Goldsmith, G. (1986) Evidence for a role of 5-hydroxytryptamine in the responses of rat raphe magnus neurones to peripheral noxious stimuli, *Neuropharmacology*, **25**, 863–868.

Dickenson, A. H., and Le Bars, D. (1987a) Supraspinal morphine and descending inhibitions acting on the dorsal horn of the rat, *J. Physiol. (Lond.)*, **384**, 81–107.

Dickenson, A. H., and Le Bars, D. (1987b) Lack of evidence for increased descending inhibition on the dorsal horn of the rat following periaqueductal grey morphine microinjections, *Brit. J. Pharmacol.*, **92**, 271–280.

Dickenson, A. H., Oliveras, J. L., and Besson, J. M. (1979) Role of the nucleus raphe magnus in opiate analgesia as studied by the microinjection technique in the rat, *Brain Res.*, **170**, 95–111.

Dickenson, A. H., Le Bars, D., and Besson, J. M. (1980a) Diffuse noxious inhibitory controls (DNIC). Effects on trigeminal nucleus caudalis neurones in the rat, *Brain Res.*, **200**, 293–305.

Dickenson, A. H., Le Bars, D., and Besson, J. M. (1980b) An involvement of nucleus raphe magnus in diffuse noxious inhibitory controls (DNIC) in the rat, *Neurosci. Lett.*, Suppl. 5, S375.

Dickenson, A. H., Rivot, J. P., Chaouch, A., Besson, J. M., and Le Bars, D. (1981) Diffuse noxious inhibitory controls (DNIC) in the rat with or without pCPA pretreatment, *Brain Res.*, **216**, 313–321.

Donzanti, B. A., and Warwick, R. O. (1979) Effect of methadone and morphine on serotonin uptake in rat periaqueductal gray slices, *Eur. J. Pharmacol.*, **59**, 107–110.

Du, H. J., and Chao, Y. F. (1976) Localization of central structures involved in descending inhibitory effect of acupuncture on viscero-somatic discharges, *Scient. Sin.*, **19**, 137–148.

Du, H. J., Kitahata, L. M., Thalhammer, J. G., and Zimmermann, M. (1984) Inhibition of nociceptive neuronal responses in the cat's spinal dorsal horn by electrical stimulation and morphine microinjection in nucleus raphe magnus, *Pain*, **19**, 249–257.

Duggan, A. W., Griersmith, B. T., Headley, P. M., and Maher, J. B. (1978) The need to control skin temperature when using radiant heat in tests of analgesia, *Exp. Neurol.*, **61**, 471–478.

Duggan, A. W., Griersmith, B. T., and North, R. A. (1980) Morphine and supra-spinal inhibition of spinal neurones: evidence that morphine decreases tonic descending inhibition in the anaesthetized cat, *Br. J. Pharmacol.*, **69**, 461–466.

Fardin, V., Oliveras, J. L., and Besson, J. M. (1984a) A reinvestigation of the analgesic effects induced by stimulation of the periaqueductal grey matter in the rat. I. The production of behavioural side effects together with analgesia, *Brain Res.*, **306**, 105–123.

Fardin, V., Oliveras, J. L., and Besson, J. M. (1984b) A reinvestigation of the analgesic effects induced by stimulation of the periaqueductal grey matter in the rat. II. Differential characteristics of the analgesia induced by ventral and dorsal PAG stimulation, *Brain Res.*, **306**, 125–139.

Fasmer, O. B., and Post, C. (1983) Behavioural responses induced by intrathecal injection of 5-hydroxytryptamine in mice are inhibited by a substance P antagonist, d-PRO^2, d-TRP^{7,9}-substance P, *Neuropharmacology*, **22**, 1397–1400.

Fasmer, O. B., Berge, O. G., and Hole, K. (1983a) Similar behavioural effects of 5-hydroxytryptamine and substance P injected intrathecally in mice, *Neuropharmacology*, **22**, 485–487.

Fasmer, O. B., Berge, O. G., Walther, B., and Hole, K. (1983b) Changes in nociception after intrathecal administration of 5,6-dihydroxytryptamine in mice, *Neuropharmacology*, **22**, 1197–1201.

Fasmer, O. B., Berge, O. G., and Hole, K. (1984) Metergoline elevates or reduces nociceptive thresholds in mice depending on test method and route of adminis-tration, *Psychopharmacology*, **82**, 306–309.

Fasmer, O. B., Berge, O. G., and Hole, K. (1985) Changes in nociception after lesions of descending serotonergic pathways induced with 5,6-dihydroxytryptamine. Different effects in the formalin and tail-flick tests, *Neuropharmacology*, **24**, 729–734.

Fasmer, O. B., Berge, O. G., Post, C., and Hole, K. (1986) Effects of the putative 5-HT$_{1A}$ receptor agonist 8-OH-2-(di-*n*-propylamino)tetralin on nociceptive sensi-tivity in mice, *Pharmacol. Biochem. Behav.*, **25**, 883–888.

Feinmann, C. (1985) Pain relief by antidepressants: possible modes of action, *Pain*, **23**, 1–8.

Fennessy, M. R., and Laska, F. J. (1979) Acute effects of opiate agonists, partial agonists and antagonists on brain monoamines in the rat, *Arch. Int. Pharmacodyn.*, **238**, 60–70.

Fennessy, M. R., and Lee, J. R. (1970) Modification of morphine analgesia by drugs affecting adrenergic and tryptaminergic mechanisms, *J. Pharm. Pharmacol.*, **22**, 930–935.

Fennessy, M. R., and Lee, J. R. (1972) Comparison of the dose-response effects of morphine on brain amines, analgesia and activity in mice, *Br. J. Pharmacol.*, **45**, 1–9.

Fibiger, H. C., and Mertz, P. H. (1972) The effect of para-chlorophenylalanine on aversion thresholds and reactivity to foot shock, *Physiol. Behav.*, **8**, 259–263.

Fields, H. L., and Anderson, S. D. (1978) Evidence that raphe-spinal neurons mediate opiate and midbrain stimulation-produced analgesia, *Pain*, **5**, 333–349.

Fields, H. L., Bry, J., Hentall, I., and Zorman, G. (1983a) The activity of neurons in the rostral medulla of the rat during withdrawal from noxious heat, *J. Neurosci.*, **3**, 2545–2552.

Fields, H. L., Vanegas, H., Hentall, I. D., and Zorman, G. (1983b) Evidence that disinhibition of brain stem neurones contributes to morphine analgesia, *Nature*, **306**, 684–686.

Fjällbrant, N., and Iggo, A. (1961) The effect of histamine, 5-hydroxytryptamine and acetycholine on cutaneous afferent fibres, *J. Physiol. (Lond.)*, **156**, 578–590.

Fock, S., and Mense, S. (1976) Excitatory effects of 5-hydroxytryptamine, histamine and potassium ions on muscular group IV afferent units: a comparison with bradykinin, *Brain Res.*, **105**, 459–469.

Foong, F. W., Terman, G., and Duggan, A. W. (1985) Methysergide and spinal inhibition from electrical stimulation in the periaqueductal grey, *Eur. J. Pharmacol.*, **116**, 239–248.

Fornal, C., Auerbach. S., and Jacobs, B. L. (1985) Activity of serotonin-containing neurons in nucleus raphe magnus in freely moving cats, *Exp. Neurol.*, **88**, 590–608.

Franklin, K. B. J., and Kelly, S. J. (1986) Sympathetic control of tryptophan uptake and morphine analgesia in stressed rats, *Eur. J. Pharmacol.*, **126**, 145–150.

Garau, L., Mulas, M. L., and Pepeu, G. (1975) The influence of raphe lesions on the effect of morphine on nociception and cortical ACh output, *Neuropharmacology*, **14**, 259–263.

Garcia-Sevilla, J. A., Ahtee, L., Magnusson, T., and Carlsson, A. (1978) Opiate-receptor mediated changes in monoamine synthesis in rat brain, *J. Pharm. Pharmacol.*, **30**, 613–621.

Gebhart, G. F., and Lorens, S. A. (1980) Attenuation of pethidine-induced antinociception by zimelidine, an inhibitor of 5-hydroxytryptamine reuptake, *Br. J. Pharmacol.*, **70**, 411–414.

Gebhart, G. F., Sandkühler, J., Thalhammer, J. G., and Zimmermann, M. (1983) Inhibition of spinal nociceptive information by stimulation in midbrain of the cat is blocked by lidocaine microinjected in nucleus raphe magnus and medullary reticular formation, *J. Neurophysiol.*, **50**, 1446–1459.

Gebhart, G. F., Sandkühler, J., Thalhammer, J. G., and Zimmermann, M. (1984) Inhibition in spinal cord of nociceptive information by electrical stimulation and morphine microinjection at identical sites in midbrain of the cat, *J. Neurophysiol.*, **51**, 75–89.

Genovese, E., Zonta, N., and Mantegazza, P. (1973) Decreased antinociceptive activity of morphine in rats pretreated intraventricularly with 5,6-dihydroxytryptamine, a long-lasting selective depletor of brain serotonin, *Psychopharmacologia (Berl.)*, **32**, 359–364.

Gerschenfeld, H. M., and Paupardin-Tritsch, D. (1974a) Ionic mechanisms and receptor properties underlying the responses of molluscan neurones to 5-hydroxytryptamine, *J. Physiol. (Lond.)*, **243**, 427–456.

Gerschenfeld, H. M., and Paupardin-Tritsch, D. (1974b) On the transmitter function of 5-hydroxytryptamine at excitatory and inhibitory monosynaptic junctions, *J. Physiol. (Lond.)*, **243**, 457–481.

Ghia, J. N., Mueller, R. A., Duncan, G. H., Scott, D. S., and Mao, W. (1981)

Serotonergic activity in man as a function of pain, pain mechanisms, and depression, *Anesth. Analg.*, **60**, 854–861.

Godefroy, F., Weil-Fugazza, J., Coudert, D., and Besson, J. M. (1980) Effect of acute administration of morphine on newly synthesized 5-hydroxytryptamine in spinal cord of the rat, *Brain Res.*, **199**, 415–424.

Godefroy, F., Butler, S. H., Weil-Fugazza, J., and Besson, J. M. (1986) Do acute or chronic tricyclic antidepressants modify morphine antinociception in arthritic rats? *Pain*, **25**, 233–244.

Godefroy, F., Weil-Fugazza, J., and Besson, J. M. (1987) Complex temporal changes in 5-hydroxytryptamine synthesis in the central nervous system induced by experimental polyarthritis in the rat, *Pain*, **28**, 223–238.

Goodlet, I., and Sugrue, M. F. (1974) Effect of acutely administered analgesic drugs on rat brain serotonin turnover, *Eur. J. Pharmacol.*, **29**, 241–248.

Görlitz, B. D., and Frey, H. H. (1972) Central monoamines and antinociceptive drug action, *Eur. J. Pharmacol.*, **20**, 171–180.

Griersmith, B. T., and Duggan, A. W. (1980) Prolonged depression of spinal transmission of nociceptive information by 5-HT administered in the substantia gelatinosa: antagonism by methysergide, *Brain Res.*, **187**, 231–235.

Griersmith, B. T., Duggan, A. W., and North, R. A. (1981) Methysergide and supraspinal inhibition of the spinal transmission of nociceptive information in the anaesthetized cat, *Brain Res.*, **204**, 147–158.

Grossman, M. L., Basbaum, A. I., and Fields, H. L. (1982) Afferent and efferent connections of the rat tail flick reflex (a model used to analyze pain control mechanisms), *J. Comp. Neurol.*, **206**, 9–16.

Guilbaud, G., Besson, J. M., Oliveras, J. L., and Liebeskind, J. C. (1973) Suppression by LSD of the inhibitory effect exerted by dorsal raphe nucleus on certain spinal cord interneurons in the cat, *Brain Res.*, **61**, 417–422.

Guilbaud, G., Peschanski, M., Gautron, M., and Binder, D. (1980) Responses of neurons of the nucleus raphe magnus to noxious stimuli, *Neurosci. Lett.*, **17**, 149–154.

Haigler, H. J. (1978) Morphine: effects on serotonergic neurons and neurons in areas with a serotonergic input, *Eur. J. Pharmacol.*, **51**, 361–376.

Hall, J. G., Duggan, A. W., Johnson, S. M., and Morton, C. R. (1981) Medullary raphe lesions do not reduce descending inhibition of dorsal horn neurones in the cat, *Neurosci. Lett.*, **25**, 25–29.

Hammond, D. L., and Yaksh, T. L. (1984) Antagonism of stimulation-produced antinociception by intrathecal administration of methysergide or phentolamine, *Brain Res.*, **298**, 329–337.

Hammond, D. L., Levy, R. A., and Proudfit, H. K. (1980) Hypoalgesia induced by microinjection of a norepinephrine antagonist in the raphe magnus: reversal by intrathecal administration of a serotonin antagonist, *Brain Res.*, **201**, 475–479.

Hammond, D. L., Tyce, G. M., and Yaksh, T. L. (1985) Efflux of 5-hydroxytryptamine and noradrenaline into spinal cord superfusates during stimulation of the rat medulla, *J. Physiol. (Lond.)*, **359**, 151–162.

Han, J. S., and Terenius, L. (1982) Neurochemical basis of acupuncture analgesia, *Ann. Rev. Pharmacol. Toxicol.*, **22**, 193–220.

Harvey, J. A., and Lints, C. E. (1965) Lesions in the medial forebrain bundle: delayed effects on sensitivity to electric shock, *Science*, **148**, 250–252.

Harvey, J. A., and Lints, C. E. (1971) Lesions in the medial forebrain bundle: relationship between pain sensitivity and telencephalic content of serotonin, *J. Comp. Physiol. Psychol.*, **74**, 28–36.

Harvey, J. A., and Yunger, L. M. (1973) Relationship between telencephalic content of serotonin and pain sensitivity, in *Serotonin and Behaviour* (Eds J. Barchas and E. Usdin), pp. 179–189, Academic Press, London.

Harvey, J. A., Schlosberg, A. J., and Yunger, L. M. (1974) Effect of p-chlorophenyl-alanine and brain lesions on pain sensitivity and morphine analgesia in the rat, *Adv. Biochem. Psychopharmacol.*, **10**, 233–245.

Haubrich, D. R., and Blake, D. E. (1973) Modification of serotonin metabolism in rat brain after acute or chronic administration of morphine, *Biochem. Pharmacol.*, **22**, 2753–2759.

Hayes, R. L., Newlon, P. G., Rosecrans, J. A., and Mayer, D. J. (1977) Reduction of stimulation-produced analgesia by lysergic acid diethylamide, a depressor of serotonergic neural activity, *Brain Res.*, **122**, 367–372.

Headley, P. M., Duggan, A. W., and Griersmith, B. T. (1978) Selective reduction by noradrenaline and 5-hydroxytryptamine of nociceptive responses of cat dorsal horn neurones, *Brain Res.*, **145**, 195–289.

Herold, M., and Cahn, J. (1968) The possible role of serotonin in affective component of pain behavioural reaction in the rat, in *Pharmacology of Pain* (Eds R. K. S. Lim, D. Armstrong, and E. G. Pardo), pp. 87–99, Pergamon Press, Oxford.

Ho, I. K., Brase, D. A., Loh, H. H., and Way, L. (1975) Influence of L-tryptophan on morphine analgesia, tolerance and physical dependence, *J. Pharmacol. Exp. Ther.*, **193**, 35–43.

Hole, K., and Lorens, S. A. (1975) Response to electric shock in rats: effects of selective midbrain raphe lesions, *Pharmacol. Biochem. Behav.*, **3**, 95–102.

Hole, K., and Marsden, C. A. (1975) Unchanged sensitivity to electric shock in L-tryptophan treated rats, *Pharmacol. Biochem. Behav.*, **3**, 307–309.

Hole, K., Fuxe, K., and Jonsson, G. (1976) Behavioral effects of 5,7-dihydroxytrypta-mine lesions of ascending 5-hydroxytryptamine pathways, *Brain Res.*, **107**, 385–399.

Hosobuchi, Y. (1978) Tryptophan reversal of tolerance to analgesia induced by central grey stimulation, *Lancet*, **2**, 47.

Hunskaar, S., Berge, O. G., Broch, O. J., and Hole, K. (1986) Lesions of the ascending serotonergic pathways and antinociceptive effects after systemic adminis-tration of p-chloroamphetamine in mice, *Pharmacol. Biochem. Behav.*, **24**, 709–714.

Hutson, P. H., Tricklebank, M. D., and Curzon, G. (1982) Enhancement of foot-shock-induced analgesia by spinal 5,7-dihydroxytryptamine lesions, *Brain Res.*, **237**, 367–372.

Hutson, P. H., Tricklebank, M. D., and Curzon, G. (1983) Analgesia induced by brief footshock: blockade by fenfluramine and 5-methoxy-*N*,*N*-dimethyltryptamine and prevention of blockade by 5-HT antagonists, *Brain Res.*, **279**, 105–110.

Hylden, J. L. K., and Wilcox, G. L. (1983) Intrathecal serotonin in mice: analgesia and inhibition of a spinal action of substance P, *Life Sci.*, **33**, 789–795.

Hynes, M. D., Lochner, M. A., Bemis, K. G., and Hymson, D. L. (1985) Fluoxetine, a selective inhibitor of serotonin uptake, potentiates morphine analgesia without altering its discriminative stimulus properties or affinity for opioid receptors, *Life Sci.*, **36**, 2317–2323.

Hyyppä, M. T., Scheinin, H., Alaranta, H., Hurme, M., Lahtela, K., and Scheinin, M. (1985) Neurotransmission and the experience of low back pain; no associates between CSF monoamine metabolites and pain, *Pain*, **21**, 57–65.

Jeftinija, S., Raspantini, C., Randic, M., Yaksh, T. L., Go, V. L. W., and Larson, A. A. (1986) Altered responsiveness to substance P and 5-hydroxytryptamine in cat dorsal horn neurons after 5-HT depletion with *p*-chlorophenylalanine, *Brain Res.*, **368**, 107–115.

Jensen, T. S., and Yaksh, T. L. (1984) Spinal monoamine and opiate systems partly mediate the antinociceptive effects produced by glutamate at brainstem sites, *Brain Res.*, **321**, 287–297.

Jensen, T. S., and Yaksh, T. L. (1986a) I. Comparison of antinociceptive action of morphine in the periaqueductal gray, medial and paramedial medulla in rat, *Brain Res.*, **363**, 99–113.

Jensen, T. S., and Yaksh, T. L. (1986b) II. Examination of spinal monoamine receptors through which brainstem opiate-sensitive systems act in the rat, *Brain Res.*, **363**, 114–127.

Johannessen, J. N., Watkins, L. R., Carlton, S. M., and Mayer, D. J. (1982) Failure of spinal cord serotonin depletion to alter analgesia elicited from the periaqueductal gray, *Brain Res.*, **237**, 373–386.

Johansson, F., and Von Knorring, L. (1979) A double-blind controlled study of a serotonin uptake inhibitor (zimelidine) versus placebo in chronic pain patients, *Pain*, **7**, 69–78.

Johnston, C. A., and Moore, K. E. (1983) The effect of morphine on 5-hydroxytryptamine synthesis and metabolism in the striatum, and several discrete hypothalamic regions of the rat brain, *J. Neural Transm.*, **57**, 65–73.

Jordan, L. M., Kenshalo, D. R. Jr., Martin, R. F., Haber, L. H., and Willis, W. D. (1978) Depression of primate spinothalamic tract neurons by iontophoretic application of 5-hydroxytryptamine, *Pain*, **5**, 135–142.

Jordan, L. M., Kenshalo, D. R. Jr., Martin, R. F., Haber, L. H., and Willis, W. D. (1979) Two populations of spinothalamic tract neurons with opposite responses to 5-hydroxytryptamine, *Brain Res.*, **164**, 342–346.

Kaergaard Nielsen, C., Magnussen, M. P., Kampmann, E., and Frey, H. H. (1967) Pharmacological properties of racemic and optically active *p*-chloroamphetamine, *Arch. Int. Pharmacodyn.*, **170**, 428–443.

Keele, C. A., and Armstrong, D. (1964) *Substances Producing Pain and Itch*, Arnold, London.

Kelly, D. D. (Ed.) (1986) Stress-induced analgesia, *Annals of the New York Academy of Sciences*, 467.

Kelly, S. J., and Franklin, K. B. J. (1984a) Evidence that stress augments morphine analgesia by increasing brain tryptophan, *Neurosci. Lett.*, **44**, 305–310.

Kelly, S. J., and Franklin, K. B. J. (1984b) Electrolytic raphe magnus lesions block analgesia induced by a stress-morphine interaction but not analgesia induced by morphine alone, *Neurosci. Lett.*, **52**, 147–152.

Kelly, S. J., and Franklin, K. B. J. (1985) An increase in tryptophan in brain may be a general mechanism for the effect of stress on sensitivity to pain, *Neuropharmacology*, **24**, 1019–1025.

King, R. B. (1980) Pain and tryptophan, *J. Neurosurg.*, **53**, 44–52.

Kiser, R. S., Lebovitz, R. M., and German, D. C. (1978) Anatomic and pharmacologic differences between two types of aversive midbrain stimulation, *Brain Res.*, **155**, 331–342.

Knapp, S., and Mandell, A. J. (1972) Narcotic drugs: effects on the serotonin biosynthetic systems of the brain, *Science*, **177**, 1209–1211.

Kostowski, W., Giacalone, E., Garattini, S., and Valzelli, L. (1968) Studies on behavioural and biochemical changes in rats after lesion of midbrain raphe, *Eur. J. Pharmacol.*, **4**, 371–376.

Kraus, E., Besson, J. M., and Le Bars, D. (1982) Behavioral model for diffuse noxious inhibitory controls (DNIC): potentiation by 5-hydroxytryptophan, *Brain Res.*, **231**, 461–465.

Kulkarni, S. K., and Robert, R. K. (1982) Reversal by serotonergic agents of reserpine-induced hyperalgesia in rats, *Eur. J. Pharmacol.*, **83**, 325–328.

Kuraishi, Y., Harada, Y., Aratani, S., Satoh, M., and Takagi, H. (1983) Separate involvement of the spinal noradrenergic and serotonergic systems in morphine analgesia: the differences in mechanical and thermal algesic tests, *Brain Res.*, **273**, 245–252.

Kuraishi, Y., Hirota, N., Satoh, M., and Takagi, H. (1985) Antinociceptive effects of intrathecal opioids, noradrenaline and serotonin in rats: mechanical and thermal algesic tests, *Brain Res.*, **326**, 168–171.

Lakoski, J. M., Mohrland, J. S., and Gebhart, G. F. (1980) The effect of morphine on the content of serotonin, 5-hydroxyindole acetic acid and substance-P in the nuclei raphe magnus and reticularis gigantocellularis, *Life Sci.*, **27**, 2639–2644.

Larsen, J. J., and Arnt, J. (1984) Spinal 5-HT or NA uptake inhibition potentiates supraspinal morphine antinociception in rats, *Acta. Pharmacol. Toxicol.*, **54**, 72–75.

Larsen, J. J., and Christensen, A. V. (1982) Subarachnoidal administration of the 5-HT uptake inhibitor citalopram points to the spinal role of 5-HT in morphine antinociception, *Pain*, **14**, 339–345.

Larsen, J. J., and Hyttel, J. (1985) 5-HT-uptake inhibition potentiates antinociception induced by morphine, pethidine, methadone and ketobemidone in rats, *Acta Pharmacol. Toxicol.*, **57**, 214–218.

Larson, A. A. (1982) Nociception is enhanced by the intrathecal injection of 5-methoxy-*N*,*N*-dimethyltryptamine in the rat, *Neurosci. Lett.*, **33**, 323–328.

Larson, A. A. (1983) Hyperalgesia produced by the intrathecal administration of tryptamine to rats, *Brain Res.*, **265**, 109–117.

Larson, A. A., and Takemori, A. E. (1977a) Effect of narcotics on the uptake of serotonin precursors by the rat brain, *J. Pharmacol. Exp. Ther.*, **200**, 216–223.

Larson, A. A., and Takemori, A. E. (1977b) Effect of fluoxetine hydrochloride (Lilly 110140), a specific inhibitor of serotonin uptake, on morphine analgesia and the development of tolerance, *Life Sci.*, **21**, 1807–1812.

Larson, A. A., and Takemori, A. E. (1978) Effect of morphine on the fate of newly transported tryptophan and 5-hydroxytryptophan in brain of rats, *J. Pharmacol. Exp. Ther.*, **205**, 265–273.

Larson, A. A., and Wilcox, G. L. (1984) Synergistic behavioral effects of serotonin and tryptamine injected intrathecally in mice, *Neuropharmacology*, **23**, 1415–1418.

Le Bars, D., and Chitour, D. (1983) Do convergent neurones in the spinal dorsal horn discriminate nociceptive from non-nociceptive information? *Pain*, **17**, 1–19.

Le Bars, D., Ménétrey, D., and Besson, J. M. (1976) Effects of morphine upon the lamina V type cells activities in the dorsal horn of the decerebrate cat, *Brain Res.*, **113**, 293–310.

Le Bars, D., Dickenson, A. H., and Besson, J. M. (1979a) Diffuse noxious inhibitory controls (DNIC). I. Effects on dorsal horn convergent neurones in the rat, *Pain*, **6**, 283–304.

Le Bars, D., Dickenson, A. H., and Besson, J. M. (1979b) Diffuse noxious inhibitory controls (DNIC). II. Lack of effect on non-convergent neurones, supraspinal involvement and theoretical implications, *Pain*, **6**, 305–327.

Le Bars, D., Dickenson, A. H., and Besson, J. M. (1980a) Microinjection of morphine within nucleus raphe magnus and dorsal horn neurone activities related to nociception in the rat, *Brain Res.*, **189**, 467–481.

Le Bars, D., Guilbaud, G., Chitour, D., and Besson, J. M. (1980b) Does systemic morphine increase descending inhibitory controls of dorsal horn neurones involved in nociception?, *Brain Res.*, **202**, 223–228.

Le Bars, D., Chitour, D., and Clot, A. M. (1981a) The encoding of thermal stimuli by diffuse noxious inhibitory controls (DNIC), *Brain Res.*, **230**, 394–399.

Le Bars, D., Chitour, D., Kraus, E., Clot, A. M., Dickenson, A. H., and Besson, J. M. (1981b) The effect of systemic morphine upon diffuse noxious inhibitory controls (DNIC) in the rat: evidence for a lifting of certain descending inhibitory controls of dorsal horn convergent neurones, *Brain Res.*, **215**, 257–274.

Le Bars, D., Dickenson, A. H., Rivot, J. P., Chitour, D., Chaouch, A., Kraus, E., and Besson, J. M. (1981c). Les systèmes sérotonergiques bulbo-spinaux jouent-ils un rôle dans la détection des messages nociceptifs?, *J. Physiol. (Paris)*, **77**, 463–471.

Le Bars, D., Dickenson, A. H., and Besson, J. M. (1982) The triggering of bulbo-spinal serotonergic inhibitory controls by noxious peripheral inputs, in *Brain Stem Control of Spinal Mechanisms* (Eds B. Sjölund and A. Björklund), pp. 381–410, Elsevier Biomedical Press, Amsterdam.

Le Bars, D., Dickenson, A. H., and Besson, J. M. (1983) Opiate analgesia and descending control systems, in *Advances in Pain Research and Therapy*, vol. 5 (Eds J. J. Bonica, U. Lindblom and A. Iggo), pp. 341–372, Raven Press, New York.

Le Bars, D., Calvino, B., Villanueva, L., and Cadden, S. (1984) Physiological approaches to counter-irritation phenomena, in *Stress-induced Analgesia* (Eds M. D. Tricklebank and G. Curzon), pp. 67–101, John Wiley & Sons Ltd., Chichester.

Le Bars, D., Dickenson, A. H., Besson, J. M., and Villanueva, L. (1986) Aspects of sensory processing through convergent neurons, in *Spinal Afferent Processing* (Ed. T. L. Yaksh), pp. 467–504, Plenum Publ. Corp., New York.

Lee, J. R., and Fennessy, M. R. (1970) The relationship between morphine analgesia and the levels of biogenic amines in the mouse brain, *Eur. J. Pharmacol.*, **12**, 65–70.

Lee, R. L., Sewell, R. D. E., and Spencer, P. S. J. (1979) Antinociceptive activity of D-ALA^2-D-LEU^5-enkephalin (BW 180C) in the rat after modification to central 5-hydroxytryptamine function, *Neuropharmacology*, **18**, 711–717.

Levine, J. D., Gordon, N. C., Smith, R., and MacBryde, R. (1986) Desipramine enhances opiate postoperative analgesia, *Pain*, **27**, 45–49.

Light, A. R., Casale, E. J., and Menétrey, D. (1986) The effects of focal stimulation in nucleus raphe magnus and periaqueductal gray on intracellularly recorded neurons in spinal laminae I and II, *J. Neurophysiol.*, **56**, 555–571.

Lin, M. T., Chi, M. L., Chandra, A., and Tsay, B. L. (1980a) Serotoninergic mechanisms of beta-endorphin and clonidine-induced analgesia in rats, *Pharmacology*, **20**, 323–328.

Lin, M. T., Chandra, A., Chi, M. L., and Kau, C. L. (1980b) Effects of increasing serotonergic receptor activity in brain on analgesic activity in rats, *Exp. Neurol.*, **68**, 548–554.

Lints, C. E., and Harvey, J. A. (1969a) Altered sensitivity to footshock and decreased brain content of serotonin following brain lesions in the rat, *J. Comp. Physiol. Psychol.*, **67**, 23–31.

Lints, C. E., and Harvey, J. A. (1969b) Drug induced reversal of brain damage in the rat, *Physiol. Behav.*, **4**, 29–31.

Liu, X., Zhu, B., and Zhang, S. X. (1986) Relationship between electroacupuncture analgesia and descending pain inhibitory mechanism of nucleus raphe magnus, *Pain*, **24**, 383–396.

Llewelyn, M. B., Azami, J., and Roberts, M. H. T. (1983) Effects of 5-hydroxytryptamine applied into nucleus raphe magnus on nociceptive thresholds and neuronal firing rate, *Brain Res.*, **258**, 59–68.

Llewelyn, M. B., Azami, J., and Roberts, M. H. T. (1984) The effect of modification of 5-hydroxytryptamine function in nucleus raphe magnus on nociceptive threshold, *Brain Res.*, **306**, 165–170.

Llewelyn, M. B., Azami, J., and Roberts, M. H. T. (1987) Brainstem mechanisms of antinociception: effects of electrical stimulation and microinjection of morphine into nucleus raphe magnus, *Neuropharmacology*, **25**, 727–735.

Long, J. B., Kalivas, P. W., Youngblood, W. W., Prange, A. J. Jr., and Kizer, J. S. (1984) Possible involvement of serotonergic neurotransmitter in neurotensin but not morphine analgesia, *Brain Res.*, **310**, 35–41.

Lorens, S. A., and Yunger, L. M. (1974) Morphine analgesia, two-way avoidance, and consummatory behavior following lesions in the midbrain raphe nuclei of the rat, *Pharmacol. Biochem. Behav.*, **2**, 215–221.

Lovick, T. A., West, D. C., and Wolstencroft, J. H. (1978) Responses of raphe-spinal and other bulbar raphe neurones to stimulation of the periaqueductal grey in the cat, *Neurosci. Lett.*, **8**, 45–49.

Lytle, L. D., Messing, R. B., Fisher, L., and Phebus, L. (1975) Effects of long-term corn consumption on brain serotonin and the response to electric shock, *Science*, **190**, 692–694.

Major, C. T., and Pleuvry, B. J. (1971) Effects of α-methyl-p-tyrosine, *p*-chlorophenylalanine, 1-β-(3,4-dihydroxyphenyl) alanine, 5-hydroxytryptophan and diethyldithiocarbamate on the analgesic activity of morphine and methylamphetamine in the mouse, *Br. J. Pharmacol.*, **42**, 512–521.

Malec, D., and Langwinski, R. (1980a) The influence of 5-HT receptor blocking agents on the behavioral effects of analgesics in rats, *Psychopharmacology*, **69**, 79–83.

Malec, D., and Langwinski, R. (1980b) Effect of quipazine and fluoxetine on analgesic-induced catalepsy and antinociception in the rat, *J. Pharm. Pharmacol.*, **32**, 71–73.

Maruyama, Y., Hayashi, G., Smits, S. E., and Takemori, A. E. (1971) Studies on the relationship between 5-hydroxytryptamine turnover in brain and tolerance and physical dependence in mice, *J. Pharmacol. Exp. Ther.*, **178**, 20–29.

Mennini, T., Pataccini, R., and Samanin, R. (1978) Effects of narcotic analgesics on the uptake and release of 5-hydroxytryptamine in rat synaptosomal preparations, *Br. J. Pharmacol.*, **64**, 75–82.

Mense, S., and Schmidt, R. F. (1974) Activation of group IV afferent units from muscle by algesic agents, *Brain Res.*, **72**, 305–310.

Merskey, H., Albe-Fessard, D. G., Bonica, J. J., Carmon, A., Dubner, R., Kerr, F. W. L., Lindblom, U., Mumford, J. M., Nathan, P. W., Noordenbos, W., Pagni, C. A., Renaer, M. J., Sternbach, R. A., and Sunderland, Sir S. (1979) Pain terms: a list with definitions and notes on usage, *Pain*, **6**, 249–252.

Messing, R. B., Phebus, L., Fisher, L., and Lytle, L. D. (1975) Analgesic effect of fluoxetine HCl (Lilly 110104), a specific uptake inhibitor for serotonergic neurons, *Psychopharm. Commun.*, **1**, 511–521.

Messing, R. B., Fisher, L. A., Phebus, L., and Lytle, L. D. (1976) Interaction of diet and drugs in the regulation of brain 5-hydroxyindoles and the response to painful electric shock, *Life Sci.*, **18**, 707–714.

Messing, R. B., Flinchbaugh, C., and Waymire, J. C. (1978a) Changes in brain tryptophan and tyrosine following acute and chronic morphine administration, *Neuropharmacology*, **17**, 391–396.

Messing, R. B., Flinchbaugh, C., and Waymire, J. C. (1978b) Tryptophan and 5-hydroxyindoles in different CNS regions following acute morphine, *Eur. J. Pharmacol.*, **48**, 137–140.

Minor, B. G., Post, C., and Archer, T. (1985) Blockade of intrathecal 5-hydroxytryptamine-induced antinociception in rats by noradrenaline depletion, *Neurosci. Lett.*, **54**, 39–44.

Miranda, F., Candelaresi. G., and Samanin, R. (1978) Analgesic effect of etorphine in rats with selective depletions of brain monoamines, *Psychopharmacology*, **58**, 105–109.

Misantone, L. J. (1976) Effects of damage to the monoamine axonal constituents of the medial forebrain bundle on reactivity to foot shock and ingestive behavior in the rat, *Exp. Neurol.*, **50**, 448–464.

Mohrland, J. S., and Gebhart, G. F. (1980) Effect of selective destruction of serotonergic neurons in nucleus raphe magnus on morphine-induced antinociception, *Life Sci.*, **27**, 2627–2632.

Monroe, P. J., Michaux, K., and Smith, D. J. (1986) Evaluation of the direct actions of drugs with a serotonergic link in spinal analgesia on the release of (^{3}H)serotonin from spinal cord synaptosomes, *Neuropharmacology*, **25**, 261–265.

Moolenaar, G. M., Holloway, J. A., and Trouth, C. O. (1976) Responses of caudal raphe neurons to peripheral somatic stimulation, *Exp. Neurol.*, **53**, 304–313.

Morton, C. R., Duggan, A. W., and Zhao, Z. Q. (1984) The effects of lesions of medullary midline and lateral reticular areas on inhibition in the dorsal horn produced by periaqueductal grey stimulation in the cat, *Brain Res.*, **301**, 121–130.

Nissbrandt, H., Yao, T., Thorén, P., and Svensson, T. H. (1982) Naloxone reversible reduction in brain monoamine synthesis following sciatic nerve stimulation, *J. Neural Trans.*, **53**, 91–100.

Ogren, S. O., and Berge, O. G. (1984) Test-dependent variations in the antinociceptive effect of *p*-chloroamphetamine-induced release of 5-hydroxytryptamine, *Neuropharmacology*, **23**, 915–924.

Ogren, S. O., and Berge, O. G. (1985) Evidence for selective serotonergic receptor involvement in *p*-chloroamphetamine-induced antinociception, *Naunyn-Schmiedeberg's Arch. Pharmacol.*, **329**, 135–140.

Ogren, S. O., and Holm, A. C. (1980) Test-specific effects of the 5-HT reuptake inhibitors alaproclate and zimelidine on pain sensitivity and morphine analgesia, *J. Neural Trans.*, **47**, 253–271.

Ogren, S. O., and Johansson, C. (1985) Separation of the associative and non-associative effects of brain serotonin released by *p*-chloroamphetamine: dissociable serotonergic involvement in avoidance learning, pain and motor function, *Psychopharmacology*, **86**, 12–26.

Ogren, S. O., Berge, O. G., and Johansson, C. (1985) Involvement of spinal serotonergic pathways in nociception but not in avoidance learning, *Psychopharmacology*, **87**, 260–265.

Oliveras, J. L., Hosobuchi, Y., Guilbaud, G., and Besson, J. M. (1978) Analgesic electrical stimulation of the feline nucleus raphe magnus: development of tolerance and its reversal by 5-HTP, *Brain Res.*, **146**, 404–409.

Oliveras, J. L., Guilbaud, G., and Besson, J. M. (1979) A map of serotonergic structures involved in stimulation producing analgesia in unrestrained freely moving cats, *Brain Res.*, **164**, 317–322.

Ossipov, M. H., Goldstein, F. J., and Malseed, R. T. (1984) Feline analgesia following central administration of opioids, *Neuropharmacology*, **23**, 925–929.

Papeschi, R., Theiss, P., and Herz, A. (1975) Effects of morphine on the turnover of brain catecholamines and serotonin in rats – acute morphine administration, *Eur. J. Pharmacol.*, **34**, 253–261.

Pepeu, G., Garau, L., and Mulas, M. L. (1974) Does 5-hydroxytryptamine influence cholinergic mechanisms in the central nervous system?, in *Advances in Biochemical Psychopharmacology*, Vol. 10, pp. 247–252, Raven Press, New York.

Pérez-Cruet, J., Thoa, N. B., and Ng, L. K. Y. (1975) Acute effects of heroin and morphine on newly synthesized serotonin in rat brain, *Life Sci.*, **17**, 349–362.

Pilowsky, P. M., Kapoor, V., Minson, J. B., West, M. J., and Chalmers, J. P. (1986) Spinal cord serotonin release and raised blood pressure after brainstem kainic acid injection, *Brain Res.*, **366**, 354–357.

Pleuvry, B. J. (1975) Mouse brain catecholamines, 5-hydroxytryptamine and the antinociceptive activity of pethidine, *Eur. J. Pharmacol.*, **34**, 351–361.

Post, C., Minor, B. G., Davies, M., and Archer, T. (1986) Analgesia induced by 5-hydroxytryptamine receptor agonists is blocked or reversed by noradrenaline-depletion in rats, *Brain Res.*, **363**, 18–27.

Prado, W. A., and Roberts, M. H. T. (1985) An assessment of the antinociceptive and aversive effects of stimulating identified sites in the rat brain, *Brain Res.*, **340**, 219–228.

Prieto, G. J., Cannon, J. T., and Liebeskind, J. C. (1983) N. raphe magnus lesions disrupt stimulation-produced analgesia from ventral but not dorsal midbrain areas in the rat, *Brain Res.*, **261**, 53–57.

Proudfit, H. K. (1980) Reversible inactivation of raphe magnus neurons: effects on nociceptive threshold and morphine-induced analgesia, *Brain Res.*, **201**, 459–464.

Proudfit, H. K. (1981) Time-course of alterations in morphine-induced analgesia and nociceptive threshold following medullary raphe lesions, *Neuroscience*, **6**, 945–951.

Proudfit, H. K., and Anderson, E. G. (1974) New-long latency bulbospinal evoked potentials blocked by serotonin antagonists, *Brain Res.*, **65**, 542–546.

Proudfit, H. K., and Anderson, E. G. (1975) Morphine analgesia: blockade by raphe magnus lesions, *Brain Res.*, **98**, 612–618.

Proudfit, H. K., and Hammond, D. L. (1981) Alterations in nociceptive threshold and morphine-induced analgesia produced by intrathecally administered amine antagonists, *Brain Res.*, **218**, 393–399.

Pycock, C. J., Burns, S., and Morris, R. (1981) In vitro release of 5-hydroxytrypta-mine and γ-amino-butyric acid from rat periaqueductal grey and raphe dorsalis region produced by morphine or an enkephalin analogue, *Neurosci. Lett.*, **22**, 313–317.

Randic, M., and Yu, H. H. (1976) Effect of 5-hydroxytryptamine and bradykinin in cat dorsal horn neurons activated by noxious stimuli, *Brain Res.*, **111**, 197–203.

Reigle, T. G., and Barker, W. L. (1983) *p*-Chlorophenylalanine antagonism of the analgesia and increase in brain noradrenaline metabolism produced by morphine, *J. Pharm. Pharmacol.*, **35**, 324–325.

Rivot, J. P., Chaouch, A., and Besson, J. M. (1980) Nucleus raphe magnus modulation of response of rat dorsal horn neurons to unmyelinated fiber inputs: partial involvement of serotonergic pathways, *J. Neurophysiol.*, **44**, 1039–1057.

Rivot, J. P., Chiang, C. Y., and Besson, J. M. (1982) Increase in serotonin metabolism within the dorsal horn of the spinal cord during nucleus raphe magnus stimulation, as revealed by in vivo electrochemical detection, *Brain Res.*, **238**, 1117–1126.

Rivot, J. P., Calvino, B., and Besson, J. M. (1987) Is there a serotonergic tonic descending inhibition on the responses of dorsal horn convergent neurons to C-fibre inputs?, *Brain Res.*, **403**, 142–146.

Rochat, C., Cervo, L., Romandini, S., and Samanin, R. (1982a) Evidence that m-chlorophenylpiperazine inhibits some nociceptive responses of rats by activating 5-hydroxytryptamine mechanisms, *J. Pharm. Pharmacol.*, **34**, 325–327.

Rochat, C., Cervo, L., Romandini, S., and Samanin, R. (1982b) Differences in the effects of d-fenfluramine and morphine on various responses of rats to painful stimuli, *Psychopharmacology*, **76**, 188–192.

Romandini, S., and Samanin, R. (1984) Methysergide and metergoline reduce morphine analgesia with no effect on the development of tolerance in rats, *Psychopharmacology*, **82**, 140–142.

Romandini, S., Pich, E. M., Esposito, E., Kruszewska, A., and Samanin, R. (1986a) The effect of different lesions of the median raphe on morphine analgesia, *Brain Res.*, **377**, 351–354.

Romandini, S., Pich, E. M., Esposito, E., Kruszewska, A. Z., and Samanin, R. (1986b) The effect of intracerebroventricular 5,7-dihydroxytryptamine on morphine analgesia is time-dependent, *Life Sci.*, **38**, 869–875.

Saarnivaara, L. (1969) Effect of 5-hydroxytryptamine on morphine analgesia in rabbits, *Ann. Med. Exp. Fenn.*, **47**, 113–123.

Sagen, J., Winker, M. A., and Proudfit, H. K. (1983) Hypoalgesia induced by the local injection of phentolamine in the nucleus raphe magnus: blockade by depletion of spinal cord monoamines, *Pain*, **16**, 253–263.

Samanin, R., and Bernasconi, S. (1972) Effects of intraventricularly injected 6-OH dopamine or midbrain raphe lesion on morphine analgesia in rats, *Psychopharmacologia (Berl.)*, **25**, 175–182.

Samanin, R., and Valzelli, L. (1971) Increase of morphine-induced analgesia by stimulation of the nucleus raphe dorsalis, *Eur. J. Pharmacol.*, **16**, 298–302.

Samanin, R., Gumulka, W., and Valzelli, L. (1970) Reduced effect of morphine in midbrain raphe lesioned rats, *Eur. J. Pharmacol.*, **10**, 339–343.

Samanin, R., Ghezzi, D., Mauron, C., and Valzelli, L. (1973) Effect of midbrain raphe lesion on the antinociceptive action of morphine and other analgesics in rats, *Psychopharmacologia (Berl.)*, **33**, 365–368.

Samanin, R., Bernasconi, S., and Quattrone, A. (1976) Antinociceptive action of quipazine: relation to central serotonergic receptor stimulation, *Psychopharmacologia (Berl.)*, **46**, 219–222.

Sandkühler, J., and Gebhart, G. F. (1984) Relative contributions of the nucleus raphe magnus and adjacent medullary reticular formation to the inhibition by stimulation in the periaqueductal gray of a spinal nociceptive reflex in the phenobarbital anaesthetized rat, *Brain Res.*, **305**, 77–87.

Sandrini, G., Alfonsi, E., De Rysky, C., Marini, S., Facchinetti, F., and Nappi, G. (1986) Evidence for serotonin-S_2 receptor involvement in analgesia in humans, *Eur. J. Pharmacol.*, **130**, 311–314.

Schlosberg, A. J., and Harvey, J. A. (1978) Diurnal changes in serotonin content of frontal pole and pain sensitivity in the rat, *Physiol. Behav.*, **20**, 117–120.

Schmauss, C., Hammond, D. L., Ochi, J. W., and Yaksh, T. L. (1983) Pharmacological antagonism of the antinociceptive effects of serotonin in the rat spinal cord, *Eur. J. Pharmacol.*, **90**, 349–357.

Sewell, R. D. E., and Spencer, P. S. J. (1975) Anti-nociceptive activity of narcotic agonists and partial agonists in mice given biogenic amines by intracerebroventricular injection, *Psychopharmacologia (Berl.)*, **42**, 67–71.

Shen, E., Tsai, T. T., and Lan, C. (1975) Supraspinal participation in the inhibitory effect of acupuncture on viscero-somatic reflex discharges, *Chim. Med. J.*, **1**, 431–440.

Shimizu, T., Koja, T., Fujisaki, T., and Fukuda, T. (1981) Effects of methysergide and naloxone on analgesia induced by the peripheral electric stimulation in mice, *Brain Res.*, **208**, 463–467.

Shiomi, H., Murakami, H., and Takagi, H. (1978) Morphine analgesia and the bulbospinal serotonergic system: increase in concentration of 5-hydroxyindoleacetic acid in the rat spinal cord with analgesics, *Eur. J. Pharmacol.*, **52**, 335–344.

Shyu, K. W., Lin, M. T., and Wu, T. C. (1984) Possible role of central serotonergic neurons in the development of dental pain and aspirin-induced analgesia in the monkey, *Exp. Neurol.*, **84**, 179–187.

Sicuteri, F.(1968) Sensitization of nociceptors by 5-hydroxytryptamine in man, in *Pharmacology of Pain* (Eds R. K. S. Lim, D. Armstrong and E. G. Pardo), pp. 57–86, Pergamon Press, Oxford.

Simansky, K. J., and Harvey, J. A. (1981) Altered sensitivity to footshock after selective serotonin depletion: comparison of electrolytic lesions and neurotoxin injections in the medial forebrain bundle of the rat, *J. Comp. Physiol. Psychol.*, **95**, 341–350.

Sinclair, J. G. (1986) The failure of morphine to attenuate spinal cord nociceptive transmission through supraspinal actions in the cat, *Gen. Pharmac.*, **17**, 351–354.

Skagerberg, G., and Björklund, A. (1985) Topographic principles in the spinal projections of serotonergic and non-serotonergic brainstem neurons in the rat, *Neuroscience*, **15**, 445–480.

Snelgar, R. S., and Vogt, M. (1980) Mapping, in the rat central nervous system, of morphine-induced changes in turnover of 5-hydroxytryptamine, *J. Physiol. (Lond.)*, **314**, 395–410.

Soja, P. J., and Sinclair, J. G. (1980) Evidence against a serotonin involvement in the tonic descending inhibition of nociceptor-driven neurons in the cat spinal cord, *Brain Res.*, **199**, 225–230.

Soja, P. J., and Sinclair, J. G. (1983) Spinal vs supraspinal actions of morphine on cat spinal cord multireceptive neurons, *Brain Res.*, **273**, 1–7.

Spampinato, U., Esposito, E., Romandini, S., and Samanin, R. (1985) Changes of serotonin and dopamine metabolism in various forebrain areas of rats injected with morphine either systemically or in the raphe nuclei dorsalis and medianus, *Brain Res.*, **328**, 89–95.

Sparkes, C. G., and Spencer, P. S. J. (1971) Antinociceptive activity of morphine after injection of biogenic amines in the cerebral ventricles of the conscious rat, *Br. J. Pharmacol.*, **42**, 230–241.

Springfield, S. A., and Moolenaar, G. M. (1983) Differences in the responses of raphe nuclei to repetitive somatosensory stimulation, *Exp. Neurol.*, **79**, 360–370.

Steinman, J. L., Komisaruk, B. R., Yaksh, T. L., and Tyce, G. M. (1983) Spinal cord monoamines modulate the antinociceptive effects of vaginal stimulation in rats, *Pain*, **16**, 155–166.

Sternbach, R. A., Janowsky, D. S., Huey, L. Y., and Segal, D. S. (1976) Effects of altering brain serotonin activity on human chronic pain, in *Advances in Pain Research and Therapy*, Vol. 1 (Eds J. J. Bonica and D. Albe-Fessard), pp. 601–606, Raven Press, New York.

Sugrue, M. F. (1979a) Effect of depletion of rat brain 5-hydroxytryptamine on morphine-induced antinociception, *J. Pharm. Pharmacol.*, **31**, 253–255.

Sugrue, M. F. (1979b) On the role of 5-hydroxytryptamine in drug-induced antinociception, *Br. J. Pharmacol.*, **65**, 677–681.

Taiwo, Y. O., Fabien, A., Pazoles, C. J., and Fields, H. L. (1985) Potentiation of morphine antinociception by monoamine reuptake inhibitors in the rat spinal cord, *Pain*, **21**, 329–337.

Takagi, H., Takashima, T., and Kimura, K. (1964) Antagonism of the analgetic effect of morphine in mice by tetrabenazine and reserpine, *Arch. Int. Pharmacodyn.*, **149**, 484–492.

Takemori, A. E., Tulunay, F. C., and Yano, I. (1975) Differential effects on morphine analgesia and naloxone antagonism by biogenic amine modifiers, *Life Sci.*, **17**, 21–28.

Tenen, S. S. (1967) The effects of p-chlorophenylalanine, a serotonin depletor, on avoidance acquisition, pain sensitivity and related behavior in the rat, *Psychopharmacologia (Berl.)*, **10**, 204–219.

Tenen, S. S. (1968) Antagonism of the analgesic effect of morphine and other drugs by p-chlorophenylalanine, a serotonin depletor, *Psychopharmacologia (Berl.)*, **12**, 278–285.

Tilson, H. A., and Rech, R. H. (1974) The effects of p-chlorophenylalanine on morphine analgesia, tolerance and dependence development in two strains of rats, *Psychopharmacologia (Berl.)*, **35**, 45–60.

Todd, A. J., and Millar, J. (1983) Receptive fields and responses to ionophoretically applied noradrenaline and 5-hydroxytryptamine of units recorded in laminae I–III of cat dorsal horn, *Brain Res.*, **288**, 159–167.

Tricklebank, M. D., and Curzon, G. (Eds) (1984) *Stress-induced Analgesia*, John Wiley and Sons, Chichester.

Tricklebank, M. D., Hutson, P. H., and Curzon, G. (1982) Analgesia induced by brief footshock is inhibited by 5-hydroxytryptamine but unaffected by antagonists of 5-hydroxytryptamine or by naloxone, *Neuropharmacology*, **21**, 51–56.

Tricklebank, M. D., Hutson, P. H., and Curzon, G. (1984) Analgesia induced by brief or more prolonged stress differs in its dependency on naloxone, 5-hydroxytryptamine and previous testing and analgesia, *Neuropharmacology.*, **23**, 417–421.

Tsubokawa, T., Yamamoto, T., Katayama, Y., and Moriyasu, N. (1981) Diencephalic modulation of activities of raphe-spinal neurons in the cat, *Exp. Neurol.*, **74**, 561–572.

Tulunay, F. C., Yano, I., and Takemori, A. E. (1976) The effect of biogenic amine modifiers on morphine analgesia and its antagonism by naloxone, *Eur. J. Pharmacol.*, **35**, 285–292.

Tyce, G. M., and Yaksh, T. L. (1981) Monoamine release from cat spinal cord by somatic stimuli: an intrinsic modulatory system, *J. Physiol. (Lond.)*, **314**, 513–529.

Uzan, A., Kabouche, M., Rataud, J., and Le Fur, G. (1980) Pharmacological evidence of a possible tryptaminergic regulation of opiate receptors by using indalpine, a selective 5-HT uptake inhibitor, *Neuropharmacology*, **19**, 1075–1079.

Vanegas, H., Barbaro, N. M., and Fields, H. L. (1984a) Midbrain stimulation inhibits tail-flick only at currents sufficient to excite rostral medullary neurons, *Brain Res.*, **321**, 127–133.

Vanegas, H., Barbaro, N. M., and Fields, H. L. (1984b) Tail-flick related activity in medullospinal neurons, *Brain Res.*, **321**, 135–141.

Vasko, M. R., and Vogt, M. (1982) Analgesia, development of tolerance, and 5-hydroxytryptamine turnover in the rat after cerebral and systemic administration of morphine. *Neuroscience*, **7**, 1215–1225.

Vasko, M. R., Pang, I. H., and Vogt, M. (1984) Involvement of 5-hydroxytryptamine-containing neurons in antinociception produced by injection of morphine into nucleus raphe magnus or into spinal cord, *Brain Res.*, **306**, 341–348.

Villanueva, L., and Le Bars, D. (1985) The encoding of thermal stimuli applied to the tail of the rat by lowering the excitability of trigeminal convergent neurones, *Brain Res.*, **330**, 245–251.

Villanueva, L., Chitour, D., and Le Bars, D. (1986) Involvement of the dorsolateral funiculus in the descending spinal projections responsible for diffuse noxious inhibitory controls in the rat, *J. Neurophysiol.*, **56**, 1185–1195.

Vogt, M. (1974) The effect of lowering the 5-hydroxytryptamine content of the rat spinal cord on analgesia produced by morphine, *J. Physiol. (Lond.)*, **236**, 483–498.

Vonvoigtlander, P. F., Lewis, R. A., and Neff, G. L. (1984) Kappa opioid analgesia is dependent on serotonergic mechanisms, *J. Pharmacol. Exp. Ther.*, **231**, 270–274.

Wang, J. K. (1977) Antinociceptive effect of intrathecally administered serotonin, *Anesthesiology*, **47**, 269–271.

Watkins, L. R., Johannessen, J. N., Kinscheck, I. B., and Mayer, D. J. (1984) The neurochemical basis of footshock analgesia: the role of spinal cord serotonin and norepinephrine, *Brain Res.*, **290**, 107–117.

Weil-Fugazza, J., Godefroy, F., and Besson, J. M. (1979) Changes in brain and spinal tryptophan and 5-hydroxyindoleacetic acid levels following acute morphine administration in normal and arthritic rats, *Brain Res.*, **175**, 291–301.

Weil-Fugazza, J., Godefroy, F., Coudert, D., and Besson, J. M. (1981a) Morphine analgesia and newly synthetized 5-hydroxytryptamine in the dorsal and the ventral halves of the spinal cord in the rat, *Brain Res.*, **214**, 440–444.

Weil-Fugazza, J., Godefroy, F., Chitour, D., and Le Bars, D. (1981b) Synthèse de la sérotonine au niveau spinal chez le rat: modifications induites par stimulation somatique nociceptive associée ou non à l'administration de morphine, *C.R. Acad. Sci. (Paris)*, **293**, 89–92.

Weil-Fugazza, J., Godefroy, F., Coudert, D., and Besson, J. M. (1981c) Changes in total and free tryptophan levels in serum following acute morphine administration in arthritic rats, *Neurochem. Int.*, **3**, 323–328.

Weil-Fugazza, J., Godefroy, F., and Le Bars, D. (1984) Increase in 5-HT synthesis in the dorsal part of the spinal cord, induced by a nociceptive stimulis: blockade by morphine, *Brain Res.*, **297**, 247–264.

Wessendorf, M. W., and Anderson, E. G. (1983) Single unit studies of identified bulbospinal serotonergic units, *Brain Res.*, **279**, 93–103.

Wessendorf, M. W., Proudfit, H. K., and Anderson, E. G. (1981) The identification of serotonergic neurons in the nucleus raphe magnus by conduction velocity, *Brain Res.*, **214**, 168–173.

West, D. C., and Wolstencroft, J. H. (1977) Electrophysiological identification of raphe-spinal neurones in the cat, *J. Physiol. (Lond.)*, **265**, 29P–30P.

Willcockson, W. S., Gerhart, K. D., Cargill, C. L., and Willis, W. D. (1983) Effects of biogenic amines on raphe-spinal tract cells, *J. Pharmacol. Exp. Ther.*, **225**, 637–645.

Willer, J. C., Roby, A., Gerard, A., and Maulet, C. (1982) Electrophysiological evidence for a possible serotonergic involvement in some endogenous opiate activity in humans, *Eur. J. Pharmacol.*, **78**, 117–120.

Willis, W. D. (1984) The raphe-spinal system, in *Brain Stem Control of Spinal Cord Function* (Ed. C. D. Barnes), pp. 141–214, Academic Press, New York.

Willis, W. D., Gerhart, K. D., Willcockson, W. S., Yezierski, R. P., Wilcox, T. K., and Cargill, C. L. (1984) Primate raphe- and reticulospinal neurons: effects of stimulation in periaqueductal gray or VPL thalamic nucleus, *J. Neurophysiol.*, **51**, 467–480.

Wood, P. L. (1985) In vivo evaluation of mu, delta and kappa opioid receptor agonists on spinal 5-HT metabolism in the rat, *Neuropeptides*, **5**, 319–322.

Woolf, C. J., Mitchell, D., and Barrett, G. D. (1980) Antinociceptive effect of peripheral segmental electrical stimulation in the rat, *Pain*, **8**, 237–252.

Yaksh, T. L. (1979) Direct evidence that spinal serotonin and noradrenaline terminals mediate the spinal antinociceptive effects of morphine in the periaqueductal gray, *Brain Res.*, **160**, 180–185.

Yaksh, T. L., and Tyce, G. M. (1979) Microinjection of morphine into the periaqueductal gray evokes the release of serotonin from spinal cord, *Brain Res.*, **171**, 176–181.

Yaksh, T. L., and Wilson, P. R. (1979) Spinal serotonin terminal system mediates antinociception, *J. Pharmacol. Exp. Ther.*, **208**, 446–453.

Yaksh, T. L., Du Chateau, J. C., and Rudy, T. A. (1976) Antagonism by methysergide and cinanserin of the antinociceptive action of morphine administered into the periaqueductal gray, *Brain Res.*, **104**, 367–372.

Yaksh, T. L., Plant, R. L., and Rudy, T. A. (1977) Studies on the antagonism by raphe lesions of the antinociceptive action of systemic morphine, *Eur. J. Pharmacol.*, **41**, 399–408.

Yarbrough, G. G., Buxbaum, D. M., and Sanders-Bush, E. (1971) Increased serotonin turnover in the acutely morphine treated rat, *Life Sci.*, **10**, 977–983.

Yarbrough, G. G., Buxbaum, D. M., and Sanders-Busch, E. (1972) Increased serotonin turnover in acutely morphine-treated mice, *Biochem. Pharmacol.*, **21**, 2667–2669.

Yarbrough, G. G., Buxbaum, D. M., and Sanders-Bush, E. (1973) Biogenic amines and narcotic effects. II. Serotonin turnover in the rat after acute and chronic morphine administration. *J. Pharmacol. Exp. Ther.*, **185**, 328–335.

Yezierski, R. P., Wilcox, T. K., and Willis, W. D. (1982) The effects of serotonin antagonists on the inhibition of primate spinothalamic tract cells produced by stimulation in nucleus raphe magnus or periaqueductal gray, *J. Pharmacol. Exp. Ther.*, **220**, 266–277.

Yoburn, B. C., Morales, R., Kelly, D. D., and Inturrisi, C. E. (1984) Constraints on the tail-flick assay: morphine analgesia and tolerance are dependent upon locus of tail stimulation, *Life Sci.*, **34**, 1755–1762.

York, J. L., and Maynert, E. W. (1978) Alterations in morphine analgesia produced by chronic deficits of brain catecholamines or serotonin: role of analgesimetric procedure, *Psychopharmacology*, **56**, 119–125.

Yunger, L. M., and Harvey, J. A. (1973) Effect of lesions in the medial forebrain bundle on three measures of pain sensitivity and noise-elicited startle, *J. Comp. Physiol. Psychol.*, **83**, 173–183.

Yunger, L. M., and Harvey, J. A. (1976) Behavioral effects of L-5-hydroxytryptophan after destruction of ascending serotonergic pathways in the rat: the role of catecholaminergic neurons, *J. Pharmacol. Exp. Ther.*, **196**, 307–315.

Zemlan, F. P., Corrigan, S. A., and Pfaff, D. W. (1980) Noradrenergic and serotonergic mediation of spinal analgesia mechanisms, *Eur. J. Pharmacol.*, **61**, 111–124.

Zemlan, F. P., Kow, L. M., and Pfaff, D. W. (1983) Spinal serotonin (5-HT) receptor subtypes and nociception, *J. Pharmacol. Exp. Ther.*, **226**, 477–485.

Zimmermann, M. (1986) Behavioural investigations of pain in animals, in *Assessing Pain in Farm Animals* (Eds I. J. H. Duncan and V. Molony), pp. 16–29, Commission of the European Communities, Luxembourg.

Neuronal Serotonin
Edited by N. N. Osborne and M. Hamon
© 1988 John Wiley & Sons Ltd

CHAPTER 8

Serotonin in Neurodegenerative Disorders

A. J. CROSS
Astra Neuroscience Research Unit
1 Wakefield Street
London WC1N 1PJ
England

INTRODUCTION

The function of the serotonergic system in the human brain has until recent years been extremely difficult to decipher. This is due to several factors, but mainly to the lack of selective serotonin receptor agonists and antagonists. Those drugs with an apparently selective action on the serotonergic system often have undesirable side-effects, and in addition as serotonergic neurones are distributed diffusely throughout many regions of the brain this adds a further complicating factor to the study of the functional effects of such drugs. Thus in comparison with, for instance, the anatomically more limited dopaminergic system (where selective drugs are available), our knowledge of the serotonergic system lags far behind.

Studies of serotonin in human neurodegenerative disorders have provided much information on the anatomy and function of serotonin-containing neurones in the human brain. It seems to be particularly the case for serotonin that marked species differences occur in the distribution and function of both serotonin neurones and receptors, and therefore care must be taken in extrapolating data from experimental animals to man. This has become even more relevant with the development of novel selective serotonergic agents, for which therapeutic applications are currently being explored in several neuropsychiatric disorders. It is thus essential that a thorough understanding be obtained of the serotonergic system in the human brain.

In the present chapter, the distribution of serotonin neurones and receptors in the normal human brain will be described. Changes in these systems will then be discussed in three main areas where the study of neurodegenerative diseases is particularly pertinent. Thus the study of Huntington's disease and Parkinson's disease has provided useful information on the role of serotonin in the basal ganglia; studies of dementia have implicated serotonin in higher cognitive functions; and finally the study of serotonin in several instances of encephalitis may have broader implications for the role of monoamine-containing neurones in CNS disease. Other illnesses in which serotonin has been implicated, such as affective disorders, psychoses and sleep disorders, are discussed elsewhere.

SEROTONIN IN THE NORMAL HUMAN BRAIN

The distribution of serotonergic neurones within the human brain has not been studied directly by histochemical techniques. In experimental animals immunohistochemical staining has been extremely useful in mapping studies (Chapter 2 of this volume); however, this technique cannot be applied to the human brain as serotonin diffuses rapidly out of neurones immediately after death. The distribution of serotonin and its metabolite 5-hydroxyindole-acetic acid (5-HIAA) have been extensively studied in the human brain, and in general the distribution is similar to that in the rat brain (Fig. 1). The distribution and properties of serotonin receptors have also been studied in the human brain, and in this instance a number of differences to the rat brain are apparent. The studies of Hoyer *et al.* (1986a, b) have demonstrated that the 5HT-2 receptor has a different pharmacological specificity in human brain compared to rat brain, particularly with respect to the ergot derivative mesulergine. In addition, the sub-types of the 5HT-1 receptor differ between rat and human brain. Interestingly, ^{3}H-mesulergine, which predominantly labels the 5HT-2 receptor in the rat cortex, appears to label the 5HT-1C receptor in the human cortex. Moreover, the 5HT-1B receptor, as defined in ligand binding studies using either ^{3}H-serotonin or ^{125}I-iodocyanopindolol, cannot be detected in human brain (Hoyer *et al.*, 1986a; Cross *et al.*, unpublished observations; see also Middlemiss *et al.*, 1986). As the 5HT-1B

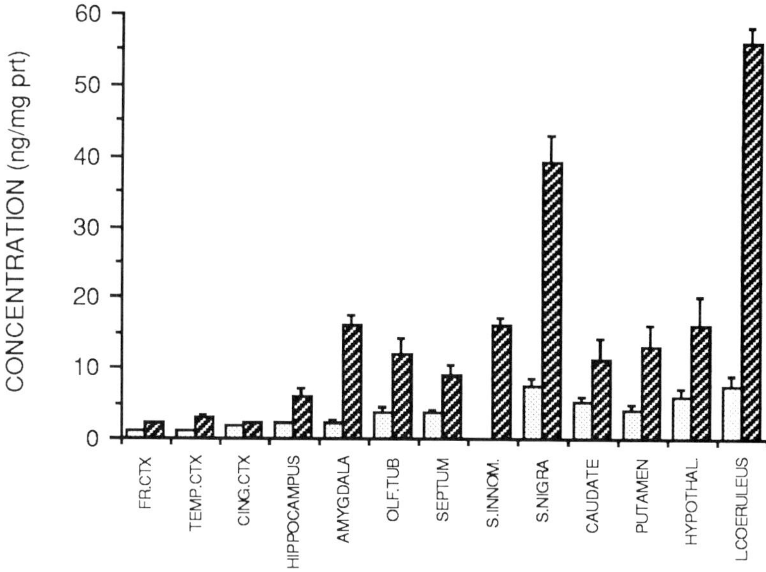

Fig. 1 Distribution of serotonin (light bars) and 5HIAA (hatched bars) concentrations in human brain. Values are the mean ±SEM of 6–12 samples

receptor probably functions as the presynaptic autoreceptor on serotonergic terminals in rat cortex, it is of some importance to determine the nature of the autoreceptor in human cortex.

In addition to these pharmacological studies, species differences also occur in the distribution of serotonin receptors, and particularly the 5HT-1A receptor (Fig. 2). Thus in human brain, ^{3}H-8-OH-DPAT binding to the 5HT-1A receptor is dense in the outer layers of the cerebral cortex whereas this is not the case in the rat (see also Hoyer et al., 1986a). As in the rat, ^{3}H-8-OH-DPAT binding is highest in the hippocampus; however, in man the CA-1 region contains the densest labelling whereas in the rat the dentate gyrus and CA-3 regions predominate. In human brain the basal forebrain (containing the ascending cholinergic neurones) is low in 5HT-1 receptors, in direct contrast to other species including monkeys (Stuart et al., 1986). In other areas 5HT-1A receptor distribution is similar in rat and man, most notably the negligible levels of ^{3}H-8-OH-DPAT binding in caudate/putamen and substantia nigra. It is also of interest that ^{3}H-serotonin labels binding sites of high density in the human globus pallidus and substantia nigra. These ^{3}H-serotonin binding sites do not have the characteristics of the 5HT-1B site, and these brain regions contain low levels of 5HT-1A, 5HT-1C and 5HT-2 receptors. In view of the high concentrations of serotonin in these regions, and the functional effects of serotonin in the basal ganglia, it will be important to determine the nature of these binding sites (see Chapter 15 of this volume).

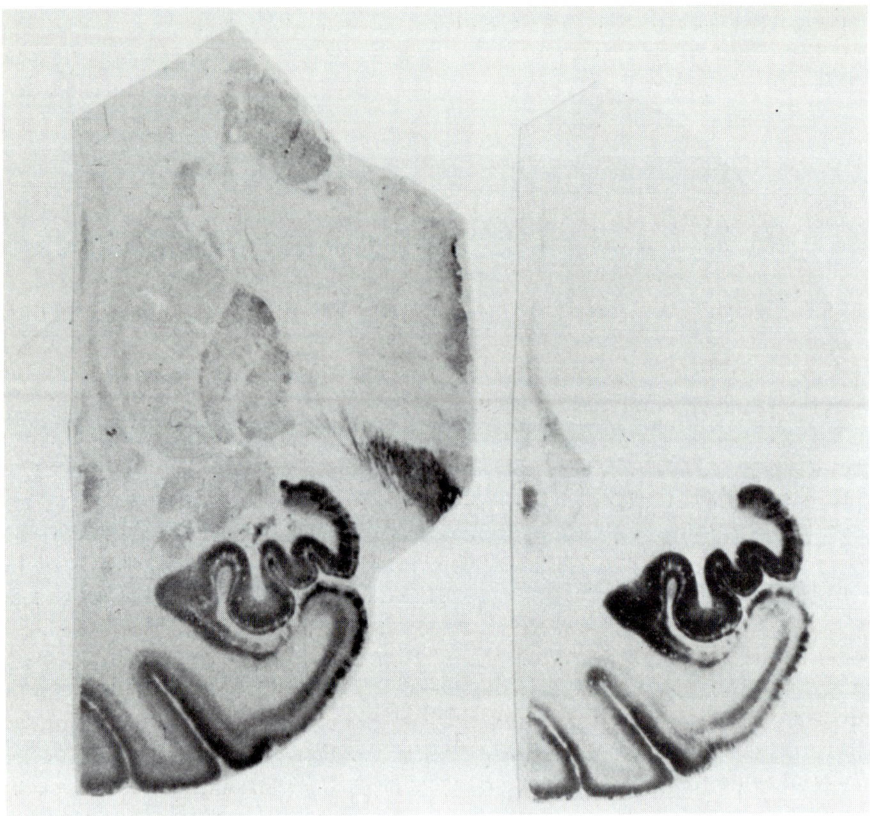

Fig. 2 Distribution of 5HT-1 receptors in human brain. The autoradiograph on the left shows ³H-serotonin binding in a section of human brain at the level of the anterior hippocampus. High densities of binding sites are present in hippocampus, globus pallidus and substantia nigra. The autoradiograph on the right shows ³H-8-OH-DPAT binding to an equivalent section. Note high binding in outer cortical layers and hippocampus and also the absence of binding in basal ganglia

SEROTONIN IN MOVEMENT DISORDERS

The serotonergic innervation arising from the dorsal raphe nucleus constitutes a major ascending projection to the striatum from the brainstem. The largest ascending innervation of the striatum is the nigro-striatal dopamine-containing pathway. As ascending serotonin neurones also innervate the dopamine-containing cells of the substantia nigra it would seem likely that serotonin might play an important role in the functioning of these brain regions, and hence in the control of motoricity. In experimental animals the involvement of serotonin in basal ganglia function has received relatively little attention in comparison with other neurotransmitters. There is, however, some evidence that serotonin is involved in some human movement disorders, although the clinical relevance of such changes is not always clear.

Parkinson's disease

Parkinson's disease is characterized by a functional imbalance in the extra-pyramidal motor system. The characteristic clinical features of the parkinsonian syndrome include akinesia, rigidity and tremor. Pathological studies have focused on losses of the pigmented dopaminergic cells of the substantia nigra, and post-mortem neurochemical studies have shown a marked decrease in dopamine concentrations in both the striatum and substantia nigra. In addition to these well-documented changes, decreased serotonin concentrations have been observed in almost all the basal ganglia regions studied by Birkmayer and Riederer (1978) and by Scatton *et al.* (1983). This has been interpreted as a loss of serotonergic terminals in the basal ganglia, and is confirmed by the marked loss of ^{3}H-imipramine binding (Raisman *et al.*, 1986) in similar brain regions (Table 1). Neuropathological studies of the brainstem in Parkinson's disease suggest that the loss of serotonergic terminals in the basal ganglia may be related to a loss of serotonergic neurones from the raphe nucleus. The widespread reductions in serotonin concentrations in the basal ganglia are not paralleled by widespread decreases in 5-HIAA concentrations. Thus 5-HIAA concentrations may be reduced in the striatum but not throughout the basal ganglia (Birkmayer and Riederer, 1978; Scatton *et al.*, 1983). The ratio of 5-HIAA/serotonin is markedly increased in caudate nucleus and putamen, possibly reflecting an increase in serotonin turnover in the remaining serotonergic terminals. It has been suggested that L-dihydroxyphenylalanine (L-DOPA), used in the treatment of Parkinson's disease, may be taken up into serotonergic neurones. Once in the neurones, DOPA is decarboxylated to dopamine and displaces serotonin from its stores, thus resulting in at least part of the depletion of brain serotonin. Subsequent studies have found no differences in serotonin concentrations in those patients whose L-DOPA treatment was discontinued

Table 1 Presynaptic serotonergic markers in Parkinson's disease putamen

	Control	Parkinson's disease
Serotonin	876 ± 80	215 ± 35
5-HIAA	1320 ± 106	618 ± 113
^{3}H-Imipramine binding	45.6 ± 3.4	34.4 ± 3.4

Values are expressed as mean ±SEM. Serotonin and 5-HIAA concentrations as ng/g tissue, ^{3}H-imipramine binding as fmol/mg protein. (Reproduced from Raisman *et al.*, 1986)

compared with those treated continuously. Moreover, if L-DOPA were taken up into serotonergic neurones, it is likely that these neurones would be the site of at least part of the decarboxylation of DOPA to produce dopamine. Studies in experimental animals suggest that this is unlikely (see Rinne *et al.*, 1974).

The relationship between changes in serotonin concentrations in basal ganglia and the appearance of motor abnormalities in Parkinson's disease is not fully understood. Administration of 5-hydroxytryptophan (5-HTP), *p*-chlorophenylalanine (PCPA), tryptophan, or the antagonist methysergide to parkinsonian patients does not generally have any effect on the motor symptoms of the disease. Moreover, in a patient with hemiparkinsonism, with a highly lateralized dopamine deficiency, serotonin concentrations were similar in both caudate nuclei (see Birkmayer and Riederer, 1978, for references). It would thus appear unlikely that changes in serotonin concentrations in the caudate nucleus are directly involved in the motor abnormalities of Parkinson's disease. This conclusion is supported by observations on MPTP-induced parkinsonism in monkeys. Monkeys treated with MPTP demonstrate most of the clinical features of Parkinson's disease, yet in these animals there may be no degeneration of basal ganglia serotonergic terminals. Furthermore, a practically complete recovery of movement control can be achieved after L-DOPA administration.

The depletion of serotonin in Parkinson's disease is not confined to the basal ganglia. Post mortem studies show that serotonin is reduced in some cortical regions and the hippocampus. Once again, the decrease in 5-HIAA is either less pronounced or not present at all. A differential loss of serotonin has also been observed in the separate raphe nuclei and it remains possible that these differential losses may relate to changes in those cells innervating the basal ganglia and some cortical areas, i.e. the dorsal and medial raphe nuclei.

Parkinson's disease is frequently associated with depressive episodes and, in a proportion of patients, with dementia. It is possible that changes in the serotonergic system may underlie some of these features of the disease. Post mortem studies suggest that serotonin and 5-HIAA concentrations may be lower in those parkinsonian patients with depression (Fahn *et al.*, 1971). However, it is not clear whether this reflects an increased severity of the disease in patients with depression, as dopamine and noradrenaline concentrations are also reduced to a greater extent in these patients.

Serotonin receptors have been studied less extensively in Parkinson's disease. ³H-serotonin binding (to 5HT-1 receptors) is reduced in the putamen but not in the caudate nucleus or globus pallidus (Reisine *et al.*, 1977). As reductions in ³H-serotonin binding were significantly correlated with reductions in choline acetyltransferase activity, it has been suggested that 5HT-1 receptors in this brain region may be present on cholinergic interneurones. Reductions in ^{3}H-serotonin binding have also been observed in the frontal cortex, whereas 5HT-2 receptors labelled with either ^{3}H-spiperone or ^{3}H-ketanserin are not reduced. The reduction of 5HT-1 receptors in Parkinson's disease patients is only observed in those patients with concomitant dementia and cortical cholinergic deficit (Table 2). Once again this suggests an association between cholinergic neurones and 5HT-1 receptors, and in experimental animals lesions of forebrain cholinergic neurones result in losses of cortical ^{3}H-serotonin binding (Table 2). It would thus seem likely that a proportion of cortical 5HT-1 receptors are present on cholinergic terminals.

Table 2 ^{3}H-Serotonin binding and ChAT activity in the ibotenate-lesioned rat and in Parkinson patients

	^{3}H-Serotonin binding (fmol/mg protein)	ChAT activity (nmol/g/h)
Control	41.8 ± 6.4	27.9 ± 8.3
NbM lesioned rat	33.4 ± 8.7*	12.4 ± 4.9**
Controls	153 ± 16	11.8 ± 4.8
Parkinson's disease (non-demented)	144 ± 22	12.9[a]
Parkinson's disease (demented)	114 ± 20*	3.9 ± 0.4**

Data from lesioned rats is for frontal cortex, data for Parkinson's disease is from parietal cortex.
NbM = nucleus basalis of Meynert.
*$p < 0.05$, **$p < 0.01$, a-mean of two samples.
(Data from Cross and Deakin, 1985; Perry *et al.*, 1984)

Huntington's disease

Huntington's disease is a genetically determined neurodegenerative disease associated with abnormal involuntary movements, psychosis and dementia. The major neuropathological abnormality in Huntington's disease is atrophy of the basal ganglia and frontal cortex. In particular the small and medium-sized neurones of the caudate nucleus and putamen degenerate, and this is reflected by marked losses of gamma-aminobutyric acid (GABA) and some neuropeptides (particularly met-enkephalin and substance P). The initial studies of Bernheimer and Hornykiewicz (1973) on four patients with Huntington's disease suggested that serotonin concentrations were either unchanged or marginally increased. More recent studies (Reynolds *et al.*, unpublished observations) suggest that the increase in serotonin concentrations may be more widespread within the basal ganglia. The binding of ^{3}H-paroxetine to the serotonin uptake site is also increased in the basal ganglia in Huntington's disease (Table 3). These changes may merely reflect the sparing of serotonergic neurones in Huntington's disease, the increases

Table 3   Presynaptic serotonergic markers in
Huntington's disease putamen

	Control	Huntington's disease
Serotonin (ng/g tissue)	337 ± 20	608 ± 47**
5-HIAA (ng/g tissue)	811 ± 70	1826 ± 47**
^{3}H-Paroxetine binding (fmol/mg protein)	108 ± 11	178 ± 10*

$^*p < 0.05$, $^{**}p < 0.01$.
(Data from Cross *et al.*, 1986b; Reynolds *et al.*, unpublished)

being the result of tissue shrinkage. It has been reported that the administration of 5-HTP to Huntington's disease patients produces a worsening of symptoms, whereas PCPA and serotonin receptor antagonists have little effect. Overall these findings suggest that increased basal ganglia serotonin concentrations may be involved in the production of abnormal movements in Huntington's disease. However, as serotonin antagonism is ineffective in relieving the movement disorder it is unlikely that serotonergic changes are of central importance.

Several reports suggest that serotonin receptors may be changed in the basal ganglia in Huntington's disease. 5HT-1 receptors labelled with ^{3}H-serotonin or ^{3}H-LSD are extensively reduced, probably reflecting the degeneration of intrinsic striatal neurones (Enna *et al.*, 1976). In contrast to Parkinson's disease, the loss of ^{3}H-serotonin binding in the putamen of Huntington's disease patients did not correlate with the loss of ChAT activity. Unlike 5HT-1 receptors, the low amounts of 5HT-2 receptors present in human basal ganglia are not reduced in Huntington's disease (Fig. 3). This suggests that 5HT-2 receptors may be located either on terminals of striatal afferents or on those neurones which are not susceptible to the degenerative process in Huntington's disease.

Serotonin concentrations and serotonin receptors are generally found to be unchanged in brain regions other than the basal ganglia. The exception to this is the hippocampus, where a loss of 5HT-1 receptors has been observed (Fig. 3). In experimental animals, cortical 5HT-1 receptors may be located in part on cholinergic terminals. As a loss of ChAT activity is also observed in the Huntington's disease hippocampus, it is possible that a similar relationship between 5HT-1 receptors and cholinergic terminals exists in man.

The biochemical changes occurring in the basal ganglia in Huntington's disease can on the whole be reproduced in experimental animals by the intrastriatal injection of excitotoxins such as kainic acid. In rats, striatal

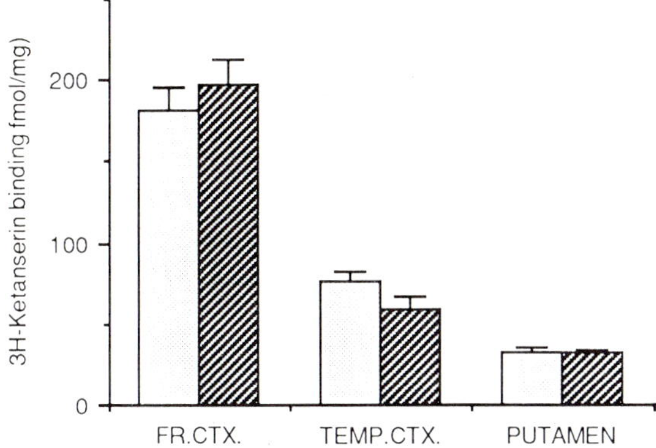

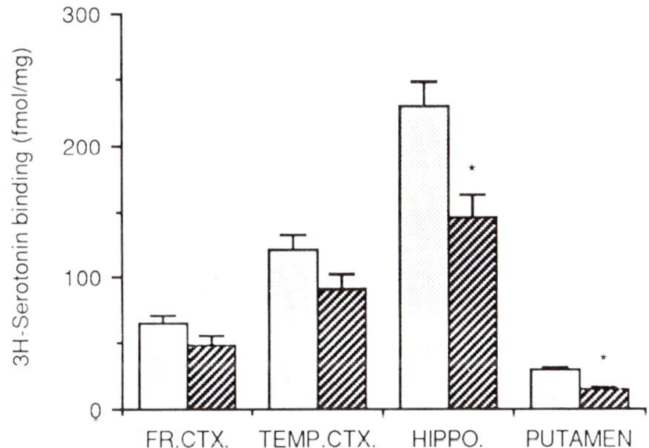

Fig. 3 Serotonin receptors in Huntington's disease. Upper: ^{3}H-ketanserin binding to 5HT-2 receptors; lower: ^{3}H-serotonin binding to 5HT-1 receptors. Light bars represent controls, hatched bars are Huntington's disease patients. A significant loss of ^{3}H-serotonin binding is observed in putamen and hippocampus of the Huntington's disease patients ($*p < 0.05$, data from Cross *et al.*, 1986b)

kainic acid lesions produce marked losses of 5HT-1 receptors with little change in 5HT-2 receptors (Schwarcz *et al.*, 1977). The concentrations of serotonin are increased in the lesioned rat striatum and serotonin turnover rises markedly after intrastriatal kainic acid injection. The studies of Berger *et al.* (1982) have shown that prior destruction of striatal serotonergic neurones can protect against the neurotoxic effects of kainic acid. Accordingly, it has been suggested that the increased serotonin concentrations following both the intrastriatal kainic acid injection and in the Huntington's disease basal ganglia may contribute to the degenerative process.

Serotonergic systems have not beeen studied in great detail in relation to other movement disorders, although abnormalities in the raphe nucleus have been noted in progressive supranuclear palsy. In conclusion, the studies to date suggest that changes in serotonergic systems in the basal ganglia do not have marked effects on motor function. This conclusion should be considered from the view that in Parkinson's disease the serotonergic deficit occurs in parallel with a large dopaminergic deficit; it may be the case that a specific serotonergic deficit in human basal ganglia could produce motor abnormalities which are 'masked' in Parkinson's disease. It also seems likely that the serotonergic deficit in Parkinson's disease may relate to some of the other features of the disease, including depression.

SEROTONIN IN DEMENTIA

Serotonin-containing nerve terminals are densely distributed throughout the cerebral cortex and related structures. The serotonergic innervation of the hippocampus and amygdala is particularly rich, and is accompanied by a dense and discrete distribution of serotonin receptor subtypes. It is therefore not surprising that recent studies suggest that serotonin may play an important role in learning and memory processes (Ogren, 1982), and that forebrain serotonergic systems may be involved in several dementing syndromes.

Dementia of the Alzheimer type

An involvement of serotonergic systems in Alzheimer type dementia (ATD) was initially suggested by the pathological studies of Ishii (1966) and Hirano and Zimmermann (1962), who described the presence of neurofibrillary tangles in the large neurones of the raphe nucleus of patients with ATD. Subsequent neuropathological studies confirmed these findings and suggested that some of the large neurones of the raphe nucleus may be lost in ATD. In addition, the histological studies of Mann *et al.* (1985) showed that the functional capacity of the remaining raphe neurones may be impaired.

Neurochemical studies on the brains from patients with ATD have demonstrated a reduction in the concentrations of serotonin. Some (but not all) studies also described reductions in the metabolite 5-HIAA, although these are variable in terms of both the extent of the reduction and the brain regions in which the reductions occur (see Palmer *et al.*, 1986, for references). The concentrations of serotonin and 5-HIAA are also reduced in biopsy samples of the temporal and frontal cortex in ATD, the reductions being greatest in the temporal cortex. As in Parkinson's disease, the ratio of 5-HIAA to serotonin may be increased, suggesting (although not conclusively) that serotonin turnover and release may increase in the surviving terminals in ATD. Studies of serotonergic parameters in cerebrospinal fluid (CSF) also point to a serotonergic deficit in ATD. Thus the concentrations of serotonin and 5-HIAA have been found to be reduced in ATD CSF in the majority of studies. It should however be noted that some evidence exists for a deficit in the clearance of 5-HIAA from CSF, and these results should be viewed with caution. Other indices of serotonergic neurones have been less extensively studied. High affinity serotonin uptake into cortical biopsies is markedly reduced in ATD, and the binding of ^{3}H-imipramine to the serotonergic uptake system is also reduced in the frontal cortex. The potassium-stimulated release of serotonin from tissue prisms from cortical biopsies is also markedly reduced in ATD (Bowen *et al.*, 1983). Taken together, this neurochemical evidence strongly favours a loss of serotonergic terminals in the cortex in ATD.

The regional distribution of the changes in serotonergic markers has not been studied fully. However, it is clear that changes do not occur evenly throughout the brain. In our own studies, reductions in 5-HIAA concentrations reached statistical significance only in the frontal and temporal cortex and the hippocampus (Fig. 4). Adolfsson *et al.* (1979) found 5-HIAA concentrations significantly reduced in the cingulate cortex and hippocampus. In a report by Gottfries *et al.* (1983), 5-HIAA concentrations in the caudate

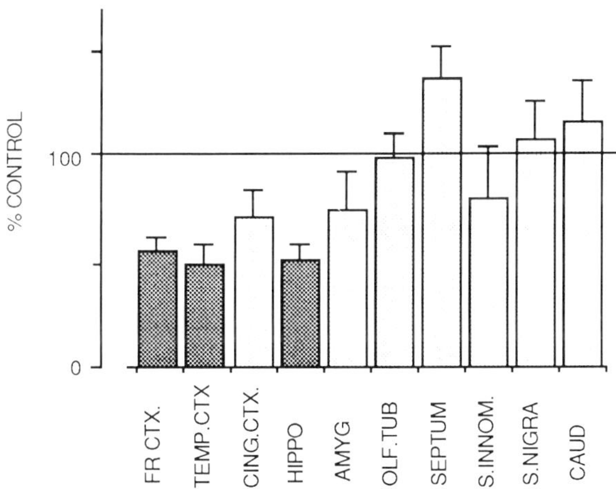

Fig. 4 Distribution of 5HIAA concentrations in ATD brain. Values are mean ±SEM of ATD patients expressed as a percentage of a control group. Hatched bars represent significant losses in the ATD group ($p < 0.05$). (Data from Cross *et al.*, 1984)

nucleus of ATD patients were not significantly different from those of controls. These findings that 5-HIAA concentrations are not reduced to a similar extent in all areas studied, and particularly within the cerebral cortex, may be of some relevance to the pathogenesis of the disease. Histological studies have demonstrated marked neurofibrillary tangle formation in the dorsal raphe nucleus. The serotonergic cell bodies in this nucleus project diffusely to the cerebral cortex and other forebrain regions. Recent neuropathological studies have described metabolic abnormalities in dorsal raphe nucleus (Mann *et al.*, 1985) comparable to those observed in other areas of pathological change. If loss of cell bodies or metabolic abnormality in the dorsal raphe nucleus were the primary change in serotonin neurones in ATD, the loss of presynaptic serotonergic markers in ATD might be expected to be distributed evenly throughout the projection areas of this serotonergic system. The suggestion that abnormalities in serotonergic terminals are highly localized to some neocortical regions and the hippocampus are consistent with the primary site of damage in serotonin neurones being at the terminal fields in the temporal cortex and hippocampus. If this is the case, the abnormalities in the raphe nucleus could be secondary to these changes. Moreover, it has been shown that changes in the neurones of the dorsal raphe nucleus in ATD correlate with the severity of pathological change in the neocortex (Mann *et al.*, 1985). Pathological studies of brainstem nuclei following cortical lesions or leucotomy are also consistent with this proposal. Thus in man large lesions of the neocortex induce retrograde degenerative changes in the nucleus basalis of Meynert, locus coeruleus and dorsal raphe nucleus (Pearson *et al.*, 1985). However, it is noteworthy that in other areas in which there is consistent and marked pathological change in ATD, such as the amygdala and septum, 5-HIAA concentrations are comparable with controls. These regions may, however, show reductions in serotonin concentration.

The losses of presynaptic serotonergic markers in ATD cortex do not correlate with the density of senile plaques (Cross *et al.*, 1983). In a recent study 5-HIAA concentrations were correlated with the density of neurofibrillary tangles and this may be interpreted as reflecting a primary cortical change associated with a serotonergic denervation (Palmer *et al.*, 1986).

Changes in postsynaptic serotonin receptors have also been described in ATD. Our initial studies using ^{3}H-LSD as ligand demonstrated a large loss of serotonin receptors in neocortex and hippocampus (Cross *et al.*, 1984a). These findings were similar to those of Bowen *et al.* (1983) who also used ^{3}H-LSD as ligand. It is apparent that ^{3}H-LSD labels at least two types of serotonin receptors in rat and human brain, and we have subsequently used ligands which are more selective for serotonin receptor subtypes. The binding of ^{3}H-serotonin to the 5HT-1 receptor is significantly reduced in ATD brains (Table 4). Further studies have suggested that the loss of ^{3}H-serotonin binding in the cortex is predominantly due to the loss of the 5HT-1A binding

Table 4 ^{3}H-Serotonin binding to 5HT-1 receptors in ATD patients

Brain region	Control	ATD
Temporal cortex	51 ± 5	27 ± 4**
Frontal cortex	136 ± 9	116 ± 13
Cingulate cortex	103 ± 14	75 ± 15
Hippocampus	54 ± 8	32 ± 6*
Amygdala	24 ± 2	11 ± 3**
Olf. tubercle	36 ± 4	29 ± 4
S. nigra	39 ± 4	23 ± 7
S. innominata	40 ± 3	22 ± 6
Septum	27 ± 4	16 ± 5
Caudate nucleus	78 ± 9	81 ± 10

Values are expressed as fmol ligand bound/mg protein, mean ±SEM.
$*p < 0.05$, $**p < 0.01$.
(Data from Cross *et al.*, 1984)

site labelled by ^{3}H-8-OH-DPAT (Middlemiss *et al.*, 1986). This is not entirely surprising as the majority of the ^{3}H-serotonin binding sites in the human cortex are of the 5HT-1A subtype. 5HT-2 receptors, labelled with ^{3}H-ketanserin, are also reduced in ATD (Table 5). The loss of 5HT-2 receptors is limited to some cortical (particularly temporal) areas and the amygdala. The loss of 5HT-2 receptors in the temporal lobe can be very large; in our studies ^{3}H-ketanserin binding was reduced to 36% of control values in the temporal cortex and a similar reduction in the entorhinal cortex has been observed in a separate study (Perry *et al.*, 1984).

In view of the marked reduction of 5HT-2 receptors in the temporal cortex,

Table 5 ^{3}H-Ketanserin binding to 5HT-2 receptors in ATD

Brain region	Control	ATD
Temporal cortex	56 ± 6	20 ± 3**
Frontal cortex	90 ± 10	49 ± 6**
Cingulate cortex	58 ± 6	30 ± 3**
Hippocampus	17 ± 2	12 ± 2
Amygdala	21 ± 2	12 ± 2**
Olf. tubercle	22 ± 3	18 ± 3
S. nigra	14 ± 1	14 ± 3
S. innominata	30 ± 8	17 ± 2
Septum	23 ± 4	12 ± 3

Values are expressed as fmol ligand bound/mg protein, mean ±SEM.
$**p < 0.01$.
(Data from Cross *et al.*, 1984)

the specificity of the receptor changes has been studied in this brain region. Of thirteen ligand binding sites studied (Fig. 5), only two were found to be significantly reduced in ATD, and of these the 5HT-2 receptor was the only monoamine receptor to be reduced. Among other binding sites, only benzodiazepine receptors were reduced, to 84% of control values. Other studies have demonstrated losses of somatostatin and corticotropin–releasing factor receptors in ATD.

A significant age-effect on 5HT-1 receptors has been noted in ATD. In our study, when the ATD group is divided on the basis of younger or older than 75 years, the significant decrease in 5HT-1 receptors is only apparent in the younger group (Fig. 6). Both younger and older ATD patients demonstrate a significant reduction in 5HT-2 receptors compared with controls. As it has been suggested that younger subjects may die with a more severe form of illness, these results suggest that the loss of 5HT-2 receptors may occur at a relatively early stage of the disease and precede the loss of 5HT-1 receptors and markers of serotonergic terminals.

Reductions in serotonin receptors in ATD do not correlate with the cholinergic deficit. In the temporal and frontal cortex, the binding of ^{3}H-ketanserin to 5HT-2 receptors was significantly correlated with somatostatin reductions (Fig. 7), but this relationship was not significant in other regions where 5HT-2 receptors were reduced in ATD. Reductions in 5HT-2 receptors did, however, correlate with GABA concentrations in the cingulate cortex of the ATD group. In experimental animals ^{3}H-ketanserin binding to 5HT-2 receptors is considered to be present on intrinsic cortical neurones. Similarly, somatostatin and GABA are thought to be present on intrinsic cortical

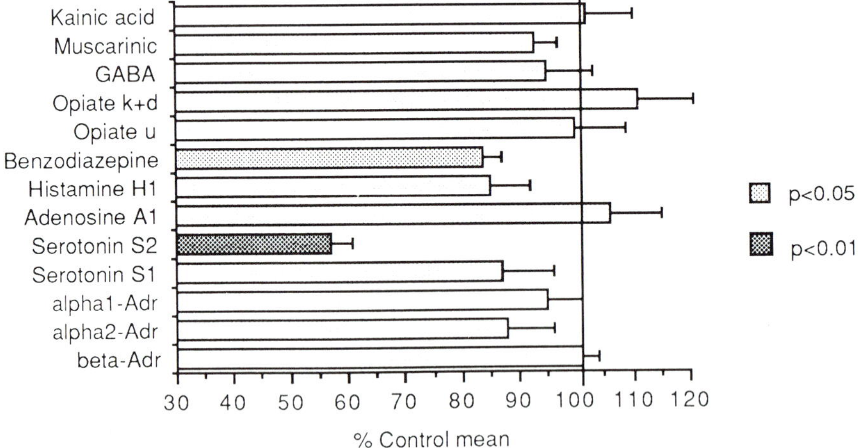

Fig. 5 Neurotransmitter receptor binding sites in ATD temporal cortex. Values are expressed as mean ±SEM of ATD patients as a percentage of a control group. (Data from Cross *et al.*, 1986a)

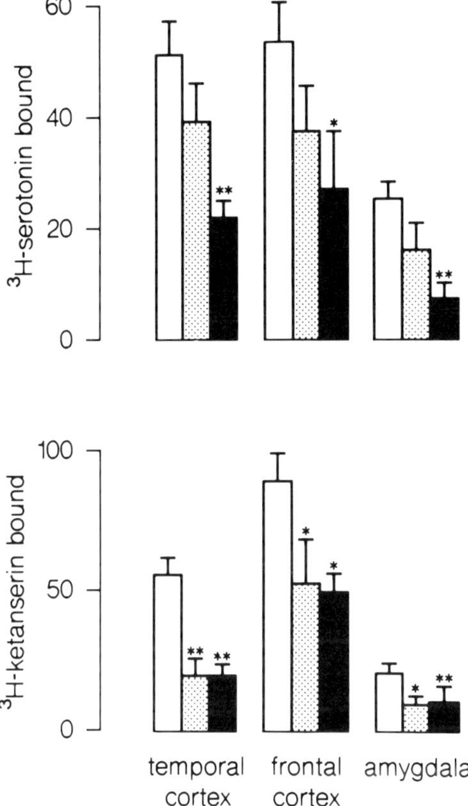

Fig. 6 Relationships between serotonin receptors and age in ATD brains. Control values are given as open columns. Shaded columns: ATD patients older than 75 y; solid columns: ATD patients younger than 75 y. Ligand binding values are fmol bound/mg protein, mean ±SEM. *$p < 0.05$; **$p < 0.01$. (Data from Cross *et al.*, 1984)

neurones in animals and man. Therefore it is possible that intrinsic cortical neurones which carry 5HT-2 receptors degenerate in ATD, and that these cells contain somatostatin in frontal and temporal cortex.

Within the group of ATD patients, no relationship is apparent between losses of 3H-LSD binding to the 5HT-2 receptor and the density of senile plaques (Fig. 8). However, a significant relationship between 3H-ketanserin binding to the 5HT-2 receptor and the density of neurofibrillary tangles has been observed in the entorhinal cortex (Perry *et al.*, 1984; Fig. 9). As previously discussed, it is unlikely that the progression of pathological

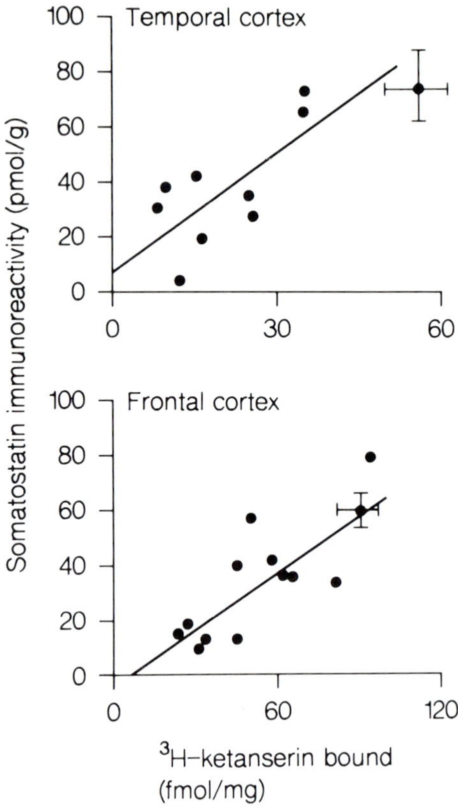

Fig. 7 Relationships between ^{3}H-ketanserin
binding and somatostatin-like immunoreactivity in temporal cortex and frontal cortex
of ATD patients. Mean control values are
included with bars representing SEM. (Data
from Cross *et al.*, 1984)

changes in the serotonergic system occurs primarily in the raphe nucleus and
spreads to the terminal fields. It would seem likely that changes in
serotonergic cell bodies in the raphe nucleus are secondary to pathological
changes in areas of terminal degeneration (i.e. neocortex and hippocampus).
It is clear that serotonin receptors, and particularly 5HT-2 receptors, are
selectively reduced in these areas. There is some evidence to suggest that
this change occurs relatively early in the disease, and it is possible to envisage
that changes in neurones expressing serotonin receptors in the neocortex and
hippocampus might lead to degeneration of serotonergic terminals and hence
to changes in the dorsal raphe nucleus. In view of the association between
5HT-2 receptors and somatostatin concentrations, and tangle-bearing

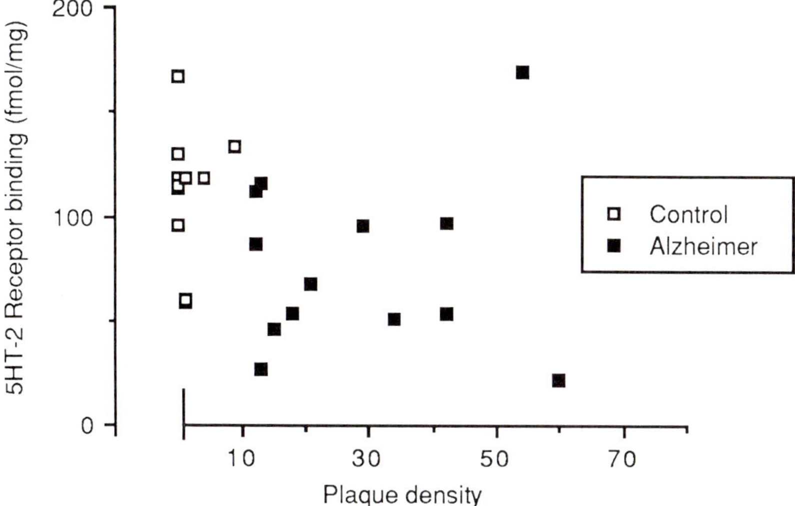

Fig. 8 Relationship between ³H-LSD binding to the 5HT-2 receptor and the density of senile plaques in control and ATD temporal cortex. No significant correlation is present. (Data from Cross *et al.*, 1983)

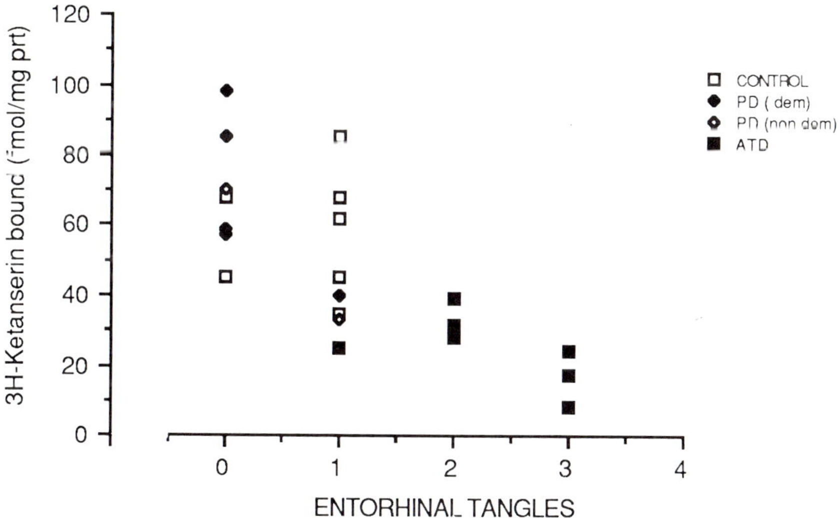

Fig. 9 Relationship between ³H-ketanserin binding to the 5HT-2 receptor and the density of neurofibrillary tangles in entorhinal cortex of control, Parkinson's disease (PD) patients and ATD patients. (Data from Perry *et al.*, 1984)

(pyramidal) neurones, it is tempting to speculate that 5HT-2 receptors are present on these neurones.

To date no association has been observed between changes in serotonergic neurones and the severity of the various symptoms of ATD. It is clear that serotonin is involved in learning and memory processes; however, no significant relationship between serotonergic changes and dementia ratings is apparent. It may well be that the serotonergic changes contribute to the dementia of ATD, but that the nature of this contribution is obscured by the cholinergic deficit. It is clear that ATD patients have a more severe and clinically distinct form of dementia to that of Parkinson's disease. As Parkinson's disease is associated with changes in presynaptic cholinergic and serotonergic markers, but not cortical 5HT-2 receptors, it is possible that the 5HT-2 receptor loss in ATD contributes to the clinical differences between the dementia of ATD and Parkinson's disease. ATD patients may also present with changes in affect, aggressive behaviour and sleep disturbances. However, the association between these features of the disease and serotonergic changes awaits detailed clinical and neurochemical analysis.

Down's syndrome

It is invariably the case that patients with Down's syndrome develop Alzheimer-like pathological changes at around 35 years of age. Pathological studies have demonstrated a loss of the large neurones of the dorsal raphe nucleus, and functional changes in the remaining neurones. Neurochemical studies have shown reductions in both serotonin and 5-HIAA concentrations in the neocortex and hippocampus, and also in basal ganglia (Yates *et al.*, 1976; Reynolds and Goodridge, 1985). The loss of serotonin in patients with Down's syndrome may be limited to older patients; in one 27-year-old patient with no Alzheimer-like pathology, brain serotonin concentrations were normal. Once again, this may be consistent with a primary pathological change in the forebrain leading to retrograde degeneration of serotonin neurones.

SEROTONERGIC SYSTEMS IN VIRAL ENCEPHALITIS

Experimentally induced virus infections of the central nervous system have been shown to result in an increase in the activity of monoamine-containing neurones. The serotonergic system appears to be particularly sensitive: mouse brain 5-HIAA concentrations are increased twofold following infection with herpes simplex virus (HSV), vaccinia, pseudorabies, encephalomyocarditis virus and influenza (Lycke and Roos, 1975). The increase in brain 5-HIAA concentrations probably reflects an increase in serotonin turnover. This change does not appear to be due to secondary factors associated with acute encephalitis as it does not occur in mice with an encephalitis resulting from infection with rabies or Semliki-forest virus. Moreover, the changes in brain 5-HIAA concentrations may be a result of infection of the serotonergic cells of the raphe nucleus. In mice infected by intraventricular injection of herpes simplex virus, the only cells in the brainstem which become infected and can be positively stained for virus antigen are those in the locus coeruleus and dorsal raphe nucleus (Neeley *et al.*, 1985) (Fig. 10). Following peripheral infection with HSV, the raphe nucleus is again infected and brain 5-HIAA concentrations are increased.

The serotonergic system has been studied in a few cases of human neuro-degenerative diseases of known viral origin. Abnormalities of serotonin metabolism have been noted in several cases of Creutzfeld–Jacob disease and some of the symptoms of the disease (e.g. myoclonus and some behavioural abnormalities) may reflect serotonergic dysfunction. Other monoamine systems appear to be equally affected.

Although very little is known about the mechanism of viral-induced increases in serotonin metabolism in experimental animals, the interest of these studies lies in the mechanism of the spread of the disease. A number of studies indicate that HSV infection of neurones involves the intra-axonal transport of the virus from nerve terminals to cell bodies. Following intraven-tricular infection, serotonergic neurones may be exposed to the virus at an early stage because of their proximity to ventricular spaces. Thus the selective infection of serotonin neurones may be due to transport of virus from fore-

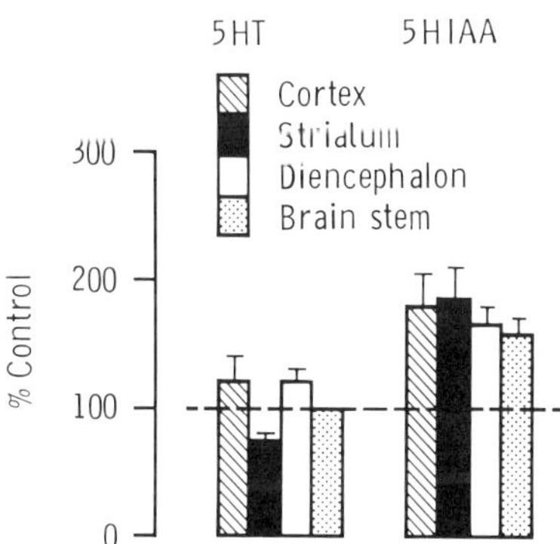

Fig. 10 Serotonin and 5HIAA concentrations in herpes simplex-infected mice. Values are expressed as a percentage of control means. 5HIAA concentrations are significantly increased in infected mice in all brain regions (*$p < 0.01$). (Data from Neeley *et al.*, 1985)

brain and ventricular terminals to brainstem cell bodies via retrograde transport mechanisms. That serotonergic cells are also infected following peripheral infection suggests that these cells might also have increased sensitivity to infection.

CONCLUSIONS AND FUTURE TRENDS

Studies of serotonergic systems in the human brain demonstrate a number of differences to the brains of experimental animals. In two regions of the brain, the basal ganglia and temporal lobe, serotonin receptors have a distinct distribution and both of these areas receive a large serotonergic innervation. Within the basal ganglia 5HT-1 receptors may be localized on the small or medium-sized neurones which degenerate in Huntington's disease. In contrast, 5HT-2 receptors are located on striatal cells (or nerve terminals) which do not degenerate. The characteristics of these basal ganglia 5HT-1 receptors have not been determined, although they are clearly distinct from 5HT-1A receptors labelled by ^{3}H-8-OH-DPAT. As 5HT-1 receptors are unchanged in the caudate nucleus and globus pallidus in Parkinson's disease, it is unlikely that they are located presynaptically on dopaminergic or serotonergic nerve terminals. Unfortunately, no motor abnormality has been identified in Huntington's disease or Parkinson's disease which correlates with changes in the striatal serotonergic system. The effects of new, selective serotonergic drugs may provide more information.

Studies on ATD have also shed light on the localization of serotonin receptors in the cortex and hippocampus. It is possible that 5HT-2 receptors may be located on somatostatin/GABA-containing neurones or glutamate-containing pyramidal neurones. In contrast, 5HT-1 receptors which are less affected in ATD may be located on different neurones. The reduction of 5HT-1 receptors in parkinsonian patients with dementia suggests that some of these receptors may be present on cholinergic terminals in cortex, and this is supported by studies in experimental animals. The relationship between serotonergic abnormalities and the clinical features of ATD have not been elucidated. However, it is interesting that Parkinson's disease and ATD are both associated with a widespread serotonergic deficit, and both are often associated with depression and changes in the sleep/wake cycle.

As serotonergic neurones degenerate in both Parkinson's disease and ATD, and are affected in some viral encephalopathies, they may be considered to be particularly sensitive to damage. It has been suggested that part of this susceptibility may be due to the structure of these neurones. As discussed by Rossor (1982), the serotonergic neurones of the raphe nucleus form part of the ascending reticular formation ('isodendritic core') which is affected in both ATD and Parkinson's disease. It is clear, however, that the degenerative process is different in these diseases. Thus in ATD serotonin neurones undergo retrograde degeneration in response to pathological changes in the forebrain, whereas in Parkinson's disease it is likely that anterograde degeneration occurs. Part of the susceptibility of serotonergic neurones may relate to their anatomy, as illustrated by studies of viral encephalitis. Thus because of their diffuse innervation of the forebrain, serotonin neurones can accumulate pathogens in their cell bodies in an apparently selective way. Whether or not transmissible agents are involved in ATD and Parkinson's disease, these studies suggest that partly due to their anatomy serotonergic neurones may be particularly sensitive to degenerative processes.

It is clear from the foregoing discussion that post-mortem neurochemical studies have been of central importance in describing the involvement of serotonergic systems in neurodegenerative disorders. It is equally clear, however, that studies of this kind are limited in their ability to provide detailed information on the consequences of these serotonergic changes. This may be due to the fact that in general, post-mortem studies are limited to examining the end-stage of the disease. Moreover, in both ATD and Parkinson's disease the serotonergic involvement accompanies other more dramatic neurochemical deficits, and any clinical correlates may be obscured. Many new selective serotonergic drugs are currently under development and it is likely that our understanding of the role of serotonin in the human brain will be enhanced by clinical studies with these drugs. Many differences in the neurochemistry and pharmacology of serotonin exist between man and other species. Therefore it seems essential that a thorough understanding be obtained of the serotonergic system in the normal human brain before such clinical studies are undertaken.

The involvement of serotonin in neurodegenerative disorders may be of relevance to new diagnostic approaches using techniques such as positron emission tomography. Thus it is possible that changes in 5HT-2 receptors (labelled with ^{11}C-methylspiperone for instance) could be determined in the living patient and possibly aid in the diagnosis of dementias. It may in addition be possible to visualize serotonin neurones with ^{11}C-5-HTP. It is in this area that neurochemical and pharmacological studies of serotonin may be most fruitful, and a greater understanding of the role of serotonin in neurodegenerative diseases may arise.

REFERENCES

Adolfsson, R., Gottfries, C. G., Roos, B. E., and Winblad, B. (1979) Changes in brain catecholamines in patients with dementia of Alzheimer type, *Br. J. Psychiat.*, **135**, 216–223.
Berger, M., Sperk, G., and Hornykiewicz, O. (1982) Serotonergic denervation partially protects rat striatum from kainic acid toxicity, *Nature*, **299**, 254–256.

Bernheimer, M., and Honrykiewicz, O. (1973) Brain amines in Huntington's chorea, *Adv. Neurol.*, **1**, 525–533.

Birkmayer, W., and Riederer, P. (1978) Serotonin and extra pyramidal disorders, in *Serotonin in Mental Abnormalities* (Ed. D. J. Boullin), pp. 273–302, John Wiley & Sons, Chichester.

Bowen, D. M., Allen, S. J., Benton, J. S., Goodhart, M. J., Haan, E. A., Palmer, A. M., Sims, N. R., Smith, C. C. T., Spillane, J. A., Esiri, M. M., Snowden, J. S., Wilcock, G. K., and Davison, A. N. (1983) Biochemical assessment of serotonergic and cholinergic dysfunction and cerebral atrophy in Alzheimer's disease, *J. Neurochem.*, **41**, 266–272.

Cross, A. J., and Deakin, J. F. W. (1985) Cortical serotonin receptor subtypes after lesion of ascending cholinergic neurones, *Neurosci. Lett.*, **60**, 261–265.

Cross, A. J., Crow, T. J., Johnson, J. A., Joseph, M. H., Perry, E. K., Perry, R. H., Blessed, G., and Tomlinson, B. E. (1983) Monoamine metabolism in senile dementia of Alzheimer type, *J. Neurol. Sci.*, **60**, 383–392.

Cross, A. J., Crow, J. J., Johnson, J. A., Perry, E. K., Perry, R. H., Blessed, G., and Tomlinson, B. E. (1984a) Studies on neurotransmitter receptor systems in neocortex and hippocampus in senile dementia of the Alzheimer type, *J. Neurol. Sci.*, **64**, 109–111.

Cross, A. J., Crow, T. J., Ferrier, I. N., Johnson, J. A,, Bloom, S. R., and Corsellis, J. A. N. (1984b) Serotonin receptor changes in dementia of the Alzheimer type, *J. Neurochem.*, **43**, 1574–1581.

Cross, A. J., Crow, T. J., Ferrier, I. N., and Johnson, J. A. (1986a) The selectivity of the reduction of serotonin S2 receptors in Alzheimer type dementia, *Neurobiol. Aging*, **7**, 3–7.

Cross, A. J., Reynolds, G. P., Hewitt, L. M., and Slater, P. (1986b) Brain serotonin receptors in Huntington's disease, *Neurochem. Int.*, **9**, 431–435.

Enna, S. J., Bennett, J. P., Bylund, D. R., Synder, S. H., Bird, E. D., and Iversen, L. L. (1976) Alterations of brain neurotransmitter receptor binding in Huntington's chorea, *Brain Res.*, **116**, 531–537.

Fahn, S., Libsch, L. R., and Cutler, R. W. (1971) Monoamines in the human neostriatum: topographic distribution in normal and in Parkinson's disease and their role in akinesia, rigidity, chorea and tremor, *J. Neurol. Sci.*, **14**, 421–426.

Gottfries, C. G., Adolfsson, R., Aquilonius, S.-M., Carlsson, A., Eckernas, S.-A., Nordberg, L., Oreland, L., and Windblad, B. (1983) Biochemical changes in dementia disorders of Alzheimer type, *Neurobiol. Aging*, **4**, 261–271.

Hirano, A., and Zimmermann, H. M. (1962) Alzheimer's neurofibrillary tangles, a topographical study, *Arch. Neurol.*, **7**, 227–242.

Hoyer, D., Pazos, A., Probst, A., and Palacios, J. M. (1986a) Serotonin receptors in the human brain I. Characterization and autoradiographic localization of 5HT-1A recognition sites. Apparent absence of 5HT-1B recognition sites, *Brain Res.*, **376**, 85–96.

Hoyer, D., Pazos, A., Probst, A., and Palacios, J. M. (1986b) Serotonin receptors in the human brain II. Characterization and autoradiographic localization of 5HT-1C and 5HT-2 recognition sites, *Brain Res.*, **376**, 97–107.

Ishii, T. (1966) Distribution of Alzheimer neurofibrillary tangles in the brain stem and hypothalamus in senile dementia, *Acta. Neuropathol.*, **6**, 181–187.

Lycke, E., and Roos, B. E. (1975) Virus infections in infant mice causing persistent impairment of turnover of brain catecholamines, *J. Neurol. Sci.*, **26**, 49–60.

Mann, D. M. A., Yates, P. O., and Marcyniuk, B. (1985) Correlation between senile plaque and neurofibrillary tangle counts in cerebral cortex and neuronal counts in cortex and subcortical structures in Alzheimer's disease, *Neurosci. Lett.*, **56**, 51–55.

Middlemiss, D. N., Palmer, A. M., Edel, N., and Bowen, D. M. (1986) Binding of the novel serotonin agonist 8-hydroxy-2-(dipropylamino) tetralin in normal and Alzheimer brain, *J. Neurochem.*, **46**, 993–996.

Neeley, S. P., Cross, A. J., Crow, T. J., Johnson, J. A., and Taylor, G. R. (1985) Herpes simplex encephalitis: Neurochemical and neuroanatomical selectivity, *J. Neurol. Sci.*, **71**, 325–337.

Ogren, S. O. (1982) Central serotonin neurones and learning in the rat, in *Biology of Serotonergic Transmission* (Ed. N. N. Osborne), pp. 317–335, John Wiley & Sons, Chichester.

Palmer, A. M., Francis, P. T., Benton, J. S., Sims, N. R., Mann, D. M. A., Neary, D., Snowden, J. S., and Bowen, D. M. (1986) Presynaptic serotonergic dysfunction in patients with Alzheimer's disease, *J. Neurochem.*, (in press).

Pearson, R. C. A., Esiri, M. M., Hiorns, R. W., Wilcock, G. K., and Powell, T. P. S. (1985) Anatomical correlates of the distribution of the pathological changes in the neocortex in Alzheimer's disease, *Proc. Natl. Acad. Sci.*, **82**, 4531–4534.

Perry, E. K., Perry, R. H., Candy, J. M., Fairbairn, A. F., Dick. D. J., Blessed, G., and Tomlinson, B. E. (1984) Cortical serotonin S2 receptor binding abnormalities in patients with Alzheimer's disease: comparisons with Parkinson's disease, *Neurosci. Lett.*, **51**, 353–357.

Raisman, R., Cash, R., and Agid, Y. (1986) Parkinson's disease: decreased density of ^{3}H-imipramine and ^{3}H-paroxetine binding sites in putamen, *Neurology*, **36**, 556–560.

Reisine, T. D., Fields, J. Z., Yamamura, M. I., Bird, E. D., Iversen, L. L., Schreiner, P. S., and Enna, S. J. (1977) Neurotransmitter receptor abnormalities in Parkinson's disease, *Life Sci.*, **21**, 335–344.

Reynolds, G. P., and Goodridge, H. (1985) Alzheimer-like brain abnormalities in adults with Down's syndrome, *Lancet* (ii), 1368–1369.

Rinne, U. K., Sonninen, V., Riekkinen, P., and Laaksonen, H. (1974) Post-mortem findings in parkinsonian patients treated with L-DOPA: Biochemical considerations, in *Current Concepts in the Treatment of Parkinsonism* (Ed. M. D. Yahr), pp. 211–233, Raven Press, New York.

Rossor, M. N. (1981) Parkinson's disease and Alzheimer's disease as disorders of the isodendritic core, *Br. Med. J.*, **283**, 1588–1590.

Scatton, B., Javoy-Agid, F., Rouquier, L., Dubois, B., and Agid, Y. (1983) Reduction of cortical dopamine, noradrenaline, serotonin and their metabolites in Parkinson's disease, *Brain Res.*, *275*, 321–328.

Schwarcz, R., Bennett, J. P., and Coyle, J. T. (1977) Loss of striatal serotonin synaptic receptor binding induced by kainic acid lesions: correlations with Huntington's disease, *J. Neurochem.*, **28**, 867–869.

Stuart, A. M., Mitchell, I. J., Slater, P., Unwin, H. L. P., and Crossman, A. R. (1986) A semi-quantitative atlas of 5-hydroxytryptamine-1 receptors in the primate brain, *Neuroscience* (in press).

Yates, C. M., Simpson, J., and Gordon, A. (1986) Regional brain 5-hydroxytryptamine levels are reduced in senile Down's syndrome as in Alzheimer's disease, *Neurosci. Lett.*, **65**, 189–192.

Neuronal Serotonin
Edited by N. N. Osborne and M. Hamon
© 1988 John Wiley & Sons Ltd

CHAPTER 9

Serotonin and Behaviour, with Special Regard to Animal Models of Anxiety, Depression and Waiting Ability

PHILIPPE SOUBRIÉ
Département de Pharmacologie
Pitié-Salpêtrière
91 Boulevard de l'Hôpital
75013 Paris
France

This article is a select review. It is not my intention to discuss the truly vast literature on the involvement of serotonin- (5-HT)-containing neurons in a large range of behavioural phenomena (aggressivity, feeding behaviour, learning and memory, etc.). In this chapter, I will discuss, in the light of recent data, the hypothesized role of 5-HT neurons in animal models of anxiety and depression and broach the question that there may be a specific working principle for the action of 5-HT neurons. The proposal will be that these systems are implicated in a dual process: control of information processing and of response emission that allows the organism to tolerate and/ or arrange a delay before acting.

This review avoids trying to provide a detailed account for how specific brain regions or restricted 5-HT pathways function. Hence, the discussion for the most part concentrates on the ascending 5-HT systems in the CNS. This does not imply that irrespective of their nucleus of origin (median vs. dorsal raphe nucleus) or of their areas of projection (extrapyramidal, limbic, cortical, etc.), 5-HT neurons may control in the same direction any kind of behaviour.

Before engaging in such a review, it must be mentioned that the research strategy which consists in isolating one system of neurotransmitter from the others and in analysing its implications in one specific function, or hypothesizing that it preferentially controls one neurobiological process, may be viewed as an oversimplification. However, investigating whether or not a given neuronal system is more directly involved in the control of one type of behaviour, or in the action of one class of psychotropic drugs, does not imply that further neuronal systems do not participate (directly or indirectly) in the same processes.

ANIMAL MODELS OF ANXIETY AND 5-HT NEURONS

Since Brodie and Shore (1957) hypothesized that behavioural arousal is a joint function of 5-HT inhibition and catecholaminergic excitation, the general idea has gradually emerged that 5-HT neurons play a role in fear- or anxiety-mediated behavioural suppression. Decreased serotonin transmission has been linked to reduced anxiety and proposed as the mechanism in the anxiolytic activity of benzodiazepines. What is known to date concerning the interplay between 5-HT neurons and animal behaviour under anxiogenic situations?

Punishment-induced blockade of ongoing behaviour

This paradigm, which seems to be a useful analogue of anxiety, consists of training laboratory animals in a task (bar pressing, licking, etc.) for reward and then of introducing transitory periods during which responding is rewarded but also punished by mild electric foot-shock. Experiments conducted with the 5-HT depletor, *para*-chlorophenylalanine (pCPA), or 5-HT receptor blockers (methysergide, cinanserin, mianserin) as well as lesion studies with specific neurotoxins such as 5,7-dihydroxytryptamine (5,7-DHT), have led to the conclusion that a depressed 5-HT transmission produces, as benzodiazepines do, a significant attenuation of punishment-induced inhibition (Iversen, 1984; Gardner, 1986; Johnston and File, 1986; Thiébot, 1986).

The involvement of 5-HT neurons is further substantiated by the fact that the administration of lysergic acid diethylamide (Schoenfeld, 1976) or the application of 5-HT to the nucleus raphe dorsalis (Thiébot *et al.*, 1982), manipulations known to cause suppression of the firing rate of 5-HT cells as well as diminished release of the indoleamine in terminal areas, attenuate the behavioural inhibition elicited by punishment or a signal paired with punishment. Likewise, intra-raphe micro-injection of chlordiazepoxide, which also depresses 5-HT neuronal activity, produces an anti-punishment effect. This effect is abolished when 5-HT neurons of the dorsal raphe are first destroyed by local infusion of 5,7-DHT (Thiébot *et al.*, 1982).

A significant number of studies, however, have failed to find any anti-punishment activity of pCPA or 5-HT receptor antagonists (Morato de Carvalho *et al.*, 1981; Commissaris and Rech, 1982; Kilts *et al.*, 1982). These discrepant results have not yet been fully explained. The type of punishment used, the pre-drug magnitude of the conditioned suppression, and the length of exposure to the conflict situation may partially account for differences in the degree of control of 5-HT systems on punishment-induced inhibition. Differences in the technical procedures used to alter 5-HT transmission may also be responsible for these discrepancies (Lorens, 1978). In particular, it might be that the lack of specificity of many such procedures for the putative 5-HT receptor subtypes creates a somewhat confused overall picture and thus masks the expression of an anxiolytic effect.

Indeed, anti-punishment effects have been reported with drugs assumed to specifically bind to 5-HT2 or 5-HT1A sites (ritanserine, 8-OH-DPAT, buspirone, etc.). Although these effects required the integrity of 5-HT pathways to be observed, they generally occurred in only a narrow dose range and failed to reach the magnitude of those produced by benzodiazepines, and greatly depended on the testing procedure used (Colpaert *et al.*, 1985; Gardner, 1986; Pich and Samanin, 1986)

Treatments which increase 5-HT availability in the synaptic cleft have been shown to reinforce response suppression but this effect seems unspecific in that non-punished behaviour is usually also reduced (Iversen, 1984; Thiébot, 1986). However, since electric stimulation of the median raphe nucleus reportedly causes a behavioural inhibition resembling fear responses to threatening events (Graeff and Silveira-Filho, 1978), the apparent unspecific effect of enhanced 5-HT transmission in punishment paradigms might be accounted for by a state of generalized anxiety.

Although conventional anxiolytics such as benzodiazepines reportedly reduce 5-HT neuronal activity and release of the indoleamine from nerve endings (see Soubrié *et al.*, 1983), there are few studies to indicate that 5-HT neurons have a central role in mediating the effects of benzodiazepines in a punishment paradigm. The doses of benzodiazepines required to modify 5-HT transmission are generally higher than those necessary to cause changes in punished behaviour. Moreover, changes in 5-HT turnover do not closely parallel the efficacy of benzodiazepines in releasing punished behaviour under acute (Sepinwall and Cook, 1978) or chronic treatment conditions (Lister and File, 1983).

Tye *et al.* (1977) reported that a lesion of ascending 5-HT pathways blocked the action of an intraperitoneal injection of chlordiazepoxide in a conflict paradigm. However, other studies clearly show that destruction or inactivation of 5-HT neurons or 5-HT receptor blockade fail to impair the release of punished behaviour induced by peripheral administration of benzodiazepines (Shephard *et al.*, 1982b; Green and Hodges, 1986). The injection into the raphe dorsalis of the benzodiazepine receptor antagonist, RO 15–1788, did not counteract the anti-punishment effect of systemic diazepam, while similar brain manipulation was able to antagonize the reduction in 5-HT release induced by the peripheral administration of benzodiazepines (Thiébot *et al.*, 1984). Finally, direct stimulation of 5-HT receptors or enhanced 5-HT availability did not antagonize, and even may potentiate, the anti-conflict effect of benzodiazepines (Kilts *et al.*, 1982; Shephard *et al.*, 1982b).

All these data give evidence that, under punishment-induced anxiety, 5-HT neurons play a role in controlling behaviour and that reduction in 5-HT transmission generally mimics the effects of anti-anxiety agents. However, this is not always the case and few data support the notion that the anti-punishment effects of benzodiazepines derive from their ability to depress 5-HT transmission.

Novelty- and frustration-induced inhibition

Novel stimuli may increase the general level of fear. Often, but not invariably, one consequence of novelty is an inhibition of behaviour: inhibition of food intake in new surroundings (hyponeophagia), reduction of ambulation or exploratory behaviour, suppression of social interactions between a pair of rats. Anti-anxiety drugs generally release these forms of behavioural inhibition (Gray, 1982).

Very few reports (Shephard *et al.*, 1982a) indicate that neophobia-induced hypophagia can be affected by decreasing 5-HT transmission. Nevertheless, indirect reduction of 5-HT activity by micro-injection of GABA-mimetics into the dorsal raphe facilitates feeding behaviour (Przewlocka *et al.*, 1979) and recent experiments with 5-HT1A agonists showed that these drugs, probably by stimulating 5-HT autoreceptors, enhance food intake (Dourish *et al.*, 1986). However, whether these behavioural changes relate to a reduced anxiety remains to be elucidated. In addition, as for the punishment procedure, most pharmacological studies did not yield evidence in support of a 5-HT mediation of the 'antineophobic' action of benzodiazepines (Shephard and Broadhurst, 1982).

Like benzodiazepines, destruction of 5-HT neurons with 5,7-DHT has been shown to preferentially increase social contacts in animals placed in an unfamiliar situation, the 5-HT innervation of lateral septum being crucially involved (Clarke and File, 1982). At variance with the negative results obtained in a punishment paradigm, the intra-raphe injection of RO 15–1788 has been reported to partially antagonize the reduction in social interactions in rat produced by an 'anxiogenic' benzodiazepine receptor inverse agonist injected peripherally (Hobbs *et al.*, 1984).

Several authors have reported increased locomotor activity or exploratory behaviour in rodents following raphe nuclei (median) lesions or pCPA treatment (see Lorens, 1978) and it has been suggested that the 5-HT inputs to the hippocampus are responsible for this phenomenon. However, other workers have questioned whether 5-HT neurons are involved in this behavioural change at all. Nevertheless, local infusion of 5,7-DHT into the medial forebrain bundle, substantia nigra, nucleus accumbens or fornix-fimbria have been shown to facilitate locomotor activity or dopamine-induced locomotor stimulation, whereas local application of exogenous 5-HT induces an opposite effect (Lyness and Moore, 1981; Williams and Azmitia, 1981). Finally, treatments which indirectly reduce 5-HT transmission, such as application of GABA-mimetics or benzodiazepines to either the median or the dorsal raphe nucleus, have been found to stimulate locomotor activity (Przewlocka *et al.*, 1979; Sainati and Lorens, 1983). In most of these studies it appears that when blockade of 5-HT transmission (electrolytic lesions excepted) stimulates locomotor activity, the effect occurs mainly when animals are placed in a familiar situation (the home cage) or in an open-field test provided that they have spent a sufficient time in the apparatus to become 'habituated' to the environment. Very few reports are available to suggest that depression of 5-HT transmission increases specifically locomotor exploratory behaviour in a novel setting, i.e. releases locomotion from novelty-induced (anxiogenic?) inhibitory influences (see also Gerson and Baldessarini, 1980). Recent data obtained with 5-HT1A agonists are in keeping with such a notion (Pellow and File, 1986).

According to Gray (1982), frustration which is assumed to be generated in animals under reward-omission or non-reward conditions, shows considerable equivalence with fear or anxiety – in particular, this state produces a benzodiazepine-sensitive behavioural suppression. Whatever the paradigm used: extinction, 'time out' or schedules of differential reinforcement of low rates (DRL), the existing evidence concerning the role of 5-HT neurons – besides being scarce – is somewhat inconsistent.

When, after a history of continuous positive reinforcement, responding is no longer followed by the expected reward, pCPA-treated rats respond at a higher rate than controls during an extinction session (Beninger and Phillips, 1979), and Rosen and Freedman (1974) have reported that *para*-chloroamphetamine increases the number of responses emitted during a signal associated with non-reward (time out), while responding during a signal associated with periods of continuous reinforcement was not altered by the drug. However, Tye *et al.* (1979) found that pCPA did not produce increases in responding during time-out periods, whereas lesions with 5,7-DHT were reported rather to depress responding during these periods (Tye *et al.*, 1977).

Omission training is a paradigm which involves the animal's ability to withhold a response (lever pressing) previously maintained by positive reinforcement in order to obtain a similar reinforcement. Again, pCPA treatment significantly attenuates behavioural suppression, this action being partially reversed by the administration of 5-hydroxytryptophan (Thornton and Goudie, 1978). DRL is a paradigm in which the animal is required to let a specified time elapse between successive responses to obtain a reward. Thus, the making of a premature response results in non-reward and delay before it is possible to gain a reward. Electrolytic lesions of the raphe nuclei have been shown to impair DRL performance, this being primarily the result of premature responses (Wing and Wirtshafter, 1982). However, the 5-HT origin of such behavioural changes remains to be established, as only a weak shift towards premature responses is caused by pCPA or cinanserin (Mele and Caplan, 1980) and is not reproduced by 5,7-DHT lesions (Wing and Wirtshafter, 1982).

In conclusion, a variety of behavioural changes have been examined in relation to the activity of 5-HT systems in anxiety-producing conditions. Changes induced by a reduced 5-HT transmission have been compared with those produced by anti-anxiety drugs. Hence, it appears that 5-HT neurons are likely to control behaviour under some anxiogenic situations (punishment in particular) but much less under conditions of frustration- or novelty-induced anxiety.

This may suggest that only certain forms of anxiety are sensitive to drugs affecting serotonin transmission, and thus would be reminiscent of the reported absence of anti-anxiety effect of buspirone in patients with a history of benzodiazepine use (Schweizer *et al.*, 1986b).

ANIMAL MODELS OF DEPRESSION AND 5-HT NEURONS

Among the behavioural procedures that have been investigated as potential animal models of depression, it would be difficult to find another model that has stimulated more research activity, theoretical interest, argument and ethical concerns than those generated by the learned helplessness paradigm. This model, originally described by Seligman and coworkers in dogs (see Maier and Seligman, 1976), is based on the finding that exposure to uncontrollable stress produces performance deficits in subsequent learning tasks, which are not seen in subjects exposed to an identical, but controllable, stressor. The appeal of the paradigm comes mainly from the attractive value of the concept of uncontrollability in relation to affective disorders and the fact that such performance deficits have been proven to be sensitive to a large variety of antidepressant therapies. It is therefore not surprising that most of the information concerning the role of 5-HT systems in controlling depressive-like behaviours has mainly issued from the learned helplessness paradigm. Consonant with the biochemical and pharmacological foundations of aminergic hypotheses of depression, much animal research implicated noradrenergic (NA) and 5-HT neurons in the induction of helpless behaviour and/or its attenuation by antidepressant drugs.

Most relevant animal data for the implication of 5-HT neurons derive from studies on 5-HT metabolism and ^{3}H-imipramine binding, suggesting a reduction in 5-HT transmission in animals that expressed helpless behaviour (Petty and Sherman, 1983; Sherman and Petty, 1984). Moreover, application of 5-HT to the septum or the frontal cortex reversed escape failure, and a relationship between reversal of learned helplessness by antidepressants and enhancement of 5-HT transmission has been proposed (Sherman and Petty, 1980).

Contradictory results have been reported, however. It has been found that alteration of brain 5-HT metabolism following uncontrollable stress did not correlate with helplessness *per se* (Weiss *et al.*, 1981). Moreover, 5-HT depletion with pCPA, unlike NA depletion, did not facilitate the induction of helpless behaviour (Anisman *et al.*, 1979; but see also Hamilton *et al.*, 1986), whereas the 5-HT precursors, L-tryptophan and 5-hydroxytryptophan, did (Brown *et al.*, 1982b).

Likewise, we found that, in rats, prior infusion of 5,7-DHT into the midbrain raphe area did not facilitate learned helplessness induction, while a 70% loss of tryptophan hydroxylase activity in the striatum, hippocampus and cerebral cortex was detected in the lesioned animals (Soubrié *et al.*, 1986).

Data obtained on additional animal models of depression such as the forced swimming test or the separation model in monkeys further support the notion that 5-HT depletion did not augment depressive-like behaviours (Kraemer *et al.*, 1984).

Moreover, although a connection between reversal of learned helplessness following antidepressant treatment and restoration of 5-HT transmission (especially in cortical and limbic areas) has been suggested, some of our findings suggest that the changes in 5-HT transmission that occur in a matter of a few days of treatment with tricyclic antidepressants or monoamine oxidase inhibitors are unlikely to underlie the effects of these drugs in the learned helplessness paradigm. Indeed, damage to 5-HT neurons associated with a 70% loss of tryptophan hydroxylase activity in the cerebral cortex, the hippocampus and the striatum did not alter the ability of any antidepressant studied (clomipramine, desipramine, imipramine and nialamide) to reverse escape failures in the shuttle-box paradigm in rats previously exposed to inescapable shocks (Soubrié *et al.*, 1986).

As suggested in the previous section, it is still conceivable that the lack of specificity of the pharmacological tools for 5-HT receptor subtypes creates a confused overall picture and masks the expression of the involvement of 5-HT neurons in depressive-like behaviours. As a matter of fact, a recent study of escape failures in the shuttle-box paradigm (Giral *et al.*, 1987) found (Fig. 1) that purported 5-HT1A agonists such as buspirone, gepirone, 8-OH-DPAT and TVXQ 7821, were all effective in eliminating helpless behaviour.

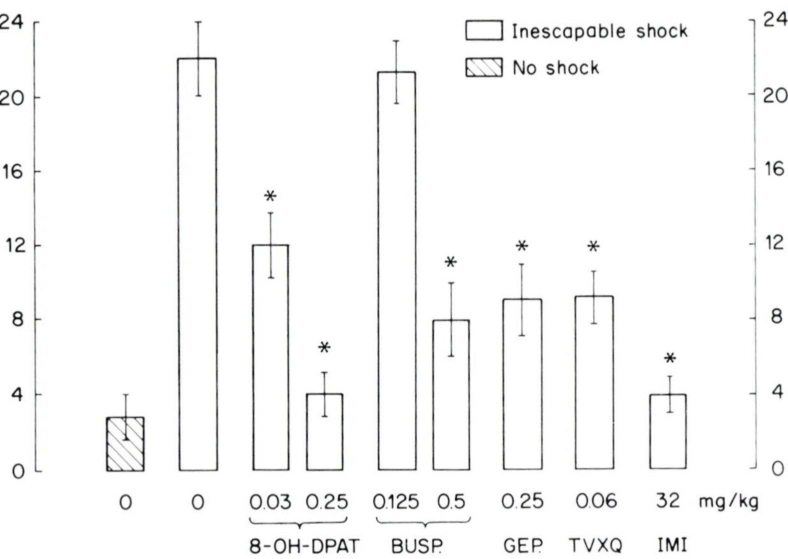

Fig. 1 Reversal by 5-HT1A agonists of escape deficits produced by inescapable shock pretreatment in rats.

Rats were first subjected to 60 inescapable, randomized electric foot shocks (15 sec duration, 0.8 mA, every minute ±15 sec). Escape deficits were estimated on three consecutive 15 min daily shuttle-box sessions (30 two-way avoidance trials/session), the first avoidance session being performed 48 h after shock pretreatment.

Drugs (BUSP = buspirone; GEP = gepirone; TVXQ = TVXQ 7821) were injected i.p. 6 h after shock pretreatment, and then, twice a day (morning, i.e. 30 min before shuttle-box sessions, and afternoon). The indicated doses correspond to the total daily dose administered. Imipramine (IMI) was used as an internal control.

Data are the mean ±SEM (n = 8 or 12 rats/group) of escape failures at the third shuttle-box session. Escape failure refers to failure of the rat to change compartments during the electric footshock. (0.8 mA; 3 sec duration).

*Significantly different (p < 0.01) from the associated saline-treated shocked rats (ANOVA)

Although several data suggest that the learned helplessness syndrome might be an animal model of depression and that conventional anxiolytics (benzodiazepines) are weakly effective in eliminating interference effects (Sherman *et al.*, 1982; Drugan *et al.*, 1984), anxiety might contribute to the development of this syndrome (Drugan *et al.*, 1984, 1985). Therefore, it might be that, in animals, 5-HT1A agonists preferentially affect anxiety associated with uncontrollable stress and oppose benzodiazepines which mainly reduce anxiety linked to controllable aversive events. Such a possibility is reminiscent of the clinical observation of Schweizer *et al.* (1986a) of an antidepressant effect of buspirone in non-melancholic depressed patients.

NA neurons might be indirectly involved in the ability of 5-HT1A agonists in reversing helpless behaviour. Elimination of the learned helplessness syndrome is reportedly accompanied by enhanced NA transmission (see Martin *et al.*, 1986), a biochemical change that is expected, as shown for some 5-HT1A agonists (Sanghera *et al.*, 1983), after blockade of the 5-HT inhibitory influence on NA neurons.

Hence, if 5-HT neurons are implicated in helpless behaviour, transitory inactivation rather than activation of these systems – perhaps by releasing NA neurons from inhibitory control – seems to protect animals from exhibiting behavioural deficits. Our failure to observe a similar protective effect with 5,7-DHT lesion can be accounted for by the fact that in this study (Soubrié *et al.*, 1986), behavioural testing was performed 21 days after the neurotoxin injection, when the activation of NA transmission might have disappeared (Renaud *et al.*, 1975).

WAITING CAPACITIES AND 5-HT NEURONS

Although based on relatively simple concepts, the animal models of anxiety or depression reviewed in this chapter require behavioural tasks which explore composite phenomena among which the salient role of a particular state or emotion (depression or anxiety) may be questioned. On this account, animal/human discrepancies (the reported association between depressed 5-HT transmission with increased anxiety in humans, for instance, Banki *et al.*, 1981; Rydin *et al.*, 1982) are not surprising. It is clear that in animals, the apparent anxiolytic or antidepressant effects observed following manipulations thought to reduce 5-HT transmission could be accounted for by alterations of intervening factors such as arousal, motivation, information processing, response selection, and data on response perseveration or latent inhibition may give some credence to this assumption (Solomon *et al.*, 1978).

In this context, and more especially for models of anxiety, the possibility that manipulating 5-HT transmission may interfere with the ability of the animals to wait for an expected reward warrants attention. Indeed, under punishment procedures, blockade of ongoing behaviour enables the animals to avoid electric shock but it also delays the possibility of obtaining food or water reward. Thus it is conceivable that delay of reward induced by behavioural suppression may be as critical a factor as punishment-induced fear or anxiety for the involvement of 5-HT neurons in controlling behaviour. Two behavioural models without any nociceptive component were designed to study the possible role of 5-HT neurons on waiting ability (Thiébot *et al.*, 1985a, b).

In model I, rats trained in a T-maze were allowed to choose between two types of reward: immediate but small (2 pellets) vs. 25 sec delayed but large (10 pellets). Given this choice, rats selected the large-but-delayed reward in less than 40% of the trials. At this stage, 5-HT uptake blockers, clomipramine, indalpine, zimelidine, increased the number of choices of the large-but-delayed reward (Table 1). In model II, rats were subjected to a FR48 schedule of food reinforcement but after completion of a series of 48 presses, and in case of no further press, free pellets were delivered according to fixed, increasing intervals (from 5 to 80 sec). Pressing during the 'free sequence' stopped the sequence and required the rat to complete 48 presses to be reinforced and to reinitiate the 'free sequence'. Waiting for free reward was lengthened by clomipramine, indalpine and zimelidine.

Some data obtained in model I, with 5-HT1A agonists or with lesion studies (Thiébot, 1986) suggest that these manipulations tend to decrease the rats' ability to wait for reward (Table 1). In general however, the effect of these 5-HT1A agonists is weaker than that of benzodiazepines or than could be expected from the data obtained with drugs assumed to enhance 5-HT transmission. The following hypothesis can tentatively be proposed in attempting to elucidate this point. Drugs which enhance NA transmission such as desipramine, maprotiline or beta-adrenoceptor stimulants, also increased the number of selections of the delayed reward (Table 1), thus suggesting a dual NA/5-HT mediation of waiting capacity. Hence, if 5-HT1A agonists release NA neurons from 5-HT inhibitory control (as shown for some of these compounds; Sanghera *et al.*, 1983) this would result in competing NA/5-HT influences on waiting behaviour.

IMPLICATIONS FOR FUTURE RESEARCH

These data on waiting might have several implications, regarding particularly the functional roles of 5-HT neurons.

In punishment models of anxiety, during the course of which animals are required to stop responding for reward in order to avoid punishment, it should be kept in mind that we are probably studying anxiety but also the ability to wait for reward. Hence, it cannot be excluded that 5-HT neurons would exert a control over punished behaviour, not by affecting anxiety *per se*, but by changing those responses (behavioural suppression) which are

Table 1 Effects of drugs on rat waiting capacity for food reward

Drug-induced changes in the number (mean $\pm$SEM) of selections of the large-but-delayed reward in rats

Enhanced 5-HT transmission (delay 25 sec)					
Citalopram	4	mg/kg	H-60	+2.60 $\pm$ 0.68	$p < 0.02$
Clomipramine	8	mg/kg	H-60	+1.38 $\pm$ 0.50	$p < 0.02$
Indalpine	2	mg/kg	H-30	+2.27 $\pm$ 0.60	$p < 0.02$
Zimelidine	16	mg/kg	H-60	+2.25 $\pm$ 0.41	$p < 0.01$
Reduced 5-HT transmission (delay 15 sec)					
Methysergide	8	mg/kg	H-30	-2.51 ± 0.67	$p < 0.01$
Buspirone	4	mg/kg	H-30	-2.33 ± 0.49	$p < 0.01$
8-OH-DPAT	0.06	mg/kg	H-60	-1.00 ± 0.47	$0.10 < p < 0.05$
Gepirone	2	mg/kg	H-30	-0.63 ± 0.25	NS
TVXQ 7821	0.5	mg/kg	H-30	-1.37 ± 0.40	$p < 0.02$
Benzodiazepines (delay 15 sec)					
Diazepam	4	mg/kg	H-30	-2.07 ± 0.80	$p < 0.01$
Chlordiazepoxide	16	mg/kg	H-30	-2.62 ± 0.68	$p < 0.01$
Enhanced NA transmission (delay 25 sec)					
Desipramine	8	mg/kg	H-60	+1.90 $\pm$ 0.23	$p < 0.01$
Maprotiline	8	mg/kg	H-30	+1.56 $\pm$ 0.60	$p < 0.05$
Clenbuterol	0.06	mg/kg	H-30	+1.29 $\pm$ 0.47	$p < 0.05$

Presented are the differences in the absolute number of selections of the large-but-delayed reward between an initial block of two 5-trial sessions performed under saline and a subsequent block of two sessions of five trials each, performed under drug. For statistical analysis, these differences were compared to those observed in control (saline–saline) rats by using ANOVA.

Rats were trained to run a T-maze (two daily sessions of five trials) for reward with left and right goal boxes provided with ten and two pellets, respectively.

Within ten sessions rats chose the arm associated with the large reward in 80–100% of the trials. Then, when the rats were detained in the arm associated with the large reward before having access to the ten pellets, the number of choices of the large reward decreased as a function of the duration of the waiting period.

Drug or vehicle injections were performed two or three sessions following the introduction of delay of reward, fixed at 15 or 25 sec depending on the expected effect of the drug. Data correspond to the doses and pretreatment times (referred to in the table as H-, in minutes) producing a maximal effect.
($n = 6$ to 8 rats/groups).

commonly taken as measures of animal anxiety. This alternative would be consonant with the fact that destruction of 5-HT innervation of brain structures, the substantia nigra for instance, assumed to be more concerned with the performance of behaviours than with the control of emotions, released behavioural suppression in two models of punishment (Thiébot *et al.*, 1983). Moreover, it can be asked whether the presence or absence of a waiting dimension, especially waiting for food reward, in a presumed model of anxiety is not critical for the apparent anti-anxiety effect of reducing 5-HT transmission to be observed.

That waiting might be a basic dimension for the involvement of 5-HT systems might open up new vistas concerning the functional roles of these systems. 5-HT neurons can be viewed as processes implicated in the control of response emission and information processing, enabling the organism to arrange or tolerate a delay before acting, so that a motor response might be triggered not only on the basis of the saliency of the information sampled but also on the basis of its relevance or significance. This suggests that 5-HT neurons do not directly govern a single behaviour nor even a set of behaviours, but that, in contrast, they govern any kind of behaviour (aggression, exploration, pressing for food, approach or avoidance) when that behaviour enters into the waiting dimension. In other words, 5-HT neurons are probably not involved in the control of specific acts as such, but rather in the way in which they are carried out. It is clear that such a contention requires further substantiation. However, it may account for the many contradictions or inconsistencies that can be noticed throughout the scientific literature concerning 5-HT systems. Moreover, it is compatible with the wide range of behaviours that are reportedly affected by 5-HT manipulation, and in particular certain forms of aggressivity and aversion-related behaviours (Copenhaver *et al.*, 1978; Katz, 1980; Schutz *et al.*, 1985).

Finally, although this same contention does not postulate any kind of human/animal behavioural isomorphism, it seems to be among those most compatible with findings from biological psychiatry or psychology. Indeed, although serious methodological criticism can be levelled against these studies, a large degree of unanimity emerges to suggest a privileged connection between indexes of reduced 5-HT transmission (low 5-HIAA in CSF, ^{3}H-imipramine binding, etc.) and impulsive conduct (Brown *et al.*, 1982a; Rydin *et al.*, 1982; Stanley *et al.*, 1982; Linnoila *et al.*, 1983). It is worthy of mention that no such relationship can be found with any classical nosographic entity.

As stated by Herrnstein (1981): 'in impulsiveness versus self-control, time is always of the essence' (p. 3), and in humans as in animals, the critical contingencies for the involvement of 5-HT neurons in controlling behaviour might (greatly though not exclusively) depend on delay of reinforcement.

ACKNOWLEDGEMENTS

I am deeply grateful to Hani Rakotondrabe for her care in preparing the manuscript.

REFERENCES

Anisman, H., Irwin, J., and Sklar, S. (1979) Deficits of escape performance following catecholamine depletion: implications for behavioural deficits induced by uncontrollable stress, *Psychopharmacology*, **64**, 163–170.

Banki, C. M., Molnar, G., and Vojnik, M. (1981) Cerebrospinal fluid amine metabolites, tryptophan and clinical parameters in depression. Part 2. Psychopathological symptoms, *J. Affective Disord.*, **3**, 91–99.

Beninger, R. J., and Phillips, A. G. (1979) Possible involvement of serotonin in extinction, *Pharmac. Biochem. Behav.*, **10**, 37–41.

Brodie, B. B., and Shore, P. A. (1957) A concept for a role of serotonin and norepinephrine as chemical mediators in the brain, *Annals of the New York Academy of Sciences*, **66**, 631–642.

Brown, G. L., Goodwin, F. K., and Bunney, W. E. (1982a) Human aggression and suicide: their relationship to neuropsychiatric diagnoses and serotonin metabolism, in *Advances in Biochemical Psychopharmacology*, Vol. 34: *Serotonin in Biological Psychiatry* (Eds B. T. Ho *et al.*), pp. 287–306, Raven Press, New York.

Brown, G. L., Rosselini, R. A., Samuels, O. B., and Riley, E. P. (1982b) Evidence for a serotonergic mechanism of the learned helplessness phenomenon, *Pharmac. Biochem. Behav.*, **17**, 877–883.

Clarke, A., and File, S. E. (1982) Selective neurotoxin lesions of the lateral septum: changes in social and aggressive behaviours, *Pharmac. Biochem. Behav.*, **17**, 623–628.

Colpaert, F. C., Meert, T. F., Niemegeers, C. J. E., and Jansen, P. A. J. (1985) Behavioural and 5-HT antagonist effects of ritanserin: a pure and selective antagonist of LSD discrimination in rat, *Psychopharmacology*, **86**, 45–54.

Commissaris, R. L., and Rech, R. H. (1982) Interactions of metergoline with diazepam, quipazine and hallucinogenic drugs on a conflict behaviour in the rat, *Psychopharmacology*, **76**, 282–285.

Copenhaver, J. H., Schalock, R. L., and Carver, M. J. (1978) *Para*-chloro-D,L-phenylalanine induced filicidal behaviour in the female rat, *Pharmac. Biochem. Behav.*, **8**, 263–270.

Dourish, C. T., Hutson, P. H., and Curzon, G. (1986) *Para*-chlorophenylalanine prevents feeding induced by the serotonin agonist 8-hydroxy-2-(di-n-propylamino)-tetralin (8-OH-DPAT), *Psychopharmacology*, **89**, 467–471.

Drugan, R. C., Maier, S. F., Skolnick, P., Paul, S. M., and Craxley, J. N. (1985) An anxiogenic benzodiazepine receptor ligand induces learned helplessness, *Eur. J. Pharmacol.*, **113**, 453–457.

Drugan, R. C., Ryan, S. M., Minor, T. H., and Maier, S. F. (1984) Librium prevents the analgesia and shuttle-box escape deficit typically observed following inescapable shock, *Pharmac. Biochem. Behav.*, **21**, 749–754.

Gardner, C. R. (1986) Recent developments in 5-HT related pharmacology of animal models of anxiety, *Pharmac. Biochem. Behav.*, **24**, 1479–1485.

Gerson, S. C., and Baldessarini, R. J. (1980) Motor effects of serotonin in the central nervous system, *Life Sci.*, **27**, 1435–1451.

Giral, Ph., Martin, P., Soubrié, Ph., and Simon, P. (1987) Reversal of helpless behaviour in rats by putative 5-HT1A agonists. *Biol. Psychiatry*, **22** (in press).

Graeff, F. G., and Silveira Filho, N. G. (1978) Behavioural inhibition induced by electrical stimulation of the median raphe nucleus of the rat, *Physiol. Behav.*, **21**, 477–484.

Gray, J. A. (1982) Précis of the neuropsychology of anxiety: an enquiry into the functions of the septo-hippocampal system, *Behav. Brain Sci.*, **5**, 469–534.

Green, S., and Hodges, H. (1986) Differential effects of dorsal raphe lesions and intraraphe GABA and benzodiazepines on conflict behavior in rats, *Behav. Neur. Biol.*, **46**, 13–29.

Hamilton, M. E., Zacharko, R. M., and Anisman, H. (1986) Influence of *p*-chloroamphetamine and methysergide on the escape deficits provoked by inescapable shock, *Psychopharmacology*, **90**, 203–206.

Herrnstein, R. J. (1981) Self control as response strength, in *Quantification of Steady-State Operant Behavior* (Eds C. M. Bradshaw, E. Szabadi and C. F. Lowe), pp. 3–20, Elsevier: North-Holland Biomedical Press.

Hobbs, A., Paterson, I. A., and Roberts, M. H. T. (1984) The effects on social interaction of microinjections of Ro–15–1788 into the nucleus raphe dorsalis of the rat, *Br. J. Pharmacol.*, **82**, 241.

Iversen, S. D. (1984) 5-HT and anxiety, *Neuropharmacology*, **23**, 1553–1560.

Johnston, A. L., and File, S. E. (1986) 5-HT and anxiety: promises and pitfalls, *Pharmac. Biochem. Behav.*, **24**, 1467–1470.

Katz, R. J. (1980) Role of serotonergic mechanisms in animal models of predation, *Prog. Neuro-Psychopharmacol.*, **4**, 219–231.

Kilts, C. D., Commissaris, R. L., Cordon, J. J., and Rech, R. H. (1982) Lack of central 5-hydroxytryptamine influence on the anticonflict activity of diazepam, *Psychopharmacology*, **78**, 156–164.

Kraemer, G. W., Ebert, M. H., Lake, C. R., and McKinney, W. T. (1984) Cerebro-spinal fluid measures of neurotransmitter changes associated with pharmacological alteration of the despair response to social separation in rhesus monkeys, *Psychiat. Res.*, **11**, 303–315.

Linnoila, M., Virkkunen, M., Scheinin, M., Nuutila, A., Rimon, R., and Goodwin, F. K. (1983) Low cerebrospinal fluid 5-hydroxyindoleacetic acid concentration differentiates impulsive form non impulsive violent behaviour, *Life Sci.*, **33**, 2609–2614.

Lister, R. G., and File, S. R. (1983) Changes in regional concentrations in the rat brain of 5-hydroxytryptamine and 5-hydroxyindoleacetic acid during the development of tolerance to the sedative action of chlordiazepoxide, *J. Pharm. Pharmacol.*, **35**, 601–603.

Lorens, S. A. (1978) Some behavioural effects of serotonin depletion depend on method: a comparison of 5,7-dihydroxytryptamine, *p*-chlorophenylalanine, *p*-chloroamphetamine and electrolytic raphe lesions, *Annals of the New York Academy of Sciences*, **305**, 522–555.

Lyness, W. H., and Moore, K. E. (1981) Destruction of 5-hydroxytryptaminergic neurons and the dynamics of dopamine in nucleus accumbens-septi and other forebrain regions of the rat, *Neuropharmacology*, **20**, 327–334.

Maier, S. F., and Seligman, M. E. P. (1976) Learned helplessness. Theory and evidence, *J. Exp. Psychol. Gen.*, **105**, 3–46.

Martin, P., Soubrié, Ph., and Simon, P. (1986) Shuttle-box deficits induced by inescapable shocks in rats: reversal by the beta-adrenoceptor stimulants clenbuterol and salbutamol, *Pharmac. Biochem. Behav.*, **24**, 117–181.

Mele, P. C., and Caplan, M. A. (1980) Effects of cinanserin and *p*-chlorophenyl-alanine and their interaction with d-amphetamine on DRL performance in rats, *Pharmac. Biochem. Behav.*, **12**, 883–891.

Morato de Carvalho, S., De Aguiar, J. C., and Graeff, F. G. (1981) Effects of minor tranquilizers, tryptamine antagonists and amphetamine on behaviour punished by brain stimulation, *Pharmac. Biochem. Behav.*, **15**, 351–356.

Pellow, S., and File, S. E. (1986) Anxiolytic and anxiogenic drug effects on explora-tory activity in an elevated plus-maze: a novel test of anxiety in the rat, *Pharmac. Biochem. Behav.*, **24**, 525–529.

Petty, F., and Sherman, A. D. (1983) Learned helplessness induction decreases in vivo cortical serotonin release, *Pharmacol. Biochem. Behav.*, **18**, 649–650.

Pich, E. M., and Samanin, R. (1986) Disinhibitory effects of buspirone and low doses of sulpiride and haloperidol in two experimental anxiety models in rats: possible role of dopamine, *Psychopharmacology*, **89**, 125–130.

Przewlocka, B., Stala, L., and Scheel-Kruger, J. (1979) Evidence that GABA in the nucleus dorsalis raphe induces stimulation of locomotor activity and eating behaviour, *Life Sci.*, **25**, 937–946.

Renaud, B., Buda, M., Lewis, B. D., and Pujol. J. F. (1975) Effects of 5,6-dihydroxytryptamine on tyrosine-hydroxylase activity in central catecholaminergic neurons of the rat, *Biochem. Pharmac.*, **24**, 1739–1742.

Rosen, A. J., and Freedman, P. E. (1974) The effects of *p*-chloroamphetamine on instrumental conditioning in the rat, *Neuropharmacology*, **13**, 585–590.

Rydin, E., Schalling, D., and Asberg, M. (1982) Rorschach ratings in depressed and suicidal patients with low levels of 5-hydroxyindoleacetic acid in cerebrospinal fluid, *Psychiat. Res.*, **7**, 229–243.

Sanghera, M. K., McMillen, B. A., and German, D. C. (1983) Buspirone, a non-benzodiazepine anxiolytic, increases locus coeruleus noradrenergic neuronal activity, *Eur. J. Pharmacol.*, **86**, 107–110.

Sainati, S. M., and Lorens, S. A. (1983) Intra-raphe benzodiazepines enhance rat locomotor activity: interactions with GABA, *Pharmac. Biochem. Behav.*, **18**, 407–414.

Schoenfeld, R. I. (1976) Lysergic acid diethylamide- and mescaline-induced attenuation of the effect of punishment in the rat, *Science*, **192**, 801–803.

Schutz, M. T. B., De Aguiar, J. C., and Graeff, F. G. (1985) Anti-aversive role of serotonin in the dorsal periaqueductal grey matter, *Psychopharmacology*, **85**, 340–345.

Schweizer, E., Amsterdam, J., Rickels, K., Kaplan, M., and Droba, M. (1986a) Open trial of buspirone in the treatment of major depressive disorder. *Psychopharmacol. Bull.*, **22**, 183–185.

Schweizer, E., Rickels, K., and Lucki, I. (1986b) Resistance to the anti-anxiety effect of buspirone in patients with a history of benzodiazepine use, *New Engl. J. Med.*, **314**, 719–720.

Sepinwall, J., and Cook, I. (1978) Behavioral pharmacology of anti-anxiety drugs, in *Handbook of Psychopharmacology* (Eds L. L. Iversen, S. D. Iversen and S. D. Snyder), Vol. 13, pp. 345–393, Plenum Press, New York.

Shephard, R. A., and Broadhurst, P. L. (1982) Effects of diazepam and of serotonin agonists on hyponeophagia in rats, *Neuropharmacology*, **21**, 337–340.

Shephard, R. A., Buxton, D. A., and Broadhurst, P. L. (1982a) Beta-adrenoreceptor antagonists may attenuate hyponeophagia in the rat through a serotonergic mechanism, *Pharmac. Biochem. Behav.*, **16**, 741–744.

Shephard, R. A., Buxton, D. A., and Broadhurst, P. L. (1982b) Drug interactions do not support reduction in serotonin turnover as the mechanism of action of benzodiazepines, *Neuropharmacology*, **21**, 1027–1032.

Sherman, A. D., and Petty, F. (1980) Neurochemical basis of the action of antidepressants on learned helplessness, *Behav. Neurol. Biol.*, **30**, 119–134.

Sherman, A. D., and Petty, F. (1984) Learned helplessness decreases ^{3}H-imipramine binding in rat cortex, *J. Affective Disord.*, **6**, 25–32.

Sherman, A. D., Sacquitne, J. L., and Petty, F. (1982) Specificity of the learned helplessness model of depression, *Pharmac. Biochem. Behav.*, **16**, 449–454.

Solomon, P. R., Kiney, C., and Scott, D. R. (1978) Disruption of latent inhibition following systemic administration of *para*-chlorophenylalanine (pCPA), *Physiol. Behav.*, **20**, 265–271.

Soubrié, Ph., Blas, C., Ferron, A., and Glowinski, J. (1983) Chlordiazepoxide reduces in vivo serotonin release in the basal ganglia of 'encéphale isolé' but not of anaesthetized cats: evidence for a dorsal raphe site of action, *J. Pharmacol. Exp. Ther.*, **226**, 526–532.

Soubrié, Ph., Martin, P., El Mestikawy, S., Thiébot, M. H., Simon, P., and Hamon, M. (1986) The lesion of serotonergic neurons does not prevent antidepressant-induced reversal of escape failures produced by inescapable shocks in rats, *Pharmac. Biochem. Behav.*, **25**, 1–6.

Stanley, M., Virgilio, J., and Gershon, S. (1982) Tritiated imipramine binding sites are decreased in the frontal cortex of suicides, *Science*, **216**, 1337–1339,

Thiébot, M. H. (1986) Are serotonergic neurons involved in the control of anxiety and in the anxiolytic activity of benzodiazepines?, *Pharmac. Biochem. Behav.*, **24**, 1471–1477.

Thiébot, M. H., Hamon, M., and Soubrié, Ph. (1982) Attenuation of induced anxiety in rats by chlordiazepoxide: role of raphe dorsalis benzodiazepine binding sites and serotonergic neurons, *Neuroscience*, **7**, 2287–2294.

Thiébot, M. H., Hamon, M., and Soubrié, Ph. (1983) The involvement of nigral serotonin innervation in the control of punishment-induced behavioral inhibition in rats, *Pharmacol. Biochem. Behav.*, **19**, 225–229.

Thiébot, M. H., Soubrié, Ph., Hamon, M., and Simon, P. (1984) Evidence against the involvement of serotonergic neurons in the antipunishment activity of diazepam in the rat, *Psychopharmacology*, **82**, 355–359.

Thiébot, M. H., Le Bihan, C., Soubrié, Ph., and Simon, P. (1985a) Benzodiazepines reduce the tolerance to reward delay in rats, *Psychopharmacology*, **88**, 147–152.

Thiébot, M. H., Soubrié, Ph., and Simon, P. (1985b) Is delay of reward mediated by shock-avoidance behavior a critical target for anti-punishment effects of diazepam in rats?, *Psychopharmacology*, **87**, 473–479.

Thornton, E. W., and Goudie, A. J. (1978) Evidence for the role of serotonin in the inhibition of specific motor responses, *Psychopharmacology*, **60**, 73–79.

Tye, N. C., Everitt, B. J., and Iversen, S. D. (1977) 5-Hydroxytryptamine and punishment, *Nature (Lond.)*, **268**, 741–743.

Tye, N. C., Iversen, S. D., and Green, A. R. (1979) The effects of benzodiazepines and serotonergic manipulations on punished responding, *Neuropharmacology*, **18**, 689–695.

Weiss, J. A., Goodman, P. A., Losito, B. G., Corrigan, S., Charry, J. M., and Bailey, W. H. (1981) Behavioral depression produced by an uncontrollable stressor: relationship to norepinephrine, dopamine and serotonin levels in various regions of rat brain, *Brain Res. Rev.*, **3**, 167–205.

Williams, J. H., and Azmitia, E. C. (1981) Hippocampal serotonin re-uptake and nocturnal locomotor activity after microinjections of 5,7-DHT in the fornix-fimbria, *Brain Res.*, **207**, 95–107.

Wing, L. L., and Wirtshafter, D. (1982) Impaired DRL performance with electrolytic median raphe lesions. *Abstract of the 12th Annual Meeting of the Society for Neuroscience*, p. 309.

Neuronal Serotonin
Edited by N. N. Osborne and M. Hamon
© 1988 John Wiley & Sons Ltd

CHAPTER 10

Serotonin and Endocrinology – The Pituitary

M. MONTANGE AND A. CALAS
Laboratoire de Physiologie des Interactions Cellulaires
Université de Bordeaux I
UA CNRS 339
Avenue des Facultés
33405 Talence Cedex
France

INTRODUCTION

The involvement of serotonin in endocrine regulations is assessed by numerous morphological, pharmacological and experimental data. At the pituitary level (which will be the scope of this review) the presence and/or the action of serotonin are demonstrated in the different parts of the gland which will be named here according to the terminology of Rioch *et al.* (1940):

adenohypophysis, which comprises the anterior (pars distalis), intermediate (pars intermedia) and tuberal (pars tuberalis) lobes; and neurohypophysis, which includes the posterior or neural lobe (pars nervosa), the infundibular stalk and the median eminence (ME).

We shall first describe the anatomical reasoning for the involvement of 5-HT and then try to evaluate the physiological evidence of serotonin in the regulation of the different pituitary secretions, principally in the rat. An attempt to integrate the morphological and physiological data will then be presented.

MORPHOLOGICAL DATA

These data will not be limited to the pituitary itself, since 5-HT can be involved in neuroendocrine regulations of the gland at higher levels and even in the whole brain. Our morphofunctional report will be both extended and confined to the area most directly involved in the pituitary regulations i.e. the neuroendocrine hypothalamus.

Morphofunctional tools

To detect the presence of 5-HT neurons and/or of 5-HT itself within the pituitary, several kinds of techniques are available: biochemical assays of the dissected gland (which however do not distinguish between the tuberal lobe and median eminence–infundibular stalk, nor the neural and intermediate lobes which are considered by biochemists as a 'neurointermediate' entity); immunocytochemical detection of serotonin at light- and electron-microscope levels; histological or high-resolution radioautography after uptake of ³H-5-HT by corresponding neurons or after binding of labelled ligands of 5-HT receptors (De Souza, 1986). All these methods can of course be linked to experimental (surgical or pharmacological) manipulations and/or associated with tracers of axonal pathways using orthograde or retrograde axonal flow. A combination of these techniques helps to localize precisely the cell bodies of origin of identified fibres and terminals.

Adenohypophysis

According to the detection of 5-HT in the anterior lobe (Björklund *et al.*, 1967) studies using the formaldehyde histofluorescence technique suggested that the amine is associated with some cells which displayed a yellow fluorescence. Pearse and MacGregor (1964) assumed that they might be adrenocorticotrophin (ACTH) cells and they classified them in the APUD (amine precursors uptake and decarboxylation) series. Björklund and Falck (1969) suggested that they contained a tryptamine-like indoleamine rather than serotonin itself. Some adenohypophyseal cells can take up and retain ³H-5-

HT introduced either intravascularly or *in vitro*, as demonstrated both in the quail (Tougard, 1970) and in the rat (Calas, 1981). In the latter, the uptake shows characteristics similar to those described for 5-HT uptake into central nervous system neurons (Johns *et al.*, 1982). Labelled cells are gonadotrophs in the rat (Johns *et al.*, 1982) and in the bat (Nunez *et al.*, 1981). In addition, in the mouse and in the bat, 5-HT immunoreactive cells are detected in the anterior lobe where they constitute a subset of gonadotrophs (Payette *et al.*, 1985) and 5-HT is colocalized with β-luteinizing hormone (LH) in some secretory granules (Payette *et al.*, 1986). However, in spite of the gonado-trophs having the ability to take up and store 5-HT, it seems likely from other findings that they do not synthesize the amine biochemically (Saavedra *et al.*, 1976; Spinedi and Negro-Vilar, 1983).

Westlund and Childs (1982) have claimed the existence of 5-HT fibres detectable by immunocytochemistry at the periphery of the pars distalis. Using the same technique, Léranth *et al.* (1983) have also described sporadic serotonergic fibres within the anterior lobe. However, this observation has not been confirmed by Payette *et al.* (1985). On the other hand, the highest density of 5-HT innervation of the whole pituitary has been described in the intermediate lobe (IL) of the pituitary both by immunocytochemistry (Westlund and Childs, 1982; Léranth *et al.*, 1983; Friedman *et al.*, 1983) and radioautography (RAG), following administration of ^{3}H-5-HT either *in vitro* (Payette *et al.*, 1985) or *in vivo* by intravascular injection (Calas, 1981, 1985). In each case, labelled fibres which seem to penetrate between intermediate and posterior lobes or from the latter, send processes and terminals into the glandular parenchyma where terminals can be detected making close contact with secretory cells or even invaginating into their cytoplasm (Figs 2, 3). Differ-entiated synaptoid-like contacts can be observed between RAG labelled fibres and secretory cells (Friedman *et al.*, 1983; Léranth *et al.*, 1983; Calas, 1985).

The main problem that arises from these findings is the following: are all or some of these fibres truly serotonergic? It is clear from the 5-HT content of the IL (4.45 ng/mg protein) that all the amine is not neuronal in origin. Half of the 5-HT biochemically detectable in the IL can be 'washed out' by saline perfusion, showing a high level to be associated with the blood. Another part of IL 5-HT which persists after stalk transection and perfusion is probably located in the mast cells on the surface of the IL (Palkovits *et al.*, 1986). The latter part corresponds to neuronal 5-HT issued from central fibres processing through the stalk (Friedman *et al.*, 1983; Holzbauer *et al.*, 1985a, b; Palkovits *et al.*, 1986). This neuronal 5-HT is abolished by *para*-chlorophenylalanine (p-CPA) treatment but 5-HT immunoreactive fibres are also destroyed by intravenous 6-hydroxydopamine (6-OH-DA), a catechol-amine neurotoxin, in the rat (Saland *et al.*, 1986) but not in the mouse (Payette *et al.*, 1985). Thus it can be hypothesized that 5-HT detected in rat IL fibres originates from uptake and/or that 5-HT coexists with dopamine in

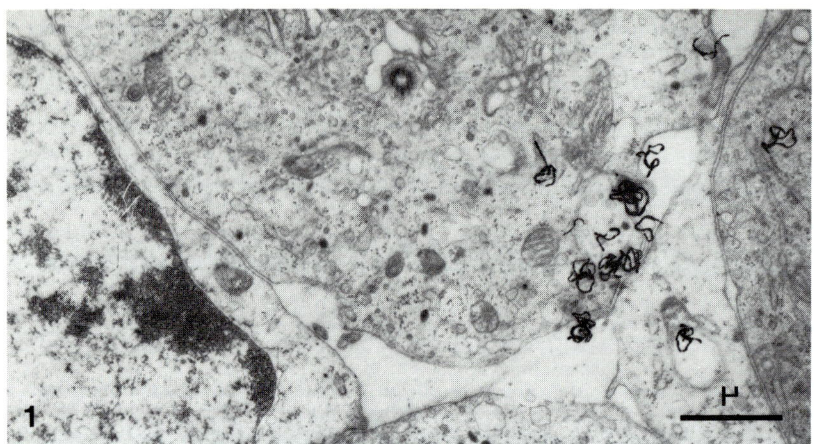

Fig. 1 Thin section of pars tuberalis after intravenous injection with ³H-5-HT and processing for radioautography (microdol developer). An intensely labelled bouton containing both clear and large granular vesicles (LGV) is invaginated within a secretory cell

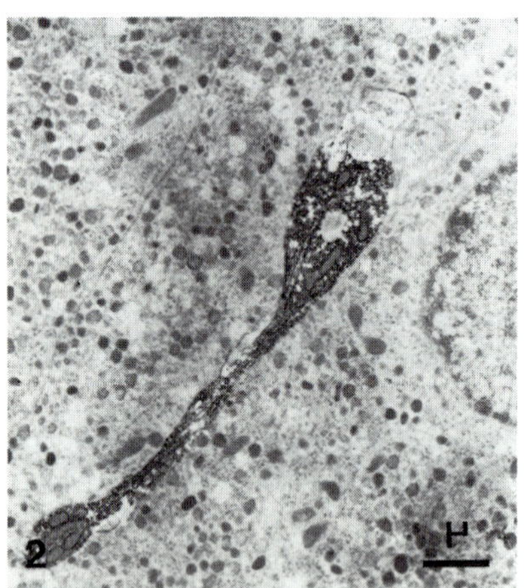

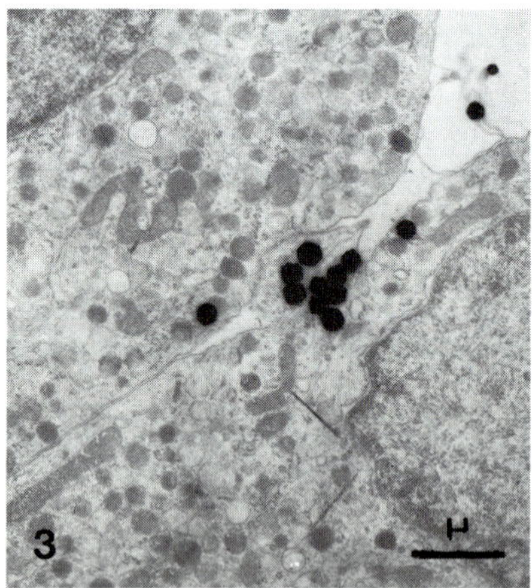

Fig. 2 Immunohistochemical localization of serotonin in a thin section of pars intermedia. A varicose fibre squeezed between secretory cells shows peroxidase labelling (by courtesy of C. Decavel)

Fig. 3 Radioautograph of thin section of pars intermedia after intravenous injection of ³H-5-HT (phenidon developer) showing a labelled bouton squeezed between two secretory cells

the nerve terminals (Saland *et al.*, 1986). Such intraneuronal colocalization could also occur with GABA, whose synthesizing enzyme, glutamic acid decarboxylase (GAD), has also been detected in IL fibres (Vincent *et al.*, 1982) and which has been claimed to coexist with 5-HT in some neuronal systems (Belin *et al.*, 1983; Gamrani *et al.*, 1984). The immunocytochemical detection of 5-HT specific biosynthetic enzyme tryptophan hydroxylase could help to resolve the problem. Such an approach would be also very useful in the third part of the adenohypophysis, the tuberal lobe, where labelled neuronal processes have been detected by radioautography after intra-vascular administration of ^{3}H-5-HT. These are thin fibres enlarged into varicosities which are in close contact with glandular cells or invaginate within their cytoplasm (Calas, 1985) as shown in Fig. 1. Their probable serotonergic nature does not exclude coexistence with other transmitters.

Neurohypophysis

The three anatomical zones of the neurohypophysis display a relatively discrete but variable serotonergic innervation (Bouchaud and Bosler, 1986).

Within the neural lobe (NL) the 5-HT immunoreactive (Steinbusch and Nieuwenhuys, 1981) or ^{3}H-5-HT radioautographically labelled (Calas, 1985; Payette *et al.*, 1985; Westlund and Childs, 1982) fibres seem to be more abundant in rat and mouse (Payette *et al.*, 1985) than in cat (Sano *et al.*, 1982). They are particularly (Saland *et al.*, 1986) or even exclusively (Léranth *et al.*, 1983) present in the areas adjacent to the IL. They are coarser in appearance than those of the pars intermedia, are depleted of immunoreac-tive 5-HT by treatment with p-CPA but not with *para*-chloroamphetamine (p-CA) and are resistant to 6-OH-DA (Saland *et al.*, 1986). At the electron microscope level RAG labelled and immunopositive fibres display clear and large granular vesicles (LGV) which can respectively predominate in some axonal profiles. They are sometimes in contact with basement membranes (Fig. 5) and, as a rule, with a pituicyte process (Calas, 1985), as shown in Figs 4 and 5. In some cases, they seem to be postsynaptic to another type of axonal process (Fig. 4). The bundle of 5-HT nerve fibres is more dense in the infundibular stalk (Payette *et al.*, 1985) than in the NL but it does not seem to ramify.

On the other hand, both 5-HT fibres and terminals are localized within the median eminence. They were first detected by radioautography following intraventricular injection of ^{3}H-5-HT (Calas *et al.*, 1974). This result has been confirmed by the same technique (Chan-Palay, 1977; Descarries and Beaudet, 1978) and also by immunohistochemistry (Steinbusch and Nieuwen-huys, 1981). These fibres can also be labelled by ^{3}H-5-HT injected into the carotid (Calas, 1985). They are particularly frequent in the external layer where they can contact the external basement membrane and/or the

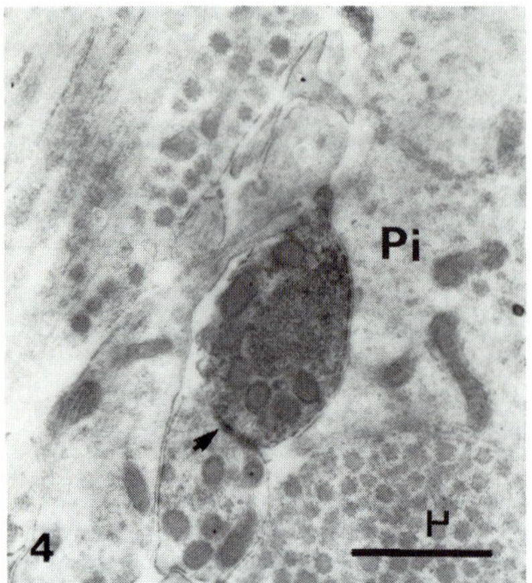

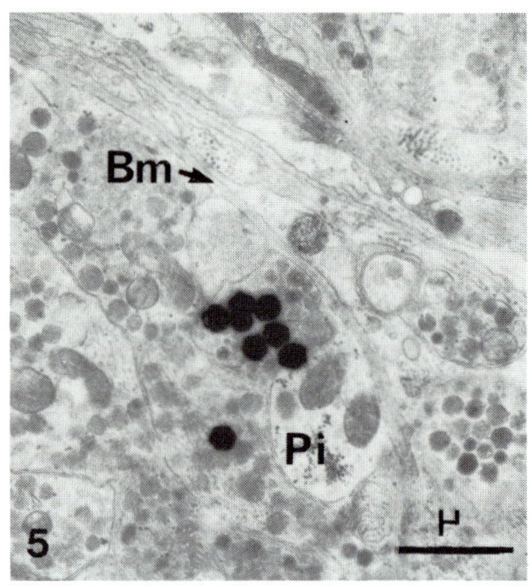

Fig. 4 Thin section of neural lobe after treatment with anti-serotonin antibodies and revealed by the peroxidase/anti-peroxidase method. The labelled varicosity displays an aggregation of small clear vesicles and some LGV. It is in close apposition with a pituicyte (Pi) and makes a postsynaptic-like contact (arrow) with an unlabelled terminal displaying clear and dense core vesicles (by courtesy of C. Decavel)

Fig. 5 Radioautograph of thin section of neural lobe after intravenous injection of ^{3}H-5-HT (phenidon developer). A labelled fibre is in close contact with basement membrane (Bm) and probably a pituicyte process (Pi). It contains numerous LGV and some clear vesicles

processes of tanycytes. In this area double labelling experiments combining radioautographic detection of ^{3}H-5-HT uptake with immunohistochemistry (of peptides or neurotransmitter biosynthetic enzymes) proved that 5-HT and catecholamine terminals are separate (Calas, 1985) as shown in Fig. 6, but may make close contact (Bosler and Beaudet, 1985b; Bouchaud and Bosler, 1986; Calas, 1985). A close relationship may exist between gonadotrophin-releasing hormone (Gn-RH) boutons and 5-HT fibres (Jennes *et al.*, 1982) as well as between thyrotropin-releasing hormone (TRH) and 5-HT nerve terminals, yet without typical synapses; in rare instances, TRH immunoreactive terminals can also take up ^{3}H-5-HT (Nakai *et al.*, 1983). This suggests the coexistence of both mediators in the same neuron, as shown in other parts of the nervous system (Hökfelt *et al.*, 1980a, b).

Hypothalamus

A detailed account of the 5-HT innervation of the different hypothalamic nuclei is beyond the scope of this review (see Descarries and Beaudet, 1978).

We shall only briefly describe the principles of this innervation (Törk, 1985) before summarizing the most recent morphofunctional data concerning the interrelationship between 5-HT and other chemically defined neurons within this region.

According to Steinbusch and Nieuwenhuys (1981) the hypothalamus as a whole is one of the highest serotonin-innervated areas within the central nervous system. The density of 5-HT fibres and terminals is maximal, at least for the rat, in the ventromedial part of the suprachiasmatic nucleus, the lateral part of the medial preoptic nucleus (Steinbusch and Nieuwenhuys, 1981), the dorsal and ventral premamillary nuclei and the mamillary nucleus itself, the nucleus ventromedialis and the ventral part of the nucleus dorsomedialis (NDM).

Remarkably, the magnocellular part of the paraventricular nucleus (but not its parvocellular part) and the supraoptic nucleus are very sparsely innervated by 5-HT fibres. A dense innervation has been detected by radioauto-

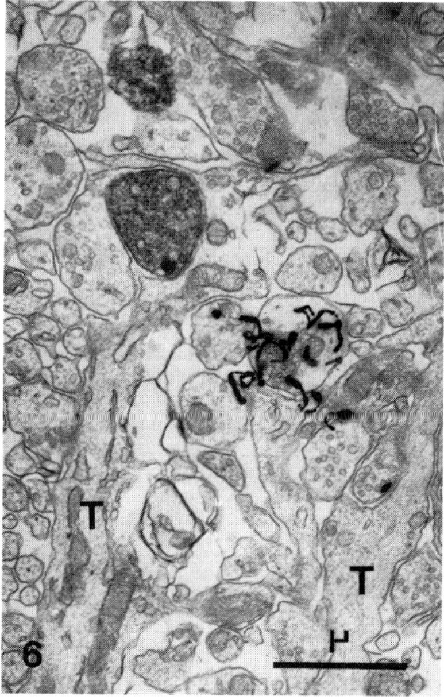

Fig. 6 Radioautograph of thin section of ME after ³H-5-HT intraventricular injection and immunocytochemical detection of tyrosine-hydroxylase (TH). Two terminals, each labelled by one procedure, are clearly distinct but in proximity. The ³H-5-HT labelled bouton displays an LGV and some clear vesicles. T: tanycyte process. By courtesy of O. Bosler and A. Beaudet

graphy in the organum vasculosum laminae terminalis (OVLT) (Bosler, 1978; Jennes *et al.*, 1982; Moore, 1977; Takeuchi and Sano, 1983) (Fig. 7).

It remains an open question whether serotonergic perikarya exist in the hypothalamus. The data described by Chan-Palay (1977) still await confirmation, while the small group of cell bodies identified by radioautography in the ventral part of the NDM (Beaudet and Descarries, 1979) has been found to display 5-HT immunoreactivity after intraventricular injection of 5-HT, dopamine (DA), noradrenaline (NA) or β-adrenergic agonist (Arezki *et al.*, 1985). Furthermore, ³H-5-HT labelled neurons are always distinct from dopaminergic, which are localized in close proximity (Arezki *et al.*, 1987).

These neurons also display 5-HT immunoreactivity after the administration of tryptophan and a monoamine oxidase (MAO) inhibitor (Frankfurt *et al.*, 1981) or after treatment with high dosages of MAO inhibitor alone (Ueda *et al.*, 1984).

However, since a high density of 5-HT terminals exists at this level it may be rationalized that the cellular labelling might be due to an uptake of 5-HT

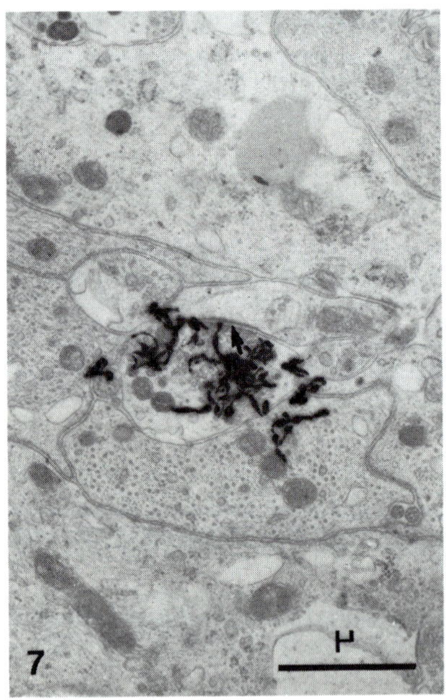

Fig. 7 Radioautograph of thin section of OVLT after intraventricular injection of ³H-5-HT. A labelled terminal containing clear, dense core and large granular vesicles constitutes a synaptoid contact with what is probably a tanycyte process (arrow). By courtesy of O. Bosler

from loaded terminals and not to a synthesis within the cell bodies. In any case, the persistence of a noticeable 5-HT content following hypothalamic differentiation (Palkovits *et al.*, 1977) argues in favour of an intrinsic hypothalamic 5-HT system.

The formaldehyde-induced fluorescence within the tanycytes (Kent and Sladek, 1978) requires confirmation and explanation by the most recent techniques of mediator identification. When continued on the same preparation, these methods allowed the description of specific interrelationships between 5-HT and other systems neurochemically identified (Calas, 1985), viz.:

(a) Serotonergic afferents upon vasoactive intestinal peptide (VIP) containing neurons in the suprachiasmatic nucleus (SCN) (Kiss *et al.*, 1984; Bosler and Beaudet, 1985a), as shown Fig. 8, and on neuropeptide

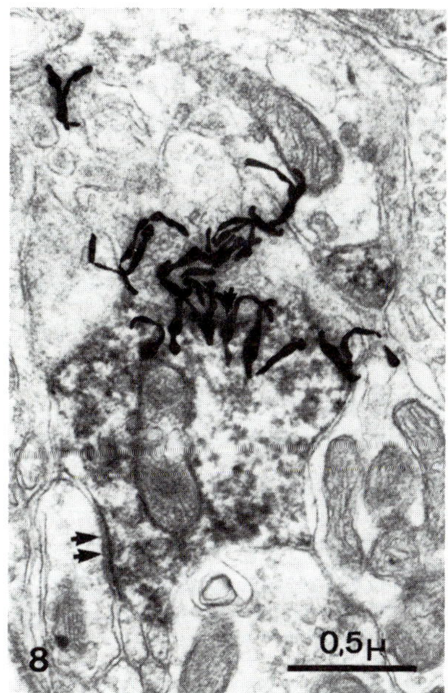

Fig. 8 Radioautograph of thin section of suprachiasmatic nucleus after ³H-5-HT intraventricular injection and immunocytochemical detection of VIP. The terminal overlaid by silver grains constitutes a synaptic contact (arrow) upon a peroxidase labelled dendrite which receives another unlabelled synaptic afferent in the plane of the section (double-arrow). By courtesy of O. Bosler and A. Beaudet

Y (NPY) immunopositive neurons in the arcuate nucleus (Guy *et al.*, 1986).

(b) 5-HT terminals upon pro-opio-melanocortin immunoreactive neurons in the arcuate and ventromedial nuclei (Bosler *et al.*, 1982; Kiss *et al.*, 1984; Bosler and Beaudet, 1985b).

(c) Serotonergic terminals in axo-axonic contact with Gn-RH boutons in the OVLT (Jennes *et al.*, 1982) and with NPY axons in the NDM (Guy *et al.*, 1986) and in the SCN (Guy *et al.*, unpublished data).

(d) Axo-somatic contacts between 5-HT terminals and Gn-RH neurons in the septo-preoptic region (Jennes *et al.*, 1982; Kiss and Halàsz, 1985) and between 5-HT fibres and dopaminergic neurons in the arcuate nucleus (Fig. 9) and the zona incerta (Bosler, 1984).

(e) Axo-axonic contacts between 5-HT and GABA terminals in the SCN (Bosler, 1984).

(f) Several types of contacts (axo-axonic, dendritic and somatic ones) between GAD immunopositive terminals and 5-HT neurons in the NDM (Arezki *et al.*, 1987).

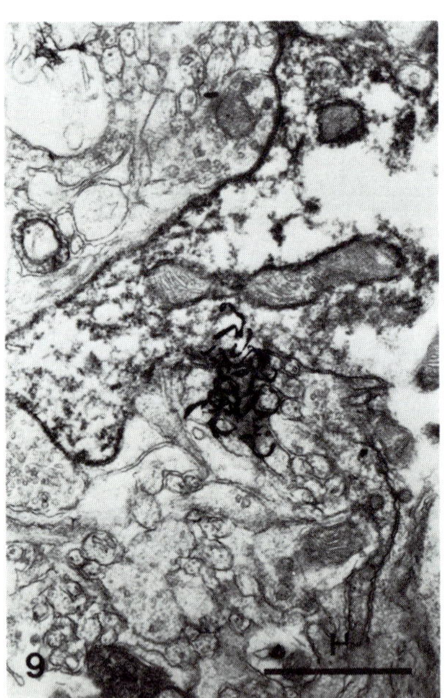

Fig. 9 Radioautograph of thin section of arcuate nucleus after ^{3}H-5-HT intraventricular injection and immunocytochemical detection of TH. A radioautographically labelled terminal is in close contact with an immunopositive dendrite. By courtesy of O. Bosler and A. Beaudet

All these interactions can also now be added to the data on the anatomical definition of axonal pathways which allows the description of serotonergic systems innervating the hypothalamus.

Using electrolytic lesions combined with biochemical assays, it has been tentatively concluded that the median raphe nucleus is the primary source of 5-HT innervation of the suprachiasmatic nucleus, anterior hypothalamus and medial preoptic area, while the 5-HT inputs to the antero-lateral hypothalamic area and arcuate nucleus appear to derive from both the dorsal and median raphe nuclei (Moore *et al.*, 1978; Van de Kar and Lorens, 1979). More precisely, the dorsal third of the dorsal raphe nucleus is a source of 5-HT innervation of the ME (Villar *et al.*, 1984). Similarly, lesions associated with immunocytochemistry showed that 5-HT fibres of the intermediate lobe of the pituitary may originate from cells both in the raphe and hypothalamus (perhaps the NDM; Mezey *et al.*, 1984). By combining 5-HT immunohistochemistry with the retrograde transport of fluorescent tracers it has also been possible to describe the origin of serotonergic projections to the supraoptic and paraventricular nuclei which appear to arise form three distinct 5-HT cell groups, B7, B8 and B9 in the mesencephalic raphe, and seem to affect preferentially the regions where oxytocinergic neurons are concentrated (Sawchenko *et al.*, 1983).

Summarizing the data described above, the following problems still have to be resolved concerning the morphological aspects of interactions between 5-HT and hypothalamo-hypophyseal system:

(a) The meaning of the 5-HT uptake and/or content in some adenohypophyseal cells (Johns *et al.*, 1982; Payette *et al.*, 1985).

(b) The existence of 5-HT fibres (and even of neuronal elements) in the anterior pituitary lobe.

(c) The possible intra-axonal coexistence of several transmitters (DA, GABA, peptides) with 5-HT in those areas to which they could be specific, and particularly in the intermediate lobe of the hypophysis.

(d) The existence of an intrahypothalamic 5-HT neuronal system arising from the nucleus dorsomedialis which could constitute a component of the serotonergic innervation of the pituitary.

(e) The precise definition of 5-HT innervation (cell bodies of origin and tracts) of the different parts of the system. Morphological tools are now available for such a study which could be particularly beneficial if combined with developmental studies (see Wallace and Lauder, 1983).

(f) The neurochemical identification of interacting neuronal elements within the hypothalamus and neurohypophysis. Such chemically defined interactions have been described above. It can be anticipated that the association of different types of labelling on the same preparations will multiply in the next few years and these data will require confirmation and interpretation from physiological studies.

PHYSIOLOGICAL DATA

As in the previous section we shall consider first the different strategies used for determining the role of 5-HT in pituitary hormone secretion before describing data obtained for each hormonal axis.

Methodological approaches

(For a review see Kordon *et al.*, 1980, 1981; Jimenez and Walker, 1985.) The use of histophysiological methods has allowed the visualization of aminergic tracts involved in the neuroendocrine function and also the correlation of the intensity of the reaction in specific brain areas with hormone secretion. Besides the fact that histochemical methods are only semi-quantitative, their main disadvantage lies in the difficulty of determining which of the two variables (amine activity or hormone secretion) is the cause or the consequence of the other. Moreover, the variations of an amine usually simultaneously affect the release and/or the synthesis of several hormones, and the correlations between amine activity and hormone secretion are difficult to establish.

The stereotaxic experiments involving lesions, surgical ablation, or electrical stimulation enable the importance of brain areas in pituitary function to be determined but do not resolve the problem of specificity and localization of the neurotransmitter–neurohormone interaction site.

In contrast, the pharmacological methods permit the localization of the causal link between neurotransmitter and hormone, but the localization of the interaction site remains difficult to determine except when local injections can be performed (e.g. of 5,6- or 5,7-DHT). The drugs used can act on the biosynthesis of the amine or at the receptor sites. This action is not specific to one area and the simultaneous changes in various systems that regulate hormone secretion directly or indirectly should be considered in interpreting the results. Moreover, the specificity of the drugs used should also be considered because some drugs are specific only when they are used in a narrow dose range.

Another methodology is the direct administration of the neurotransmitter into the brain ventricles. Pharmacological doses (several hundred times their endogenous concentration) are very often used to obtain an effect by direct infusion and can induce side-effects on other brain structures and unspecific activation of other amine systems. This kind of approach can produce contradictory data, and pretreatment with transmitter antagonists should be performed before the final interpretation.

In vitro incubation methods of pituitaries, hypothalamic fragments, brain slices or synaptosomal fractions are the only techniques which permit the characterization and localization of the neurotransmitter–neurohormone interaction at the cellular level. These *in vitro* studies are limited by the survival of the cells. More recently, the use of pituitary or hypothalamic isolated cell cultures or organotypic hypothalamic explant cultures tends to overcome this disadvantage. However, the extension of such results to physiological and *in vivo* conditions may be sometimes hazardous.

In conclusion, the use of complementary approaches is necessary to give more confidence in interpreting each kind of data.

Regulation of anterior pituitary hormone secretion

Thyrotropin secretion

Conflicting results are reported concerning the involvement of 5-HT in the control of thyrotropin function (see Jordan, 1985). Jordan *et al.* (1978) reported that serotonin injection into the third ventricle caused a rapid increase in serum thyroid-stimulating hormone (TSH) (with a dose-related stimulatory effect from 1 to 10 µg) and that the effect was completely prevented by pretreatment with cyproheptadine (a specific 5-HT receptor blocker at the dose range used). In contrast, Krulich *et al.* (1979) reported a decrease in TSH concentration in rats after intraventricular injection of serotonin (4–20 µg) or systemic administration of quipazine, a serotonin agonist; the effect of quipazine was antagonized by cyproheptadine but not by methysergide. Di Renzo *et al.* (1979) further found that the inhibitory effect of fenfluramine was antagonized by raphe lesions and by intraventricular injection of 5,7-DHT. Ruzsas *et al.* (1979) reported that p-CPA or surgical lesions of serotonergic systems increased blood and pituitary levels of TSH. Jordan *et al.* (1978) reported that 5-HT appears not to have a direct pituitary effect on thyrotropin *in vitro*. Grimm and Reichlin (1973) have shown, also *in vitro*, that 5-HT has an inhibitory effect on TRH synthesis but this was not confirmed by others who further found no effect of 5-HT on TRH release (Joseph-Bravo *et al.*, 1979; Maeda and Frohman, 1980). More recently, Chen and Ramirez (1981), using a superfusion model of the hypothalamus, reported that low concentrations of 5-HT stimulate TRH release.

The injection of 5-hydroxytryptophan (5-HTP), the precursor of 5-HT, has been shown to induce either an increase of TSH release (Chen and Meites, 1975; Sakoda *et al.*, 1979) or to have no effect (Mueller *et al.*, 1976) and to inhibit the TSH increase following cold exposure (Tuomisto *et al.*, 1975; Onaya and Hashizume, 1976). Tryptophan has been reported to inhibit (Mueller *et al.*, 1976) or to have no effect (Mattila and Männisto, 1981) on basal TSH release, but it has been shown to inhibit TSH response to cold exposure.

The data concerning the effects of p-CPA on basal TSH release are conflicting (inhibitory, stimulatory or without effect; Tuomisto *et al.*, 1975; Ruzsas *et al.*, 1979, Mattila and Männisto, 1981, respectively). All reports agree, however, on the absence of the effect of p-CPA on TSH response to cold (Tuomisto *et al.*, 1975; Onaya and Hashizume, 1976) and its ability to abolish the diurnal circadian peak of TSH (Scapagnini *et al.*, 1978; Jordan *et al.*, 1979; Fukuda *et al.*, 1980) which can be reversed by 5-HTP. The implication of 5-HT in the circadian rhythm of TSH was also confirmed by the intracerebroventricular injection of the 5-HT neurotoxin, 5,6-DHT, which decreased or abolished the diurnal peak of TSH (Scapagnini *et al.*, 1978; Jordan *et al.*, 1979) and electrolytic lesion of raphe nuclei which also induced a decrease in the amplitude of the TSH peak (Jordan *et al.*, 1979).

In conclusion, the role of serotonin neurons in TSH secretion remains controversial in relation to the basal TSH release, but the amine is implicated as a modulating factor in the TSH circadian rhythm.

Prolactin secretion

Serum prolactin (PRL) concentration increased after the administration of the serotonin precursor 5-HTP, particularly when it was given with a monoamine oxidase inhibitor (Lu and Meites, 1973; Fuller and Clemens, 1981) or after treatment with serotonin agonists (quipazine; Meltzer *et al.*, 1976; Clemens *et al.*, 1977) or a serotonin 'releaser' (p-CA; Van de Kar and Bethea, 1982; or fenfluramine; Van de Kar *et al.*, 1985). Prolactin concentrations also increased after the intraperitoneal or third ventricle injection of 5-HT (Delita *et al.*, 1980; Beccu and Libertin, 1982; Krulich *et al.*, 1979; Lawson and Gala, 1975). A stimulatory role of serotonin has also been demonstrated in the suckling- or oestrogen-induced release of prolactin (Kordon *et al.*, 1973; Mena *et al.*, 1976; Caligaris and Taleisnik, 1974). Serotonergic stimulation of prolactin release could be abolished by a discrete lesion of the dorsal raphe (Van de Kar *et al.*, 1982). Moreover, an intact neurointermediate pituitary lobe (NIL) is required to observe the 5-HTP induced increase in plasma PRL (Johnston *et al.*, 1986). Some factors present in the NIL could be responsible for mediating the 5-HTP induced increase in PRL.

No effect of 5-HT on prolactin secretion was reported using an *in vitro* system with rat anterior pituitary cells (Delita *et al.*, 1980) and it seems that serotonergic stimuli act by enhancing the secretion of a prolactin-releasing factor (PRF) (Clemens *et al.*, 1978). Vasoactive intestinal polypeptide (VIP) could be implicated in the positive effect of 5-HT on PRL secretion since it acts as a PRF *in vivo* (Kato *et al.*, 1978) and *in vitro* (Ruberg *et al.*, 1978;

Enjalbert *et al.*, 1980; Hagen *et al.*, 1986). Moreover, an intraventricular injection of 5-HT induced an increase of VIP concentration into hypophyseal portal blood followed by an increased release of pituitary PRL (Shimatsu *et al.*, 1982).

The diurnal rhythm of PRL could be abolished by pretreatment with p-CPA (Scapagnini *et al.*, 1978) or destruction of midbrain raphe nuclei (Dunn *et al.*, 1980). This cyclic secretion of PRL can be potentiated by reconstitution of serotonin biosynthetic pathways by 5-HTP administration after pharmacological inhibition of tryptophan hydroxylase (Quattrone *et al.*, 1981).

In conclusion, the involvement of 5-HT in the basal or rhythmic prolactin secretion seems to be clearly demonstrated.

Gonadotrophin secretion

Intraventricular administration of 5-HT or increasing the levels of endogenous serotonin by systemic administration of 5-HTP, or a monoamine oxidase inhibitor, or both, have been reported as decreasing luteinizing hormone (LH) secretion (Schneider and McCann, 1970; Kordon and Glowinski, 1972; Vijayan *et al.*, 1978), indicating a role for this amine in gonadotrophin release.

Electrical stimulation of the dorsal raphe nucleus suppresses the pulse-release of LH (Gallo and Moberg, 1977; Arendash and Gallo, 1978), indicating that endogenous 5-HT could inhibit LH release. These effects were prevented by prior treatment with p-CPA and were restored by 5-HTP administration. However, the administration of p-CPA failed to modify gonadotrophin release in castrated animals.

Pharmacological inhibition of 5-HT biosynthesis or blockade of the receptors completely abolished the circadian variation in LH (Héry *et al.*, 1976) and lesion of the serotonin innervation of the SCN resulted in a marked decrease in the amplitude of the circadian hormone rhythm (Héry *et al.*, 1978).

Serotonergic tone may be needed for the expression of the preovulatory discharge of LH since administration of p-CPA or lesions of raphe nuclei resulted in an inhibition of oestrogen-induced LH release (Héry *et al.*, 1976; Iyengar and Rabbi, 1983). The contribution of serotonin to a normal expression of the LH positive feedback response to oestrogen in female rats can be masked by inhibitory effects on LH secretion which sometimes accompany experimental stimulation of 5-HT synthesis. Using a sequential double chamber perfusion system, it has been shown *in vitro* that serotonin stimulated the gonadotrophin release from incubated median eminence (Vitale *et al.*, 1985) but 5-HT does not affect directly LH or FSH release (Schneider and McCann, 1970). Furthermore, daily variations in serotonin metabolism in the SCN are modified by oestradiol and are required for ovulation in the rat (Héry *et al.*, 1982). Ovarian steroids could modulate the rate of serotonin uptake and partially coordinate the serotonergic component of LH release (Meyer and Quay, 1976; Meyer and Eadens, 1985).

The action of the drugs modifying serotonin function could be different depending on the time of the administration with respect to cyclic LH release. The facilitatory effect of 5-HT occurs only during a 'critical period' and the daily pattern of hypothalamic 5-HT metabolism influences the pattern of the secretion during the oestrogen positive feedback response (Walker, 1983). The differential effects of 5-HT may be due to important temporal or hormonal factors (Coen and MacKinnon, 1979; Chen *et al.*, 1981) as well as to the existence of anatomically and functionally separate serotonergic neuronal systems influencing LH secretion. Johnson and Crowley (1983) reported that 5-HT innervation to the preoptic area may stimulate LH release in female rats while the medio-basal hypothalamus may contain an inhibitory 5-HT system. They also reported that central 5-HT neurons innervating the preoptic area are involved in the LH surge induced by progesterone but not in the increase in LH occurring after treatment with oestradiol alone or after blockade of opiate receptors (Johnson and Crowley, 1986). Furthermore, it has been shown that ovarian steroids could stimulate 5-HT synthesis in the preoptic area–anterior hypothalamus but they do not seem to influence 5-HT synthesis in the median eminence or the medio-basal hypothalamus (King *et al.*, 1986).

It has been reported recently that the serotonergic system has an inhibitory effect on the development of the positive feedback of ovarian steroids on LH secretion; this may be an example for 5-HT having a regulatory role in the onset of puberty (Moguilevsky *et al.*, 1985).

In conclusion, the possible existence of anatomically separate excitatory and inhibitory systems involving 5-HT, temporal factors and the steroid hormone environment all may contribute to determining the direction of 5-HT effects on gonadotrophin secretion. Further studies are required to establish the nature of 5-HT interaction with LH-releasing hormone (LH-RH) cells and fibres, the mechanisms by which ovarian hormones alter 5-HT activity (stimulation of 5-HT synthesis, King *et al.*, 1986; and/or change of 5-HT receptors, Biegon *et al.*, 1983) and the interaction of 5-HT activity with other regulators such as dopamine, norepinephrine, epinephrine, and opiates.

Growth hormone secretion

Intraventricular injection of serotonin, systemic injection of 5-HT or serotonin agonist administration (quipazine) have all been shown to increase serum levels of growth hormone (GH) (Collu *et al.*, 1972; Smythe, 1977; Vijayan *et al.*, 1978), suggesting a stimulating role for serotonin neurons in growth hormone release in rats. It is however questionable as to whether quipazine is a specific serotonergic drug. The stimulatory effect of 5-HTP and quipazine can be blocked respectively by cyproheptadine (Smythe, 1977) and methysergide (Vijayan *et al.*, 1978). Controversy exists concerning the effect of p-CPA, which can lower GH secretion in the rat as does methysergide (Martin *et al.*, 1978) or have no effect on episodic GH secretion, as in the case of p-CA (Clemens, 1978; Eden *et al.*, 1979). However, p-CPA can decrease GH responsiveness to some stimuli, more particularly hypoglycaemia (Bluet-Pajot *et al.*, 1980).

To investigate the precise mechanism by which serotonergic stimuli influence GH secretion, Murakami *et al.* (1986) recently showed that passive immunization of rats with antiserum specific for rat growth hormone-releasing hormone (GRH) blunts the plasma GH increase induced by 5-HTP or 5-HT, suggesting that GH secretion induced by serotonergic mechanisms is mediated at least in part by hypothalamic GRH. On the other hand, Lopez *et al.* (1986), using two experimental models, hypophysectomized autografted rats and perfused pituitaries, recently showed that 5-HT may stimulate GH secretion through a direct pituitary action.

The spontaneous rise in GH that occurs in rats about the time of onset of darkness is inhibited by certain serotonin antagonists, cyproheptadine and metergoline (Arnold and Fernstrom, 1978) or pizotifen (Clarenbach *et al.*, 1980). In conclusion, despite some controversies about the effect of p-CPA, it seems that 5-HT has essentially a stimulating effect on the physiological and experimentally modified GH secretion in rat.

Adrenocorticotrophin secretion

Many reports suggest that serotonergic mechanisms are involved in the regulation of the function of the pituitary adrenocortical axis. There is considerable pharmacological evidence consistent with the hypothesis that the central release of serotonin has a stimulating influence on this function.

In vivo injection of precursors, releasers, reuptake inhibitors or receptor agonists of 5-HT has been shown to increase plasma corticosterone levels (Popova *et al.*, 1972; Meyer *et al.*, 1978; Fuller *et al.*, 1976; Fuller and Snoddy, 1980; Fuller, 1981) and plasma immunoreactive ACTH and beta-endorphin (Bruni *et al.*, 1982; Sapun *et al.*, 1981). Intraventricular administration of serotonin also increased ACTH secretion in rats (Rose and Ganong, 1976) and serotonin has been shown to stimulate corticotrophin-releasing factor (CRF) secretion *in vitro* (Jones *et al.*, 1976; Nagakami *et al.*, 1986). Gibbs and Vale (1983) also reported that administration of the serotonin uptake inhibitor fluoxetine to anaesthetized rats caused increased release of CRF into hypophyseal portal blood.

Evidence for a direct dose-dependent stimulatory effect of 5-HT on ACTH release from dispersed anterior pituitary cells incubated *in vitro* has been shown by Spinedi and Negro-Vilar (1983), which contradicts previous data reporting no direct effect of the amine on ACTH secretion from the anterior pituitary *in vitro* (Rose and Ganong, 1976).

In contrast, other groups have reported inhibitory effects of 5-HT on ACTH release, particularly on the stress-induced secretion of ACTH (Telegdy *et al.*, 1976; Vernikos-Danellis *et al.*, 1977; Ixart *et al.*, 1980) or even no effect at all (Dixit and Buckley, 1969; Bhattacharya and Marks, 1970); the explanation for these conflicting results is unclear. Drugs used to deplete serotonin are sometimes given at high doses and could influence other transmitters in the brain and subsequently different physiological processes which can be responsible for the different effects observed.

Serotonin could also modulate ACTH secretion by altering its circadian rhythm rather than directly stimulating or inhibiting it. There have been studies showing that pharmacological inhibition of 5-HT biosynthesis abolished or attenuated the normal diurnal rhythm of plasma corticosterone (Scapagnini *et al.*, 1971; Van Delft *et al.*, 1973; Rotsztejn *et al.*, 1977) and the circadian variation in ACTH (Szafarczyk *et al.*, 1979). The action of 5-HT may only be permissive to those hormone responses which also need other non-serotonergic signals to be expressed.

Extensive studies have also attempted to localize 5-HT neurons involved in these effects. Like pharmacological depletion of whole-brain 5-HT, lesions of the raphe nuclei have been reported to induce disruption of the circadian rhythm of corticosterone secretion (Scapagnini and Preziosi, 1972) but later, with more precise histological control, it has been shown that the plasma corticosterone rhythm persists in rats bearing lesions of the nuclei raphe centralis superior and raphe dorsalis (Rotsztejn *et al.*, 1977) and that the basal corticosterone secretion is not modified after lesions of either the dorsal or median raphe nucleus (Van de Kar *et al.*, 1982). On the other hand, lesions of the SCN or of the 5-HT inputs to the SCN with 5,7-DHT result in a marked decrease in the amplitude of the circadian rhythm (Szafarczyk *et al.*, 1979) or abolition of the diurnal corticoid rhythm (Moore and Eichler, 1972; Williams *et al.*, 1983).

Thus, the suprachiasmatic nucleus may be an anatomical site at which serotonergic systems act to control the corticoid rhythm. Additional findings that may support serotonergic involvement in the corticoid rhythm are that diurnal changes in serotonin concentration in discrete brain regions seem to correlate with diurnal changes in corticosterone secretion (Scapagnini and Preziosi, 1972).

In conclusion, data on the relation of brain serotonin neurons to ACTH secretion are somewhat conflicting, while its involvement in adrenocorticotropic rhythms is well established.

In an attempt to summarize the data described above it appears that serotonin exerts an influence on the secretion of the different anterior pituitary hormones (Tuomisto and Männisto, 1985). This effect is very often mediated through the control of either hypothalamic releasing or inhibiting factors. A direct effect at the anterior pituitary level is only demonstrated for ACTH and growth hormone release. An effect of 5-HT, which seems well documented, is the control of hormonal rhythms contributing in the regulation of the amplitude for certain hormonal cyclic secretions such as TSH, ACTH and PRL (for a review see Szafarczyk, 1986).

Regulation of the pars intermedia hormone secretion

The pars intermedia (PI) of the adenohypophysis synthesizes pro-opio-melanocortin (POMC) which, after post-translational processing, produces different peptides including αMSH (melanotrophin), ACTH, CLIP (corticotrophin-like intermediate lobe peptide), LPHs (lipotrophins) and endorphins. Using an *in vitro* system, Randle *et al.* (1983) have shown that serotonin stimulated the release of ACTH bioactivity in a dose-related manner, but 5-HT had a smaller stimulatory effect on the release of αMSH and LPHs immunoreactivity. This confirms previous data showing that 5-HT stimulated ACTH release in a dose-related manner from dispersed PI cells, although only with high concentrations of 5-HT (Kraicer and Morris, 1976), but is in disagreement with data from Fischer and Moriarty (1977) who found no effect of 5-HT on ACTH release from the isolated neurointermediate lobe. Cells of the PI preferentially secrete beta-endorphin *in vivo* and it seems that serotonin drugs do not influence this release either *in vivo* or *in vitro* (Sapun-Malcolm *et al.*, 1983, 1986).

In conclusion, serotonin appears to exert a differential control on the release of some of the peptide activities from the pars intermedia.

Regulation of the hypothalamo-posthypophyseal axis

Serotonergic neuronal systems have been implicated in the control of prolactin release (described earlier) and of milk secretion in the rat, but there is less definite evidence of the involvement of 5-HT in release of oxytocin (OT) induced by suckling. Systemic administration of 5-HT, 5-HTP or monoamine oxidase inhibitors before suckling in conscious rats has been reported as inhibiting milk ejections by a central effect with an inhibition of OT release (Mizuno *et al.*, 1967). Suckling acts also on the release and synthesis of 5-HT in the hypothalamus (Mena *et al.*, 1976).

More recently, opposing effects for 5-HT have been reported for conscious and anaesthetized rats (Moos *et al.*, 1983). Neither serotonin nor its precursor affected the milk ejection reflex in conscious animals but had an inhibitory effect under urethane anaesthesia. Furthermore, the activity of paraventricular oxytocinergic neurones was inhibited after microelectrophoretic application of 5-HT in urethane-anaesthetized lactating rats (Honda *et al.*, 1985). The inhibition of 5-HT synthesis or the use of 5-HT receptor antagonists prevents OT release during suckling in conscious rats, and 5-HTP partially reverses this effect. In contrast, these drugs fail to prevent the milk ejection in anaesthetized animals (Moos *et al.*, 1983). An explanation for the opposing results obtained with conscious and anaesthetized animals still has to be elucidated. Clarke and Wright (1985) have shown that 5-HT can suppress or stimulate OT release during suckling in anaesthetized rats depending on the method of administration (spinal or intraventricular injection).

There is also some evidence that 5-HT plays a role in regulating vasopressin (VP) secretion from the neural lobe in rats. Serotonin has been reported to stimulate VP release from rat NL *in vitro* (Lemay *et al.*, 1979). Fenfluramine or quipazine increased plasma VP, an effect prevented by pretreatment with p-CPA. The 5-HT induced elevation of VP is not due to an activation of the peripheral renin–angiotensin system since quipazine and p-CA have been shown to stimulate VP secretion in rats pretreated with a specific antagonist of angiotensin II receptors or in rats with lesions of the subfornical organ which contains high concentrations of angiotensin II receptors (Steardo and Iovino, 1986).

Conversely, intracerebroventricular injection of 5,7-DHT blocked the VP elevation induced by water deprivation (Iovino and Steardo, 1985; Steardo *et al.*, 1985).

In conclusion, it is generally agreed that 5-HT is involved in the control of the hypothalamo-posthypophyseal axis; however, the conflicting data described above clearly demonstrate that further experiments are needed to clarify both the sites and the modes of 5-HT action at this level. Nevertheless, this regulation constitutes a good model for elucidating some aspects of neuroendocrine effects of serotonin since it is possible to monitor directly the release of hypothalamic hormones.

CONCLUDING REMARKS AND FUTURE TRENDS

If we attempt to integrate the morphological and experimental data described above, we can define three possible levels where serotonin is implicated in the regulation of the pituitary. The first one is the gland itself or, more precisely, the adenohypophysis since the neural lobe is physiologically the homologue of the median eminence which is the second level of integration. A third level, more multiple and diffuse, is constituted by the other sites of interrelationship between serotonin and neuroendocrine pathways. We shall limit our comments to those described in the hypothalamus, but we should emphasize that the different levels of interactions could be a factor in the discrepancies between results of experimental studies on the neuroendocrine role of 5-HT as discussed above.

In the anterior lobe we should note that 5-HT uptake and/or storage sites defined by morphology do not seem to correspond with physiological data since they concern gonadotrophs while the only direct regulation of glandular cells postulated at this level affects corticotrophs and somatotrophs. However, since the significance of the 5-HT cellular labelling is not established and since the morphological tools are still not sensitive enough to detect 5-HT on its targets, this discrepancy between both kinds of data is more a source for future experimentation than an aspersion on their validity. In fact, 5-HT$_2$ receptors have been recently identified in the different lobes of the pituitary (De Souza, 1986). Furthermore, the belief that 5-HT acts directly on the anterior lobe is supported by the existence of 5-HT terminals within the ME close to the external basement membrane at the proximity of primary portal vessels.

On the other hand, the pituitary intermediate lobe displays a firm morphological basis for physiological interactions between 5-HT and hormone secretion with the direct and differentiated innervation of glandular cells by 5-HT terminals. These morphological data may be assumed as giving a new impetus to physiological investigations in this area and/or could display a useful model for studying the morphofunctional implications of the possible intraneuronal colocalization of several neurotransmitters.

In the median eminence as well as in the neural lobe the lack of differentiated interneuronal contacts does not exclude a possible interrelationship between 5-HT and hypophysiotropic factors. Axo-axonic contacts may be sites where one factor can influence the liberation of another, while a control can also be exerted on distal terminals after release of the factors in the intercellular space (Bouchaud and Bosler, 1986). For example, *in vivo* studies suggest that 5-HT acts positively upon LH-RH terminals (Vitale *et al.*, 1985). Thus, the options remain open to all types of distal and reciprocal interreactions between 5-HT, neurotransmitters and/or neuropeptides. However, the differentiated axo-axonic contacts that we have illustrated for the first time in the neural lobe (see Fig. 4) suggest more specific interactions between nerve terminals than was previously believed.

Finally, the numerous appositions mentioned between 5-HT terminals and processes of tanycytes and pituicytes may support the still controversial hypothesis that the amine acts on the function of these cells. This projects a new light on the histofluorescence results, indicating the presence of 5-HT within ME tanycytes (Kent and Sladek, 1978).

Most interesting convergences between morphological and physiological data can be found at the third, proximal level of integration, the dendrites and cell bodies of hypothalamic neurons. It should perhaps be noted that while morphological data are ever increasing, experimental advances are slower. While morphology suggests a direct link between the 5-HT inner-vation of dopamine neurons in the arcuate nucleus and its subsequent stimu-latory effect on prolactin secretion, experimental data are still lacking. Where VIP has been postulated as being an intermediate link between serotonin and prolactin secretion, morphological data support such a 5-HT-VIP inter-relationship without proving that these structures are the site(s) where this control is exerted. Furthermore, when the serotonin neurons are postsynaptic to other types of afferents as in the nucleus dorsomedialis, it is clear that they constitute only one element in a complex chain of regulation where their function is probably not prominent and will thus be difficult to investi-gate. Finally, the recent morphological evidence that serotonin is involved in the sexually dimorphic development of the preoptic area in the rat brain (Handa *et al.*, 1986) suggests a morphogenetic influence of the amine which could involve numerous behavioural and neuroendocrine consequences.

Obviously, the main obstacles for a useful dialogue between morphologists and physiologists lie in the different approaches and scales. Quantitative measurements for the morphologist always present a problem, while for the physiologist/biochemist the precise neurochemical and/or anatomical speci-ficity of the interventions is often difficult to characterize. However, the specificity, sensitivity and resolution of the different approaches are now drawing close together and we can anticipate a development of interactions between different methodological approaches which will become truly morphofunctional. A decisive breakthrough can also be anticipated from the application of recent methodological advances to neuroendocrine regulations. For example, the grafts of serotonergic nuclei in the hypothalamus might help to define both their morphological and physiological implications at the different levels described above. Furthermore, the systematic application of *in situ* hybridization to neurosecretory cells might allow one to determine whether an increased amount of secretory material detected by morphology originates from synthesis or accumulation. Finally, the development of new techniques for the detection of the electrical activity of neurones and of the *in vivo* release of their transmitters may fill a gap between the different approaches for the possible interrelationship between 'informative' molecules and their targets at a cellular level. It is possible that future reviews on 'serotonin and endocrinology' will focus more on specific, neurochemically defined neuronal interactions which will not be artificially limited to the pituitary level but will involve other and higher levels of integration.

ACKNOWLEDGEMENTS

The authors thank M. Héry and O. Bosler for their comments and C. Fargues and I. Dratch for typing the manuscript. The original data described in this review were supported by INSERM (grant 86 6002).

REFERENCES

Arendash, G. W., and Gallo, R. V. (1978) Serotonin involvement in the inhibition of episodic luteinizing hormone release during electrical stimulation of the midbrain dorsal raphe nucleus in ovariectomized rats, *Endocrinology*, **102**, 1199–1206.

Arezki, F., Bosler, O., and Steinbusch, H. W. M. (1985) Morphological evidence that serotonin-immunoreactive neurons in the nucleus dorsomedialis hypothalami could be under catecholaminergic influence, *Neurosci. Lett.*, **56**, 161–166.

Arezki, F., Afailal, I., Bosler, O., Steinbusch, H. W. M., and Calas, A. (1987) Serotonin and serotonergic neurons. A radioautographic and immunocytochemical study on the nucleus raphe dorsalis and nucleus dorsomedialis hypothalami, *La Cellule* (in press).

Arnold, M. A., and Fernstrom, J. D. (1978) Serotonin receptor antagonists block a natural, short term surge in serum growth hormone levels, *Endocrinology*, **103**, 1159–1163.

Beaudet, A., and Descarries, L. (1979) Radioautographic characterization of a serotonin-accumulating nerve cell group in adult rat hypothalamus, *Brain Res.*, **160**, 231–243.

Beccu, D., and Libertun, C. (1982) Comparative maturation of the regulation of prolactin and thyrotropin by serotonin and thyrotropin-releasing hormone in male and female rats, *Endocrinology*, **110**, 1879–1884.

Belin, M. F., Nanopoulos, D., Didier, M., Aguera, M., Steinbusch, H., Verhofstad, A., Maitre, M., and Pujol, J. F. (1983) Immunohistochemical evidence for the presence of γ-aminobutyric acid and serotonin in one nerve cell. A study on the raphe nuclei of the rat using antibodies to glutamate decarboxylase and serotonin, *Brain Res.*, **275**, 329–339.

Bhattacharya, A. N., and Marks, B. H. (1970) Effects of alpha methyl tyrosine and *p*-chlorophenylalanine on the regulation of ACTH secretion, *Neuroendocrinology*, **6**, 49–55.

Biegon, A., Reches, A., Snyder, L., and McEwen, B. S. (1983) Serotonergic and noradrenergic receptors in the rat brain: modulation by chronic exposure to ovarian hormones, *Life Sci.*, **32**, 2015–2021.

Björklund, A., and Falck, B. (1969) Pituitary monoamines of the cat with special reference to the presence of an unidentified monoamine-like substance in the adenohypophysis, *Z. Zellforsch. Mikrosk. Anat.*, **93**, 254–264.

Björklund, A., Falck. B., and Rosengren, E. (1967) Monoamines in the pituitary gland of the pig, *Life Sci.*, **6**, 2103–2110.

Bluet-Pajot, M. T., Schaub, C., Mounier, F., Segalen, A., Duhault, J., and Kordon, C. (1980) Monoaminergic regulation of growth hormone in the rat, *J. Endocr.*, **86**, 387–396.

Bosler, O. (1978) Radioautographic identification of serotonin axon terminals in the rat organum vasculosum laminae terminalis, *Brain Res.*, **150**, 177–181.

Bosler, O. (1984) Interrelations entre systèmes monoaminergiques, GABA-ergiques et peptidergiques dans l'hypothalamus. Données ultrastructurales récentes, *Ann. Endocrinol.*, **45**, 7N.

Bosler, O., and Beaudet, A. (1985a) VIP neurons as prime synaptic targets for serotonin afferents in rat suprachiasmatic nucleus: a combined radioautographic and immunocytochemical study, *J. Neurocytol.*, **14**, 749–763.

Bosler, O., and Beaudet, A. (1985b) Relations ultrastructurales entre systèmes monoaminergiques et peptidergiques dans l'hypothalamus. Approche radioautographique et immunocytochimique couplée dans le noyau arqué et le noyau supra-chiasmatique du rat, *Ann. Endocrinol. (Paris)*, **46**, 19–26.

Bosler, O., Bloch, B., BBugnon, C., and Calas, A. (1982) Bases morphologiques des interactions monoamines-peptides dans l'hypothalamus neuroendocrine du rat. Données radioautographiques et immunocytochimiques, *Coll. INSERM*, **110**, 17–36.

Bosler, O., Joh, T. H., and Beaudet, A. (1984) Ultrastructural relationships between serotonin and dopamine neurons in the rat arcuate nucleus and medial zona incerta: a combined radioautographic and immunocytochemical study, *Neurosci. Lett.*, **48**, 279–285.

Bouchaud, C., and Bosler, O. (1986) The circumventricular organs of the mammalian brain with special reference to monoaminergic innervation, *Int. Rev. Cytol.*, **105**, 283–307.

Bruni, J. F., Hawkins, R. L., and Yen, S. S. (1982) Serotoninergic mechanism in the control of beta endorphin and ACTH release in male rats, *Life Sci.*, **30**, 1247–1254.

Calas, A. (1981) L'innervation monoaminergique de l'hypophyse. Approche radio-autographique chez le rat, *Bull. Soc. Neuroendocr. Exp.*, **1**, 15.

Calas, A. (1985) Morphological correlates of chemically specified neuronal inter-actions in the hypothalamo-hypophyseal area, *Neurochem. Int.*, **7**, 927–940.

Calas, A., Alonso, G., Arnauld, E., and Vincent, J. D. (1974) Demonstration of indolaminergic fibres in the median eminence of the duck, the rat and the monkey, *Nature*, **250**, 241–243.

Caligaris, L., and Taleisnik, S. (1974) Involvement of neurons containing 5-hydroxy-tryptamine in the mechanism of prolactin release induced by estrogen, *J. Endo-crinol.*, **62**, 25–33.

Chan-Palay, V. (1977) Indoleamine neurons and their processes in the normal rat brain and in chronic diet-induced thiamine deficiency demonstrated by uptake of ^{3}H-serotonin, *J. Comp. Neurol.*, **176**, 467–494.

Chan-Palay, V., Jonsson, G., and Palay, S. L. (1978) Serotonin and substance P coexist in neurons of the rat's central nervous system, *Proc. Natl. Acad. Sci. USA*, **75**, 1582–1586.

Chen, H. T., Sylvester, P. W., Ieri, T., and Meites, J. (1981) Potentiation of lutein-izing hormone release by serotonin agonists in ovariectomized steroid primed rats, *Endocrinology*, **108**, 948–952.

Chen, M. J., and Meites, J. (1975) Effects of biogenic amines and TRH on release of prolactin and TSH in the rat, *Endocrinology*, **96**, 10–14.

Chen, Y. F., and Ramirez, V. D. (1981) Serotonin stimulates thyrotropin-releasing hormone release from superfused rat hypothalami, *Endocrinology*, **108**, 2359–2366.

Clarenbach, P., Delpozo, E., Brownell, J., Heredia, E., Spiegel, R., and Cramer, H. (1980) Characterization of ergot and non-ergot serotonin antagonists by prolactin and growth hormone profiles during wakefulness and sleep, *Brain Res.*, **202**, 357–363.

Clarke, G., and Wright, D. M. (1985) Effect of 5-hydroxytryptamine on oxytocin release in anaesthetized lactating rats, *J. Physiol.*, **369**, 128 P.

Clemens, J. A. (1978) Effects of serotonin neurotoxins on pituitary hormone release, *Ann. N.Y. Acad. Sci.*, **305**, 399–410.

Clemens, J. A., Sawyer, B. D., and Cerimelle, B. (1977) Further evidence that serotonin is a neurotransmitter involved in the control of prolactin secretion, *Endocrinology*, **100**, 692–698.

Clemens, J. A., Roush, M. E., and Fuller, R. W. (1978) Evidence that serotonin neurons stimulate secretion of prolactin releasing factor, *Life Sci.*, **22**, 2209–2213.

Coen, C. W., and MacKinnon, P. C. B. (1979) Serotonin involvement in the control of phasic luteinizing hormone release in the rat: evidence for a critical period, *J. Endocrinol.*, **82**, 105–113.

Collu, R., Fraschini, F., Visconti, P., and Martini, L. (1972) Adrenergic and serotoninergic control of growth hormone secretion in adult male rats, *Endocrinology*, **90**, 1231–1237.

Delita, G., Yeo, T., Stubbs, W. A., Jones, A., and Besser, G. M. (1980) Effect of serotonin on prolactin secretion, *Adv. Biochem. Psychopharm.*, **24**, 443–444.

Descarries, L., and Beaudet, A. (1978) The serotonin innervation of adult rat hypothalamus, in 'Cell biology of hypothalamic neurosecretion', *Coll. Internat. C.N.R.S.* (Eds J. D. Vincent and C. Kordon), **280**, 135–153, CNRS, Paris.

De Souza, E. B. (1986) Serotonin and dopamine receptors in the rat pituitary gland: autoradiographic identification, characterization and localization, *Endocrinology*, **119**, 1534–1542.

Di Renzo, G. F., Quattrone, A., Schettini, G., and Preziosi, P. (1979) Effect of selective lesioning of serotonin-containing neurons on the TSH-inhibiting action of d-fenfluramine in male rats, *Life Sci.*, **24**, 489–494.

Dixit, B. N., and Buckley, J. P. (1969) Brain 5-hydroxytryptamine and anterior pituitary activation by stress, *Neuroendocrinology*, **4**, 32–41.

Dunn, J. D., Johnson, D. C., Castro, A. J., Swenson, R. (1980) Twenty four hours pattern of prolactin levels in female rats subjected to transection of the mesencephalic raphe or ablation of the suprachiasmatic nuclei, *Neuroendocrinology*, **31**, 85–91.

Eden, S., Bolle, P., and Modigh, K. (1979) Monoaminergic control of episodic growth hormone secretion in the rat: effect of reserpine, alpha-methyl-ptyrosine, *p*-chlorophenylalanine and haloperidol, *Endocrinology*, **105**, 523–529.

Enjalbert, A., Arancibia, S., Ruberg, M., Priam, M., Bluet-Pajot, M. T., Rotsztejn, W. H., and Kordon, C. (1980) Stimulation of *in vitro* prolactin release by vasoactive intestinal peptide, *Neuroendocrinology*, **31**, 200–204.

Fischer, J. I., and Moriarty, C. II. (1977) Control of bioactive corticotropin release from the neurointermediate lobe of the rat pituitary in vitro, *Endocrinology*, **100**, 1047–1054.

Frankfurt, M., Lauder, J. M., and Azmitia, E. C. (1981) The immunocytochemical localization of serotonergic neurons in the rat hypothalamus, *Neurosci. Lett.*, **24**, 227–232.

Friedman, E., Krieger, D. T., Léranth, C., Mezey, E., Brownstein, M. J., and Palkovits, M. (1983) Serotonergic innervation of the rat pituitary intermediate lobe: decrease after stalk section, *Endocrinology*, **112**, 1943–1947.

Fukuda, H., Mori, M., Ohshima, K., and Kobayashi, I. (1980) The role of central serotonergic and noradrenergic neurons in the regulation of nyctohemeral rhythm of plasma thyrotropin, *J. Endocrinol. Invest.*, **3**, 243–249.

Fuller, R. W. (1981) Serotonergic stimulation of pituitary adrenocortical function in rats, *Neuroendocrinology*, **32**, 118–127.

Fuller, R. W., and Clemens, J. A. (1981) Role of serotonin in the hypothalamic regulation of pituitary function, in *Serotonin: current aspects of neurochemistry and function. Advances in experimental medicine and biology* (Eds B. Haber, S. Gabay, M. R. Issidorides, and S. G. A. Alivisatos), Vol. 133, pp. 431–444.

Fuller, R. W., and Snoddy, H. D. (1980) Effect of serotonin releasing drugs on serum corticosterone concentration in rats, *Neuroendocrinology*, **31**, 96–100.

Fuller, R. W., Snoddy, H. D., and Molloy, B. B. (1976) Pharmacologic evidence for a serotonin neural pathway involved in hypothalamus pituitary adrenal function in rats, *Life Sci.*, **19**, 337–345.

Gallo, R. V., and Moberg, G. P. (1977) Serotonin mediated inhibition of episodic luteinizing hormone release during electrical stimulation of the arcuate nucleus in ovariectomized rats, *Endocrinology*, **100**, 945–954.

Gamrani, H., Harandi, M., Belin, M. F., Dubois, M. P., and Calas, A. (1984) Direct electron microscopic evidence for the coexistence of GABA uptake and endogenous serotonin in the same rat central neurons by coupled radioautographic and immunocytochemical procedures, *Neurosci. Lett.*, **48**, 25–30.

Gibbs, D. M., and Vale, W. (1983) Effect of the serotonin reuptake inhibitor fluoxetine on corticotropin-releasing factor and vasopressin secretion into hypophyseal portal blood, *Brain Res.*, **280**, 176–179.

Grimm, Y., and Reichlin, S. (1973) Thyrotropin-releasing hormone (TRH): neurotransmitter regulation of secretion by hypothalamic tissue in vitro, *Endocrinology*, **93**, 626–631.

Guy, J., Bosler, O., Dusticier, G., Pelletier, G., and Calas, A. (1986) Morphological correlates of NPY/serotonin interactions in rat hypothalamus, *Neurosci. Lett.*, suppl. **26**, S608.

Hagen, T. C., Arnaout, M. A., Scherzer, W. J., Martinson, D. R., and Garthwaite, T. L. (1986) Antisera to vasoactive intestinal polypeptide inhibit basal prolactin release from dispersed anterior pituitary cells, *Neuroendocrinology*, **43**, 641–645.

Handa, R. J., Hines, M., Schoonmaker, J. N., Shryne, J. E., and Gorski, R. A. (1986) Evidence that the serotonin is involved in the sexually dimorphic development of the preoptic area in the rat brain, *Dev. Brain Res.*, **30**, 278–282.

Héry, M., Laplante, E., and Kordon, C. (1976) Participation of serotonin in the phasic release of LH. I. Evidence from pharmacological experiments, *Endocrinology*, **99**, 496–503.

Héry, M., Laplante, E., and Kordon, C. (1978) Participation of serotonin in the phasic release of LH. II. Effects of lesions of serotonin containing pathways in the central nervous system, *Endocrinology*, **102**, 1019–1025.

Héry, M., Dusticier, G., and Calas, A. (1982) Application of the 2-deoxy-(1-^{14}C) glucose method to the study of suprachiasmatic nucleus activity and functions: phasic luteinizing hormone secretion and serotonin innervation, *Exp. Brain Res.*, **47**, 465–468.

Hökfelt, T., Johansson, O., Ljungdahl, A., and Lundberg, J. M. (1980a) Peptidergic neurones, *Nature*, **284**, 515–521.

Hökfelt, T., Lundberg, J. M., Schultzberg, M., Johansson, O., Ljungdahl, A., and Rehfeld, J. (1980b) Coexistence of peptides and putative transmitters in neurons, in *Neural Peptides and Neuronal Communication* (Eds E. Costa and M. Trabucchi), pp. 1–23, Raven Press, New York.

Holzbauer, M., Racké, K., and Sharman, D. F. (1985a) Release of endogenous 5-hydroxytryptamine from the neural and the intermediate lobe of the rat pituitary gland evoked by electrical stimulation of the pituitary stalk, *Neuroscience*, **15**, 723–728.

Holzbauer, M., Sharman, D. F., Cohen, G., and Cooper, T. R. (1985b) Pituitary 5-hydroxytryptamine nerves – A possible link with pituitary hormone secretion, *J. Neural Transm.*, **63**, 53–71.

Honda, K., Negoro, H., Fukuoka, T., Higuchi, T., and Uchide, K. (1985) Effect of microelectrophoretically applied acetylcholine, noradrenaline, dopamine and serotonin on the discharge of paraventricular oxytocinergic neurones in the rat, *Endocrinol. Japon*, **32**, 127–133.

Iovino, M., and Steardo, L. (1985) Effect of substances influencing brain serotonergic transmission on plasma vasopressin levels in the rat, *Eur. J. Pharmacol.*, **113**, 99–103.

Ixart, G., Szafarczyk, A., Malaval, F., Nouguier-Soulé, J., and Assenmacher, I. (1980) Facteurs de régulations des réponses plasmatiques de l'ACTH et de LH au stress chez le rat, *Ann. Endocr.*, **42**, 10c.

Iyengar, S., and Rabbi, J. (1983) Role of serotonin in estrogen-progesterone induced luteinizing hormone release in ovariectomized rats, *Brain Res. Bull.*, **10**, 339–343.

Jennes, L., Beckman, W. C., Stumpf, W. E., and Grzanna, R. (1982) Anatomical relationship of serotonergic and noradrenalinergic projections with the GnRH system in septum and hypothalamus, *Exp. Brain Res.*, **46**, 331–338.

Jimenez, A., and Walker, R. F. (1985) The serotonergic system, in *Handbook of Pharmacologic Methodologies for the Study of the Neuroendocrine System* (Eds R. W. Steger and A. Jones), pp. 109–154, CRC Press.

Johns, M. A., Azmitia, E. C., and Krieger, D. T. (1982) Specific in vitro uptake of serotonin by cells in the anterior pituitary of the rat, *Endocrinology*, **110**, 754–759.

Johnson, M. D., and Crowley, W. R. (1983) Acute effects of estradiol on circulating luteinizing hormone and prolactin concentrations and on serotonin turnover in individual brain nuclei, *Endocrinology*, **113**, 1935–1941.

Johnson, M. D., and Crowley, W. R. (1986) Role of central serotonin systems in the stimulatory effects of ovarian hormones and naloxone on luteinizing hormone release in female rats, *Endocrinology*, **118**, 1180–1186.

Johnston, C. A., Fagin, K. D., Alper, R. H., and Negro-Vilar, A. (1986) Prolactin release after 5-hydroxytryptophan treatment requires an intact neurointermediate pituitary lobe, *Endocrinology*, **118**, 805–810.

Jones, M. T., Millhouse, E. W., and Burden, J. (1976) Effect of various putative neurotransmitters on the secretion of corticotrophin-releasing hormone from the rat hypothalamus *in vitro* – a model of the neurotransmitters involved, *J. Endocrinol.*, **69**, 1–10.

Jordan, D. (1985) Neurotransmitter control of thyrotropin functions, *Progr. Neuroendocrinol.*, **1**, 75–98.

Jordan, D., Poncet, C., Mornex, R., and Ponsin, G. (1978) Participation of serotonin on thyrotropin release. I. Evidence for the action of serotonin on thyrotropin releasing hormone release, *Endocrinology*, **103**, 414–419.

Jordan, D., Pigeon, M., McRae-Degueurce, A., Pujol, J. F., and Mornex, R. (1979) Participation of serotonin in thyrotropin release. II. Evidence for the action of serotonin on the phasic release of thyrotropin, *Endocrinology*, **105**, 975–979.

Joseph-Bravo, P., Charli, J. L., Palacios, J. M., and Kordon, C. (1979) Effect of neurotransmitters on the *in vitro* release of immunoreactive thyrotropin-releasing hormone from rat mediobasal hypothalamus, *Endocrinology*, **104**, 801–806.

Kato, Y., Iwasaki, Y., Iwasaki, J., Abe, H., Yanaihara, N., and Imura, H. (1978) Prolactin release by vasoactive intestinal polypeptide in rats, *Endocrinology*, **103**, 554–558.

Kent, D. L., and Sladek, J. R. (1978) Histochemical, pharmacological and microspectrofluorometric analysis of new sites of serotonin localization in the rat hypothalamus, *J. Comp. Neurol.*, **180**, 221–236.

King, T. S., Steger, R. W., and Morgan, W. W. (1986) Effect of ovarian steroids to stimulate region-specific hypothalamic 5-hydroxytryptamine synthesis in ovariectomized rats, *Neuroendocrinology*, **42**, 344–350.

Kiss, J. Z., Léranth, C., and Halasz, B. (1984) Serotoninergic endings on VIP-neurons in the suprachiasmatic nucleus and on ACTH-neurons in the arcuate nucleus of the rat hypothalamus. A combination of high resolution autoradiography and electron microscopic immunocytochemistry, *Neurosci. Lett.*, **44**, 119–124.

Kiss, J., and Halasz, B. (1985) Demonstration of serotonergic axons terminating on luteinizing hormone-releasing hormone neurons in the preoptic area of the rat using a combination of immunocytochemistry and high resolution autoradiography, *Neuroscience*, **14**, 69–78.

Kordon, C., and Glowinski, J. (1972) Role of hypothalamic monoaminergic neurons in the gonadotrophin release regulating mechanisms, *Neuropharmacology*, **11**, 153–162.

Kordon, C., Blake, C. A., Terke, L. J., and Sawyer, C. H. (1973) Participation of serotonin-containing neurons in the suckling-induced rise in plasma prolactin levels in lactating rats, *Neuroendocrinology*, **13**, 213–223.

Kordon, C., Enjalbert, A., Héry, M., Joseph-Bravo, P. I., Rotsztejn, W., and Ruberg, M. (1980) Role of neurotransmitters in the control of adenohypophyseal secretion, in *Handbook of the Hypothalamus* (Eds P. J. Morgane and J. Panknepp), pp. 253–308, M. Dekker, New York.

Kordon, C., Héry, M., Szafarczyk, A., Ixart, G., and Assenmacher, I. (1981) Serotonin and the regulation of pituitary hormone secretion and of neuroendocrine rhythms, *J. Physiol. (Paris)*, **77**, 489–496.

Kraicer, J., and Morris, A. R. (1976) In vitro release of ACTH from dispersed rat pars intermedia cells. II. Effects of neurotransmitter substances, *Neuroendocrinology*, **21**, 175–192.

Krulich, L., Vijayan, E., Coppings, R. J., Giachetti, A., MacCann, S. M., and Mayfield, M. A. (1979) On the role of the central serotonergic system in the regulation of the secretion of thyrotropin and prolactin: thyrotropin-inhibition and prolactin releasing effects of 5-hydroxytryptamine and quipazine in the male rat, *Endocrinology*, **105**, 276–283.

Lawson, D., and Gala, P. R. (1975) The influence of adrenergic, dopaminergic, cholinergic and serotoninergic drugs on plasma prolactin levels in ovariectomized estrogen treated rats, *Endocrinology*, **96**, 313–318.

Lemay, A., Browllette, A., Denizeau, F., and Lawie, M. (1979) Melatonin and serotonin stimulated release of vasopressin from rat neurohypophysis in vitro, *Mol. Cell. Endocrinol.*, **14**, 157–161.

Léranth, C., Palkovits, M., and Krieger, D. T. (1983) Serotonin immunoreactive nerve fibers and terminals in the rat pituitary–light and electron-microscopic studies, *Neuroscience*, **9**, 289–296.

Lopez, F., Gonzalez, D., and Aguilar, E. (1986) Serotonin stimulates GH secretion through a direct pituitary action: studies in hypophysectomized autografted animals and in perifused pituitaries, *Acta Endocrinol.*, **113**, 317–322.

Lu, K. H., and Meites, J. (1973) Effects of serotonin precursors and melatonin on serum prolactin release in rats, *Endocrinology*, **93**, 152–155.

Maeda, K., and Frohman, L. A. (1980) Release of somatostatin and thyrotropin-releasing hormone from rat hypothalamic fragments *in vitro*, *Endocrinology*, **106**, 1837–1842.

Martin, J. B., Durand, D., Gurd, W., Faille, G., Audet, J., and Brazeau, P. (1978) Neuropharmacological regulation of episodic growth hormone and prolactin secretion in the rat, *Endocrinology*, **102**, 106–113.

Mattila, J., and Mannisto, P. T. (1981) Complex role of 5-HT in the regulation of TSH secretion in the male rat, *Hormone Res.*, **14**, 165–179.

Meltzer, H. Y., Fang, V. S., Paul, S. M., and Kaluskar, R. (1976) Effect of quipazine on rat plasma prolactin levels, *Life Sci.*, **19**, 1073–1078.

Mena, F., Enjalbert, A., Carbonel, L., Priam, M., and Kordon, C. (1976) Effect of suckling on plasma prolactin and hypothalamic monoamine levels in the rat, *Endocrinology*, **99**, 445–451.

Meyer, D. C., and Quay, W. B. (1976) Hypothalamic and suprachiasmatic uptake of serotonin *in vitro*: twenty-four-hour changes in male and proestrous female rats, *Endocrinology*, **98**, 1160–1165.

Meyer, D. C., and Eadens, D. J. (1985) The role of endogenous serotonin in phasic LH release, *Brain Res. Bull.*, **15**, 283–286.

Meyer, J. S., Buckholtz, N. S., and Boggan, W. A. (1978) Serotonergic stimulation of pituitary adrenal activity in the mouse, *Neuroendocrinology*, **26**, 312–324.

Mezey, E., Léranth, C., Brownstein, M. J., Friedman, E., Krieger, D. T., and Palkovits, M. (1984) On the origin of the serotonergic input to the intermediate lobe of the rat pituitary, *Brain Res.*, **294**, 231–237.

Mizuno, H., Talwalker, P. K., and Meites, J. (1967) Central inhibition by serotonin of reflex release of oxytocin in response to suckling stimulus in the rat, *Neuroendocrinology*, **2**, 222–231.

Moguilevsky, J. A., Faigon, M. R., Scacchi, P., and Szwarcfarb, B. (1985) Effect of the serotonergic system on luteinizing hormone secretion in prepubertal female rats, *Neuroendocrinology*, **40**, 135–138.

Moore, R. Y. (1977) Organum vasculosum laminae terminalis: innervation by serotonin neurons of the midbrain raphe, *Neurosc. Lett.*, **5**, 297–302.

Moore, R. Y., and Eichler, V. B. (1972) Loss of a circadian adrenal corticosterone rhythm following suprachiasmatic lesions in the rat, *Brain Res.*, **42**, 201–206.

Moore, R. Y., Halaris, A. E., and Jones, B. E. (1978) Serotonin neurons of the midbrain raphe: ascending projections, *J. Comp. Neurol*, **180**, 417–438.

Moos, F., and Richard, P. (1983) Serotonergic control of oxytocin release during suckling in the rat: opposite effects in conscious and anesthetized rats, *Neuroendocrinology*, **36**, 300–306.

Mueller, G. P., Twohy, C. P., Chen, H. T., Advis, J. P., and Meites, J. (1976) Effects of L-tryptophan and restraint stress on hypothalamic and brain serotonin turnover, and pituitary TSH and prolactin release in rats, *Life Sci.*, **18**, 715–724.

Murakami, Y., Kato, Y., Kobayama, Y., Tojo, K., Inoné, T., and Imura, H. (1986) Involvement of growth hormone (GH)-releasing factor in GH secretion induced by serotonergic mechanism in conscious rats, *Endocrinology*, **119**, 1089–1092.

Nakagami, Y., Suda, T., Yajima, F., Ushiyama, T., Tomori, N., Sumitomo, T., Demura, H., and Shizume, K. (1986) Effects of serotonin, cyproheptadine and reserpine on corticotropin-releasing factor release from the rat hypothalamus *in vitro*, *Brain Res.*, **386**, 232–236.

Nakai, Y., Shioda, S., Ochiai, H., Kudo, J., and Hashimoto, A. (1983) Ultrastructural relationship between monoamine and TRH-containing axons in the rat median eminence as revealed by combined autoradiography and immunocytochemistry in the same tissue section, *Cell Tissue Res.*, **230**, 1–14.

Nunez, E. A., Gershon, M. D., and Silverman, A.-J. (1981) Uptake of 5-hydroxytryptamine by gonadotrophs of the bat's pituitary: a combined immunocytochemical radioautographic analysis, *J. Histochem.*, **29**, 1336–1346.

Onaya, T., and Hashizume, K. (1976) Effects of drugs that modify brain biogenic amine concentrations on thyroid activation induced by exposure to cold, *Neuroendocrinology*, **20**, 47–58.

Palkovits, M., Saavedra, J. M., Jacobowitz, D. M., Kizer, J. S., Zaborszky, L., and

Brownstein, M. J. (1977) Serotonergic innervation of the forebrain: effect of lesions on serotonin and tryptophan hydroxylase levels, *Brain Res.*, **130**, 121–134.

Palkovits, M., Mezey, E., Chiueh, C. G., Krieger, D. T., Gallatz, K., and Brownstein, M. J. (1986) Serotonin-containing elements of the rat pituitary intermediate lobe, *Neuroendocrinology*, **42**, 522–525.

Payette, R. F., Gershon, M. D., and Nunez, E. A. (1985) Serotonergic elements of the mammalian pituitary, *Endocrinology*, **116**, 1933–1942.

Payette, R. F., Gershon, M. D., and Nunez E. A. (1986) Colocalization of luteinizing hormone and serotonin in secretory granules of mammalian gonadotrophs, *Anat. Rec.*, **215**, 51–58.

Pearse, A. G. E., and MacGregor, M. M. (1964) Functional cytology of the pituitary gland, *Ann. Rep. Brit. Emp. Cancer, Comp*, **22**, 665–666.

Popova, N. K., Maslova, L. N., and Naumenko, E. V. (1972) Serotonin and the regulation of the pituitary adrenal system after deafferentation of the hypothalamus, *Brain Res.*, **47**, 61–67.

Quattrone, A., Schettini, G., Annunziato, L., and Di Renzo, G. (1981) Pharmacological evidence of supersensitivity of central serotonergic receptors involved in control of prolactin secretion, *Eur. J. Pharmacol.*, **76**, 9–14.

Randle, J. C. R., Moor, B. C., and Kraicer, J. (1983) Differential control of the release of proopiomelanocortin derived peptides from the pars intermedia of the rat pituitary, *Neuroendocrinology*, **37**, 131–140.

Rioch, D. M. E., Wislocki, G. B., and O'Leary, T. L. (1940) A precis of preoptic, hypothalamic and hypophysial terminology, with atlas, in the hypothalamus and central levels of autonomic function, *Proc. Ass. Res. Nerv. Ment. Dis.*, **20**, 1–30.

Rose, J. C., and Ganong, W. F. (1976) Neurotransmitter regulation of the pituitary gland secretion, in *Current Developments in Psychopharmacology* (Eds W. F. Essman, L. Walzelli and N.Y. Holliswood), pp. 87–123, Spectrum.

Rotsztejn, W. H., Beaudet, A., Roberge, A. G., Lalonde, J., and Fortier, C. (1977) Role of brain serotonin in the circadian rhythm of corticosterone and the corticotropic response to adrenalectomy in the rat, *Neuroendocrinology*, **23**, 157–170.

Ruberg, M., Rotsztejn, W. H., Arancibia, S., Besson, J., and Enjalbert, A. (1978) Stimulation of prolactin release by vasoactive intestinal peptide (VIP), *Eur. J. Pharmacol*, **51**, 319–320.

Ruzsas, C., Jozsa, R., and Mess, B. (1979) Inhibitory role of brain stem serotonergic neurons system on thyroid function in rat, *Endocrinol. Exp.*, **13**, 9–15.

Saavedra, J. M., Brownstein, M. J., Kizer, J. S., and Palkovits, M. (1976) Biogenic amines and related enzymes in the circumventricular organs of the rat, *Brain Res.*, **107**, 412–417.

Sakoda, M., Kusaka, T., Takeiwa, M., Mori, H., and Baba, S. (1979) Neuroendocrine control of the pituitary TSH secretion. VII. Effect of intraventricular and intra-arterial administration of 5-hydroxytryptophan on rat pituitary TSH release, *Kobe J. Med. Sci.*, **212**, 53–59.

Saland, L. C., Wallace, J. A., and Comunas, F. (1986) Serotonin-immunoreactive nerve fibers of the rat pituitary: effects of anticatecholamine and antiserotonin drugs on staining patterns, *Brain Res.*, **368**, 310–318.

Sano, Y., Takeuchi, Y., Matsuura, T., Kawata, M., and Yamada, H. (1982) Immunohistochemical demonstration of serotonin nerve fibers in the cat neurohypophysis, *Histochemistry*, **75**, 293–299.

Sapun, D. I., Farah, J. M., and Mueller, G. B. (1981) Evidence that a serotonergic mechanism stimulates the secretion of pituitary beta endorphin-like immunoreactivity in the rat, *Endocrinology*, **109**, 421–426.

Sapun-Malcolm, D., Farah, J. M., and Mueller, G. P. (1983) Evidence for serotonergic stimulation of pituitary β-endorphin release: preferential release from the anterior lobe *in vivo*, *Life Sci.*, **33**, 95–102.

Sapun-Malcolm, D., Farah, J. M., and Mueller, G. P. (1986) Serotonin and dopamine independently regulate pituitary β-endorphin release in vivo, *Neuroendocrinology*, **42**, 191–196.

Sawchenko, P. E., Swanson, L. W., Steinbusch, H. W. M., and Verhofstad, A. A. J. (1983) The distribution and cells of origin of serotonergic inputs to the paraventricular and supraoptic nuclei of the rat, *Brain Res.*, **277**, 355–360.

Scapagnini, U., and Preziosi, P. (1972) Role of brain norepinephrine and serotonin in the tonic and phasic regulation of hypothalamic hypophyseal adrenal axis, *Arch. Int. Pharmacodyn. Ther.*, **196**, suppl., 205–220.

Scapagnini, U., Morberg, G. P., Van Loon, G. R., De Groot, J., and Ganong, W. F. (1971) Relation of brain 5-hydroxytryptamine content to the diurnal variation in plasma corticosterone in the rat, *Neuroendocrinology*, **7**, 90–96.

Scapagnini, U., Gerendai, I., Clementi, G., Fiore, L., Marchetti, B., and Prato, A. (1978) Role of brain monoamines in the regulation of the circadian variations of activity of some neuroendocrine axis, *Proc. Life Sci. Environmental Endocrinol.*, (Eds I. Assenmacher and D. S. Farner), pp. 137–143, Springer-Verlag, Berlin.

Schneider, H. P. G., and McCann, S. M. (1970) Mono- and indoleamines and control of LH secretion, *Endocrinology*, **86**, 1127–1133.

Shimatsu, A., Kato, Y., Matsushita, N., Katakami, H., Yanaihara, N., and Imura, H. (1982) Stimulation by serotonin of vasoactive intestinal polypeptide release into rat hypophyseal portal blood, *Endocrinology*, **111**, 338–344.

Smythe, G. A. (1977) The role of serotonin and dopamine in hypothalamic pituitary function, *Clin. Endocrinol.*, **7**, 325–342.

Spinedi, E., and Negro-Vilar, A. (1983) Serotonin and adrenocorticotropin (ACTH) release: direct effects at the anterior pituitary level and potentiation of arginine vasopressin-induced ACTH release, *Endocrinology*, **112**, 1217–1223.

Steardo, L., and Iovino, M. (1986) Vasopressin release after enhanced serotonergic transmission is not due to activation of the peripheral renin-angiotensin system, *Brain Res.*, **382**, 145–148.

Steardo, L., Iovino, M., Hunnicutt, E. (1985) Evidence for a 5-HT involvement in vasopressin release from neurohypophysis of rats, *Neurosci. Abstracts*, **11**, 44.

Steinbusch, H. W. M., and Nieuwenhuys, R. (1981) Localization of serotonin-like immunoreactivity in the central nervous system and pituitary of the rat, with special references to the innervation of the hypothalamus, in *Serotonin, Current Aspects of Neurochemistry and Function* (Eds B. Haber, S. Gabay, M. R. Issidorides and S. G. A. Alivisatos), *Adv. Expl. Med. Biol.*, **133**, 7–36, Plenum Press, New York.

Szafarczyk, A. (1986) Le système sérotoninergique dans les régulations neuroendocriniennes, in *Le système sérotoninergique et sa régulation, Les entretiens du Carla* (Ed. M. Briley), pp. 47–77, Centre de Recherche P. Fabre, Castres.

Szafarczyk, A., Ixart, G., Malaval, F., Nouguier-Soulé, J., and Assenmacher, I. (1979) Effects of lesions of the suprachiasmatic nuclei and of *p*-chlorophenylalanine on the circadian rhythms of adrenocorticotrophic hormone and corticosterone in the plasma and on locomotor activity of rats, *J. Endocrinol.*, **83**, 1–16.

Takeuchi, Y., and Sano, Y. (1983) Serotonin distribution in the circumventricular organs of the rat. An immunohistochemical study, *Anat. Embryol.*, **167**, 311–320.

Telegdy, G., Vermes, I., and Kovacs, G. L. (1976) Effect of drug induced changes in brain monoamines on neuroendocrine and behavioural processes, in *Second Congr. Hungarian Pharmacol. Soc. Symp. Pharmacology of catecholaminergic and serotonergic mechanisms* (Ed. K. Magyar), pp. 101–105, Aked Kiado, Budapest.

Törk, I. (1985) Raphe nuclei and serotonin containing systems, in *The Rat Nervous System, Hindbrain and Spinal Cord* (Ed. G. Paxinos), Vol. 2, pp. 43–69.

Tougard, C. (1970) Distribution et sites de fixation des monoamines dans le parenchyme adénohypophysaire, in *Neuroendocrinologie* (Eds J. Benoit and C. Kordon), 101, CNRS, Paris.

Tuomisto, J., and Mannisto, P. (1985) Neurotransmitter regulation of anterior pituitary hormones, *Pharmacol. Rev.*, **37**, 249–332.

Tuomisto, J., Ranta, T., Mannisto, P., Saarinen, A., and Leppaluoto, J. (1975) Neurotransmitter control of thyrotropin secretion in the rat, *Eur. J. Pharmacol.*, **30**, 221–229.

Ueda, S., Nishida, K., Nojyo, Y., Takeuchi, Y., and Sano, Y. (1984) Serotonin-containing neurons in the rat and cat brain, especially in the hypothalamus, following monoamine oxidase inhibitor pretreatment: An immunohistochemical study using anti-serotonin antiserum, *Arch. Histol. Jap.*, **47**, 405–410.

Van De Kar, L. D., and Bethea, C. L. (1982) Pharmacological evidence that serotonergic stimulation of prolactin secretion is mediated via the dorsal raphe nucleus, *Neuroendocrinology*, **35**, 225–230.

Van De Kar, L. D., and Lorens, S. A. (1979) Differential serotonergic innervation of individual hypothalamic nuclei and other forebrain regions by the dorsal and median midbrain raphe nuclei, *Brain Res.*, **162**, 45–54.

Van De Kar, L. D., Wilkinson, C. W., Skrobik, K., Brownfield, M. S., and Ganong, W. F. (1982) Evidence that serotonergic neurons in the dorsal raphe nucleus exert a stimulatory effect on the secretion of renin but not of corticosterone, *Brain Res.*, **235**, 233–243.

Van De Kar, L. D., Urban, J. H., Richardson, K. D., and Bethea, C. L. (1985) Pharmacological studies on the serotonergic and non-serotonin mediated stimulation of prolactin and corticosterone secretion by fenfluramine, *Neuroendocrinology*, **41**, 283–288.

Van Delft, A. M. L., Kaplanski, J., and Smelik, P. G. (1973) Circadian periodicity of pituitary adrenal function after *p*-chlorophenylalanine administration in the rat, *J. Endocr.*, **59**, 469–474.

Vernikos-Danellis, J., Kellar, K. J., Kent, D., Gonzales, C., Berger, P. A., and Barchas, J. D. (1977) Serotonin involvement in pituitary adrenal function, *Ann. N.Y. Acad. Sci.*, **297**, 518–526.

Vijayan, E., Krulich, L., and McCann, S. M. (1978) Stimulation of growth hormone release by intraventricular administration of 5-HT or quipazine in unanaesthetized male rats, *Proc. Soc., Exp. Biol. Med.*, **159**, 210–212.

Villar, M. J., Chiocchio, S. R., and Tramezzani, J. H. (1984) Origin and termination of dorsal raphe-median eminence projection, *Brain Res.*, **324**, 165–170.

Vincent, S. R., Hökfelt, T., and Wu, J. Y. (1982) GABA neuron systems in hypothalamus and the pituitary gland, *Neuroendocrinology*, **34**, 117–125.

Vitale, M. L., De Las Nieves Parisi, M., Chiocchio, S. R., and Tramezzani, J. (1985) Serotonin stimulates gonadotrophin release by acting directly on the median eminence, *Acta Physiol. Pharmacol. Latinoam.*, **35**, 473–479.

Wallace, J. A., and Lauder, J. M. (1983) Development of the serotonergic system in the rat embryo: an immunocytochemical study, *Brain Res.*, **10**, 459–479.

Walker, R. F. (1983) Quantitative and temporal aspects of serotonin's facilitatory action on phasic secretion of luteinizing hormone in female rats, *Neuroendocrinology*, **36**, 468–474.

Westlund, K. N., and Childs, G. V. (1982) Localization of serotonin fibers in the rat adenohypophysis, *Endocrinology*, **111**, 1761–1763.

Williams, J. H., Miall Allen, V. M., Klinowski, M., and Azmitia, E. (1983) Effects of microinjections of 5,7-dihydroxytryptamine in the suprachiasmatic nuclei of the rat on serotonin reuptake and the circadian variation of corticosterone levels, *Neuroendocrinology*, **36**, 431–435.

Neuronal Serotonin
Edited by N. N. Osborne and M. Hamon
© 1988 John Wiley & Sons Ltd

CHAPTER 11

Physiological and Behavioral Correlates of Serotonergic Single-Unit Activity

CASIMIR A. FORNAL AND BARRY L. JACOBS
Program in Neuroscience
Department of Psychology
Princeton University
Princeton
NJ 08544
USA

INTRODUCTION

In the mammalian brain, cell bodies of serotonergic neurons are located on or near the midline, primarily within the brainstem raphe nuclei (Dahlström and Fuxe, 1964; Wiklund *et al.*, 1981; Jacobs *et al.*, 1984). These cells project widely throughout the central nervous system (Azmitia, 1978; Moore, 1981) and have been implicated in a great variety of physiological and behavioral processes (see Jacobs and Gelperin, 1981; and review by Vogt, 1982). Most of the data that have implicated central serotonergic neurons in physiological and behavioral control have been derived from experiments employing methods such as systemic drug administration, brain stimulation, electrolytic or chemical lesions, and measurement of neurotransmitter release or metabolism. However, before any function can be ascribed unequivocally to central serotonergic neurons, an understanding of the cellular activity of these neurons in relation to physiology and behavior is necessary.

Our laboratory has been engaged in the systematic analysis of the electrophysiological activity of serotonin-containing neurons in various raphe nuclei in unanesthetized and unrestrained cats. These studies have examined the activity of these neurons under a variety of conditions without the confounding influences inherent in anesthetized or immobilized preparations. The importance of conducting single-unit studies in unanesthetized animals is underscored by the observation that anesthesia can drastically alter the response of serotonergic neurons (Heym *et al.*, 1984). This approach also combines neurochemical specificity, precision of localization, and temporal continguity between unit activity and the environmental, behavioral, or physiological variable under study. An additional advantage of this approach is that it produces minimal pertubation of the system under investigation.

To date, our laboratory has examined the activity of serotonergic neurons in two mesencephalic raphe nuclei, raphe dorsalis (DRN) and raphe centralis superior (NCS), and in two medullary raphe nuclei, raphe magnus (NRM) and raphe pallidus (NRP). These raphe nuclei represent the four largest aggregations of serotonin-containing cell bodies in the cat brain (Wiklund *et al.*, 1981; Jacobs *et al.*, 1984). The aims of this review are two-fold: (1) to provide a general description of the basic characteristics of serotonergic neurons in various raphe nuclei; and (2) to examine several specific hypotheses regarding the role of these neurons in physiology and behavior.

RECORDING TECHNIQUE

Electrophysiological recording of single-unit activity in behaving cats is accomplished using a microwire technique. This procedure has been described in detail elsewhere (Fornal *et al.*, 1985a; Heym and Jacobs, in 1986). Briefly, a microdrive is sterotaxically implanted above the particular raphe nucleus under study. Two microelectrode bundles, each consisting of six flexible insulated nichrome wires (three each of 32 μm and 64 μm diameter) are then lowered through the microdrive unit so that their tips are positioned 1 mm above the target nucleus. The microelectrodes are permanently anchored to the microdrive which can then be advanced slowly through the brain by means of an attached screw. A schematic representation of the microdrive assembly is shown in Fig. 1.

Cats are also routinely implanted with gross electrodes for recording the electroencephalogram (EEG), electrooculogram (EOG), and nuchal electromyogram (EMG). Following a two-week recovery period, cats are placed in

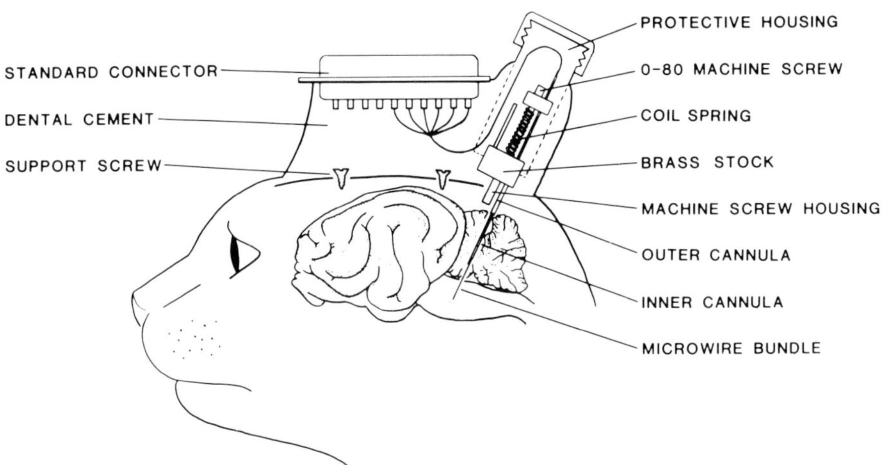

Fig. 1 Schematic diagram of a single-cannula version of our microdrive implant assembly. (Reprinted from Jacobs, 1987)

a chamber where behavior, physiological variables, and brain electrical activity are monitored continuously during each experimental session. Electrical potentials are led from the cat by means of a counter-weighted low-noise cable system that passes through a slip-ring assembly at the top of the chamber. Only recordings that display a signal-to-noise ratio of at least 3:1 are used for an experiment. Data are displayed on-line on an oscilloscope and polygraph, and stored on magnetic tape for off-line analysis. In order to control for alterations in unit activity that are associated with spontaneous or induced changes in the level of behavioral arousal, unit activity before and after each manipulation is measured during comparable behavioral states defined on the basis of polygraphic and behavioral criteria.

NEUROCHEMICAL IDENTIFICATION

Since single-unit studies of serotonergic neurons examine extracellularly recorded action potentials, the identification of neurons as 'serotonergic' is necessarily based on indirect criteria. Our identification of serotonergic neurons is based largely on extrapolation from studies by Aghajanian and coworkers that were carried out in anesthetized rats. These investigators recorded from neurons within the midbrain raphe nuclei and found a sub-population of cells that: (1) discharged in a slow (~1 2 spikes/sec) and regular manner; (2) had biphasic action potentials of relatively long duration (~2 msec); and (3) were selectively and completely suppressed by low doses of serotonin agonist drugs (e.g. D-lysergic acid diethylamide, LSD; 5-methoxy-N,N-dimethyltryptamine, 5-MeODMT) (Aghajanian *et al.*, 1968;

DeMontigny and Aghajanian, 1977; Aghajanian *et al.*, 1978). In subsequent studies, these investigators demonstrated that neurons which displayed these characteristics were serotonergic in nature (Aghajanian and Haigler, 1974; Wang and Aghajanian, 1977; Aghajanian *et al.*, 1978). Perhaps the best evidence for the serotonergic identity of these neurons comes from double labeling experiments in which slow and regular DRN cells were labeled intracellularly with a fluorescent dye, localized, and subsequently examined for the presence of serotonin by fluorescence histochemistry (Aghajanian and Vandermaelen, 1982). All cells that displayed slow and regular activity were found to contain serotonin. Other investigators have similarly found slow and regular firing cells in the DRN, as well as in other raphe nuclei including pontis, magnus, obscurus, and pallidus (Bramwell, 1974; Trulson and Jacobs, 1976; Dickenson, 1977; Haigler, 1978). The activity of neurons in these other raphe groups was also found to be suppressed by low doses of LSD. Cells displaying these electrophysiological properties have only been found in areas where anatomical studies have confirmed the presence of serotonergic neurons (Dahlström and Fuxe, 1964; Wiklund *et al.*, 1981; Jacobs *et al.*, 1984). The slow and rhythmic discharge pattern of serotonergic neurons may be related to intrinsic autoregulatory or pacemaker properties (see review by Aghajanian, 1982), and appears to be a distinguishing feature of serotonergic neurons regardless of their location within the CNS. In fact, serotonergic neurons recorded *in vitro* from midbrain tissue slices also display slow and regular activity, which further supports the idea that these cells are autoactive (Mosko and Jacobs, 1976; Trulson *et al.*, 1982).

In our single-unit studies in behaving animals, serotonergic neurons were initially identified on-line by their slow and regular activity and characteristic action potential waveform and duration. Neuronal identification was additionally confirmed by the observation that the activity of presumed serotonergic neurons was completely suppressed after systemic administration of the specific serotonin agonist 5-MeODMT. The inhibitory effect of 5-MeODMT is believed to be mediated via a direct action at autoreceptors on serotonergic cell bodies (DeMontigny and Aghajanian, 1977; Mosko and Jacobs, 1977). Furthermore, the location of each recording site was verified histologically. In brain regions where we have already characterized the activity of serotonergic neurons, the observation of an almost complete suppression of neuronal activity during REM sleep was used as an additional criterion for the identification of these cells (see below).

COMPARISON OF MESENCEPHALIC AND MEDULLARY SEROTONERGIC NEURONS

We have studied the activity of serotonergic neurons in awake, freely moving cats in the DRN, NCS, NRM, and NRP (Trulson and Jacobs, 1979a; Heym *et al.*, 1982a; Rasmussen *et al.*, 1984; Fornal *et al.*, 1985). Because serotonergic neurons in the midbrain (DRN and NCS) and medulla (NRM and NRP) give rise to predominantly ascending and descending projections, respectively (Dahlström and Fuxe, 1964; Bobillier *et al.*, 1976), it was of interest to compare the activity of serotonergic neurons in these two areas, since they are thought to subserve different functions.

Initially, the activity of serotonergic neurons was examined across the sleep–wake cycle. The sleep–wake continuum, from active waking (AW) to REM sleep, was divided into eight categories based on polygraphic and behavioral criteria which have been described previously (Trulson and Jacobs, 1979a). The relationship between discharge rate and behavioral state for serotonergic neurons in the DRN, NCS, NRM, and NRP is shown in Fig. 2. The following characteristics were observed for serotonergic neurons in all four raphe nuclei: (1) they displayed slow (~2–5 spikes/sec) and regular spontaneous activity when animals were in a quiet, resting state (see Fig. 3); (2) their discharge rates increased slightly during AW (~10–30%), steadily declined during the transition from quiet waking through the stages of slow-wave sleep (SWS), and were lowest during REM sleep, when most cells showed a complete or near complete cessation of activity; and (3) their activity was unrelated to gross body movements or phasic changes in the nuchal EMG.

Although these similarities were found, several distinct differences in activity between serotonergic neurons in mesencephalic and medullary raphe nuclei were noted. For example, NRM and NRP neurons generally displayed a higher spontaneous firing rate than DRN or NCS neurons during comparable behavioral states (Fig. 2). Also, the discharge of serotonergic neurons in the DRN and NCS exhibited a strong inverse relationship to the occurrence of sleep spindles and ponto-geniculo-occipital (PGO) spikes, which are two phasic events associated with SWS and REM sleep, respectively. In contrast, the discharge of NRM and NRP serotonergic neurons was unrelated to sleep spindles and PGO spikes. Finally, the activity of NRP neurons, as well as a small subset of NCS neurons (22%), was not as strongly related to behavioral state as DRN, NRM, and the majority of NCS cells. The NRP and some NCS neurons did not show as great a decline in activity across the sleep–wake cycle, and their activity, although significantly reduced, was not completely suppressed during REM sleep (Fig. 2). Moreover, NRP neurons were generally unresponsive to arousing stimuli, whereas DRN, NCS, and NRM cells were highly responsive to such stimuli. These results are in agreement with the findings of several other laboratories (McGinty and Harper, 1976; Lydic *et al.*, 1983; Sakai *et al.*, 1983; Trulson *et al.*, 1984).

The response of serotonergic neurons in the DRN, NCS, NRM, and NRP to phasic sensory stimulation was also determined (Heym *et al.*, 1982a; Heym *et al.*, 1982c; Rasmussen *et al.*, 1984; Fornal *et al.*, 1985; Rasmussen *et al.*,

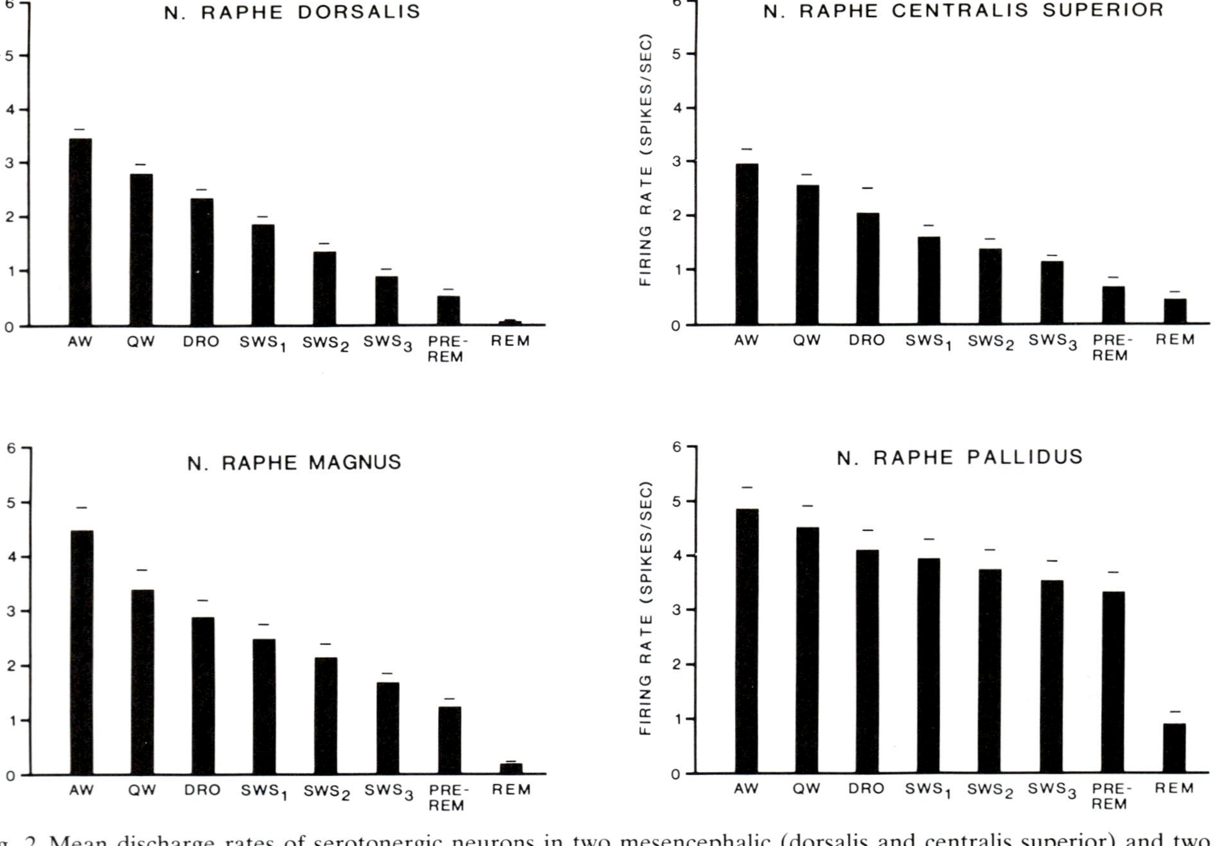

Fig. 2 Mean discharge rates of serotonergic neurons in two mesencephalic (dorsalis and centralis superior) and two medullary (magnus and pallidus) raphe nuclei across the sleep–waking continuum in the cat. Each bar represents the mean $\pm$SEM, $n = 18$–47 cells/nucleus. The sleep–waking continuum was divided into the following categories: AW, active waking; QW, quiet waking; DRO, drowsy; SWS_1, SWS_2, SWS_3, beginning, middle, and end of a slow-wave sleep epoch, respectively; PRE-REM, 60 sec period prior to REM onset; REM, rapid-eye-movement sleep. (Data

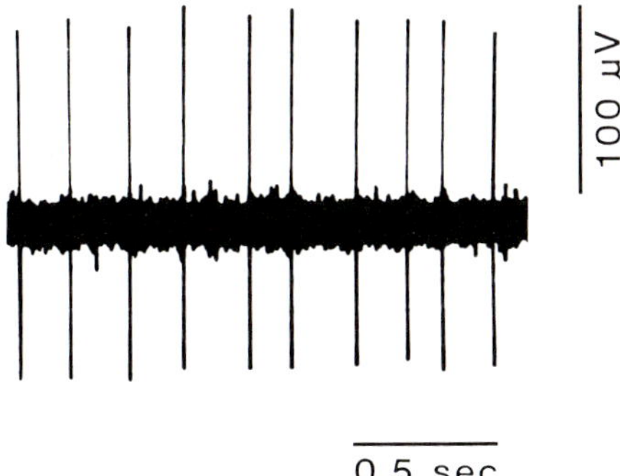

Fig. 3 Oscilloscope trace of extracellularly recorded action potentials from a typical serotonergic NRM neuron in a freely moving cat. The recording was obtained during quiet waking. (Modified from Fornal *et al.*, 1985)

1986). Cats were exposed to a series of clicks or light flashes delivered once every 2 sec. Figure 4 shows a peri-stimulus time histogram for the response of a typical serotonergic DRN neuron to the repeated presentation of an auditory stimulus. Serotonergic neurons in the DRN were highly responsive to both auditory and visual stimulation. Typically, one spike was elicited by each stimulus presentation (latency ~40–130 msec), and the next spike occurred at the normal interspike interval for that particular cell. This response pattern was also observed for the majority of NCS neurons, although a small subset of these cells displayed only inhibitory responses to the presentation of identical visual and auditory stimuli. In contrast to DRN and NCS cells, the activity of NRP cells was virtually unaffected by the presentation of either visual or auditory stimuli. The response of NRM cells was intermediate between the strong excitatory response of DRN and NCS cells and the weak excitatory response of NRP cells. Other investigators have reported similar findings (Trulson *et al.*, 1984).

Serotonergic neurons in all four raphe nuclei studies showed significant decreases in spontaneous unit activity in response to systemic injections of LSD or 5-MeODMT (Trulson and Jacobs, 1979b; Heym *et al.*, 1982a, 1982b; Rasmussen *et al.*, 1984; Fornal *et al.*, 1985). However, serotonergic neurons in the medullary raphe nuclei were substantially less responsive to serotonin agonists than serotonergic neurons in the mesencephalic raphe nuclei. For example, a 50 μg/kg (i.m.) dose of 5-MeODMT produced a large decrease in the activity of DRN and NCS cells, but had relatively little or no effect on the activity of NRM and NRP cells. However, the activity of serotonergic neurons in all four raphe nuclei was greatly inhibited by a larger dose of 5-MeODMT (250 μg/kg). In each of the raphe nuclei examined, the degree

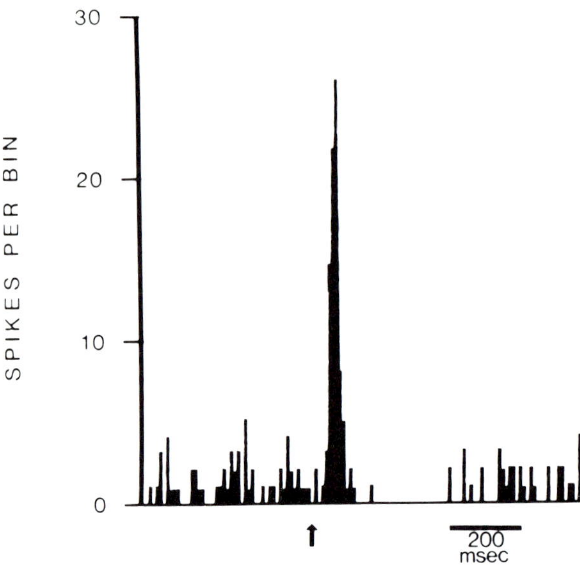

Fig. 4 Peri-stimulus time histogram of the response of a serotonergic DRN neuron, in an awake cat, to the repeated presentation of a click (113 dB, 1 msec). The histogram was constructed from 64 stimulus presentations using a bin width of 10 msec. The arrow marks the occurrence of the stimulus. (Modified from Heym *et al.*, 1982c)

of neuronal suppression produced by systemic injections of 5-MeODMT or LSD was negatively correlated with the pre-injection firing rate (Jacobs *et al.*, 1983; Fornal *et al.*, 1985). We have previously suggested that differences in autoreceptor density or sensitivity between medullary and mesencephalic serotonergic neurons may account for both the differential sensitivity to serotonin agonists, as well as differences in baseline firing rate (Heym *et al.*, 1982b; Jacobs *et al.*, 1983).

An example of the marked sensitivity of serotonergic neurons to the inhibitory action of specific serotonin autoreceptor agonists is shown in Fig. 5. This figure illustrates the rapid and complete suppression of unit activity for a single serotonergic DRN neuron following intravenous administration of 8-hydroxy-2-(di-*n*-propylamino) tetralin (8-OH-DPAT; 5 μg/kg) (Fornal *et al.*, 1987b), a potent and highly selective serotonin$_{1A}$ agonist (Leysen, 1985). The suppression of unit activity produced by 8-OH-DPAT was accompanied by mild behavioral activation.

In summary, medullary serotonergic neurons were found to be different from mesencephalic serotonergic neurons in that they were less responsive to changes in behavioral state (NRP cells only), sensory stimuli, and serotonin autoreceptor agonists, and their activity was unrelated to the occurrence of

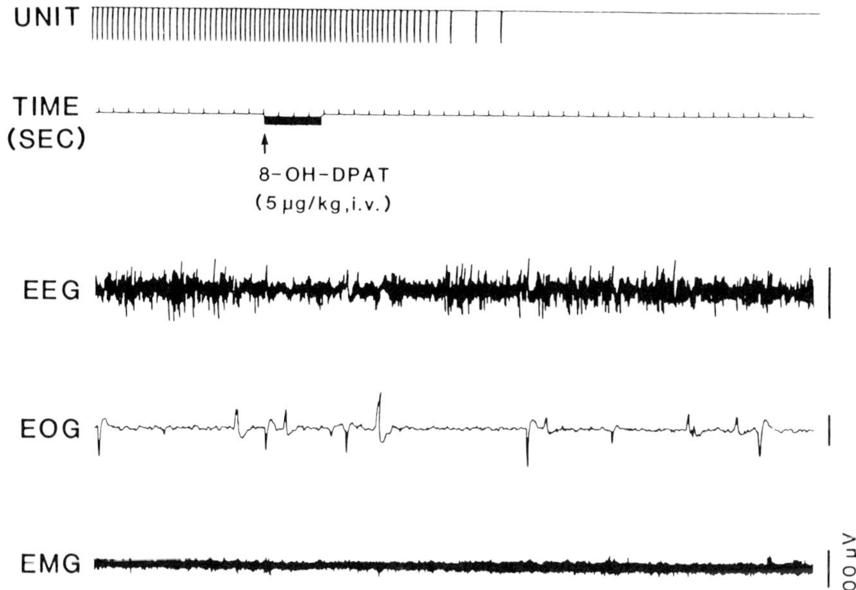

Fig. 5 Polygraph recording of the response of a serotonergic DRN unit in an awake cat to an intravenous injection of the serotonin agonist drug 8-hydroxy-2-(di-*n*-propylamino) tetralin hydrobromide (8-OH-DPAT; 5 μg/kg, i.v.). Complete suppression of unit activity occurred within 15 sec following 8-OH-DPAT administration. Unit activity returned to pre-injection levels within 20–30 min

sleep spindles or PGO spikes. The greater responsiveness of serotonergic neurons in the midbrain to behavioral state-related variables is consistent with the concept that these neurons, by virtue of their predominant ascending projections to the forebrain, are involved in higher-order behavioral processes. In contrast, serotonergic neurons in the medulla project mainly to brainstem and spinal cord sites, and are thought to be primarily involved in the regulation of somatosensory, autonomic and motor functions. Thus, the differences observed between mesencephalic and medullary raphe neurons may reflect a functional differentiation of ascending versus descending serotonergic neuronal systems.

SEROTONERGIC NEURONS AND PAIN MODULATION

Of the various groups of serotonergic neurons, the NRM has been most strongly implicated in the control of nociception. A large percentage of NRM cells respond to noxious stimuli and opiate administration (Moolenaar *et al.*, 1976; Anderson *et al.*, 1977; Deakin *et al.*, 1977; Fields and Anderson, 1978; Guilbaud *et al.*, 1980; Toda, 1982; Wessendorf and Anderson, 1983). Electrical stimulation of the NRM produces analgesia (Oliveras *et al.*, 1975;

Proudfit and Anderson, 1975), whereas lesioning of this nucleus produces hypersensitivity to pain and attenuates the analgesic action of morphine (Proudfit and Anderson, 1975; Yaksh *et al.*, 1977). Whether serotonergic neurons of the NRM are specifically involved in these effects remains controversial. However, the following evidence suggests that serotonergic cells are specifically involved.

1. Serotonergic neurons of the NRM provide the major source of serotonin to the dorsal horn of the spinal cord (Oliveras *et al.*, 1977; Basbaum *et al.*, 1978), an area where iontophoresis of serotonin inhibits neuronal discharge evoked by noxious stimuli (Belcher *et al.*, 1978; Headley *et al.*, 1978).
2. Direct application of serotonin to the spinal cord produces analgesia (Yaksh and Wilson, 1979) whereas the opposite occurs with serotonin antagonists (Proudfit and Hammond, 1981).
3. Interruption of serotonergic neurotransmitter in the spinal cord partially prevents opiate-induced analgesia (see review by Gebhart, 1982).
4. The analgesia produced by electrical stimulation of the NRM or the midbrain periaqueductal gray matter is reduced by disruption of serotonergic neurotransmitter in the spinal cord (Akil and Mayer, 1972; Basbaum *et al.*, 1977; Hammond and Yaksh, 1984).
5. Morphine increases serotonin release and 5-hydroxyindoleacetic acid levels in the spinal cord (Shiomi *et al.*, 1978; Yaksh and Tyce, 1979; Commissiong, 1983).
6. Pain (or stress), which has been shown to produce analgesia (Watkins *et al.*, 1984), also increases serotonin release in the spinal cord (Tyce and Yaksh, 1981).

Taken together these data suggest that serotonergic neurons in the NRM may play a critical role in mediating various forms of analgesia.

It has been hypothesized (see review by Basbaum and Fields, 1984) that painful or stressful stimuli activate endogenous opioid-containing neurons, which, in turn, activate NRM serotonergic cells to release serotonin in the spinal cord, thus inhibiting pain transmission. If opiate-induced and nociceptor-induced analgesia are indeed dependent on activation of serotonergic NRM neurons, then both noxious stimuli and opiates should increase the activity of these neurons. We have directly tested this hypothesis in behaving animals by examining the single-unit activity of identified serotonergic neurons in the NRM in response to painful stimuli, and following systemic administration of morphine (Auerbach *et al.*, 1985).

Our objective in these initial studies was to characterize the response of serotonergic NRM neurons to painful stimuli of several different modalities, i.e., thermal, mechanical, chemical, and electrical. A major issue that we wanted to address in these studies was whether serotonergic NRM neurons respond specifically to noxious stimuli.

Initially, we examined the activity of serotonergic NRM neurons in response to radiant heating of the tail. For these studies, cats were first habituated to a loose canvas bag from which the head and tail protruded. Radiant heat from a lamp was focused on a distal, shaved portion of the tail. The heat was adjusted to produce baseline tail-flick latencies of 2 to 5 sec. Termination of the stimulus was therefore under the cat's control. However, if a cat did not respond within 10 sec, the lamp was turned off to prevent possible tissue damage. Cats usually appeared unstressed by this procedure but any signs of pain such as struggling or vocalization during a particular test were noted.

Our results clearly show that the discharge of serotonergic NRM neurons was not consistently increased during the interval when the tail was being heated, or at the time of the tail-flick response. Figure 6 shows the response of a typical serotonergic NRM neuron during two tail-flick trials. In one trial,

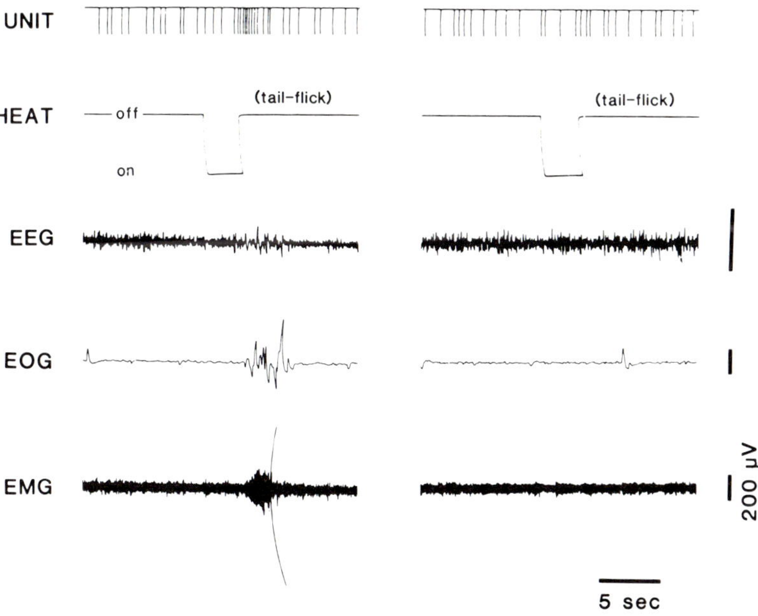

Fig. 6 Polygraph records of the activity of a single serotonergic NRM neuron during two successive tail-flick trials in an awake cat. The first tail-flick trial produced an elevation of unit activity, as well as behavioral arousal indicated by EEG desynchronization and an increase in EOG and EMG activity. No behavioral arousal or change in unit activity were noted during the second trial. (Reprinted by permission of Academic Press from Auerbach *et al.*, 1985)

heating produced an increase in unit discharge at the time of the tail-flick; however, this was accompanied by behavioral arousal, as indicated by EEG desynchronization, and an increase in EOG and EMG activity. By contrast, in another trial studying the same cell, heating did not cause behavioral arousal and failed to substantially increase unit discharge. It should be noted that in both trials, the tail-flick latencies were nearly identical. In about half of all trials, cats were behaviorally aroused by the heat stimulus. For these trials, unit activity was increased by about 70% from baseline at the time of the tail-flick response. This contrasts with the 23% increase in unit activity observed for those trials not resulting in behavioral arousal. Thus, these data indicate that the increase in unit activity associated with noxious heating was correlated with behavioral arousal, and not with the tail-flick response.

We have also examined the activity of serotonergic NRM neurons in response to mildly painful pinches. A hemostat, with rubber tubing slipped over its jaws, was applied to various body regions (e.g., ears, toes, flanks, or tail). The force exerted by the clamp was adjusted to produce escape behavior, orienting, and vocalization, without eliciting signs of more intense pain such as aggressive responses. The clamp was applied for a period lasting no longer than 30 sec. The data show that about half of the cells tested were activated during the pinch. However, the increase in unit activity was most consistently produced when the cats were quiet immediately before the trial and were clearly aroused by the pinch. Figure 7 displays the activation of one serotonergic NRM neuron to an ear pinch that produced vocalization.

In another study, we examined the response of serotonergic NRM neurons to a more prolonged painful stimulus. This was accomplished by monitoring unit activity before and after a subcutaneous injection of dilute Formalin into the forepaw. Pain intensity was rated according to the following numerical scale previously developed for cats (Dubuisson and Dennis, 1977): 0 – no difference in use of the forepaws; 1 – injected paw touches floor but full weight not placed on it; 2 – injected paw held elevated; and 3 – cat licks, shakes, or bites injected paw. Figure 8 displays the mean firing rate and simultaneously recorded pain ratings for a group of serotonergic NRM neurons after the injection of Formalin. Immediately after the injection, cats displayed strong behavioral reactions and signs of pain such as licking, biting and shaking the injected paw (behavioral score = 3). During this period, the discharge rate of serotonergic NRM neurons increased, but never exceeded the pre-injection, normal active waking baseline levels. Following the injection, pain intensity and behavioral arousal gradually decreased over time, as did unit activity. By 30 min after the injection, the pain score approached 0–level and the discharge of these neurons was back to the quiet waking baseline level.

These data suggest that the increase in serotonergic unit discharge may be related more closely to arousal than to the perception of pain. This is

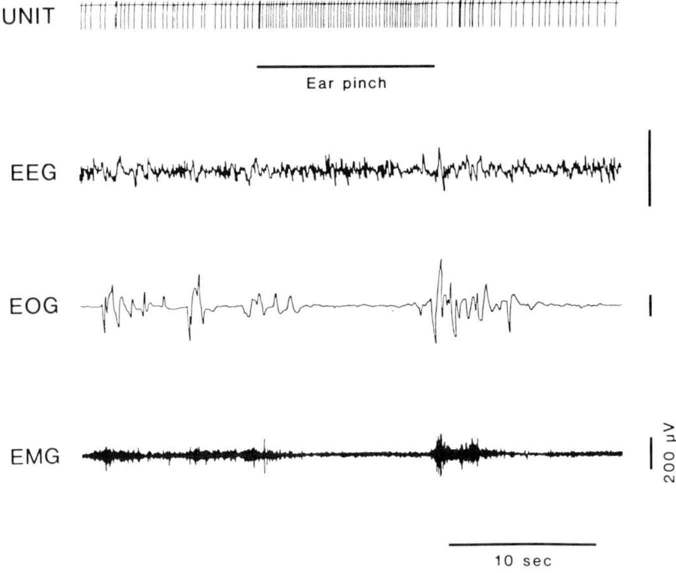

Fig. 7 Polygraph record of the activation of a serotonergic NRM neuron to an ear pinch that produced escape behavior and vocalization in a cat. (Reprinted by permission of Academic Press from Auerbach *et al.*, 1985)

supported by our observation that serotonergic unit activity is also increased by non-noxious, but arousing, stimuli. Figure 9 shows the activation of a typical serotonergic NRM neuron when the experimenter tapped on the door and later entered the recording room. The magnitude of the increase in unit activity is approximately the same as that observed in response to painful stimulation. The point to be made here is that these neurons are activated during any period of behavioral arousal, whether or not arousal was the result of exposure to aversive stimuli.

In follow-up experiments, we conducted a series of studies to obtain more precise quantification of the response of serotonergic NRM neurons to pain using inferior alveolar nerve stimulation. The alveolar nerve carries Aδ and C afferent fibers which are thought to convey nociceptive information from the tooth pulp, and lower threshold Aβ afferent fibers which probably mediate non-noxious sensations and the jaw-opening reflex (JOR) (Cadden *et al.*, 1983; Tal, 1984). This system provides an opportunity to examine the response of serotonergic NRM neurons to both noxious (high-intensity) and non-noxious (low-intensity) electrical stimulation of the alveolar nerve. For these experiments, a pair of stimulating electrodes were implanted into the ventral surface of the mandibular bone, and the inferior alveolar nerve was

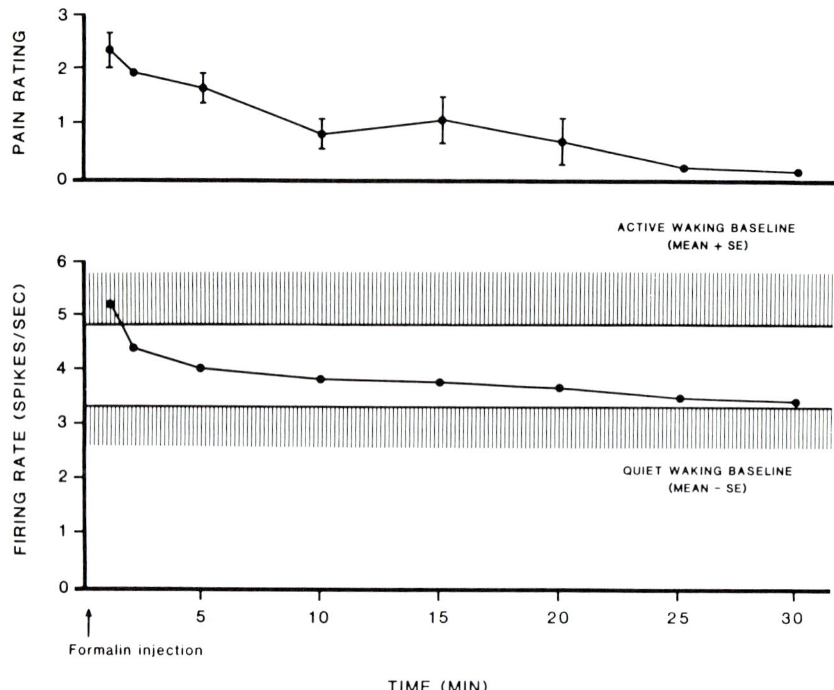

Fig. 8 Mean discharge rate of serotonergic NRM neurons ($n = 4$) and pain ratings after subcutaneous injection of 0.1 ml of 2% formaldehyde into the forepaw. The horizontal lines and striped areas represent the pretreatment mean ±SEM, respectively, for active waking (upper) and quiet waking (lower) baseline discharge rates for these cells. Note that the mean discharge rate following formaldehyde injection was not above the normal waking range. (Modified from Auerbach *et al.*, 1985)

stimulated by constant-current monophasic pulses using a stimulus intensity range from 50 to 1200 μA.

Our data indicate that high-intensity, noxious stimulation of the alveolar nerve was no more effective than low-intensity, non-noxious stimulation in activating serotonergic NRM neurons. Figure 10 displays peri-stimulus time histograms of the response of one serotonergic NRM neuron to the repetitive stimulation of the alveolar nerve at several different current intensities. The discharge of this neuron was increased at a current of 100 μA which was the threshold for eliciting the JOR, but this current did not produce any overt signs of arousal or distress. In fact, during repetitive trials at relatively low current intensities, cats often became drowsy and would even fall asleep, as indicated by the appearance of slow waves in the cortical EEG. A current level of 400 μA elicited a strong JOR without producing any signs of pain (e.g. escape behavior, flinch, or vocalization). Increasing the current intensity

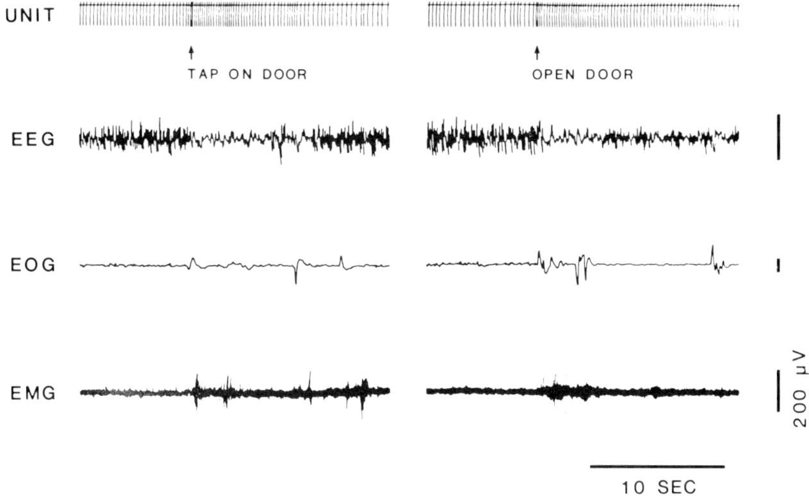

Fig. 9 Polygraph records showing the activation of a serotonergic NRM neuron, in an awake cat, in response to a knock on the recording booth door (left panel), and to opening the door (right panel). Both of these manipulations produced an increase in behavioral arousal in this animal. (Reprinted by permission of Academic Press from Auerbach *et al.*, 1985)

to 800 µA was clearly painful as judged by the cat's reaction to the stimulus, but did not further increase the unit response above levels produced at the lower current intensities, i.e. 100 and 400 µA. Stimulation with the 800 µA current was repeated 30 min after the injection of 2 mg/kg morphine. This dose of morphine abolished overt behavioral signs of pain associated with high-intensity alveolar nerve stimulation but did not affect the magnitude of the unit response, or abolish the JOR. Thus, these data indicate that serotonergic NRM neurons are not specifically responsive to pain. Also shown in the same figure is the response of a non-serotonergic neuron recorded simultaneously on an adjacent electrode. In contrast to serotonergic neurons, the activity of this cell was dramatically increased when the stimulus reached a noxious level, i.e. 800 µA, but showed a minimal response at non-noxious levels, i.e. 100 and 400 µA. Furthermore, the excitatory response of this cell to high-intensity alveolar nerve stimulation was blocked by morphine. These neurochemically unidentified neurons will be discussed in greater detail later in this section.

The ability of non-nociceptive afferents to influence serotonergic NRM unit activity is further supported by our auditory and visual stimulation data (Fornal *et al.*, 1985a). About half of all serotonergic NRM neurons tested were excited by the repetitive presentation of clicks and flashes. Of special significance is the fact that the neuronal activation produced by these noxious

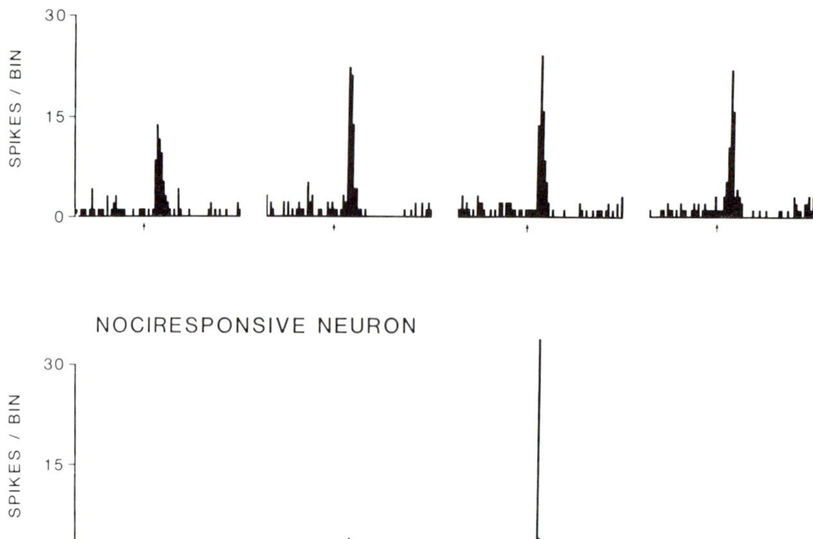

Fig. 10 Peri-stimulus time histograms showing the response of a serotonergic NRM neuron (top panel), and a non-serotonergic NRM neuron (bottom panel), to repeated alveolar nerve stimulation at various current intensities above and below pain threshold (constant-current, monophasic pulses of 1.4 msec duration once every 2 sec). Each histogram was constructed from 64 alveolar nerve stimulations using a bin width of 10 msec. The arrow marks the occurrence of the stimulation. See text for details. (Reprinted by permission of Academic Press from Auerbach *et al.*, 1985)

stimuli was of a similar magnitude to that elicited by painful alveolar nerve stimulation.

Since opiate-induced analgesia may depend on increased release of serotonin in the spinal cord, we also examined the effects of morphine on the discharge of serotonergic NRM neurons. Single-unit activity was monitored before and after the systemic injection of various doses of morphine (0.5, 2.0 and 4.0 mg/kg, i.p.). Most of our data were obtained using the 2 mg/kg dose. We have determined, using the tail-flick test, that this dose of morphine exerts a strong analgesic action in cats that is evident within 10 min and lasts for several hours. The analgesia was naloxone-reversible and was not accompanied by hyperexcitability (i.e., feline mania) which may occur with higher doses of morphine.

Figure 11 shows the mean firing rate for a group of serotonergic NRM

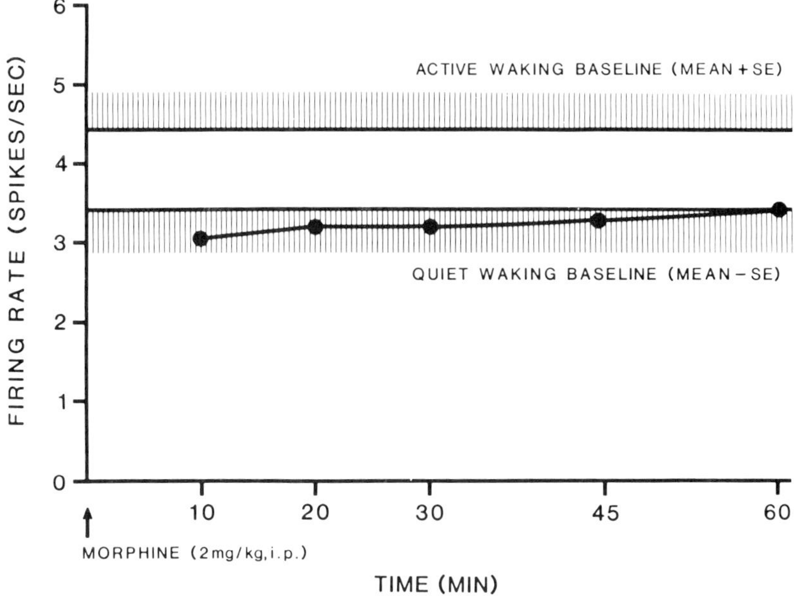

Fig. 11 Mean firing rate of serotonergic NRM neurons ($n = 16$) after systemic administration of morphine sulfate. The horizontal lines and striped areas represent the pretreatment mean ±SEM, respectively, for active waking (upper) and quiet waking (lower) baseline discharge rates for these cells. Note that the mean discharge rate after morphine administration was not above the normal waking range. (Modified from Auerbach *et al.*, 1985)

neurons after 2 mg/kg of morphine. The discharge rate of these neurons was unaffected by morphine during the 60-min observation period, and was always within the pre-injection, normal quiet waking range. Furthermore, no change in unit activity was observed following the other doses of morphine.

The activity of a less frequently encountered class of non-serotonergic NRM neurons was more specifically and consistently related to noxious stimuli. In contrast to serotonergic neurons, these cells showed little or no spontaneous activity during waking behavior. However, they discharged in bursts during REM sleep, and their activity was unaffected by systemic administration of 5-MeODMT. The discharge of these neurons was dramatically increased at the time of the tail-flick response and throughout the period of noxious pinch (see Figure 9 in Auerbach *et al.*, 1985). Furthermore, these neurochemically unidentified neurons were generally unresponsive to non-noxious auditory, visual, or tactile stimuli. Morphine also activated some of these neurons (see Figure 10 in Auerbach *et al.*, 1985). Thus, it is possible that these non-serotonergic NRM neurons are more directly involved in mediating analgesia associated with opiate administration, pain and/or stress.

These results indicate that morphine-induced analgesia is not dependent on an increase in serotonergic NRM neuronal activity. Chiang and Pan (1985), in a study using anesthetized rats, have recently provided evidence in support of this idea. They found that systemic administration of morphine activated only 1 of 10 raphe-spinal neurons classified as being serotonergic on the basis of their slow conduction velocity (<12 m/sec) and low spontaneous discharge rate (<3 spikes/sec). In contrast, they observed that of 38 raphe-spinal neurons classified as non-serotonergic, 20 were activated by systemic morphine administration. Furthermore, Haigler (1978) reported that ionto-phoretic application of morphine had no effect on the firing rate of a majority (90%) of brainstem raphe neurons, identified as serotonergic on the basis of their spontaneous discharge pattern. Although activation of NRM neurons by morphine has been reported by several investigators (Anderson *et al.*, 1977; Deakin *et al.*, 1977; Fields and Anderson, 1978; Toda, 1982), none of these studies attempted to distinguish between serotonergic and non-serotonergic cell types. In light of the above findings, it is important to consider the cell type under study, since in the NRM, serotonergic neurons comprise only a small proportion of the cell population (less than 15% of all neurons in the cat NRM; Wiklund *et al.*, 1981). Therefore, any interpretation of these studies is confounded by the uncertainty of neuronal identity. The cells activated by morphine are most likely non-serotonergic in nature. These results, however, do not exclude a role for serotonin in the mediation of morphine analgesia, since opiates may modulate serotonin release at the axon terminal (Bineau-Thurotte *et al.*, 1984).

It is also worth noting that several recent studies have questioned the involvement of descending serotonergic neurons in the mediation of opiate-induced analgesia. Selective destruction of serotonergic projections to the spinal cord had no effect on analgesia following either the systemic or central administration of morphine (see Anderson and Proudfit, 1981; Johannessen *et al.*, 1982). Likewise, pharmacological blockade of spinal serotonin receptors had no effect on morphine-induced analgesia (Proudfit and Hammond, 1981). Furthermore, the release of endogenous serotonin into spinal cord superfusates was not increased after the systemic administration of an analgesic dose of morphine (Proudfit *et al.*, 1985).

Overall, the following conclusions can be drawn from these studies: (1) morphine-induced analgesia does not appear to be dependent on the activation of serotonergic NRM neurons, since systemic administration of analgesic doses of morphine fail to affect either the spontaneous discharge rate or pain-induced activation of these neurons; and (2) serotonergic NRM neurons do not appear to be specifically involved in pain modulation, since they respond to both noxious and non-noxious stimuli to the same degree. This suggests that they are not involved in a closed-loop, negative feedback analgesic mechanism. However, the activation of these neurons in response to arousing stimuli could conceivably be part of a more general pain modulatory system operating during periods of increased motor activity and behavioral arousal. This conceptualization is also compatible with the notion that serotonergic neurons exert modulatory control over a large variety of sensory, motor and endocrine functions.

SEROTONERGIC NEURONS AND TEMPERATURE REGULATION

A long-standing and large literature has implicated brain serotonin in the regulation of body temperature (see reviews by Hellon, 1975; Myers, 1980; Cox and Lee, 1982; Dascombe, 1985). However, no clear picture regarding the role of serotonin in thermoregulation has emerged. A brief review of this literature follows. The anterior hypothalamic-preoptic area (AH/POA), the primary thermoregulatory center of the CNS, receives a substantial serotonergic innervation from the mescencephalic raphe nuclei (Bobillier *et al.*, 1976; van de Kar *et al.*, 1980). Local application of serotonin to the AH/POA produces alterations in body temperature in several mammalian species (see reviews by Hellon, 1975; Myers, 1980). In cats, Feldberg and Myers (1965) reported that intrahypothalamic injection of serotonin produced a rise in body temperature that was accompanied by shivering, piloerection, and cutaneous vasoconstriction. Another study in cats (Komiskey and Rudy, 1977) further delineated both rostral and caudal sites within the AH/POA where microinjection of serotonin elicited hyperthermia and hypothermia, respectively. Destruction of serotonergic nerve terminals in the AH/POA impaired the ability of animals to defend against heat and cold stress (Myers, 1975; Waller *et al.*, 1976). Similarly, cats were unable to thermoregulate against exposure to a hot environment following inhibition of serotonin synthesis (Harvey and Milton, 1976). Furthermore, neurochemical studies have shown increased brain serotonin metabolism in animals exposed to either heat or cold (Squires, 1974; Okuda *et al.*, 1986), as well as increased serotonin release in the AH/POA in animals exposed to peripheral cooling (Myers and Beleslin, 1971). These studies indicate that an intact brain serotonergic system may be necessary for the physiological processes underlying thermoregulation.

Central serotonergic neurons have also been implicated in fever. Administration of serotonin-depleting agents or serotonin antagonists has been shown to alter the febrile response in several species (see reviews by Cox and Lee, 1982; Dascombe, 1985). In cats, inhibition of serotonin synthesis or blockade of postsynaptic serotonin receptors attenuated the febrile response to bacterial pyrogen administration (Milton and Harvey, 1975). Consistent with these findings, Myers (1977) showed that the release of radiolabeled serotonin in the anterior hypothalamus in cats was increased during the development of fever. These results suggest that an activation of the central serotonergic system may be involved in fever development.

Because of the presumed involvement of brain serotonin in thermoregulation, a number of studies have examined the electrophysiological response of serotonergic neurons to thermal stimuli. All of the studies to date have been conducted in anesthetized animals. Weiss and Aghajanian (1971) reported that the activity of serotonergic neurons in the dorsal and median raphe nuclei of rats was increased in response to elevated body temperature induced by radiant heating. However, other investigators failed to confirm these findings (Bramwell, 1974; Trulson and Jacobs, 1976). In another study, Dickenson (1977) examined the effects of changes in skin temperature on the activity of serotonergic neurons in the DRN, NRM, raphe medianus, and raphe pontis. He reported that many serotonergic cells in the NRM responded with an increase in unit activity to either peripheral warming (27%) or cooling (16%), whereas only a few cells in the other raphe nuclei were influenced by changes in skin temperature. Since these effects were independent of core temperature, the unit responses were ascribed to afferent inputs from cutaneous thermoreceptors. Studies in the cat (Cronin and Baker, 1976) and rat (Hori and Harada, 1976) have shown that serotonergic neurons in the midbrain respond to a rise in local brain temperature with an increase in unit discharge. Consistent with these results, Keenan and Chu (1987) reported that rat DRN neurons recorded *in vitro* were inherently temperature-sensitive, with the majority of cells (61%) showing an increase in firing rate as the bath temperature was elevated. These results raise questions regarding the possible manner in which serotonergic neurons may participate in thermoregulation under physiological conditions (i.e. do they function as central thermoreceptors, relay neurons transmitting information from peripheral thermoreceptors, or comprise a component of effector mechanisms in thermoregulation?).

In this section we describe our studies which were directed at examining some of these issues at the single-unit level in behaving animals. Because of the important role attributed to hypothalamic serotonin in thermoregulation, we studied the serotonergic neurons of the DRN, since they provide a major input to the AH/POA (Bobillier *et al.*, 1976; van de Kar *et al.*, 1980). The effects of both increased ambient temperature and pyrogen-induced fever were examined (Fornal *et al.*, 1987a).

The febrile response to pyrogen administration is particularly interesting in relation to environmental heating because both manipulations are characterized by an increase in body temperature, but opposite thermoregulatory mechanisms are evoked. Environmental heating produces compensatory heat loss responses (e.g. panting, vasodilation) which resist the increase in body temperature, while pyrogen administration causes a regulated rise in body temperature associated with heat gain responses (e.g. shivering, piloerection, vasoconstriction). During the falling phase of the febrile response to pyrogen administration, heat-loss mechanisms are activated to restore normal body temperature.

In general, the following predictions can be made based on the data already discussed. If serotonergic DRN neurons are involved in heat-gain mechanisms, their unit activity should be increased during the thermogenic phase following pyrogen administration. Conversely, if these neurons are involved in heat-loss mechanisms, their unit activity should be increased both during environmental heating and during defervescence following pyrogen administration. An additional possibility is that serotonergic DRN neurons may be inherently temperature-sensitive, i.e. they may respond directly to changes in brain temperature. If this is the case, then serotonergic DRN unit activity should be similarly affected following both manipulations regardless of the thermoregulatory response elicited.

For the environmental heating studies, a fan and a variable-resistance heating coil were connected to the recording chamber via a small duct. The temperature in the chamber was monitored continuously throughout the experiment by means of a thermistor probe mounted inside the chamber. For each trial, heating was preceded by a short acclimatization period (e.g. 10 min) during which room air was blown into the chamber. Following this, the desired experimental temperature (43 $\pm$ 1°C) was typically achieved within 15–30 min, and was maintained until each cat displayed continuous panting (respiratory rate >150/min). Cats were maintained at this elevated temperature for periods up to 2 hrs. Body temperature for each cat was monitored by either obtaining rectal temperatures before and after each heating trial, or by recording brain temperature on-line throughout the heating trial via a supradurally implanted thermistor. Respiratory rate was monitored using a bellows pneumograph before, during, and after heating. The sessions were terminated by room air being blown into the chamber until the environmental temperature returned to baseline levels.

The activity of serotonergic DRN neurons remained unaffected during the interval when ambient temperature was being raised from approximately 25° to 43°C. During this initial phase of heating, cats displayed no appreciable behavioral or physiological responses to the elevation in ambient temperature. Following prolonged heat exposure, however, cats displayed intense panting, relaxation of posture, and a progressive rise in body temperature (range = 0.5 to 2.0°C). However, once again, no change in unit activity was observed. The discharge rate of a typical serotonergic DRN neuron, as well as respiratory rate and body temperature following prolonged heating, are shown in Fig. 12. The discharge rate of this neuron did not change during the 60-min heating trial, although respiratory rate increased from 28/min to 160/min and brain temperature was elevated 1.7°C above baseline (38.3°C). These findings are consistent with the previously mentioned studies in

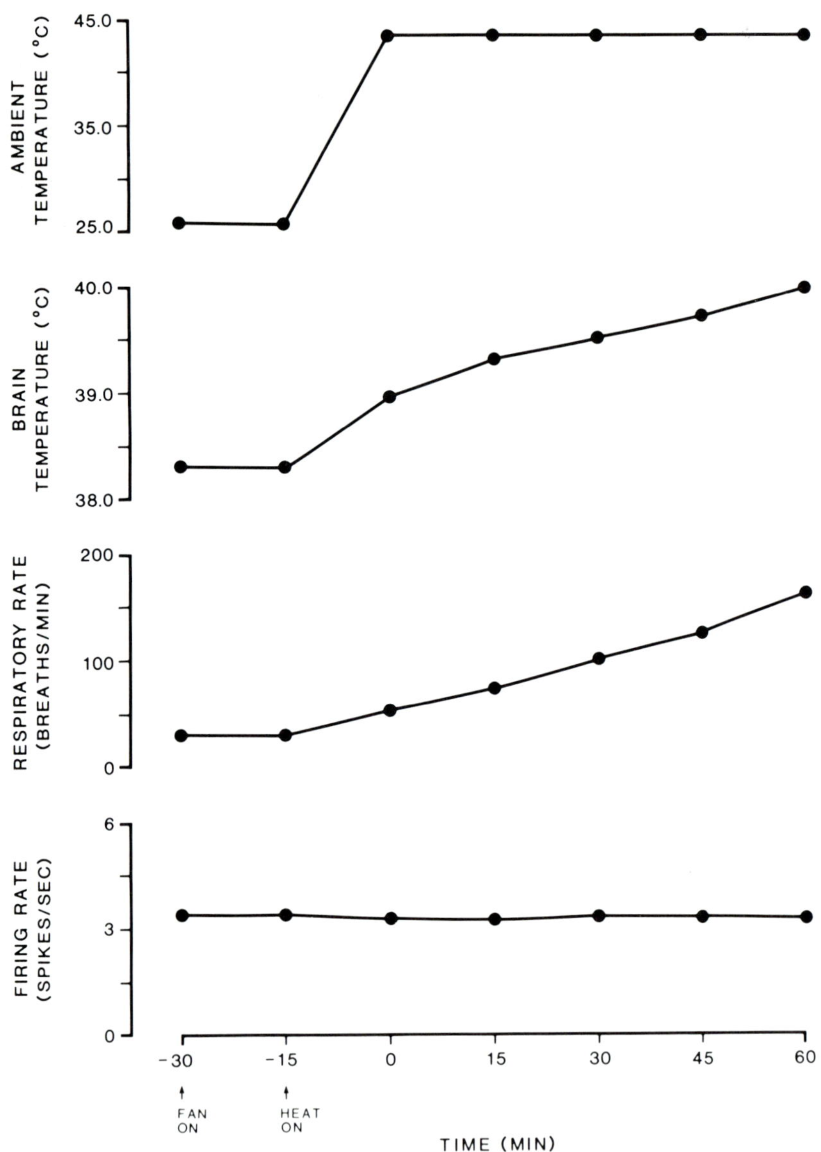

Fig. 12 Effect of environmental heating on brain temperature, respiratory rate, and single-unit activity of a serotonergic DRN neuron in an awake cat. Environmental heating had no effect on serotonergic unit activity despite producing marked increases in respiratory rate and brain temperature

anesthetized rats which showed that serotonergic DRN neurons were unaffected by elevations in either skin temperature or body temperature.

In our pyrogen studies, fever was induced by a single intravenous dose (50 μg/kg) of the synthetic pyrogen muramyl dipeptide (MDP). Body temperature was monitored by either obtaining rectal temperatures at predetermined intervals both before and after MDP administration, or by recording brain temperature on-line. Single-unit activity and behavior were monitored continuously throughout the pyrogenic response. All pyrogen studies were carried out at an ambient temperature of 25 ± 1°C. Following MDP administration, body temperature typically began to rise within 30 min, reached a peak at 1–2 hrs, and returned to normal by about 6 hrs. Peak elevations in body temperature of about 1.5 to 2.5°C above baseline were routinely produced.

Contrary to the hypothesized role of serotonin in heat gain (Feldberg and Myers, 1965), we found no change in the activity of serotonergic DRN neurons in response to pyrogen-induced fever (Fornal *et al.*, 1987a). Figure 13 shows body temperature and unit responses of a typical serotonergic DRN neuron in a cat administered MDP. The firing rate of this neuron was unchanged during both the onset, or the 'chill phase', of the pyrogenic response, when cats displayed overt piloerection and shivering (evident in the EMG), and during the declining temperature phase. During the peak febrile response, cats typically became drowsy and spent a large portion of their time sleeping. Although body temperature was considerably higher than normal in these animals, the cells still showed decreased unit activity during SWS and were off during REM sleep (although occurrences of REM sleep tended to be suppressed). Thus, the characteristic pattern of serotonergic DRN unit activity across the sleep–wake cycle was the same for both febrile and afebrile states.

In some instances, we examined the activity of a serotonergic DRN neuron in response to both environmental heating and pyrogen administration. Figure 14 shows the activity of one such neuron during the time of peak brain temperature produced by both manipulations. This figure illustrates the lack of effect of increased brain temperature *per se* on serotonergic DRN unit activity, and does not support the idea that these neurons are directly temperature sensitive under physiological conditions (Cronin and Baker, 1976; Hori and Harada, 1976; Keenan and Chu, 1987).

Our studies support the following conclusions: (1) serotonergic DRN neurons do not appear to be directly responsive to elevations in ambient temperature, suggesting that they do not receive afferent input from heat-sensitive cutaneous receptors; (2) the thermoregulatory responses associated with both environmental heating and pyrogen-induced fever are not accompanied by changes in serotonergic DRN unit discharge; and (3) serotonergic DRN neurons do not appear to be directly thermosensitive. Overall, these results do not support a specific role for serotonergic DRN

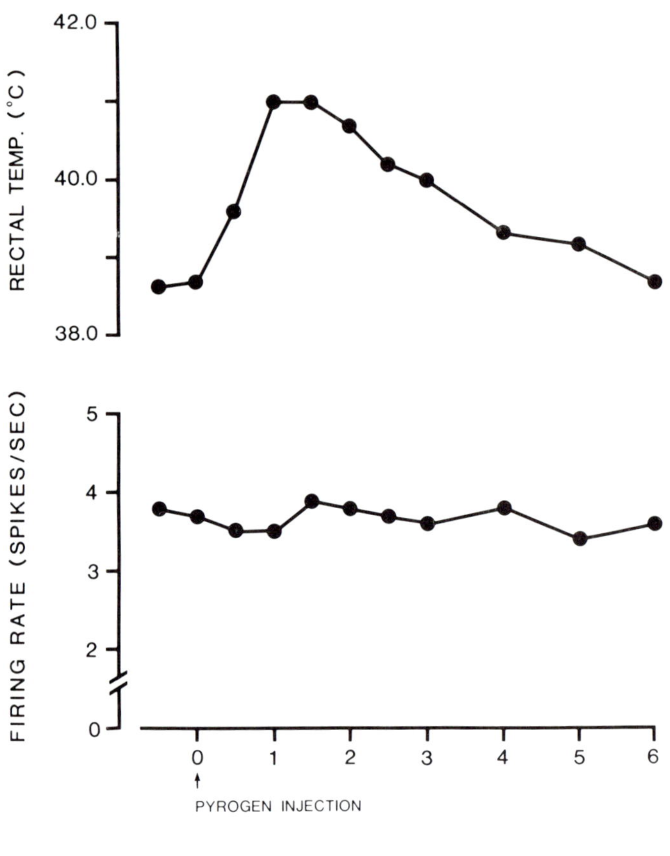

Fig. 13 Activity of a serotonergic DRN neuron during the
febrile response following pyrogen administration (muramyl
dipeptide; 50 μg/kg, i.v.). All unit measures were obtained
when the animal was in a quiet waking state. There were no
significant changes in unit activity following pyrogen
administration

neurons in thermoregulation. However, they do not rule out the possibility
that serotonergic neurons in other raphe nuclei may be more directly involved
in thermoregulation. In addition, since the activity of serotonergic DRN
neurons is strongly related to behavioral state, they may act in a more general
way by modulating state-dependent changes in central thermoregulation.
For example, serotonergic neurons, by modulating the gain or threshold of
thermoregulatory responses, may be involved in the downward resetting
of the thermoregulatory set-point during SWS and/or in the suppression of
thermoregulatory effector mechanisms during REM sleep.

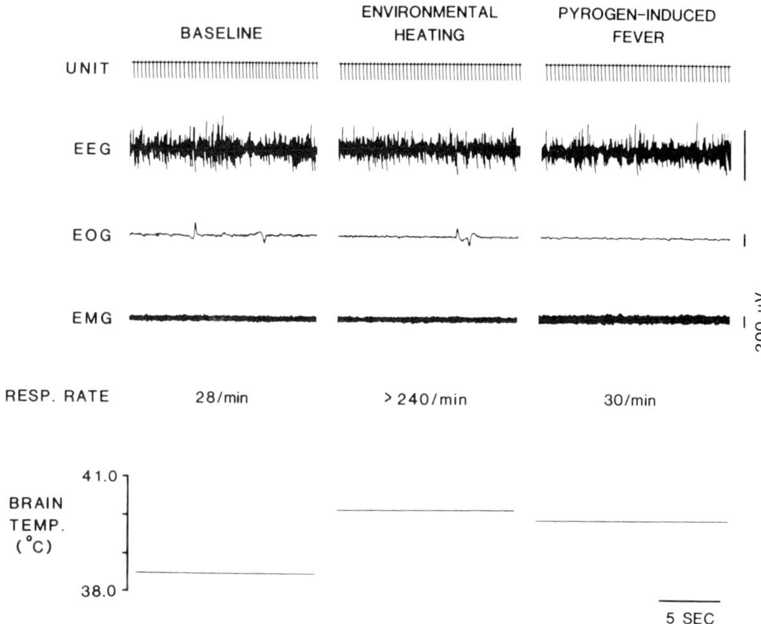

Fig. 14 Polygraph records showing the activity of a single serotonergic DRN neuron, as well as physiological responses during the peak elevation of brain temperature produced in response to both environmental heating and pyrogen administration. This figure illustrates the lack of effect of elevated brain temperature on DRN unit activity regardless of the thermoregulatory mechanisms invoked. See text for further details

SEROTONERGIC NEURONS AND CARDIOVASCULAR REGULATION

Brain serotonergic neurons have been implicated in the regulation of cardiovascular functions (see reviews by Kuhn *et al.*, 1980a; Antonaccio, 1984; Wolf *et al.*, 1985). Descending serotonergic pathways innervate many areas of the brainstem and spinal cord known to be involved in the regulation of sympathetic activity and in the mediation of cardiovascular reflexes (e.g. intermediolateral cell column, nucleus tractus solitarius, dorsal motor nucleus of the vagus) (Loewy and McKellar, 1981; Moore, 1981). Likewise, ascending serotonergic pathways provide a major input to the hypothalamus which has long been implicated in cardiovascular control and sympathetic responses (Azmitia, 1978; Moore, 1981).

Electrical stimulation of the dorsal and median raphe nuclei in anesthetized rats (Smits *et al.*, 1978; Kuhn *et al.*, 1980b; Robinson *et al.*, 1985), as well as chemical or electrical stimulation of the DRN in anesthetized cats (Piper and Goadsby, 1985), have been shown to consistently produce increases in arterial blood pressure with little or no change in heart rate. Pharmacological evidence presented in these studies suggests that serotonin is involved in the pressor responses to raphe stimulation, although the results are not in total agreement. In addition, Kuhn *et al*., (1980b) implicated an ascending serotonergic pathway to the AH/POA in the pressor response to DRN stimulation in rats. Furthermore, an increase in sympathetic outflow appears to mediate the pressor response to raphe stimulation, since pharmacological inhibition of postganglionic adrenergic nerves (Kuhn *et al*., 1980b) or blockade of α-adrenergic receptors (Piper and Goadsby, 1985) attenuates the pressor response. These results support a pressor role for midbrain serotonergic neurons.

The role that serotonergic neurons in the medulla play in cardiovascular regulation, however, remains controversial. Electrical stimulation within each of the medullary raphe nuclei has been shown to produce both an increase and decrease in arterial blood pressure (Adair *et al*., 1977; Yen *et al*., 1983). The variability of the response (i.e. pressor or depressor effect) following stimulation of a particular nucleus may be due to the fact that the medullary raphe nuclei, in contrast to the mesencephalic groups, are comprised predominantly of non-serotonergic cells (Moore, 1981; Wiklund *et al*., 1981). Another possibility is that serotonergic neurons in the medullary raphe nuclei may subserve both pressor and depressor functions. However, more recent studies combining electrical or chemical stimulation with pharmacological manipulations of serotonergic neurons have provided evidence in favor of a pressor role for medullary serotonergic neurons (Howe *et al*., 1983; McCall, 1984; Pilowsky *et al*., 1986). This pressor action is presumably mediated through sympathetic activation. Consistent with this idea, several studies have shown that iontophoretic application of serotonin predominantly activates sympathetic preganglionic neurons in the intermediolateral cell column of the spinal cord (Coote *et al*., 1981; Kadzielawa, 1983; McCall, 1983). Therefore, serotonergic neurons in both the midbrain and medulla may serve to elevate arterial blood pressure by enhancing sympathetic activity.

Although the central serotonergic system is thought to play an important role in cardiovascular regulation, so far only two studies have examined the single-unit activity of serotonergic neurons in relation to cardiovascular variables, and both of these studies were conducted in anesthetized rats. Foote *et al*. (1969) found that intravenous administration of L-norepinephrine had no effect on the activity of serotonergic neurons in the DRN despite producing an increase in blood pressure. Furthermore, Bramwell (1974) reported that there was no relationship between the spontaneous discharge rate of serotonergic DRN neurons and baseline blood pressure. These studies suggest that serotonergic neurons are unresponsive to changes in blood pressure in anesthetized animals. Whether or not serotonergic unit activity in unanesthetized animals is affected by alterations in blood pressure is an important point that remains to be determined.

Recently, we have begun studying the role of serotonergic neurons in the regulation of the cardiovascular system. Our strategy has consisted of examining the activity of serotonergic neurons in relation to the cardiac cycle and also in response to pharmacological manipulations of arterial blood pressure in behaving cats. Since many cardiovascular receptor systems undergo rhythmic fluctuations in their activity during the cardiac cycle, the demonstration of a cardiac-related periodicity in the discharge of serotonergic neurons would be one indication that they receive afferent information from the cardiovascular system. One possible pathway by which this information may reach the raphe nuclei is via the nucleus tractus solitarius (NTS), the major site of termination of primary afferent fibers from arterial and cardio-pulmonary receptors. Fibers from the NTS have been shown to project to the DRN (Aghajanian and Wang, 1977). In addition, the demonstration that serotonergic neurons respond to changes in blood pressure would provide further evidence for their involvement in cardiovascular regulation. Our studies focused primarily on serotonergic DRN neurons; however, a small group of serotonergic NRM neurons was also included, since they may be more directly involved in cardiovascular regulation by virtue of their descending projections to the intermediolateral cell column.

In the cardiac studies, the electrocardiogram (ECG) signal was fed through a variable threshold Schmitt-trigger set to deliver a pulse at the peak of the R-wave. Peri-stimulus time histograms of unit spike occurrence were then constructed around the triggered R-wave pulse using a microcomputer. The probability of unit discharge of serotonergic neurons in both the DRN and NRM was found to be unrelated to the phase of the cardiac cycle (Morilak *et al.*, 1986). Data from a representative serotonergic neuron, illustrating the lack of cardiac-related periodicity in unit discharge, is shown in Fig. 15. Serotonergic neurons exhibited a uniform distribution of neuronal spikes throughout the cardiac cycle under quiet, resting conditions. In addition, we considered the possibility that the discharge of these neurons may exhibit a relationship to the cardiac cycle under conditions other than the quiet, resting state, i.e. when the cardiovascular system is mobilized. However, additional studies have also failed to demonstrate any cardiac-related periodicity in the discharge of these neurons during either the presentation of an arousing exteroceptive stimulus (100 dB white noise), or during a tonic cardiovascular manipulation (15% blood volume expansion with isotonic saline).

We have also examined the activity of serotonergic neurons in the DRN and NRM in response to peripherally induced changes in blood pressure. In these experiments, phenylephrine, a direct-acting vasoconstrictor, was used to induce dose-dependent increases in mean arterial pressure (MAP) over a range of 20–70 mmHg, and sodium nitroprusside, a direct-acting vasodilator,

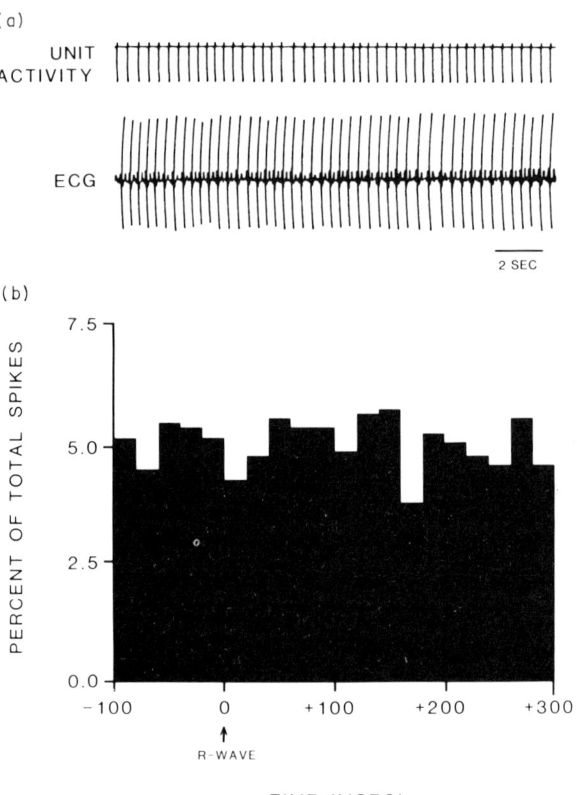

Fig. 15 Lack of cardiac-related periodicity in the discharge of a typical serotonergic DRN neuron. A: Simultaneous recording of single-unit activity and the electrocardiogram (ECG) in a freely moving cat during quiet waking. B: Computer-generated peri-R-wave time histogram produced from the activity of the same cell during quiet waking. The histogram was constructed from 1017 spikes using a bin width of 20 msec and represents approximately 800 cardiac cycles. Percentages were obtained by dividing each bin count by the total number of recorded spikes. The obtained spike distribution was not significantly different from a uniform rectangular distribution

was used to induce dose-dependent decreases in MAP in a range of 10–50 mmHg. The discharge of serotonergic DRN neurons was unaffected by changes in blood pressure in either direction. The activity of one serotonergic DRN neuron during both phenylephrine-induced hypertension and nitroprusside-induced hypotension is shown in Fig. 16. During a slow infusion of

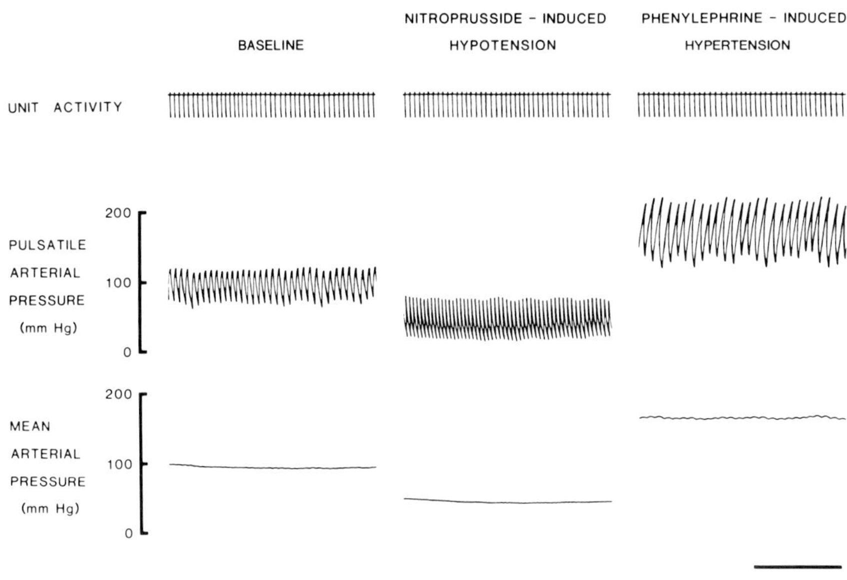

Fig. 16 Activity of a single serotonergic DRN neuron during both phenylephrine-induced hypertension and nitroprusside-induced hypotension. Phenylephrine hydrochloride (100 μg/ml) and sodium nitroprusside (100 μg/ml) were infused intravenously to achieve and maintain the alterations in blood pressure

phenylephrine (100 μg/ml), the discharge of this neuron did not change even though MAP was raised 65 mmHg from baseline (~100 mmHg). During the hypertensive period following phenylephrine infusion, heart rate decreased from a baseline rate of 175 to 130 beats/min, indicating strong activation of arterial baroreceptors. The activity of this same neuron was also unchanged following a slow infusion of nitroprusside (100 μg/ml), although MAP was lowered to approximately 50 mmHg. During the hypotensive period, heart rate increased to 265 beats/min, indicating marked unloading of arterial baroreceptors. Initial evidence also indicates that the activity of serotonergic NRM neurons is unaltered during similar manipulations of blood pressure.

Overall, these data demonstrate that the activity of both serotonergic DRN and NRM neurons is unaffected by either the normal beat-to-beat changes in cardiovascular afferent activity during the cardiac cycle, or by drug-induced alterations of primary cardiovascular variables. Thus, these neurons do not appear to be affected by afferent information from the cardiovascular system. This suggests that any involvement of serotonergic DRN and NRM neurons in cardiovascular regulation is mediated independently of afferent activity. Furthermore, as in the case of thermoregulation, serotonergic neurons may act in a more general way by modulating state-dependent changes in central cardiovascular regulation. For example, serotonergic neurons may affect blood pressure, heart rate, and sympathetic activity by modulating the gain or threshold of certain cardiovascular reflexes across the sleep–wake cycle.

DISCUSSION

The purpose of this chapter was to review the literature concerning the role of central serotonergic neurons in physiology and behavior, and to describe our recent studies on the electrophysiological response of these neurons to manipulations of specific physiological variables. Previous studies in mammals have strongly implicated brain serotonergic neurons in the control of physiological processes such as nociception, cardiovascular function, and thermoregulation. Studies in our laboratory, conducted in behaving cats, indicate that serotonergic neurons are not primarily involved in these physiological processes, since manipulations of these systems had no effect on serotonergic unit activity independent of changes in behavioral arousal. For example, our studies on serotonergic NRM neurons demonstrate that the activation of these neurons by noxious stimuli was similar to that elicited by non-noxious, but arousing stimuli. Thus, NRM serotonergic neurons do not appear to be specifically responsive to nociceptive input, and, therefore, it is unlikely that they play a preferential role in the control of nociception. Similarly, in our studies on serotonergic DRN neurons, we found that the activity of these neurons was unaffected by manipulations of arterial blood pressure or body temperature, unless there was an accompanying change in behavioral state. For example, following the administration of vasoactive drugs, changes in unit activity always paralleled behavioral state changes, and there was also no difference in the absolute level of unit activity for a particular behavioral state, regardless of the absolute level of arterial blood pressure. Likewise, the correlation between serotonergic DRN unit activity and behavioral state, as well as the absolute level of unit activity for a particular behavioral state, were unaffected by elevations in body temperature induced by either pyrogen administration, or environmental heating. Overall, these studies suggest that the activity of central serotonergic neurons is strongly related to behavioral state, and not to changes in specific physiological variables. Additional studies in our laboratory have shown that alterations in serotonergic DRN unit activity in cats, in response to a variety of interoceptive and exteroceptive stimuli (e.g. insulin-induced hypoglycemia, emesis, blood volume expansion, hypertonic saline administration, acute restraint, exposure to 100 dB white noise, or presentation of a dog) paralleled changes in behavioral state induced by these stimuli (Fornal *et al.*, 1986; Wilkinson *et al.*, 1986; Fornal, Morilak, and Jacobs, unpublished observations). Thus, the unit responses to these stimuli can be accounted for by the concomitant changes in behavioral activation. To date, no physiological or behavioral manipulation has been found to influence the activity of serotonergic neurons independent of changes in behavioral arousal. Thus, serotonergic neurons appear to respond in a stereotypical fashion to stimuli of various modalities, i.e. any stimulus that elicits behavioral activation will evoke a concomitant increase in serotonergic unit activity, and vice versa. Since the tonic activity of serotonergic neurons appears to vary only in association with behavioral state, the primary role of these neurons in physiology and behavior may be to coordinate the activity of target structures with the level of behavioral arousal.

Several unique anatomical and electrophysiological properties of central serotonergic neurons support a modulatory role for serotonin in the CNS. First, serotonergic neurons have long axons that project diffusely throughout the CNS to innervate areas which subserve diverse physiological and behavioral functions (Dahlström and Fuxe, 1964; Azmitia, 1978; Moore, 1981). Second, the purported non-synaptic nature of many serotonergic terminals suggests that upon release, serotonin may potentially interact with several postsynaptic neurons (Chan-Palay, 1975; Descarries et al., 1975). Third, individual serotonergic neurons have been shown to project to more than one target site (van der Kooy and Kuypers, 1979). In addition, the slow and regular activity of serotonergic neurons, coupled with their slow conduction velocity and long-duration synaptic actions (Segal, 1975; Wang and Aghajanian 1977; Wessendorf et al., 1981), is also compatible with a tonic maintenance function, as opposed to a rapid and more phasic function. Therefore, the design of this system would appear ideal for supporting a global modulatory function, i.e. to coordinate the activity of several heterogeneous physiological and behavioral systems.

Recent studies have lent support to the hypothesis that brain serotonin serves primarily a modulatory role, rather than a conventional excitatory or inhibitory transmitter action. For example, it has been shown that local application of serotonin exerts a long-lasting facilitatory action on both glutamate-induced and afferent stimulation-induced excitation of facial and spinal motoneurons, but alone does not elicit action potentials in these neurons (McCall and Aghajanian, 1979; White and Neuman, 1980). These results suggest that serotonin can modulate the sensitivity of neurons to afferent inputs. Serotonin is thought to mediate this facilitatory action on motoneurons by producing a decrease in potassium conductance, which leads to a slow and moderate depolarization of the postsynaptic membrane (Vandermaelen and Aghajanian, 1982). This action would then presumably result in a generalized increase in motoneuron excitability.

Based on the anatomy of this system, serotonin could potentially modulate several regulatory systems, and thus influence a wide range of sensory, autonomic, and endocrine functions. Furthermore, our data, as well as studies from other laboratories, suggest that the modulatory action of serotonin may vary as a function of the general level of behavioral arousal. According to this hypothesis, the activity of serotonergic neurons would have to be precisely controlled in order to match the excitability of target neurons to the behavioral state of the animal. In this respect, the activity of serotonergic neurons for a given behavioral state has been found to remain virtually constant when monitored for periods lasting from one day to several weeks (Trulson and Jacobs, 1983; Trulson *et al.*, 1984; Fornal, Wilkinson and Jacobs, unpublished observation.) The physiological basis of the remarkable constancy of unit activity over time is thought to involve local negative feedback mechanisms mediated by axon collaterals, and possibly via connections with neighboring serotonergic neurons (see review by Aghajanian, 1982). The pacemaker-like properties of serotonergic neurons probably also contribute to their constant firing rate. Collectively, these mechanisms would then operate to maintain the tonic activity of serotonergic neurons within a relatively narrow range. Influences exerted during state changes apparently override the above autoregulatory mechanisms to produce the alterations in unit activity that are observed across the sleep–wake continuum. We found that the spontaneous activity of serotonergic neurons was highest during active waking (AW), a time when animals are most likely to interact with their environment. The firing rate observed during AW was near-maximum, since strong arousing stimuli (e.g. pain, exposure to loud white noise, barking dog, restraint) did not further increase the firing rate of these neurons beyond AW levels (Auerbach *et al.*, 1985; Wilkinson *et al.*, 1986).

The importance of tightly regulating the level of serotonergic neurotransmitter is exemplified by the finding that pharmacological stimulation of postsynaptic serotonin receptors produces a classical motor syndrome characterized by resting tremor, rigidity, myoclonus, and hyperreflexia (Jacobs, 1976). This syndrome presumably results from a hyperfacilitation of spinal and brainstem motoneuron excitability produced by overstimulation of postsynaptic serotonin receptors.

The overall level of many physiological systems has been shown to vary across the sleep–wake cycle (see Orem and Barnes, 1980; Zanchetti *et al.*, 1982). Several autonomic reflexes are considerably dampened or suspended during REM sleep. For example, the baroreceptor-mediated control of arterial blood pressure, heart rate, and sympathetic activity is diminished. Likewise, there is muscle atonia and suspension of thermoregulatory capacity. Since the activity of serotonergic neurons is at its lowest level during REM sleep (i.e. most cells are completely silent), these findings indicate that central serotonergic neurons may be involved in state-related changes in physiological regulation. Furthermore, the reappearance of serotonergic unit activity several seconds prior to animals awakening from REM sleep may serve to re-establish the sensitivity of target neurons to a level compatible with the waking state. Thus, central serotonergic neurons may coordinate physiological adjustments (i.e. reflex gain and excitability thresholds) which underlie behavioral state transitions.

Finally, our results underscore the importance of controlling for behavioral state when studying the effects of physiological and behavioral manipulations, or drug actions on the serotonergic system. This would be especially important in studies involving neurochemical measurements of neurotransmitter release or turnover. For example, following the administration of a pyrogen, the febrile response in animals is accompanied by behavioral depression and an increase in total sleep time. Since the discharge of serotonergic neurons parallels changes in behavioral state, the overall activity of these neurons should be reduced in response to pyrogen-induced fever; and this indeed appears to be the case. Therefore, unless appropriate measures are taken to control for the influence of behavioral state, a manipulation that produces changes in serotonin release or turnover could be mistakenly interpreted as having a direct effect on serotonergic neurons. It is quite possible, however, that such a manipulation could have induced changes in various serotonergic parameters indirectly by producing an overall change in behavioral state. Furthermore, an additional factor to consider when interpreting studies of the central serotonergic system is that the release of serotonin at terminal areas may be locally altered by neuromodulators or neurotransmitters, independent of changes in unit activity. In this respect we are currently investigating the relationship between the release of serotonin in target areas, measured by an *in vivo* dialysis technique, and unit activity recorded from cell bodies in the brainstem (Wilkinson *et al.*, 1987). A primary aim of these studies is to determine under what conditions the terminal release of serotonin may be dissociated from unit activity.

In summary, the ultimate role of central serotonergic neurons may be to coordinate various physiological processes in concert with arousal level to produce smooth, integrated behavioral outputs.

FUTURE TRENDS

Future study of serotonergic single-unit activity in behaving animals should address the following major issues: (1) the relationship between unit activity and terminal release of serotonin in target areas under a variety of physiological and behavioral conditions; (2) the relationship between unit activity and functional changes in target systems; (3) examination of the local control of serotonergic unit activity by neurotransmitter/neuromodulatory substances; and (4) analysis of the single-unit activity of serotonergic neurons located off the midline in the ventrolateral medulla (these cells have not been studied previously and it would be of interest to determine whether their activity conforms to the general pattern displayed by serotonergic neurons located in the midline raphe nuclei). Along these lines we plan to investigate the functional consequences of changes in serotonergic unit activity, using the masseteric reflex as an index of serotonergic modulation of motoneuron excitability. A major aim of these studies will be to establish a cause-and-effect relationship between unit activity and reflex amplitude by combining these studies with pharmacological manipulations, including neurotoxin-induced destruction of specific neurochemical inputs. Furthermore, we have recently developed a technique which allows the infusion of substances into the area where single-unit activity is being recorded (Jacobs and Abercrombie, 1986). Using this technique, we plan to investigate the local influence of various neurotransmitter/neuromodulatory substances on central serotonergic neurons which may be involved in mediating the state-related changes in the activity of these neurons. This technique will also be utilized to investigate the direct action of a variety of drugs on the activity of serotonergic neurons in behaving animals.

ACKNOWLEDGEMENTS

Work from the authors' laboratory was supported by National Institute of Mental Health Grant MH 23433. The authors wish to thank Ms Arlene Kronewitter and Ms Daphne Flowers for typing the manuscript, and would also like to acknowledge the contributions of Drs David Morilak, Joseph Litto and Sidney Auerbach, as well as Ms Lynn Wilkinson.

REFERENCES

Adair, J. R., Hamilton, B. L., Scappaticci, K. A., Helke, C. J., and Gillis, R. A. (1977) Cardiovascular responses to electrical stimulation of the medullary raphe area of the cat, *Brain Res.*, **128**, 141–145.

Akil, H., and Mayer, D. J. (1972) Antagonism of stimulation-produced analgesia by p-CPA, a serotonin synthesis inhibitor, *Brain Res.*, **44**, 692–697.

Aghajanian, G. K. (1982) Regulation of serotonergic neuronal activity: autoreceptors and pacemaker potentials, in *Advances in Biochemical Psychopharmacology*, Vol. 34, *Serotonin in Biological Psychiatry* (Eds B. T. Ho, J. C. Schoolar, and E. Usdin), pp. 173–181, Raven Press, New York.

Aghajanian, G. K., and Haigler, H. J. (1974) L-Tryptophan as a selective histochemical marker for serotonergic neurons in single-cell recording studies, *Brain Res.*, **81**, 364–372.

Aghajanian, G. K., and Vandermaelen, C. P. (1982) Intracellular identification of central noradrenergic and serotonergic neurons by a new double labeling procedure, *J. Neurosci.*, **2**, 1786–1792.

Aghajanian, G. K., and Wang, R. Y. (1977) Habenular and other midbrain raphe afferents demonstrated by a modified retrograde tracing technique, *Brain Res.*, **122**, 229–242.

Aghajanian, G. K., Foote, W. E., and Sheard, M. H. (1968) Lysergic acid diethylamide: sensitive neuronal units in the midbrain raphe, *Science*, **161**, 706–708.

Aghajanian, G. K., Wang, R. Y., and Baraban, J. (1978) Serotonergic and non-serotonergic neurons of the dorsal raphe: reciprocal changes in firing induced by peripheral nerve stimulation, *Brain Res.*, **153**, 169–175.

Anderson, S. D., Basbaum, A. I., and Fields, H. L. (1977) Response of medullary raphe neurons to peripheral stimulation and to systemic opiates, *Brain Res.*, **123**, 363–368.

Anderson, E. G., and Proudfit, H. K. (1981) The functional role of the bulbospinal serotonergic nervous system, in *Serotonin Neurotransmission and Behavior* (Eds B. L. Jacobs and A. Gelperin), pp. 307–338, MIT Press, Cambridge, Mass.

Antonaccio, M. J. (1984) Central transmitters: physiology, pharmacology, and effects on the circulation, in *Cardiovascular Pharmacology* (Ed. M. J. Antonaccio), pp. 155–195, Raven Press, New York.

Auerbach, S., Fornal, C., and Jacobs, B. L. (1985) Response of serotonin-containing neurons in nucleus raphe magnus to morphine, noxious stimuli, and periaqueductal gray stimulation in freely moving cats, *Exp. Neurol.*, **88**, 609–628.

Azmitia, E. C. (1978) The serotonin-producing neurons of the midbrain medial and dorsal raphe nuclei, in *Handbook of Psychopharmacology*, Vol. 9 (Eds L. L. Iversen, S. D. Iversen, and S. H. Snyder), pp. 233–314, Plenum Press, New York.

Basbaum, A. I., and Fields, H. L. (1984) Endogenous pain control systems: brainstem spinal pathways and endorphin circuitry, *Annu. Rev. Neurosci.*, **7**, 309–338.

Basbaum, A. I., Clanton, C. H., and Fields, H. L. (1978) Three bulbospinal pathways from the rostral medulla of the cat: an autoradiographic study of pain modulating systems, *J. Comp. Neurol.*, **178**, 209–224.

Basbaum, A. I., Marley, N. J. E., O'Keefe, J., and Clanton, C. H. (1977) Reversal of morphine and stimulus-produced analgesia by subtotal spinal cord lesions, *Pain*, **3**, 43–56.

Belcher, G., Ryall, R. W., and Schaffner, R. (1978) The differential effects of 5-hydroxytryptamine, noradrenaline and raphe stimulation on nociceptive and non-nociceptive dorsal horn interneurones in the cat, *Brain Res.*, **151**, 307–321.

Bineau-Thurotte, M., Godefroy, F., Weil-Fugazza, J., and Besson, J.-M. (1984) The effect of morphine on the potassium evoked release of tritiated 5-hydroxytryptamine from spinal cord slices in the rat, *Brain Res.*, **291**, 293–299.

Bobillier, P., Seguin, S., Petitjean, F., Salvert, D., Touret, M., and Jouvet, M. (1976) The raphe nuclei of the cat brainstem: a topographical atlas of their efferent projections as revealed by autoradiography, *Brain Res.*, **113**, 449–486.

Bramwell, G. J. (1974) Factors affecting the activity of 5-HT-containing neurones, *Brain Res.*, **79**, 515–519.

Cadden, S. W., Lisney, S. J. W., and Matthews, B. (1983) Thresholds to electrical stimulation of nerves in cat canine tooth-pulp with Aβ-, Aδ- and C-fibre conduction velocities, *Brain Res.*, **261**, 31–41.

Chan-Palay, V. (1975) Fine structure of labelled axons in the cerebellar cortex and nuclei of rodents and primates after intraventricular infusions with tritiated serotonin, *Anat. Embryol.*, **148**, 235–265.

Chiang, C. Y., and Pan, Z. Z. (1985) Differential responses of serotonergic and non-serotonergic neurons in nucleus raphe magnus to systemic morphine in rats, *Brain Res.*, **337**, 146–150.

Commissiong, J. W. (1983) Mass fragmentographic analysis of monoamine metabolites in the spinal cord of rat after the administration of morphine, *J. Neurochem.*, **41**, 1313–1318.

Coote, J. H., Macleod, V. H., Fleetwood-Walker, S., and Gilbey, M. P. (1981) The response of individual sympathetic preganglionic neurones to microelectrophoretically applied endogenous monoamines, *Brain Res.*, **215**, 135–145.

Cox, B., and Lee, T. F. (1982) Role of central neurotransmitters in fever, in *Handbook of Experimental Pharmacology*, Vol. 60, *Pyretics and Antipyretics* (Ed. A. S. Milton), pp. 125–150, Springer-Verlag, Berlin.

Cronin, M. J., and Baker, M. A. (1976) Heat-sensitive midbrain raphe neurons in the anesthetized cat, *Brain Res.*, **110**, 175–181.

Dahlström, A., and Fuxe, K. (1964) Evidence for the existence of monoamine-containing neurons in the central nervous system. I. Demonstration of monoamines in the cell bodies of brain stem neurons, *Acta Physiol. Scand.*, **62** (Suppl. 232), 1–55.

Dascombe, M. J. (1985) The pharmacology of fever, *Progress in Neurobiology*, **25**, 327–373.

Deakin, J. F. W., Dickenson, A. H., and Dostrovsky, J. O. (1977) Morphine effects on rat raphe magnus neurones, *J. Physiol. (Lond.)*, **267**, 43–45P.

DeMontigny, C., and Aghajanian, G. K. (1977) Preferential action of 5-methoxytryptamine and 5-methoxydimethyltryptamine on presynaptic serotonin receptors: a comparative iontophoretic study with LSD and serotonin, *Neuropharmacology*, **16**, 811–818.

Descarries, L., Beaudet, A., and Watkins, K. C. (1975) Serotonin nerve terminals in adult rat neocortex, *Brain Res.*, **100**, 563–588.

Dickenson, A. H. (1977) Specific responses of rat raphe neurones to skin temperature, *J. Physiol. (Lond.)*, **273**, 277–293.

Dubuisson, D., and Dennis, S. G. (1977) The Formalin test: a quantitative study of the analgesic effects of morphine, meperidine, and brainstem stimulation in rats and cats, *Pain*, **4**, 161–174.

Feldberg, W., and Myers, R. D. (1965) Changes in temperature produced by microinjections of amines into the anterior hypothalamus of cats, *J. Physiol. (Lond.)*, **177**, 239–245.

Fields, H. L., and Anderson, S. D. (1978) Evidence that raphe-spinal neurons mediate opiate and midbrain stimulation-produced analgesias, *Pain*, **5**, 333–349.

Foote, W. E., Sheard, M. H., and Aghajanian, G. K. (1969) Comparison of effects of LSD and amphetamine on midbrain raphe units, *Nature*, **222**, 567–569.

Fornal, C., Auerbach, S., and Jacobs, B. L. (1985) Activity of serotonin-containing neurons in nucleus raphe magnus in freely moving cats, *Exp. Neurol.*, **88**, 590–608.

Fornal, C. A., Litto, W. J., Morilak, D. A., and Jacobs, B. L. (1987a) Single-unit responses of serotonergic dorsal raphe nucleus neurons to environmental heating and pyrogen administration in freely moving cats, *Exp. Neurol.*, **98**, 388–403.

Fornal, C., Morilak, D. A., and Jacobs, B. L. (1986) Physiological stress and the activity of dorsal raphe serotonergic neutrons in freely moving cats, *Soc. Neurosci. Abstr.*, **12**, 1134.

Fornal, C. A., Wilkinson, L. O., and Jacobs, B. L. (1987b) Effects of intravenous buspirone and 8-OH-DPAT on the activity of serotonergic dorsal raphe neurons in freely moving cats, *Soc. Neurosci. Abstr.*, **13**, 456.

Gebhart, G. F. (1982) Opiate and opioid peptide effects on brainstem neurons: relevance to nociception and antinociceptive mechanisms, *Pain*, **12**, 93–140.

Guilbaud, G., Peschanski, M., Gautron, M., and Binder, D. (1980) Responses of neurons of the nucleus raphe magnus to noxious stimuli, *Neurosci. Lett.*, **17**, 149–154.

Haigler, H. J. (1978) Morphine: effects on brainstem raphe neurons, in *Iontophoresis and Transmitter Mechanisms in the Mammalian Central Nervous System* (Eds R.

W. Ryall and J. S. Kelly), pp. 326–328, Elsevier/North-Holland Biomedical Press, Amsterdam.

Hammond, D. L., and Yaksh, T. L. (1984) Antagonism of stimulation-produced antinociception by intrathecal administration of methysergide or phentolamine, *Brain Res.*, **298**, 329–337.

Harvey, C. A., and Milton, A. S. (1976) The effects of parachlorophenylalanine and 6-hydroxydopamine on thermoregulatory responses to heat and cold stress, *J. Physiol. (Lond.)*, **263**, 208–209P.

Headley, P. M., Duggan, A. W., and Griersmith, B. T. (1978) Selective reduction by noradrenaline and 5-hydroxytryptamine of nociceptive responses of cat dorsal horn neurones, *Brain Res.*, **145**, 185–189.

Hellon, R. F. (1975) Monoamines, pyrogens and cations: their actions on central control of body temperature, *Pharmac. Rev.*, **26**, 289–321.

Heym, J., and Jacobs, B. L. (1986) Recording single unit activity of neurons in freely moving cats, in *Modern Methods in Pharmacology: Electrophysiological Techniques* (Ed. H. M. Geller), pp. 17–34, Alan R. Liss, Inc., New York.

Heym, J., Steinfels, G. F., and Jacobs, B. L. (1982a) Activity of serotonin-containing neurons in the nucleus raphe pallidus of freely moving cats, *Brain Res.*, **251**, 259–276.

Heym, J., Steinfels, G. F., and Jacobs, B. L. (1982b) Medullary serotonergic neurons are insensitive to 5-MeODMT and LSD, *Eur. J. Pharmacol.*, **81**, 677–680.

Heym, J., Steinfels, G. F., and Jacobs, B. L. (1984) Chloral hydrate anesthesia alters the responsiveness of central serotonergic neurons in the cat, *Brain Res.*, **291**, 63–72.

Heym, J., Trulson, M. E., and Jacobs, B. L. (1982c) Raphe unit activity in freely moving cats: effects of phasic auditory and visual stimuli, *Brain Res.*, **232**, 29–39.

Hori, T., and Harada, Y. (1976) Responses of midbrain raphe neurons to local temperature, *Pflügers Arch.*, **364**, 205–207.

Howe, P. R. C., Kuhn, D. M., Minson, J. B., Stead, B. H., and Chalmers, J. P. (1983) Evidence for a bulbospinal serotonergic pressor pathway in the rat brain, *Brain Res.*, **270**, 29–36.

Jacobs, B. L. (1976) An animal behavior model for studying central serotonergic synapses, *Life Sci.*, **19**, 777–785.

Jacobs, B. L. (1987) Brain monoaminergic unit activity in behaving animals, in *Progress in Psychobiology and Physiological Psychology* (Eds A. N. Epstein and A. R. Morrison), pp. 171–206, Academic Press, New York.

Jacobs, B. L., and Abercrombie, E. D. (1986) Simultaneous single unit recording and local drug injection in the locus coeruleus of freely moving cats, *Soc. Neurosci. Abstr.*, **12**, 1390.

Jacobs, B. L., and Gelperin, A. (Eds) (1981) *Serotonin Neurotransmission and Behavior*, MIT Press, Cambridge, Mass.

Jacobs, B. L., Gannon, P. J., and Azmitia, E. C. (1984) Atlas of serotonergic cell bodies in the cat brainstem: an immunocytochemical analysis, *Brain Res. Bull.*, **13**, 1–31.

Jacobs, B. L., Heym, J., and Rasmussen, K. (1983) Raphe neurons: firing rate correlates with size of drug response, *Eur. J. Pharmacol.*, **90**, 275–278.

Johannessen, J. N., Watkins, L. R., Carlton, S. M., and Mayer, D. J. (1982) Failure of spinal cord serotonin depletion to alter analgesia elicited from the periaqueductal gray, *Brain Res.*, **237**, 373–386.

Kadzielawa, K. (1983) Antagonism of the excitatory effects of 5-hydroxytryptamine on sympathetic preganglionic neurones and neurones activated by visceral afferents, *Neuropharmacology*, **22**, 19–27.

Keenan, C. L., and Chu, N.-S. (1987) Thermosensitivity of dorsal raphe neurons in vitro, *Brain Res.*, **410**, 189–194.

Komiskey, H. L., and Rudy, T. A. (1977) Serotonergic influences on brainstem thermoregulatory mechanisms in the cat, *Brain Res.*, **134**, 297–315.

Kuhn, D. M., Wolf, W. A., and Lovenberg, W. (1980a) Review of the role of the central serotonergic neuronal system in blood pressure regulation, *Hypertension*, **2**, 243–255.

Kuhn, D. M., Wolf, W. A., and Lovenberg, W. (1980b) Pressor effects of electrical stimulation of the dorsal and median raphe nuclei in anesthetized rats, *J. Pharmacol. Exp. Ther.*, **214**, 403–409.

Leysen, J. E. (1985) Characterization of serotonin receptor binding sites, in *Neuropharmacology of Serotonin* (Ed. A. R. Green), pp. 79–116, Oxford University Press, Oxford.

Loewy, A. D., and McKellar, S. (1981) Serotonergic projections from the ventral medulla to the intermediolateral cell column in the rat, *Brain Res.*, **211**, 146–152.

Lydic, R., McCarley, R. W., and Hobson, J. A. (1983) The time-course of dorsal raphe discharge, PGO waves, and muscle tone averaged across multiple sleep cycles, *Brain Res.*, **274**, 365–370.

McCall, R. B. (1983) Serotonergic excitation of sympathetic preganglionic neurons: a microiontophoretic study, *Brain Res.*, **289**, 121–127.

McCall, R. B. (1984) Evidence for a serotonergically mediated sympathoexcitatory response to stimulation of medullary raphe nuclei, *Brain Res.*, **311**, 131–139.

McCall, R. B., and Aghajanian, G. K. (1979) Serotonergic facilitation of facial motoneuron excitation, *Brain Res.*, **169**, 11–27.

McGinty, D. J., and Harper, R. M. (1976) Dorsal raphe neurons: depression of firing during sleep in cats, *Brain Res.*, **101**, 569–575.

Milton, A. S., and Harvey, C. A. (1975) Prostaglandins and monoamines in fever, in *Temperature Regulation and Drug Action* (Eds P. Lomax, E. Schönbaum, and J. Jacob), pp. 133–142, Karger, Basel.

Moolenaar, G.-M., Holloway, J. A., and Trouth, C. O. (1976) Responses of caudal raphe neurons to peripheral somatic stimulation, *Exp. Neurol.*, **53**, 304–313.

Moore, R. Y. (1981) The anatomy of central serotonin neuron systems in the rat brain, in *Serotonin Neurotransmission and Behavior* (Eds B. L. Jacobs and A. Gelperin), pp. 35–71, MIT Press, Cambridge, Mass.

Morilak, D. A., Fornal, C., and Jacobs, B. L. (1986) Single unit activity of noradrenergic neurons in locus coeruleus and serotonergic neurons in the nucleus raphe dorsalis of freely moving cats in relation to the cardiac cycle, *Brain Res.* **399**, 262–270.

Mosko, S. S., and Jacobs, B. L. (1976) Recording of dorsal raphe unit activity in vitro, *Neurosci. Lett.*, **2**, 195–200.

Mosko, S. S., and Jacobs, B. L. (1977) Electrophysiological evidence against negative neuronal feedback from the forebrain controlling midbrain raphe unit activity, *Brain Res.*, **119**, 291–303.

Myers, R. D. (1975) Impairment of thermoregulation, food and water intakes in the rat after hypothalamic injections of 5,6-dihydroxytryptamine, *Brain Res.*, **94**, 491–506.

Myers, R. D. (1977) New aspects of the role of hypothalamic calcium ions, 5-HT and PGE during normal thermoregulation and pyrogen fever, in *Drugs, Biogenic*

Amines and Body Temperature (Eds K. E. Cooper, P. Lomax, and E. Schönbaum), pp. 51–53, Karger, Basel.

Myers, R. D. (1980) Hypothalamic control of thermoregulation, in *Handbook of the Hypothalamus*, Vol. 3 (Eds P. J. Morgane and J. Panksepp), pp. 83–210, Marcel Dekker, New York.

Myers, R. D., and Beleslin, D. B. (1971) Changes in serotonin release in hypothalamus during cooling or warming of the monkey, *Am. J. Physiol.*, **220**, 1746–1754.

Okuda, C., Saito, A., Miyazaki, M., and Kuriyama, K. (1986) Alteration of the turnover of dopamine and 5-hydroxytryptamine in rat brain associated with hypothermia, *Pharmacol. Biochem. Behav.*, **24**, 79–83.

Oliveras, J. L., Redjemi, F., Guilbaud, and Besson, J. M. (1975) Analgesia induced by electrical stimulation of the inferior centralis nucleus of the raphe in the cat, *Pain*, **1**, 139–145.

Oliveras, J. L., Bourgoin, S., Héry, F., Besson, J. M., and Hamon, M. (1977) The topographical distribution of serotonergic terminals in the spinal cord of the cat: biochemical mapping by the combined use of microdissection and microassay procedures, *Brain Res.*, **138**, 393–406.

Orem, J., and Barnes, C. D. (Eds) (1980) *Physiology in Sleep*, Academic Press, New York.

Pilowsky, P. M., Kapoor, V., Minson, J. B., West, M. J., and Chalmers, J. P. (1986) Spinal cord serotonin release and raised blood pressure after brainstem kainic acid injection, *Brain Res.*, **366**, 354–357.

Piper, R. D., and Goadsby, P. J. (1985) Pressor response to electrical and chemical stimulation of nucleus raphe dorsalis in the cat, *Stroke*, **16**, 307–312.

Proudfit, H. K., and Anderson, E. G. (1975) Morphine analgesia: blockade by raphe magnus lesions, *Brain Res.*, **98**, 612–618.

Proudfit, H. K., Arai, A., and Monsen, M. (1985) Effect of morphine on the release of endogenous monoamines into spinal cord superfuates, *Soc. Neurosci. Abstr.*, **11**, 131.

Proudfit, H. K., and Hammond, D. L. (1981) Alterations in nociceptive threshold and morphine-induced analgesia produced by intrathecally administered amine antagonists, *Brain Res.*, **218**, 393–399.

Rasmussen, K., Heym, J., and Jacobs, B. L. (1984) Activity of serotonin-containing neurons in nucleus centralis superior of freely moving cats, *Exp. Neurol.*, **83**, 302–317.

Rasmussen, K., Strecker, R. E., and Jacobs, B. L. (1986) Single unit response of noradrenergic, serotonergic and dopaminergic neurons in freely moving cats to simple sensory stimuli, *Brain Res.*, **369**, 336–340.

Robinson, S. E., Austin, M. J. F., and Gibbens, D. M. (1985) The role of serotonergic neurons in dorsal raphe, median raphe and anterior hypothalamic pressor mechanisms, *Neuropharmacology*, **24**, 51–58.

Sakai, K., Vanni-Mercier, G., and Jouvet, M. (1983) Evidence for the presence of PS-OFF neurons in the ventromedial medulla oblongata of freely moving cats, *Exp. Brain Res.*, **49**, 311–314.

Segal, M. (1975) Physiological and pharmacological evidence for a serotonergic projection to the hippocampus, *Brain Res.*, **94**, 115–131.

Shiomi, H., Murakami, H., and Takagi, H. (1978) Morphine analgesia and the bulbospinal serotonergic system: increase in concentration of 5-hydroxyindoleacetic acid in the rat spinal cord with analgesics, *Eur. J. Pharmacol.*, **52**, 335–344.

Smits, J. F. M., van Essen, H., and Struyker-Boudier, H. A. J. (1978) Serotonin-mediated cardiovascular responses to electrical stimulation of the raphe nuclei in the rat, *Life Sci.*, **23**, 173–178.

Squires, R. F. (1974) Hyperthermia- and L-tryptophan-induced increases in serotonin turnover in rat brain, in *Advances in Biochemical Psychopharmacology*, Vol. 10, *Serotonin: New Vistas* (Eds E. Costa, G. L. Gessa, and M. Sandler), pp. 207–211, Raven Press, New York.

Tal, M. (1984) The threshold for eliciting the jaw opening reflex in rats is not increased by neonatal capsaicin, *Behav. Brain Res.*, **13**, 197–200.

Toda, K. (1982) Responses of raphe magnus neurons to systemic morphine in rats, *Brain Res. Bull.*, **8**, 101–103.

Trulson, M. E., and Jacobs, B. L. (1976) Serotonin-containing neurons: lack of response to changes in body temperature in rats, *Brain Res.*, **116**, 523–530.

Trulson, M. E., and Jacobs, B. L. (1979a) Raphe unit activity in freely moving cats: correlation with level of behavioral arousal, *Brain Res.*, **163**, 135–150.

Trulson, M. E., and Jacobs, B. L. (1979b) Effects of 5-methoxy-*N*,*N*-dimethyltryptamine on behavior and raphe unit activity in freely moving cats, *Eur. J. Pharmacol.*, **54**, 43–50.

Trulson, M. E., and Jacobs, B. L. (1983) Raphe unit activity in freely moving cats: lack of diurnal variation, *Neurosci. Lett.*, **36**, 285–290.

Trulson, M. E., Crisp, T., and Trulson, V. M. (1984) Activity of serotonin-containing nucleus centralis superior (raphe medianus) neurons in freely moving cats, *Exp. Brain Res.*, **54**, 33–44.

Trulson, M. E., Howell, G. A., Brandstetter, J. W., Frederickson, M. H., and Frederickson, C. J. (1982) *In vitro* recording of raphe unit activity: evidence for endogenous rhythms in presumed serotonergic neurons, *Life Sci.*, **31**, 785–790.

Tyce, G. M., and Yaksh, T. L. (1981) Monoamine release from cat spinal cord by somatic stimuli: an intrinsic modulatory system, *J. Physiol. (Lond.)*, **314**, 513–529.

van de Kar, L. D., Lorens, S. A., Vodraska, A., Allers, G., Green, M., Van Orden, D. E., Van Orden, L. S., III. (1980) Effect of selective midbrain and diencephalic 5,7-dihydroxytryptamine lesions on serotonin content in individual preopticohypothalamic nuclei and on serum luteinizing hormone level, *Neuroendocrinology*, **31**, 309–315.

van der Kooy, D., and Kuypers, H. G. J. M. (1979) Fluorescent retrograde double labeling: axonal branching in the ascending raphe and nigral projections, *Science*, **204**, 873–875.

Vandermaelen, C. P., and Aghajanian, G. K. (1982) Serotonin-induced depolarization of rat facial motoneurons in vivo: comparison with amino acid transmitters, *Brain Res.*, **239**, 139–152.

Vogt, M. (1982) Some functional aspects of central serotonergic neurons, in *Biology of Serotonergic Transmission* (Ed. N. N. Osborne), pp. 299–315, John Wiley & Sons, Chichester.

Waller, M. B., Myers, R. D., and Martin, G. E. (1976) Thermoregulatory deficits in the monkey produced by 5,6-dihydroxytryptamine injected into the hypothalamus, *Neuropharmacology*, **15**, 61–68.

Wang, R. Y., and Aghajanian, G. K. (1977) Antidromically identified serotonergic neurons in the rat midbrain raphe: evidence for collateral inhibition, *Brain Res.*, **132**, 186–193.

Watkins, L. R., Johannessen, J. N., Kinscheck, I. B., and Mayer, D. J. (1984) The neurochemical basis of footshock analgesia: the role of spinal cord serotonin and norepinephrine, *Brain Res.*, **290**, 107–117.

Weiss, B. L., and Aghajanian, G. K. (1971) Activation of brain serotonin metabolism by heat: role of midbrain raphe neurons, *Brain Res.*, **26**, 37–48.

Wessendorf, M. W., and Anderson, E. G. (1983) Single unit studies of identified bulbospinal serotonergic units, *Brain Res.*, **279**, 93–103.

Wessendorf, M. W., Proudfit, H. K., and Anderson, E. G. (1981) The identification of serotonergic neurons in the nucleus raphe magnus by conduction velocity, *Brain Res.*, **214**, 168–173.

White, S. R., and Neuman, R. S. (1980) Facilitation of spinal motoneurone excitability by 5-hydroxytryptamine and noradrenaline, *Brain Res.*, **188**, 119–127.

Wiklund, L., Léger, L., and Persson, M. (1981) Monoamine cell distribution in the cat brainstem. A fluorescence histochemical study with quantification of indolaminergic and locus coeruleus cell groups, *J. Comp. Neurol.*, **203**, 613–647.

Wilkinson, L. O., Abercrombie, E. D., and Jacobs, B. L. (1986) Environmental stress and activity of dorsal raphe serotonergic neurons in the freely moving cat, *Soc. Neurosci. Abstr.*, **12**, 1134.

Wilkinson, L. O., Martin, K. M., Auerbach, S. A., Marsden, C. A., and Jacobs, B. L. (1987) Simultaneous analysis of serotonin neuronal activity and release in behaving animals, *Soc. Neurosci. Abstr.*, **13**, 662.

Wolf, W. A., Kuhn, D. M., and Lovenberg, W. (1985) Serotonin and central regulation of arterial blood pressure, in *Serotonin and the Cardiovascular System* (Ed. P. M. Vanhoutte), pp. 63–73, Raven Press, New York.

Yaksh, T. L., Plant, R. L., and Rudy, T. A. (1977) Studies on the antagonism by raphe lesions of the antinociceptive action of systemic morphine, *Eur. J. Pharmacol.*, **41**, 399–408.

Yaksh, T. L., and Tyce, G. M. (1979) Microinjection of morphine into the periaqueductal gray evokes the release of serotonin from spinal cord, *Brain Res.*, **171**, 176–181.

Yaksh, T. L., and Wilson, P. R. (1979) Spinal serotonin terminal system mediates antinociception, *J. Pharmacol. Exp. Ther.*, **208**, 446–453.

Yen, C.-T., Blum, P. S., and Spath, J. A., Jr. (1983) Control of cardiovascular function by electrical stimulation within the medullary raphe region of the cat, *Exp. Neurol.*, **79**, 666–679,

Zanchetti, A., Baccelli, G., and Mancia, G. (1982) Cardiovascular regulation during sleep, *Arch. Ital. Biol.*, **120**, 120–137.

Neuronal Serotonin
Edited by N. N. Osborne and M. Hamon
© 1988 John Wiley & Sons Ltd

CHAPTER 12

Serotonin Uptake into Astrocytes and its Implications

H. K. KIMELBERG
Division of Neurosurgery and Departments of Biochemistry and Pharmacology/Toxicology
Albany Medical College
Albany
NY 12208
USA

INTRODUCTION

An important factor in allowing serotonin to effectively fulfill the many roles described by other authors in this book is the ability to control both the initiation and termination of its action. In the mammalian nervous system, initiation of action is thought to be due to release of serotonin from presynaptic nerve endings by the exocytosis of vesicles in which this transmitter is stored. Termination of its action after serotonin has combined with postsynaptic receptors and achieved its effects, is thought to be mainly by re-uptake mechanisms located in the same nerve terminals from which it was released. This scheme is now an established textbook tenet (e.g. McGeer *et al.*, 1978;

Cooper *et al.*, 1982; Feldman and Quenzer, 1984) and is well accepted in the literature, e.g. 'there is strong evidence for the existence of a high-affinity transport of 5-HT into peripheral and central serotonergic neurons' (Descarries and Beaudet, 1983). However, many synapses in the CNS are surrounded by the process of astroglial cells (astrocytes) (Lugaro, 1907; Wolff, 1970; Palay and Chan-Palay, 1974; Peters *et al.*, 1976; Varon and Somjen, 1979). These cells, together with oligodendroglia, constitute the macroglia which represent the major portion of the non-neuronal cells in the CNS. This perisynaptic location of astrocytes has long led neuroscientists to speculate that these cells may have a role in transmitter uptake (e.g. Lugaro, 1907).

In this chapter I will review recent evidence indicating that there may be significant uptake of serotonin into astrocytes, in addition to the already established evidence for neuronal re-uptake of serotonin. I will also discuss possible reasons why such uptake has only been seen in a few studies done *in situ*, and speculate on the possible roles for astrocytic uptake based on the limited data available.

LOCALIZATION OF SEROTONIN UPTAKE IN NERVOUS TISSUE *IN SITU*

The major evidence to date from studies *in situ* indicates that concentrative uptake of added, exogenous serotonin is localized to neurons, and specifically to serotonergic neurons synthesizing and releasing serotonin. Such uptake is thought to occur principally by a high-affinity uptake system (see Chapter 13 of this volume). This uptake system shows K_m values for serotonin of around 10^{-7}M, is highly dependent on the presence of Na^+ in the medium and is inhibited specifically by several classes of compounds. In turn, as discussed in several other chapters in this volume, it is considered highly significant that some of these inhibitory compounds are clinically successful drugs for psychiatric disorders (e.g. Green and Costain, 1981; Feldman and Quenzer, 1984).

Evidence for neuronal uptake includes observations that high-affinity uptake into synaptosome preparations is quite intense (Iversen, 1974; Koide and Uyemura, 1980). Also, high-affinity uptake is inhibited by prior specific surgical or chemical lesioning of neuronal tracts in the intact animal. Surgical lesions of the midbrain raphe nuclei, containing most of the serotonin cells which project to other parts of the brain, or chemical lesions of serotonergic neurons by injection of 5,6- or 5,7-dihydroxytryptamine, which appear to selectively destroy serotonergic neurons (Breese and Cooper, 1975), all inhibit high-affinity uptake of serotonin by brain (Kuhar *et al.*, 1972; Björklund *et al.*, 1973). However, it is not clear whether these lesion studies are completely specific. Thus, destruction of nerve terminals may have indirect effects on the astroglial processes surrounding them.

Histochemical and autoradiographic studies have shown that the pattern of localization of the serotonin taken up corresponds to the pattern for endogenous levels of this transmitter in serotonergic neurons (Fuxe *et al.*, 1968). Autoradiography at the electron microscopic level has also localized uptake of exogenous, labeled serotonin to nerve endings. Quantitation of the results showed that after intraventricular injection of [³H]serotonin in rat brain (Aghajanian and Bloom, 1967), about 80% of the grain clusters were over nerve endings and axons, and only about 5% over glia and blood vessels.

What then is the evidence for any significant extraneuronal uptake of serotonin and why should this question even be raised? Ruda and Gobel (1980), using electron microscopy, localized grains to astrocytic processes in layers I and II of the dorsal horn of cat medulla after topical application of [³H]serotonin and pretreatment of the animal with a monoamine oxidase (MAO) inhibitor, in addition to finding uptake of [³H]serotonin into serotonergic nerve endings of different types. Also, in the filum terminale of the frog, which contains only glial cell bodies, increased grain density over such cell bodies and their processes was seen after incubation of slices of this tissue with [³H]serotonin (Ritchie *et al.*, 1981). However, the major evidence for astroglial uptake comes from work showing that glial preparations isolated from brain tissue (Henn and Hamberger, 1971) and several types of glial cell cultures (Suddith *et al.*, 1978; Tardy *et al.*, 1982; Whitaker *et al.*, 1983; Katz and Kimelberg, 1985; Kimelberg and Katz, 1985) show high-affinity uptake of [³H]serotonin. One possible reason for the general failure to observe concentrative uptake into extraneuronal sites in the CNS *in situ* is that such uptake may well be followed by metabolism. Also, even when metabolism is inhibited, astrocytes may be much less able than neurons to concentrate the transmitters because they do not have a vesicular storage system such as occurs in nerve terminals (Heuser and Reese, 1977). Thus, the astrocytic high-affinity uptake system may be, as has been described for extraneuronal norepinephrine uptake at peripheral synapses, an 'uptake-followed-by-metabolism' process as opposed to the intraneuronal 'uptake-and-retention' process (Iversen, 1973).

At concentrations greater than 1 μM, serotonin is probably taken up predominantly by a non-saturating system, which shows no dependence on external Na^+ (Bogdanski *et al.*, 1970). Uptake of serotonin by astrocytes described in this chapter will be predominantly of the high-affinity type, usually characterized by maintaining serotonin at a concentration range of 10^{-7}M, by omitting Na^+ or by adding specific inhibitors, especially in *in vitro* studies.

UPTAKE OF SEROTONIN IN ISOLATED BRAIN FRACTIONS AND GLIAL CULTURES

Uptake into brain slices, total homogenates or crude, predominantly mito-chondrial (10,000–50,000 g) membrane fractions (Fonnum *et al.*, 1980) *in vitro* could be into neurons, glia, or membrane vesicles of varying size. It would not be expected, for example, that shearing forces caused by homogen-ization would differentiate between nerve or glial cell processes. Thus, frac-tions obtained by simple differential centrifugation could be expected to contain a mixture of vesiculated synaptic nerve terminals, axonal and dendritic processes, glial processes and other vesiculated plasma membrane fractions. Further purification of the 10,000–50,000 g fraction by centrifug-ation on density gradients will enable predominantly synaptosome-containing fractions to be obtained, but it is still likely that such fractions will contain other vesiculated components. Indeed, when membrane fractions obtained from homogenization of the C_6 glioma line were mixed with brain homogen-ates, they were found to be present in the 'synaptosome' band after density gradient centrifugation (Cotman *et al.*, 1971). Although this experiment does not necessarily show the fate of glial cells after density gradient centrifugation of actual brain fractions, it illustrates the problems of obtaining pure brain components by subfractionation procedures.

The earliest report of concentrative uptake of serotonin by glia appears to have been that by Henn and Hamberger (1971) using glial fractions isolated by gradient centrifugation from rabbit cerebral cortices or whole brain. At 10^{-7}M and 5 minutes of incubation time, [^{3}H]serotonin showed a sixfold cell-to-medium concentration ratio. By comparison, synaptosomal fractions showed a 49-fold cell-to-medium ratio after 5 minutes of incubation with 10^{-7}M [^{3}H]serotonin. In contrast, at 10^{-7}M, a 53-fold concentrative uptake of gamma-aminobutyric acid by the glial fraction was found. It should be noted that although 80–90% of the isolated fractions was reported to be glia, the proportion of the fraction that was astrocytic is unclear (Henn and Hamberger, 1971).

Uptake of [^{3}H]serotonin by the C_6 glial tumor cell line in the presence of the MAO inhibitor nialamide appears to have been the first report of sero-tonin uptake by glial cultures (Suddith *et al.*, 1978). One component of this uptake was markedly Na^+ dependent. It had a somewhat high K_m of 1–2×10^{-6}M and was inhibited by uptake blockers such as chlorimipramine (Cl-IMI) and desipramine (DMI) at relatively high concentrations, with IC_{50} values of around 10^{-5}M. The effect of the more specific serotonin uptake inhibitor fluoxetine was not examined. Similar results for [^{3}H]serotonin uptake by C_6 cells were reported by Whitaker *et al.* (1983). Again, relatively high K_m values of 2.2 μM for serotonin and IC_{50} values for Cl-IMI and DMI of 28 and > 1000 μM, respectively, were reported. Zimelidine and mepyramine had IC_{50} values of 19 and 25 μM respectively. In this study it was not stated whether an MAO inhibitor was added. Tardy *et al.* (1982) reported uptake of [³H]serotonin by primary astrocyte cultures from mouse brain with a K_m of 0.17 μM and a V_{max} of 0.6 pmol/mg protein/min. Uptake was inhibited by Cl-IMI and fluoxetine, but again only at very high concentrations of around $10^{-4}M$. Also, it was not mentioned whether an MAO inhibitor was added in this study.

We have also recently observed significant uptake of [³H]serotonin by rat primary astrocyte cultures, that is increased by the presence of an MAO inhibitor (Katz and Kimelberg, 1985; Kimelberg and Katz, 1985, 1986). This uptake showed a high affinity for the Na^+-sensitive component of [³H]serotonin uptake with a K_m of 0.40 μM, and [³H]serotonin uptake was also very sensitive to specific uptake inhibitors such as the clinically effective antidepressants. The order of effectiveness of inhibition for the antidepressants tested was chlorimipramine > fluoxetine > imipramine = amitriptyline > desmethylimipramine > iprindole > mianserin. The IC_{50} values are shown in Table 1 and these values are very close to those found for inhibition of [³H]serotonin uptake in various brain preparations (Shaskan and Snyder, 1970; Iversen, 1974; Koide and Uyemura, 1980; Green and Costain, 1981). Inhibition of this astrocytic uptake by the specific inhibitor fluoxetine, with an IC_{50} value of $2.3 \times 10^{-8}M$, is close to the value of $6 \times 10^{-8}M$ reported for inhibition of serotonin uptake in rat brain synaptosome (Wong *et al.*, 1975). Combined autoradiography (after uncubation with [³H]serotonin) and immunostaining for the specific astroglial marker glial fibrillary acidic protein

Table 1 IC_{50} values for inhibition of Na^+-sensitive [³H]serotonin uptake by rat primary astrocyte cultures

Compound	IC_{50} (μM)
Cl-imipramine	0.009
Fluoxetine	0.023
Imipramine	0.14
Amitryptyline	0.14
Desmethylimipramine	0.62
Iprindole	2.8
Mianserin	4.9

IC_{50} values (concentrations needed to inhibit Na^+-dependent [³H]serotonin uptake at $10^{-7}M$, constituting approximately 80% of total uptake, by 50%) are given for a number of clinically effective antidepressants. All the antagonists shown were present during a 20-min preincubation period. Uptake of [³H]serotonin was then measured for 4 min. Homogeneous astrocyte cultures were prepared from neonatal rat cerebral cortex and were used after 22–23 days growth. (See Katz and Kimelberg, 1985, for further details)

(GFAP) showed increased grain density over GFAP(+) cells (Kimelberg and Katz, 1985). Moreover, this grain density was reduced to background or close to background levels when Na$^+$ was omitted or potent inhibitors such as fluoxetine or chlorimipramine were added. An example of such an auto-radiograph, including the effects of fluoxetine is shown in Fig. 1.

We estimated that the high-affinity serotonin uptake system enabled the astroglial cells in primary culture to concentrate [^{3}H]serotonin up to 40–50 fold at an external serotonin concentration of 10^{-7}M (Katz and Kimelberg, 1985). Also, we estimated (Katz and Kimelberg, 1985) that the rate of uptake by the primary astrocyte cultures was about 10% of that reported for rat brain striatal or hypothalamic slices (Shaskan and Snyder, 1970). These

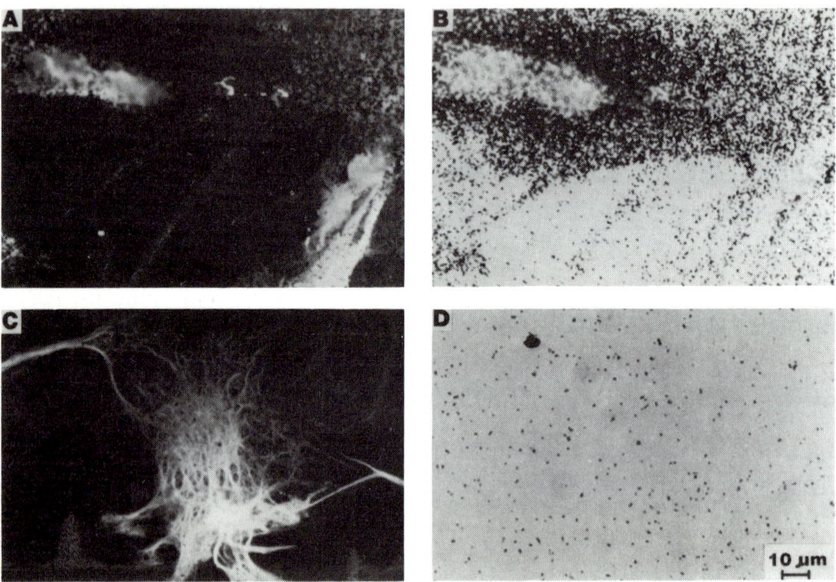

Fig. 1 Autoradiographs of uptake of 3 × 10^{-7}M [^{3}H]serotonin (B and D) and staining of same fields for GFAP (A and C) in primary rat astrocyte cultures by immunofluorescence. A and B show localization of grains over GFAP(+) astro-cytes. Note that in some areas the grain density is sufficiently intense so that it partially obscures the fluorescence. For C and D, 10^{-7}M fluoxetine was added and this culture showed no grain density above background over the GFAP(+) cells. Six-week cultures growing on glass coverslips were exposed to 3 × 10^{-7}M [^{3}H]ser-otonin ± 10^{-7}M fluoxetine in the presence of 10^{-4}M pargyline and 10^{-5}M Na ascorbate. After 30 min, the cells were rapidly washed, fixed in 4% paraformal-dehyde for 30 min at room temperature, washed again and then stained for GFAP using a mouse monoclonal primary antibody and a rhodamine conjugated secondary antibody and then dipped for autoradiography in Kodak NTB2 nuclear track emulsion. After 26 days at 4°C, the cultures were developed and photographed for GFA immunofluorescence and autoradiography. (See Kimelberg and Katz, 1985, for further experimental details)

lower values may reflect an actual lower transport rate for astrocytes *in situ* compared to neurons, or a lower uptake into the cells when grown in culture. In addition, uptake in astrocytes may be localized to cells in specific brain regions and/or to localized membrane sites.

In contrast to these findings, Hansson (1983) reported 'weak' autoradiographic grain localization over primary astrocyte cultures after exposure of cultures to [³H]serotonin. We found that virtually all the cells had a grain density that was above background after incubation with [³H]serotonin in Na⁺-containing medium in the presence of an MAO inhibitor and after fixation with 4% glutaraldehyde (Katz and Kimelberg, 1985). However, the intensity of uptake did vary in different areas within the same culture as well as from culture to culture for cells that we would expect all to be GFAP(+). Examples of this autoradiographic localization and variation within a single culture are shown in Fig. 2 for a standard culture prepared from the neocortex of 1-day-old rat brains. Cells showing the most intense uptake are depicted in Fig. 2A, while the weakest uptake seen is shown in Fig. 2E. The corresponding phase micrographs are on the right-hand side of the figure. Figure 2G shows that in Na⁺-free media the density over the cells (see corresponding phase micrograph in Fig. 2H) is reduced close to background levels, i.e., the grain density that is found in the cell-free area on the right-hand half of the micrograph. Also, in Fig. 1A and 1B it can be seen that some GFAP(+) areas have more intense uptake than others. This variation in uptake suggests some degree of specialization of astrocytes within different regions and even from the same region. These standard cultures are grown from a broad region of the brain, the neocortex of one day old rats. The meninges are removed as completely as possible. In more recent studies we have found that astrocyte cultures prepared from different brain regions do indeed show different levels of uptake of [³H]serotonin as measured in pmol/mg protein (Kimelberg and Katz, 1986). The relative magnitude of uptake for the different regions was corpus striatum ≫ cerebral cortex > hippocampus. The actual uptake rates were 15.42, 14.98 and 8.06 pmol/mg protein/30 minutes for one series of cultures. In other cultures, uptake into the corpus striatum exceeded uptake into the cerebral cortex. [³H]serotonin uptake in all the regional cultures was inhibited 75 to 87% by removal of Na⁺, and inhibited to a similar extent when 10⁻⁷M fluoxetine was present. 10⁻⁷ Desmethylimipramine had essentially no effect (Kimelberg and Katz, 1986).

The projections of the serotonergic cell bodies radiate widely throughout the CNS, and the cerebral cortex, corpus striatum, and hippocampus all receive serotonergic innervation. Such innervation occurs early during fetal and early postnatal life in the rat (McGeer *et al.*, 1978). Thus the occurrence of high-affinity uptake of serotonin in 2–3 week old astrocyte cultures prepared from 1-day-old rat pups is not unexpected. The differences we found

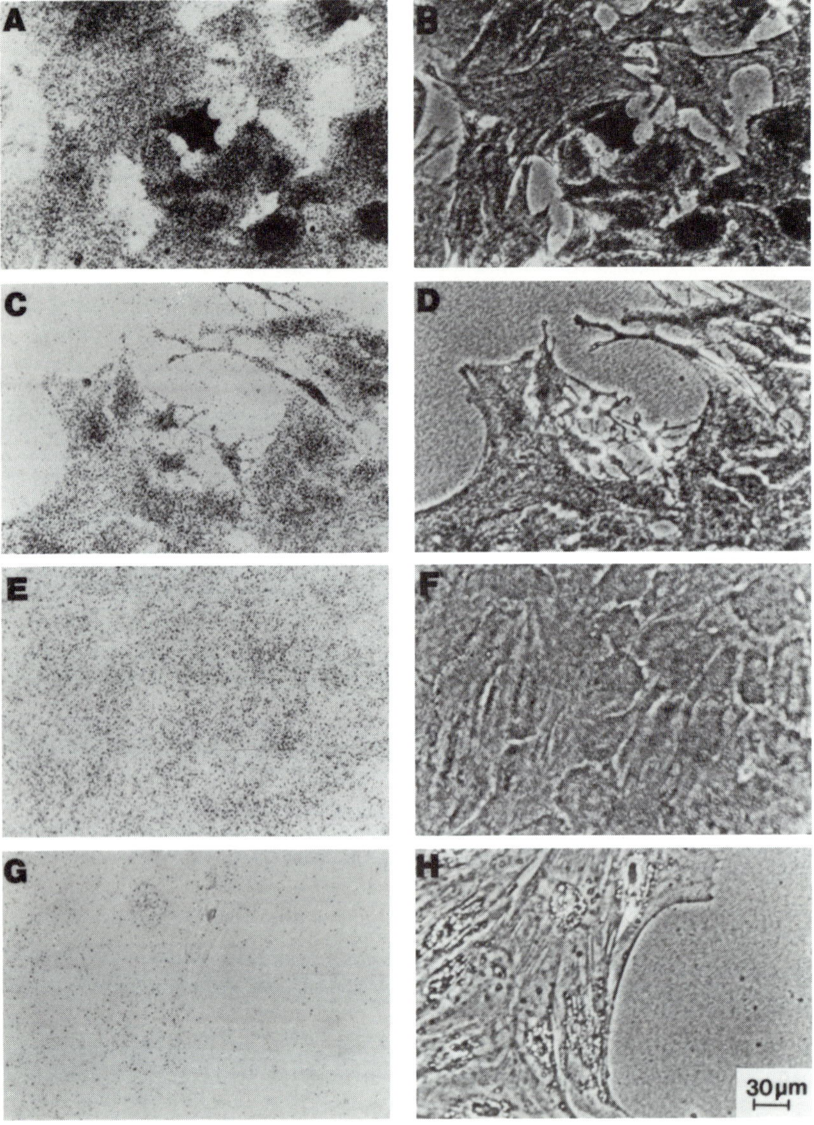

Fig. 2 Autoradiographs of cells growing on glass coverslips after incubation with [3H]serotonin. Uptake in presence of 3×10^{-7}M [3H]serotonin was allowed to proceed for 30 minutes and the cells were then fixed in 4% glutaraldehyde. (A) brightfield photomicrograph showing uptake in an area of a culture containing predominantly flat cells; (B) the corresponding phase contrast photomicrograph of the same field. As can be seen, all the cells show a high grain density compared to virtually no silver grains in the cell-free areas between cells. One cell with an

in high-affinity serotonin uptake for astrocyte cultures from the different regions (Kimelberg and Katz, 1986) may reflect quantitative differences in the degree of innervation.

INVOLVEMENT OF MONOAMINE OXIDASE IN SEROTONIN UPTAKE BY ASTROCYTES

Omission of the MAO inhibitor pargyline markedly reduced the Na^+-dependent component of [^{3}H]serotonin uptake, but had a negligible effect on the Na^+-independent component (Katz and Kimelberg, 1985). This suggests that there is a significant oxidative deamination by MAO of serotonin after it is taken up by the high-affinity system, and subsequent release of its metabolite 5-hydroxyindoleacetic acid (5-HIAA). Since the steady-state uptake level is a balance between the rate of uptake and the rate of release, we can speculate that the carboxylate anion metabolite of serotonin is more permeant than the primary amino parent compound. Thus, the absence of MAO inhibitors would not be expected to affect the rate of uptake, but the rate of efflux would be faster, and a lower steady-state content of [^{3}H]serotonin plus [^{3}H]5-HIAA would result. We have studied the efflux of label after loading cells with [^{3}H]serotonin to a steady-state level in the presence of pargyline, or to a lower steady-state level in the absence of pargyline, and observed a two-fold higher rate of efflux of tritium label in the latter case (Kimelberg and Herrick, unpublished observations).

MAO exists as two isozymes, MAO-A and MAO-B. We have therefore

---

extremely high grain density is present in the middle of the field in A Examination of regions from all areas of the coverslip did not reveal any cells which did not show grain density above background. (C) represents an area with an intermediate grain density and (D) the corresponding phase contrast micrograph. In this figure both process-bearing cells and flat cells were present, both showing a grain density clearly greater than background. E shows an area which represents the least intense uptake seen, and F is the corresponding phase contrast micrograph showing that the area contains a confluent monolayer of cells with the characteristic flat morphology of GFAP(+) cells in primary astrocyte cultures (Kimelberg, 1983). When Na^+ was omitted from the medium (replaced with choline$^+$) during the initial incubation with [^{3}H]5-HT, all the cells showed a grain density almost identical to background levels as shown in G. H is the corresponding phase micrograph showing the area of the field which contained cells and the area which was cell-free. This field was chosen so that a direct comparison can be made of the grain density in the cell-containing and cell-free areas. Since our cultures stain 95% or more for GFAP and include all the morphological types which were seen as taking up [^{3}H]serotonin label in the autoradiographs, these results strongly suggest that GFAP(+) cells are taking up [^{3}H]serotonin, and this is also directly shown in Fig. 1. Cells were 5 weeks old and were left for 9 days before development of the emulsion. (From Katz and Kimelberg, 1985)

studied the relative effectiveness of clorgyline, an inhibitor specific for the MAO-A isozyme for which serotonin is a preferred substrate (McGeer *et al.*, 1978) in promoting uptake of [^{3}H]serotonin, and compared it with the effectiveness of pargyline, which at lower concentrations is a more effective inhibitor of MAO-B and only inhibits MAO-A at higher concentrations (McGeer *et al.*, 1978). These results are shown in Fig. 3 and, as can be seen, clorgyline causes a maximum increase in Na$^+$-dependent uptake of [^{3}H]serotonin at 10^{-9} to 10^{-7}M, whereas pargyline is only effective at 10^{-5} to 10^{-4}M. In contrast, neither clorgyline nor pargyline had any effect on Na$^+$-independent uptake.

These results fit with serotonin being a preferred substrate for MAO-A (McGeer *et al.*, 1978). Also, primary mouse astrocyte cultures have been shown to possess predominantly MAO-A (Yu and Hertz, 1982). These cultures at 23 days were found to contain 92% MAO-A and 8% MAO-B. With continued growth in culture and/or a treatment with dibutyryl cyclic AMP, however, the proportion of MAO-B increases, with a maximum of 29% MAO-B and 71% MAO-A being obtained after 85 days plus treatment with dibutyryl cyclic AMP (Yu and Hertz, 1982). Immunocytochemical

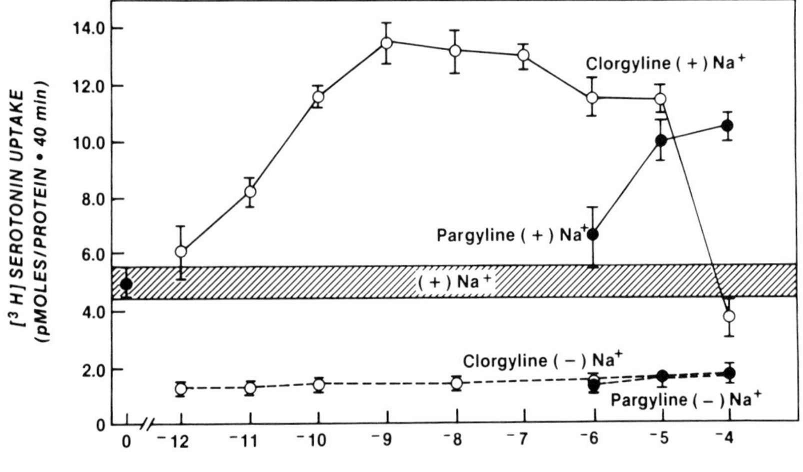

Fig. 3 Effects of clorgyline and pargyline on the uptake of [^{3}H]serotonin expressed as pmol/mg protein. Uptake of [^{3}H]serotonin at a final concentration of 10^{-7}M was measured at 37°C for 40 min in a buffered physiological salt solution as described previously (Katz and Kimelberg, 1985). Uptake was in the presence of varying concentrations of clorgyline and pargyline with ((+) Na$^+$) and without ((−) Na$^+$) Na$^+$ present as indicated. Na$^+$ was replaced by choline$^+$. The hatched area shows the level of uptake representing the mean ±SEM (see data point at the zero point on the abscissa), obtained in the presence of Na$^+$ but without any MAO inhibitor. Cells were 28 days old.

Values are means ±SEM, *n* (number of culture wells per data point) = 4

studies have shown that MAO-B is localized in the brain to astrocytes and serotonergic neurons (Levitt *et al.*, 1982; Westlund *et al.*, 1985). This latter localization is unexpected since serotonin is specifically oxidized by the A rather than the B isozyme. Currently, one study has shown that serotonin in synaptosomes is preferentially metabolized by MAO-A (Williams *et al.*, 1986). Such metabolism, however, would presumably only occur for serotonin leaked intraneuronally from vesicles. Similarly, the existence of A as well as B isozymes in astroglial cells *in situ* has not yet been clearly established. However, studies by Schoepp and Azzaro (1983) showed that injection of kainic acid into the striatum in rat leads to a gliosis and a moderate decrease in MAO-A, but a delayed increase in MAO-B levels associated with increasing astrogliosis. It was suggested that there is both MAO-A and B in the normal and proliferating astrocyte population, but that the astroglial component of MAO-A is only a small proportion of the total MAO-A, so that for this isozyme a small decrease was observed after kainic acid-induced neuronal loss. Remarkably, an increased uptake of [³H]dopamine and increased production of both deaminated (dihydroxyphenylacetic acid, DOPAC) and deaminated plus *o*-methylated (homovanillic acid, HVA) products was also observed in this study, consistent with increased extra-neuronal or glial uptake and metabolism of dopamine after kainic acid treatment. This is consistent with our previous observations of considerable uptake of [³H]dopamine and its metabolism to both DOPAC and HVA in rat primary astrocyte cultures (Pelton *et al.*, 1981). Based on our data on [³H]serotonin uptake in astrocyte cultures, we propose that similar astroglial uptake and metabolism can occur for serotonin. It will be of interest to study the production of the oxidative metabolite of serotonin, 5-HIAA, by these cultures, to correlate its production with the existence of both A and B isozymes of MAO, and also to study how 5-HIAA gets out the cell.

The cellular localization of the MAO isozymes and the uptake and release of monoamines have recently been the objects of renewed interest in relation to Parkinsonism symptoms induced by inadvertent administration of 1-methyl-4-phenyl-1,2,3,6-tetrahydropyridine (MPTP). It appears that MPTP itself is not toxic but is converted to a toxic, oxidized derivative MPP⁺ by MAO-B. MPP⁺ is selectively toxic to neurons of the substantia nigra, the degeneration of which then causes the symptoms of Parkinsonism (Burns *et al.*, 1983; Langston *et al.*, 1984). It has been suggested that MPP⁺ is taken up by the high-affinity dopamine (DA) uptake system in these cells, while one site of MPTP oxidation may be via MAO-B action in astrocytes (Singer *et al.*, 1987). More recently, Brooks *et al.* (1986) reported that fluoxetine protects against the major dopamine depletion as well as a much smaller serotonin depletion in mouse striatum, after intraperitoneal injection of MPTP. It was suggested that a major conversion of MPTP to MPP⁺ occurs in serotonergic neurons and that MPTP is taken up into such neurons by the fluoxetine-sensitive high-affinity serotonin uptake system. More recent data (Brooks, Jarvis and Wagner, personal communication) have shown that when the serotonergic neurons were destroyed by prior treatment with 5,7-dihydroxytryptamine, considerable fluoxetine-sensitive depletion of dopamine after injection of MPTP still occurred. This could clearly be due to uptake of MPTP into astrocytes by the fluoxetine-sensitive serotonin uptake system we have described, and its subsequent conversion to MPP$^+$ in astrocytes by MAO-B.

IMIPRAMINE BINDING SITES IN ASTROCYTE CULTURES

It has been shown that high-affinity binding of [^{3}H]imipramine in the nanomolar range to membranes from the brain is associated with serotonin uptake by the high-affinity uptake system (Langer *et al.*, 1980; Paul *et al.*, 1981; Sette *et al.*, 1981; Reith *et al.*, 1983). Consequently, we examined [^{3}H]imipramine binding to both intact rat astrocytes in primary culture and membranes prepared from these cultures, to see whether a high-affinity binding site could be identified (Waniewski *et al.*, 1986). No evidence for the existence of a high-affinity binding site was detected in either intact astrocyte cultures or membranes prepared from the cells. However, a very dense population of low-affinity binding sites was observed for both intact cells (K_d = 959 nM, B_{max} = 672 pmol/mg protein) and a membrane fraction prepared from these cells (K_d = 606 nM, B_{max} = 1610 pmol/mg protein). This binding was Na$^+$ independent in both cases. These data were in contrast with results obtained in the same study using brain membranes, where two binding sites with K_d values of 8.9 nM and 1.79 μM and B_{max} values of 0.31 and 53.2 pmol/mg protein respectively, were obtained. These values presumably correspond to the high- and low-affinity components, respectively, seen in brain (e.g., see Reith *et al.*, 1983, with K_d values for mouse brain membranes of 13 and 769 nM and B_{max} values of 0.82 and 6.3 pmol/mg protein respectively). At least two possible explanations can be offered for the discrepancy between the association between high-affinity [^{3}H]imipramine binding and high-affinity [^{3}H]serotonin uptake in brain, and the absence of high-affinity [^{3}H]imipramine binding but the presence of high-affinity [^{3}H]serotonin uptake in primary astrocyte cultures. One is that the high-affinity [^{3}H]imipramine binding site is not on the high-affinity serotonin uptake site, a viewpoint which has been put forward by Laduron *et al.* (1982) based on the lack of correspondence between the distribution of [^{3}H]imipramine binding and [^{3}H]serotonin uptake in sub-cellular fractions from rat brains. Alternatively, the very high density of low-affinity binding sites in astrocytes could have obscured detection of a much smaller amount of a high-affinity, Na$^+$-dependent binding site also present on the cells. Also, the finding of such a high level of low-affinity [^{3}H]imipramine binding in astrocyte cultures suggests that a significant proportion of the comparable sites in the brain may be present on astrocytes.

High-affinity [3H]imipramine binding has, however, been obtained in homogenates of C_6 glioma cells (Whitaker *et al.*, 1983) with a single high-affinity binding site with a K_d of 1.7 nM and a B_{max} of 0.202 pmol/mg protein. However, this binding did not correlate with the serotonin uptake also seen in these cells since antidepressants which displaced [3H]imipramine binding at low concentrations (e.g. chlorimipramine, IC_{50} = 79 nM) were very weak inhibitors of [3H]serotonin uptake (e.g. chlorimipramine, IC_{50} = 27.6 μM). Tardy *et al.* (1982) found a K_d for [3H]serotonin binding of 15 nM and a B_{max} of 0.18 pmol/mg protein in primary astrocyte cultures from mice. A high-affinity [3H]serotonin uptake system with a K_m of 0.17 to 0.27 μM was also present in these cells. However, a rather high concentration of 10^{-4}M fluoxetine or chlorimipramine was needed to inhibit [3H]serotonin uptake by 45 and 32% respectively.

Clearly, controversy exists on the relation between high-affinity binding sites for 5-HT uptake blockers and the presence of a high-affinity serotonin uptake system, so that the absence of a detectable high-affinity [3H]imipramine binding site in rat primary astrocyte cultures, if it is simply not being seen for technical reasons as discussed above, need not be contradictory to the clearly demonstrated presence of a high-affinity [3H]serotonin uptake system in these cells.

SPECULATION ON ROLES FOR ASTROGLIAL UPTAKE OF SEROTONIN, AND FUTURE TRENDS

As can be appreciated from the foregoing review, at the present time the evidence for uptake of serotonin by astrocytes comes mainly from studies on isolated preparations and glial cultures. The failure to routinely localize this uptake into astrocytes *in situ* may be simply due to the lack of precise definition of grain localization at the cellular level in light-microscopic studies coupled with the fact that astrocytes do not appear to concentrate serotonin to anywhere near the same extent as do neurons or neuronal nerve endings. In *in vitro* preparations containing only astrocytes, such lower uptake can be detected using quantitative measurements of uptake of radioactive label. In the case of autoradiographic localization, the fixed cultures can also be developed for a time sufficient to obtain an adequate grain density. In tissue, conditions are usually optimized for sites showing the greatest uptake, namely neuronal nerve endings, and regions of lower uptake, such as glia, may then be missed. Chemical or surgical lesion studies to eliminate specific nerve terminals cannot be considered definitive because of the possibility of indirect effects of neuronal destruction on surrounding astrocytes. Also, the chemicals used may be toxic or affect transmitter uptake in glial cells themselves. For example, we have found that treatment of primary astrocyte cultures with the serotonergic-specific toxin 5,7-dihydroxytryptamine affects serotonin uptake. After exposure of cultures to 10^{-4}M, 5,7-dihydroxytryptamine for 3 days in the absence of serum, there was only a 10% reduction in cell protein relative to serum-free controls, but a 47% reduction in the Na^+-dependent component of [^{3}H]serotonin uptake, expressed on a per mg protein basis (Kimelberg, Katz and Waniewski, unpublished experiments). However, serotonergic neurons may indeed be more sensitive to the toxic and inhibitory effects of this compound.

Alternatively, the finding of uptake into primary astrocyte cultures may represent uptake by immature astrocytes that express this property transiently at an early stage in development. However, the existence of this system in glial cells isolated from adult or near-adult animals argues against this possibility. Further studies on uptake of [^{3}H]serotonin as a function of age in culture may provide useful information on this topic.

Accepting that the work described in this article indicates that astrocytes *in situ* can take up serotonin by a high-affinity system and thus compete with neuronal terminals for uptake of the released transmitter, the question of the function of such uptake remains an interesting topic for speculation. Studies so far indicate that the high-affinity serotonin uptake system in astrocytes *in vitro* behaves pharmacologically like the same system in various brain preparations, so that at present it does not seem possible to selectively inhibit astrocytic uptake and determine its effect on neuronal function *in situ*. Further pharmacological studies on this point should be made, and if selective inhibition can be obtained it might enable this question to be addressed *in situ*. Mianserin and iprindole do block uptake of [^{3}H]serotonin at relatively high concentrations (see Table 1), but inhibition of serotonin uptake in synaptosomes by 8 μM mianserin has also been reported (Green and Costain, 1981).

The likely fate of serotonin taken up into astrocytes appears to be metabolism and removal – a pass-through, uptake and metabolism system, rather than uptake and storage in presynaptic vesicles for re-release, as seems to occur in nerve endings. Astrocytes are usually closely apposed to the axonal varicosities which represent the presynaptic specialization of serotonergic neurons in the mammalian CNS (Peters *et al.*, 1976; Heuser and Reese, 1977; Gershon *et al.*, 1981). Sometimes postsynaptic, high-density regions, thought to represent receptor regions, are also apposed to the varicosities. However, in many cases, this junctional membrane differentiation characteristic of chemical synapses is absent (Descarries *et al.*, 1975), suggesting action at a distance or non-synaptic or humoral type effects of serotonin. In such cases re-uptake by the same varicosity which released the serotonin would not occur, and uptake by the many small astrocytic processes which richly ramify within the neuropil would serve a useful function.

In the astrocyte, no other fate for the serotonin taken up other than inactivation by metabolism seems likely at the present time. In preliminary studies, we obtained no evidence for depolarization-induced release of [^{3}H]serotonin when the cells were exposed to elevated [K$^+$] in the medium. It appears from our studies on astrocyte cultures (Katz and Kimelberg, 1985) and other studies (Trendelenburg, 1980) that metabolites such as 5-HIAA permeate membranes more readily than the parent amines, and thus can exit by diffusion through the cell membrane anywhere within the neuropil. Alternatively, there could be specific membrane transport systems located at the perivascular surface of an astrocytic process or end-foot (Peters *et al.*, 1976) which closely appose capillaries, and also in astrocytic processes located beneath the pia mater and subependymal zone facing CSF, which would preferentially direct such metabolites into the blood and CSF respectively, for ultimate removal from the CNS. Some of the intramembraneous assemblies preferentially localized in astrocytic membranes at the perivascular and CSF-facing sites (Landis and Reese, 1981) might represent the molecular basis of such transport processes.

Recently it has been hypothesized that transmitter uptake may be due to internalization of receptors, as seen in receptor down-regulation (LaBella, 1985). Thus, a separate specific uptake system for transmitters was proposed to be unnecessary, and such a dual role for receptors would explain why some drugs have effects on both receptors and uptake systems. Since astrocytes, particularly in culture, are now known to have a large variety of receptors for neurotransmitters (van Calker and Hamprecht, 1980; Kimelberg, 1983), including serotonin receptors (Tardy *et al.*, 1982), such a route of uptake, if it does prove to be significant, could also apply to astrocytes.

Would uptake by astrocytes be likely to represent a first or second line of defense for removing released serotonin from the synaptic cleft and extracellular space, thus terminating its actions? The close proximity of astrocytic processes to the synapse (Lugaro, 1907; Wolff, 1970; Palay and Chan-Palay, 1974; Peters *et al.*, 1976; Heuser and Reese, 1977), as discussed above, plus the possible existence of a high-affinity uptake system in such cells, as documented in this article, suggest that astrocytic uptake could compete equally with re-uptake into neurons. We have described (Pelton *et al.*, 1981; Kimelberg and Pelton, 1983; Semenoff and Kimelberg, 1985) similar high-affinity uptake systems for norepinephrine and dopamine, so that astrocytes may have uptake systems for catecholamines as well as serotonin. Similar considerations in terms of localization *in situ* and in culture, the effect of specific inhibitors and the role of chemical and surgical lesioning on uptake *in situ* that have been discussed for serotonin also apply to the catecholamine uptake systems (Kimelberg, 1986a, b). Since the serotonin uptake system in astrocytes is sensitive to clinically effective antidepressants, the therapeutic effects of such agents may thus be mediated to some extent by their action on astrocytes. Long-term treatment with the antidepressant amitriptyline has been reported to decrease CSF levels of 5-HIAA (Charney *et al.*, 1984), and this could be a consequence of inhibition of serotonin uptake and deamination in astrocytes, as well as in nerve endings. Such inhibition by the tricyclic antidepressants of serotonin uptake and its subsequent inactivation through metabolism, would contribute to increasing the levels of released serotonin, and alleviate at least some types of depression which are considered to be a consequence of low, effective levels of serotonin (Green and Costain, 1981).

The involvement of astrocytic uptake as a significant means of terminating the action of serotonin in the CNS seems rarely to have been considered, but it may well play an important role. Future studies in this area, particularly to define the occurrence and possible regional localization of such uptake *in situ*, could shed light on this important aspect of neuron–astroglia interrelationships and its role in both normal and abnormal brain functions.

ACKNOWLEDGEMENTS

Work reported in this article from the author's laboratory was supported by grant NS19492 from NINCDS. I thank D. Katz, D. Semenoff and R. A. Waniewski for their excellent collaboration in some aspects of these studies, N. Brzyski and K. Herrick for expert technical assistance and E. P. Graham for typing the manuscript.

REFERENCES

Aghajanian, G. K., and Bloom, F. E. (1967) Localization of tritiated serotonin in rat brain by electron-microscopic autoradiography, *J. Pharmacol. Exp. Ther.*, **156**, 23–30.

Björklund, A., Nobin, A., and Steveni, M. (1973) Effects of 5,6-dihydroxytryptamine on nerve terminal serotonin and serotonin uptake in the rat brain, *Brain Res.*, **53**, 117–127.

Bogdanski, D. F., Tissari, A. H., and Brodie, B. B. (1970) Mechanism of transport and storage of biogenic amines. III. Effects of sodium and potassium on kinetics of 5-hydroxytryptamine and norepinephrine transport by rabbit synaptosomes, *Biochim. Biophys. Acta.*, **219**, 189–199.

Breese, G. R., and Cooper, B. R. (1975) Behavioral and biochemical interreactions of 5,7-dihydroxytryptamine with various drugs when administered intracisternally to adult and developing rats, *Brain Res.*, **98**, 517–527.

Brooks, W. J., Jarvis, M. F., and Wagner, G. C. (1986) Mediation of MPTP-induced dopaminergic neurotoxicity by serotonergic neurons, *Soc. Neurosci. Abst.*, **12**, 88.

Burns, R. S., Chiueh, C. C., Markey, S. P., Ebert, M. H., Jacobowitz, D. M., and Kopin, I. J. (1983) A primate model of Parkinsonism: selective destruction of dopaminergic neurons in the pars compacta of the substantia nigra by N-methyl-4-phenyl-1,2,3,6-tetrahydropyridine, *Proc. Natl. Acad. Sci. USA*, **80**, 4546–4550.

Charney, D. S., Heninger, G. R., and Sternberg, D. E. (1984) Serotonin function and mechanism of action of antidepressant treatment, *Arch. Gen. Psychiat.*, **41**, 359–365.

Cooper, J. R., Bloom, F. E., and Roth, R. H. (1982) *The Biochemical Basis of Neuropharmacology*, 4th edn., Oxford University Press, New York and Oxford.

Cotman, C. W., Herschman, H., and Taylor, D. (1971) Subcellular fractionation of cultured glial cells, *J. Neurobiol.*, **2**, 169–180;

Descarries, L., and Beaudet, A. (1983) The use of radioautography for investigating transmitter-specific neurons, in *Handbook of Chemical Neuroanatomy* (Eds A. Björklund and T. Hokfelt), Vol. I., pp. 286–328, Elsevier, Amsterdam.

Descarries, L., Beaudet, A., and Watkins, K. C. (1975) Serotonin nerve terminals in adult rat neocortex, *Brain Res.*, **100**, 563–588.

Feldman, R. S., and Quenzer, L. F. (1984) *Fundamentals of Neuropsychopharmacology*, Sinauer Associates, Sunderland, Ma.

Fonnum, F., Karlsen, R. L., Malthe-Sorenssen, D., Sterri, S., and Walaas, I. (1980) High affinity transport systems and their role in transmitter action, in *The Cell Surface and Neuronal Function* (Eds C. W. Cotman, G. Poste and G. L. Nicolson), pp. 455–504, North-Holland, Amsterdam.

Fuxe, K., Hokfelt, T., Ritzen, M., and Ungerstedt, U. (1968) Studies on uptake of intraventricularly administered tritiated noradrenaline and 5-hydroxytryptamine with combined fluorescence, histochemical and autoradiographic techniques, *Histochemie*, **16**, 186–194.

Gershon, M. D., Schwartz, J. H., and Kandel, E. R. (1981) Morphology of chemical synapses and patterns of interconnection, in *Principles of Neural Science* (Eds E. R. Kandel and J. H. Schwartz), p. 91, Elsevier, Amsterdam.

Green, A. R., and Costain, D. W. (1981) *Pharmacology and Biochemistry of Psychiatric Disorders*, Wiley, New York, pp. 71–88.

Hansson, E. (1983) Accumulation of putative amino acid neurotransmitters, monoamines and D-Ala^2-Met-Enkephalinamide in primary astroglial cultures from various brain areas, visualized by autoradiography, *Brain Res.*, **289**, 189–196.

Henn, F. A., and Hamberger, A. (1971) Glial cell function: Uptake of transmitter substances, *Proc. Natl. Acad. Sci. USA*, **68**, 2686–2690.

Heuser, J. E., and Reese, T. S. (1977) Structure of the synapse, in *Handbook of Physiology – The Nervous System* I (Ed. E. R. Kandel), pp. 261–293, Williams and Wilkins, Baltimore.

Iversen, L. L. (1973) Catecholamine uptake processes, *Br. Med. Bull.*, **29**, 130–135.

Iversen, L. L. (1974) Uptake mechanisms for neurotransmitter amines, *Biochem. Pharmac.*, **23**, 1927–1935.

Katz, D., and Kimelberg, H. K. (1985) Kinetics and autoradiography of high affinity uptake of serotonin by primary astrocyte cultures, *J. Neurosci.*, **5**, 1901–1908.

Kimelberg, H. K. (1983) Primary astrocyte cultures – a key to astrocyte function, *Cell. Molec. Neurobiol.*, **3**, 1–16.

Kimelberg, H. K. (1986a) Occurrence and functional significance of serotonin and catecholamine uptake by astrocytes, *Biochem. Pharmacol.*, **35**, 2273–2281.

Kimelberg, H. K. (1986b) Catecholamine and serotonin uptake in astrocytes, in *Astrocytes* (Eds S. Fedoroff and A. Vernadakis), Academic Press, Vol. 2, pp. 107–131.

Kimelberg, H. K., and Katz, D. (1985) Identification of high affinity serotonin uptake into immunocytochemically identified astrocytes, *Science*, **228**, 889–891.

Kimelberg, H. K., and Katz, D. M. (1986) Regional differences in serotonin and catecholamine uptake in primary astrocyte cultures, *J. Neurochem.*, **47**, 1647–1652.

Kimelberg, H. K., and Pelton, E. W. (1983) High-affinity uptake of [^{3}H]norepinephrine by primary astrocyte cultures and its inhibition by tricyclic antidepressants, *J. Neurochem.*, **40**, 1265–1270.

Koide, T., and Uyemura, K. (1980) A comparison of the inhibitory effects of new non-tricyclic amine uptake inhibitors on the uptake of norepinephrine and 5-hydroxytryptamine into synaptosomes of the rat brain, *Neuropharmacology*, **19**, 349–354.

Kuhar, M. J., Roth, R. H., and Aghajanian, G. K. (1972) Synaptosomes from forebrains of rats with midbrain raphe lesions: selective reduction of serotonin uptake, *J. Pharmac. Exp. Ther.*, **181**, 36–45.

LaBella, F. S. (1985) Neurotransmitter uptake and receptor-ligand internalization – are they two distinct processes?, *Trends Pharmacol. Sci.*, **6**, 319–322.

Laduron, P. M., Robbyns, M., and Schotte, A. (1982) [^{3}H]desipramine and [^{3}H]imipramine binding are not associated with noradrenaline and serotonin uptake in the brain, *Eur. J. Pharmacol.*, **78**, 491–493.

Landis, D. M. D., and Reese, T. S. (1981) Membrane structure in mammalian astrocytes: a review of freeze-fracture studies on adult, developing, reactive and cultured astrocytes, *J. Exp. Biol.*, **95**, 35–48.

Langer, S. Z., Moret, C., Raisman, R., Dubocovich, M. L., and Briley, M. (1980) High-affinity [^{3}H]imipramine binding in rat hypothalamus: Association with uptake of serotonin but not of norepinephrine, *Science*, **210**, 1133–1135.

Langston, J. W., Forno, L. D., Robert, C. S., and Irwin, I. (1984) Selective nigral toxicity after systemic administration of 1-methyl-4-phenyl-1,2,3,6-tetrahydropyrine (MPTP) in the squirrel monkey, *Brain Res.*, **292**, 390–394.

Levitt, P., Pintar, J. E., and Breakefield, X. O. (1982) Immunocytochemical demonstration of monoamine oxidase B in brain astrocytes and serotonergic neurons, *Proc. Natl. Acad. Sci. USA*, **79**, 6385–6389.

Lugaro, E. (1907) Sulle funzioni della neuroglia, *Rivista di Patologia Nervosa e Mentale*, **12**, 225–233.

McGeer, P. L., Eccles, S. C., and McGeer, E. G. (1978) *Molecular Neurobiology of the Mammalian Brain*, Plenum Press, New York.

Palay, S., and Chan-Palay, V. (1974) *Cerebellar Cortex – Cytology and Organization*, Springer, Berlin.

Paul, S. M., Rehavi, M., Rice, K. C., Ittah, Y., and Skolnick, P. (1981) Does high-affinity [^{3}H]imipramine binding label serotonin reuptake sites in brain and platelet?, *Life Sci.*, **28**, 2753–2760.

Pelton, E. W., Kimelberg, H. K., Shipherd, S. V., and Bourke, R. S. (1981) Dopamine and norepinephrine uptake and metabolism by astroglial cells in culture, *Life Sci.*, **28**, 1655–1663.

Peters, A., Palay, S. L., and Webster, H. de F. (1976) *The Fine Structure of the Nervous System: The Neurons and Supporting Cells*, pp. 231–263, W. B. Saunders, Philadelphia, London, Toronto.

Reith, M. E. A., Sershen, H., Allen, D., and Lajtha, A. (1983) High- and low-affinity binding of [^{3}H]imipramine in mouse cerebral cortex, *J. Neurochem.*, **40**, 389–395.

Ritchie, T., Glusman, S., and Haber, B. (1981) The filum terminale of the frog spinal cord, a nontransformed glial preparation. II. Uptake of serotonin, *Neurochem. Res.*, **6**, 441–452.

Ruda, M. A., and Gobel, S. (1980) Ultrastructural characterization of axonal endings in the substantia gelatinosa which take up [^{3}H]serotonin, *Brain Res.*, **184**, 57–83.

Schoepp, D. D., and Azzaro, A. J. (1983) Effects of intrastriatal kainic acid injection on [^{3}H]dopamine metabolism in rat striatal slices: evidence for postsynaptic glial cell metabolism by both the type A and B forms of monoamine oxidase, *J. Neurochem.*, **40**, 1340–1348.

Semenoff, D., and Kimelberg, H. K. (1985) Autoradiography of high affinity uptake of catecholamines by primary astrocyte cultures, *Brain Res.*, **348**, 125–136.

Sette, M., Raisman, R., Briley, M., and Langer, S. Z. (1981) Localization of tricyclic antidepressant binding sites on serotonin nerve terminals, *J. Neurochem.*, **37**, 40–42.

Singer, T. P., Trevor, A. J. and Castagnoli, Jr., N. (1987) Biochemistry of the neurotoxic action of MPTP. *Trends Biochem. Sci.* **12**, 266–270.

Shaskan, E. G., and Snyder, S. H. (1970) Kinetics of serotonin accumulation into slices from rat brain: relationship to catecholamine uptake, *J. Pharmacol. Exp. Ther.*, **175**, 404–418.

Suddith, R. L., Hutchison, H. T., and Haber, B. (1978) Uptake of biogenic amines by glial cells in culture. I. A neuronal-like transport system for serotonin, *Life Sci.*, **22**, 2179–2188.

Tardy, M., Costa, M. F. D., Fages, C., Bardakdjian, J., and Gonnard, P. (1982) Uptake and binding of serotonin by primary cultures of mouse astrocytes, *Dev. Neurosci.*, **5**, 19–26.

Trendelenburg, U. (1980) A kinetic analysis of the extraneuronal uptake and metabolism of catecholamines, *Rev. Physiol. Biochem. Pharmacol.*, **87**, 31–115.

van Calker, D., and Hamprecht, B. (1980) Effects of neurohormones on glial cells, in *Advances in Cellular Neurobiology* (Eds S. Fedoroff and L. Hertz), pp. 31–67, Academic Press, New York.

Varon, S. S., and Somjen, G. G. (1979) Neuron-glia interactions, *Neurosci. Res. Prog. Bull.*, **17**, 129–174.

Waniewski, R. A., Katz, D. M., and Kimelberg, H. K. (1987) Low-affinity binding of [3H]imipramine to primary astrocyte cultures, *Biochem. Pharmacol.* **36**, 639–646.

Westlund, K. N., Denney, R. M., Kochersperger, R. M., Rose, R. M., and Abell, C. W. (1985) Distinct monoamine oxidase A and B populations in primate brain, *Science*, **230**, 181–183.

Whitaker, P. M., Vint, C. K., and Morin, R. (1983) [3H]imipramine labels sites on brain astroglial cells not related to serotonin uptake, *J. Neurochem.*, **41**, 1319–1323.

Williams, G. M., Brown, L. M., Amedro, J. B., Azzaro, A. J., and Smith, D. J. (1986) The type of monoamine oxidase isozyme associated with the metabolism of serotonin in serotonergic nerve terminals, *Soc. Neurosci. Abst.*, **12**, 1293.

Wolff, J. R. (1970) The astrocyte as link between capillary and nerve cell, *Triangle*, **9**, 153–164.

Wong, D. T., Bymaster, F. P., Horng, J. S., and Molloy, B. B. (1975) A new selective inhibitor for uptake of serotonin into synaptosomes of rat brain: 3-(p-trifluoromethylphenoxy)-N-3-phenylpropylamine, *J. Pharmac. Exp. Ther.*, **193**, 804–811.

Yu, P. H., and Hertz, L. (1982) Differential expression of type A and type B monoamine oxidase of mouse astrocytes in primary cultures, *J. Neurochem.*, **39**, 1492–1495.

Neuronal Serotonin
Edited by N. N. Osborne and M. Hamon
© 1988 John Wiley & Sons Ltd

CHAPTER 13

The Neuronal Sodium-Dependent Serotonin Transporter: Studies with [³H]Imipramine and [³H]Paroxetine

DAVID GRAHAM AND SALOMON Z. LANGER
Department of Biology
Laboratoires d'Etudes et de Recherches Synthélabo (LERS)
58 rue de la Glacière
75013 Paris
France

INTRODUCTION

Neurotransmitter transporter systems have a key role to play in mono-aminergic (noradrenaline, dopamine, adrenaline, serotonin) and amino acid

(GABA, glycine, glutamate, aspartate) synaptic transmission. These transporter systems are located in presynaptic terminals or in associated glial cells, and function to inactivate released transmitter by reducing high synaptic cleft concentrations of neurotransmitter. As such, whenever neurotransmitter is released from presynaptic terminals and diffuses across the synaptic cleft to interact with its postsynaptic receptors, these transporter systems act to terminate the trans-synaptic signal and recycle the neurotransmitter molecule itself. Acetylcholine is one exception, in that this neurotransmitter is inactivated by acetylcholinesterase cleavage in the synaptic cleft, and choline is subsequently taken up by the nerve terminal through a presynaptic transporter system.

In the process of neurotransmitter re-uptake, two different classes of transporter systems have been described. One class of transporter functions to translocate neurotransmitter across the neuronal or glial plasma membrane, whereas the other is associated with the intracellular transport of neurotransmitter into synaptic storage vesicles. This review discusses the neuronal plasma membrane transporter for the neurotransmitter, serotonin (Fig. 1).

UPTAKE STUDIES

Using brain slice and synaptosomal preparations, transmitter re-uptake across the plasma membrane has been demonstrated to be an active process which occurs with relatively high affinity. This uptake is inhibited by cyanide and dinitrophenol which block adenosine triphosphate (ATP) synthesis, and also by ouabain, an inhibitor of the enzyme Na^+/K^+-ATPase which generates transmembrane ion gradients. A need for external sodium ions is also an absolute requirement for inward transport.

For serotonin, many studies on the plasma membrane transport of this neurotransmitter have been performed on platelets. This is because blood platelets are considered to be a model for certain aspects of serotonergic neuronal function as they are thought to accumulate and store serotonin in a manner similar to that occurring in serotonergic nerve terminals (Sneddon, 1973; Stahl, 1977). In an elegant series of studies, Rudnick and coworkers have analyzed the transport of serotonin across the platelet plasma membrane using plasma membrane vesicle preparations. Since plasma membrane vesicles do not contain intracellular binding compartments or metabolizing enzymes, such preparations have the advantage that, in contrast to brain slices or synaptosomes, one can impose particular transmembrane gradients across the plasma membrane. With this system, it was thus possible to obtain direct evidence that serotonin was accumulated only if appropriate Na^+/K^+ transmembranal gradients were constructed (Rudnick, 1977; Talvenheimo *et al.*, 1979; Nelson and Rudnick, 1982). In this artificial system, ouabain did not inhibit the uptake process, demonstrating that the transport of serotonin

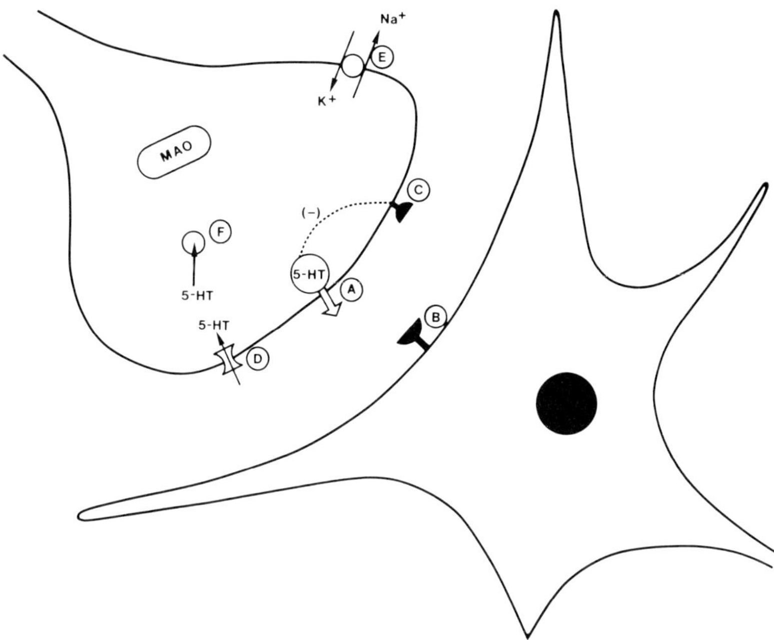

Fig. 1 Schematic representation of a serotonergic synapse. Following exocytotic release of serotonin from the presynaptic terminal (A) the transmitter crosses the synaptic cleft to bind to its postsynaptic receptor sites (5HT$_1$, 5HT$_2$ or 5HT$_3$; B). Subsequently, presynaptic mechanisms serve to terminate this signal, and recycle the neurotransmitter. The binding of serotonin to the presynaptic autoreceptor (C) leads to a reduction in the amount of neurotransmitter released. Also, the high synaptic cleft concentrations of serotonin are reduced by active uptake of neurotransmitter across the neuronal plasma membrane by the sodium-dependent serotonin transporter (D). This uptake process is dependent on a Na$^+$/K$^+$ transmembrane gradient generated by the enzyme Na$^+$/K$^+$ ATPase (E). Intracellularly translocated serotonin is then taken up into storage vesicles (F) by a pharmacologically distinct serotonin transporter

is not directly coupled to the Na$^+$/K$^+$-ATPase, but rather depends on the Na$^+$/K$^+$ transmembranal gradients that this enzyme generates. In addition, the presence of external sodium and chloride ions was shown to be needed for the serotonin uptake process to occur. Moreover, in these platelet plasma membrane vesicle preparations, serotonin transport was inhibited in the presence of serotonin uptake inhibitors such as imipramine.

Similar models of plasma membrane neurotransmitter transport have also been reported from studies on plasma membrane vesicles for other neurotransmitters including γ-aminobutyric acid (Kanner, 1978) and L-glutamate (Kanner and Sharon, 1978). Thus, it is considered that, in general, neurotransmitter re-uptake across the plasma membrane occurs through co-transport with sodium ions which move down their electrochemical potential gradient into the cell.

THE NEURONAL SODIUM-DEPENDENT SEROTONIN TRANSPORTER AND RADIOLIGAND BINDING ASSAYS

A number of potent and, in some cases, very selective inhibitors of sodium-dependent serotonin uptake have been described. Radiolabelled forms of these inhibitors have been used in *in vitro* binding assays to try and derive more information at the molecular level on the sodium-dependent serotonin transporter.

[3H]Imipramine

[³H]Imipramine was used initially for this purpose, and high-affinity [³H]imipramine binding sites on mammalian platelet and brain membrane preparations have been described (Briley *et al.*, 1979; Raisman *et al.*, 1979, 1980; Paul *et al.*, 1980; Rehavi *et al.*, 1980). In the brain, these [³H]imipramine binding sites were shown to be distributed heterogeneously, displaying a regional localization that corresponded closely with the regional density of serotonergic, but not noradrenergic innervation (Palkovits *et al.*, 1981). Moreover, denervation of serotonergic neurons using the neurotoxin 5,7-dihydroxytryptamine or by electrolytic lesions resulted in a substantial decrease in the density of [³H]imipramine binding (Sette *et al.*, 1981, 1983b; Gross *et al.*, 1981). The pharmacological profile of inhibition of [³H]imipramine binding to platelet or brain membranes by various drugs was reported to correlate significantly with the rank order of potency of these drugs to inhibit serotonin uptake into platelets or presynaptic serotonergic terminals (Langer *et al.*, 1980a, b; Paul *et al.*, 1980). These findings have therefore led to the proposal that [³H]imipramine binding is associated with the sodium-dependent serotonin transporter (Langer and Briley, 1981; Langer *et al.*, 1981).

[³H]Imipramine binding to platelet or brain membranes requires the presence of sodium ions, and chloride ions have also been reported to enhance this binding (Talvenheimo *et al.*, 1979; Briley and Langer, 1981). However, although [³H]imipramine binds to the transporter molecule, Talvenheimo *et al.* (1979) have reported that it is not transported. It is therefore of interest to note that the K_d values reported in initial studies on [³H]imipramine binding to both platelet and brain membranes were of much higher affinity than might have been expected from the K_i values of imipramine for inhibition of sodium-dependent serotonin uptake.

In all initial studies, [³H]imipramine binding was studied at 0°C. It has subsequently been reported that the binding of [³H]imipramine to brain (Plenge and Mellerup, 1984; Reith *et al.*, 1984) and platelet membranes (Davis *et al.*, 1983; Davis, 1984) is temperature-dependent and is of progressively lower affinity at increasing incubation temperatures (Davis *et al.*, 1983; Segonzac *et al.*, 1980). This thermodynamic aspect of imipramine binding to the serotonin transporter could therefore explain the differences noted between the K_d of [³H]imipramine binding at 0°C and the K_i of this compound to inhibit serotonin uptake at 37°C. Moreover, recent studies reveal that, although imipramine binds to the sodium-dependent transporters for serotonin and noradrenaline with similar potency at 37°C, the affinity of imipramine for noradrenaline is temperature-independent (Segonzac *et al.*, 1980). This could therefore account for the paradox that imipramine is an equipotent inhibitor of both noradrenaline and serotonin uptake at 37°C, whereas at 0°C, [³H]imipramine labels the serotonin transporter preferentially.

A similar phenomenon of temperature-dependent interaction of other drugs with the sodium-dependent serotonin transporter has also been reported. For example, 5-methoxytryptoline displays fivefold higher affinity for the serotonin transporter at 0°C than at 20°C, as measured by the inhibition of [³H]imipramine binding to platelet membranes (Segonzac *et al.*, 1980). Therefore, although [³H]imipramine has been shown to be a useful marker for the sodium-dependent serotonin transporter, the need to conduct these assays at 4°C for routine purposes can give rise to misleading results regarding the affinity of certain compounds at physiological temperatures for this transporter.

There have been several reports recently indicating heterogeneity of [³H]imipramine binding to brain membrane preparations (Reith *et al.*, 1983; Hrdina, 1984; Marcusson *et al.*, 1985, 1986; Moret and Briley, 1986). For example, in one report, two classes of imipramine binding sites in the brain have been described, one of which is reported to be of high affinity and protease-sensitive, whereas the other is insensitive to protease attack (Marcusson *et al.*, 1985). It has now been proposed that only this high-affinity component of [³H]imipramine binding to cerebral cortical membranes is associated with the sodium-dependent serotonin transporter. The nature of the lower-affinity [³H]imipramine binding component of cerebral cortical membranes remains to be determined. This lower-affinity [³H]imipramine binding component of cerebral cortical membranes is of relatively high capacity, and, as such, poses a serious constraint to the use of [³H]-imipramine as a marker of the neuronal sodium-dependent serotonin transporter. To some extent, the use of desipramine to define nonspecific [³H]imipramine binding in [³H]imipramine binding assays performed in the past might not have been the most appropriate choice for displacing ligand. Recently, Marcusson *et al.* (1986) have suggested that this methodological problem associated with [³H]imipramine labelling of the neuronal sodium-dependent serotonin transporter might be resolved by using norzimelidine or serotonin instead of desipramine to define nonspecific [3H]imipramine binding. It must be pointed out, however, that irrespective of the ligand chosen to define nonspecific imipramine binding, there still remains the constraint of performing [3H]imipramine binding assays at 4°C. As alluded to earlier, [3H]imipramine binding assays performed at this temperature can give rise in some instances to discordant results when compared to sodium-dependent serotonin uptake data performed at 37°C. In considering *in vitro* binding assays, a more appropriate choice would therefore be to label the sodium-dependent serotonin transporter at a more physiological temperature with a much more selective ligand.

[3H]Paroxetine

In comparison to imipramine, paroxetine (α(-)*trans*-4-(*p*-fluorophenyl)-3-[3,4-methylenedioxyphenoxy-methyl]piperidine) is a very potent and selective inhibitor of sodium-dependent serotonin uptake *in vitro* (Buus-Lassen, 1978). For this reason, we chose to examine the properties of [3H]paroxetine binding to rat cerebral cortical membrane preparations (Habert *et al.*, 1985). Using this radioligand in *in vitro* binding assays at either 22°C or 37°C, a single class of high-affinity [3H]paroxetine binding sites was demonstrated in rat cerebral cortical membranes (Habert *et al.*, 1985). Fluoxetine, another selective sodium-dependent serotonin uptake inhibitor, was used to define nonspecific binding.

Lesioning experiments indicated that these [3H]paroxetine binding sites were located on serotonergic neurons. Following 5,7-dihydroxytryptamine lesions, endogenous serotonin levels were decreased by 94% compared to sham controls, and this resulted in a 90% reduction in [3H]paroxetine binding to cerebral cortical membranes prepared from lesioned rats (Habert *et al.*, 1985). On the other hand, lesions performed with the neurotoxin, 6-hydroxy-dopamine, did not significantly affect the parameters of [3H]paroxetine binding (Graham and Langer, unpublished observations). It is interesting to contrast the effect of 5,7-dihydroxytryptamine lesions on [3H]imipramine and [3H]paroxetine binding. In an earlier report from our laboratory, [3H]imipramine binding to rat cerebral cortical membranes prepared from 5,7-dihydroxytryptamine lesioned rats showed only a 42% reduction in binding compared to sham controls, although endogenous serotonin levels were decreased by 92% (Sette *et al.*, 1983b). This therefore suggests heterogeneity of [3H]imipramine binding to rat cerebral cortical membranes, as [3H]imipramine binding would appear to be labelling additional sites which are not located on serotonergic neurons.

Several lines of evidence indicate that the [3H]paroxetine binding sites located on serotonergic neurons are associated with the sodium-dependent serotonin transporter. First, the equilibrium dissociation constant of [³H]paroxetine binding to rat cerebral cortical membranes (K_d = 0.15 nM; Fig. 2) was shown to be close to the potency of paroxetine to inhibit sodium-dependent serotonin uptake into rat whole brain synaptosomes (K_i = 0.26 nM; Hytell, 1982). Second, [³H]paroxetine binding to rat cerebral cortical membranes could be potently inhibited by sodium-dependent serotonin uptake inhibitors. Drugs which label other uptake systems or receptors gave K_i values > 1 μM at inhibiting [³H]paroxetine binding. Moreover, an excellent correlation was shown between the potency of various compounds to inhibit [³H]paroxetine binding to rat cerebral cortical membranes and sodium-dependent serotonin uptake into rat brain synaptosomes (r = 0.985; p < 0.001; Habert et al., 1985). Among the various neurotransmitters tested, only serotonin and tryptamine, both substrates for the sodium-dependent serotonin transporter in platelets (Segonzac et al., 1984), gave K_i values of less than 1 μM at inhibiting [³H]paroxetine binding to rat cerebral cortical membranes (Habert et al., 1985).

In these competition inhibition experiments, the sodium-dependent serotonin uptake inhibitors, both tricyclic and non-tricyclic, and serotonin itself inhibited [³H]paroxetine binding to rat cerebral cortical membranes with Hill coefficients close to unity.

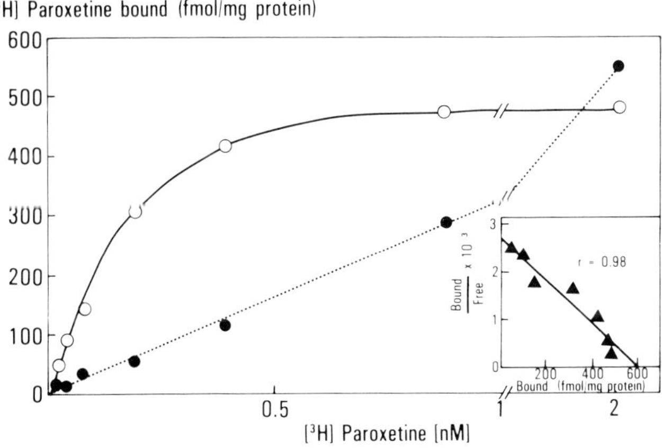

Fig. 2 Specific binding of [³H]-paroxetine to rat cerebral cortical membranes. Rat cerebral cortical membranes were incubated for 60 min at 22°C in the presence and absence of 10 μM fluoxetine with seven concentrations of [³H]paroxetine between 0.03 and 2 nM (Habert et al., 1985). Each point represents the mean of duplicate determinations. The saturation binding isotherm shows specific binding (open circles) and nonspecific binding (solid circles). Inset: transformation of the specific binding data according to Scatchard. (Reproduced from Habert et al. (1985), by permission of Elsevier Science Publishers B.V., Amsterdam)

The potency of a number of tryptamine derivatives and analogues to inhibit the binding of [³H]paroxetine to rat cerebral cortical membranes was investigated to obtain more information on the structural requirements of the substrate recognition site of the sodium-dependent serotonin transporter (Table 1). Using the *in vitro* [³H]paroxetine binding assay, tryptamine displayed an affinity for the transporter which was almost equipotent to serotonin. Substitutions at position 5 of serotonin with fluoro, methyl and methoxy groups, led to decreases in affinity of 2.7-fold, 24-fold and >43-fold, respectively. However, substitution of tryptamine at position 7 with a methyl group produced the most active tryptamine derivative tested. Curiously, although *N*-methyltryptamine and 5-methoxytryptamine were 9.4-fold and >34-fold less potent than tryptamine, 5-methoxy-*N,N*-dimethyltryptamine was only 3.5-fold less potent than tryptamine. This finding with 5-

Table 1 Inhibition of [³H]paroxetine binding to rat cerebral cortical membranes by various tryptamine derivatives and analogues

Substitution	K_i (µM)
None	0.88 ± 0.05
5-hydroxy	0.70 ± 0.09
5-hydroxy *N*-methyl	1.17 ± 0.06
5-hydroxy *N,N*-dimethyl	0.91 ± 0.11
5-hydroxy *N*-acetyl	29.05 ± 4.52
N-methyl	8.32 ± 1.64
7-methyl	0.55 ± 0.11
5-methyl	16.58 ± 1.87
5-fluoro	1.88 ± 0.20
5-methoxy	> 30.00
5-methoxy *N,N*-dimethyl	3.10 ± 0.27
5-methoxy *N*-acetyl	> 30.00
Rigid analogues	
Tryptoline	4.07 ± 0.14
5-hydroxy tryptoline	1.53 ± 0.17
5-methoxy tryptoline	0.42 ± 0.01

The inhibition of [³H]paroxetine binding to rat cortical membranes at 22°C was determined at 0.2 nM [³H]paroxetine for the various compounds listed. The concentration of displacer that inhibited specific binding by 50% (IC$_{50}$ value) was determined from a Hofstee plot, in which the line of best fit was calculated by linear regression using the method of least squares. K_i values for each displacer were subsequently calculated using the formula $K_i = IC_{50}/(1 + c/K_d)$, where c is the concentration of [³H]paroxetine employed, and K_d the equilibrium dissociation constant. The values shown in the table represent the means ±SEM of triplicate experiments.

methoxy-N,N-dimethyltryptamine is at variance with the structure–activity relationship that one might predict from N-methyltryptamine and 5-methoxy-tryptamine. However, it must be noted that it confirms a previous study in which 5-methoxy-N,N-dimethyltryptamine was reported to be a fairly potent inhibitor of sodium-dependent serotonin uptake into synaptosomes (Ubelhack *et al.*, 1978). Melatonin (5-methoxy-N-acetyl-tryptamine), on the other hand, did not inhibit [³H]paroxetine binding (Table 1).

Among the other aromatic hydroxylated tryptamine derivatives tested, N-methyl and N,N-dimethyl substitutions on serotonin led to only slight decreases in potency. This result agrees with an earlier report that 5-hydroxy-N,N-dimethyltryptamine is a potent inhibitor of sodium-dependent serotonin uptake into synaptosomes (Horn, 1973). N-Acetyl substitution of serotonin, on the other hand, proved to be very detrimental, with a 41.5 fold decrease in potency compared to serotonin (Table 1).

The rigid analogue of tryptamine, tetrahydro-β-carboline (tryptoline), was 4.6 fold less potent than tryptamine in inhibiting [³H]paroxetine binding to rat cerebral cortical membranes. Interestingly, substitution of the tryptoline molecule at position 5 with hydroxy and methoxy groups led to pronounced increases in affinity, with 5-methoxytryptoline being slightly more potent than the most potent tryptamine derivative tested, 7-methyltryptamine. This increase in tryptoline potency by the 5-methoxy substitution was opposite to that noted with tryptamine, where 5-methoxy substitution proved to be detrimental to its affinity for [³H]paroxetine binding.

In addition to the above tryptamine derivatives and analogues, several phenylethylamine derivatives were tested, as *para*-substituted phenylethyla-mines and amphetamines have been reported to be potent inhibitors of serotonin accumulation in brain slices and synaptosomes (Carlsson, 1970; Ross *et al.*, 1977; Ross and Kelder, 1977; Chuang and Van Woert, 1980). DL-*p*-Chloroamphetamine, fenfluramine and *p*-chlorophenylethylamine gave K_i values on [³H]paroxetine binding to rat cerebral cortical membranes of 6.53 μM, 3.86 μM and 20.32 μM, respectively. Therefore, although a functional sodium-dependent serotonin transporter is needed in order for these compounds to exert their central effects (Ross *et al.*, 1977; Ross and Kelder, 1977; Chuang and Van Woert, 1980), the [³H]paroxetine binding results indicate that the compounds themselves have low affinity for the transporter. This lends support to the theory that these lipophilic compounds enter serotonergic neurons by diffusion. Once inside the neuron, these compounds might bring about the release of serotonin from intracellular storage sites, triggering the translocation by the serotonin transporter of free serotonin across the neuronal membrane in an outward direction (Ross and Kelder, 1977).

As discussed earlier, a change to higher temperatures can reveal important thermodynamic information on drug interactions with proteins. In this respect, it is interesting to note that at 0°C, chlorimipramine and desme-thylchlorimipramine inhibit [3H]imipramine binding to rat cerebral cortical membranes with similar potencies (Graham, unpublished observations). However, chlorimipramine has been reported to be about 15 fold more potent than desmethylchlorimipramine as an inhibitor of sodium-dependent serotonin uptake at 37°C (Hytell, 1982). Interestingly, the potencies of these compounds to inhibit [3H]paroxetine binding carried out at 22°C were noted to be in accord with their affinities as serotonin uptake inhibitors (Habert *et al.*, 1985).

Thus, [3H]paroxetine is a new and highly selective ligand with which to study the sodium-dependent neuronal serotonin transporter using *in vitro* binding assays.

[3H]Indalpine and [3H]norzimelidine

In addition to [3H]paroxetine, two other selective inhibitors of serotonin uptake, [3H]norzimelidine and [3H]indalpine, have recently been reported to label the neuronal sodium-dependent serotonin transporter. In a study of [3H]indalpine binding to slide-mounted sections of rat brain, pharmacological characterization and regional distribution of [3H]indalpine binding indicated that this binding was associated with the sodium-dependent serotonin trans-porter (Bénavidès *et al.*, 1985; Savaki *et al.*, 1985). [3H]Norzimelidine was also reported to label the sodium-dependent serotonin transporter in rat brain membrane preparations (Hall, 1984). Interestingly, the binding of [3H]norzimelidine to the neuronal sodium-dependent serotonin transporter could be potently inhibited by serotonin (K_i value = 24 nM), which contrasts with the much reduced potency of serotonin to inhibit the binding of [3H]imi-pramine (K_i value = 300 nM; Marcusson *et al.*, 1986), [3H]paroxetine (K_i value = 698 nM; Habert *et al.*, 1985) and [3H]indalpine (K_i value = 280 nM; Bénavidès *et al.*, 1985).

COMPARISON BETWEEN THE PLATELET AND NEURONAL SODIUM-DEPENDENT SEROTONIN TRANSPORTERS

Serotonin transport across platelet and neuronal plasma membranes exhibits many similarities. For example, serotonin is transported by a sodium-depen-dent high-affinity uptake system in both tissues (Stahl, 1977; Ross, 1982). Also, [3H]imipramine and [3H]paroxetine display similar binding properties to platelet and neuronal membranes (Langer *et al.*, 1981; Habert *et al.*, 1985; Segonzac *et al.*, 1980). Moreover, in uptake and binding studies, the tricyclic and non-tricyclic serotonin uptake inhibitors exhibit similar rank orders of inhibitory potency in both tissues. This has therefore led many workers to utilize platelets as an appropriate system to derive information about neuronal sodium-dependent serotonin uptake. In this regard, studies on the platelet sodium-dependent serotonin transporter have provided useful information for our understanding of the mechanism of action of its neuronal counterpart.

It must be stated, however, that subtle differences between these two transporter moieties have been reported. For example, serotonin and its non-hydroxylated form, tryptamine, both act as substrates for the platelet transporter (Segonzac et al., 1984), whereas only serotonin (and not tryptamine) is taken up by the neuronal transporter (Ross and Ask, 1980). Interestingly, although tryptamine and serotonin are almost equipotent at inhibiting [³H]paroxetine binding to rat cerebral cortical membranes (Habert et al., 1985), subsequent solubilization of the neuronal sodium-dependent serotonin transporter has revealed that the potency of serotonin to inhibit [³H]paroxetine binding to the solubilized transporter remains unchanged, whereas the potency of tryptamine is decreased approximately tenfold (Habert et al., 1986).

Studies with the tryptolines have revealed a dissociation in their rank order of potency on inhibition of sodium-dependent serotonin uptake into platelets and brain. In both rabbit and human platelets, the rank order of inhibitory potency on serotonin uptake is 5-MeO-tryptoline > 5-hydroxytryptoline > tryptoline (Table 2; Segonzac et al., 1985b). In rat hypothalamic slices, however, 5-hydroxytryptoline is more potent than 5-methoxytryptoline and the unsubstituted tryptoline (Table 2), in agreement with an earlier report that 5-hydroxytryptoline is a potent inhibitor of serotonin uptake in rat brain (Kellar et al., 1976). The reason for this variance in tryptoline structure–activity relationships between the platelet and neuronal serotonin transporters remains to be elucidated. However, since the rank order of potency of these tryptolines to inhibit [³H]paroxetine binding to both human platelet and rat cerebral cortical membranes is 5-methoxy- > 5-hydroxy- > unsubstituted (Table 2), the differences noted on neuronal and platelet uptake studies would not appear to reflect a change in the affinity of these tryptolines to bind to the transporter.

PLASTICITY OF [³H]IMIPRAMINE AND [³H]PAROXETINE BINDING SITES

Many studies have reported that chronic tricyclic antidepressant administration produces a decrease in the number of [³H]imipramine binding sites (without affecting the K_d of [³H]imipramine binding) in various regions of mammalian brain (Raisman et al., 1980; Kinnier et al., 1980; Briley et al., 1982; Barbaccia et al., 1983; Racagni et al., 1983). However, these findings remain controversial as there exist other reports where no changes in [³H]imipramine binding parameters upon similar drug administration could be

Table 2 Inhibition by various tryptolines of [³H]paroxetine binding to human platelet and rat cerebral cortical membranes and of serotonin uptake into human platelets and rat hypothalamic slices

Drugs	K_i values (μM)			
	[³H]paroxetine binding		[³H]serotonin uptake	
	Human platelets	Rat cerebral cortex	Human platelets	Rat hypothalamus
5-MeO-tryptoline	0.87 ± 0.14	0.42 ± 0.01	0.33 ± 0.04	0.251 ± 0.115
5-hydroxy-tryptoline	4.54 ± 1.25	1.53 ± 0.17	0.72 ± 0.23	0.014 ± 0.006
Tryptoline	6.98 ± 0.79	4.07 ± 0.14	1.97 ± 0.45	1.180 ± 0.330

Inhibition of [³H]paroxetine binding to human platelet membranes was determined at 37°C using 0.1 nM [³H]paroxetine (Segonzac *et al.*, 1980), and to rat cerebral cortical membranes at 22°C using 0.2 nM [³H]paroxetine (Habert *et al.*, 1985). Inhibition of serotonin uptake into fresh human platelets was determined at 37°C using 0.6 μM [³H]serotonin (Segonzac *et al.*, 1980). Inhibition of serotonin uptake into rat hypothalamic slices was determined at 37°C using 50 nM [³H]serotonin (Galzin and Langer, unpublished observations). K_i values were calculated from IC$_{50}$ values, as described in the legend to Table 1. The values shown in the table represent means ±SEM of at least three experiments.

demonstrated (Lee *et al.*, 1983; Kinnier *et al.*, 1980; Barbaccia *et al.*, 1982; Abel *et al.*, 1985). This discrepancy in results between different groups might have arisen because the heterogeneity of [³H]imipramine binding sites in neuronal membrane preparations (Reith *et al.*, 1983; Hrdina, 1984; Marcusson *et al.*, 1985) imposed methodological constraints on the labelling of the sodium-dependent serotonin transporter by [³H]imipramine. To overcome this problem, we therefore chose to reinvestigate some of these studies using [³H]paroxetine, instead of [³H]imipramine, as a marker of the sodium-dependent serotonin transporter.

Chronic administration of the selective serotonin uptake inhibitors, chlorimipramine and citalopram, was found not to alter the parameters of [³H]paroxetine binding to the neuronal sodium-dependent serotonin transporter of rat cerebral cortical or hippocampal membranes (Graham *et al.*, 1986). In this context, it is interesting to note that chronic citalopram also failed to affect the kinetic parameters of sodium-dependent serotonin uptake into rat brain synaptosomes (Hytell *et al.*, 1984). Moreover, changes in serotonin concentrations produced by other chronic pharmacological treatments such as the monoamine oxidase inhibitor, clorgyline (Graham *et al.*, 1980), or the tryptophan hydroxylase inhibitor, parachlorophenylalanine (Galzin *et al.*, 1986) did not modify the parameters of [³H]paroxetine binding to rat cerebral cortical membranes. These findings therefore indicate that, under the experimental conditions that we tested, the neuronal sodium-dependent serotonin transporter does not undergo substrate-induced plasticity changes.

The inability to demonstrate [³H]paroxetine binding site plasticity is at variance with the results of some of the [³H]imipramine binding studies. One reason for this discrepancy could be that the down-regulation of [³H]imipramine binding sites noted in these studies as a result of chronic imipramine or desipramine administration might have been effected through a pharmacological action of these tricyclic antidepressants other than serotonin uptake inhibition, for instance, inhibition of neuronal uptake of noradrenaline. If this were the case, the use of selective serotonin uptake inhibitors in the [³H]paroxetine binding study could not be expected to produce the same effect. It must be pointed out, however, that equivocal results on the plasticity of [³H]imipramine binding sites in the same brain region have still been observed in different studies in which the same tricyclic, desipramine, was administered chronically (Barbaccia *et al.*, 1983; Gentsch *et al.*, 1984; Abel *et al.*, 1985).

In one study in which chronic tricyclic antidepressant administration produced a decrease in the number of [³H]imipramine binding sites in rat hippocampal membranes, a concomitant increase in the V_{max} of sodium-dependent serotonin uptake into hippocampal slices was reported (Barbaccia *et al.*, 1983). This has raised the hypothesis that the tricyclic antidepressant binding site associated with the transporter could be under a separate control mechanism and might exert a modulatory influence on sodium-dependent serotonin uptake. It could be that [³H]imipramine and [³H]paroxetine bind to different sites associated with the transporter, and it is only the [³H]imipramine binding site that is subjected to down-regulation upon chronic tricyclic antidepressant treatment. However, it must be stated that we have observed that chlorimipramine inhibits [³H]paroxetine binding to rat cerebral cortical membranes upon chronic chlorimipramine treatment, with a Hill coefficient close to unity, and with a K_i value similar to that observed in sham controls (Graham and Langer, unpublished observations). Also, upon solubilization of the neuronal sodium-dependent serotonin transporter, we have reported that imipramine inhibits [³H]paroxetine binding with the same potency as in the membrane preparation, and with a Hill coefficient close to unity (Habert *et al.*, 1986). These latter observations therefore indicate that the imipramine and paroxetine binding sites on the neuronal serotonin transporter are closely associated on the same protein. Furthermore, a down-regulation of [³H]imipramine upon chronic tricyclic antidepressant administration to laboratory animals is the opposite paradigm that might be envisaged from studies of depressed patients, in that many studies have described a lower density of [³H]imipramine binding sites in the platelets of untreated depressed patients compared to healthy controls (Briley *et al.*, 1980; Raisman *et al.*, 1981; Poirier *et al.*, 1986; Lewis and McChesney, 1985; Langer *et al.*, 1986). In addition, although the administration of imipramine was reported not to affect the parameters of platelet [³H]imipramine binding in healthy volunteers (Suranyi-Cadotte *et al.*, 1985b), the low B_{max} value of [³H]imipramine binding in platelets of depressed patients was shown to increase during chronic treatment with imipramine (Suranyi-Cadotte *et al.*, 1985a).

The importance of an adequate washout period after chronic tricyclic antidepressant treatment has been pointed out by Poireir *et al.* (1984). In that study, the administration of chlorimipramine for one week in healthy volunteers resulted in a reduction in the B_{max} of platelet [³H]imipramine binding, which recovered to control values two weeks after washout (Poirier *et al.*, 1984).

In conclusion, the present results using [³H]paroxetine favour the view that the neuronal sodium-dependent serotonin transporter from normal rats does not exhibit plasticity upon chronic antidepressant treatment. Nevertheless, it is of interest to note that serotonin uptake has been reported to exhibit a circadian rhythm in the rat suprachiasmatic nucleus (Meyer and Quay, 1976). Since a circadian rhythm in [³H]imipramine binding has also been noted in rat suprachiasmatic nucleus and raphe (Wirz-Justice *et al.*, 1983; Kraeuchi *et al.*, 1986), the possibility exists that a circadian control over serotonin re-uptake is exerted through the binding of a regulatory substance or an endocoid to the sodium-dependent serotonin transporter.

IS THERE AN ENDOCOID REGULATING SODIUM-DEPENDENT SEROTONIN UPTAKE?

Uptake studies have, for the most part, indicated that the tricyclic antidepressants inhibit sodium-dependent serotonin uptake with Hill coefficients close to unity (Stacey, 1961; Long and Lessin, 1962; Tuomisto, 1974; Rudnic, 1977; Lingjaerde, 1977, 1979; Talvenheimo *et al.*, 1979), although there have been several reports of non-competitive type inhibition by imipramine (Wielosz *et al.*, 1976; Le Fur and Uzan, 1977; Wood and Wyllie, 1981).

In addition, competition binding experiments reveal that serotonin inhibits [³H]imipramine and [³H]paroxetine binding to rat brain and platelet membranes with a Hill coefficient close to unity (Segonzac *et al.*, 1984, 1985a; Habert *et al.*, 1985; Marcusson *et al.*, 1986). An earlier report that serotonin inhibited [³H]imipramine binding to rat cerebral cortical membranes with a Hill coefficient much less than unity (Sette *et al.*, 1983a) probably arose because the methodology employed at that time did not take into consideration the heterogeneity of [³H]imipramine binding to cerebral cortical preparations (Reith *et al.*, 1983; Hrdina, 1984; Marcusson, 1985, 1986).

Nevertheless, although [³H]imipramine and [³H]paroxetine binding assays suggest a competitive interaction between imipramine, paroxetine and serotonin, dissociation kinetic experiments indicate that serotonin, imipramine and paroxetine might bind to different sites on the macromolecular complex of the serotonin transporter (Wennogle and Meyerson, 1983; Segonzac *et al.*, 1985a; Plenge and Mellerup, 1985; Arbilla *et al.*, 1987). For example, serotonin produces a pronounced decrease in the rate of dissociation of [3H]imipramine binding to platelet membranes, compared to the rate of [3H]imipramine dissociation in the presence of fluoxetine (Segonzac *et al.*, 1985a). Moreover, possible allosterism in serotonin and imipramine binding is also suggested by a study in which chronic antidepressant administration was reported to produce a decrease in the B_{max} of [3H]imipramine binding in rat hippocampal membranes, but a concomitant increase in the V_{max} of sodium-dependent serotonin uptake into hippocampal slices (Barbaccia *et al.*, 1983). In order to explain these findings, the concept of an endogenous regulator or endocoid that would function to modulate serotonin uptake by binding to a regulatory site associated with the serotonin transporter has been evoked (Sette *et al.*, 1983a; Langer and Raisman, 1983; Barbaccia *et al.*, 1983; Langer *et al.*, 1984). This hypothesis has stimulated research to try and identify this endogenous ligand.

Several groups have utilized various fractionation procedures from the brain and body fluids of several mammalian species, in an attempt to isolate a possible endocoid (different from serotonin) associated with the sodium-dependent serotonin transporter (Barbaccia *et al.*, 1983, 1986; Barbaccia and Costa, 1984; Angel and Paul, 1984; Rehavi *et al.*, 1985; Brusov *et al.*, 1985; Lee *et al.*, 1980). By deproteinizing rat brain homogenates in aqueous acetic acid, followed by reverse-phase HPLC of the concentrated extracts, Costa and coworkers have described two peaks of [3H]imipramine-displacing activity, one of which was due to serotonin itself (Barbaccia *et al.*, 1983; Barbaccia and Costa, 1984). Interestingly, these fractions of peak [3H]imipramine-displacing activity also inhibited serotonin uptake, which has led these authors to conclude that the two peaks of activity corresponded to the endocoid and serotonin, respectively (Barbaccia *et al.*, 1983; Barbaccia and Costa, 1984). However, in a subsequent study, Lee *et al.* (1980), in replicating the protocol of Barbaccia *et al.* (1983), have found that the first peak of [3H]imipramine-displacing activity, which eluted from the HPLC column and which corresponded to the endocoid activity of Barbaccia *et al.* (1983), was of extremely high osmolarity. Since this first peak of [3H]imipramine-displacing activity corresponded to the void volume of the HPLC column and, as such, contained a high concentration of salts, this inhibitory activity on [3H]imipramine binding and [3H]serotonin uptake was concluded to be non-specific in nature (Lee *et al.*, 1980).

The above example serves to illustrate the need to exert stringent precautions in attempting to demonstrate the existence of a novel endogenous ligand. In such studies, controls should be run to ensure that substances have not been inadvertently introduced and/or concentrated during the isolation protocol, such that they interfere with the assay system. In view of this, the presence in brain or plasma of an endogenous modulator of serotonin uptake, other than serotonin itself, remains to be convincingly demonstrated.

DO THE [3H]IMIPRAMINE AND [3H]PAROXETINE RECOGNITION SITES FULFIL THE ROLE OF A PHARMACOLOGICAL RECEPTOR?

According to the original definition of Langley, a receptor was described as the specific component of the organism with which a chemical reagent (or drug) reacted. Clearly, this is a very broad definition as many cellular proteins such as structural proteins or enzymes of crucial metabolic or regulatory pathways could be considered as specific targets for drug action. For this reason, the term 'pharmacological receptor' has been coined to refer to a group of cellular proteins whose function is to act as receptors for hormones, neurotransmitters and endocoids.

The binding of specific [3H]ligands, such as [3H]antagonists, to these pharmacological receptors, and [3H]imipramine or [3H]paroxetine to the sodium-dependent serotonin transporter, share a number of common properties. For example, each displays saturable, stereospecific and, in most cases, high-affinity binding. Also, the binding sites for these ligands show an asymmetrical distribution in regional, tissue and cellular localization. Moreover, competition inhibition experiments with these [3H]ligands using drugs belonging to different chemical and pharmacological classes, illustrate a pharmacological selectivity that can be correlated with a functional response (serotonin uptake).

One important difference exists, however, between these pharmacological receptors and the sodium-dependent serotonin transporter in their functional properties, in that the former serve to sense extracellular regulatory signals and translate them into intracellular messengers which trigger physiological or metabolic responses. The sodium-dependent serotonin transporter, in contrast, functions to internalize the neurotransmitter molecule itself. In this respect, it might be perhaps more appropriate to consider this transporter as belonging to a large family of membrane-bound carrier proteins, such as the so-called LDL and transferrin receptors. At the level of the sodium-dependent transporter for noradrenaline, dopamine and serotonin, there are clear pharmacological differences, as well as structure–activity relationships, for drugs which act as uptake inhibitors and for the respective substrates and their analogues.

There are many examples in the literature illustrating that pharmacological receptors are subject to chronic regulatory and homeostatic control. In an earlier section it was reported that the [3H]paroxetine binding sites associated with the sodium-dependent serotonin transporter do not exhibit plasticity, although in some reports, but not all, modulation of [3H]imipramine binding sites has been described as a result of various chronic treatments. Nevertheless, it must be noted that regulation of synthesis is not *per se* a criterion for classification as a pharmacological receptor. For example, many enzymes are under feedback control mechanisms. Also, the so-called LDL receptor has been shown to be regulated by the cell's demand for cholesterol.

The possibility that an endocoid exists which, by binding to an allosteric site associated with the sodium-dependent serotonin transporter, would modulate serotonin uptake, has been discussed in the previous section. An appropriate isolation of such an endocoid has yet to be demonstrated, let alone its possible interaction with an allosteric site such as the recognition site(s) for paroxetine and imipramine on the sodium-dependent serotonin transporter. Consequently, the concept that the binding site(s) for paroxetine and imipramine are possible endocoid receptor moieties, which function to modulate serotonin uptake, still remains a speculative hypothesis.

In conclusion, available evidence to date must be interpreted as indicating that the [³H]imipramine and [³H]paroxetine binding sites are recognition sites on the sodium-dependent serotonin transporter which fulfil most but not all of the criteria to be considered as pharmacological receptors. Nevertheless, the use of a selective ligand such as [³H]paroxetine to label a high-affinity binding site on the sodium-dependent serotonin transporter can help to provide information regarding the properties of this transporter in much the same way as [³H]antagonists have been utilized to examine classical pharmacological receptors.

SOLUBILIZATION OF THE NEURONAL SODIUM-DEPENDENT SEROTONIN TRANSPORTER

We have recently started a programme to solubilize and purify the sodium-dependent serotonin transporter from rat cerebral cortex.

Crude synaptic membrane preparations from rat cerebral cortex were gently agitated in the presence of the non-ionic detergent, digitonin, and then the suspension was centrifuged at $125,000g$ for 1 h. The presence of solubilized sodium-dependent serotonin transporter in the supernatant detergent extracts was tested using a [³H]paroxetine binding assay in which bound ligand was separated from free ligand by filtration over glass-fibre filters pretreated with 0.3% polyethylenimine (Habert *et al.*, 1986). This method of filtering detergent-solubilized membrane proteins was first described by Bruns *et al.* (1983).

Using the above solubilization procedure, the detergent extracts were shown to contain a single class of high-affinity, non-interacting [³H]paroxetine binding sites. The K_d of [³H]paroxetine binding to these solubilized extracts gave a value of 0.140 ± 0.015 nM, which was very similar to the affinity of [³H]paroxetine binding to the membrane-bound form of the sodium-dependent serotonin transporter of rat cerebral cortex (Table 3). Between different preparations, the amount of solubilized [³H]paroxetine binding sites corre-

Table 3 Equilibrium parameters derived from Scatchard analysis of [³H]paroxetine binding to solubilized and parent membrane preparations. Values are means ±SEM of four independent experiments.
(Reproduced by permission of Elsevier Science Publishers from Habert *et al.*, 1986)

	Membrane preparation	Solubilized preparation
K_d (nM)	0.150 ± 0.010	0.140 ± 0.015
B_{max}		
(pmol/g tissue)	19.2 ± 1.2	8.0 ± 1.1
(fmol/mg protein)	549.0 ± 36.2	321.0 ± 40.0
Solubilization yield (%)	–	42

sponded to 40–50% of those recognition sites present in the parent membrane preparations.

A variety of sodium-dependent serotonin uptake inhibitors and other drugs were used to characterize pharmacologically the solubilized high-affinity [³H]paroxetine binding sites. The non-tricyclic inhibitors of sodium-dependent serotonin uptake, paroxetine, citalopram, fluoxetine, indalpine and fluvoxamine, inhibited [³H]paroxetine binding with K_i values in the low nanomolar range (Habert *et al.*, 1986). Chlorimipramine, imipramine and serotonin gave K_i values of 1.7 nM, 57.8 nM and 466 nM, respectively. The potency of these various compounds to inhibit [³H]paroxetine binding to the detergent extracts and serotonin uptake into rat brain synaptosomes showed a highly significant correlation ($p < 0.001$; Fig. 3). Also, of numerous other neurotransmitter receptor agonists and antagonists tested, all gave K_i values greater than 1 μM. Thus, the pharmacological profile of the high-affinity [³H]paroxetine binding sites present in the detergent extracts indicates that these sites are associated with the sodium-dependent serotonin transporter.

We are now attempting to purify these solubilized [³H]paroxetine binding sites by an affinity chromatographic step utilizing a citalopram derivative coupled to an agarose support. Between 70 and 85% of the [³H]paroxetine binding sites can be removed from the detergent extracts by adsorption to the support. Experiments are currently in progress to optimize the subsequent elution of the purified transporter.

CONCLUSIONS AND FUTURE TRENDS

[³H]Imipramine has been utilized extensively in *in vitro* binding assays to label the neuronal sodium-dependent serotonin transporter. Nevertheless, [³H]imipramine as a ligand poses certain problems, such as the need to conduct routine binding assays at 4°C and the presence of binding site heterogeneity. These limitations can now be circumvented by utilizing the highly selective ligand, [³H]paroxetine, as a marker of the neuronal sodium-dependent serotonin transporter.

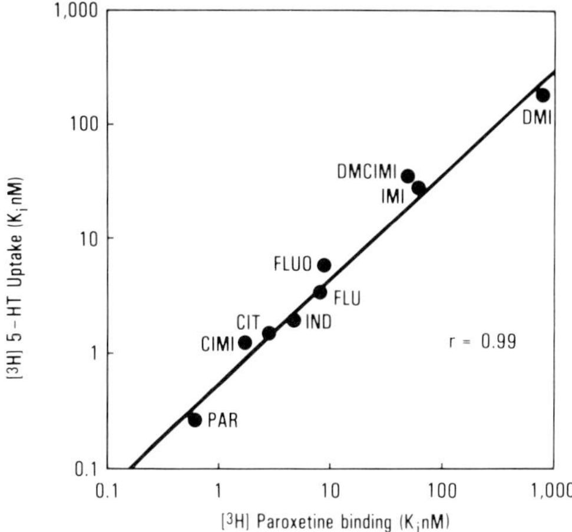

Fig. 3 Correlation between potency for inhibition of [3H]-paroxetine binding to the solubilized transporter from the rat brain cortex, and potency for inhibition of [3H]-serotonin uptake into rat brain synaptosomes. The K_i values for inhibition of uptake and binding are taken from Habert *et al.* (1986). PAR, paroxetine; CIT, citalopram; IND, indalpine; FLU, fluvoxamine; FLUO, fluoxetine; CIMI, chlorimipramine; DMCIMI, desmethylchlorimipramine; IMI, imipramine; DMI, desipramine. (Reproduced from Halbert *et al.*, 1986 by permission of Elsevier Science Publishers B.V., Amsterdam)

Studies performed in the past on [3H]imipramine binding to human post-mortem brain tissue have yielded equivocal results. [3H]Paroxetine could well offer a more appropriate ligand to investigate changes in neuronal serotonin transporter properties in certain psychiatric disorders such as depression. The use of [3H]paroxetine for those clinical studies which have measured platelet [3H]imipramine binding to the serotonin transporter might also be considered.

Radioligand binding assays and uptake studies on the neuronal sodium-dependent serotonin transporter have so far provided information on the structural requirements which are important for binding and transport. These techniques have also aided in the development of specific inhibitors that act upon this transporter. However, although it is known that sodium and chloride ions are co-transported upon uptake, the molecular details by which serotonin is bound and translocated across the neuronal plasma membrane by the transporter remain to be elucidated. In this respect, analysis of the purified transporter protein might prove beneficial. The successful solubilization of this membrane-bound transporter, as evidenced by retention of the substrate and inhibitor binding site(s) of this macromolecule, now permits a starting point towards purification. In this purification procedure, the [³H]paroxetine binding assay should provide a useful tool. Eventually, it would be hoped to apply molecular biological techniques to clone the purified transporter. Information regarding the primary and tertiary structures of this protein and the effects of site-directed mutagenesis on carrier function will ultimately permit a detailed insight into the molecular mechanisms operating in this transport process under normal as well as under pathological conditions.

ACKNOWLEDGEMENTS

We thank Miss Colette Villebeuf for excellent secretarial assistance in the preparation of this article.

REFERENCES

Abel, M. S., Clody, M. E., Wennogle, L. P., and Meyerson, L. R. (1985) Effect of chronic desmethylimipramine or electroconvulsive shock on selected brain and platelet neurotransmitter recognition sites, *Biochem. Pharmacol.*, **34**, 679–683.

Angel, I., and Paul, S. M. (1984) Inhibition of synaptosomal 5-[³H]hydroxytryptamine uptake by endogenous factor(s) in human blood, *FEBS Letters*, **171**, 280–284.

Arbilla, S., Bigg, D., Claustre, Y., Graham, D., Habert, E., Langer, S. Z., Pimoule, C., Scatton, B., and Schoemaker, H. (1987) SL 81.0385, a new selective and potent 5-hydroxytryptamine uptake inhibitor, *Br. J. Pharmacol.* (in press).

Barbaccia, M. L., and Costa, E. (1984) Autacoids for drug receptors: a new approach in drug development, *Ann. New York Acad. Sci.*, **430**, 103–114.

Barbaccia, M. L., Brunello, N., Chuang, D. M., and Costa, E. (1982) On the mode of action of imipramine: relationship between serotonergic axon terminal function and down regulation of β-adrenergic receptors, *Neuropharmacol.*, **22**, 373–383.

Barbaccia, M. L., Gandolfi, O., Chuang, D. M., and Costa, E. (1983) Modulation of neuronal serotonin uptake by a putative endogenous ligand of imipramine recognition sites, *Proc. Natl. Acad. Sci. USA*, **80**, 5134–5138.

Barbaccia, M. L., Melloni, P., Pozzi, O., and Costa, E. (1986) In rat brain, 5-methoxytryptamine may not be the endocoid for the [³H]imipramine recognition sites, *Eur. J. Pharmacol.*, **132**, 45–52.

Bénavidès, J., Savaki, H. E., Malgouris, C., Laplace, C., Margelidon, C., Daniel, M., Courteix, J., Uzan, A., Guérémy, G., and Le Fur, G. (1985) Quantitative autoradiography of [³H]indalpine binding sites in the rat brain. I. Pharmacological characterization, *J. Neurochem.*, **45**, 514–520.

Briley, M., and Langer, S. Z. (1981) Sodium-dependency of [³H]-imipramine binding to rat cerebral cortex, *Eur. J. Pharmacol.*, **72**, 377–380.

Briley, M., Langer, S. Z., Raisman, R., Sechter, D., and Zarifian, E. (1980) Tritiated imipramine binding sites are decreased in platelets of untreated depressed patients, *Science*, **209**, 303–305.

Briley, M. S., Raisman, R., and Langer, S. Z. (1979) Human platelets possess high-affinity binding sites for [³H]imipramine, *Eur. J. Pharmacol.*, **58**, 347–358.

Briley, M. S., Raisman, R., Arbilla, S., Casadamont, M., and Langer, S. Z. (1982) Concomitant decrease in [³H]imipramine binding in cat brain and platelets after chronic treatment with imipramine, *Eur. J. Pharmacol.*, **81**, 309–314.

Bruns, R. F., Lawson-Wendling, K., and Pugsley, T. A. (1983) A rapid filtration assay for soluble receptors using polyethylenimine-treated filters, *Anal. Biochem.*, **132**, 74–81.

Brusov, O. S., Fomenko, A. M., and Katasonov, A. B. (1985) Human plasma inhibitors of platelet serotonin uptake and imipramine receptor binding: extraction and heterogeneity, *Biol. Psychiatry*, **20**, 235–244.

Buus-Lassen, J. (1978) Potent and long-lasting potentiation of two 5-hydroxytryptophan-induced effects in mice by three selective 5-HT uptake inhibitors, *Eur. J. Pharmacol.*, **47**, 351–358.

Carlsson, A. (1970) Structural specificity for inhibition of [¹⁴C]-5-hydroxytryptamine uptake by cerebral slices, *J. Pharm. Pharmacol.*, **22**, 729–732.

Chuang, H. E., and Van Woert, M. H. (1980) Comparative effects of substituted phenylethylamines on brain serotonergic mechanisms, *J. Pharmacol. Exp. Ther.*, **213**, 254–260.

Davis, A. (1984) Temperature-sensitive conformational changes in [³H]imipramine binding sites and the involvement of sulphur-containing bonds, *Eur. J. Pharmacol.*, **102**, 341–347.

Davis, A., Morris, J. M., and Tang, S. W. (1983) Temperature-sensitive conformational changes in membrane-bound and solubilized [³H]-imipramine binding sites, *Eur. J. Pharmacol.*, **88**, 407–410.

Galzin, A. M., Langer, S. Z., and Passarelli, F. (1986) Effects of 6-nitroquipazine on [³H]-5-HT overflow in hypothalamic slices from normal and PCPA treated rats, *Br. J. Pharmacol.*, **87**, 17P.

Gentsch, C., Lichtsteiner, M., and Feer, H. (1984) [³H]Imipramine and [³H-cyano-imipramine binding in rat brain tissue: effect of long-term antidepressant administration, *J. Neural Transm.*, **59**, 257–264.

Graham, D., Tahraoui, L., and Langer, S. Z. (1980) Effects of chronic treatments with selective monoamine oxidase inhibitors and specific 5-hydroxytryptamine uptake inhibitors on [³H]paroxetine binding to rat cerebral cortical membranes, *Neuropharmacol*, **26**, 1087–1092.

Gross, G., Göthert, M., Ender, H. C., and Schümann, H. J. (1981) [³H]Imipramine binding sites in the rat brain: selective localization on serotonergic neurons, *Naunyn-Schmiedeberg's Arch. Pharmacol.*, **317**, 310–314.

Habert, E., Graham, D., Tahraoui, L., Claustre, Y., and Langer, S. Z. (1985) Characterization of [³H]paroxetine binding to rat cortical membranes, *Eur. J. Pharmacol.*, **118**, 107–114.

Habert, E., Graham, D., and Langer, S. Z. (1986) Solubilization and characterization of the 5-hydroxytryptamine transporter complex from rat cerebral cortical membranes, *Eur. J. Pharmacol.*, **122**, 197–204.

Hall, H. (1984) Characterization of the binding of [³H]norzimelidine, a 5-HT uptake inhibitor, to rat brain homogenates, *Acta. Pharmacol. Toxicol.*, **55**, 33–40.

Horn, A. S. (1973) Structure-activity relations for the inhibition of 5-HT uptake into rat hypothalamic homogenates by serotonin and tryptamine analogues, *J. Neurochem.*, **21**, 883–888.

Hrdina, P. D. (1984) Differentiation of two components of specific [³H]imipramine binding in rat brain, *Eur. J. Pharmacol.*, **102**, 481–488.

Hytell, J. (1982) Citalopram: pharmacological profile of a specific serotonin uptake inhibitor with antidepressant activity, *Prog. Neuropsychopharmacol. Biol. Psychiat.*, **6**, 277–295.

Hytell, J., Overo, K. F., and Arnt, J. (1984) Biochemical effects and drug levels in rats after long-term treatment with the specific 5-HT uptake inhibitor, citalopram, *Psychopharmacol.*, **83**, 20–27.

Kanner, B. I. (1978) Active transport of γ-aminobutyric acid by membrane vesicles isolated from rat brain, *Biochemistry*, **17**, 1207–1211.

Kanner, B. I., and Sharon, I. (1978) Active transport of L-glutamate by membrane vesicles isolated from rat brain, *Biochemistry*, **17**, 3949–3953.

Kellar, K. J., Elliot, G. R., Holman, R. B., Vernikoss-Danellis, J., and Barchas, J. D. (1976) Tryptoline inhibition of serotonin uptake in rat forebrain homogenates, *J. Pharmacol. Exp. Ther.*, **198**, 619–625.

Kinnier, W. J., Chuang, D. M., and Costa, E. (1980) Down-regulation of dihydro-alprenolol and imipramine binding sites in brains of rats repeatedly treated with imipramine, *Eur. J. Pharmacol.*, **67**, 289–294.

Kraeuchi, K., Wirz-Justice, A., Morimasa, T., Suetterlin-Willener, R., and Feer, H. (1986) Temporal distribution of [³H]imipramine binding in rat brain regions is not changed by chronic metamphetamine, *Chronobiology International*, **3**, 127–133.

Langer, S. Z., and Briley, M. S. (1981) High affinity [³H]imipramine binding: a new biological tool for studies in depression, *Trends Neurosci.*, **4**, 28–31.

Langer, S. Z., and Raisman, R. (1983) Binding of [³H]imipramine and [³H]desipramine as biochemical tools for studies in depression, *Neuropharmacology*, **22**, 407–413.

Langer, S. Z., Moret, C., Raisman, R., Dubocovich, M. L., and Briley, M. S. (1980a) High affinity [³H]imipramine binding in rat hypothalamus is associated with the uptake of serotonin but not norepinephrine, *Science*, **210**, 1133–1135.

Langer, S. Z., Briley, M., Raisman, R., Henry, J.-F., and Morselli, P. L. (1980b) Specific [³H]imipramine binding in human platelets: influence of age and sex, *Naunyn-Schmiedeberg's Arch. Pharmacol.*, **313**, 189–194.

Langer, S. Z., Zarifian, E., Briley, M. S., Raisman, R., and Sechter, D. (1981) High affinity binding of [³H]imipramine in brain and platelets and its relevance to the biochemistry of affective disorders, *Life Sci.*, **29**, 211–218.

Langer, S. Z., Raisman, R., Tahraoui, L., Scatton, B., Niddam, R., Lee, C. R., and Claustre, Y. (1984) Substituted tetrahydro-β-carbolines are possible candidates as endogenous ligand of the [³H]imipramine recognition site, *Eur. J. Pharmacol.*, **98**, 153–154.

Langer, S. Z., Sechter, D., Loo, H., Raisman, R., and Zarifian, E. (1986) Electroconvulsive shock therapy and maximum binding of platelet [³H]imipramine binding in depression, *Arch. Gen. Psychiatry*, **43**, 949–952.

Lee, C. M., Javitch, J. A., and Snyder, S. H. (1983) Recognition sites for norepinephrine uptake: regulation by neurotransmitter, *Science*, **22**, 626–629.

Lee, C. R., Galzin, A. M., Taranger, M. A., and Langer, S. Z. (1980) Pitfalls in demonstrating an endogenous ligand of imipramine recognition sites, *Biochem. Pharmacol*, **6**, 945–949.

Le Fur, G., and Uzan, A. (1977) Effects of 4-(3-indolyl-alkyl)piperidine derivatives on uptake and release of noradrenaline, dopamine and 5-hydroxytryptamine in rat brain synaptosomes, rat heart and human blood platelets, *Biochem. Pharmacol.*, **26**, 497–503.

Lewis, D. A., and McChesney, C. (1985) Tritiated imipramine binding distinguishes among subtypes of depression, *Arch. Gen. Psychiatry*, **42**, 485–488.

Lingjaerde, O. (1977) Platelet uptake and storage of serotonin, in *Serotonin in Health and Disease* (Ed. W. B. Essman), Vol. 4, pp. 139–199, Spectrum Publications Inc.

Lingjaerde, O. (1979) Inhibitory effect of clomipramine and related drugs on serotonin uptake in platelets more complicated than previously thought, *Psychopharmacol.*, **61**, 245–249.

Long, R. F., and Lessin, A. W. (1962) Inhibition of 5-hydroxytryptamine uptake by platelets *in vitro* and *in vivo*, *Biochem. J.*, **82**, 5P.

Marcusson, J., Fowler, C. J., Hall, H., Ross, S. B., and Winblad, B. (1985) Specific binding of [^{3}H]imipramine to protease-sensitive and protease-resistant sites, *J. Neurochem.*, **44**, 705–711.

Marcusson, J. O., Bäckström, I. T., and Ross, S. B. (1986) Single-site model of the neuronal 5-hydroxytryptamine uptake and imipramine-binding site, *Mol. Pharmacol.*, **30**, 121–128.

Meyer, D. C., and Quay, W. B. (1976) Hypothalamic and suprachiasmatic uptake of serotonin *in vitro*: twenty-four hour changes in male and proestrous female rats, *Endocrinology*, **98**, 1160–1165.

Moret, C., and Briley, M. S. (1986) High- and low-affinity binding of [^{3}H]imipramine in rat hypothalamus, *J. Neurochem.*, **47**, 1609–1613.

Nelson, P. J., and Rudnick, G. (1982) The role of chloride ion in platelet serotonin transport, *J. Biol. Chem.*, **257**, 6151–6155.

Palkovits, M., Raisman, R., Briley, M. S., and Langer, S. Z. (1981) Regional distribution of [^{3}H]imipramine binding in rat brain, *Brain Res.*, **210**, 493–498.

Paul, S. M., Rehavi, M., Skolnick, P., and Goodwin, F. K. (1980) Demonstration of specific high-affinity binding sites for [^{3}H]imipramine on human platelets, *Life Sci.*, **26**, 953–959.

Plenge, P., and Mellerup, E. T. (1984) Imipramine binding site. Temperature-dependence of the binding of [^{3}H]-labelled paroxetine to human platelet membrane, *Biochim. Biophys. Acta*, **770**, 22–28.

Plenge, P., and Mellerup, E. T. (1985) Antidepressive drugs can change the affinity of [^{3}H]imipramine and [^{3}H]paroxetine binding to platelet and neuronal membranes, *Eur. J. Pharmacol.*, **119**, 1–8.

Poirier, M. F., Loo, H., Benkelfat, C., Sechter, D., Zarifian, E., Galzin, A. M., Schoemaker, H., Segonzac, A., and Langer, S. Z. (1984) [^{3}H]Imipramine binding and [^{3}H]5-HT uptake in human blood platelets: changes after one week chlorimipramine treatment, *Eur. J. Pharmacol.*, **106**, 629–633.

Poirier, M. F., Benkelfat, C., Loo, H., Sechter, D., Zarifian, E., Galzin, A. M., and Langer, S. Z. (1986) Reduced B_{max} of [^{3}H]imipramine binding to platelets of depressed patients free of previous medication with 5-HT uptake inhibitors, *Psychopharmacol.*, **89**, 456–461.

Racagni, G., Mocchetti, I., Calderini, G., Battistella, A., and Brunello, N. (1983) Temporal sequence of changes in central noradrenergic system of rat after prolonged antidepressant treatment: receptor desensitization and neurotransmitter interactions, *Neuropharmacol.*, **22**, 415–424.

Raisman, R., Briley, M. S., and Langer, S. Z. (1979) Specific tricyclic antidepressant binding sites in rat brain, *Nature*, **281**, 148–150.

Raisman, R., Briley, M. S., and Langer, S. Z. (1980) Specific tricyclic antidepressant binding sites in rat brain characterized by high-affinity [^{3}H]imipramine binding, *Eur. J. Pharmacol.*, **61**, 373–380.

Raisman, R., Sechter, D., Briley, M. S., Zarifian, E., and Langer, S. Z. (1981) High affinity [^{3}H]imipramine binding in platelets from untreated and treated depressed patients compared to healthy volunteers, *Psychopharmacol.*, **75**, 368–371.

Rehavi, M., Paul, S. M., Skolnick, P., and Goodwin, F. K. (1980) Demonstration of specific high-affinity binding sites for [³H]imipramine in human brain, *Life Sci.*, **26**, 2273–2279.

Rehavi, M., Ventura, I., and Sarne, Y. (1985) Demonstration of endogenous 'imipramine-like' material in rat brain, *Life Sci.*, **36**, 687–693.

Reith, M. E. A., Sershen, H., Allen, D., and Lajtha, A. (1983) High- and low-affinity binding of [³H]imipramine in mouse cerebral cortex, *J. Neurochem.*, **40**, 389–395.

Reith, M. E. A., Sershen, H., and Lajtha, A. (1984) Thermodynamics of the interactions of tricyclic drugs with binding sites for [³H]imipramine in mouse cerebral cortex, *Biochem. Pharmacol.*, **33**, 4101–4104.

Ross, S. B. (1982) The characteristics of serotonin uptake systems, in *Biology of Serotonergic Transmission* (Ed. N. N. Osborne), pp. 159–195, Wiley, London.

Ross, S. B., and Ask, A. L. (1980) Structural requirements for uptake into serotonergic neurons, *Acta Pharmacol. Toxicol.*, **46**, 270–277.

Ross, S. B., and Kelder, D. (1977) Efflux of 5-hydroxytryptamine from synaptosomes of rat cerebral cortex, *Acta Physiol. Scand.*, **99**, 27–36.

Ross, S. B., Ogren, S. O., and Renyi, A. L. (1977) Substituted amphetamine derivatives. I. Effect on uptake and release of biogenic monoamines and on monoamine oxidase in the mouse brain, *Acta Pharmacol. Toxicol.*, **41**, 337–352.

Rudnick, G. (1977) Active transport of 5-hydroxytryptamine by plasma membrane vesicles isolated from human blood platelets, *J. Biol. Chem.*, **252**, 2170–2174.

Savaki, H., Malgouris, C., Bénavidès, J., Laplace, C., Uzan, A., Guérémy, C., and Le Fur, G. (1985) Quantitative autoradiography of [³H]imipramine binding sites in the rat brain. II. Regional distribution, *J. Neurochem.*, **45**, 521–526.

Segonzac, A., Schoemaker, H., and Langer, S. Z. (1980) Temperature-dependence of drug interaction with the platelet 5HT transporter: a clue to the imipramine selectivity paradox, *J. Neurochem*, **48**, 331–339.

Segonzac, A., Raisman, R., Tateishi, T., Schoemaker, H., Hicks, P. E., and Langer, S. Z. (1985a) Tryptamine, a substrate for the serotonin transporter in human platelets, modifies the dissociation kinetics of [³H]imipramine binding: possible allosteric interaction, *J. Neurochem.*, **44**, 349–356.

Segonzac, A., Schoemaker, H., Tateishi, T., and Langer, S. Z. (1985b) 5-Methoxytryptoline, a competitive endocoid acting at [³H]-imipramine recognition sites in human platelets, *J. Neurochem.*, **45**, 249–256.

Segonzac, A., Tateishi, T., and Langer, S. Z. (1984) Saturable uptake of [³H]tryptamine in rabbit platelets is inhibited by 5-hydroxytryptamine uptake blockers, *Naunyn-Schmiedeberg's Arch. Pharmacol.*, **328**, 33–37.

Sette, M., Raisman, R., Briley, M. S., and Langer, S. Z. (1981) Localization of tricyclic antidepressant binding sites on serotonin nerve terminals in rat hypothalamus, *J. Neurochem.*, **37**, 40–42.

Sette, M., Briley, M. S., and Langer, S. Z. (1983a) Complex inhibition of [³H]imipramine binding by 5-HT and non-tricyclic 5-HT uptake blockers, *J. Neurochem.*, **40**, 622–628.

Sette, M., Ruberg, M., Raisman, R., Scatton, B., Zivkovic, B., Agid, Y., and Langer, S. Z. (1983b) [³H]Imipramine binding in subcellular fractions of the rat cerebral cortex after chemical lesion of serotonergic neurones, *Eur. J. Pharmacol.*, **95**, 41–51.

Sneddon, J. M. (1973) Blood platelets as a model for monoamine-containing neurons, *Prog. in Neurobiol.*, **1**, 153–198.

Stacey, R. D. (1961) Uptake of 5-hydroxytryptamine by platelets, *Br. J. Pharmacol.*, **16**, 284–295.

Stahl, S. M. (1977) The human platelet: a diagnostic and research tool for the study of biogenic amines in psychiatric and neurologic disorders, *Arch. Gen. Psychiat.*, **34**, 509–516.

Suranyi-Cadotte, B. E., Quirion, R., Nair, N. P. V., Lafaille, F., and Schwartz, G. (1985a) Imipramine treatment differentially affects platelet [3H]imipramine binding and serotonin uptake in depressed patients, *Life Sci.*, **36**, 795–799.

Suranyi-Cadotte, B. E., Lafaille, F., Schwartz, G., Nair, N. P. V., and Quirion, R. (1985b) Unchanged platelet [3H]imipramine binding in normal subjects after imipramine administration, *Biol. Psychiatry*, **20**, 1237–1240.

Talvenheimo, J., Nelson, P. J., and Rudnick, G. (1979) Mechanism of imipramine inhibition of platelet 5-hydroxytryptamine transport, *J. Biol. Chem.*, **254**, 4631–4635.

Tuomisto, J. (1974) A new modification for studying 5-HT uptake by blood platelets: a re-evaluation of tricyclic antidepressants as uptake inhibitors, *J. Pharm. Pharmacol.*, **26**, 92–100.

Ubelhack, R., Franke, L., and Seidel, K. (1978) Wirkung von LSD und 5-methoxy-N-N-dimethyltryptamine auf die aktive Aufnahme von [3H]-serotonin in Synaptosomenfraktionen aus Rattenhirn, *Acta Biol. Med. Germ.*, **37**, 1611–1614.

Wennogle, L. P., and Meyerson, L. R. (1983) Serotonin modulates the dissociation of [3H]imipramine from human platelet recognition sites, *Eur. J. Pharmacol.*, **86**, 303–307.

Wielosz, M., Salmona, M., De Gaetono, G., and Garattini, S. (1976) Uptake of 14C-5-hydroxytryptamine by human and rat platelets and its pharmacological inhibition, *Naunyn-Schmiedeberg's Arch. Pharmacol.*, **296**, 59–65.

Wirz-Justice, A., Kraeuchi, K., Morimasa, T., Willener, R., and Feer, H. (1983) Circadian rhythm of [3H]imipramine binding in the rat suprachiasmatic nuclei, *Eur. J. Pharmacol.*, **87**, 331–333.

Wood, M. D., and Wyllie, M. C. (1981) The inhibitory action of imipramine on monoamine transport, *Br. J. Pharmacol.*, **74**, 890P.

Neuronal Serotonin
Edited by N. N. Osborne and M. Hamon
© 1988 John Wiley & Sons Ltd

CHAPTER 14

Biochemical Properties of Central Serotonin Receptors

M. Hamon, H. Gozlan, S. El Mestikawy, M. B. Emerit, J. M. Cossery
and O. Lutz
INSERM U. 288
Neurobiologie Cellulaire et Fonctionnelle
Faculté de Médecine Pitié-Salpêtrière
91 Boulevard de l'Hôpital
75634 Paris Cédex 13
France

INTRODUCTION

Since the first pharmacological characterization of serotonin (5-HT) receptors 30 years ago (Gaddum and Picarelli, 1957), considerable progress has been made in their identification in the central nervous system owing principally to the development of the *in vitro* binding technique with appropriate radio-active ligands (see Hamon, 1984). Today, it is generally accepted that at least four classes of specific 5-HT receptor binding sites exist in brain membranes: the three 5-HT_1 classes called 5-HT_{1A}, 5-HT_{1B} and 5-HT_{1C}, which are characterized by a nM affinity for 5-HT and selected agonists, but only a μM affinity for most antagonists (Hoyer *et al.*, 1985); and the 5-HT_2 class which in contrast has a nM affinity for selective antagonists and a μM affinity for 5-HT and related agonists (Peroutka and Synder, 1979). In addition, a fifth category of 5-HT receptors, called 5-HT_3, has been clearly identified on peripheral nerves (Richardson and Engel, 1986), and recent data on the 5-HT-induced facilitation of norepinephrine release from hippo-campal slices strongly suggest that this effect involves central receptors with the same pharmacological properties as those of the 5-HT_3 class (Feuerstein and Hertting, 1986).

Although the pharmacological data in support of this classification are well established, they are obviously not enough for asserting that the various 5-HT binding sites identified exclusively by pharmacological approaches really correspond to different 5-HT receptors, i.e. to different protein complexes in membranes. Biochemical investigations are absolutely needed at this stage. In spite of the relatively recent development of such investigations, important data have already been obtained about the physico-chemical characteristics of central 5-HT receptors. As will be shown in this review, these data are far from being complete, but seem to confirm the actuality of the heterogeneity of central 5-HT receptors previously established by the pharmacologists.

BIOCHEMICAL EVIDENCE OF THE IDENTITY OF 5-HT BINDING SITES WITH TRUE 5-HT RECEPTORS IN THE CNS

If a binding site actually corresponds to a receptor in membranes, its occupancy by a putative agonist must trigger well-defined cellular events such as the synthesis and/or release of a hormone, the transmembrane flux of ions, etc. In the case of neuromodulators such as monoamines in the brain, these events are in fact generated by second messengers which are synthesized in target cells as a result of the agonist binding to the specific sites. This occurs notably for catecholamines which can trigger the synthesis of several second messengers such as cyclic AMP (by acting on β-adrenoceptors) or diacyl-glycerol and inositol triphosphate (by acting on alpha$_1$ adrenoceptors). Similarly, we will see that 5-HT acting on 5-HT_{1A} receptors can affect the production of cyclic AMP in target cells, whereas the binding of the indole-amine to 5-HT$_{1C}$ or 5-HT$_2$ receptors results in the production of diacylglycerol and inositol triphosphate.

Coupling of 5-HT$_{1A}$ receptor binding sites to adenylate cyclase in the rat brain

Enhanced cyclic AMP production by 5-HT in rat brain membranes was first reported by Von Hungen et al. (1975). The demonstration that this effect involves a postsynaptic 5-HT receptor coupled to adenylate cyclase was then made by Enjalbert et al. (1978a, b). Although the apparent affinity of this receptor for 5-HT is much lower (K_A app. = 0.5 μM) than that of 5-HT$_1$ sites (K_d = 1–5 nM; see Nelson et al., 1980a, b), the relative potency of a series of 5-HT agonists and antagonists to affect adenylate cyclase in the rat hippocampus is positively correlated to their apparent affinity for 5-HT$_1$ binding sites, which led Barbaccia et al. (1983) to conclude that the receptor coupled to adenylate cyclase in the rat brain belongs to the 5-HT$_1$ class. More recently, a similar pharmacological study has been carried out by Markstein et al. (1986), who used newly developed 5-HT agonists and ant-agonists acting selectively (or preferentially) on a given class of 5-HT recep-tors. According to these authors, the 5-HT receptor responsible for the activation of adenylate cyclase in homogenates of rat hippocampus has a K_A app. of 0.1 μM and pharmacological properties resembling those of the 5-HT$_{1A}$ class. In contrast, no correlation was found between the pharmaco-logical profile of the 5-HT receptor positively coupled to adenylate cyclase and those of 5-HT$_{1B}$, 5-HT$_{1C}$ or 5-HT$_2$ binding sites in the rat brain (Markstein et al., 1986). That 5-HT$_{1C}$ sites (in the choroid plexus) are not linked in a stimulatory or inhibitory way to adenylate cyclase has been recently confirmed by Palacios et al. (1986).

In addition to stimulating adenylate cyclase activity, 5-HT has also been found to exert an inhibitory effect on the forskolin-activated enzyme in guinea pig and rat hippocampal membranes (De Vivo and Maayani, 1986). This effect occurs in the same concentration range as the 5-HT-induced stimulation and, furthermore, exhibits the pharmacological profile typical of the involvement of 5-HT$_{1A}$ receptor (De Vivo and Maayani, 1986). Similarly, Weiss et al. (1986) noted that cyclic AMP production stimulated by vasoactive intestinal polypeptide (VIP) in striatal or cortical neurones in primary culture could be reduced by 5-HT through an action on 5-HT$_1$ receptors, probably of the 5-HT$_{1A}$ subtype (recognized by the selective agonist 8-OH-DPAT; see Arvidsson et al., 1981, 1984).

Therefore, convincing evidence exists for the coupling of 5-HT$_{1A}$ binding sites to adenylate cyclase in brain membranes. However, this coupling can result in the activation as well as in the inhibition of the enzyme, which indicates that the 5-HT$_{1A}$ receptor binding site might be associated with different GTP-binding protein complexes; N_s or N_i (see Taylor and Merritt, 1986). GTP modulation of ^{3}H-8-OH-DPAT (Gozlan *et al.*, 1983; Hall *et al.*, 1986) or ^{3}H-WB 4101 (Norman *et al.*, 1985) specific binding to 5-HT$_{1A}$ sites confirms the existence of such associations, but provides no information on the nature (N_s and/or N_i) of the GTP-binding proteins involved. Further biochemical studies, notably on solubilized 5-HT$_{1A}$ receptors, must be carried out in order to answer this question.

In case of 5-HT-induced increase in cyclic AMP production, subsequent events in target cells are the activation of cyclic AMP-dependent protein kinase, and the phosphorylation of specific proteins. Evidence for 5-HT-induced phosphorylation of the synaptic vesicle protein I involving cyclic AMP-dependent protein kinase has been reported by Dolphin and Green-gard (1981a, b) in slices of the facial nucleus of the rat brainstem. In addition, Lemos *et al.* (1984) found specifically in brain regions densely innervated by serotonergic projections in the rat a 48-kD protein which is a substrate of cyclic AMP-dependent protein kinase. However, the most convincing demonstration of cyclic AMP-dependent phosphorylation of proteins due to 5-HT receptor stimulation has been obtained in invertebrates. For instance, 5-HT-induced increase in cyclic AMP levels (Deterre *et al.*, 1982), and subsequent cyclic AMP-dependent phosphorylation of specific proteins (Lemos *et al.*, 1982) could be observed in *single* identified neurones in molluscs (*Aplysia, Helix*). In these cells, 5-HT-induced protein phosphory-lation finally triggers the closing of K$^+$ channels (Deterre *et al.*, 1982; Shuster *et al.*, 1985), leading to slow modulatory synaptic potential that contributes to presynaptic facilitation by 5-HT.

Possible coupling of 5-HT receptors to cyclic GMP production

Although an increased cerebellar production of cyclic guanosine monophos-phate (GMP) was reported several years ago by Lykouras *et al.* (1980) after the i.p. injection of the potent 5-HT agonist 5-methoxy-*N,N*-dimethyltrypta-mine (5-MeO-DMT) to rats, the possible involvement of specific 5-HT recep-tors could not be proven, as the 5-HT antagonists cyproheptadine and methy-sergide failed to prevent the 5-MeO-DMT effect. In contrast, more recent studies on a clonal cell of glial origin (Ogura *et al.*, 1986) have shown that the stimulatory effect of 5-HT (K_A app. $= 0.1$ μM) on cyclic GMP production in these cells really involves specific receptors since it can be prevented by 5-HT antagonists such as methysergide, metergoline and cyproheptadine.

As noted above regarding cyclic AMP production, the cellular events which follow the synthesis of cyclic GMP have also been mainly analysed in invertebrates. Thus Paupardin-Tritsch *et al.* (1986) have clearly demonstrated that 5-HT-induced cyclic GMP production leads to the phosphorylation by cyclic GMP-dependent protein kinase of specific proteins which finally control Ca^{2+} conductance in snail neurones. However, the pharmacological properties of the 5-HT receptor(s) responsible for cyclic GMP accumulation have not been explored in detail, and it cannot yet be concluded whether this receptor is directly coupled to some enzyme(s) involved in cyclic GMP metabolism, or corresponds to one of those which control the phosphatidylinositol pathway (see next subsection). For glioma cells, Ogura *et al.* (1986) proposed that 5-HT-induced stimulation of guanylate cyclase activity results from the initial activation of phospholipase C; indeed, arachidonic acid, which can be considered as a product of this enzyme, is well known as a potent direct activator of guanylate cyclase (see Ogura *et al.*, 1986). Since 5-HT$_2$ sites are functionally coupled to phospholipase C in membranes (see below), including in glial cells (Ananth *et al.*, 1987), their possible identity with the 5-HT receptors controlling cyclic GMP production in glioma cells remains a working hypothesis for further investigations.

Coupling of 5-HT$_{1C}$ and 5-HT$_2$ receptors to phosphatidylinositol turnover

Evidence that 5-HT receptors are coupled not only to adenylate cyclase but also to phospholipase C responsible for the hydrolysis of phosphatidylinositol biphosphate, was first reported by Berridge and colleagues (see Berridge and Heslop, 1981), who studied the stimulatory effect of the indoleamine on the fluid secretion by salivary gland of the blowfly. Subsequently, improved assay conditions for the measurement of phosphatidylinositol turnover have allowed studies in mammalian brain.

Two groups (Conn and Sanders-Bush, 1985; Sanders-Bush and Conn, 1986; Kendall and Nahorski, 1985) succeeded in demonstrating a stimulatory action of 5-HT and related agonists on phospholipase C activity in slices of the rat cerebral cortex. Furthermore, extensive investigations on the pharmacological characteristics of the putative receptor involved in this action revealed a highly significant correlation with those of ^{3}H-ketanserin binding sites, indicating that this receptor belongs to the 5-HT$_2$ subtype (Conn and Sanders-Bush, 1985; Kendall and Nahorski, 1985). The same conclusion has been reached by several groups who demonstrated that 5-HT$_2$ receptors also trigger phosphatidylinositol breakdown in platelets (De Chaffoy de Courcelles *et al.*, 1985), in C$_6$ glioma cells (Ananth *et al.*, 1987), in rat aorta (Roth *et al.*, 1986), in rat aortic myocytes (Cory *et al.*, 1986), in smooth muscle cells (Doyle *et al.*, 1986), etc.

However, all 5-HT-induced phosphatidylinositol responses cannot be attributed to the stimulation of 5-HT$_2$ receptors. This is notably the case in the choroid plexus where the pharmacological characteristics of the receptor mediating the 5-HT-induced activation of phospholipase C are strikingly different from those of 5-HT$_2$ binding sites (Conn *et al.*, 1986). In particular, the potent 5-HT$_2$ antagonist spiperone is only poorly active against the 5-HT effects in rat choroid plexus. Furthermore, ten times less 5-HT are required to trigger phosphatidylinositol breakdown in the choroid plexus than in the cerebral cortex where 5-HT$_2$ receptors predominate (Conn *et al.*, 1986).

It is now clearly established that another class of 5-HT receptors, called 5-HT$_{1C}$, is responsible for the activation of phospholipase C by the indoleamine in the choroid plexus. This receptor class is in fact that which has been previously characterized by Pazos *et al.* (1985) with ^{3}H-mesulergine as radioligand.

Conclusions – the problem of 5-HT$_{1B}$ sites

As indicated in Table 1, which briefly summarizes the present knowledge on 5-HT receptors and/or binding sites in mammals, convincing evidence exists for the coupling of 5-HT$_{1A}$ sites with adenylate cyclase, and of 5-HT$_{1C}$ and 5-HT$_2$ sites with phospholipase C in brain membranes. Therefore, biochemical studies on the cellular transduction mechanisms associated to 5-HT receptors confirm the receptor heterogeneity previously established by pharmacological studies.

However, in addition to those sites already mentioned, the pharmacologists have described a 5-HT$_{1B}$ site which is characterized by a nM affinity for 5-HT and a poor affinity for preferential 5-HT$_{1A}$ and 5-HT$_{1C}$ ligands (Hoyer *et al.*, 1985; Hamon *et al.*, 1986). Evidence for the correspondence of this site with a true 5-HT receptor derives from *in vitro* release studies which

Table 1 Main characteristics of the various classes of 5-HT receptors or binding sites presently identified in mammals

Receptor type	Main location	Selective radioligand	Coupling mechanism
5-HT$_{1A}$	Postsynaptic Limbic areas	^{3}H-8-OH-DPAT	Adenylate cyclase
5-HT$_{1B}$	Post and pre(?) synaptic Extrapyramidal areas	[^{125}I]iodocyanopindolol	Adenylate cyclase
5-HT$_{1C}$	Postsynaptic Choroid plexus	^{3}H-mesulergine	Phosphatidylinositol
5-HT$_2$	Postsynaptic Frontal cortex	^{3}H-ketanserin	Phosphatidylinositol
5-HT$_3$	Peripheral nerves	none	(Neuronal depolarization)
5-HT$_{Pre}$	Presynaptic Striatum	^{3}H-8-OH-DPAT	5-HT carrier? (high-affinity uptake)

demonstrated that the 5-HT-induced inhibition of acetylcholine release from striatal slices (Gillet *et al.*, 1985) and hippocampal synaptosomes (Maura and Raiteri, 1986), and of its own release from various brain regions (see Engel *et al.*, 1986) could be reproduced or prevented by drugs acting preferentially as 5-HT_{1B} agonists or antagonists respectively. On account of the Ca^{2+} dependence of neurotransmitter release from brain slices, it can be assumed that 5-HT_{1B} receptors probably participate in the control of Ca^{2+} flux through target cell membranes. However, the cellular mechanisms involved in this putative control have yet to be explored.

Another problem about 5-HT_{1B} sites concerns their postulated presynaptic location on serotonergic terminals, since they apparently correspond to the presynaptic autoreceptors responsible for a local inhibitory feedback control of 5-HT release (Engel *et al.*, 1986). Indeed, extensive degeneration of serotonergic terminals resulting from an intracerebral administration of 5,7-dihydroxytryptamine is not associated with any decrease in the density of 5-HT_{1B} sites in areas such as the cerebral cortex, the striatum and the hippocampus where 5-HT autoreceptors controlling 5-HT release have been clearly identified (Vergé *et al.*, 1985, 1986). In contrast, the same treatment produces a significant reduction of $^3\text{H-8-OH-DPAT}$ binding to 5-HT_{Pre} sites in striatal membranes (Gozlan *et al.*, 1983; Hall *et al.*, 1985). but these presynaptic sites are more probably related to the 5-HT carrier involved in the neurotransmitter reuptake than to the 5-HT autoreceptor controlling 5-HT release (see Schoemaker and Langer, 1986; Hamon *et al.*, 1987).

PHYSICO-CHEMICAL CHARACTERISTICS OF CENTRAL 5-HT RECEPTOR BINDING SITES

Since the same 5-HT binding protein may theoretically account for the heterogeneity of 5-HT receptors by means of coupling to different transduction mechanisms, interaction with various phospholipids in membranes, etc., direct biochemical investigations on the receptor binding protein(s) must also be considered for a possible further proof of actual 5-HT receptor heterogeneity in the CNS. In this section, we review the numerous studies which have revealed differences and similarities in the sensitivity of putative 5-HT receptor binding protein(s) to physical parameters and chemical reagents.

Effects of temperature

Temperature is a critical parameter since it determines the thermodynamics of receptor–ligand interactions, and affects the receptor environment, notably by changing the membrane viscosity. The measurement of ^3H-ligand binding to specific sites at various temperatures can therefore provide interesting information, especially for a comparison between the various 5-HT receptors already identified in brain membranes. Thus, Hall *et al.* (1986) reported that ^{3}H-8-OH-DPAT binding to 5-HT$_{1A}$ sites increases between 0°C and 23°C and then decreases progressively at higher temperatures, down to the non-specific binding value at 55°C because of thermal denaturation of specific sites. In contrast, ^{3}H-5-HT binding to 5-HT$_{1B}$ sites in striatal membranes is maximal at 0°C, and decreases regularly as a function of temperature increase, to finally disappear completely at temperatures higher than 55°C (Hall *et al.*, 1986). Therefore the thermodynamics of receptor–agonist interactions are clearly distinct for 5-HT$_{1A}$ and 5-HT$_{1B}$ sites.

Temperature dependence has also been described for ^{3}H-antagonist binding to 5-HT$_2$ sites (Kendall and Nahorski, 1983) but it differs markedly from that found for 5-HT$_1$ sites. Indeed, temperature-dependent modifications of ^{3}H-5-HT and ^{3}H-8-OH-DPAT binding are always associated with changes in K_d but not in B_{max} (Hall *et al.*, 1986), whereas Kendall and Nahorski (1983) noted an increase in B_{max} with no change in K_d when ^{3}H-spiperone binding to cortical 5-HT$_2$ sites was assayed at 22°C (for 60 min) rather than at 37°C (for 15 min).

Differences between 5-HT$_1$ and 5-HT$_2$ receptor binding sites also concern the heat denaturation process since cortical membranes heated at 45°C for 10 min have a 50% decrease in ^{3}H-5-HT binding capacity, but keep almost completely their capacity to bind ^{3}H-spiperone (Foller *et al.*, 1981).

Effects of membrane viscosity

An increased ^{3}H-5-HT binding capacity of mouse brain membranes treated with cholesteryl hemisuccinate or stearic acid to enhance their viscosity was first reported by Heron *et al.* (1980). Conversely, the same authors observed that membrane fluidization by incorporation of lecithin or linoleic acid causes a small but significant decrease in ^{3}H-5-HT binding. Convergent reports have been published by other groups (Shih and Ohsawa, 1983; Wesemann *et al.*, 1986) who used other chemicals for altering *in vitro* membrane viscosity.

To our knowledge, only one group (Shih and Ohsawa, 1983) examined the possible changes in 5-HT$_2$ binding sites resulting from altered membrane viscosity. The failure of stearic acid incorporation in rat cortical membranes to affect ^{3}H-spiperone binding to 5-HT$_2$ sites contrasts with the enhanced ^{3}H-5-HT binding to 5-HT$_1$ sites in the same membrane preparation (Shih and Ohsawa, 1983).

Effects of pH

Although binding assays are usually performed at the physiological pH, 7.4, this value is not optimum for the specific binding of ^{3}H-5-HT or ^{3}H-8-OH-

DPAT to respective 5-HT$_1$ sites. Maximum ^{3}H-5-HT binding to 5-HT$_{1B}$ sites in striatal membranes occurs around pH 8.0, and that of ^{3}H-8-OH-DPAT to hippocampal 5-HT$_{1A}$ sites is highest between pH 8.5 and 8.75 (Hall *et al.*, 1986). In both cases, the pH-dependent increase is associated with a significant enhancement of the apparent affinity of 5-HT$_1$ sites for their respective ligands.

Similarly, Battaglia *et al.* (1983a) have noted a dramatic increase in the affinity of cortical 5-HT$_2$ binding sites for agonists when ^{3}H-ketanserin binding was assayed at pH 8.2 instead of 7.0. In contrast, antagonists demonstrate little or no affinity changes between these two pH values. Such differential effects of pH probably reflect the critical role played by the charge of amino- or sulfhydryl-groups for the binding of agonists to 5-HT$_2$ receptor.

Effects of ions

On account of the possible charge of important chemical groups at the level of the ligand binding site on the receptor, it is not surprising that ions can affect the ligand–receptor interaction.

The first report on ionic modulation of ^{3}H-5-HT binding to 5-HT$_1$ sites was made by Bennett and Snyder (1976), who described an inhibitory effect of 120 mM NaCl or KCl, but a stimulatory effect of 4 mM CaCl$_2$. Similar findings were then published by several other groups who subsequently showed that Mn^{2+} is even more potent than Ca^{2+} in enhancing the affinity of 5-HT$_{1A}$ and 5-HT$_{1B}$ sites for selective ^{3}H-agonists (see Fig. 1; Hamon *et al.*, 1983; Fujita *et al.*, 1986; Norman *et al.*, 1985). Although the stimulatory effect of divalent cations can be probably ascribed to some interaction with negative charges (of sialic acid, for instance) in the immediate vicinity of the binding site for the ligand, no experimental proof of this mechanism has yet been reported. Apparently such cation-induced effects only concern receptor–agonist interactions since the affinity (calculated from displacement curves) of 5-HT$_1$ sites for antagonists is not shifted in the presence of Ca^{2+} or Mn^{2+} (Hamon *et al.*, 1983; Fujita *et al.*, 1986; Norman *et al.*, 1985).

Similarly Li$^+$, Na$^+$ and K$^+$ have been shown to affect selectively the agonist binding to cortical 5-HT$_2$ sites (Battaglia *et al.*, 1983b). Thus a significant reduction of the apparent affinity of 5-HT$_2$ sites for 5-HT but not for spiperone has been noted when 120 mM NaCl, KCl and particularly LiCl are included in the ^{3}H-ketanserin binding assay mixture. Conversely, the divalent cations Mn^{2+}, Ca^{2+} and Mg^{2+} increase the apparent affinity of 5-HT$_2$ sites for 5-HT and related agonists, but not for antagonists (Battaglia *et al.*, 1984), exactly as previously described in the case of 5-HT$_1$ sites.

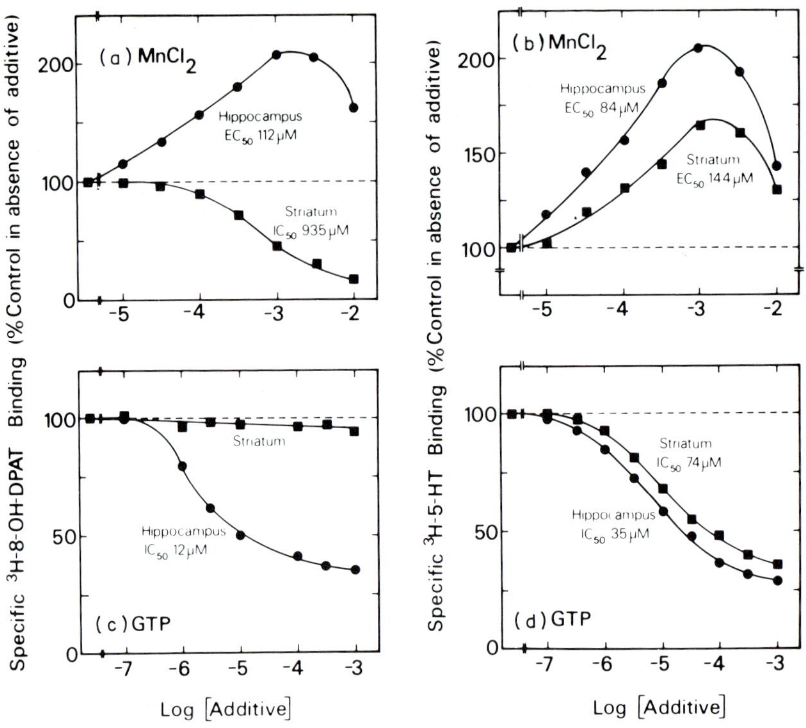

Fig. 1 Effects of MnCl₂ and GTP on specific ³H-8-OH-DPAT (a,c) and ³H-5-HT (b,d) binding to hippocampal and striatal membranes. Hippocampal (circles) and striatal (squares) membranes were incubated with either ³H-8-OH-DPAT (0.96 nM) or ³H-5-HT (1.06 nM) in the absence or presence of increasing concentrations of MnCl₂ (a,b) or GTP (c,d). Specific binding was defined as that displaced by 10 μM 5-HT (see Hall *et al.*, 1985). Data (means of four separate experiments) are expressed as the percentage specific binding compared to that observed in the absence of MnCl₂ and GTP. EC₅₀ and IC₅₀ values correspond to concentrations producing half maximal effects (stimulation by MnCl₂; inhibition by GTP)

Effects of guanine nucleotides

As for all receptors coupled to adenylate cyclase, GTP and its poorly metabolizable analogue GppNHp (Hamon *et al.*, 1982), reduce ³H-agonist binding to 5-HT₁ₐ and 5-HT₁ᵦ sites (Fig. 1). In contrast, the affinity of these sites for antagonists remains unaltered in the presence of guanine nucleotides (Mallat and Hamon, 1982; Hamon *et al.*, 1983; Norman *et al.*, 1985; Hall *et al.*, 1986). Further indirect evidence of the close relationship between GTP regulation of agonist binding and receptor coupling with adenylate cyclase has been obtained by recent investigations on 5-HT$_{1C}$ sites: no GTP modulation of 5-HT binding to these sites could be found by Palacios *et al.* (1986), as expected from their complete independence of adenylate cyclase in membranes (see earlier section on 5-HT$_{1A}$ binding sites).

Similarly, Peroutka *et al.* (1979) mentioned that 5-HT displacement of ^{3}H-spiperone bound to cortical 5-HT$_2$ receptors cannot be affected by GTP and other guanine nucleotides. However, this statement has been challenged by many other groups (Hamon *et al.*, 1980; Rosenfeld and Makman, 1981; Kendall and Nahorski, 1983; Battaglia *et al.*, 1984), who demonstrated a significant reduction in the affinity of 5-HT$_2$ sites for 5-HT and other agonists in the presence of GTP or GppNHp. Although 5-HT$_2$ receptors are not coupled to adenylate cyclase, such a GTP modulatory effect is not surprising as membrane receptors (such as 5-HT$_2$ sites) coupled to phospholipase C are also associated with GTP-binding proteins (Taylor and Merritt, 1986). Therefore similar affinity changes to those occurring in the case of receptors coupled to adenylate cyclase may well be induced by the dissociation of putative 5-HT$_2$ receptor-N protein complexes in the presence of GTP or GppNHp.

Conclusions

Examination of the modulatory effects of physical parameters and chemicals indicates generally striking differences between the respective sensitivity of 5-HT$_{1A}$, 5-HT$_{1B}$, 5-HT$_{1C}$ and 5-HT$_2$ sites, which confirms the heterogeneity previously inferred from pharmacological investigations. In addition, marked differences also concern the effects of these agents on the binding of agonists and antagonists. In most cases, the agonist–receptor interaction is altered by these agents, but not the antagonist–receptor interaction. Such differential modulations are particularly useful for the development of *in vitro* tests aiming at the distinction between 5-HT agonists and antagonists (see Hamon *et al.*, 1983). They also illustrate the fundamental differences between the receptor recognition of agonists and antagonists. In particular, the recognition of a 5-HT agonist, but not that of an antagonist, largely depends on the receptor association with GTP-binding proteins (in the case of 5-HT$_{1A}$, 5-HT$_{1B}$ and 5-HT$_2$ sites).

THE CHEMICAL NATURE OF CENTRAL 5-HT RECEPTORS

The differential sensitivity to various physical parameters and chemicals noted in the previous section suggests important differences in the chemical nature of the various types of central 5-HT receptors. Evidence of such differences is reviewed in the present section.

Amino acid residues possibly involved in ligand binding to active site on the 5-HT receptor protein(s)

As early as 1976, Bennett and Snyder demonstrated that the 5-HT$_1$ receptors are proteins since ^{3}H-5-HT specific binding is markedly reduced in membranes pretreated with trypsin. SH groups (of cysteine residues) apparently play a crucial role for the specific binding of the radioactive ligand since membrane treatment with the SH-modifying agents *N*-ethylmaleimide and 5,5'-dithiobis(2-nitrobenzoic acid) results in the irreversible disappearance of ^{3}H-5-HT (Bennett and Snyder, 1976) and ^{3}H-8-OH-DPAT (Table 2; see also Hall *et al.*, 1986) specific binding capacities. In contrast, ^{3}H-ketanserin binding to cortical 5-HT$_2$ sites is markedly less affected by *N*-ethylmaleimide (unpublished observations).

Table 2 Effects of protein-modifying reagents on ^{3}H-8-OH-DPAT binding to 5-HT$_{1A}$ sites in rat hippocampal membranes

Modifying reagent		Amino acid residue preferentially altered	^{3}H-8-OH-DPAT specifically bound (%)
None (control)		–	100
N-Ethylmaleimide	1 mM	cysteine	12
5,5'-Dithiobis(2-nitrobenzoic acid)	1 mM	cysteine	25
p-Bromo-phenacylbromide	0.1 mM	{ histidine cysteine }	30
2-Hydroxy-5-nitrobenzylbromide	1 mM	{ tryptophan cysteine }	60
N-Bromosuccinimide	10 μM	tryptophan	83
	0.1 mM		38
Diethylpyrocarbonate	0.1 mM	histidine	89
1,2-Cyclohexane dione	1 mM	{ lysine	100
2,3-Butane dione	1 mM	arginine }	100

Hippocampal membranes were treated with each reagent at the indicated concentration, washed, and then used for binding assays with 1 nM ^{3}H-8-OH-DPAT. Specific ^{3}H-8-OH-DPAT binding (mean of three independent determinations) is expressed as a percentage of that found with untreated membranes.

Other protein-modifying agents have been used for the possible identification of amino acid residues functionally important for the ligand binding to specific sites (Table 2). However, neither 1,2-cyclohexane dione nor butane dione altered ^{3}H-8-OH-DPAT binding, suggesting that lysyl and/or arginyl residues do not participate in the ligand recognition process of 5-HT$_{1A}$ receptors. Similarly, histidyl and tryptophan residues are probably not involved in ^{3}H-8-OH-DPAT binding since selective modifying reagents such as diethylpyrocarbonate and N-bromosuccinimide (at 10 μM) respectively exerted only minor effect (Table 2; see also Bennett and Snyder, 1976). In contrast, p-bromo-phenacylbromide, and to a lesser extent 2-hydroxy-5-nitrobenzylbromide, affected ^{3}H-8-OH-DPAT binding to hippocampal 5-HT$_{1A}$ sites, but their effects resulted more probably from the blockade of cysteine-SH groups than from the modification of histidyl or tryptophan residues. A marked reduction in ^{3}H-5-HT binding has been noted by Bennett and Snyder (1976) in cortical membranes treated with the carboxyl group reagent 1-ethyl-3-(3-dimethylaminopropyl)carbodiimide, suggesting that acidic residues (i.e. acidic amino acids, sialic acid, etc.) are critical for 5-HT$_1$ sites. However, it is not known yet whether only one, two or the three 5-HT$_1$ subtypes can be altered by this reagent.

Overall, these data regarding the chemical groups possibly involved in the ligand binding to the receptor proteins are only partial.

Involvement of lipids in the ligand binding to central 5-HT receptors

In the 1960s Woolley and Gommi (1964, 1965, 1966) reported the inactivation of peripheral 5-HT receptors by neuraminidase, which led to the postulate that these receptors are in fact gangliosides. Although this speculation has only a historical value at present, the possible role of gangliosides in the regulation of 5-HT$_1$ receptors must not be neglected, as Berry-Kravis and Dawson (1985) have shown recently that the incorporation of selected gangliosides in cultured NCB-20 hybrid cells results in a tenfold increase in receptor affinity for ^{3}H-5-HT. Nevertheless, Bennett and Snyder (1976) found some increase (+45%) of ^{3}H-5-HT binding to specific 5-HT$_1$ sites following ganglioside alteration in cortical membranes treated by neuraminidase (which removes sialic acid residues).

Membrane treatment with lipid-modifying reagents such as Azure A (which affects preferentially sulphatides and phosphatidylinositol), arylsulphatase (which catalyses the desulphatation of sulfolipids to lipids and inorganic sulphate), and various lipases, has confirmed the critical role of lipids for the binding characteristics of 5-HT$_1$ sites. Azure A (10 μM) and arylsulphatase produce a marked reduction of ^{3}H-5-HT (Yoshikawa and Ishitani, 1985) and ^{3}H-8-OH-DPAT (Gozlan et $al.$, unpublished observations) binding to 5-HT$_1$ sites without affecting ^{3}H-spiperone binding to 5-HT$_2$ sites. Similarly, membrane treatment with phospholipase A$_2$ results in a dramatic fall of ^{3}H-5-HT (Yoshikawa and Ishitani, 1985) and ^{3}H-8-OH-DPAT (Table 3) binding, with only partial reduction of ^{3}H-spiperone binding. As shown in Table 3, ^{3}H-8-OH-DPAT binding is only marginally recovered when the products of phospholipase A$_2$ (lysophosphatides and fatty acids) are removed from membranes by serum albumin, which confirms that the deleterious effects of

Table 3   Effects of phospholipase A_2 on ^{3}H-8-OH-DPAT binding to 5-HT$_{1A}$ sites in hippocampal membranes

Phospholipase A_2 (U/ml)	^{3}H-8-OH-DPAT specifically bound (%)	
	Without serum albumin	With serum albumin
0	100	100
0.005	73	81
0.010	23	38
0.025	13	25
0.050	5	13
0.100	0	0

Hippocampal membranes (1 mg protein/ml) were incubated in 0.05 M Tris-HCl, pH 7.4, at 37°C for 30 min in the presence of various amounts of phospholipase A_2 (from 0.005 Unit/ml to 0.100 Unit/ml) and 1 mM $CaCl_2$. They were then extensively washed with ice-cold 0.05 M Tris-HCl, pH 7.4, with and without 1% bovine serum albumin, and used for the measurement of ^{3}H-8-OH-DPAT (1 nM) binding to 5-HT$_{1A}$ sites. Results (means of three independent experiments) are expressed in percentages of ^{3}H-8-OH-DPAT binding to hippocampal membranes incubated in the absence of phospholipase A_2.

this enzyme are due principally to intrinsic alterations of 5-HT$_{1A}$ (and also 5-HT$_{1B}$) sites produced by the loss of the specified lipid moieties.

In contrast to phospholipase A_2, phospholipases C and D have been reported to be either inactive (Yoshikawa and Ishitani, 1985) or to enhance slightly (+22% with phospholipase D; Bennett and Snyder, 1976) ^{3}H-5-HT binding to 5-HT$_1$ sites in rat cortical membranes.

Another important observation which indirectly confirms the implication of lipids in 5-HT receptor functioning concerns the effects of ascorbate. Indeed, Muakkassah-Kelly *et al.* (1982, 1983) have shown that lipid peroxidation in cortex membranes incubated with ascorbate is significantly correlated with a loss of 5-HT$_1$ and 5-HT$_2$ binding sites. However, not only lipids but also proteins and membrane fluidity are deeply altered following the peroxidation of unsaturated fatty acids in the presence of ascorbic acid (Muakkassah-Kelly *et al.*, 1982), and further investigations are necessary before concluding which of these alterations really accounts for the inactivation of 5-HT receptor binding sites.

Although much has yet to be done to understand the relevance of lipid–protein interactions for the integrity of 5-HT receptors in brain membranes, the data presently reported clearly indicate that lipids play a major role in the binding characteristics of these receptors. This observation must be taken into account, particularly for the purification of central 5-HT receptors.

THE PURIFICATION OF CENTRAL 5-HT RECEPTORS

Two series of approaches have been attempted for the purification of central 5-HT receptors: (1) the solubilization of the native receptors, followed by various chromatographic procedures, (2) the irreversible labelling of the receptors in membranes, and their subsequent solubilization and extraction.

The solubilization of central 5-HT receptor binding sites

Successful solubilization of 5-HT$_{1A}$ (Gozlan *et al.*, 1987), 5-HT$_{1B}$ (Allgren *et al.*, 1985), 5-HT$_{1C}$ (Yagaloff and Hartig, 1986) and 5-HT$_2$ (Ilien *et al.*, 1980, 1982; Schotte *et al.*, 1984; Wouters *et al.*, 1985a) receptor binding sites has been achieved using various detergents. Thus, digitonin (1%) and sodium cholate (1%) allowed Rousselle *et al.*, (1985) to solubilize ^{3}H-5-HT binding sites from the rat brain. Van den Berg *et al.* (1983) used a mixture of Triton X-100, Tween 80 and octyl-beta-glucopyranoside (see also Allgren *et al.*, 1985), but high-affinity ^{3}H-5-HT binding could be detected in the solubilized material only after extensive removal of the detergents on Biobeads SM-2, since Triton X-100 exerts a direct inhibitory effect on solubilized 5-HT$_1$ sites (see Bennett and Snyder, 1976; El Mestikawy *et al.*, unpublished observations). Under such conditions, the characteristics of ^{3}H-5-HT binding to the solubilized fraction suggest that 5-HT$_{1B}$ sites are preferentially solubilized, but also considerably altered. In particular, Allgren *et al.* (1985) noted that solubilized 5-HT$_{1B}$ sites bound D- and L-LSD with comparable high affinities, in contrast to functional 5-HT$_1$ receptors in membranes which recognize only the D isomer.

In the case of 5-HT$_{1A}$ receptors, we found that a mixture of the zwitterionic detergent 3-(3-cholaminopropyl)dimethyl-ammonio-l-propane sulfonate (CHAPS, 10 mM) and NaCl (0.2 M) allows the solubilization of half the membrane-bound sites in the rat hippocampus (Gozlan *et al.*, 1987). The solubilized 5-HT$_{1A}$ receptor exhibits the same high affinity for ^{3}H-8-OH-DPAT and 5-HT as its membrane-bound form (Fig. 2). Furthermore, ^{3}H-8-OH-DPAT binding to soluble sites is still sensitive to GTP (Gozlan *et al.*, 1987), which indicates that the solubilization procedure does not dissociate the receptor-N complex. However, some changes in the ligand binding characteristics do occur upon solubilization since for instance, the pH-dependent increase of ^{3}H-8-OH-DPAT binding to 5-HT$_{1A}$ sites in membranes is markedley less pronounced with soluble receptors (Table 4). Whether these pH effects involve the membrane receptor environment (phospholipids) more than the receptor itself *sensu stricto* is an interesting possibility, which deserves further investigations.

The 5-HT$_{1C}$ site has also been successfully solubilized from pig choroid plexus using 10 mM CHAPS (Yagaloff and Hartig, 1986) with 45% recovery. The pharmacological characteristics of the solubilized 5-HT$_{1C}$ sites are generally identical to those of the membrane-bound receptors, except for 5-HT which shows a 20-fold increase in affinity for the solubilized receptors.

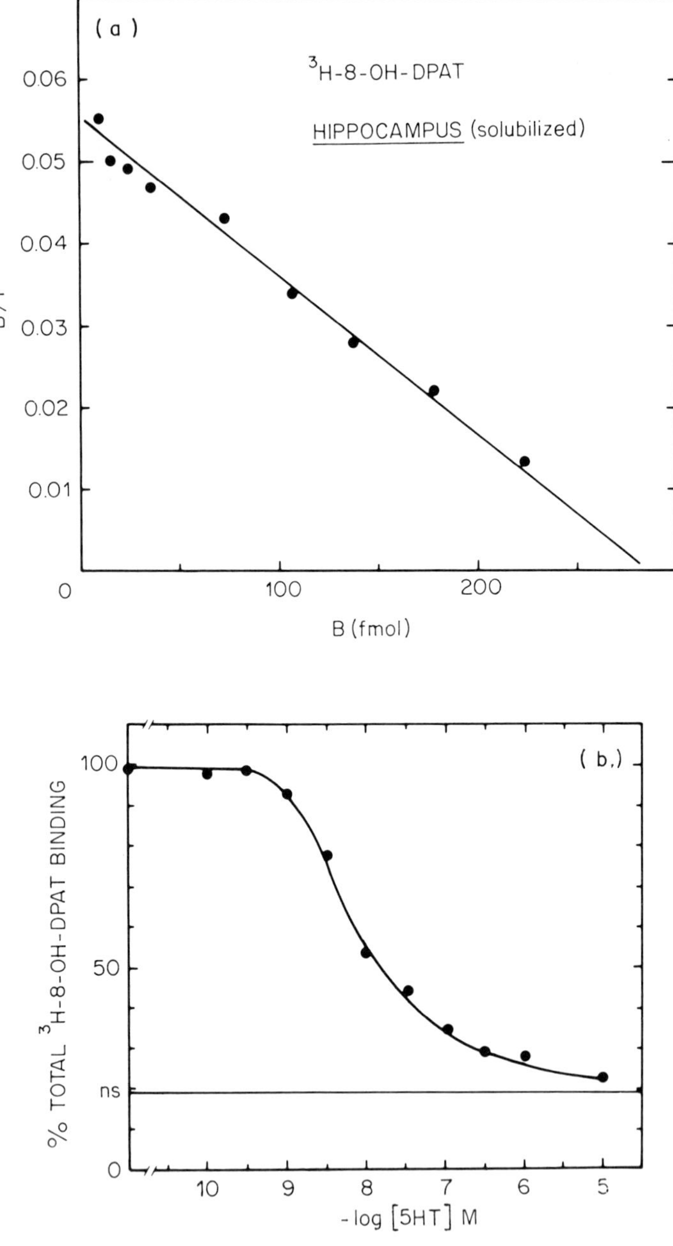

Fig. 2 Characteristics of ^{3}H-8-OH-DPAT binding to solubilized hippocampal 5-HT$_{1A}$ sites.
(a) Scatchard plot. Binding assays were carried out with 0.2–7.0 nM ^{3}H-8-OH-DPAT as described by Gozlan *et al.* (1987). *B* is

Table 4 pH-dependence of ^{3}H-8-OH-DPAT binding to 5-HT$_{1A}$ sites in rat hippocampal membranes and in the corresponding solubilized fraction

	^{3}H-8-OH-DPAT specifically bound (pmol/g)			
pH	Membranes (%)		Soluble fraction (%)	
6.9	6.18	(100)	3.57	(100)
7.2	7.73	(125)	3.72	(104)
7.5	10.39	(168)	3.84	(108)
7.9	11.25	(182)	3.90	(109)
8.3	14.86	(240)	4.60	(129)
8.6	15.49	(251)	5.38	(159)

Solubilization of hippocampal 5-HT$_{1A}$ sites was achieved using 10 mM CHAPS and 0.2 M NaCl (Gozlan *et al.*, 1987). Binding assays were carried out using 1 nM ^{3}H-8-OH-DPAT as described by Gozlan *et al.* (1987), except that the final pH in the assay mixture varied between 6.9 and 8.6. Each value (mean of triplicate determinations) corresponds to pmol ^{3}H-8-OH-DPAT specifically bound per g of starting fresh tissue. Figures in parentheses are the percentages compared to binding values measured at pH 6.9.

According to Yagaloff and Hartig (1986), this shift suggests the loss of a receptor modulatory component by treatment with the detergent.

Originally, the solubilization of cortical 5-HT$_2$ sites was achieved using 0.25% lysolecithin (Ilien *et al.*, 1980, 1982) or 1% digitonin (Chan and Madras, 1982). However, much higher recovery (40%) of 5-HT$_2$ sites in soluble form could be obtained by a mixture of 6.8 mM CHAPS and 1.4 M NaCl (Wouters *et al.*, 1985a). The solubilized 5-HT$_2$ receptor has a high affinity for the selective ligand ^{3}H-ketanserin (see Leysen *et al.*, 1982), and exhibits a pharmacological profile identical to that of membrane bound 5-HT$_2$ sites: in particular, ^{3}H-ketanserin binding can be inhibited by nanomolar concentrations of most 5-HT antagonists and by micromolar concentrations of 5-HT agonists (Wouters *et al.*, 1985a).

---

expressed in fmol ^{3}H-8-OH-DPAT bound/0.1 ml of solubilized membranes. *B/F* represents the ratio of ^{3}H-8-OH-DPAT specifically bound to that free in the assay medium. Each point is the mean of triplicate determinations.

(b) Displacement by 5-HT of ^{3}H-8-OH-DPAT bound to solubilized 5-HT$_{1A}$ sites. Binding assays were carried out with 1 nM ^{3}H-8-OH-DPAT. Specific binding in the absence of 5-HT was equal to 44 fmol/0.1 ml of solubilized membranes (100%). Each point is the mean of triplicate determinations

Physico-chemical properties of solubilized 5-HT receptor binding sites

The 5-HT- (probably 5-HT$_{1B}$) binding site solubilized by Van den Berg *et al.* (1983) could be filtered through a Sephacryl S300 column without losing its high affinity for ^{3}H-5-HT, which allowed an estimate of its molecular weight (MW) according to gel filtration standard procedure. The value obtained, 98 kD, is about twice as high as that inferred from sedimentation velocity through glycerol gradient (58 kD; Van den Berg *et al.*, 1983). The latter value agrees remarkably well with that calculated from the inactivation of 5-HT$_{1B}$ sites in frozen tissues irradiated by high-energy electrons (Gozlan *et al.*, 1986): ~60 kD.

Much higher MWs have been reported by Rousselle *et al.* (1985) who used also the gel filtration procedure (through ACA 22 Ultrogel) for their estimation. According to these authors, two major peaks of proteins corresponding to 235 kD and 438 kD could be separated from the main bulk of solubilized proteins owing to their ^{3}H-5-HT binding capacity. The discrepancy between Rousselle's findings and those of Van den Berg *et al.* (1983) further illustrates that the size of neurotransmitter receptor-containing micelles in solubilized fractions is highly dependent on the detergent used. Indeed, Rousselle *et al.* (1985) used sodium cholate whereas Van den Berg *et al.*(1983) used Triton X-100 and Tween 80.

In the case of 5-HT$_{1C}$ binding sites, Yagaloff and Hartig (1986) reported a MW of 850 kD on account of the sedimentation characteristics of solubilized receptors through a Ficoll gradient. However, aggregation phenomena and the presence of tightly bound detergent and lipids in the solubilized complex very probably contributed to this large apparent molecular size.

Similarly, the sedimentation characteristics of solubilized 5-HT$_2$ binding site through sucrose gradient largely depend on its association with the detergents and lipids since membrane solubilization by lysophosphatidyl-choline gave a Svedberg coefficient of 11.5 S, whereas CHAPS led to a much lower value of 5 S (see Wouters *et al.*, 1985a). Interestingly, Wouters *et al.* (1985a) reported that the latter value may correspond to the 5-HT$_2$ receptor with little or no phospholipid attached to it. Since the pharmacological properties of '5 S' soluble 5-HT$_2$ receptors are remarkably similar to those of the membrane-bound receptors (Wouters *et al.*, 1985a), it can be inferred that lipids play probably only a minor role, if any, in the pharmacological characteristics of this particular class of central 5-HT receptors. Indeed, as already mentioned, lipid-modifying reagents and enzymes do not significantly alter the ligand binding characteristics of 5-HT$_2$ sites in rat brain membranes (Yoshikawa and Ishitani, 1985).

Other estimates of the MW of 5-HT$_2$ receptor have been made using the radiation inactivation technique, but no agreement has been reached, since MWs of 55–60 kD (Gozlan *et al.*, 1986), 145–152 kD (Nishino and Tanaka,

1985) and 200–210 kD (Brann, 1985) have been reported so far. Different technical conditions probably explain these discrepancies, which further emphasize the limits of the radiation inactivation technique for MW determination of membrane receptors.

In addition to MW estimates, a few studies have been conducted on the physico-chemical properties of solubilized binding sites. It has been noted for instance that thermal denaturation is much faster for the solubilized 5-HT$_{1A}$ (El Mestikawy *et al.*, unpublished observations) and 5-HT$_{1C}$ (Yagaloff and Hartig, 1986) binding sites than for the corresponding receptors inside membranes. In addition, Chan and Madras (1983) reported an isoelectric point (pI) of 5.03 for the solubilized 5-HT$_2$ receptor from frontal cortex membranes.

The labelling of central 5-HT receptor proteins by irreversible ligands

Two types of ligands have been developed for the irreversible labelling of 5-HT receptors: alkylating agents and photosensitive probes.

Alkylating agents

Emerit *et al.* (1985) synthesized a chloroamine derivative of 8-OH-DPAT, 8-methoxy-2′-chloro-PAT, which is rapidly converted into a highly reactive aziridinium ion in solution. At 1 μM this compound can block up to 60% of 5-HT$_{1A}$ sites in hippocampal membranes. This effect really originates in an alkylating reaction following prior occupancy of 5-HT$_{1A}$ receptors by the drug, since preincubation of hippocampal membranes with an excess (10 μM) of 5-HT or other 5-HT$_{1A}$ ligands protects 5-HT$_{1A}$ sites from subsequent irreversible blockade by 8-methoxy-2′-chloro-PAT (Emerit *et al.*, 1985).

Unfortunately, the apparent affinity of 8-methoxy-2′-chloro-PAT for 5-HT$_{1A}$ sites is too low (IC$_{50}$ = 80 nM) for the possible use of a radioactive derivative as a selective probe of these sites *in vitro*. Nevertheless, its high reactivity allowed us to use 8-methoxy-2′-chloro-PAT as an irreversible blocker of 5-HT$_{1A}$ receptors *in vivo*. As shown in Table 5, a long-lasting reduction of ^{3}H-8-OH-DPAT binding capacity was found in the ventral hippocampus after the local microinjection of a low dose of 8-methoxy-2′-chloro-PAT. Calculations (see Nelson *et al.*, 1979) based on the time course recovery of ^{3}H-8-OH-DPAT binding (Table 5) allow an estimate of 2.30 days for the half-life of hippocampal 5-HT$_{1A}$ sites in the adult rat. This value corresponds to a synthesis rate of ~120 fmol of 5-HT$_{1A}$ receptor binding sites per mg membrane protein per day, i.e. ~720 fmol per day for both hippocampi. Using this approach, it should be possible to measure the turn-over rate of 5-HT$_{1A}$ receptors not only in the ventral hippocampus but in any brain region after various pharmacological treatments (see Hamon *et al.*,

Table 5   Recovery of 5-HT$_{1A}$ sites in the ventral hippocampus following the local microinjection of 8-methoxy-2′-chloro-PAT

Treatment	^{3}H-8-OH-DPAT specifically bound (fmol/mg prot.)
Control	170.8 ± 13.0
8-methoxy-2′-chloro-PAT (2.25 nmol)	
1 day	88.1* ± 13.1
2 days	87.4* ± 5.1
4 days	119.7* ± 10.3
6 days	145.4 ± 14.3
9 days	158.6 ± 15.7

8-Methoxy-2′-chloro-PAT (2.25 nmol/2 µl) was injected bilaterally into the ventral hippocampus of adult male rats, and animals were sacrificed 1–9 days after the treatment. ^{3}H-8-OH-DPAT binding to 5-HT$_{1A}$ sites was measured using 1.0 nM of the labelled ligand. Each value is the mean ±SEM of 4 independent determinations.
*$p < 0.05$ when compared to ^{3}H-8-OH-DPAT binding to ventral hippocampal membranes from untreated rats.

1984), during ontogenesis (Daval *et al.*, 1987), etc., i.e. under any conditions associated with alterations in the density of these receptors in rats.

Similarly, Battaglia *et al.* (1986) have demonstrated that *N*-ethoxy-carbonyl-2-ethoxy-1,2-dihydroquinoline (EEDQ) can be used as an irreversible blocker of 5-HT$_2$ sites in the rat cortex *in vitro* and *in vivo*. Calculations based on the time course recovery of 5-HT$_2$ binding sites after *in vivo* EEDQ administration as reported by Battaglia *et al.* (1986), allowed us to estimate a half-life of 3.5 days, apparently longer than for 5-HT$_{1A}$ binding sites (see above). As in the case of 8-methoxy-2′-chloro-PAT for the latter sites, the use of EEDQ would be particularly relevant for measuring the rate of 5-HT$_2$ receptor turnover following treatments known to affect their density (for instance, the chronic administration of anti-depressant drugs; see Hamon *et al.*, 1984).

Photosensitive ligands

Cheng and Shih (1979) were the first to develop a photoaffinity ligand for the irreversible labelling of 5-HT receptor binding sites. Unfortunately, their compound ^{3}H-nitroaryl-azidophenyl-serotonin (^{3}H-NAP-5-HT) only exhibited a µM affinity for 5-HT$_1$ sites, and a high degree of non-specific binding. Nevertheless, these authors reported that three protein bands corresponding to MWs of 80 kD, 49 kD and 38 kD could be specifically labelled by ^{3}H-NAP-5-HT in membranes from the whole rat brain. However, no inference can be made about their possible identity with 5-HT receptors, as the pharmacological characteristics of ^{3}H-incorporation into these proteins have not been investigated (Cheng and Shih, 1979).

In contrast, Ransom *et al.* (1986) and Emerit *et al.* (1986, 1987) developed new photosensitive radioligands, ^{3}H-*p*-azido-PAPP and ^{3}H-8-methoxy-3'-NAP-amino-PAT respectively, with much higher affinity for 5-HT$_{1A}$ sites in rat brain membranes. Both groups demonstrated that ^{3}H irreversible labelling could be prevented by 5-HT$_{1A}$ agonists (such as 8-OH-DPAT; Fig. 3) and antagonists but neither 5-HT$_{1B}$ nor 5-HT$_2$ related drugs, which confirms the selective occupancy of 5-HT$_{1A}$ sites by ^{3}H-photosensitive ligands. SDS-polyacrylamide gel electrophoresis of ^{3}H-labelled material revealed a major peak corresponding to 55 kD (Ransom *et al.*, 1986) or 63 kD (Emerit *et al.*, 1987). Interestingly, these MWs are very close to that previously reported for the 5-HT$_{1A}$ receptor binding site: ~60 kD by Gozlan *et al.* (1986), who used the radiation inactivation procedure. Further confirmation of the identity of the ^{3}H-labelled ~60 kD protein and the 5-HT$_{1A}$ receptor binding subunit has been reported by Emerit *et al.* (1987), who noted that this protein has the same regional and subcellular distributions as 5-HT$_{1A}$ receptors in the rat brain.

In addition to these newly synthesized photoaffinity probes, ^{3}H-5-HT itself has been successfully used by Rousselle *et al.* (1985) for the irreversible labelling of specific sites in brain membranes. Following UV irradiation in the presence of 5 nM ^{3}H-5-HT, brain membranes were solubilized and ^{3}H-labelled proteins separated by SDS-polyacrylamide gel electrophoresis. Thus, a major radioactive band also corresponding to ~60 kD was identified by Rousselle *et al.* (1985). Since ^{3}H-5-HT labelled both 5-HT$_{1A}$ and 5-HT$_{1B}$ sites under the assay conditions used by Rousselle *et al.* (1985), these results would suggest that both binding sites correspond to protein(s) of ~60 kD. In agreement with this inference, Gozlan *et al.* (1986) reported that both 5-HT$_{1A}$ and 5-HT$_{1B}$ sites have the same NW, close to 60 kD, as calculated from their radiation-induced inactivation in frozen brain tissues. Therefore, it cannot be settled at present whether the pharmacologically distinct 5-HT$_{1A}$ and 5-HT$_{1B}$ receptors correspond to the same protein-binding subunit in different membrane environments (involving notably phospholipids), or to different ~60 kD proteins.

Photoaffinity ligands derived from the selective 5-HT$_2$ antagonist ketanserin have been recently synthesized by Wouters *et al.* (1985b, c). Among these molecules, 7-azidoketanserin was shown to block irreversibly 5-HT$_2$ receptor binding sites in membranes from the rat prefrontal cortex (Wouters *et al.*, 1985c). The development of a radioactive derivative of 7 azidoketanserin would be particularly useful for the identification of the 5-HT$_2$ receptor binding subunit. In particular, it should allow a more accurate determination of its MW than the radiation inactivation technique.

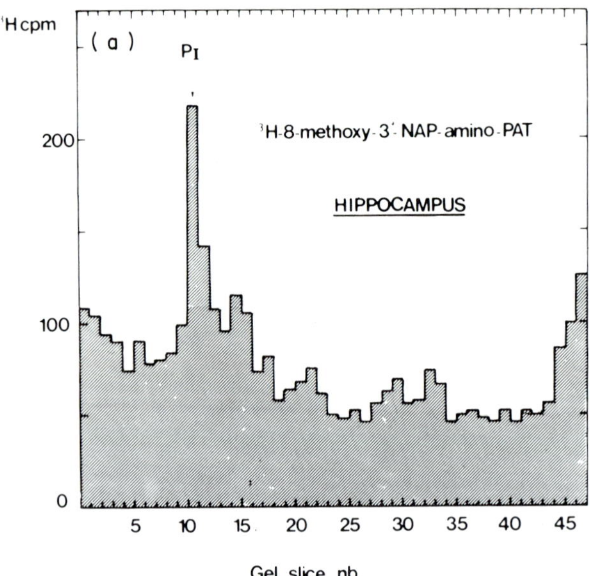

³Hcpm
(a)
P_I
³H-8-methoxy-3'-NAP-amino-PAT
HIPPOCAMPUS
200
100
0
5   10   15   20   25   30   35   40   45
Gel slice nb

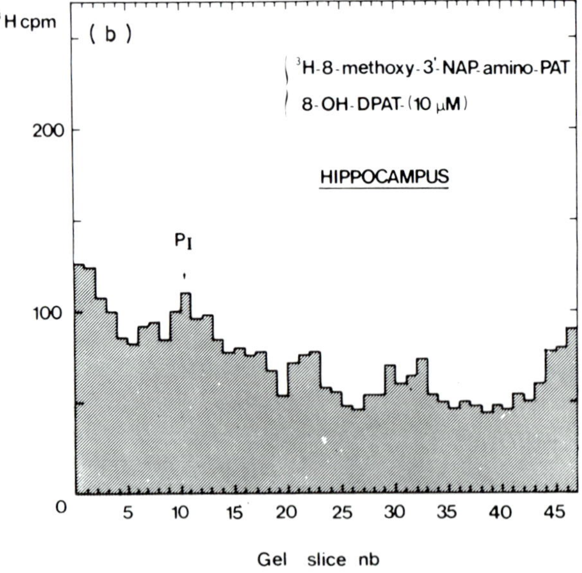

³Hcpm
(b)
³H-8-methoxy-3'-NAP-amino-PAT
8-OH-DPAT-(10 µM)
HIPPOCAMPUS
200
P_I
100
0
5   10   15   20   25   30   35   40   45
Gel slice nb

CONCLUSIONS AND FUTURE TRENDS

Although considerable progress has been made in the pharmacological characterization of central 5-HT receptors in mammals, including humans, numerous important questions concerning their biochemical properties are still unanswered at present. Thus, the mechanism(s) transducing 5-HT messages at the cellular level are known for $5-HT_{1A}$ sites (coupled to adenylate cyclase) and $5-HT_{1C}$ and $5-HT_2$ sites (coupled to phospholipase C), but not for $5-HT_{1B}$ and $5-HT_3$ receptors (mainly at the periphery). The binding proteins themselves are only partially characterized. The best known is probably the $5-HT_{1A}$ receptor binding subunit which has a MW of ~60 kD, and possesses SH group(s) of critical importance for the recognition of selective $5-HT_{1A}$ ligands. However, the pharmacological properties of the $5-HT_{1A}$ receptor depend not only on the intrinsic characteristics of the protein binding subunit but also on its lipidic environment, as membrane treatment with azure A and phospholipase A_2 results in marked alterations of specific ^{3}H-8-OH-DPAT binding. Therefore, in addition to further investigations on the physico-chemical properties of receptor proteins, there is an obvious need for detailed studies on the possible relevance of protein–lipid interactions to the specific pharmacological profile of a given 5-HT receptor.

To be performed, such studies imply notably the purification of 5-HT receptor protein(s) and the reconstitution of active receptors in artificial lipid membranes. More than ten years ago, Shih *et al.* (1974) developed an affinity column for the purification of 5-HT binding proteins from the rat brain. Elution by LSD of the material bound to the column (5-HT-Sepharose) gave only one protein band as revealed by subsequent gel electrophoresis. However, this protein has not been further characterized and its possible correspondence with 5-HT receptors is therefore quite speculative. Nevertheless, such data illustrate that affinity chromatography with selective ligands bound to various supports is undoubtedly one of the major strategies for the purification of the various types of central 5-HT receptors.

From the purified receptor(s), it will then be possible to develop monoclonal and/or polyclonal antibodies, notably for the ultrastructural localization of 5-HT receptors. Alternatively, 5-HT receptor antibodies can be theoretically produced via the anti-idiotype paradigm. Thus, antibodies

---

Fig. 3 SDS-polyacrylamide gel electrophoretic pattern of the irreversible labelling of hippocampal membranes by [^{3}H]8-methoxy-3'-NAP-amino-PAT. Membranes were incubated and UV irradiated with 20 nM [^{3}H]8-methoxy-3'-NAP-amino-PAT in the absence (a) or in the presence (b) of 10 μM 8-OH-DPAT (see Emerit *et al.*, 1987). Labelled membranes were then processed for SDS-polyacrylamide gel electrophoresis, and gels divided into 2 mm slices which were counted by liquid scintillation spectrometry. The $5-HT_{1A}$ binding subunit corresponds to P_1 (MW = 63 kD)

against a selective 5-HT ligand can be used as antigens for the production
of anti-idiotypic antibodies possibly recognizing the 5-HT binding site(s) on
the receptor(s). Apparently, this approach should be successful, since Todd
and Ciaranello (1985) discovered natural antibodies which bound to 5-HT_{1A}
receptors in an autistic child. Using these antibodies, these authors already
established that 5-HT_{1A}, 5-HT_{1B} and 5-HT_2 receptor binding sites are immu-
nologically distinct (Todd and Ciaranello, 1985).

Finally, in light of the recent spectacular progress in the knowledge of
other neurotransmitter receptors owing to molecular genetics technology
(Kubo *et al.*, 1986; Dixon *et al.*, 1986), it can be reasonably predicted that
the most complete information on the biochemical characteristics of 5-HT
receptors will be provided by this type of approach. Two groups (Gundersen
et al., 1983; Sakai *et al.*, 1986) have already shown that the synthesis of
functional 5-HT receptors can be induced in *Xenopus* oocytes by the injection
of total mRNA extracted from rat brain. These preliminary data suggest that
it should be possible to purify the specific mRNA(s) of 5-HT receptors using
Xenopus oocyte as a translation system. The sequencing of corresponding
cDNA will then give the final answer to the possible molecular heterogeneity
of central 5-HT receptors.

REFERENCES

Allgren, R. L., Kyncl, M. M. A., and Ciaranello, R. D. (1985) Pharmacological
    characterization of solubilized 5-HT_1 serotonin binding sites from bovine brain,
    Brain Research, **348**, 77–85.
Ananth, U. S., Leli, U., and Hauser, G. (1987) Stimulation of phosphoinositide
    hydrolysis by serotonin in C_6 glioma cells, *J. Neurochem.*, **48**, 253–261.
Arvidsson, L.-E., Hacksell, U., Johansson, A., Nilsson, J. L. G., Lindberg, P.,
    Sanchez, D., Wikström, H., Svensson, K., Hjorth, S., and Carlsson, A. (1984) 8-
    Hydroxy-2-alkylamino-tetralins and related compounds as central 5-hydroxytrypta-
    mine receptor agonists, *J. Med. Chem.*, **27**, 45–51.
Arvidsson, L.-E., Hacksell, U., Nilsson, J. L. G., Hjorth, S., Carlsson, A., Sanchez,
    D., Lindberg, P., and Wikström, H. (1981) 8-Hydroxy-2-(di-*n*-propylamino)te-
    tralin, a new centrally acting 5-HT receptor agonist, *J. Med. Chem.*, **24**, 921–923.
Barbaccia, M. L., Brunello, N., Chuang, D. M., and Costa, E. (1983) Serotonin
    elicited amplification of adenylate cyclase activity in hippocampal membranes from
    adult rat, *J. Neurochem*, **40**, 1671–1679.
Battaglia, G., Norman, A. B., Newton, P. L., and Creese, I. (1986) *In vitro* and *in
    vivo* irreversible blockade of cortical S_2 serotonin receptors by *N*-ethoxycarbonyl-
    2-ethoxy-1,2-dihydroquinoline: a technique for investigating S_2 serotonin receptor
    recovery, *J. Neurochem.*, **46**, 589–593.
Battaglia, G., Shannon, M., Borgundvaag, B., and Titeler, M. (1983a) pH-dependent
    modulation of agonist interactions with [^{3}H]ketanserin-labelled S_2 serotonin recep-
    tors. *Life Sci.*, **33**, 2011–2016.
Battaglia, G., Shannon, M., and Titeler, M. (1983b) Modulation of brain S_2 serotonin
    receptors by lithium, sodium and potassium chloride, *Life Sci.*, **32**, 2597–2601.

Battaglia, G., Shannon, M., and Titeler, M. (1984) Guanyl nucleotide and divalent cation regulation of cortical S_2 serotonin receptors, *J. Neurochem.*, **45**, 1213–1219.

Bennett, J. P. Jr, and Snyder, S. H. (1976) Serotonin and lysergic acid diethylamide binding in rat brain membranes: relationship to postsynaptic serotonin receptors, *Mol. Pharmacol.*, **12**, 373–389.

Berridge, M. J., and Heslop, J. P. (1981) Separate 5-hydroxytryptamine receptors on the salivary gland of the blowfly are linked to the generation of either cyclic adenosine 3′,5′-monophosphate or calcium signals, *Brit. J. Pharmacol.*, **73**, 729–738.

Berry-Kravis, E., and Dawson, G. (1985) Possible role of gangliosides in regulating an adenylate cyclase-linked 5-hydroxytryptamine (5-HT_1) receptor, *J. Neurochem.*, **45**, 1739–1747.

Brann, M. R. (1985) Serotonin-S_2 and dopamine-D_2 receptors are the same size in membranes, *Biochem. Biophys. Res. Commun.*, **133**, 1181–1186.

Chan, B., and Madras, B. K. (1982) [^{3}H]mianserin binding to solubilized membranes of frontal cortex, *Eur. J. Pharmacol.*, **83**, 1–10.

Chan, B., and Madras, B. K. (1983) Partial purification of [^{3}H]mianserin binding sites, *Eur. J. Pharmacol.*, **87**, 357–365.

Cheng, S. H., and Shih, J. C. (1979) Photoaffinity labeling of serotonin binding proteins, *Life Sci.*, **25**, 2197–2203.

Conn, P. J., and Sanders-Bush, E. (1985) Serotonin-stimulated phosphoinositide turnover: mediation by the S_2 binding site in rat cerebral cortex but not in subcortical regions, *J. Pharmacol. Exp. Ther.*, **234**, 195–203.

Conn, P. J., Sanders-Bush, E., Hoffman, B. J., and Hartig, P. R. (1986) A unique serotonin receptor in choroid plexus is linked to phosphatidylinositol turnover, *Proc. Natl. Acad. Sci. USA*, **83**, 4086–4088.

Cory, R. N., Berta, P., Haiech, J., and Bockaert, J. (1986) 5-HT_2 receptor-stimulated inositol phosphate formation in rat aortic myocytes, *Eur. J. Pharmacol.*, **131**, 153–157.

Daval, G., Vergé, D., Becerril, A., Gozlan, H., Spampinato, U., and Hamon, M. (1987) Transient expression of 5-HT_{1A} receptor binding sites in some areas of the rat CNS during postnatal development, *Int. J. Develop. Neurosci.* **5**, 171–180.

De Chaffoy de Courcelles, D., Leysen, J. E., De Clerck, F., Van Belle, H., and Janssen, P. A. J. (1985) Evidence that phospholipid turnover is the signal transducing system coupled to serotonin-S_2 receptor sites, *J. Biol. Chem.*, **260**, 7603–7608.

Deterre, P., Paupardin-Tritsch, D., Bockaert, J., and Gerschenfeld, H. M. (1982) cAMP-mediated decrease in K^+ conductance evoked by serotonin and dopamine in the same neuron: a biochemical and physiological single-cell study, *Proc. Natl. Acad. Sci. USA*, **79**, 7934–7938.

De Vivo, M., and Maayani, S. (1986) Characterization of the 5-hydroxytryptamine$_{1A}$ receptor-mediated inhibition of forskolin-stimulated adenylate cyclase activity in guinea pig and rat hippocampal membranes, *J. Pharmacol. Exp. Ther.*, **238**, 248–253.

Dixon, R. A. F., Kobilka, B. K., Strader, D. J., Benovic, J. L., Dohlman, H. G., Frielle, T., Bolanowski, M. A., Bennett, C. D., Rands, E., Diehl, R. E., Mumford, R. A., Slater, E. E., Sigal, I. S., Caron, M. G., Lefkowitz, R. J., and Strader, C. D. (1986) Cloning of the gene and cDNA for mammalian β-adrenergic receptor and homology with rhodopsin, *Nature (Lond.)*, **321**, 75–79.

Dolphin, A. C., and Greengard, P. (1981a) Serotonin stimulates phosphorylation of protein I in the facial motor nucleus of rat brain, *Nature (Lond.)*, **289**, 76–79.

Dolphin, A. C., and Greengard, P. (1981b) Neurotransmitter- and neuromodulator-dependent alterations in phosphorylation of protein I in slices of rat facial nucleus, *J. Neurosci.*, **1**, 192–203.

Doyle, V. M., Creba, J. A., Rüegg, U. T., and Hoyer, D. (1986) Serotonin increases the production of inositol phosphates and mobilizes calcium via the 5-HT$_2$ receptor in A$_7$r$_5$ smooth muscle cells, *Naunyn-Schmiedeberg's Arch. Pharmacol.*, **333**, 98–103.

Emerit, M. B., El Mestikawy, S., Gozlan, H., Cossery, J. M., Besselièvre, R., Marquet, A., and Hamon, M. (1987) Identification of the 5-HT$_{1A}$ receptor binding subunit in rat brain membranes using the photoaffinity probe [^{3}H]8-methoxy-3'-NAP-amino-PAT, *J. Neurochem.* **49**, 373–380.

Emerit, M. B., Gozlan, H., Hall, M. D., Hamon, M., and Marquet, A. (1985) Irreversible blockade of central 5-HT binding sites by 8-methoxy-2'-chloro-PAT, *Biochem. Pharmacol.*, **34**, 883–892.

Emerit, M. B., Gozlan, H., Marquet, A., and Hamon, M. (1986) Irreversible blockade of central 5-HT$_{1A}$ receptor binding sites by the photoaffinity probe 8-methoxy-3'-NAP-amino-PAT, *Eur. J. Pharmacol.*, **127**, 67–81.

Engel, G., Göthert, M., Hoyer, D., Schlicker, E., and Hillenbrand, K. (1986) Identity of inhibitory presynaptic 5-hydroxytryptamine (5-HT) autoreceptors in the rat brain cortex with 5-HT$_{1B}$ binding sites, *Naunyn-Schmiedeberg's Arch. Pharmacol.*, **332**, 1–7.

Enjalbert, A., Bourgoin, S., Hamon, M., Adrien, J., and Bockaert, J. (1978a) Postsynaptic serotonin-sensitive adenylate cyclase in the central nervous system. I. Development and distribution of serotonin- and dopamine-sensitive adenylate cyclases in rat and guinea pig brain, *Mol. Pharmacol.*, **14**, 2–10.

Enjalbert, A., Hamon, M., Bourgoin, S., and Bockaert, J. (1978b) Postsynaptic serotonin-sensitive adenylate cyclase in the central nervous system. II. Comparison with dopamine- and isoproterenol-sensitive adenylate cyclases in rat brain, *Mol. Pharmacol.*, **14**, 11–23.

Feuerstein, T. J., and Hertting, G. (1986) Serotonin (5-HT) enhances hippocampal noradrenaline (NA) release: evidence for facilitatory 5-HT receptors within the CNS, *Naunyn-Schmiedeberg's Arch. Pharmacol.*, **333**, 191–197.

Fujita, M., Seo, T., Nishio, H., and Segawa, T. (1986) Effects of bicarbonate ion on serotonin binding to rat frontal cortex membranes, *Neurochem. Int.*, **8**, 235–242.

Gaddum, J. H., and Picarelli, Z. P. (1957) Two kinds of tryptamine receptor, *Brit. J. Pharmacol. Chemother.*, **12**, 323–328.

Gillet, G., Ammor, S., and Fillion, G. (1985) Serotonin inhibits acetylcholine release from rat striatum slices: evidence for a presynaptic receptor-mediated effect, *J. Neurochem.*, **45**, 1687–1691.

Gozlan, H., El Mestikawy, S., Pichat, L., Glowinski, J., and Hamon, M. (1983) Identification of presynaptic serotonin autoreceptors using a new ligand: ^{3}H-PAT, *Nature (Lond.)*, **305**, 140–142.

Gozlan, H., Emerit, M. B., El Mestikawy, S., Cossery, J. M., Marquet, A., Besselièvre, R., and Hamon, M. (1987) Photoaffinity labelling and solubilization of the central 5-HT$_{1A}$ receptor binding site, *J. Recept. Res.* **7**, 195–221.

Gozlan, H., Emerit, M. B., Hall, M. D., Nielsen, M., and Hamon, M. (1986) *In situ* molecular sizes of the various types of 5-HT binding sites in the rat brain, *Biochem. Pharmacol.*, **35**, 1891–1897.

Gundersen, C. B., Miledi, R., and Parker, I. (1983) Serotonin receptors induced by exogenous messenger RNA in *Xenopus* oocytes, *Proc. Roy. Soc. Lond.*, Ser. B, **219**, 103–109.

Hall, M. D., Gozlan, H., Emerit, M. B., El Mestikawy, S., Pichat, L., and Hamon, M. (1986) Differentiation of pre- and post-synaptic high affinity serotonin receptor binding sites using physico-chemical parameters and modifying agents, *Neurochem. Res.*, **11**, 891–912.

Hall, M. D., El Mestikawy, S., Emerit, M. B., Pichat, L., Hamon, M., and Gozlan, H. (1985) [^{3}H]8-hydroxy-2-(di-*n*-propylamino) tetralin binding to pre- and postsynaptic 5-hydroxytryptamine sites in various regions of the rat brain, *J. Neurochem.*, **44**, 1685–1696.

Hamon, M. (1984) Radioactive ligand binding studies: identification of central serotonin receptors, in *Brain Receptor Methodologies, Part A* (Eds P. J. Marangos *et al.*), pp. 309–337, Academic Press Inc.

Hamon, M., Bourgoin S., El Mestikawy, S., and Goetz, C. (1984) Central serotonin receptors, in *Handbook of Neurochemistry*, Vol. 6, 2nd edn. (Ed. A. Lajtha), pp. 107–143, Plenum Publ. Corp., New York.

Hamon, M., Cossery, J. M., Spampinato, U., and Gozlan, H. (1986) Are there selective ligands for 5-HT$_{1A}$ and 5-HT$_{1B}$ receptor binding sites in brain? *Trends Pharmacol. Sci.*, **7**, 336–338.

Hamon, H., Emerit, M. B., El Mestikawy, S., Vergé, D., Daval, G., Marquet, A., and Gozlan, H. (1987) Pharmacological, biochemical and functional properties of 5-HT$_{1A}$ receptor binding sites labelled by [^{3}H]8-hydroxy-2-(di-*n*-propylamino)-tetralin in the rat brain, in *Brain Serotonergic Mechanisms: The Pharmacological, Biochemical and Potential Therapeutic Actions of 8-OH-DPAT* (Eds C. T. Dourish, S. Ahlenius and P. H. Hutson), Ellis Horwood Ltd (Health Science series) (in press).

Hamon, M., Goetz, C., and Gozlan, H. (1983) Reciprocal modulations of central 5-HT receptors by GTP and cations, in *CNS Receptors – From Molecular Pharmacology to Behavior* (Eds P. Mandel and F. V. De Feudis), pp. 349–359, Raven Press, New York.

Hamon, M., Mallat, M., El Mestikawy, S., and Pasquier, A. (1982) Ca^{2+}-guanine nucleotide interactions in brain membranes. II. Characteristics of [^{3}H]guanosine triphosphate and [^{3}H]β,γ-imidoguanosine 5′-triphosphate binding and catabolism in the rat hippocampus and striatum, *J. Neurochem.*, **38**, 162–172.

Hamon, M., Nelson, D. L., Herbet, A., and Glowinski, J. (1980) Multiple receptors for serotonin in the rat brain, in *Receptors for Neurotransmitters and Peptide Hormones* (Eds G. Pepeu, M. J. Kuhar and S. J. Enna), pp. 223–233, Raven Press, New York.

Heron, D., Shinitzky, M., Hershkowitz, M., and Samuel, D. (1980) Lipid fluidity markedly modulates the binding of serotonin to mouse brain membranes, *Proc. Natl. Acad. Sci. USA*, **77**, 7463–7467.

Hoyer, D., Engel, G., and Kalkman, H. O. (1985) Molecular pharmacology of 5-HT$_1$ and 5-HT$_2$ recognition sites in rat and pig brain membranes: radioligand binding studies with [^{3}H]5-HT, [^{3}H]8-OH-DPAT, (−)[^{125}I]iodocyanopindolol, [^{3}H]mesulergine and [^{3}H]ketanserin, *Eur. J. Pharmacol.*, **118**, 13–23.

Ilien, B., Gorissen, H., and Laduron, P. (1980) Solubilization of serotonin receptors from rat frontal cortex, *Biochem. Pharmacol.*, **29**, 3341–3344.

Ilien, B., Gorissen, H., and Laduron, P. (1982) Characterization of solubilized serotonin (5-HT$_2$) receptors in rat brain, *Mol. Pharmacol.*, **22**, 243–249.

Kendall, D. A., and Nahorski, S. R. (1983) Temperature-dependent 5-hydroxytryptamine (5-HT)-sensitive [^{3}H]spiperone binding to rat cortical membranes: Regulation by guanine nucleotide and antidepressant treatment, *J. Pharmacol. Exp. Ther.*, **227**, 429–434.

Kendall, D. A., and Nahorski, S. R. (1985) 5-Hydroxytryptamine-stimulated inositol phospholipid hydrolysis in rat cerebral cortex slices: pharmacological characterization and effects of antidepressants, *J. Pharmacol. Exp. Ther.*, **233**, 473–479.

Kubo, T., Fukuda, K., Mikami, A., Maeda, A., Takahashi, H., Mishina, M., Haga, T., Haga, K., Ichiyama, A., Kangawa, K., Kojima, M., Matsuo, H., Hirose, T., and Numa, S. (1986) Cloning, sequencing and expression of complementary DNA encoding the muscarinic acetylcholine receptor, *Nature (Lond.)*, **323**, 411–416,

Lemos, J. R., Barberis, C., Tassin, J. P., and Bockaert, J. (1984) Topographical distributions of 32 K and 48 K cAMP-regulated phosphoproteins: relationships to dopamine and serotonin innervations in striatum, substantia nigra and cerebral cortex, *Brain Res.*, **323**, 47–54.

Lemos, J. R., Novak-Hofer, I., and Levitan, I. B. (1982) Serotonin alters the phosphorylation of specific proteins inside a single living nerve cell, *Nature (Lond.)*, **298**, 64–65.

Leysen, J. E., Niemegeers, C. J. E., Van Nueten, J. M., and Laduron, P. M. (1982) [^{3}H]ketanserin (R 41468), a selective ^{3}H-ligand for serotonin$_2$ receptor binding sites. Binding properties, brain distribution, and functional role, *Mol. Pharmacol.*, **21**, 301–314.

Lykouras, E., Eccleston, D., and Marshall, E. F. (1980) The effect of a 5-HT agonist on cyclic guanosine monophosphate in rat cerebellum, *Biochem. Pharmacol.*, **29**, 827–828.

Mallat, M., and Hamon, M. (1982) Calcium-guanine nucleotide interactions in brain membranes. I. Modulation of central 5-HT receptors in the rat, *J. Neurochem.*, **38**, 151–161.

Markstein, R., Hoyer, D., and Engel, G. (1986) 5-HT$_{1A}$-receptors mediate stimulation of adenylate cyclase in rat hippocampus, *Naunyn-Schmiedeberg's Arch. Pharmacol.*, **333**, 335–341.

Maura, G., and Raiteri, M. (1986) Cholinergic terminals in rat hippocampus possess 5-HT$_{1B}$ receptors mediating inhibition of acetylcholine release, *Eur. J. Pharmacol.*, **129**, 333–337.

Muakkassah-Kelly, S. F., Andresen, J. W., Shih, J. C., and Hochstein, P. (1982) Decreased [^{3}H]serotonin and [^{3}H]spiperone binding consequent to lipid peroxidation in rat cortical membranes, *Biochem. Biophys. Res. Commun.*, **104**, 1003–1010.

Muakkassah-Kelly, S. F., Andresen, J. W., Shih, J. C., and Hochstein, P. (1983) Dual effects of ascorbate on serotonin and spiperone binding in rat cortical membranes, *J. Neurochem.*, **41**, 1429–1439.

Nelson, D. L., Herbet, A., Adrien, J., Bockaert, J., and Hamon, M. (1980a) Serotonin-sensitive adenylate cyclase and [^{3}H]serotonin binding sites in the CNS of the rat. II. Respective regional and subcellular distributions and ontogenetic developments, *Biochem. Pharmacol.*, **29**, 2455–2463.

Nelson, D. L., Herbet, A., Enjalbert, A., Bockaert, J., and Hamon, M. (1980b) Serotonin-sensitive adenylate cyclase and [^{3}H]serotonin binding sites in the CNS of the rat. I. Kinetic parameters and pharmacological properties, *Biochem. Pharmacol.*, **29**, 2445–2453.

Nelson, D. L., Herbet, A., Glowinski, J., and Hamon, M. (1979) [^{3}H]Harmaline as a specific ligand of MAO A. II. Measurement of the turnover rates of MAO A during ontogenesis in the rat brain, *J. Neurochem.*, **32**, 1829–1836.

Nishino, N., and Tanaka, C. (1985) Target size analysis of serotonin 5-HT$_1$ and 5-HT$_2$ receptors in bovine brain membranes, *Life Sci.*, **37**, 1167–1174.

Norman, A. B., Battaglia, G., and Creese, I. (1985) [^{3}H]WB 4101 labels the 5-HT$_{1A}$

serotonin receptor subtype in rat brain. Guanine nucleotide and divalent cation sensitivity, *Mol. Pharmacol.*, **28**, 487–494.

Ogura, A., Ozaki, K., Kudo, Y., and Amano, T. (1986) Cytosolic calcium elevation and cGMP production induced by serotonin in a clonal cell of glial origin, *J. Neurosci.*, **6**, 2489–2494.

Palacios, J. M., Markstein, R., and Pazos, A. (1986) Serotonin-1C sites in the choroid plexus are not linked in a stimulatory or inhibitory way to adenylate cyclase, *Brain Res.*, **380**, 151–154.

Paupardin-Tritsch, D., Hammond, C., Gerschenfeld, H. M., Nairn, A. C., and Greengard, P. (1986) cGMP-dependent protein kinase enhances Ca^{2+} current and potentiates the serotonin-induced Ca^{2+} current increase in snail neurones, *Nature (Lond.)*, **323**, 812–814.

Pazos, A., Hoyer, D., and Palacios, J. M. (1985) The binding of serotonergic ligand to the porcine choroid plexus. Characterization of a new type of serotonin recognition site, *Eur. J. Pharmacol.*, **106**, 539–546.

Peroutka, S. J., and Snyder, S. H. (1979) Multiple serotonin receptors: differential binding of [^{3}H]5-hydroxytryptamine, [^{3}H]lysergic acid diethylamide and [^{3}H]spiro-peridol, *Mol. Pharmacol.*, **16**, 687–699.

Peroutka, S. J., Lebovitz, R. M., and Snyder, S. H. (1979) Serotonin receptor binding sites affected differentially by guanine nucleotides, *Mol. Pharmacol.*, **16**, 700–708.

Poller, D., Carroll, J. A., and Middlemiss, D. N. (1981) The [^{3}H]5-HT and [^{3}H]spiro-peridol binding sites in the rat frontal cortex are thermally inactivated at different rates, *Eur. J. Pharmacol.*, **72**, 121–123.

Ransom, R. W., Asarch, K. B., and Shih, J. C. (1986) Photoaffinity labelling of the 5-hydroxytryptamine$_{1A}$ receptor in rat hippocampus, *J. Neurochem.*, **47**, 1066–1072.

Richardson, B. P., and Engel, G. (1986) The pharmacology and function of 5-HT$_3$ receptors, *Trends Neurosci.*, **9**, 424–428,

Rosenfeld, M. R., and Makman, M. H. (1981) The interaction of lisuride, an ergot derivative, with serotonergic and dopaminergic receptors in rabbit brain, *J. Pharmacol. Exp. Ther.*, **216**, 526–531.

Roth, B. L., Nakaki, T., Chuang, D.-M., and Costa, E. (1986) 5-Hydroxytryptamine$_2$ receptors coupled to phospholipase C in rat aorta: modulation of phosphoinositide turnover by phorbolester, *J. Pharmacol. Exp. Ther.*, **238**, 480–485.

Rousselle, J. C., Gillet, G., and Fillion, G. (1985) Solubilization and characterization of [^{3}H]5-HT high affinity binding sites (5-HT$_1$ and 5-HT$_3$), *J. Pharmacol. (Paris)*, **16**, 421–438.

Sakai, Y., Kimura, H., and Okamoto, K. (1986) Pharmacological characterization of serotonin receptor induced by rat brain messenger RNA in *Xenopus* oocytes, *Brain Res.*, **362**, 199–203.

Sanders-Bush, E., and Conn, P. J. (1986) Effector systems coupled to serotonin receptors in brain: serotonin stimulated phosphoinositide hydrolysis, *Psychopharmacol. Bull.*, **22**, 829–836.

Schoemaker, H., and Langer, S. Z. (1986) [^{3}H]8-OH-DPAT labels the serotonin transporter in the rat striatum, *Eur. J. Pharmacol.*, **124**, 371–373.

Schotte, A., Maloteaux, J. M., and Laduron, P. M. (1984) Solubilization of serotonin S$_2$-receptors from human brain, *Eur. J. Pharmacol.*, **100**, 329–333.

Shih, J. C., Eiduson, S., Geller, E., and Costa, E. (1974) Serotonin-binding proteins isolated by affinity chromatography, *Adv. Biochem. Psychopharmacol.*, **11**, 101–104.

Shih, J. C., and Ohsawa, R. (1983) Differential effect of cholesterol on two types of 5-hydroxytryptamine binding sites, *Neurochem. Res.*, **8**, 701–710.

Shuster, M. J., Camardo, J. S., Siegelbaum, S. A., and Kandel, E. R. (1985) Cyclic AMP-dependent protein kinase closes the serotonin-sensitive K^+ channels of *Aplysia* sensory neurones in cell-free membrane patches, *Nature (Lond.)*, **313**, 392–395.

Taylor, C. W., and Merrit, J. E. (1986) Receptor coupling to polyphosphoinositide turnover: a parallel with the adenylate cyclase system, *Trends Pharmacol. Sci.*, **7**, 238–242.

Todd, R. D., and Ciaranello, R. D. (1985) Demonstration of inter- and intraspecies differences in serotonin binding sites by antibodies from an autistic child, *Proc. Natl. Acad. Sci. USA*, **82**, 612–616.

Van den Berg, S. R., Allgren, R. L., Todd, R. D., and Ciaranello, R. D. (1983) Solubilization and characterization of high affinity [^{3}H]serotonin binding sites from bovine cortical membranes, *Proc. Natl. Acad. Sci. USA*, **80**, 3508–3512.

Vergé, D., Daval, G., Marcinkiewicz, M., Patey, A., El Mestikawy, S., Gozlan, H., and Hamon, M. (1986) Quantitative autoradiography of multiple 5-HT$_1$ receptor subtypes in the brain of control or 5,7-dihydroxytryptamine-treated rats, *J. Neurosci.* **6**, 3474–3482.

Vergé, D., Daval, G., Patey, A., Gozlan, H., El Mestikawy, S., and Hamon, M. (1985) Presynaptic 5-HT autoreceptors on serotonergic cell bodies and/or dendrites but not terminals are of the 5-HT$_{1A}$ subtype, *Eur. J. Pharmacol.*, **113**, 463–464.

Von Hungen, K., Roberts, S., and Hill, D. F. (1975) Serotonin-sensitive adenylate cyclase activity in immature rat brain, *Brain Res.*, **84**, 257–267.

Weiss, S., Sebben, M., Kemp, D. E., and Bockaert, J. (1986) Serotonin 5-HT$_1$ receptors mediate inhibition of cyclic AMP production in neurons, *Eur. J. Pharmacol.*, **120**, 227–230.

Wesemann, W., Weiner, N., and Hoffmann-Bleihauer, P. (1986) Modulation of serotonin binding in rat brain by membrane fluidity, *Neurochem. Int.*, **9**, 447–454.

Woolley, D. W., and Gommi, B. W. (1964) Serotonin receptors. V. Selective destruction by neuraminidase plus EDTA and reactivation with tissue lipids, *Nature (Lond.)*, **202**, 1074–1075.

Woolley, D. W., and Gommi, B. W. (1965) Serotonin receptors. VII. Activities of pure gangliosides as the receptors. *Proc. Natl. Acad. Sci. USA*, **53**, 959–963.

Woolley, D. W., and Gommi, B. W. (1966) Serotonin receptors. VI. Methods for the direct measurement of isolated receptors, *Arch. Int. Pharmacodyn. Ther.*, **159**, 8–17.

Wouters, W., Van Dun, J., Leysen, J. E., and Laduron, P. M. (1985a) Solubilization of rat brain serotonin-S$_2$ receptors using CHAPS/salt, *Eur. J. Pharmacol.*, **115**, 1–9.

Wouters, W., Van Dun, J., Leysen, J. E., and Laduron, P. M. (1985b) Photoaffinity labelling of serotonin-S$_2$ receptors with 7-azidoketanserin, *Eur. J. Pharmacol.*, **107**, 399–400.

Wouters, W., Van Dun, J., Leysen, J. E., and Laduron, P. M. (1985c) Photoaffinity probes for serotonin and histamine receptors. Synthesis and characterization of two azide analogues of ketanserin, *J. Biol. Chem.*, **260**, 8423–8429.

Yagaloff, K. A., and Hartig, P. R. (1986) Solubilization and characterization of the serotonin 5-HT$_{1C}$ site from pig choroid plexus, *Mol. Pharmacol.*, **29**, 120–125.

Yoshikawa, S., and Ishitani, R. (1985) Implication of acidic lipids in 5-hydroxytryptamine receptor mechanisms, *Life Sci.*, **36**, 485–492.

CHAPTER 15

Functional Correlates of Central 5-HT Binding Sites

STEPHEN J. PEROUTKA
Departments of Neurology and Pharmacology
Stanford University Medical Center
Stanford
CA 94305
USA

INTRODUCTION

Receptor site analysis has been a productive approach to the understanding of the actions of 5-hydroxytryptamine (5-HT) in the central and peripheral nervous systems. In the thirty years since the differentiation of M and D receptors (Gaddum and Picarelli, 1957), it has become clear that multiple 5-HT receptors exist. The heterogeneity of 5-HT receptors has become even more apparent in the past decade. To a significant degree, the appreciation of this fact is a direct result of the development and application of radioligand binding techniques (Snyder, 1983). At the present time, at least five 5-HT binding site subtypes have been differentiated by ^{3}H-radioligand techniques in brain homogenates (Peroutka and Synder, 1979; Pedigo *et al.*, 1981; Fillion, 1983; Fuller, 1984; Hamon *et al.*, 1984; Hoyer *et al.*, 1985b; Peroutka, 1986; Heuring and Peroutka, 1986). Anatomic studies using autoradiographic techniques have also confirmed that a variety of serotonergic recognition sites, with distinct regional localizations, exist in the central nervous system (Biegon *et al.*, 1982; Marcinkiewicz *et al.*, 1984; Pazos and Palacios, 1985; Pazos *et al.*, 1985a; Hoyer *et al.*, 1986a, b).

This chapter will focus on the pharmacologic data derived from radioligand binding studies. These data will then be compared with the effects of 5-HT and related agents in a variety of physiologic systems. Thus, an attempt will be made to define functional correlates of 5-HT binding sites in order that these membrane recognition sites might be considered 5-HT 'receptors'. Peripheral 5-HT$_3$ receptors (Fozard, 1984; Richardson *et al.*, 1985; Bradley *et al.*, 1986) will not be discussed in this chapter (see Chapter 17 of this volume).

RADIOLIGAND BINDING STUDIES OF 5-HT RECEPTORS

Preliminary binding studies

The first successful radioligand analysis of 5-HT receptors was reported by Bennett and Aghajanian (1974). The binding of ^{3}H-D-lysergic acid diethylamide (D-LSD) was saturable, reversible, stereoselective and displayed high affinity (K_D = 7.5 nM) for its membrane recognition site. The binding sites also displayed appropriate regional variations since brain regions with the highest density of receptors were areas known to receive a dense projection of 5-HT neuronal terminals. These findings were soon extended and confirmed by other laboratories (Bennett and Snyder, 1975; Lovell and Freedman, 1976).

The second radioligand used to label 5-HT receptors was ^{3}H-5-HT (Bennett and Snyder, 1976; Fillion *et al.*, 1978; Nelson *et al.*, 1978). Like ^{3}H-LSD binding, the ^{3}H-5-HT binding was saturable, stereoselective and displayed appropriate regional variations. However, important discrepancies were noted between the binding of ^{3}H-LSD and ^{3}H-5-HT. At the time, Bennett and Snyder (1976) suggested that ^{3}H-5-HT and ^{3}H-LSD did not label the same membrane recognition site but rather two different 'states' of the same receptor.

Radioligand analysis of 5-HT receptors was considerably advanced by the finding that ^{3}H-spiperone could also be used to label presumed 5-HT recognition sites (Leysen *et al.*, 1978). Previously, ^{3}H-spiperone had been considered a pure dopaminergic ligand. However, in the rat frontal cortex, where dopamine receptors are sparse, ^{3}H-spiperone was found to label a receptor that appeared to be 'serotonergic' in the sense that 5-HT antagonists were the most potent displacers of the ligand. However, 5-HT and related tryptamines were weak displacers of ^{3}H-spiperone. The ability of ^{3}H-spiperone to label 5-HT receptors was quickly confirmed by other laboratories (Creese and Snyder, 1978; Quik *et al.*, 1978).

Differentiation of 5-HT$_1$ and 5-HT$_2$ receptors

Thus, marked differences were noted between the binding characteristics of ^{3}H-5-HT, ^{3}H-LSD and ^{3}H-spiperone (Peroutka and Snyder, 1979). If each ligand labeled the same membrane recognition site, then unlabeled drugs should be equipotent in displacing ^{3}H-5-HT, ^{3}H-LSD and ^{3}H-spiperone. This pattern is observed with D-LSD displacement of the three ligands. A K_i value of approximately 10 nM is observed with D-LSD competition studies against each of these three ligands. In marked contrast, 5-HT is approximately three orders of magnitude more potent in displacing ^{3}H-5-HT (3.8 nM) than ^{3}H-spiperone (2,700 nM). Its apparent K_i for ^{3}H-LSD binding is 110 nM, a value which is intermediate between its affinity for ^{3}H-5-HT and ^{3}H-spiperone labeled sites. Furthermore, the Hill slope for 5-HT displacement of ^{3}H-LSD, but not of ^{3}H-5-HT or ^{3}H-spiperone, binding is significantly less than unity (Peroutka and Snyder, 1979). The converse pattern is observed with spiperone displacement of the three radioligands. For example, spiperone is extremely potent against ^{3}H-spiperone binding (0.51 nM) yet has 1400 fold less affinity for total ^{3}H-5-HT labeled sites (730 nM). In addition, its apparent K_i for ^{3}H-LSD binding is 18 nM but the displacement curve is biphasic, with a 'plateau' occurring between 10 and 30 nM spiperone (Peroutka and Snyder, 1979).

Given the results outlined above and the fact that no correlation exists between drug potencies for ^{3}H-5-HT and ^{3}H-spiperone labeled 'serotonergic' receptors, Peroutka and Snyder (1979) concluded that at least two distinct 5-HT membrane recognition sites are present in the central nervous system. The sites labeled by ^{3}H-5-HT were designated '5-HT$_1$ receptors' and those labeled by ^{3}H-spiperone were designated '5-HT$_2$ receptors'. Since ^{3}H-LSD had equal affinity for both sites, it was proposed that this ligand could be used to label both 5-HT$_1$ and 5-HT$_2$ receptors.

Characterization of 5-HT$_1$ binding site subtypes

However, 5-HT$_1$ binding sites labeled by ^{3}H-5-HT were soon shown to be heterogeneous. Non-sigmoidal displacement of ^{3}H-5-HT by spiperone led to the suggestion that sites with high affinity ($K_i = 2$–13 nM) for spiperone should be designated 5-HT$_{1A}$ sites while sites with relatively low affinity for spiperone (K_i = 34 μM) should be designated 5-HT$_{1B}$ sites (Pedigo *et al.*, 1981). These two subtypes have different regional localizations and have been identified in many species (Schnellmann *et al.*, 1984). Autoradiographic studies have also confirmed the ability of spiperone to differentiate at least two types of sites labeled by ^{3}H-5-HT (Deshmukh *et al.*, 1983). A third subtype of total 5-HT$_1$ recognition sites (the 5-HT$_{1C}$ site) has been identified in the choroid plexus and cortex of various species (Pazos *et al.*, 1984b; Yagaloff and Hartig, 1985; Peroutka, 1986). Most recently, a fourth subtype site labeled by ^{3}H-5-HT, the 5-HT$_{1D}$ site, has been identified in bovine brain (Heuring and Peroutka, 1987). In the past three years, the availability of selective and novel agents has greatly facilitated the analysis of 5-HT$_1$ binding site subtypes. A summary of the currently accepted 5-HT receptor classification system based on radioligand data is presented in Table 1. A summary of drug potencies at each of the five known 5-HT binding site subtypes is provided in Table 2. These data are discussed in greater detail in the following section.

5-HT$_{1A}$ binding sites

A number of radioligands have been shown to label the 5-HT$_{1A}$ binding site. Based on spiperone displaceable ^{3}H-5-HT binding, 5-HT$_{1A}$ sites were found to predominate in the hippocampus, septum and cerebral cortex (Deshmukh *et al.*, 1983). The 5-HT$_{1A}$ site can be more directly labeled with ^{3}H-8-hydroxy-2-(di-*n*-propylamino)tetralin (8-OH-DPAT) (Gozlan *et al.*, 1983; Hall *et al.*, 1985; Peroutka, 1985; Hoyer *et al.*, 1985b), ^{3}H-ipsapirone (formerly called TVX Q 7821) (Dompert *et al.*, 1985), ^{3}H-buspirone (Moon and Taylor, 1985) and ^{3}H-1-(2-(4-aminophenyl) ethyl) -4-(3-trifluromethylphenyl) piperazine (PAPP) (Asarch *et al.*, 1985). In addition, ^{3}H-WB 4101, previously

Table 1 Characteristics of 5-HT$_{1A}$, 5-HT$_{1B}$, 5-HT$_{1C}$, 5-HT$_{1D}$ and 5-HT$_2$ binding sites

	5-HT$_{1A}$	5-HT$_{1B}$	5-HT$_{1C}$	5-HT$_{1D}$	5-HT$_2$
Radiolabeled by	^{3}H-5-HT	^{3}H-5-HT	^{3}H-5-HT	^{3}H-5-HT	^{3}H-Spiperone
	^{3}H-8-OH-DPAT	^{125}I-CYP	^{3}H-Mesulergine		^{3}H-Mesulergine
	^{3}H-Ipsapirone	(Rat and	^{125}I-LSD		^{125}I-LSD
	^{3}H-WB 4101	mouse only)			^{3}H-Ketanserin
	^{3}H-Buspirone				^{3}H-DOB
	^{3}H-PAPP				^{3}H-Mianserin
					^{125}I-Methyl-LSD
High-density regions	Raphe nuclei	Substantia nigra	Choroid plexus	Basal ganglia	Layer IV cortex
	Hippocampus	Globus pallidus			

Table 2 Drug affinities for 5-HT$_{1A}$, 5-HT$_{1B}$, 5-HT$_{1C}$, 5-HT$_{1D}$ and 5-HT$_2$ receptors

Drug potencies (K_i, nM)	5-HT$_{1A}$	5-HT$_{1B}$	5-HT$_{1C}$	5-HT$_{1D}$	5-HT$_2$
< 10 nM	5-CT 8-OH-DPAT 5-HT RU 24969 D-LSD	RU 24969 5-CT 5-HT	Mesulergine Metergoline Methysergide	5-CT 5-HT Metergoline	Spiperone Mesulergine Methysergide Metergoline Mianserin
10–1000 nM	Metergoline Methysergide Spiperone Mesulergine	Metergoline Methysergide D-LSD	Mianserin 5-HT RU 24969 5-CT D-LSD	Methysergide Mianserin 8-OH-DPAT D-LSD RU 24969	D-LSD
> 1000 nM	Mianserin	Mianserin Spiperone Mesulergine 8-OH-DPAT	Spiperone 8-OH-DPAT	Mesulergine Spiperone	RU 24969 5-HT 8-OH-DPAT

Data given are derived from Peroutka and Snyder (1979), Peroutka (1986), Hoyer *et al.* (1985b), Heuring and Peroutka (1987) and unpublished observations.

considered a selective alpha$_1$-adrenergic radioligand, has been demonstrated to label the 5-HT$_{1A}$ site (Norman *et al.*, 1985). Regardless of the ^{3}H-ligand used to label the site, it displays high and selective affinity for 8-OH-DPAT, 5-methoxydimethyltryptamine, ipsapirone and buspirone. The 5-HT$_{1A}$ site is densely present in the CA1 region and dentate gyrus of the hippocampus and in the raphe nuclei (Deshmukh *et al.*, 1983; Marcinkiewicz *et al.*, 1984; Pazos and Palacios, 1985; Glaser *et al.*, 1985; Hoyer *et al.*, 1986a). In addition, the fact that 5, 7-dihydroxytryptamine-induced lesions of the forebrain serotonergic projections cause a loss of ^{3}H-8-OH-DPAT binding in the striatum but not hippocampus has led Hamon and colleagues to hypothesize that ^{3}H-8-OH-DPAT also labels a presynaptic 5-HT autoreceptor (Gozlan *et al.*, 1983; Hall *et al.*, 1985, 1986). ^{3}H-8-OH-DPAT has also been reported to label the presynaptic '5-HT transporter' in the rat striatum (Schoemaker and Langer, 1986). In the absence of ascorbate, Peroutka and Demopulos (1986) found that ^{3}H-8-OH-DPAT could label glass fiber filter paper under special conditions (see Demopulos and Peroutka, 1987). The 5-HT$_{1A}$ binding site is the only postsynaptic site labeled by ^{3}H-8-OH-DPAT that has, to date, been pharmacologically characterized in the presence of ascorbate (Peroutka, 1985, 1986; Hoyer *et al.*, 1985b; 1986a; Vergé *et al.*, 1986).

5-HT$_{1B}$ binding sites

The putative 5-HT$_{1B}$ site has been more difficult to characterize. Sills *et al.* (1984) defined 5-HT$_{1B}$ binding as specific ^{3}H-5-HT binding in the presence of 1 mM guanosine triphosphate (GTP) and 2000 nM spiperone. They concluded that RU 24969 and TFMPP were selective 5-HT_{1B} agents. The 5-HT_{1B} site has been more directly labeled in rat brain with ^{125}I-cyanopindolol (Pazos *et al.*, 1985b; Hoyer *et al.*, 1985a,b). The ^{125}I-cyanopindolol site has high affinity for 5-HT and RU 24969 and relatively low affinity for D-LSD and 8-OH-DPAT. The highest densities of 5-HT_{1B} sites in rat brain are found in the globus pallidus, dorsal subiculum and substantia nigra (Pazos and Palacios, 1985). Recent data have demonstrated that this site can also be labeled with ^{3}H-5-HT in rat frontal cortex (Peroutka, 1986; Blurton and Wood, 1986). Moreover, the 5-HT_{1B} site appears to be species specific in that it is present in rat and mouse brain but not in guinea pig, cow, chicken, turtle, frog or human brain membranes (Heuring *et al.*, 1986).

5-HT_{1C} binding sites

The 5-HT_{1C} site was first characterized in membranes from pig choroid plexus and cortex (Pazos *et al.*, 1984a,b; Hoyer *et al.*, 1985b). The site was labeled by both ^{3}H-5-HT and ^{3}H-mesulergine. Independently, Yagaloff and Hartig (1985) labeled the site with ^{125}I-LSD in the rat choroid plexus. The site has high affinity for 5-HT, methysergide and mianserin and relatively low affinity for RU 24969. These pharmacological characteristics are also shared with the putative 5-HT_{1C} site identified in rat cortex (Hoyer *et al.*, 1985b; Peroutka, 1986).

5-HT_{1D} binding sites

Recently, a fourth subtype of 5-HT_1 site labeled by ^{3}H-5-HT has been identified in bovine brain membranes (Heuring and Peroutka, 1987). The addition of either the 5-HT_{1A} selective drug 8-OH-DPAT (100 nM) or the 5-HT_{1C} selective drug mesulergine (100 nM) to the assay results in a 5–10% decrease in specific ^{3}H-5-HT binding in bovine caudate. Scatchard analysis reveals that the simultaneous addition of both drugs decreases the B_{max} of ^{3}H-5-HT binding by 10–15% without affecting the K_D value (1.8 ± 0.3 nM). Competition studies using a series of pharmacologic agents reveal that the sites labeled by ^{3}H-5-HT in bovine caudate in the presence of 100 nM 8-OH-DPAT and 100 nM mesulergine appear to be homogeneous. 5-HT_{1A} selective agents such as 8-OH-DPAT, ipsapirone and buspirone display micromolar affinities for these sites. RU 24969 and ($-$)pindolol are approximately two orders of magnitude less potent at these sites than at 5-HT_{1B} sites which have been identified in rat brain. Agents which display nanomolar potencies for 5-HT_{1C} sites such as mianserin and mesulergine are two to three orders of magnitude less potent at the ^{3}H-5-HT binding sites in bovine caudate. In addition, both 5-HT_2 and 5-HT_3 selective agents are essentially inactive at these binding sites. These ^{3}H-5-HT sites display nanomolar affinity for 5-

carboxyamidotryptamine (5-CT), 5-methoxytryptamine, metergoline and 5-HT. Apparent K_i values of 10–100 nM are obtained for D-LSD, RU 24969, methiothepin, tryptamine, methysergide and yohimbine whereas L-LSD and corynanthine are significantly less potent. Regional studies demonstrate that this class of sites is most dense in the basal ganglia but exists in all regions of bovine brain. These data therefore demonstrate the presence of a homogeneous class of 5-HT$_1$ binding sites in bovine caudate which is pharmacologically distinct from previously defined 5-HT$_{1A}$, 5-HT$_{1B}$, 5-HT$_{1C}$ and 5-HT$_2$ binding site subtypes. We therefore suggested that this class of sites be designated the 5-HT$_{1D}$ subtype of binding sites labeled by ^{3}H-5-HT (Heuring and Peroutka, 1987).

5-HT$_2$ binding sites

In contrast to the subtypes of the 5-HT$_1$ sites, 5-HT$_2$ sites appear to be homogeneous. ^{3}H-spiperone, ^{3}H-LSD, ^{3}H-mianserin, ^{3}H-ketanserin, ^{3}H-mesulergine, ^{125}I-LSD and N$_1$-methyl-2-^{125}I-LSD can be used to label the 5-HT$_2$ binding site (Leysen, 1981; Leysen et al., 1978, 1982; Hartig et al., 1983; Closse, 1983; Pazos et al., 1984a; Engel et al., 1984b; Kadan et al., 1984; Hoffman et al., 1985; Hartig et al., 1985). Serotonergic antagonists have high affinity for this site, while 5-HT and related tryptamines are markedly less potent. The number of 5-HT$_2$, but not 5-HT$_1$, binding sites can be decreased by chronic treatment with antidepressant drugs (Peroutka and Snyder, 1980a, b). The highest level of 5-HT$_2$ binding is in the cerebral cortex and caudate, with all other brain regions having substantially fewer binding sites (Schotte et al., 1983). The 5-HT$_2$ site has been extensively studied using ^{3}H-spiperone and ^{3}H-ketanserin. Using cinanscrin displaceable ^{3}H-spiperone to define binding to 5-HT$_2$ sites, the highest density of specific binding was observed in lamina IV of the neocortex (Palacios et al., 1981). Similarly, a moderate amount of 5-HT$_2$ binding using ^{3}H-ketanserin was found in lamina IV of the cortex while a high density of specific binding was identified in the striatum (Slater and Patel, 1983). The most extensive study to date of 5-HT$_2$ sites by autoradiographic techniques has utilized ^{3}H-ketanserin (Pazos et al., 1985a; Hoyer et al., 1986b). Recently, the hallucinogen ^{3}H-DOB has been shown to label an apparent 5-HT$_2$ recognition site (Titeler et al., 1987). Studies are in progress to determine the relevance of this site to 5-HT$_2$ sites labeled by ^{3}H-antagonists.

PHYSIOLOGICAL MODELS OF CENTRAL 5-HT RECEPTORS

Although radioligand techniques provide a rapid and sensitive measure of drug potencies at specific membrane recognition sites, they cannot differentiate the agonist versus antagonist properties of drugs, the relative efficacy of agents or drug interactions with other receptor sites involved in a specific physiologic response. This type of pharmacologic information can only be derived from physiologic experiments. In this section, a number of physiological models will be analyzed for possible relationships to 5-HT$_{1A}$, 5-HT$_{1B}$, 5-HT$_{1C}$, 5-HT$_{1D}$ and 5-HT$_2$ binding sites. As shown in Table 3, a number of investigators have proposed that a relationship does indeed exist between various physiologic systems and radioligand binding sites. As will be described below, the suggested relationships have been based on the similarities between drug potencies in various physiologic systems and their affinities for radioligand binding sites.

Table 3 Proposed functional correlates of 5-HT binding site subtypes

5-HT$_{1A}$	Adenylate cyclase stimulation Raphe cell and CA1 hippocampal cell inhibition Canine basilar artery contractions Forepaw treading, tremor, head-weaving Facilitation of ejaculation and/or seminal emissions Thermoregulation Hypotensive effects
5-HT$_{1B}$	'Autoreceptor'
5-HT$_{1C}$	Phosphatidylinositol turnover
5-HT$_{1D}$	? Rat kidney perfusion ? Rat stomach fundus
5-HT$_2$	Phosphatidylinositol turnover Contraction of vascular smooth muscle 5-Hydroxytryptophan or mescaline-induced head twitches Tryptamine-induced seizures Rat forepaw edema Discriminative cue properties Contraction of bronchial smooth muscle Platelet shape changes and aggregation Smooth muscle prostacyclin synthesis

5-HT$_{1A}$ receptors

Neurophysiological effects of 5-HT in raphe and hippocampus

The 5-HT receptor mediating inhibition of raphe nuclei has been extensively studied using a variety of pharmacologic agents. In the 1970s, neurophysiological analysis focused mainly on the effects of 5-HT, tryptamine analogs and D-LSD on raphe cell firing (Rogawski and Aghajanian, 1981). More recently, the effects of 5-HT binding site subtype selective agents have been analyzed in this model system using more advanced intracellular techniques. VanderMaelen and Wildmeran (1984) first showed that buspirone, a 5-HT$_{1A}$ selective agent, caused complete inhibition of dorsal raphe neuronal firing in the rat. Buspirone also has identical effects in mouse brain slices (Trulson and Arasteh, 1986). The ability of other 5-HT$_{1A}$ selective agents to produce this effect was soon demonstrated by several laboratories. For example, ipsapirone, 8-OH-DPAT and 5-CT were found to mimic the effect of 5-HT on raphe cell firing (Sprouse and Aghajanian, 1985; Sinton and Fallon, 1986; VanderMaelen *et al.*, 1986). By contrast, (−)propranolol, a beta-adrenergic agent which also displays high affinity for the 5-HT$_{1A}$ receptor (Hiner *et al.*, 1986), was shown to reversibly block the inhibitory effects of ipsapirone and 8-OH-DPAT on raphe cell inhibition (Sprouse and Aghajanian, 1986). Thus, inhibition of dorsal raphe cells appears to be mediated by 5-HT$_{1A}$ receptors (Dourish *et al.*, 1986).

The hippocampus may be another ideal region in which to study 5-HT$_{1A}$ receptor function at the cellular level. The structure has a high density of 5-HT$_{1A}$ sites, particularly in the dentate gyrus and CA1 region (Marcinkiewicz *et al.*, 1984; Pazos and Palacios, 1985) and is particularly amenable to neurophysiological analysis. Bath application of 5-HT$_{1A}$ selective agents such as 8-OH-DPAT, ipsapirone and buspirone decreases the amplitude of the field potential recorded in the pyramidal cell layer of CA1 after stimulation of Schaffer collaterals in the stratum radiatum (Peroutka *et al.*, 1986b). The 5-HT$_2$ selective antagonist, ketanserin, has no independent effect and does not affect the action of the 5-HT$_{1A}$ selective drugs. The effect of 5-HT and 8-OH-DPAT on the population spike is partially blocked by spiperone, a finding which also supports a role for 5-HT$_{1A}$ receptors in the CA1 region of the hippocampus (Beck *et al.*, 1985). Intracellular studies have shown that 5-HT markedly hyperpolarizes CA1 pyramidal cells (15–20mV). By contrast, buspirone causes only a modest (1–2 mV) hyperpolarization of CA1 pyramidal neurons (Andrade and Nicoll, 1985). These preliminary studies suggest that analysis of the actions of selective 5-HT$_{1A}$ receptor subtype agents may yield important information on the physiological effects of this receptor in the hippocampus.

5-HT sensitive adenylate cyclase

A 5-HT sensitive adenylate cyclase in mammalian brain was first detected in newborn rats (Von Hungen *et al.*, 1974, 1975). In a study using guinea pig hippocampal membranes, a 5-HT-sensitive adenylate cyclase could be stimulated by nanomolar concentrations of two 5-HT$_{1A}$ selective agents, 5-CT and 8-OH-DPAT (Shenker *et al.*, 1985). In addition, it has also been demonstrated recently that 5-HT can inhibit forskolin-stimulated adenylate cyclase in guinea pig and rat hippocampal membranes (De Vivo and Maayani, 1986). The pharmacologic data derived from this system appeared to be consistent with a single, homogeneous population of receptors. 8-OH-DPAT, d-LSD and buspirone were similar to 5-HT in their ability to inhibit cyclase activity. By contrast, spiperone was a competitive antagonist at this receptor whereas ketanserin had no effect on either cyclase activity or on 5-HT-induced inhibition of the cyclase activity. Thus, inhibition of forskolin-stimulated adenylate cyclase activity in hippocampal membranes appears to be mediated by the 5-HT$_{1A}$ receptor.

Radioligand binding data also support an association between 5-HT$_{1A}$ receptors and a 5-HT-sensitive adenylate cyclase. Regulation of neuro-transmitter binding by guanine nucleotides often reflects an association of the agonist binding site with an adenylate cyclase (Maguire et al., 1977; Rodbell, 1980; Limbird, 1981). Thus, GTP and GDP (guanosine diphos-phate), but not GMP (guanosine monophosphate), inhibit the binding of ^{3}H-8-OH-DPAT to brain homogenates (Hall et al., 1985; Schlegel and Peroutka, 1986). In addition, guanine nucleotides significantly reduce agonist potencies for ^{3}H-8-OH-DPAT binding sites whereas antagonist potencies are not affected by nucleotides. In summary, an association between the 5-HT$_{1A}$ binding site and a high-affinity adenylate cyclase system is supported by both the pattern of pharmacological interactions at the cyclase in certain systems as well as by the effects of guanine nucleotides on specific ^{3}H-8-OH-DPAT binding.

Canine basilar artery contractions

The canine basilar artery has been proposed as a functional correlate of the 5-HT$_{1A}$ receptor. In a study using 5-HT and ten putative 5-HT$_{1A}$ agonists, each agent was shown to induce a contraction of the canine basilar artery (Peroutka et al., 1986a). However, each of the putative 5-HT agonists elicited a less forceful contraction than did 5-HT. The K_{ED50} value in the basilar artery system correlated significantly with drug affinity for 5-HT$_{1A}$ binding sites labeled by ^{3}H-8-OH-DPAT. Importantly, no significant correlation was found between artery K_{ED50} values and agonist affinities for total 5-HT$_1$ sites labeled by ^{3}H-5-HT. In an independent study of the canine basilar artery, Taylor et al. (1986) determined agonist association constants (K_A values) using the method of 'partial irreversible blockade'. With this method, no unwarranted assumptions are made concerning either the function relating receptor occupancy to effect or to the amount of 'spare receptors' present (Taylor et al., 1986). The values obtained in the basilar artery correlated with K_i values for the 5-HT$_{1A}$ binding site subtype. Taken together, these two independent studies strongly suggest that 5-HT induced contractions of the canine basilar artery are mediated, at least in part, by 5-HT$_{1A}$ receptors.

At the same time, evidence exists that 5-HT$_2$ receptors may also be present in the canine basilar artery (Peroutka, 1984).

Behavioral studies

A complex behavioral syndrome results after central 5-HT stimulation with drugs such as 5-hydroxytryptophan (Jacobs, 1976; Green, 1984). The syndrome includes resting tremor, forepaw treading, flattened body posture, head-weaving, hindlimb abduction, Straub tail and head-twitching. Attempts have been made to correlate each of the behavioral components with 5-HT$_1$ and/or 5-HT$_2$ binding sites. Only metergoline and methysergide block the entire set of behaviors, while relatively high concentrations of ketanserin and pipamperone are needed to block 5-HTP-induced forepaw treading, head-weaving, tremor, hindlimb abduction, extended posture or Straub tail (Lucki *et al.*, 1984). These findings suggest that the many components of the 5-HT behavioral syndrome may be related to activation of 5-HT$_1$ receptors.

The recent development and characterization of 5-HT$_{1A}$ selective agents has clarified the role of 5-HT receptor subtypes in the mediation of specific behaviors. The results of behavioral studies with 8-OH-DPAT and 5-MeODMT indicate that forepaw treading and flattened body posture may be regarded as behavioral correlates of 5-HT$_{1A}$ receptor activation (Tricklebank *et al.*, 1984, 1985; Tricklebank, 1985). The effects of 8-OH-DPAT, 5-MeODMT, buspirone and ipsapirone were recently examined on the 5-HT behavioral syndrome (Smith and Peroutka, 1986). 8-OH-DPAT, 5-MeODMT and buspirone induce hindlimb abduction, flattened body posture and Straub's tail. Ipsapirone induces only a slight flattening of body posture. By contrast, 8-OH DPAT and 5-MeODMT, but not buspirone and ipsapirone, also induce forepaw treading, head-weaving and tremor. However, both buspirone and ipsapirone antagonize the induction of these three behaviors by 8-OH-DPAT or 5-MeODMT. These data show that 8-OH-DPAT and 5-MeODMT are 'full agonists' in relation to these six components of the 5-HT behavioral syndrome. Buspirone and ipsapirone, on the other hand, act as 'antagonists' in relation to forepaw treading, head-weaving and tremor. Therefore, these data suggest that specific components of the 5-HT behavioral syndrome are mediated by 5-HT$_{1A}$ receptors. In addition, preliminary studies in male rats suggest that 5-HT$_{1A}$ selective agonists facilitate seminal emissions and/or ejaculations (Ahlenius *et al.*, 1981; Kwong *et al.*, 1986).

Other systems

In a study utilizing tryptamine derivatives, the hypotensive potencies in pentobarbitone-anaesthesized rats correlated with their affinity for total 5-

HT$_1$ sites (Kalkman *et al.*, 1983b, 1984). Furthermore, similar studies using 8-OH-DPAT and RU 24969 have suggested that the 5-HT$_{1A}$ site may mediate these hypotensive effects (Doods *et al.*, 1985). The thermoregulatory effects of 8-OH-DPAT and RU 24969 also have been attributed to 5-HT$_{1A}$ receptor activation (Gudelsky *et al.*, 1985; Tricklebank *et al.*, 1986).

5-HT$_{1B}$ binding sites: 'autoreceptors' in rat

Neurotransmitters are stored in synaptic vesicles located in the nerve terminal. Depolarization of the neuron by potassium or electrical stimulation causes release of the pre-stored neurotransmitter. The release can be inhibited by 5-HT and related agonists, presumably through a 'presynaptic autoreceptor'. Initial studies analyzed ^{3}H-5-HT release from rat brain synaptosomes or slices (Farnebo and Hamberger, 1974; Hamon *et al.*, 1974; Cerrito and Raiteri, 1979; Göthert, 1980; Martin and Sanders-Bush, 1982a,b). In both systems, nanomolar concentrations of 5-HT inhibited ^{3}H-5-HT release. Slightly higher concentrations of methiothepin and metergoline were able to block the action of 5-HT. However, micromolar concentrations of cyproheptadine, methysergide and mianserin had no effect on the action of exogenous 5-HT. Metergoline and methiothepin were also the most potent inhibitors of ^{3}H-5-HT release following potassium-evoked depolarization of rat raphe nuclei and hypothalamic slices (Cox and Ennis, 1982; Ennis and Cox, 1982).

Direct comparisons have been made between the pharmacologic interactions at 'autoreceptors' and 5-HT$_1$ binding sites. Engel and co-workers initially compared the potencies of fourteen indole derivatives at 'autoreceptors' with their affinity for 5-HT$_1$ binding sites (Engel *et al.*, 1983). A significant correlation between the 5-HT$_1$ binding site and the 5-HT 'autoreceptor' was found using either electrical stimulation to induce release of ^{3}H-5-HT from rat cortical slices, or overflow of ^{3}H-norepinephrine from strips of canine saphenous vein. Furthermore, agonist interactions in each of these two systems were significantly correlated ($r = 0.83$). More recently, Engel *et al.* (1986) have expanded their analysis in view of recent data on 5-HT$_1$ binding site subtypes. No significant correlations were observed between drug potencies at the 5-HT 'autoreceptor' and their affinities for 5-HT$_{1C}$ or 5-HT$_2$ binding sites. Significant correlations were obtained with drug affinities for both 5-HT$_{1A}$ and 5-HT$_{1B}$ sites. However, selective 5-HT$_{1A}$ drugs such as 8-OH-DPAT and ipsapirone had no effect on the autoreceptor and therefore were not used in the calculation of the correlation coefficient. As a result, extensive pharmacologic analyses of 5-HT 'autoreceptors' support the hypothesis that the receptor-mediating release of 5-HT and certain other neurotransmitters from presumed nerve terminals is mediated by the 5-HT$_{1B}$ binding site. A similar conclusion was reached from an analysis of 5-HT and related drug effects on the release of ^{3}H-5-HT induced by depolarization of rat cerebellum synaptosomes (Raiteri *et al.*, 1986). By contrast, the 5-HT heteroreceptor-mediating release of glutamate did not conform to the previously described $5\text{-}HT_{1A}$, $5\text{-}HT_{1B}$ or $5\text{-}HT_{1C}$ binding sites.

$5\text{-}HT_{1C}$ receptors: phosphatidylinositol turnover

Besides adenylate cyclase, phosphoinositide turnover is believed to be a common 'second messenger' in the transduction of neurotransmitter signals to the cell interior (Berridge, 1984; Nishizuka, 1984). 5-HT has been shown to increase phosphoinositide turnover in the mammalian central nervous system (Brown *et al.*, 1984; Conn and Sanders-Bush, 1984; Kendall and Nahorski, 1985). The hydrolysis of phosphoinositides may modulate a number of intracellular processes including calcium flux, increased arachidonate metabolism, increased cyclic GMP and protein kinase C production.

Preliminary studies have shown that the $5\text{-}HT_{1C}$ site appears to mediate 5-HT-induced phosphoinositide turnover in the rat choroid plexus (Conn *et al.*, 1986). The pharmacology of 5-HT stimulated phosphatidylinositol hydrolysis in choroid plexus was compared to the potency of the drugs versus ^{125}I-LSD binding to $5\text{-}HT_{1C}$ receptors. Mianserin and ketanserin were potent antagonists of 5-HT-induced changes whereas spiperone was more than an order of magnitude less potent in this system. The authors concluded that the $5\text{-}HT_{1C}$ site in choroid plexus is functionally linked to phosphatidylinositol turnover.

$5\text{-}HT_{1D}$ receptors

Isolated perfused rat kidney

To date, a specific physiologic effect of 5-HT has not been correlated with the recently defined $5\text{-}HT_{1D}$ binding site. However, Charlton *et al.* (1986) described an inhibitory prejunctional '5-HT₁-like' receptor in the isolated perfused rat kidney. This 5-HT receptor inhibited the stimulus-induced release of ^{3}H-noradrenaline following sympathetic periarterial nerve stimulation to the kidney. 5-CT was the most potent agonist (IC_{30} = 1.8 nM) whereas 5-HT (IC_{30} = 45 nM) and RU 24969 (IC_{30} = 250 nM) were slightly less potent agonists. 8-OH-DPAT was totally inactive in this system. The effects of these agonists could be antagonized by methiothepin, metergoline and methysergide but not by cyproheptadine, ketanserin, mesulergine, (−)propranolol or (±)pindolol. The authors concluded that the 5-HT receptor in this system conformed to the general criteria of a '5-HT₁-like' receptor but could not be correlated to either the $5\text{-}HT_{1A}$, $5\text{-}HT_{1B}$ or $5\text{-}HT_{1C}$ binding site subtype. The potencies of 5-HT and related agents in the isolated perfused rat kidney system are therefore quite similar to their potencies at the putative 5-HT$_{1D}$ binding site.

Rat stomach fundus

5-HT-induced contraction of the rat stomach fundus is a second physiologic effect that does not appear to be mediated by 5-HT$_{1A}$, 5-HT$_{1B}$, 5-HT$_{1C}$ or 5-HT$_2$ binding site subtypes (Leysen and Tollenaere, 1982; Cohen and Wittenauer, 1986). Interestingly, in the first major study of the system (Vane, 1959), it was demonstrated that 5-HT and 5-methoxytryptamine were equipotent agents in the rat stomach fundus while tryptamine and *N*, *N*-dimethyltryptamine were 32- and 110-fold, respectively, less active than 5-HT. Similar rank-order potencies exist for these analogs at the putative 5-HT$_{1D}$ site (Heuring and Peroutka, 1986). More recently, Clineschmidt *et al.* (1986) concluded that the fundic 5-HT receptor resembled the '5-HT$_1$' binding site but that the heterogeneous nature of the radioligand binding precluded a more definitive correlation. Of note is the fact that yohimbine was a potent antagonist in this system, in agreement with its high affinity for the putative 5-HT$_{1D}$ site. Clearly, more detailed future studies are needed to define functional correlates of the putative 5-HT$_{1D}$ binding site subtype.

5-HT$_2$ binding sites

Phosphoinositol turnover

Analysis of 5-HT induced inositol phospholipid hydrolysis in rat cerebral cortex (Kendall and Nahorski, 1985; Conn and Sanders-Bush, 1985) suggests that this event may occur as a result of 5-HT$_2$ receptor activation. For example, the response to 5-HT is blocked by nanomolar concentrations of ketanserin, and phosphoinositide turnover is not affected by 8-OH-DPAT. The rat thoracic aorta (Roth *et al.*, 1984; Nakaki *et al.*, 1985), cultured bovine aortic smooth muscle cells (Coughlin *et al.*, 1981) and platelets (de Chaffoy de Courcelles *et al.*, 1985) are three additional systems in which 5-HT appears to modulate phosphoinositide turnover via a receptor which is similar to the 5-HT$_2$ binding site.

Vascular contractions

In addition to 5-HT-induced contractions of the canine basilar artery, 5-HT and related agents also interact with a number of other vascular tissues. For example, the 5-HT K_{ED50} for induced contractions of the rabbit and rat aorta, canine femoral artery and rat caudal artery range from 150 to 6000 nM (Peroutka, 1984). Antagonist interactions with these vessels are almost exclusively competitive in nature. In the rabbit aorta, for example, methysergide, cyproheptadine, spiperone and ketanserin all produce parallel shifts of the 5-HT dose response curve (Maayani *et al.*, 1984). Moreover, classical 5-HT antagonists are extremely potent agents in many vascular tissues, with antagonist potencies in the nanomolar range (Apperley *et al.*, 1976, 1980; Peroutka, 1984; Cory *et al.*, 1986). Furthermore, a similarity between the pattern of inhibition at these receptor sites and classical 'D receptors' has been noted (Apperley *et al.*, 1980). Thus, 5-HT$_2$ receptors mediate 5-HT-induced contractions in a wide variety of vascular tissues.

Behavioral studies

Drug antagonism of the 'head-shake' or 'head-twitch component' of the 5-HT behavioral syndrome has clearly been related to a blockade of the 5-HT$_2$ receptors (Peroutka *et al.*, 1981; Leysen *et al.*, 1982; Leysen, 1983; Yap and Taylor, 1983; Colpaert and Janssen, 1983). Likewise, tryptamine-induced seizure activity could be prevented by antagonists in a manner consistent with antagonism of the 5-HT$_2$ receptor (Leysen *et al.*, 1978). Other behavioral effects have also been attributed to the 5-HT$_2$ receptor. For example, two independent behavioral studies of the discriminative cue properties of 5-HT agonists have concluded that this behavioral response may be mediated by 5-HT$_2$ receptors (Glennon *et al.*, 1983; Friedman *et al.*, 1983).

Other systems

Antagonism of 5-HT-induced forepaw edema in the rat by 22 antagonists correlated with drug affinity for 5-HT$_2$ sites labeled by ^{3}H-spiperone (Ortmann *et al.*, 1982). Similarly, drug antagonism of tracheal smooth muscle contraction and *in vivo* bronchoconstriction is consistent with mediation by 5-HT$_2$ sites (Leysen *et al.*, 1984), as is contraction of guinea pig ileum (Engel *et al.*, 1984a). The dose-dependent pressor response which occurs after administration of 5-HT to pithed normotensive rats can also be blocked in a manner consistent with 5-HT$_2$ receptor mediation (Kalkman *et al.*, 1983a). 5-HT$_2$ receptors have also been implicated in the regulation of aldosterone production (Matsuoka *et al.*, 1985). 5-HT induction of platelet shape changes and aggregation may also be mediated by 5-HT$_2$ receptors (DeClerck and Herman, 1983; Pletscher and Affolter, 1983; Geaney *et al.*, 1984; DeClerck *et al.*, 1984). Leysen *et al.* (1984) have reviewed the role of 5-HT$_2$ receptors in platelet function.

FUTURE TRENDS

In the present report, evidence is provided from a variety of biochemical, physiologic and behavioral studies that multiple 5-HT receptors exist in the central nervous system. At the present time, the differentiation of 5-HT receptors into 5-HT_{1A}, 5-HT_{1B}, 5-HT_{1C}, 5-HT_{1D} and 5-HT_2 subtypes appears to be the most relevant classification system. Indeed, a number of functional correlates of these '5-HT binding sites' have been proposed (Table 3). In many of these systems, significant correlations have been documented between physiologic drug potencies and drug affinities for specific binding site subtypes.

The development of selective 5-HT_1 agonists such as 8-OH-DPAT (Hjorth *et al.*, 1982; Middlemiss and Fozard, 1983) and RU 24969 (Euvrard and Boisser, 1980) has greatly facilitated research into the analysis of 5-HT_1 binding site subtypes. To a significant degree, the recent advancements in the analysis of 5-HT receptor subtypes are a direct result of radioligand binding techniques. Although binding techniques have been criticized in terms of their relevance to functional receptor sites, the current data strongly support an association between the majority of serotonergic binding sites defined in radioligand studies and distinct functional responses.

Further advancements in the understanding of the physiologic effects of 5-HT would have important implications. To the basic scientist, characterization of all specific 5-HT receptors would greatly clarify the role of 5-HT in the central nervous system. Clinically, 5-HT has been implicated in a number of human disorders such as anxiety, depression, migraine, vasospasm and epilepsy, although the exact role of 5-HT in these disorders remains speculative. Analysis of 5-HT receptor subtypes and their functional role in the central nervous system should greatly elucidate the pathophysiological basis of many of these human diseases.

ACKNOWLEDGEMENTS

This work was supported in part by the John A. and George L. Hartford Foundation, the Alfred P. Sloan Foundation, the McKnight Foundation and NIH Grant 12151–12.

REFERENCES

Ahlenius, S., Larsson, K., Svensson, L., Hjorth, S., Carlsson, A., Lindberg, P., Wikström, H., Sanchez, D., Arvidsson, L.-E., Hacksell, U., and Nilsson, J. L. G. (1981) Effects of a new type of 5-HT receptor agonist on male rat sexual behavior, *Pharmacol. Biochem. Behav.*, **15**, 785–792.

Andrade, R., and Nicoll, R. A. (1985) The novel anxiolytic buspirone elicits a small hyperpolarization and reduces serotonin responses at putative 5-HT_1 receptors on hippocampal CA1 pyramidal cells, *Soc. Neurosci. Abs.*, **11**, 597.

Apperley, E., Feniuk, W., Humphrey, P. P. A., and Levy, G. P. (1980) Evidence for two types of excitatory receptor for 5-hydroxytryptamine in dog isolated vasculature, *Br. J. Pharmac.*, **68**, 215–224.

Apperley, E., Humphrey, P. P. A., and Levy, G. P. (1976) Receptors for 5-hydroxytryptamine and noradrenaline in rabbit isolated ear artery and aorta, *Br. J. Pharmac.*, **58**, 211–221.

Asarch, K. B., Ransom, R. W., and Shih, J. C. (1985) [³H]-1-(2-4-aminophenyl)ethyl-4-(3-trifluoromethylphenyl) piperazine: a selective radioligand for 5-HT₁ₐ receptors in rat brain, *Soc. Neurosci.*, **11**, 1257.

Beck, S. G., Clarke, W. P., and Goldfarb, J. (1985) Spiperone differentiates multiple 5-hydroxytryptamine responses in rat hippocampal slices in vitro, *Eur. J. Pharmacol.*, **116**, 195–197.

Bennett, J. L., and Aghajanian, G. K. (1974) D-LSD binding to brain homogenates: Possible relationship to serotonin receptors, *Life Sci.*, **15**, 1935–1944.

Bennett, Jr., J. P., and Snyder, S. H. (1975) Stereospecific binding of d-lysergic acid diethylamide (LSD) to brain membranes: Relationship to serotonin receptors, *Brain Res.*, **94**, 523–544.

Bennett, Jr., J. P., and Snyder, S. H. (1976) Serotonin and lysergic acid diethylamide binding in rat brain membranes: Relationship to postsynaptic serotonin receptors, *Mol. Pharmacol.*, **12**, 373–389.

Berridge, M. J. (1984). Inositol trisphosphate and diacylglycerol as second messengers, *Biochem. J.*, **220**, 345–360.

Biegon, A., Rainbow, T. C., and McEwen, B. S. (1982) Quantitative autoradiography of serotonin receptors in the rat brain, *Brain Res.*, **242**, 197–204.

Blurton, P. A., and Wood, M. D. (1986) Identification of multiple binding sites for [³H]5-hydroxytryptamine in the rat CNS, *J. Neurochem.*, **46**, 1392–1398.

Bradley, P. B., Engel, G., Feniuk, W., Fozard, J. R., Humphrey, P. P. A., Middlemiss, D. N., Mylecharane, E. J., Richardson, B. P., and Saxena, P. R. (1986) Proposals for the classification and nomenclature of functional receptors for 5-hydroxytryptamine, *Neuropharmacol.*, **25**, 563–576.

Brown, E., Kendall, D. A., and Nahorski, S. R. (1984) Inositol phospholipid hydrolysis in rat cerebral cortical slices: I. receptor characterization, *J. Neurochem.*, **42**, 1379–1387.

Cerrito, F., and Raiteri, M. (1979) Serotonin release is modulated by presynaptic autoreceptors, *Eur. J. Pharmacol.*, **57**, 427–430.

Charlton, K. G., Bond, R. A., and Clarke, D. E. (1986) An inhibitory prejunctional 5-HT₁-like receptor in the isolated perfused rat kidney, *Naunyn-Schmiedeberg's Arch. Pharmacol.*, **332**, 8–15.

Clineschmidt, B. V., Reiss, D. R., Pettibone, J., and Robinson, J. L. (1986) Characterization of 5-hydroxytryptamine receptors in rat stomach fundus, *J. Pharmacol. Exp. Ther.*, **235**, 696–708.

Closse, A. (1983) [3H]Mesulergine, a selective ligand for serotonin-2 receptors, *Life Sci.*, **32**, 2485–2495.

Cohen, M. L., and Wittenauer, L. A. (1986) Further evidence that the serotonin receptor in the rat stomach fundus is not 5-HT₁ₐ or 5-HT₁ᵦ, *Life Sci.*, **38**, 1–5.

Colpaert, F. C., and Janssen, P. A. J. (1983) The head-twitch response to intraperitoneal injection of 5-hydroxytryptophan in the rat: Antagonist effects of purported 5-hydroxytryptamine antagonists and of pirenperone, an LSD antagonist, *Neuropharmacol.*, **22**, 993–1000.

Conn, P. J., and Sanders-Bush, E. (1984) Selective 5HT-2 antagonists inhibit serotonin stimulated phosphatidylinositol metabolism in cerebral cortex, *Neuropharmacol.*, **8**, 993–996.

Conn, P. J., and Sanders-Bush, E. (1985) Serotonin-stimulated phosphoinositide turnover: Mediation by the S_2 binding site in rat cerebral cortex but not in subcortical regions, *J. Pharmacol. Exp. Ther.*, **234**, 195–203.

Conn, P. J., Sanders-Bush, E., Hoffman, B. J., and Hartig, P. R. (1986) A unique serotonin receptor in choroid plexus is linked to phosphatidylinositol turnover, *Proc. Natl. Acad. Sci. USA*, **83**, 4086–4088.

Cory, R. N., Osman, R., and Maayani, S. (1986) Kinetic definition of agonist efficacy at a 5-hydroxytryptamine (5-HT_2) receptor in the isolated rabbit aorta, *J. Pharmacol. Exp. Ther.*, **236**, 48–54.

Coughlin, S. R., Moskowitz, M. A., Antoniades, H. N., and Levine, L. (1981) Serotonin receptor-mediated stimulation of bovine smooth muscle cell prostacyclin synthesis and its modulation by platelet-derived growth factor. *Proc. Natl. Acad. Sci. USA*, **78**, 7134–7138.

Cox, B., and Ennis, C. (1982) Characterization of 5-hydroxytryptaminergic autoreceptors in the rat hypothalamus, *J. Pharm. Pharmacol.*, **34**, 438–441.

Creese, I., and Snyder, S. H. (1978) ^{3}H-Spiroperidol labels serotonin receptors in rat cerebral cortex and hippocampus, *Eur. J. Pharmacol.*, **49**, 201–202.

DeClerck, F. F., and Herman, A. G. (1983) Hydroxytryptamine and platelet aggregation, *Fed. Proc.*, **42**, 228–232.

DeClerck, F., Xhonneux, B., Leysen, J., and Janssen, P. A. (1984) The involvement of 5-HT_2-receptor sites in the activation of cat platelets, *Thrombosis Res.*, **33**, 305–321.

de Chaffoy de Courcelles, D., Leysen, J. E., De Clerck, F., Van Belle, H., and Janssen, P. A. (1985) Evidence that phospholipid turnover is the signal transducing system coupled to serotonin-S_2 receptor sites, *J. Biol. Chem.*, **260**, 7603–7608.

Demopulos, C. M., and Peroutka, S. J. (1987) 'Specific' ^{3}H-8-OH-DPAT binding to glass fiber filter paper: Implications for the analysis of serotonin binding site subtypes, *Neurochem. Int.* **10**, 371–376.

Deshmukh, P. P., Yamamura, H. I., Woods, L., and Nelson, D. L. (1983) Computer-assisted autoradiographic localization of subtypes of serotonin$_1$ receptors in rat brain, *Brain Res.*, **288**, 338–343.

De Vivo, M., and Maayani, S. (1986) Characterization of the 5-hydroxytryptamine$_{1A}$ receptor-mediated inhibition of forskolin-stimulated adenylate cyclase activity in guinea pig and rat hippocampal membranes, *J. Pharmacol. Exp. Ther.*, **238**, 248–253.

Dompert, W. U., Glaser, T., and Traber, J. (1985) ^{3}H-TVX Q 7821: Identification of 5-HT_1 binding sites as target for a novel putative anxiolytic, *Naunyn-Schmiedeberg's Arch. Pharmacol.*, **328**, 467–470.

Doods, H. N., Kalkman, H. O., De Jonge, A., Thoolen, M., Wilffert, B., Timmermans, P., and Van Zwieten, P. A. (1985) Differential selectivities of RU 24969 and 8-OH-DPAT for the purported 5-HT_{1A} and 5-HT_{1B} binding sites. Correlation between 5-HT_{1A} affinity and hypotensive activity, *Eur. J. Pharmacol.*, **112**, 363–370.

Dourish, C. T., Hutson, P. H., and Curzon, G. (1986) Putative anxiolytics 8-OH-DPAT, buspirone and TVX Q 7821 are agonists at 5-HT_{1A} autoreceptors in the raphe nuclei, *Trends Pharmacol. Sci.*, **7**, 212–214.

Engel, G., Gothert, M. K., Hoyer, D., Schlicker, E., and Hillenbrand (1986) Identity of inhibitory presynaptic 5-hydroxytryptamine (5-HT) autoreceptors in the rat brain cortex with 5-HT_{1B} binding sites, *Naunyn-Schmiedeberg's Arch. Pharmacol.*, **357**, 1–7.

Engel, G., Gothert, M., Muller-Schweinitzer, E., Schlicker, E., Sistonen, L., and Stadler, P. A. (1983) Evidence for common pharmacological properties of [^{3}H]5-

hydroxytryptamine binding sites, presynaptic 5-hydroxytryptamine autoreceptors in CNS and inhibitory presynaptic 5-hydroxytryptamine receptors on sympathetic nerves, *Naunyn Schmiedeberg's Arch. Pharmacol.*, **324**, 116–124.

Engel, G., Hoyer, G., Kalkman, H. O., and Wick, M. B. (1984a) Identification of 5-HT$_2$ receptors on longitudinal muscle of the guinea pig ileum, *J. Recep. Res.*, **4**, 113–126.

Engel, G., Muller-Schweinitzer, E., and Palacios, J. M. (1984b) 2-[^{125}Iodo] LSD, a new ligand for the characterization and localisation of 5HT$_2$ receptors, *Naunyn Schmiedeberg's Arch. Pharmacol.*, **325**, 328–336.

Ennis, C., and Cox, B. (1982) Pharmacological evidence for the existence of two distinct serotonin receptors in rat brain, *Neuropharmacol.*, **21**, 41–44.

Euvrard, C., and Boissier, J. R. (1980) Biochemical assessment of the central 5-HT activity of RU 24969 (A piperidinyl indole), *Eur. J. Pharmacol.*, **63**, 65–72.

Farnebo, L. O., and Hamberger, B. (1974) Regulation of [^{3}H] 5-hydroxytryptamine release from rat brain slices, *J. Pharmacol.*, **26**, 642–644.

Fillion, G. (1983) 5-Hydroxytryptamine receptors in brain, *Handbook Psychopharmacol.*, **17**, 139–166.

Fillion, G. M. B., Rousselle, J., Fillion, M., Beaudoin, D. M., Goiny, M. R., Deniau, J., and Jacob, J. J. (1978) High-affinity binding of [^{3}H]5-hydroxytryptamine to brain synaptosomal membranes: Comparison with [^{3}H]lysergic acid diethylamide binding, *Mol. Pharmacol.*, **14**, 50–59.

Fozard, J. R. (1984) Neuronal 5-HT receptors in the periphery, *Neuropharmacol.*, **23**, 1473–1486.

Friedman, R. L., Barrett, R. J., and Sanders-Bush, E. (1983) Discriminative cue properties of quipazine: mediation by serotonin-2 binding sites, *Soc. Neurosci. Abs.*, **9**, 335.

Fuller, R. W. (1984) Serotonin receptors, *Monogr. Neural Sci.*, **10**, 158–181.

Gaddum, J. H., and Picarelli, Z. P. (1957) Two kinds of tryptamine receptor, *Br. J. Pharmacol. Chemother.*, **12**, 323–328.

Geaney, D. P., Schachter, M., Elliot, J. M., and Grahame-Smith, D. G. (1984) Characterisation of [^{3}H]lysergic acid diethylamide binding to a 5-hydroxytryptamine receptor on human platelet membranes, *Eur. J. Pharmacol.*, **97**, 87–93.

Glaser, T., Rath, M., Traber, J., Zilles, K., and Schleicher, A. (1985) Autoradiographic identification and topographical analyses of high affinity serotonin receptor subtypes as a target for the novel putative anxiolytic TVX Q 7821, *Brain Res.*, **358**, 129–136.

Glennon, R. A., Young, R., and Rosecrans, J. A. (1983) Antagonism of the effects of the hallucinogen DOM and the purported 5-HT agonist quipazine by 5-HT$_2$ antagonists, *Eur. J. Pharmacol.*, **91**, 189–196.

Göthert, M. (1980) Serotonin-receptor-mediated modulation of Ca^{2+} dependent 5-hydroxytryptamine release from neurones of the rat brain cortex, *Naunyn-Schmiedeberg's Arch. Pharmacol.*, **314**, 223–230.

Gozlan, H., El Mestikawy, S., Pichat, L., Glowinski, J., and Hamon, M. (1983) Identification of presynaptic serotonin autoreceptors using a new ligand: ^{3}H-PAT, *Nature*, **305**, 140–142.

Green, A. R. (1984) 5-HT mediated behaviour, *Neuropharmacol.*, **23**, 1521–1528.

Gudelsky, G. A., Koenig, J. I., and Meltzer, H. Y. (1985) Serotonin receptor subtypes and thermoregulation, *Abs. Am. Soc. Neuropsychopharmacol.*, **24**, 64.

Hall, M. D., El Mestikawy, S., Emerit, M. B., Pichat, L., Hamon, M., and Gozlan, H. (1985) [^{3}H]8-hydroxy-2-(di-n-propyl-amino)tetralin binding to pre- and postsynaptic 5-hydroxytryptamine sites in various regions of the rat brain, *J. Neurochem.*, **44**, 1685–1696.

Hall, M. D., Gozlan, H., Emerit, M. B., El Mestikawy, S., Pichat, L., and Hamon, M. (1986) Differentiation of pre- and post-synaptic high affinity serotonin receptor binding sites using physico-chemical parameters and modifying agents, *Neurochem. Res.*, **11**, 891–912.

Hamon, M., Bourgoin, S., El Mestikawy, S., and Goetz, C. (1984) Central serotonin receptors, in *Handbook of Neurochemistry* 2nd edn, Vol. 6 (Ed. H. Lajtha), pp. 107–143, Plenum Press, New York.

Hamon, M., Bourgoin, S., Jagger, J., and Glowinski, J. (1974) Effects of LSD on synthesis and release of 5-HT in rat brain slices, *Brain Res.*, **69**, 265–280.

Hartig, P. R., Kadan, M. J., Evans, J. J., and Krohn, A. M. (1983) ^{125}I-LSD: A high sensitivity ligand for serotonin receptors, *Eur. J. Pharmacol.*, **89**, 321–322.

Hartig, P. R., Scheffel, U., Frost, J. J., and Wagner, H. N. (1985) *In vivo* binding of ^{125}I-LSD to serotonin 5-HT$_2$ receptors in mouse brain, *Life Sci.*, **37**, 657–664.

Heuring, R. E., and Peroutka, S. J. (1986) Characterization of ^{3}H-5-HT binding in bovine caudate, *J. Neurosci.* (in press).

Heuring, R. E., Schlegel, J. R., and Peroutka, S. J. (1986) Species variations in 5-HT$_{1B}$ and 5-HT$_{1C}$ binding sites defined by RU 24969 competition studies, *Eur. J. Pharmacol.*, **122**, 279–282.

Hiner, B. C., Roth, H. L., and Peroutka, S. J. (1986) Antimigraine drug interactions with 5-hydroxytryptamine$_{1A}$ receptors, *Ann. Neurol.*, **19**, 511–513.

Hjorth, S., Carlsson, A., Lindberg, P., Sanchez, D., Wikstrom, H., Arvidsson, L. E., Hacksell, U., and Nilsson, J. L. G. (1982) 8-Hydroxy-(di-*n*-propylamino) tetralin, 8-OH-DPAT, a potent and selective simplified ergot congener with central 5-HT receptor stimulating activity, *J. Neural Trans.*, **55**, 169–188.

Hoffman, B. J., Karpa, M. D., Lever, J. R., and Hartig, P. R. (1985) N$_1$-methyl-2-I^{125} LSD (I^{125} MIL), a preferred ligand for serotonin 5-HT$_2$ receptors, *Eur. J. Pharmacol.*, **110**, 147–148.

Hoyer, D., Engel, G., and Kalkman, H. O. (1985a) Characterization of the 5-HT$_{1B}$ recognition site in rat brain: Binding studies with ($-$) [^{125}I] iodocyanopindolol, *Eur. J. Pharmacol.*, **118**, 1–12.

Hoyer, D., Engel, G., and Kalkman, H. O. (1985b) Molecular pharmacology of 5-HT$_1$ and 5-HT$_2$ recognition sites in rat and pig brain membranes: radioligand binding studies with [^{3}H]5-HT, [^{3}H]8OH-DPAT, ($-$)[^{125}I]-iodocyanopindolol, [^{3}H]mesulergine and [^{3}H]ketanserin, *Eur. J. Pharmacol.*, **118**, 13–23.

Hoyer, D., Pazos, A., Probst, A., and Palacios, J. M. (1986a) Serotonin receptors in the human brain: I. Characterization and autoradiographic localization of 5-HT$_{1A}$ recognition sites. Apparent absence of 5-HT$_{1B}$ recognition sites, *Brain Res.*, **376**, 85–96.

Hoyer, D., Pazos, A., Probst, A., and Palacios, J. M. (1986b) Serotonin receptors in the human brain: II. Characterization and autoradiographic localization of 5-HT$_{1C}$ and 5-HT$_2$ recognition sites. *Brain Res.*, **376**, 97–107.

Jacobs, B. L. (1976) An animal behavioral model for studying central serotonergic synapses, *Life Sci.*, **19**, 777–786.

Kadan, M. J., Krohn, A. M., Evans, M. J., Waltz, R. L., and Hartig, P. R. (1984) Characterization of ^{125}I-lysergic acid diethylamide binding to serotonin receptors in rat frontal cortex, *J. Neurochem.*, **43**, 601–606.

Kalkman, H. O., Batink, H. D., Thoolen, M. J. M. C., Timmermans, P. B. M. W. M., and Van Swieten, P. A. (1983a) Correlation between the affinity for

[³H]mianserin-labelled receptors in brain and antagonism of the serotonin pressor response in pithed rats, *Biochem. Pharmacol.*, **32**, 2111–2113.

Kalkman, H. O., Boddeke, H. W. G. M., Doods, H. N., Timmermans, P. B. M. W. M., and Van Zwieten, P. A. (1983b) Hypotensive activity of serotonin receptor agonists in rats is related to their affinity for 5-HT$_1$ receptors, *Eur. J. Pharmacol.*, **91**, 155–156.

Kalkman, H. O., Engel, G., and Hoyer, G. (1984) Three distinct subtypes of serotonergic receptors mediate the triphasic blood pressure response to serotonin in rats, *J. Hypertension*, **2**, 143–145.

Kendall, D. A., and Nahorski, S. R. (1985) 5-hydroxytryptamine-stimulated inositol phospholipid hydrolysis in rat cerebral cortex slices: Pharmacological characterization and effects of antidepressants. *J. Pharmacol. Exp. Ther.*, **233**, 473–479.

Kwong, L. L., Smith, E. R., Davidson, J. M., and Peroutka, S. J. (1986) Differential interactions of 'prosexual' drugs with 5-hydroxytryptamine$_{1A}$ and alpha$_2$-adrenergic receptors, *Behavioral Neurosci.*, **100**, 664–668.

Leysen, J. E. (1981) Serotonergic receptors in brain tissue: Properties and identification of various ³H-ligand binding studies in vitro, *J. Physiol. (Paris)*, **77**, 351–362.

Leysen, J. (1983) Serotonin receptor binding sites: Is there pharmacological and clinical significance?, *Med. Biol.*, **61**, 139–143.

Leysen, J. E., de Courcelles, D. C., De Clerck, F., Niemegeers, J. E., and Van Nueten, J. M. (1984) Serotonin-S$_2$ receptor binding sites and functional correlates, *Neuropharmacol.*, **23**, 1493–1501.

Leysen, J. E., Niemegeers, C. J. E., Tollenaere, J. P., and Laduron, P. M. (1978) Serotonergic component of neuroleptic receptors, *Nature*, **272**, 163–166.

Leysen, J. E., Niemegeers, C. J. E., Van Nueten, J. M., and Laduron, P. M. (1982) ³H-Ketanserin (R 41 468), a selective ³H-ligand for receptor binding sites, *Mol. Pharmacol.*, **21**, 301–314.

Leysen, J. E., and Tollenaere, J. P. (1982) Biochemical models for serotonin receptors, *Ann. Rev. Med. Chem.*, **17**, 1–10.

Limbird, L. E. (1981) Activation and attenuation of adenylate cyclase, *Biochem. J.*, **195**, 1–13.

Lovell, R. A., and Freedman, D. X. (1976) Stereospecific receptor sites for d-lysergic diethylamide in rat brain: Effects of neuro-transmitters, amine antagonists, and other psychotropic drugs, *Mol. Pharmacol.*, **12**, 620–630.

Lucki, I., Nobler, M. S., and Frazer, A. (1984) Differential actions of serotonin antagonists on two behavioral models of serotonin receptor activation in the rat, *J. Pharmacol. Exp. Ther.*, **228**, 133–139.

Maayani, S., Wilkinson, C. W., and Stollak, J. S. (1984) 5-Hydroxytryptamine receptor in rabbit aorta: Characterization by butyrophenone analogs, *J. Pharmacol. Exp. Ther.*, **229**, 346–350.

Maguire, M. E., Ross, E. M., and Gilman, A. G. (1977) Beta-adrenergic receptor: ligand binding properties and the interaction with adenyl cyclase, *Adv. Cyclic Nucleotide Res.*, **8**, 1–83.

Marcinkiewicz, M., Verge, D., Gozlan, H., Pichat, L., and Hamon, M. (1984) Autoradiographic evidence for the heterogeneity of 5-HT$_1$ sites in the rat brain, *Brain Res.*, **291**, 159–163.

Martin, L. L., and Sanders-Bush, E. (1982a) The serotonin autoreceptor: antagonism by quipazine, *Neuropharmacol.*, **21**, 445–450.

Martin, L. L., and Sanders-Bush, E. (1982b) Comparison of the pharmacological characteristics of 5 HT$_1$ and 5 HT$_2$ binding sites with those of serotonin autoreceptors which modulate serotonin release, *Naunyn-Schmiedeberg's Arch. Pharmacol.*, **321**, 165–170.

Matsuoka, H., Ishii, M., Goto, A., and Sugimoto, T. (1985) Role of serotonin type 2 receptors in regulation of aldosterone production, *Am. J. Physiol.*, **249**, E234–E238.

Middlemiss, D. N., and Fozard, J. R. (1983) 8-Hydroxy-2-(di-*n*-propylamino)-tetralin discriminates between subtypes of the 5-HT$_1$ recognition site, *Eur. J. Pharmacol.*, **90**, 151–153.

Moon, S. L., and Taylor, D. P. (1985) *In vitro* autoradiography of ^{3}H-buspirone and ^{3}H-2-deoxyglucose after buspirone administration, *Soc. Neurosci. Abs.*, **11**, 114.

Nakaki, T., Roth, B. L., Chuang, D. M., and Costa, E. (1985) Phasic and tonic components in 5-HT$_2$ receptor-mediated rat aorta contraction: Participation of Ca^{++} channels and phospholipase, C, *J. Pharmacol. Exp. Ther.*, **234**, 442–447.

Nelson, D. L., Herbet, A., Bourgoin, S., Glowinski, J., and Hamon, M. (1978) Characteristics of central 5-HT receptors and their adaptive changes following intracerebral 5, 7-dihydroxytryptamine administration in the rat, *Mol. Pharmacol.*, **14**, 983–995.

Nishizuka, Y. (1984) Turnover of inositol phospholipids and signal transduction, *Science*, **225**, 1365–1369.

Norman, A. B., Battaglia, G., Morrow, A. L., and Creese, I. (1985) [^{3}H]WB4101 labels S$_1$ serotonin receptors in rat cerebral cortex, *Eur. J. Pharmacol.*, **106**, 461–462.

Ortmann, R., Bischoff, S., Radeke, E., Buech, O., and Delini-Stula, A. (1982) Correlations between different measures of antiserotonin activity of drugs, *Naunyn-Schmiedeberg's Arch. Pharmacol.*, **321**, 265–270.

Palacios, J. M., Niehoff, D. L., and Kuhar, M. J. (1981) [^{3}H]Spiperone binding sites in brain: autoradiographic localization of multiple receptors, *Brain Res.*, **213**, 277–289.

Pazos, A., and Palacios, J. M. (1985) Quantitative autoradiographic mapping of serotonin receptors in the rat brain. I. Serotonin-1 receptors, *Brain Res.*, **346**, 205–230.

Pazos, A., Cortes, R., and Palacios, J. M. (1985a) Quantitative autoradiographic mapping of serotonin receptors in the rat brain. II. Serotonin-2 receptors, *Brain Res.*, **346**, 231–249.

Pazos, A., Engel, G., and Palacios, J. M. (1985b) Beta-adrenoceptor blocking agents recognize a subpopulation of serotonin receptors in brain, *Brain Res.*, **343**, 403–408.

Pazos, A., Hoyer, D., and Palacios, J. M. (1984a) Mesulergine, a selective serotonin-2 ligand in the rat cortex, does not label these receptors in porcine and human cortex: evidence for species differences in brain serotonin-2 receptors, *Eur. J. Pharmacol.*, **106**, 531–538.

Pazos, A., Hoyer, D., and Palacios, J. M. (1984b) The binding of serotonergic ligands to the porcine choroid plexus: characterization of a new type of serotonin recognition site, *Eur. J. Pharmacol.*, **106**, 539–546.

Pedigo, N. W., Yamamura, H. I., and Nelson, D. L. (1981) Discrimination of multiple [^{3}H]5-hydroxytryptamine binding sites by the neuroleptic spiperone in rat brain, *J. Neurochem.*, **36**, 220–226.

Peroutka, S. J. (1984) Vascular serotonin receptors: Correlation with 5-HT$_1$ and 5-HT$_2$ binding sites, *Biochem. Pharmacol.*, **33**, 2349–2353.

Peroutka, S. J. (1985) Selective labeling of 5-HT$_{1A}$ and 5-HT$_{1B}$ binding sites in bovine brain, *Brain Res.*, **344**, 167–171.

Peroutka, S. J. (1986) Pharmacological differentiation and characterization of 5-

HT$_{1A}$, 5-HT$_{1B}$ and 5-HT$_{1C}$ binding sites in rat frontal cortex, *J. Neurochem*, **47**, 529–540.

Peroutka, S. J., and Demopulos, C. M. (1986) ^{3}H-8-OH-DPAT 'specifically' labels glass fiber filter paper, *Eur. J. Pharmacol.*, **129**, 199–200.

Peroutka, S. J., and Snyder, S. H. (1979) Multiple serotonin receptors: Differential binding of ^{3}H-serotonin, ^{3}H-lysergic acid diethylamide and ^{3}H-spiroperidol, *Mol. Pharmacol.*, **16**, 687–699.

Peroutka, S. J., and Snyder, S. H. (1980a) Long-term antidepressant treatment decreases spiroperidol-labeled serotonin receptor binding, *Science*, **210**, 88–90.

Peroutka, S. J., and Snyder, S. H. (1980b) Regulation of serotonin$_2$ (5-HT$_2$) receptors labeled with ^{3}H-spiroperidol by chronic treatment with the antidepressant amitripty-line, *J. Pharmacol. Exp. Ther.*, **215**, 582–587.

Peroutka, S. J., Lebovitz, R. M., and Snyder, S. H. (1981) Two distinct central serotonin receptors with different physiological functions, *Science*, **212**, 827–829.

Peroutka, S. J., Huang, S., and Allen, G. S. (1986a) Canine basilar artery contrac-tions mediated by 5-hydroxytryptamine$_{1A}$ receptors, *J. Pharmacol. Exp. Ther.*, **237**, 901–906.

Peroutka, S. J., Mauk, M. D., and Kocsis, J. D. (1986b) Modulation of hippocampal neuronal activity by 5-hydroxytryptamine and 5-hydroxytryptamine$_{1A}$ selective drugs, *Neuropharmacol.* (in press).

Pletscher, A., and Affolter, H. (1983) The 5-hydroxytryptamine receptor of blood platelets, *J. Neural Transmission*, **57**, 233–242.

Quik, M., Iversen, L. L., Lardner, A., and Mackay, A. V. P. (1978) Use of ADTN to define specific ^{3}H-spiperone binding to receptors in brain, *Nature*, **274**, 513–514.

Raiteri, M., Maura, G., Bonanno, G., and Pittaluga, A. (1986) Differential pharma-cology and function of two 5-HT$_1$ receptors modulating transmitter release in cerebellum, *J. Pharmacol. Exp. Ther.*, **237**, 644–648.

Richardson, B. P., Engel, G., Donatsch, P., and Stadler, P. A. (1985) Identification of serotonin M-receptor subtypes and their specific blockade by a new class of drugs, *Nature*, **336**, 126–131.

Rodbell, M. (1980) The role of hormone receptors and GTP regulatory proteins in membrane transduction, *Nature*, **284**, 17–21.

Rogawski, M. A., and Aghajanian, G. K. (1981) Serotonin autoreceptors on dorsal raphe neurons: Structure-activity relationships of tryptamine analogs, *J. Neurosci.*, **1**, 10–1148.

Roth, B. L., Nakaki, T., Chuang, D. M., and Costa, E. (1984) Aortic recognition sites for serotonin (5-HT) are coupled to phospholipase C and modulate phospha-tidylinositol turnover. *Neuropharmacol.*, **23**, 1223–1225.

Schlegel, J. R., and Peroutka, S. J. (1986) Nucleotide interactions with 5-HT$_{1A}$ binding sites directly labeled by [^{3}H]-8-hydroxy-2-(di-*n*-propylamino)tetralin ([^{3}H]-8-OH-DPAT), *Biochem. Pharmacol.*, **35**, 1943–1949.

Schnellmann, R. G., Waters, S. J., and Nelson, D. L. (1984) [^{3}H]5-hydroxytrypta-mine binding sites: Species and tissue variation, *J. Neurochem.*, **42**, 65–70.

Schoemaker, H., and Langer, S. Z. (1986) [^{3}H]8-OH-DPAT labels the serotonin transporter in the rat striatum, *Eur. J. Pharmacol.*, **124**, 371–373.

Schotte, A., Maloteaux, J. M., and Laduron, P. M. (1983) Characterization and regional distribution of serotonin S$_2$-receptors in human brain, *Brain Res.*, **276**, 231–235.

Shenker, A., Maayani, S., Weinstein, H., and Green, J. P. (1985) Two 5-HT recep-tors linked to adenylate cyclase in guinea pig hippocampus are discriminated by 5-carboxamidotryptamine and spiperone, *Eur. J. Pharmacol.*, **109**, 427–429.

Sills, M. A., Wolfe, B. B., and Frazer, A. (1984) Determination of selective and nonselective compounds for the 5-HT$_{1A}$ and 5-HT$_{1B}$ receptor subtypes in rat frontal cortex, *J. Pharmacol. Exp. Ther.*, **231**, 480–487.

Sinton, C. M., and Fallon, S. L. (1986) Differences in response of dorsal and median raphe serotonergic neurons to 5-HT1 receptor ligands, *Soc. Neurosci. Abs.*, **12**, 1239.

Slater, P., and Patel, S. (1983) Autoradiographic distribution of serotonin$_2$ receptors in rat brain, *Eur. J. Pharmacol.*, **92**, 297–298.

Smith, L. M., and Peroutka, S. J. (1986) Differential effects of 5-hydroxytryptamine$_{1A}$ selective drugs on the 5-HT behavioral syndrome, *Pharmacol. Biochem. Behav.*, **24**, 1513–1519.

Snyder, S. H. (1983) Molecular aspects of neurotransmitter receptors: An overview, *Handbook Psychopharmacol.*, **17**, 1–12.

Sprouse, J. S., and Aghajanian, G. K. (1985) Serotonergic dorsal raphe neurons: Electrophysiological responses in rats to 5-HT$_{1A}$ and 5-HT$_{1B}$ receptor subtype ligands, *Soc. Neurosci. Abs.*, **11**, 47.

Sprouse, J. S., and Aghajanian, G. K. (1986) Inhibition of serotonergic dorsal raphe cell firing by 5-HT$_{1A}$ agonists: Stereoselective blockage by propranolol, *Soc. Neurosci. Abs.*, **12**, 1240.

Taylor, E. W., Duckles, S. P., and Nelson, D. L. (1986) Dissociation constants of serotonin agonists in the canine basilar artery correlate to K_i values at the 5-HT$_{1A}$ binding site, *J. Pharmacol. Exp. Ther.*, **236**, 118–125.

Titeler, M., Lyon, R. A., Davis, K. H. and Glennon, R. A. (1987) Selectivity of serotonergic drugs for multiple brain serotonin receptors. *Biochem. Pharmacol.* **36**, 3265–3271.

Tricklebank, M. D. (1985) The behavioral response to 5-HT receptor agonists and subtypes of the central 5-HT receptor, *Trends Pharmacol. Sci.*, **6**, 403–407.

Tricklebank, M. D., Forler, C., and Fozard, J. R. (1984) The involvement of subtypes of the 5-HT$_1$ receptor and of catecholaminergic systems in the behavioral response to 8-hydroxy-2-(di-*n*-propylamino)tetralin in the rat, *Eur. J. Pharmacol.*, **106**, 271–282.

Tricklebank, M. D., Forler, C., Middlemiss, D. N., and Fozard, J. R. (1985) Subtypes of the 5-HT receptor mediating the behavioural responses to 5-methoxy-*N*,*N* dimethyltryptamine in the rat, *Eur. J. Pharmacol.*, **117**, 15–24.

Tricklebank, M. D., Middlemiss, D. N., and Neill, J. (1986) Pharmacological analysis of the behavioural and thermoregulatory effects of the putative 5-HT$_1$ receptor agonist, RU 24969, in the rat, *Neuropharmacol.*, **25**, 877–886.

Trulson, M. E., and Arasteh, K. (1986) Buspirone decreases the activity of 5-hydroxytryptamine-containing dorsal raphe neurons in-vitro, *J. Pharm. Pharmacol.*, **38**, 380–382.

VanderMaelen, C. P., and Wilderman, R. C. (1984) Buspirone, a non-benzodiazepine anxiolytic drug, causes inhibition of serotonergic dorsal raphe neurons in the rat, *Soc. Neurosci. Abs.*, **10**, 259.

VanderMaelen, C. P., Gehlbach, G., Yocca, F. D., and Mattson, R. J. (1986) Inhibition of serotonergic dorsal raphe neurons in rat brain slice by the 5-HT$_1$ agonist 5-carboxyamidotryptamine, *Soc. Neurosci. Abs.*, **12**, 1239.

Vane, J. R. (1959) The relative activities of some tryptamine analogues on the isolated rat stomach strip preparation, *Brit. J. Pharmacol.*, **14**, 87–98.

Verge, D., Daval, G., Marcinkiewicz, M., Patey, A., El Mestikawy, S., Gozlan, H and Hamon, M. (1986) Quantitative autoradiography of multiple 5-HT; receptor subtypes in the brain of control or 5, 7-dihydroxytryptamine-treated rats. *J. Neurosci.* **6,** 3474–3482.

Von Hungen, K., Roberts, S., and Hill, D. F. (1974) Developmental and regional variations in neurotransmitter-sensitive adenylate cyclase systems in cell-free preparations from rat brain, *J. Neurochem.*, **22,** 811–819.

Von Hungen, K., Roberts, S., and Hill, D. F. (1975) Serotonin sensitive adenylate cyclase activity in immature rat brain, *Brain Res.*, **84,** 257–267.

Yagaloff, K. A., and Hartig, P. R. (1985) ^{125}I-LSD binds to a novel serotonergic site on rat choroid plexus epithelial cells, *J. Neurosci.*, **5,** 3178–3183.

Yap, C. Y., and Taylor, D. A. (1983) Involvement of 5-HT$_2$ receptors in the wet-dog shake behaviour induced by 5-hydroxytryptophan in the rat, *Neuropharmacol.*, **22,** 801–804.

Neuronal Serotonin
Edited by N. N. Osborne and M. Hamon
© 1988 John Wiley & Sons Ltd

CHAPTER 16

5HT Receptors: Transmembrane Signaling Mechanisms

ELAINE SANDERS-BUSH

Department of Pharmacology and Psychiatry
School of Medicine
Vanderbilt University
Nashville
TN 37232
USA

INTRODUCTION

Radioligand binding studies and functional studies suggest that there are multiple 5HT receptors. The current, most widely accepted classification is based on radioligand binding studies in the central nervous system, in which the two major categories are 5HT-1 and 5HT-2 sites, with the former being subdivided into 5HT-1a, 5HT-1b and 5HT-1c sites. How these sites are related to the multiple actions of 5HT and indeed whether all four binding sites do in fact function as receptors for mediating specific actions of 5HT is the object of intense investigation. Further, how and if other proposed functional classes of 5HT receptors, such as 5HT-M sites (Richardson *et al.*, 1985), will fit into this classification scheme is unclear. The current review focuses on the rapidly developing field of the biochemical transducing mechanisms which link 5HT recognition sites on the cell surface with intracellular responses. An underlying premise of these studies is that the demonstration of the coupling of a specific 5HT binding site to a specific transducing system provides direct evidence that that site is a functional receptor. Furthermore, characterization of the transmembrane signaling pathway is a first, important step in elucidation of the molecular mechanisms of serotonergic transmission. Two major transmembrane transducing systems which link cell surface receptors with intracellular responses are the adenylate cyclase/cyclic AMP signal cascade and the phospholipase C/phosphoinositide hydrolysis pathway. As illustrated schematically in Figs 1 and 2, these two signaling pathways have many features in common, including a final common mechanism involving protein phosphorylation. The current review focuses on those studies of 5HT-activated adenylate cyclase and phosphoinositide hydrolysis which have attempted to relate these responses to the 5HT binding sites as defined in the mammalian brain (Table 1).

Table 1   Proposed serotonin receptor/effector coupling systems

Tissue	References
5HT-1 receptors positively coupled to adenylate cyclase	
Guinea pig hippocampus	Barbaccia *et al.*, 1983
Rat hippocampus	Shenker *et al.*, 1983
5HT-1a receptors negatively coupled to adenylate cyclase	
Guinea pig and rat hippocampus	Devivo and Maayani, 1986
Cultured neurons	Weiss *et al.*, 1986
5HT-1c receptors positively coupled to phospholipase C	
Rat choroid plexus	Conn *et al.*, 1986b
Pig choroid plexus	Hoffman *et al.*, 1986
5HT-2 receptors positively coupled to phospholipase C	
Rat cerebral cortex	Conn and Sanders-Bush, 1985
Rat aorta	Roth *et al.*, 1986
Human platelets	de Chaffoy *et al.*, 1985
Rabbit platelets	Schachter *et al.*, 1985

Stimulation of specific cell surface receptors (R_s, Fig. 1) results in the activation of adenylate cyclase via a regulatory coupling protein (G_s, also referred to as N_s) that binds guanine nucleotides. The activation of adenylate cyclase leads to an increase in the intracellular concentration of cyclic AMP, which via cyclic AMP-dependent protein kinases and protein phosphorylation, mediates the physiological effects of receptor activation. Other receptors (R_i) are negatively coupled to adenylate cyclase, so that stimulation of

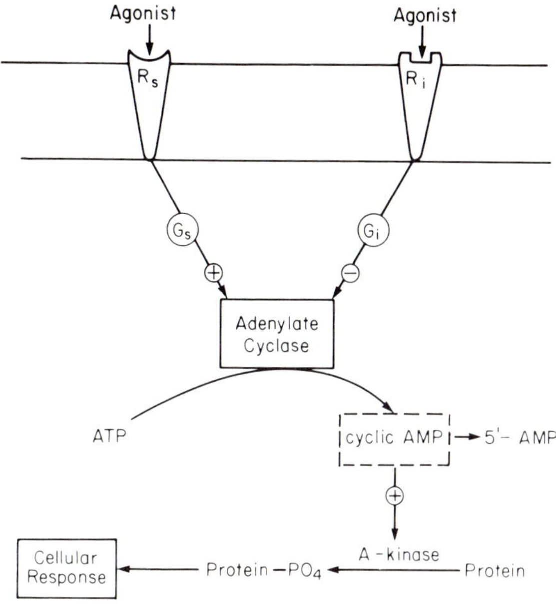

Fig. 1 Schematic illustration of the major components of the adenylate cyclase/cyclic AMP signal cascade. Activation of specific cell surface receptors (R_s or R_i) leads to the activation or inhibition, respectively, of adenylate cyclase. The receptors are linked to adenylate cyclase by different, specific guanine nucleotide binding proteins, G_s and G_i. Adenylate cyclase converts ATP to cyclic AMP, which activates a protein kinase, kinase A, to regulate cellular function via phosphorylation of specific proteins. The action of cyclic AMP is terminated by its metabolism to 5'-AMP by cAMP phosphodiesterase

these sites inactivates adenylate cyclase and decreases intracellular cyclic AMP levels. The coupling of these inhibitory receptors to adenylate cyclase also involves a guanine nucleotide binding protein, G_i (also referred to as N_i).

Other cell surface receptors are linked to a specific phosphodiesterase in the plasma membrane, phospholipase C. Activation of this enzyme leads to the hydrolysis of membrane inositol lipids and the initiation of a multifunctional transmembrane signaling process (Fig. 2). Three different membrane inositol lipid substrates are hydrolyzed by phospholipase C; phosphatidylinositol (PI), phosphatidylinositol-4-phosphate (PIP) and phosphatidylinositol-4,5-bisphosphate (PIP2). It is generally thought that PIP2 is the primary substrate of receptor-activated phospholipase C, although in some tissues the other two

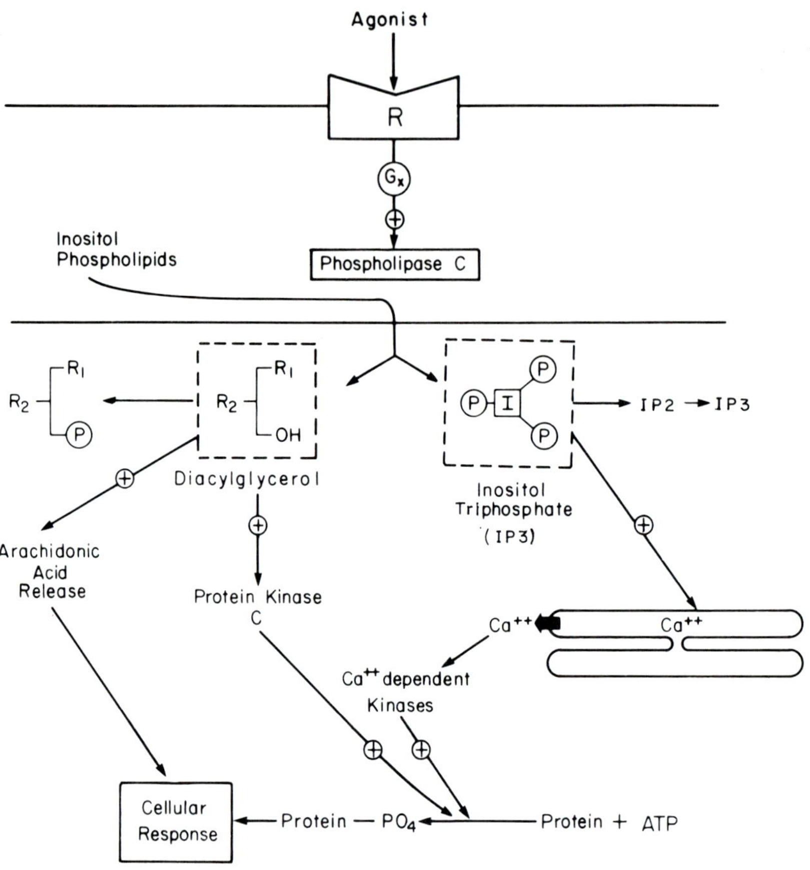

Fig. 2 Schematic illustration of the phospholipase C/phosphoinositide hydrolysis signal pathway. The interaction of an agonist with specific cell surface receptors (R) leads to the activation of phospholipase C. Phospholipase C catalyzes the hydrolysis of membrane inositol lipids, in particular phosphatidylinositol-4,5-bisphosphate, leading to the formation of at least two second messengers, inositol-1,4.5-triphosphate (IP3) and diacylglycerol (DAG). IP3 is metabolized to the inactive product IP2 and then IP by sequential dephosphorylation. The action of IP3 is mediated by calcium ions, released from intracellular storage sites (probably the endoplasmic reticulum). Calcium regulates cellular function in numerous ways, including the activation of calcium dependent protein kiinases. The other second messenger product DAG activates srotein kinase C, which catalyzes protein phosphorylation. DAG is metabolized to an inactive phosphorylated product or to arachidonic acid which leads to an increase in prostaglandins, prostacyclins and leukotrienes

lipids may serve as substrates for the activated phospholipase C (Dixon and Hokin, 1985). Recent evidence suggests that, analogous to the receptor-activated adenylate cyclase system, the coupling of cell surface receptors to phospholipase C is mediated by a guanine nucleotide binding protein. In some systems, the G-protein is similar to G_i, but in the majority of studies, the phosphoinositide hydrolysis-linked G-protein is different from both G_i or G_s (Taylor and Merritt, 1986) and is therefore referred to as G_x or N_x. The hydrolysis of PIP2 leads to the formation of two second messengers, inositol-1,4,5-trisphosphate (IP3) and diacylglycerol (DAG). Both products are rapidly metabolized, as would be required for an intracellular mediator. IP3 is metabolized by sequential dephosphorylation to IP2, then to IP and finally to inositol, all of which are inactive. The most common assay of phosphoinositide hydrolysis involves inhibition of the further metabolism of IP by the addition of lithium and measuring the accumulation of IP as an index of IP3 formation (Berridge *et al.*, 1982). DAG is metabolized by DAG kinase to the inactive product, phosphatidic acid, and also by a specific lipase (DAG lipase) to release glycerol and fatty acids, including arachidonic acid, a major component of phospholipase C derived DAG. The released arachidonic acid is subsequently metabolized via lipoxygenase and cyclooxygenase pathways to form prostaglandins and leukotrienes. These products may have local effects within the same cell (an autocrine function) and may also be released from the cell and alter the function of neighboring cells (i.e., have a paracrine action). The two primary second messengers, IP3 and DAG, alter cellular function by activation of specific protein kinases with subsequent protein phosphorylation – DAG activates protein kinase C and IP3 releases intracellular stores of calcium which then has numerous potential cellular effects including activation of calcium-dependent protein kinases. Additional products of phosphoinositide hydrolysis which are formed in some tissues include inositol-1,3,4-trisphosphate, an inactive isomer of IP3 (Irvine *et al.*, 1984; Burgess *et al.*, 1985), inositol-1,3,4,5-tetrakisphosphate (Batty *et al.*, 1985) and inositol-1,2-cyclic-4,5-trisphosphate (Wilson *et al.*, 1985). For clarity, these products have been omitted from the scheme in Fig. 2, but evidence is emerging that the latter two substances may also regulate cellular function.

5HT-1 RECEPTORS LINKED TO ADENYLATE CYCLASE

Positively coupled sites

5HT-induced cAMP formation was first demonstrated in tissue preparations of the liver fluke *Fasciola hepatica* (Mansour *et al.*, 1960). Since then, 5HT-sensitive adenylate cyclase systems have been described in a number of invertebrate tissues and in various cell lines (see Conn and Sanders-Bush,

1987, for a review), but the classification of these receptors and their relationship to the various 5HT binding sites described in mammalian systems has,
for the most part, not been investigated. One such study (Berry-Kravis and
Dawson, 1983) in which receptor classification was attempted involves the
NCB-20 hybrid cell line. The 5HT receptor linked to adenylate cyclase in
NCB-20 cells has properties in common with the [^{3}H]-5HT (5HT-1) binding
sites in those cells, but the pharmacology of these sites is not identical to the
5HT-1 sites in brain homogenates. 5HT-stimulated adenylate cyclase has
been extensively studied in homogenates of superior and inferior colliculi of
immature rats, but evidence suggests that the receptor which mediates this
response is distinct from the 5HT binding sites described in radioligand
binding studies (Nelson *et al.*, 1980). 5HT also stimulates adenylate cyclase
in a crude membrane fraction of striatum and the properties of this system
resemble, but are not identical to, the pharmacology of [^{3}H]-5HT binding
sites (Fillion *et al.*, 1979). Since all of these early studies were done prior to
the recognition of multiple 5HT-1 binding sites, which are labeled with [^{3}H]-
5HT, the possible mediation of these responses by one of the specific subtypes
should be re-evaluated. More recent studies have described a 5HT-activated
adenylate cyclase in hippocampus of adult rats (Barbaccia *et al.*, 1983) and
guinea pigs (Shenker *et al.*, 1983), which is markedly potentiated by depletion
of 5HT stores. The finding that nonselective 5HT antagonists, but not selective 5HT-2 antagonists, block the effect of 5HT suggests that a 5HT-1 site
mediates these effects. However, [^{3}H]-5HT binding and 5HT-stimulated
adenylate cyclase do not change correspondingly after lesions with kainic
acid or with 5,7-dihydroxytryptamine (5,7-DHT), leading to the suggestion
that the adenylate cyclase response is mediated by a subpopulation of 5HT-
1 recognition sites (Barbaccia *et al.*, 1983). More recently, Shenker *et al.*
(1985) found that spiperone, which has a greater affinity for the 5HT-1a than
for the 5HT-1b and 5HT-1c sites, partially blocks the adenylate cyclase
activation by 5-carboxyamidotryptamine, a 5HT agonist that has equal
affinity for the 5HT-1 subsites. Based on these results, they suggested that
the 5HT-1a receptor is positively linked adenylate cyclase. Other investigations have shown that 5HT-1c receptor sites are not linked to adenylate
cyclase (Palacios *et al.*, 1986).

Negatively coupled sites

Recent evidence suggest that 5HT receptors are also negatively coupled to
adenylate cyclase. For these studies, adenylate cyclase activity is stimulated
with a receptor agonist or with forskolin, which directly activates adenylate
cyclase, and then receptor-mediated inhibition of this response is measured.
In guinea pig hippocampal membranes, 5HT and 5HT agonists inhibit
forskolin-activated adenylate cyclase activity and their potencies are roughly

consistent with their binding affinities for the 5HT-1a site (Devivo and Maayani, 1985). Furthermore, the potencies of antagonists at preventing the 5HT inhibition of adenylate cyclase agree with their potencies at competing for the 5HT-1a binding site (Devivo and Maayani, 1986). 5HT-1a-like receptors which are negatively coupled to adenylate cyclase have also been found in cultured striatal and cortical neurons (Weiss *et al.*, 1986). Although many neurotransmitter substances may be positively or negatively coupled to adenylate cyclase at different receptors, the claim that the opposite effects of 5HT on adenylate cyclase are mediated by the same membrane receptor is unprecedented. Additional studies of this possibility are clearly needed. The co-existence of pharmacologically similar 5HT receptors in brain tissue, which are linked to adenylate cyclase in a stimulatory and an inhibitory way, may explain why studies of 5HT-stimulated adenylate cyclase in mammalian brain have not been very successful.

5HT-1 RECEPTORS LINKED TO PHOSPHOLIPASE C

5HT-stimulated phosphoinositide hydrolysis has been extensively studied in the insect salivary gland. In this organ, 5HT mobilizes calcium and increases saliva secretion, and evidence suggest that the phosphoinositide hydrolysis signal pathway mediates these responses (Fain and Berridge, 1979). The pharmacological profile of 5HT-stimulated phosphoinositide hydrolysis in the insect salivary gland resembles, but is not identical to, the 5HT-2 binding site in mammalian brain. It would be interesting to re-evaluate the effects of 5HT on phosphoinositide hydrolysis in the salivary gland in light of the recent evidence (as discussed below) that the 5HT-1c site is linked to phosphoinositide hydrolysis in mammalian tissues, since the pharmacology of the 5HT-1c site resembles the 5HT-2 site. In the salivary gland, 5HT also activates adenylate cyclase and a comparison of this response to the phosphoinositide hydrolysis response provided the first indication that different 5HT receptors are linked to these separate signaling pathways (Berridge and Heslop, 1981).

In another secretory tissue, the mammalian choroid plexus, 5HT induces a strong phosphoinositide hydrolysis response, while other neurotransmitters which are known to be coupled to phosphoinositide hydrolysis are weak or inactive (Fig. 3). The weak effect of norepinephrine is apparently mediated by activation of the 5HT receptors in this tissue (Conn and Sanders-Bush, 1986a). Radioligand binding studies in choroid plexus have led to the discovery of the latest 5HT binding site, the 5HT-1c site, which is highly localized in this tissue (Pazos *et al.*, 1984; Yagaloff and Hartig, 1985). Since other 5HT binding sites are apparently not present (Hoyer *et al.*, 1986), pharmacological studies of the 5HT phosphoinositide hydrolysis response in the choroid plexus have centered around the possible role of the 5HT-1c binding site (Conn *et al.*, 1986a). In addition to 5HT, a number of other

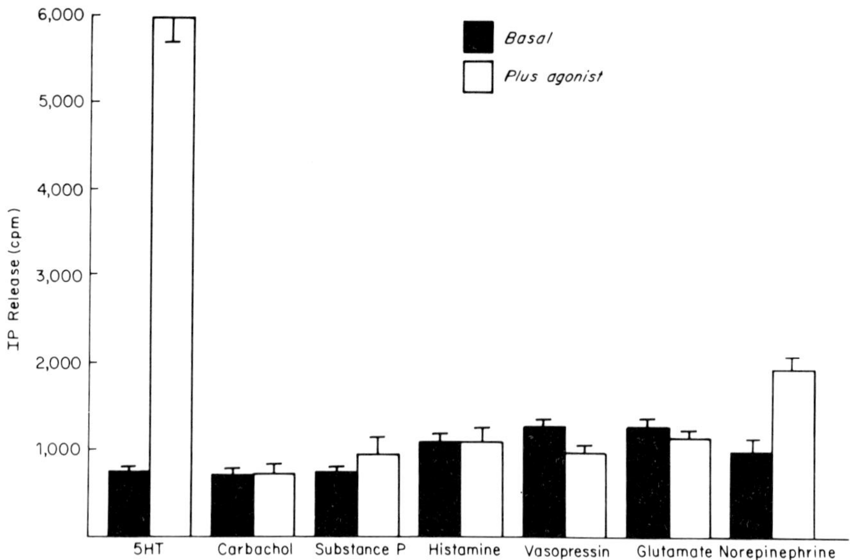

Fig. 3 Stimulation of phosphoinositide hydrolysis in the choroid plexus. The formation of [³H]-IP is measured in the presence of lithium. The concentration of 5HT is 50 nM. Other potential agonists are added in concentrations (0.1 to 1 mM) which give a maximum effect in other tissues

presumed 5HT agonists stimulate phosphoinositide hydrolysis in the choroid plexus (Table 2). Agonists potencies in the choroid plexus are roughly consistent with their binding affinities at the 5HT-1c site. Furthermore, the rank order and the absolute potencies of antagonists at blocking 5HT-stimulated phosphoinositide hydrolysis in the choroid plexus agrees with their affinities at the 5HT-1c binding site (Table 3). In choroid plexus of pigs, 5HT induces a 30-fold increase in phosphoinositide hydrolysis with a pharmacology consistent with the 5HT-1c site (Hoffman *et al.*, 1986). These data are consistent with the hypothesis that the 5HT-1c binding site in choroid plexus utilizes phosphoinositide hydrolysis as a signal-transducing mechanism. The 5HT-1c binding sites are localized on the choroidal epithelial cells (Yagaloff and Hartig, 1985; Yagaloff *et al.*, 1986) and, similarly, the 5HT receptors linked to phosphoinositide hydrolysis are apparently found on these cells (unpublished results). Given that phosphoinositide hydrolysis-linked receptors in other secretory tissues (including a 5HT receptor in salivary gland) regulate fluid secretion, it seems likely that the 5HT-1c sites in choroid plexus have a similar role, e.g. they regulate the production of cerebrospinal fluid. Given this possibility, it is important to know if these sites are in fact innervated by 5HT neurons. Neuroanatomical studies suggest that the choroid plexus receives serotonergic innervation; however, it is not clear that these fibers make synaptic contact with the epithelial cells rather than, or in

Table 2 Agonist profiles at phosphoinositide hydrolysis-linked receptors in cortex and choroid plexus

	Cerebral cortex			Choroid plexus		
	EC50* (nM)	Relative efficacy*	K_i† (nM)	EC50* (nM)	Relative efficacy*	K_i‡ (nM)
5-HT	3000	1.0	6000	30	1.0	30
5-HT	7000	1.1	4600	—	—	—
LSD-25	4	0.2	4	8	0.3	4
Quipazine	10000	0.6	—	800	0.4	1700
MK-212	18000	0.8	35000	830	0.9	—
TFMPP	Antagonist		—	550	0.4	—
MCPP	Antagonist		—	—	—	—

* Relative efficacy is the maximal release of [^{3}H]-IP relative to the maximal effect of 5-HT (220% and 600% of basal in cortex and choroid plexus, respectively) which is assigned a value of 1. EC50 refers to that concentration of agonist that elicits a half-maximal response. From Conn and Sanders-Bush (1985); Conn *et al.* (1986a) and unpublished results.

† Apparent K_i values at the 5HT-2 binding site were determined in frontal cortex in competition type experiments using [^{3}H]-ketanserin to label the 5HT-2 site. From Conn and Sanders-Bush (1985).

‡ Apparent K_i values at the 5HT-1c binding site were determined in choroid plexus in competition type experiments using [^{125}I]-LSD to label the 5HT-1c site. From Yagaloff and Hartig (1985) and Hoyer *et al.* (1986).

Table 3 Antagonists of choroid plexus 5-HT-stimulated phosphoinositide hydrolysis

	PI hydrolysis (K_i)	5HT-1c site (K_i)
Mianserin	10	5
Ketanserin	130	190
Imipramine	380	100
Chlorpheneramine	2600	3100
Propranolol	3600	750
Spiperone	6200	4600
8-OH-DPAT	>100000	20000
Haloperidol	>100000	>100000

Apparent K_i values in inhibiting the effect of 5HT (50 nM) on phosphoinositide hydrolysis in choroid plexus were estimated from dose–response curves. Apparent K_i values at the 5HT-1c site were determined in competition experiments using [^{125}I]-LSD as a radioligand in membranes of choroid plexus. From Conn *et al.* (1986a), Yagaloff and Hartig (1985), and unpublished results.

addition to, the vasculature in this tissue (Nakamura and Mariyasu, 1978; Moskowitz *et al.*, 1979; Napoleone *et al.*, 1982).

Since neuronal denervation of a structure that receives tonic input from a given neurotransmitter often results in a supersensitive response to that neurotransmitter, indirect evidence that the 5HT-1c receptor is under the tonic influence of 5HT is provided by the finding that lesions of 5HT neurons induce a supersensitive 5HT-1c phosphoinositide hydrolysis response in the choroid plexus (Conn *et al.*, 1986b). Given the high density of 5HT-1c receptor sites in the choroid plexus, the strong and apparently unique coupling of these receptors to phosphoinositide hydrolysis, and the finding that these receptors receive tonic serotonergic input, it may be possible for the first time to develop a specific correlate of 5HT receptor activation (fluid secretion) in a mammalian tissue that can be dissected at the molecular and cellular level with respect to the mechanisms of receptor regulation and the role of specific second messengers in the action of 5HT.

5HT-2 RECEPTORS LINKED TO PHOSPHOLIPASE C

Berridge *et al.* (1982) were the first to demonstrate that 5HT stimulates phosphoinositide hydrolysis in mammalian brain. The effect of 5HT and other agonists on [³H]-IP formation in cerebral cortical slices was markedly potentiated by the addition of lithium, which amplifies the signal by inhibiting the metabolism of IP. The first indication that the cerebral cortical phosphoinositide hydrolysis-linked 5HT receptor is related to one of the previously described 5HT binding sites was the finding that two potent 5HT-2 antagonists, ketanserin and pizotyline, block 5HT-stimulated phosphoinositide hydrolysis in cortical slices (Conn and Sanders-Bush, 1984). This finding has been confirmed and extended in a number of investigations of cerebral cortex (Conn and Sanders-Bush, 1985; Kendall and Nahorski, 1985), rat aorta (Roth *et al.*, 1984), and human (de Chaffoy de Courcelles *et al.*, 1985) and rabbit (Schachter *et al.*, 1985) platelets. Furthermore, as illustrated in Fig. 4 for the 5HT receptors in cerebral cortex, a strong correlation is found between the K_d values at blocking 5HT-stimulated phosphoinositide hydrolysis and at competing for the 5HT-2 binding site (Conn and Sanders-Bush, 1986b). A similar correlation has been found in human blood platelets (de Chaffoy de Courcelles *et al.*, 1985) and in rat aorta (Roth *et al.*, 1986). 5HT agonists stimulate phosphoinositide hydrolysis in cerebral cortex with affinities that are consistent with their apparent K_i values at inhibiting [³H]-ketanserin binding to the 5HT-2 binding site (Table 2). Furthermore, the effects of 5HT in cortical brain slices is apparently not a secondary response mediated by release of another neurotransmitter (Conn and Sanders-Bush, 1986c). These studies have led to a general agreement that the 5HT-2 recognition site employs phosphoinositide hydrolysis for signal transduction.

Studies of 5HT-stimulated phosphoinositide hydrolysis in the cerebral cortex, a cellular response closely linked to 5HT-2 receptor activation, have provided a direct assessment of 5HT-2 receptor sensitivity after *in vivo* treatments which modulate 5HT transmission (Conn and Sanders-Bush, 1986b). Previous studies of the 5HT-2 binding site and of *in vivo* functional

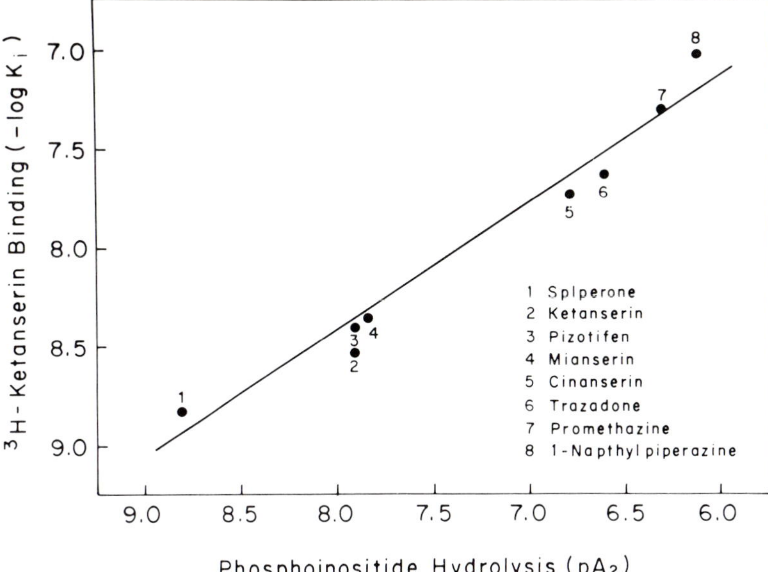

Fig. 4 Comparison of affinities at blocking 5HT-stimulated phosphoinositide hydrolysis and binding to the 5HT-2 site. The pA2 values (-log K_d) were determined by Schild analyses of 5HT concentration–effect curves in cortical slices as described by Conn and Sanders-Bush (1986b). Apparent K_i values were determined in competition binding studies of [^{3}H]-ketanserin in membranes prepared from frontal cortex. The correlation coefficient determined by linear regression analysis is 0.985

(behavioral) measures of the 5HT-2 receptor found an unexplained dissociation between binding and function. Chronic administration of 5HT-2 antagonists causes an apparent down-regulation of the 5HT-2 binding site (Blackshear and Sanders-Bush, 1982; Blackshear *et al.*, 1983; Leysen *et al.*, 1986), but a behavioral supersensitivity has been reported after these treatments (Mogilnicka and Klimek, 1979; Friedman *et al.*, 1983; Stoltz *et al.*, 1983). Denervation supersensitivity is suggested by behavioral studies (Nakamura and Fukushima, 1978; Yamamoto and Ueki, 1981), but the 5HT-2 binding sites are not up-regulated after lesions of 5HT neurons (Seeman *et al.*, 1980; Blackshear *et al.*, 1981; Quik and Azmita, 1983). Destruction of noradrenergic neurons has a similar lack of effect upon alpha-1 adrenergic receptor density, but a supersensitive alpha-1-mediated phosphoinositide hydrolysis response is found (Akhtar and Abdel-Latif, 1986; Kendall *et al.*, 1985; Janowsky *et al.*, 1984; Zatz, 1985). This suggests that the alpha-1 adrenergic receptor is regulated by alterations in receptor–effector coupling rather than by alterations in receptor density. This apparently is not the case for 5HT-2 neurons, since neither the phosphoinositide hydrolysis response nor the 5HT-2 binding site is changed after chronic depletion of 5HT stores by a chemical lesion of 5HT neurons or by the chronic administration of *p*-chlorophenylalanine (PCPA), an inhibitor of the rate-limiting step in 5HT synthesis (Table 4). Indeed, the phosphoinositide hydrolysis response mirrors precisely changes in the density of the 5HT-2 binding site (Table 4). For example, the chronic administration of mianserin, a 5HT-2 antagonist, causes a 48% loss of sites and a corresponding 52% decrease in the maximal phosphoinositide hydrolysis response to 5HT. Apparently, the mechanism of the 5HT-2 behavioral supersensitivity after chronic antagonist administration or after denervation occurs distal to the 5HT-2 receptor/effector complex. One possible postmembrane mechanism has been characterized in 1321N1 astrocytoma cells, where sensitivity to muscarinic agonists is regulated by changes in the sensitivity of the cell to the second messenger, IP3, without accompanying changes in muscarinic binding properties or in muscarinic-induced phosphoinositide hydrolysis (Masters *et al.*, 1985). It must also be recognized that the behavioral assays used to evaluate 5HT-2 receptor sensitivity measure motor output and hence are not specific. Thus, the *in vivo* changes in 5HT-2 mediated behavior may reflect changes in some other neurotransmitter system or alterations in intracellular metabolism distal to the mechanism of coupling of receptors to initial membrane responses.

Table 4 Serotonin-stimulated phosphoinositide hydrolysis and 5HT-2 receptor density after chronic treatments

	EC50 (μM)	Maximal response (% above basal)	B_{max} (fmol/mg protein)
Control	1.3 ± 0.4	159 ± 8	209 ± 25
Mianserin	2.4 ± 1.0	$81 \pm 7^*$	$100 \pm 10^*$
Control	1.9 ± 0.2	99 ± 21	172 ± 17
PCPA	2.2 ± 1.0	107 ± 12	191 ± 26
Control	0.6 ± 0.1	141 ± 13	227 ± 19
5,7-DHT	0.4 ± 0.1	146 ± 15	256 ± 18

EC50 value refers to the concentration of 5HT that elicits a half-maximal response. The maximal effect of 5HT is reported as the % increase above the basal level of [^{3}H]-IP. The density (B_{max}) of [^{3}H]-ketanserin binding sites was determined by Scatchard analysis.

Values are means $\pm$SEM. Adapted from Conn and Sanders-Bush (1986b)
$^*p < 0.001$ versus control

SUMMARY

Cell surface receptors are linked to intracellular responses by two major transmembrane transducing systems, the adenylate cyclase/cyclic AMP signal cascade and the phospholipase C/phosphoinositide hydrolysis signaling pathway. This review presents the evidence that specific 5HT receptor subtypes are linked to these transducing systems. Studies of 5HT-stimulated adenylate cyclase in mammalian brain suggest that the 5HT-1a site mediates this response, and recent studies suggest that a similar site may be negatively coupled to adenylate cyclase. Two 5HT receptor sites, 5HT-1c and 5HT-2, are positively coupled to phosphoinositide hydrolysis. So far, there is no evidence that 5HT receptor sites are negatively coupled to phospholipase C. Given that specific 5HT receptors are now known to be linked to specific transducing systems, future studies should focus on the intracellular biochemical changes elicited by these signals and their relation to the physiological functions of 5HT.

REFERENCES

Akhtar, R. A., and Abdel-Latif, A. A. (1986) Surgical sympathetic denervation increases alpha-1-mediated accumulation of myo-inositol trisphosphate and muscle contraction in rabbit iris dilator smooth muscle, *J. Neurochem.*, **46**, 96–104.

Barbaccia, M. L., Brunello, N., Chuang, D. M., and Costa, E. (1983) Serotonin-elicited amplification of adenylate cyclase activity in hippocampal membranes from adult rat, *J. Neurochem.*, **40**, 1671–1679.

Batty, I. R., Nahorski, S. R., and Irvine, R. F. (1985) Rapid formation of inositol 1,3,4,5-tetrakisphosphate following muscarinic receptor stimulation of rat cerebral cortical slices, *Biochem. J.*, **232**, 211–215.

Berridge, M. J., and Heslop, J. P. (1981) Separate 5-hydroxytryptamine receptors on the salivary gland of the blowfly are linked to the generation of either cyclic adenosine 3′,5′-monophosphate or calcium signals, *Brit. J. Pharmacol.*, **73**, 729–738.

Berridge, M. J., Downes, C. P., and Hanley, M. R. (1982) Lithium amplifies agonist-dependent phosphatidylinositol responses in brain and salivary glands, *Biochem. J.*, **206**, 587–595.

Berry-Kravis, E., and Dawson, G. (1984) Characterization of an adenylate cyclase-linked serotonin (5-HT$_1$) receptor in a neuroblastoma × brain explant hybrid cell line (NCB-20), *J. Neurochem.*, **40**, 977–985.

Blackshear, M. A., and Sanders-Bush, E. (1982) Serotonin receptor sensitivity after acute and chronic treatment with mianserin, *J. Pharmacol. Exp. Ther.*, **221**, 303–308.

Blackshear, M. A., Steranka, L. R., and Sanders-Bush, E. (1981) Multiple serotonin receptors: regional distribution and effect of raphe lesions, *Eur. J. Pharmacol.*, **76**, 325–334.

Blackshear, M. A., Friedman, R. L., and Sanders-Bush, E. (1983) Acute and chronic effects of serotonin (5HT) antagonists on serotonin binding sites, *Naunyn-Schmiedeberg's Arch. Pharmacol.*, **324**, 125–129.

Burgess, G. M., McKinney, J. S., Irvine, R. F., and Putney, Jr., J. W. (1985) Inositol 1,4,5-trisphosphate and inositol 1,3,4-trisphosphate formation in Ca^{2+}-mobilizing-hormone activating cells, *Biochem. J.*, **232**, 237–243.

Conn, P. J., and Sanders-Bush, E. (1984) Selective 5HT-2 antagonists inhibit serotonin stimulated phosphatidylinositol metabolism in cerebral cortex, *Neuropharmacology*, **23**, 993–996.

Conn, P. J., and Sanders-Bush, E. (1985) Serotonin-stimulated phosphoinositide turnover: mediation by the S_2 binding site in rat cerebral cortex but not in subcortical regions, *J. Pharmacol. Exp. Ther.*, **234**, 195–203.

Conn, P. J., and Sanders-Bush, E. (1986a) Agonist-induced phosphoinositide hydrolysis in choroid plexus, *J. Neurochem.*, **47**, 1754–1760.

Conn, P. J., and Sanders-Bush, E. (1986b) Regulation of serotonin-stimulated phosphoinositide hydrolysis: relation to the serotonin 5HT-2 binding site, *J. Neurosci.* **6**, 3669–3675.

Conn, P. J., and Sanders-Bush, E. (1986c) Biochemical characterization of serotonin stimulated phosphoinositide turnover, *Life Sci.*, **38**, 663–669.

Conn, P. J., and Sanders-Bush, E. (1987) Central serotonin receptors: Effector systems, physiological roles and regulation, *Psychopharmacology*, **92**, 267–277.

Conn, P. J., Janowsky, A., and Sanders-Bush, E. (1986b) Denervation supersensitivity of 5HT-1C receptors in rat choroid plexus, *Brain Res.*, **400**, 396–398.

Conn, J. P., Sanders-Bush, E., Hoffman, B. J., and Hartig, P. R. (1986a) A unique serotonin receptor in choroid plexus is linked to phosphatidylinositol turnover, *Proc. Natl. Acad. Sci.*, **83**, 4086–4088.

De Chaffoy de Courcelles, D., Leysen, J. E., De Clerck, F., Van Belle, H., and Janssen, P. A. J. (1985) Evidence that phospholipid turnover is the signal transducing system coupled to serotonin-S_2 receptor sites, *J. Biol. Chem.*, **260**, 7603–7608.

Devivo, M., and Maayani, S. (1985) Inhibition of forskolin-stimulated adenylate cyclase activity by 5-HT receptor agonists, *Eur. J. Pharmacol.*, **119**, 231–234.

Devivo, M., and Maayani, S. (1986) Characterization of the 5-hydroxytryptamine-1a receptor-mediated inhibition of forskolin-stimulated adenylate cyclase activity in guinea pig and rat hippocampal membranes, *J. Pharmacol. Exp. Therap.*, **238**, 248–253.

Dixon, J. K., and Hokin, L. E. (1985) The formation of inositol-1,2-cyclic phosphate on agonist stimulation of phosphoinositide breakdown in mouse pancreatic mini-lobules, *J. Biol. Chem.*, **260**, 16068–16071.

Fain, J. N., and Berridge, M. J. (1979) Relationship between hormonal activation of phosphotidylinositol hydrolysis, fluid secretion and calcium flux in the blowfly salivary gland, *Biochem. J.*, **178**, 45–58.

Fillion, G., Rousselle, J. C., Beaudoin, D., Pradelles, P., Goiny, M., Dray, F., and Jacob, J. (1979) Serotonin sensitive adenylate cyclase in horse brain synaptosomal membranes, *Life Sci.*, **24**, 1813–1822.

Friedman, E., Cooper, T. B., and Dallob, A. (1983) Effects of chronic antidepressant treatment on serotonin receptor activity in mice, *Eur. J. Pharmacol.*, **89**, 69–76.

Hoffman, B. J., Hartig, P. R., Conn, P. J., and Sanders-Bush, E. (1986) Thirty-fold stimulation of phosphoinositide hydrolysis by serotonin 5-HT-1c receptors in pig choroid plexus, *Neurosci. Abs.*, **12**, part 1, 576.

Hoyer, D., Srivatsa, S., Pazos, A., Engel, G., and Palacios, J. M. (1986). [125I]-LSD labels 5-HT-1c recognition sites in pig choroid plexus membranes. Comparison with [3H]-mesulergine and [3H]-5-HT binding, *Neurosci. Lett.*, **69**, 269–274.

Irvine, R. F., Letcher, A. J., Lander, D. J., and Downes, C. P. (1984) Inositol trisphosphates in carbachol-stimulated rat parotid glands, *Biochem. J.*, **223**, 237–243.

Janowsky, A., Labarca, R., and Paul, S. M. (1984) Noradrenergic denervation increases alpha-1 adrenoceptor-mediated inositol-phosphate accumulation in the hippocampus, *Eur. J. Pharmacol.*, **102**, 193–194.

Kendall, D. A., and Nahorski, S. R. (1985) 5-Hydroxytryptamine-stimulated inositol phospholipid hydrolysis in rat cerebral cortex slices: pharmacological characterization and effects of antidepressants, *J. Pharmacol. Exp. Ther.*, **233**, 473–479.

Kendall, D. A., Brown, E., and Nahorski, S. R. (1985) Alpha-1-adrenoceptor mediated inositol phospholipid hydrolysis in rat cerebral cortex: Relationship between receptor occupancy and response and effects of denervation, *Eur. J. Pharmacol.*, **114**, 41–52.

Leysen, J. E., Van Gompel, P., Gommeren, W., Woestenborghs, R., and Janssen, P. A. J. (1986) Down regulation of serotonin-S_2 receptor sites in rat brain by chronic treatment with the serotonin-S_2 antagonists: ritanserin and setoperone, *Psychopharmacology*, **88**, 434–444.

Mansour, T. E., Sutherland, E. W., Rall, T. W., and Bueding, E. (1960) The effect of serotonin (5-hydroxytryptamine) on the formation of adenosine $3',5'$-phosphate by tissue particles from the liver fluke, *Fasciola hepatica*, *J. Biol. Chem.*, **235**, 466–470.

Masters, S. B., Quinn, M. T., and Brown, J. H. (1985) Agonist induced desensitization of muscarinic receptor-mediated calcium efflux without concommitant desensitization of phosphoinositide hydrolysis, *Mol. Pharmacol.*, **27**, 325–332.

Mogilnicka, E., and Klimek, V. (1979) Mianserin, danitracen and amitriptyline withdrawal increases the behavioral responses of rats to L-5HTP, *Comm. J. Pharmacol.*, **31**, 704–705.

Moskowitz, M. A., Liebman, J. F., Reinhard, Jr., J. F., and Schlosberg, A. (1979) Raphe origin of serotonin-containing neurons within choroid plexus of the rat, *Brain Res.*, **169**, 590–594.

Nakamura, M., and Fukushima, H. (1978) Effects of reserpine, para-chlorophenylalanine, 5,6-dihydroxytryptamine and fludiazepam on the head twitches induced by 5-hydroxytryptamine or 5-methoxytryptamine in mice, *J. Pharm. Pharmacol.*, **30**, 254–256.

Nakamura, S., and Mariyasu, N. (1978) Nerve fibers and nerve endings in the choroid plexus: electron microscopic study, *Brain and Nerve*, **30**, 259–266.

Napoleone, P., Sancesario, G., and Amenta, F. (1982) Indoleaminergic innervation of rat choroid plexus: a fluorescence histochemical study, *Neurosci. Lett.*, **34**, 143–147.

Nelson, D. L., Herbet, A., Enjalbert, A., Bockaert, J., and Hamon, M. (1980) Serotonin-sensitive adenylate cyclase and [^{3}H]serotonin binding sites in the CNS of the rat. I. Kinetic parameters and pharmacological properties, *Biochem. Pharmacol.*, **29**, 2445–2453.

Palacios, J. J., Markstein, R., and Pazos, A. (1986) Serotonin 5-HT-1c sites in the choroid plexus are not linked in a stimulatory or inhibitory way to adenylate cyclase, *Brain Res.* (in press).

Pazos, A., Hoyer, D., and Palacios, J. M. (1984) The binding of serotonergic ligand to the porcine choroid plexus: characterization of a new type of serotonin recognition site, *Eur. J. Pharmacol.*, **106**, 539–546.

Quik, M., and Azmita, E. (1983) Selective destruction of the serotonergic fibers of the fornix-fimbria and cingulum bundle increases 5HT-1 but not 5HT-2 receptors in rat midbrain, *Eur. J. Pharmacol.*, **90**, 377–384.

Richardson, B. P., Engel, G., Donatsch, P., and Stadler, P. A. (1985) Identification of serotonin M-receptor subtypes and their specific blockade by a new class of drugs, *Nature*, **316**, 126–131.

Roth, B. L., Nakaki, T., Chuang, D. M., and Costa, E. (1984) Aortic recognition sites for serotonin (5HT) are coupled to phospholipase C and modulate phosphatidylinositol turnover, *Neuropharmacology*, **23**, 1223–1225.

Roth, B. L., Nakaki, T., Chuang, D., and Costa, E. (1986) 5-Hydroxytryptamine-2 receptors coupled to phospholipase C in rat aorta: Modulation of phosphoinositide turnover by phorbol ester, *J. Pharmacol. Exp. Ther.*, **238**, 480–486.

Schachter, M., Godfrey, P. P., Minchin, M. C. W., McClue, S. J., and Young, M. M. (1985) Serotonergic agonists stimulate inositol lipid metabolism in rabbit platelets, *Life Sci.*, **37**, 1641–1647.

Seeman, P., Westman, K., Coscina, D., and Warsh, J. J. (1980) Serotonin receptors in hippocampus and frontal cortex, *Eur. J. Pharmacol.*, **66**, 179–191.

Shenker, A., Maayani, S., Weinstein, H., and Green, J. P. (1983) Enhanced serotonin-stimulated adenylate cyclase activity in membranes from adult guinea pig hippocampus, *Life Sci.*, **32**, 2335–2342.

Shenker, A., Maayani, S., Weinstein, H., and Green, J. P. (1985) Two 5-HT receptors linked to adenylate cyclase in guinea pig hippocampus are discriminated by 5-carboxamidotryptamine and spiperone, *Eur. J. Pharmacol.*, **109**, 427–429.

Stoltz, J. F., Marsden, C. A., and Middlemiss, D. N. (1983) Effect of chronic antidepressant treatment and subsequent withdrawal on [3H]-5-hydroxytryptamine and [3H]-spiperone binding in rat frontal cortex and serotonin mediated behavior, *Psychopharmacology*, **80**, 150–155.

Taylor, C. W., and Merritt, J. E. (1986) Receptor coupling to polyphosphoinositide turnover: a parallel with the adenylate cyclase system, *TIPS*, **7**, 238–242.

Weiss, S., Sebben, M., Kemp, D. E., and Bockaert, J. (1986) Serotonin 5-HT$_1$ receptors mediate inhibition of cyclic AMP production in neurons, *Eur. J. Pharmacol.*, **120**, 227–230.

Wilson, D. B., Connolly, T. M., Bross, T. E., Majerus, P. W., Sherman, A. N. T., Rubin, L. J., and Brown, J. E. (1985) Isolation and characterization of the inositol cyclic phosphate products of polyphosphoinositide cleavage by phospholipase C, *J. Biol. Chem.*, **260**, 13496–13501.

Yagaloff, K. A., and Hartig, P. R. (1985) ^{125}I-Lysergic acid diethylamide binds to a novel serotonergic site on rat choroid plexus epithelial cells, *J. Neurosci.*, **5**, 3178–3183,

Yagaloff, K. A., Lozano, G., Van Dyke, T., Levine, A. J., and Hartig, P. R. (1986) Serotonin 5HT-1c receptors are expressed at high density on choroid plexus tumors from transgenic mice, *Brain Res.* (in press).

Yamamoto, T., and Ueki, S. (1985) The role of central serotonergic mechanisms in head-twitch and backward locomotion induced by hallucinogenic drugs, *Pharmacol. Biochem. Behav.*, **14**, 95–100.

Zatz, M. (1986) Denervation supersensitivity of the rat pineal to norepinephrine-stimulated [3H]inositide turnover revealed by lithium and a convenient procedure, *J. Neurochem.*, **45**, 95–100.

Neuronal Serotonin
Edited by N. N. Osborne and M. Hamon
© 1988 John Wiley & Sons Ltd

CHAPTER 17

The Pharmacology, Distribution and Function of 5-HT₃ Receptors

BRIAN P. RICHARDSON AND KARL-HEINZ BUCHHEIT
Preclinical Research Department
Sandoz Ltd
CH-4002 Basle
Switzerland

INTRODUCTION

Serotonin (5-hydroxytryptamine, 5-HT) is a neurotransmitter and inflammatory mediator which is stored and released from neurones, blood platelets, enterochromaffin cells of the gastrointestinal tract and – in some species – mast cells. Once released, serotonin produces an array of biological effects subsequent to binding and activating specific receptors located on the surface of its target cells. Evidence for the existence of different subtypes of serotonin receptor was first obtained more than 30 years ago (Gaddum and Picarelli, 1957), and it is the aim of this chapter to review recent progress which has been made specifically regarding the pharmacology and function of one of these subtypes – the 5-HT$_3$ receptor. Before doing this, however, it will be necessary to define the term 5-HT$_3$ receptor precisely, since only recently has it been used to describe a particular subtype of serotonin receptor found on various elements of the peripheral nervous system (Bradley *et al.*, 1986; Richardson and Engel, 1986). Thus it is a term which will probably not be familiar to all readers.

A new classification and nomenclature for serotonin receptor subtypes has been proposed (Bradley *et al.*, 1986) and is based on the action of several highly potent and selective agonist and antagonist drugs that have recently become available. These compounds possess the ability to mimic or antagonize a restricted number of the biological actions of serotonin in the body selectively and their use has led to the conclusion that three general categories of serotonin receptor can be distinguished. These have accordingly been designated 5-HT$_1$, 5-HT$_2$ and 5-HT$_3$ receptors (Bradley *et al.*, 1986). Details concerning this new nomenclature are given in the next section, which also provides a brief outline of the key historical events in serotonin receptor pharmacology that have paved the way for the new proposal. This background should enable the reader to understand exactly what is being referred to by the term 5-HT$_3$ receptor, and to appreciate how it differs from the 5-HT$_1$ and 5-HT$_2$ receptor subtypes. Subsequently a detailed account of the pharmacology, distribution and function of 5-HT$_3$ receptors will be given.

CLASSIFICATION OF 5-HT RECEPTOR SUBTYPES

As indicated in the introduction, the first evidence for heterogeneity of serotonin receptors was obtained more than 30 years ago by two groups working independently. In Rocha e Silva's laboratory in São Paulo and in Gaddum's laboratory in Edinburgh it was shown that serotonin contracts the isolated guinea pig ileum by two different mechanisms: indirectly through excitation of intramural nerves which then release acetylcholine, and through the direct stimulation of smooth muscle cells (Rocha e Silva *et al.*, 1953; Gaddum and Hameed, 1954). Working together, a representative from each of these groups later performed a cardinal series of experiments which showed that the serotonin receptors located on the cholinergic nerves in this preparation are pharmacologically distinguishable from those on the smooth muscle. Using morphine and dibenzyline – the latter nowadays being more commonly referred to as phenoxybenzamine – these investigators showed that the ability of serotonin to release acetylcholine in the guinea pig ileum could be blocked by morphine, whereas its direct action on the smooth muscle cells could not. Conversely, dibenzyline blocked the action of serotonin on the smooth muscle cells, but left its ability to release acetylcholine unaffected. Because of the differential sensitivity of these serotonin receptors to morphine and dibenzyline, the neuronal and smooth muscle receptors were designated M- and D-receptors respectively (Gaddum and Picarelli, 1957).

Although the general concept of different types of receptors mediating the contractile action of serotonin in the guinea pig ileum is still entirely valid today, morphine and dibenzyline are clearly unsuitable tools for generally classifying serotonin receptors since they are neither selective nor potent at these sites. Dibenzyline has well-recognized α-adrenoreceptor blocking properties, antagonizes responses to histamine and acetylcholine and inhibits the uptake of catecholamines into adrenergic nerve terminals and extra-neuronal tissue (Day and Vane, 1963; Furchgott, 1966; Cubeddu et al., 1974). Morphine, on the other hand, acts as a physiological antagonist of serotonin in the guinea pig ileum and reduces the acetylcholine release caused by a variety of different stimuli in this tissue (Rocha e Silva et al., 1953; Kosterlitz and Robinson, 1958).

Subsequent to these initial attempts to classify serotonin receptor subtypes, several other proposals have been made, but these too have suffered from the severe handicap that, until very recently, specific agonist and antagonist drugs for the various putative 5-HT receptor subtypes described were simply not available (see, for example, Drakontides and Gershon, 1968; Wallis, 1981; Humphrey, 1983). This resulted in considerable confusion in this area, a situation which has been further complicated by the proposal that serotonin receptors in the central nervous system be classified by the nature of the neuronal response elicited by serotonin in a particular preparation rather than by the sensitivity of such responses to specific agonist and antagonist drugs (Aghajanian, 1981). As a result, a parallel and entirely independent physiological classification for central serotonin receptors has emerged using the appellation S_1, S_2 and S_3 -receptors.

Fortunately several recent events have considerably clarified the function of serotonin receptor subtypes and enabled their improved pharmacological classification (see Fig. 1 and Table 1). The first of these was data published by Peroutka and Snyder showing that [³H]-5-HT and [³H]-spiperone bind in nanomolar concentrations to two different sites in rat brain cortex membranes (Peroutka and Snyder, 1979). That with high affinity for [³H]-5-HT was called the 5-HT₁ site, and the other – with high affinity for [³H]-spiperone – was called the 5-HT₂ site. The same group subsequently suggested that stimu-lation of 5-HT₁ sites produced an increase in neuronal adenylate cyclase activity, whereas 5-HT₂ sites are involved in mediating some of the central behavioural effects of serotonin in rats (Peroutka et al., 1981).

Using a large number of agonist and antagonist drugs it has recently been shown that the affinities of these compounds for the 5-HT₂ binding site in rat cerebral cortex membranes correlate very closely with those obtained in functional studies of the D-receptor in vascular and intestinal smooth muscle

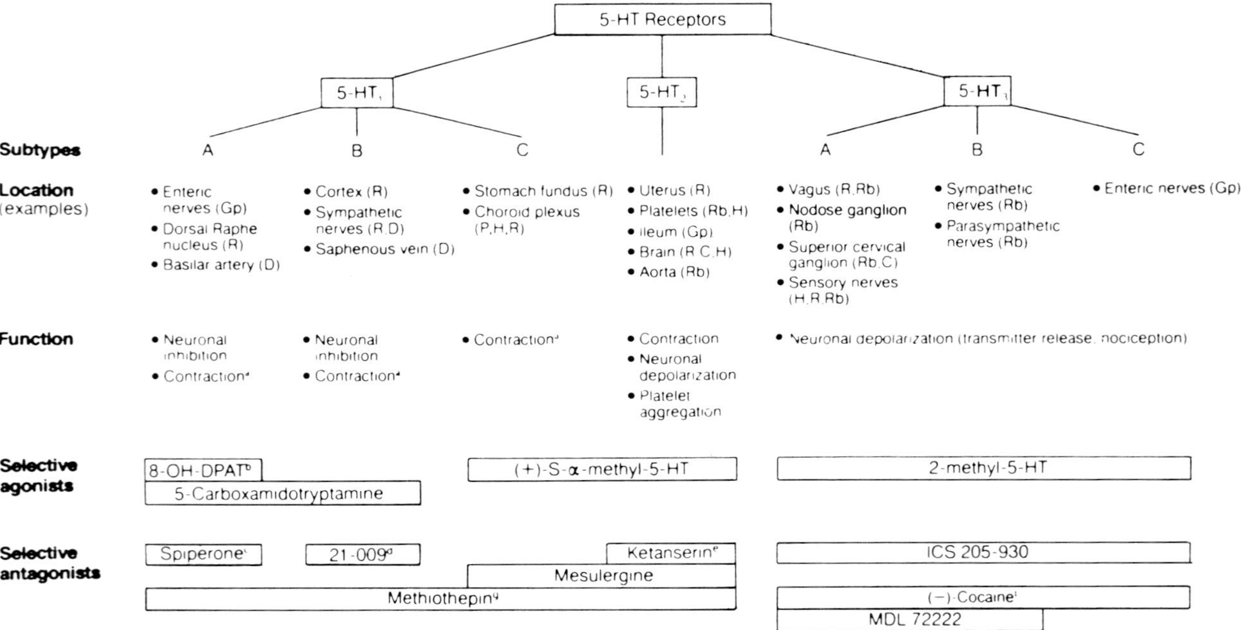

Fig. 1 Classification of 5-HT receptors. Abbreviations: C, cat; D, dog; Gp, guinea pig; H, human; P, pig; R, rat; Rb, rabbit. *Notes*: [a]5-HT₁ receptors also mediate relaxant responses in the precontracted guinea pig ileum and cat saphenous vein, but the exact subtype of 5-HT₁ receptor involved is still unclear; [b]8-OH-DPAT is 8-hydroxy-2(di-*n*-propylamino)tetralin. [c]Spiperone also has affinity for 5-HT₂ receptors, dopamine D₂ receptors and α-adrenoreceptors. [d]21-009 (4-(3-terbutylamino-2-hydroxypropoxy)-indole-2-carbonic acid isopropylester) also has affinity for β-adrenoreceptors. [e]Ketanserin also has affinity for α₁-adrenoreceptors and histamine H₁ receptors; [f](−)-Cocaine also has local anaesthetic action and inhibits neuronal catecholamine reuptake. [g]Methiothepin has affinity for histamine H₁ receptors, α₁-adrenoreceptors and dopamine D₂ receptors. (From Richardson and Engel, 1986; reproduced by permission)

(Humphrey *et al.*, 1982; Engel *et al.*, 1984, 1985; Maayani *et al.*, 1984). Thus it is now clear that the $5\text{-}HT_2$ binding sites originally described in the central nervous system are identical with the peripheral 5-HT D-receptors first described some 30 years ago by Gaddum and Picarelli in the guinea pig ileum (Engel *et al.*, 1985; Bradley *et al.*, 1986).

$5\text{-}HT_2$ receptors are widely distributed throughout the body and invariably mediate excitatory actions of serotonin such as neuronal depolarization, smooth muscle contraction and platelet activation (Leysen *et al.*, 1984; Fozard, 1985a; Bradley *et al.*, 1986). They can be specifically blocked by the drug ketanserin (see Fig. 1). There is no evidence to date to suggest heterogeneity within the $5\text{-}HT_2$ receptor subtype.

In contrast, $5\text{-}HT_1$ receptors are responsible for mediating inhibitory actions of serotonin, both in the central nervous system and in the periphery. Thus this receptor subtype elicits the presynaptic and prejunctional inhibitory actions of serotonin on central and peripheral neurones respectively (Feniuk *et al.*, 1979; Engel *et al.*, 1983, 1986; Charlton *et al.*, 1985; Fozard, 1985a) as well as the relaxant action of serotonin on a variety of smooth muscle preparations (Feniuk *et al.*, 1983). Although $5\text{-}HT_1$ receptors are involved in the inhibitory actions of serotonin, its contractile actions in some smooth muscle preparations can also be mediated by this receptor subtype (Humphrey, 1984; Bradley *et al.*, 1986). These receptors are insensitive to the $5\text{-}HT_2$ receptor antagonist ketanserin, but can be blocked by methiothepin, although the latter compound possesses considerable affinity for several other neurotransmitter receptors (see Fig. 1).

Recent studies employing radioligand binding and autoradiography have demonstrated heterogeneity within the $5\text{-}HT_1$ receptor category (Pedigo *et al.*, 1981; Pazos *et al.*, 1984; Hoyer *et al.*, 1985). Currently these receptors are subdivided into $5\text{-}HT_{1A}$, $5\text{-}HT_{1B}$ and $5\text{-}HT_{1C}$ sites, functional receptor correlates having now been established for each of these (Middlemiss, 1986; Buchheit *et al.*, 1986; see also Fig. 1). Thus $5\text{-}HT_{1A}$ sites mediate the stimulant effects of serotonin on neuronal adenylate cyclase (Markstein *et al.*, 1986) which presumably accounts for the inhibition of neuronal firing caused by serotonin in the dorsal raphe nucleus and rat hippocampal slice preparation (Fallon *et al.*, 1983; Beck *et al.*, 1985). On the other hand, $5\text{-}HT_{1B}$ sites mediate the inhibition of evoked serotonin and noradrenaline release from nerve terminals (Engel *et al.*, 1983, 1986) and $5\text{-}HT_{1C}$ sites the contractile actions of serotonin in the rat stomach fundus strip (Buchheit *et al.*, 1986). Unfortunately no specific antagonists currently exist for any of these $5\text{-}HT_1$ receptor subtypes, which makes their definitive characterization difficult.

In addition to the $5\text{-}HT_1$ and $5\text{-}HT_2$ subtypes of serotonin receptor, a third category clearly exists, since the 'M' receptors originally described by Gaddum and Picarelli are not blocked by methiothepin or ketanserin. These receptors, which are responsible for mediating many of the excitatory actions of serotonin in the peripheral nervous system (Fozard, 1985a; Richardson and Engel, 1986), have been designated the 5-HT$_3$ subtype in the recently proposed new classification (Bradley *et al.*, 1986). The remainder of this chapter will provide a detailed account of the pharmacology, distribution and function of these receptors.

THE PHARMACOLOGY OF 5-HT$_3$ RECEPTORS

Only in the last couple of years have selective agonist and antagonist drugs for 5-HT$_3$ receptors become available. Some of them are among the most potent drugs known to man. The background concerning their discovery and development has recently been reviewed in some detail (Richardson *et al.*, 1985; Richardson and Engel, 1986).

Since this is a relatively new field, the various assay systems which can be used to evaluate the activity of drugs at 5-HT$_3$ receptors will not be generally known and so will be described before the structure–activity relationships of 5-HT$_3$ receptor agonist and antagonist drugs are reviewed.

Bioassay systems

In vitro assays

The agonistic or antagonistic potency of compounds at 5-HT$_3$ receptors located on peripheral neurones can be conveniently assessed in three different *in vitro* bioassay systems: the rabbit vagus nerve, the rabbit heart and the guinea pig ileum. In all three systems compounds with activity at 5-HT$_1$ or 5-HT$_2$ receptors are totally inactive under the conditions employed. Thus neither methiothepin nor ketanserin influence the effects of 5-HT$_3$ agonists in these preparations (see Fig. 1 and Table 1).

In contrast to the situation with 5-HT$_1$ and 5-HT$_2$ receptors, no radioligand binding assay is presently available for determinating the affinity constants of compounds for 5-HT$_3$ receptors. This is because 5-HT$_3$ receptors are largely located on the terminals of peripheral neurones, making the preparation of membranes rich in these receptors difficult. Moreover, no suitable radioligand for these receptors yet exists.

Desheathed rabbit vagus nerve The isolated rabbit vagus nerve, where the epineurium has been removed (i.e. a 'desheathed' preparation), can be used for the determination of the agonistic or antagonistic potency of compounds at 5-HT$_3$ receptors on sensory nerves (Neto, 1978). Extracellular recordings are made such that the amplitude of the compound action potential carried by the myelinated (A) and non-myelinated (C) fibres can be recorded separately

Table 1   *In vitro* potencies of selective agonists and antagonists for the various 5-HT receptor subtypes

	5-HT receptor subtype selectivity	5-HT$_1$ receptors[a]			5-HT$_2$ receptors[b]	5-HT$_3$ receptors[c]		
		A	B	C		A	B	C
		pEC$_{50}$	pEC$_{30}$	pD$_2$	pD$_2$	pD$_2$	EC$_{50}$ (nmol)	pD$_2$
Agonists								
5-HT	None	6.9	6.8	8.0	6.8	6.0	11	5.9
5-CONH$_2$-T[f]	5-HT$_1$	7.2	7.7	6.9	5.9	<4.0[e]	>1000[e]	<4.0
α-CH$_3$-5-HT	5-HT$_2$	5.5	4.6	8.0[d]	7.1	<4.0	250	4.6
2-CH$_3$-5-HT	5-HT$_3$	4.4	4.2	6.2[d]	<4.0	5.7	18	5.4
		pK$_1$	pA$_2$	pA$_2$	pA$_2$	pA$_2$	pA$_2$	pA$_2$
Antagonists								
Methiothepin	5-HT$_1$/5-HT$_2$	8.5	7.0	N.D.	N.D.	<6.0	<6.0	<6.0
Ketanserin	5-HT$_2$	<5.0	<5.7	<6.0[d]	8.5	<6.0	<6.0	<6.0
ICS 205–930	5-HT$_3$	<5.0	<5.0	<5.0[d]	<6.0	10.2	10.6	7.9
206-792[g]	5-HT$_3$	N.D.	N.D.	N.D.	N.D.	8.9	9.8	7.7
206-830[g]	5-HT$_3$	N.D.	N.D.	N.D.	N.D.	13.1	10.1	9.1
(−)-Cocaine	5-HT$_3$	<4.0	N.D.	N.D.	N.D.	6.8	6.2	6.1
MDL 72222	5-HT$_3$	N.D.	N.D.	<4.3	<4.5	7.9	8.9	<6.0

The following functional assays were used to generate the data presented in this table (details of methods and sources of data are given in parenthesis):

[a] 5-HT$_1$ receptors: 5-HT$_{1A}$ = serotonin-sensitive adenylate cyclase in rat hippocampal homogenates (Markstein *et al.*, 1986); 5-HT$_{1B}$ = serotonin autoreceptor in rat cerebral cortex (Engel *et al.*, 1983); 5-HT$_{1C}$ = rat stomach fundus strip (Buchheit *et al.*, 1986; Fozard, 1984b).
[b] 5-HT$_2$ receptors: rabbit aorta (Apperley *et al.*, 1976; Thompson, C., unpublished observations).
[c] 5-HT$_3$ receptors: 5-HT$_{3A}$ = rabbit vagus nerve (Donatsch *et al.*, 1984b; Richardson *et al.*, 1985); 5-HT$_{3B}$ = rabbit heart (Fozard and Mobarok Ali, 1978b; Donatsch *et al.*, 1984b; Richardson, B. P., unpublished data); 5-HT$_{3C}$ = guinea pig ileum (Fozard *et al.*, 1979; Buchheit *et al.*, 1985b)
[d] Kalkman, H. O. unpublished data
[e] antagonist
[f] 5-carboxamide-tryptamine
[g] see Fig. 11
N.D. = no data available

(Oakley and Schafer, 1978). Serotonin and other 5-HT$_3$ agonists cause a concentration-dependent reduction in the amplitude of the C-fibre action potential (Fig. 2) without affecting that carried in the faster conducting myelinated fibres. The maximal reduction in the C-fibre action potential produced by serotonin is about 45%, indicating that a subpopulation of C-fibres is preferentially affected. To assess the agonistic potency of test compounds at the 5-HT$_3$ receptors in this preparation, cumulative concentration response curves are constructed. Subsequent to establishing a concentration response curve for serotonin itself, a curve for the putative agonist is determined. To confirm that the effect of this putative 5-HT$_3$ agonist is really mediated by 5-HT$_3$ receptors and not by any of the other receptors (e.g. nicotinic or muscarinic receptors) which are also present in this preparation, a 5-HT$_3$ antagonist like ICS 205-930 should be used to block the action of the agonist. ICS 205-930 is a highly selective competitive antagonist and, at concentrations above 10^{-10}M, causes a progressive rightward shift in the serotonin concentration response curve (see Fig. 3). The pA$_2$ value obtained with ICS 205-930 should be about 10.2 (Richardson *et al.*, 1985; see also Table 1).

Isolated rabbit heart The isolated heart of the rabbit perfused by the Langendorff technique in the presence of muscarinic antagonists responds to the injection of boluses of serotonin with an increase in rate and contractile force (Fig. 4; see also Jacob and Poite-Bevierre, 1960). This effect is caused by a specific receptor-mediated release of catecholamines from sympathetic neurones (Fig. 5; see also Fozard and Mwaluko, 1976). Serotonin agonists and antagonists with selectivity for these 5-HT$_3$ receptors mimic and inhibit respectively the serotonin-induced tachycardia and inotropic responses (Fozard and Mobarok Ali, 1978b). The action of 5-HT$_3$ receptor agonists is dose-dependent and reproducible and can be inhibited in a competitive manner by compounds such as ICS 205–930 which has a pA$_2$ value of 10.6 in this preparation (Fig. 6, Table 1; see also Richardson *et al.*, 1985).

Isolated guinea pig ileum Both the whole ileum and the longitudinal muscle-myenteric plexus preparation are contracted by small quantities of serotonin in a concentration-dependent manner. Three different mechanisms contribute to this contraction (see below; and Buchheit *et al.*, 1985b). To assess the action of agonists it is thus necessary to eliminate contributions by non-5-HT$_3$ receptor-mediated events. This can be achieved by including atropine and a 5-HT$_2$ receptor antagonist in the buffer solution. Under these conditions the residual serotonin-induced contractions are totally abolished by 5-HT$_3$ receptor antagonists (Buchheit *et al.*, 1985b) and the system can be used to determine the potency of agonists and antagonists at 5-HT$_3$

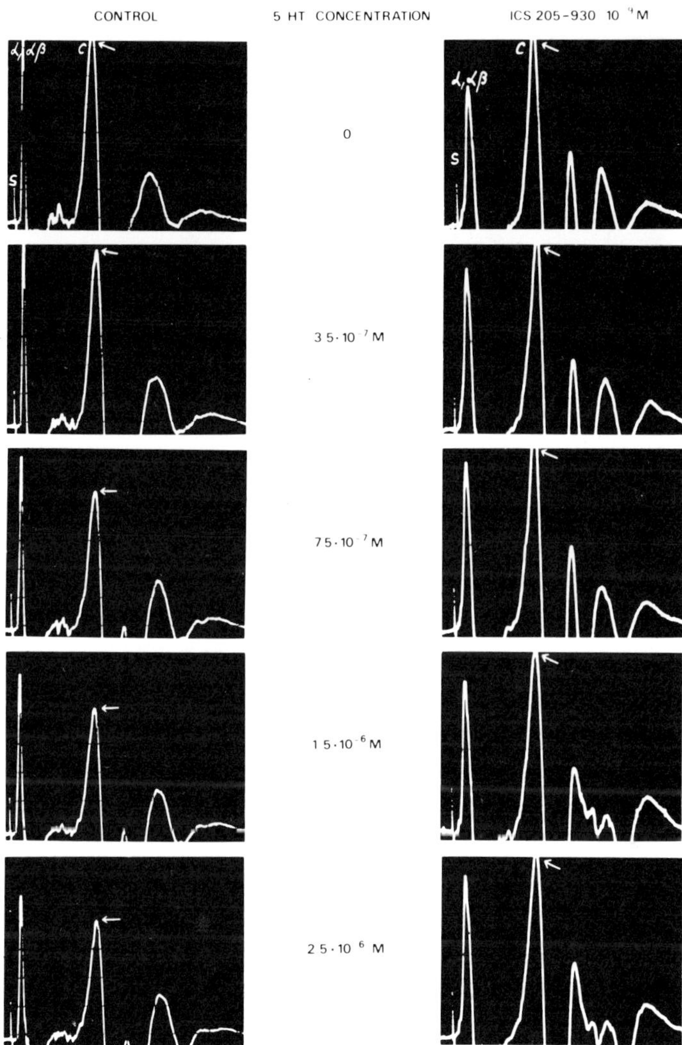

Fig. 2 Oscilloscope tracings of extracellularly recorded action potentials in desheathed rabbit vagus nerve. S = stimulus artefact; α, αβ = action potential carried by fast conducting A-fibres; C = action potential carried by non-myelinated C-fibres. Arrow indicates amplitude of C-fibre action potentials in the presence of increasing concentrations (top to bottom) of serotonin (5-HT). Left panels were recorded from a nerve bathed in control buffer solution (Locke's solution); the right panels from the same nerve incubated in the presence of the 5-HT₃ receptor antagonist ICS 205-930 (10^{-9}M). Note blockade of serotonin's ability to reduce the amplitude of the C-fibre action potential by ICS 205-930 (see also Fig. 3)

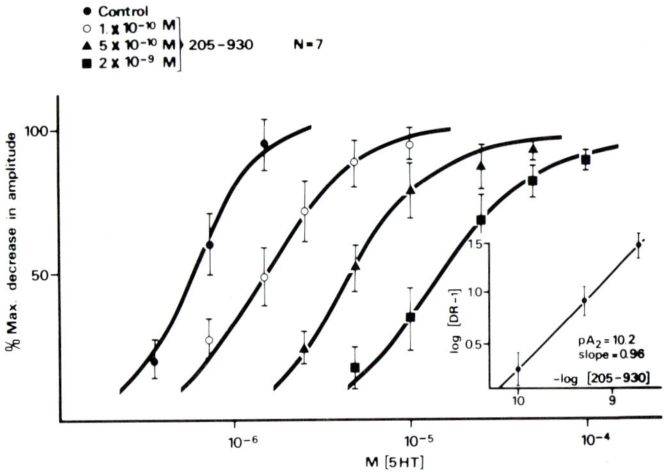

Fig. 3 Concentration–response curves for the effect of serotonin to reduce the C-fibre action potential in desheathed rabbit vagus nerves and competitive antagonism of this effect by ICS 205-930: solid circles, serotonin alone; open circles, serotonin + 10^{-10}M ICS 205-930; triangles, serotonin + 5×10^{-10}M ICS 205-930; squares, serotonin + 2×10^{-9}M ICS 205-930. $n = 7$ for all points. *Inset*: The regression of $\log_{10}$ of the dose ratio $(DR - 1)$ on-$\log_{10}$ of the antagonist concentration, where DR is the factor by which the agonist (i.e. serotonin) concentration has to be increased in the presence of the antagonist to obtain a response identical in magnitude to that recorded in the absence of the antagonist. For a competitive antagonist the slope of this regression line should not be significantly different from unity. The intercept on the abscissa gives the pA_2 value which provides a convenient measure of the potency of the antagonist under investigation (Arunlakshana and Schild, 1959)

receptors within the enteric nervous system. The action of 5-HT$_3$ receptor agonists is concentration-dependent and reproducible and can be inhibited in a competitive manner by compounds such as ICS 205-930 which has a pA_2 value of 7.9 in this system (Fig. 7 and Table 1).

In vivo *assays*

Besides *in vitro* assays it is possible to evaluate the agonist or antagonist activity of compounds on 5-HT$_3$ receptors in rats and even in man. Again, drugs with selective action at 5-HT$_1$ or 5-HT$_2$ receptors are inactive in these systems (Richardson *et al.*, 1985; Richardson and Engel, 1986).

Von Bezold–Jarisch reflex in rats The von Bezold–Jarisch reflex is a vagally

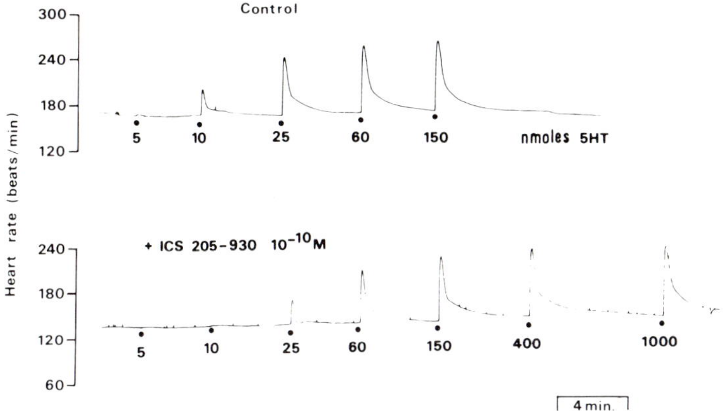

Fig. 4 Isolated rabbit heart preparation perfused by the Langendorff technique. Boluses of serotonin injected into the perfusion fluid cause a dose-dependent increase in heart rate (upper trace). After establishing a full dose–response curve for serotonin, ICS 205-930 (10^{-10}M) was added to the perfusion fluid and the procedure repeated. Note surmountable antagonism caused by ICS 205-930 (see also Fig. 6)

mediated bradycardia and consequent hypotension of short duration which occurs when afferent nerve endings in the right ventricle are depolarized by various means (Krayer, 1961; Paintal, 1973). Serotonin is one of the agents that is able to elicit this reflex when injected intravenously as a bolus into the jugular vein of anaesthetized rats (Fozard and Host, 1982). Selective 5-HT$_3$ receptor agonists produce a similar dose-dependent reflex bradycardia which is blocked by pretreatment of the rats with 5-HT$_3$ receptor antagonists like ICS 205-930 or MDL 72222 (see Figs 8 and 9; and Fozard, 1984a; Richardson *et al.*, 1985). Thus this model can be used to estimate the potency of 5-HT$_3$ receptor agonists or antagonists *in vivo*. The former are given into the jugular vein and the latter can be given either orally or parenterally (Fozard and Host, 1982; Fozard, 1984a; Richardson *et al.*, 1985).

Cantharidine blister base in humans The application of cantharidine to the medial aspect of the forearm of humans for a few hours results in the development of a blister. Serotonin and analogues with selectivity for 5-HT$_3$ receptors cause pain when applied to the base of such a blister and markedly potentiate the pain produced by bradykinin (Keele and Armstrong, 1964; Sicuteri *et al.*, 1965; Richardson *et al.*, 1985). The pain experienced while the blister base is exposed to the test substance is registered by the subject, who moves a lever on a continuous graduated scale and which is coupled to a linear integrator. By means of this integrator the area under the pain–time curve is recorded. 5-HT$_3$ receptor antagonists, but not antagonists at 5-HT$_1$ or 5-HT$_2$ receptors, inhibit the painful effects of serotonin and cause a parallel

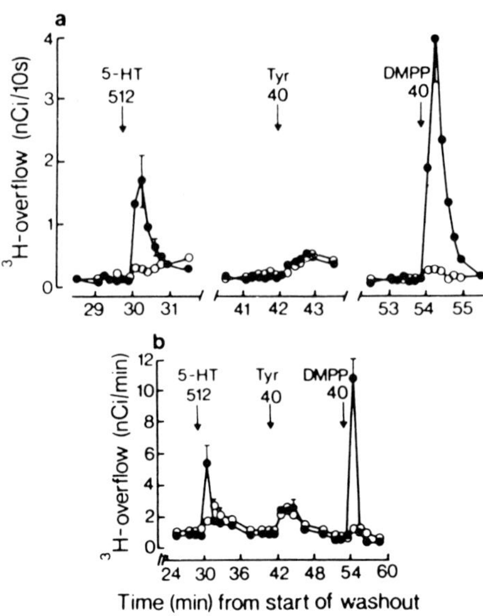

Fig. 5 Spontaneous and evoked ³H-overflow from rabbit hearts whose neuronal transmitter stores had been labelled by perfusion with [³H]-[−]-noradrenaline (10 ng/ml, 43nCi/ml), for 12 minutes. Atropine (0.5 μg/ml) was present throughout the experiments. Zero time marks the end of the loading period with [³H]-noradrenaline. Washout was with either Tyrode solution containing its normal calcium ion concentration (solid circles), or Tyrode solution containing 6% of its normal calcium ion concentration (open circles). The points represent the means of four observations; vertical lines show SEM. (a) Effect of bolus injections of 5-HT (512 μg), tyramine (Tyr, 40 μg), and dimethylphenylpiperazinium iodide (DMPP, 40 μg); results presented as nCi/10 s. (b) Results presented as nCi/minute. (From Fozard and Mwaluko, 1976; reproduced by permission)

rightward shift of the dose–response curve (Fig. 10; Richardson *et al.*, 1985). This bioassay allows an estimation of the potency of agonist and antagonist drugs at 5-HT$_3$ receptors on sensory nerves in humans after their topical application. Such a system might also permit the evaluation of 5-HT$_3$ receptor antagonists after oral or parenteral administration.

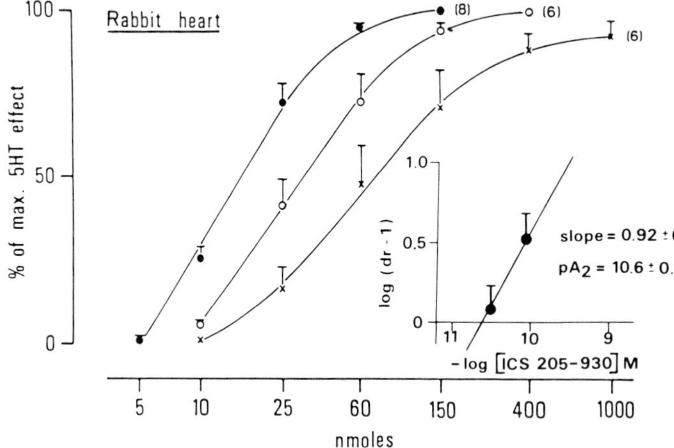

Fig. 6 Concentration–response curves for the ability of bolus injections of serotonin to cause increases in the rate of contraction of isolated rabbit hearts perfused by the Langendorff technique and antagonism of this effect by ICS 205-930. Solid circles, serotonin alone (n = 8); open circles, serotonin + 3 × 10^{-11}M ICS 205-930 (n = 6); crosses, serotonin + 10^{-10}M ICS 205-930 (n = 6). *Inset*: transformation of data according to Arunlakshana and Schild (1959)

Structure–activity relationship of drugs active at 5-HT$_3$ receptors

This section summarizes current knowledge regarding the structure–activity relationships of 5-HT$_3$ receptor agonists and antagonists. For the sake of brevity, only compounds with selectivity for 5-HT$_3$ receptors are considered.

Antagonists

The potent and selective 5-HT$_3$ receptor antagonists which have recently become available display no structural similarities to the older antagonists which are active at 5-HT$_1$ or 5-HT$_2$ receptors. This reflects the fact that they were not merely derived from pre-existing compounds by slight structural modification, but that entirely new chemical starting points were used to generate them. An analysis of the currently available 5-HT$_3$ receptor antagonists reveals that there were three different starting points for the chemistry programmes from which they resulted (Fig. 11). In the late 1970s it was reported that both the dopamine antagonist metoclopramide (Fozard and Mobarok Ali, 1978a) and the alkaloid cocaine (Fozard et al., 1979) block neuronal serotonin receptors present on sympathetic nerve endings in the isolated rabbit heart. Subsequently scientists at the Merrell Dow Research Institute in Strasbourg synthesized cocaine analogues (e.g. MDL 72222),

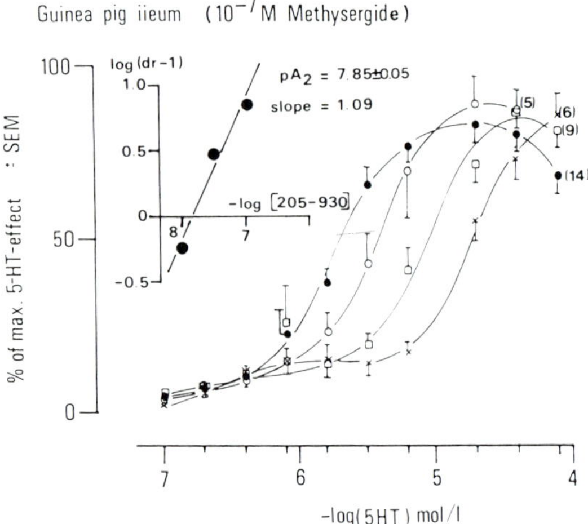

Fig. 7 Concentration–response curves for the contractile action of serotonin on the isolated longitudinal muscle of the guinea pig ileum. 10^{-7}M methysergide and 10^{-7}M atropine were present in the Tyrode solution throughout. Solid circles, serotonin alone ($n = 14$); open circles, serotonin + 10^{-8}M ICS 205-930 ($n = 5$); squares, serotonin + 3×10^{-8}M ICS 205-930 ($n = 6$); crosses, serotonin + 10^{-7}M ICS 205-930 ($n = 9$). *Inset*: transformation of the data according to Arunlakshana and Schild (1959)

whereas the Beecham's group in England based their chemical programme around metoclopramide (e.g. BRL 20627). A third approach, adopted by the Sandoz scientists in Basle, Switzerland, used the serotonin (5-HT) molecule itself as the chemical starting point. The three prototype antagonists which emerged all closely resemble the parent lead compound (see Fig. 11), and consist of an aromatic moiety (indole or benzene nucleus), an ester or amide bridge, and a more or less rigid ring system containing a basic nitrogen atom.

ICS 205-930 and MDL 72222, which were the first selective 5-HT$_3$ receptor antagonists to be reported, were developed independently at Sandoz and Merrell Dow respectively and appeared at about the same time. Both compounds, which are tropine esters, differ from cocaine in that the carboxymethyl group in position 2 of the tropine ring is missing and the benzoyloxy moiety in position 3 has been moved from the exo- to the endo- position. In the case of ICS 205-930, which was derived from 5-HT itself, the indole nucleus has replaced the benzene ring of cocaine, and with MDL 72222 a 3,5-dichlorobenzene has been used. BRL 20627, on the other hand, is derived

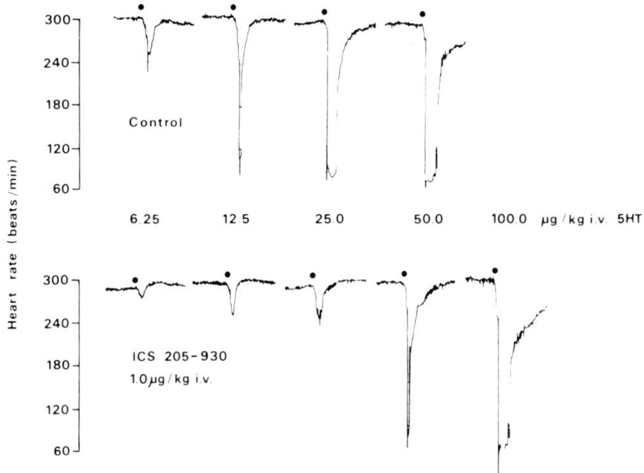

Fig. 8 The von Bezold–Jarisch reflex induced by bolus injections of serotonin into the jugular vein of anaesthetized rats. Increasing doses of serotonin (5-HT) applied at 5 minute intervals cause reflex reductions in heart rate. The area under the heart rate–time curve is used to quantitate the effect. Upper trace: serotonin alone; lower trace: serotonin 5 minutes after the intravenous administration of ICS 205-930 (1 µg/kg). Recordings have been taken from the same animal. Note surmountable antagonism caused by ICS 205-930

from metoclopramide by incorporating the aminoethyl side chain into a rigid ring system.

Table 2 compares the 5-HT₃ antagonistic potencies of these three prototypes with those of metoclopramide and cocaine in the isolated rabbit heart. The data clearly demonstrate that moving the ester-linked aromatic nucleus

Table 2 Affinity of various antagonists for 5-HT₃ receptors in the isolated rabbit heart

Compound	pA₂ value	Source
(−)-Cocaine	6.2	Fozard *et al.*, 1979
Benzoyltropine	7.2	Fozard *et al.*, 1979
Metoclopramide	7.2	Fozard and Mobarok Ali, 1978a
ICS 205-930	10.6	Richardson *et al.*, 1985
206-830	10.1	Richardson *et al.*, 1985
206-792	9.8	Richardson *et al.*, 1985
MDL 72222	9.3	Fozard, 1984a
MDL 72422	9.0	Fozard and Gittos, 1983
BRL 20627	7.2	Buchheit and Richardson (unpublished results)
BRL 24924	8.6	Buchheit and Richardson (unpublished results)

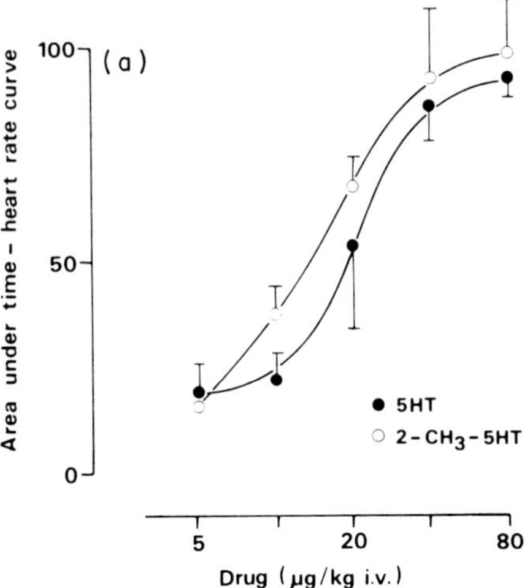

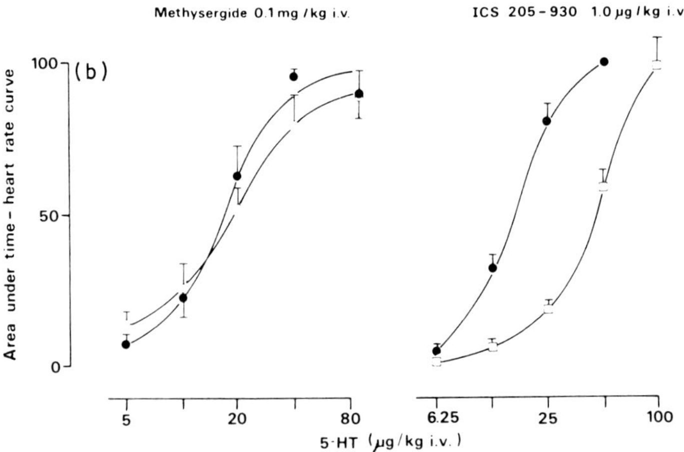

Fig. 9 Dose–response curves for the von Bezold–Jarisch reflex in anaesthetized rats. (a) Dose–response curves for 5-HT (solid circles; $n = 5$) and 2-methyl-5-HT (open circles; $n = 4$). The selective 5-HT$_2$ receptor agonist S(+)-α-methyl-5-HT was totally inactive up to a dose of 1,000 μg per kg, the highest dose tested ($n = 5$, data not shown). (b) Dose–response curves for 5-HT before (solid circles; $n = 5$) and 5 min after the intravenous administration of the 5-HT$_1$ and 5-HT$_2$ receptor antagonist methysergide (open circles) (100 μg per kg; left panel; $n = 4$) or the 5-HT$_3$ receptor antagonist ICS 205-930 (squares) (1 μg per kg; right panel; $n = 5$). Vertical bars indicate SEM in each case. (From Richardson *et al.*, 1985; reproduced by permission)

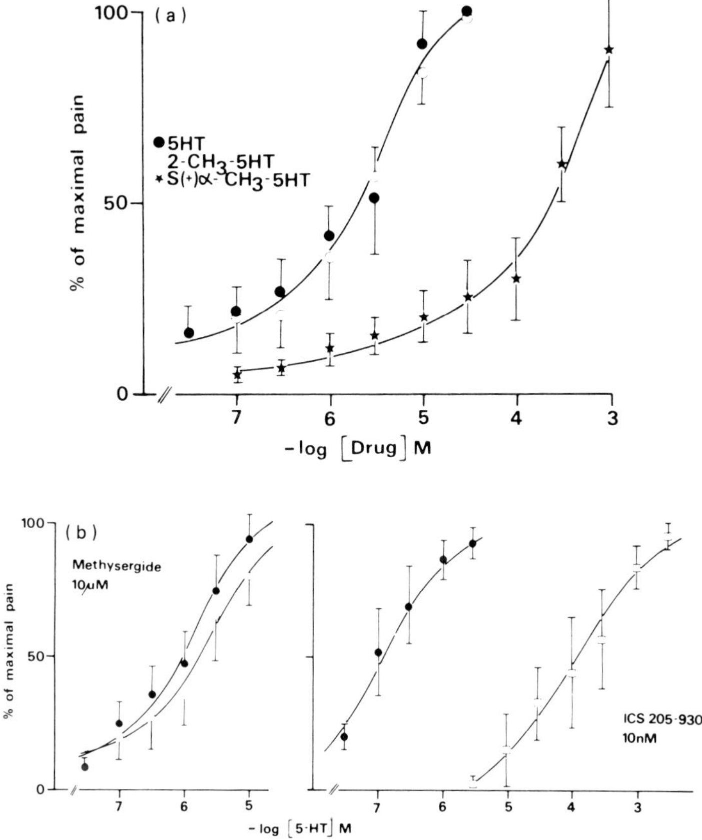

Fig. 10 Painful effects of serotonin and methylated analogues when applied to a blister base on the human forearm. (a) Effect of serotonin (solid circles; $n = 5$), 2-methyl-5-HT (open circles; $n = 3$) and S($+$)-α-methyl-5-HT (asterisks; $n = 3$) in causing pain. (b) Concentration--response curves for 5-HT in the absence (solid circles; $n = 4$) or presence of 10^{-5}M methysergide (open circles; $n = 3$; left panel) or 10^{-8}M ICS 205-930 (squares; $n = 4$; right panel). Antagonists were applied to the blister base for 20 min before applying 5-HT in their continued presence. Vertical bars indicate SEM in each case. (From Richardson *et al.*, 1985, with permission)

of cocaine from the exo- to the endo-position (i.e. benzoyltropine) leads to a modest increase in affinity. Replacing the benzene ring by an indole nucleus (ICS 205-930) or a 3,5-dichlorobenzene ring (MDL 72222), however, substantially increases the affinity of these antagonists for the 5-HT$_3$ receptors found in the rabbit heart. The compound BRL 20627, on the other hand, shows that replacing the flexible side chain of metoclopramide by a conformationally more constrained system does not necessarily result in higher affinity for these

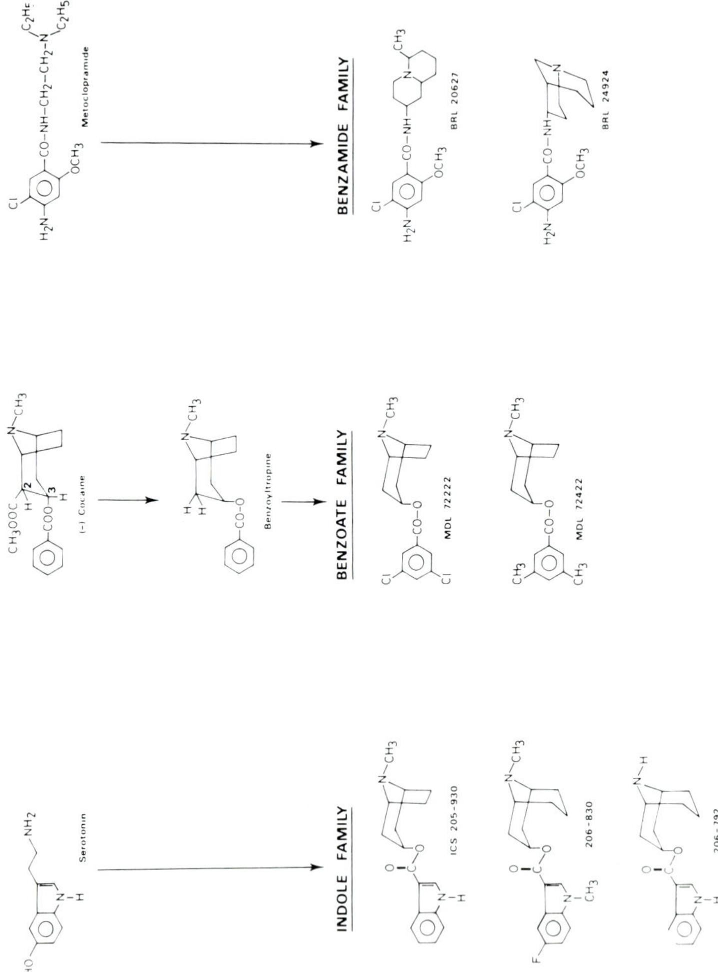

Fig. 11 New 5-HT$_3$ receptor antagonists and the parent lead compounds from which they were derived

5-HT$_3$ receptors. The only advantage of such compounds is their increased specificity, since the dopamine receptor blocking activity of substituted benzamides can be considerably reduced by such a manoeuvre.

Within each family of compounds the affinity for 5-HT$_3$ receptors in general, as well as the selectivity for each of the putative subtypes (see below) can be modified by changing the substitution pattern. Thus within the indole family, slight variation of the prototype molecule ICS 205-930 affects the affinity for 5-HT$_3$ receptors on enteric, sympathetic and sensory neurones differently, making the identification of at least three different subtypes of 5-HT$_3$ receptor possible (Richardson *et al.*, 1985; see also Fig. 1 and Table 1).

Within the benzoic ester group of antagonists, the chlorine atoms in position 3 and 5 of MDL 72222 can be replaced by methyl groups without substantial loss in affinity for the 5-HT$_3$ receptors found in the rabbit heart (Fozard and Gittos, 1983). Mono-chlorine or mono-methyl derivatives, however, are less active, independent of the position of the substituent.

Substitution of the condensed nitrogen-containing 2-ring system of BRL 20627 by a bridged 2-ring system (BRL 24924) results in a modest increase in affinity for 5-HT$_3$ receptors (Dunbar *et al.*, 1986).

Besides the three distinct families of 5-HT$_3$ receptor antagonists already discussed, there are other compounds which display blocking properties: for example, both cisapride and R 53434 (Janssen Pharmaceuticals) inhibit contractions produced by activation of neuronal serotonin receptors in the isolated guinea pig ileum (Schuurkes *et al.*, 1985). Whereas cisapride possesses some structural features of the benzamide group of 5-HT$_3$ antagonists, R 53434 shows no resemblance to any of the three groups at all. However, neither of these compounds is selective for 5-HT$_3$ receptors.

Agonists

In contrast to the highly potent and selective antagonists for 5-HT$_3$ receptors that are now available and which are suitable for both *in vivo* and *in vitro* studies, no equivalent agonist yet exists. In fact there is only one compound, 2-methyl-5-HT, that has been established as a selective 5-HT$_3$ receptor agonist (Richardson *et al.*, 1985; see below).

Most compounds that have been investigated for agonist activity at 5-HT$_3$ receptors and which have been described in the literature are close derivatives of serotonin itself. Thus serotonin has been systematically methylated in various positions. Despite such a relatively small structural change, the resulting analogues show large differences in agonistic potency at 5-HT$_2$ and 5-HT$_3$ receptors (Fig. 12). Among these methylated analogues, compounds with selectivity for 5-HT$_2$ and 5-HT$_3$ receptors can be found. Unfortunately,

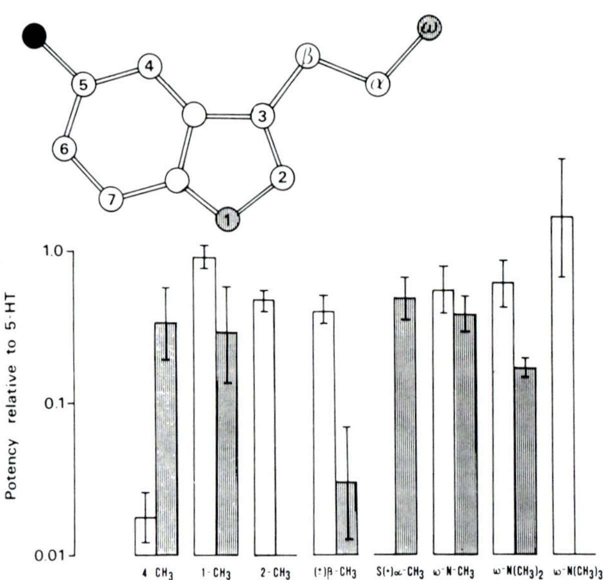

Fig. 12 Agonist potencies of methylated serotonin (5-HT) analogues relative to 5-HT (potency = 1.0) on 5-HT$_3$ receptors in the rabbit vagus nerve (open columns) and 5-HT$_2$ receptors in the rat uterus (striped columns). Each column indicates the geometric mean value obtained from at least three independent experiments. Vertical lines on each column indicate 90% confidence limits. (From Richardson *et al.*, 1985; reproduced by permission)

no published data exists on the activity of most of these analogues at 5-HT$_1$ receptors (but see Engel *et al.*, 1986; Markstein *et al.*, 1986; and Table 1).

Methylation of the indole nucleus of serotonin in position 2 gives 2-methyl-5-HT, which is about half as active as 5-HT itself at 5-HT$_3$ receptors in the rabbit heart, vagus nerve and guinea pig ileum preparations. At 5-HT$_2$ receptors in the rat uterus, however, 2-methyl-5-HT is 2000 times less potent than 5-HT, whereas its affinity for 5-HT$_1$ receptors, as determined by binding studies using a rat cortex membrane preparation (Richardson *et al.*, 1985) or functional studies (Table 1) is 70 to 400 times lower than that of 5-HT, depending on 5-HT$_1$ receptor subtype (Markstein *et al.*, 1986). Thus 2-methyl-5-HT has considerable selectivity for 5-HT$_3$ receptors and is especially useful for *in vitro* experiments. However, with regard to its ability to penetrate the blood–brain barrier, 2-methyl-5-HT suffers the same disadvantage as 5-HT itself: it does not enter the brain, making the *in vivo* study of central 5-HT$_3$ receptors, if they exist, still very difficult (see later). Another problem is that, like 5-HT itself, 2-methyl-5-HT is not very stable *in vivo*, and this obviously limits its value for whole animal studies.

Substitution of the indole NH– in position 1 by an N–CH$_3$– group does not markedly influence potency or selectivity for 5-HT$_2$ or 5-HT$_3$ receptors – the agonistic activity of 1-methyl-5-HT at 5-HT$_2$ receptors being only slightly reduced compared with 5-HT, and its activity at 5-HT$_3$ receptors being indistinguishable from that of the natural ligand. In contrast, methylation in position 4 of the indole system almost completely abolishes agonistic action at 5-HT$_3$, but not at 5-HT$_2$ receptors.

Improved selectivity of serotonin analogues as agonists for 5-HT$_3$ receptors can be achieved by introducing an increasing number of methyl groups at the terminal side chain nitrogen atom. The first member of this series, ω-N-methyl-5-HT, is not very different from 5-HT itself, although its agonistic potency at both 5-HT$_2$ and 5-HT$_3$ receptors is somewhat reduced, its potency at the latter being slightly the more affected. Bufotenin, the N,N-dimethyl compound, is as active as the mono-methyl derivative at 5-HT$_3$ receptors, but shows a further and marked reduction in potency at 5-HT$_2$ receptors. This selectivity is even more pronounced with the quaternary trimethyl compound (bufotenidin), which retains full activity at 5-HT$_3$ receptors but is virtually inactive at 5-HT$_2$ receptors. Its potency for the former is roughly the same as that of 5-HT. Unfortunately, this compound is not entirely suitable as a selective 5-HT$_3$ agonist, since it also possesses potent nicotinic agonist properties (Wallis and Nash, 1980). Moreover, no information regarding its affinity for 5-HT$_1$ receptors is presently available.

Methylation in the alpha-position of 5-HT leads to racemic alpha-methyl-5-HT. The S-($+$)-enantiomer shows considerable selectivity for 5-HT$_2$ receptors. This compound is 10 times, 20 times and 700 times less potent as a 5-HT$_3$ agonist than 5-HT in the guinea pig ileum, the rabbit heart and the rabbit vagus respectively (Table 1). In the rat uterus S-($+$)-α-methyl-5-HT is one-half as active as 5-HT (Richardson $et\ al.$, 1985). Thus, although the selectivity of this compound for 5-HT$_2$ receptors is less pronounced than that of 2-methyl-5-HT for 5-HT$_3$ receptors, it can nevertheless be useful when used in conjunction with the latter to discriminate between 5-HT$_2$ and 5-HT$_3$ receptor mediated effects. It should also be remembered that, although this compound has greatly reduced potency at 5-HT$_{1A}$ and 5-HT$_{1B}$ receptors compared with 5-HT, it has a similar potency at 5-HT$_{1C}$ receptors in the rat fundus preparation and is thus not absolutely selective for 5-HT$_2$ receptors (Table 1).

Introduction of a methyl group into the beta position of the side chain results in a racemic mixture that does not show selectivity for 5-HT$_2$ or 5-HT$_3$ receptors although its agonistic effects at 5-HT$_3$ receptors are slightly more pronounced than those at 5-HT$_2$ receptors.

One striking feature of the structure–activity relationships of compounds active as agonists at 5-HT$_3$ receptors is that substitution of the hydroxy group of 5-HT in position 5 of the indole ring causes a substantial loss of activity. For example, 5-methoxytryptamine (Fozard, 1984b), 5-methyltryptamine

(Fozard and Mobarok Ali, 1978b), and tryptamine itself (Buchheit *et al.*, unpublished), all of which retain agonist activity at 5-HT$_2$ receptors, are totally inactive as 5-HT$_3$ agonists. In fact, replacement of the 5-OH group by a 5-carboxamido moiety produces a weak antagonist at 5-HT$_3$ receptors (Giger *et al.*, 1985). This clearly indicates that, unlike the case with 5-HT$_1$ or 5-HT$_2$ receptors, the hydroxy group in the 5 position plays a crucial role in the interaction between 5-HT and the 5-HT$_3$ receptor.

What can be concluded about the active conformation of 5-HT at 5-HT$_3$ receptors? Some information is provided by the fact that both 4-methyl- and S-(+)-α-methyl-5-HT are almost devoid of 5-HT$_3$ agonistic effects, whereas 2-methyl-5-HT is highly selective at this receptor. This suggests that introducing a methyl group in position 2 of the indole ring favours adoption of a conformation active at 5-HT$_3$ receptors, and that this occurs at the expense of the alternative conformations necessary for 5-HT$_1$ or 5-HT$_2$ receptor activation. In contrast, methyl groups in position 4 or at the alpha carbon atom of the side chain do not allow the conformation necessary for 5-HT$_3$ activation to occur but do permit that required for 5-HT$_2$ receptor activation. These hints may in future permit the synthesis of rigid serotonin analogues where the active conformation for 5-HT$_3$ receptors is 'frozen'. This could provide potent and highly selective 5-HT$_3$ agonists stable enough for both *in vitro* and *in vivo* use.

DISTRIBUTION AND FUNCTION OF 5-HT$_3$ RECEPTORS

The criteria for defining whether a particular effect of serotonin is mediated through 5-HT$_3$ receptor activation are as follows:

1. The response should be inhibited by ICS 205-930 or MDL 72222. For *in vitro* studies these antagonists should yield pA$_2$ values of 8.0–11.0 and 7.5–9.5 respectively, as indicated in Table 1. In whole animal studies the effects of exogenously applied serotonin are usually blocked by doses of these antagonists below 1 mg/kg body weight, providing the dose of serotonin given is not massively supramaximal. The blockade produced by these two compounds is surmountable when the concentration or dose of serotonin is increased.
2. The response should be resistant to antagonism by 5-HT$_1$ receptor (e.g. methiothepin) or 5-HT$_2$ receptor (e.g. ketanserin) antagonists.
3. The response should be mimicked by 2-methyl-5-HT, with a potency only slightly less than that of serotonin itself. In contrast the 5-HT$_2$ receptor agonist S-(+)-α-methyl-5-HT should be a factor of at least 20 less potent than serotonin. 5-carboxamidotryptamine, the 5-HT$_1$ receptor agonist, should be inactive.

Responses that have been shown to fulfil these criteria are summarized in

Table 3, from which it can be seen that 5-HT$_3$ receptors are widely distributed over the peripheral nervous system and mediate exclusively excitatory actions of serotonin. At present there is no convincing evidence for their occurrence either in the central nervous system or in non-neuronal tissue. However, as previously indicated, selective 5-HT$_3$ receptor agonist and antagonist drugs to study these possibilities have only just become available, making any definitive conclusions in this regard premature.

Some analogues of ICS 205-930 that are also potent 5-HT$_3$ receptor antagonists yield clearly differing pA$_2$ values when tested against serotonin in the isolated rabbit vagus nerve, the rabbit heart or the guinea pig ileum. This indicates the presence of at least three different subtypes of 5-HT$_3$ receptor (Donatsch *et al.*, 1984a; Richardson *et al.*, 1985; Richardson and Engel, 1986; see also Fig. 1, Table 1; and below).

Postganglionic autonomic neurones

Sympathetic neurones

Serotonin is capable of depolarizing the cell bodies, axons and terminals of postganglionic sympathetic neurones (reviewed by Fozard, 1984b), thus causing release of noradrenaline (Jacob and Poite-Bevierre, 1960; Fozard and Mwaluko, 1976; Göthert and Duehrsen, 1979). The two systems which have been most extensively studied from a pharmacological point of view are the isolated rabbit heart (see section on bioassay systems) and the superior cervical ganglion. In both tissues the stimulatory effects of serotonin can be antagonized by very low concentrations of ICS 205–930 or MDL 72222 in a competitive manner, whereas methiothepin and ketanserin are completely inactive (see Tables 1–3, and Richardson and Engel, 1986). The rank order of agonist potency for serotonin analogues on the rabbit heart is also consistent with the actions of serotonin being mediated by a 5-HT$_3$ receptor in this preparation (Table 1).

A point of considerable interest is the fact that MDL 72222 has different affinity constants in the two systems (Fozard, 1984a; Azami *et al.*, 1985; Round and Wallis, 1985; Richardson and Engel, 1986; and Table 1). This indicates the existence of different subtypes of 5-HT$_3$ receptors. Moreover, although ICS 205-930 possesses a similar pA$_2$ value in the rabbit heart and vagus nerve preparations, compounds 206-792 and 206-830 (Fig. 11) are analogues which, like MDL 72222, can also clearly distinguish between the different types of 5-HT$_3$ receptors in these two tissues (Richardson *et al.*, 1985). Based on the affinity constants obtained with MDL 72222 in different systems, it would appear that the 5-HT$_3$ receptors present on the rabbit superior cervical ganglion are identical with those found on sensory fibres of the vagus nerve and nodose ganglion but different from those present on the

Table 3 Distribution of 5-HT$_3$ receptors and the various responses they mediate

Location	Species	Organ/tissue	Response to 5-HT	Key references
Postganglionic autonomic neurones				
Sympathetic	Rabbit Rabbit, Cat, Rat	Heart Superior cervical ganglion	Depolarization and noradrenaline release	Fozard and Mwaluko, 1976; Fozard, 1984b; Azami *et al.*, 1985; Round and Wallis, 1986
Parasympathetic	Cat Rabbit Calf	Bladder Heart Trachea	Depolarization and acetylcholine release	Saxena *et al.*, 1985; Fozard, 1984c Offermeier and Ariëns, 1966
Enteric nervous system	Guinea pig, Rat	Small intestine	Depolarization and neurotransmitter release	Buchheit *et al.*, 1985a; Cooper *et al.*, 1986
	Guinea pig	Small intestine		Gaddum and Picarelli, 1957; Fozard *et al.*, 1979; Buchheit *et al.*, 1985b; Richardson *et al.*, 1985
Sensory neurones	Human	Skin, Veins	Pain, wheal and flare	Keele and Armstrong, 1964; Sicuteri *et al.*, 1965; Richardson *et al.*, 1985; Orwin and Fozard, 1986
	Cat	Carotid body	Neuronal depolarization	Kirby and McQueen, 1984
	Cat, Rat, Human	Heart	Neuronal excitation	Fozard and Host, 1982; Fozard, 1984b; Richardson *et al.*, 1985
	Rabbit Rabbit, Rat, Cat	Nodose ganglion Vagus Nerve	Depolarization	Neto, 1978; Wallis *et al.*, 1982; Ireland *et al.*, 1982; Fozard, 1984b; Azami *et al.*, 1985 Round and Wallis, 1986; Richardson *et al.*, 1985
	Rabbit	Lung	Neuronal excitation	Armstrong *et al.*, 1986

sympathetic nerve terminals in the rabbit heart (Round and Wallis, 1985, 1986; Richardson and Engel, 1986; and see below).

Parasympathetic neurones

Serotonin releases acetylcholine from postganglionic parasympathetic nerve fibres in the rat bronchus (Aas, 1983), the calf trachea (Offermeier and Ariëns, 1966), the rabbit heart (Fozard, 1984b, c) and the cat urinary bladder (Saxena *et al.*, 1985). Only in the case of the rabbit heart and the cat bladder have attempts been made to inhibit this action of serotonin with the selective 5-HT₃ receptor antagonists now available. The affinity constant for MDL 72222 at 5-HT₃ receptors on parasympathetic nerve endings in the rabbit heart is very similar to that obtained for inhibition of serotonin-induced noradrenaline release in the same preparation. This suggests that the 5-HT₃ receptors on cardiac parasympathetic and sympathetic nerve endings are identical (see Fig. 1).

In anaesthetized cats, intra-arterial injections of serotonin induce a biphasic contraction of the urinary bladder. The initial spike phase of this contracture is blocked by low intravenous doses of MDL 72222 whereas the second phase of contraction is unaffected. Conversely, ketanserin blocks the second phase of contraction but not the first. Thus the early phase of contracture is elicited by 5-HT₃ receptors and the late phase by 5-HT₂ receptors (Saxena *et al.*, 1985).

Enteric nervous system (ENS)

Vast amounts of serotonin are present in the mammalian gastrointestinal tract (GIT). For example, in humans, about 90% of the body's total serotonin content is found in the GIT (Erspamer, 1966), a majority of which is stored in the mucosal enterochromaffin cells that are scattered throughout its length. Although only a comparatively minor portion of serotonin resides within the neural elements of the GIT, it is functionally the more important and has consequently attracted much attention recently (for review, see Gershon, 1982).

Different types of serotonin receptors exist in the GIT and are targets for the serotonin released from neuronal sources. They occur both on neuronal and on non-neuronal tissue, as demonstrated by biochemical (autoradiography, binding studies), electrophysiological, and pharmacological techniques (see Gershon, 1982; Gershon *et al.*, 1985).

Unfortunately, a systematic pharmacological characterization of enteric serotonergic receptors has not been performed to date, due to the lack of specific agonist and antagonist drugs. Only recently have attempts been made to define the location of serotonin receptor subtypes in the gastrointestinal tract using combined autoradiography and electrophysiological techniques. These studies have demonstrated that an excitatory 5-HT receptor exists on type II/AH cells of the myenteric plexus (Gershon *et al.*, 1985; Branchek *et al.*, 1984). Although this neuronally located serotonin receptor is unaffected by either $5-HT_1$ or $5-HT_2$ antagonists and could thus be a $5-HT_3$ receptor, formal demonstration of this fact using selective $5-HT_3$ receptor agonists and antagonists has not been made.

Thus detailed information concerning the exact location of $5-HT_3$ receptors in the GIT is still lacking. Despite this their presence in the ENS, as defined by the strict criteria listed above, has been demonstrated in a limited number of functional pharmacological experiments (see Table 3). Thus $5-HT_3$ receptors have been conclusively demonstrated in the guinea pig myenteric plexus (Gaddum and Picarelli, 1957; Buchheit *et al.*, 1985b), and there is also good evidence for their presence on neuronal elements of the guinea pig stomach (Buchheit *et al.*, 1985a; and see below). It seems likely that $5-HT_3$ receptors will be demonstrated in other parts of the gastrointestinal tract in the near future. For instance, in the mouse duodenum there is a neuronal serotonin receptor that is insensitive to methysergide or hyoscine and which mediates a tetrodotoxin-sensitive relaxation (Drakontides and Gershon, 1968). The new pharmacological tools now available will certainly show whether this is a $5-HT_3$ receptor.

Regarding the subtypes of $5-HT_3$ receptors found in the GIT, there is good reason to believe that those located within the myenteric plexus of the guinea pig ileum are different from those found either on sensory or on sympathetic neurones. Thus ICS 205-930 is more than 100 times less potent in the guinea pig ileum than in the rabbit vagus nerve or heart, and MDL 72222 is completely devoid of any specific inhibitory action at intestinal $5-HT_3$ receptors (see Table 1). These observations support the idea that, in addition to the two different subtypes of $5-HT_3$ receptor found on sympathetic and sensory neurones and which have already been discussed, a third subtype of $5-HT_3$ receptor exists in the ENS.

The neurones forming the intrinsic ENS can be subdivided into afferent sensory and efferent motor elements. Different types of neurones using a variety of neurotransmitters are found in each part. This poses the question as to whether the $5-HT_3$ receptor subtype present in the myenteric plexus of the guinea pig ileum is the only one found within the ENS. Results obtained using muscle strips from the guinea pig stomach in fact suggest this is not the case. Contractions induced by electrical field stimulation of such preparations are enhanced by metoclopramide, ICS 205-930 and MDL 72222 at concentrations similar to those previously shown to antagonize the action of serotonin in the rabbit vagus nerve or rabbit heart (Fig. 13; Buchheit *et al.*, 1985a). Although the currently available data do not yet permit a definitive conclusion as to which type of $5-HT_3$ receptor is involved, this effect is clearly

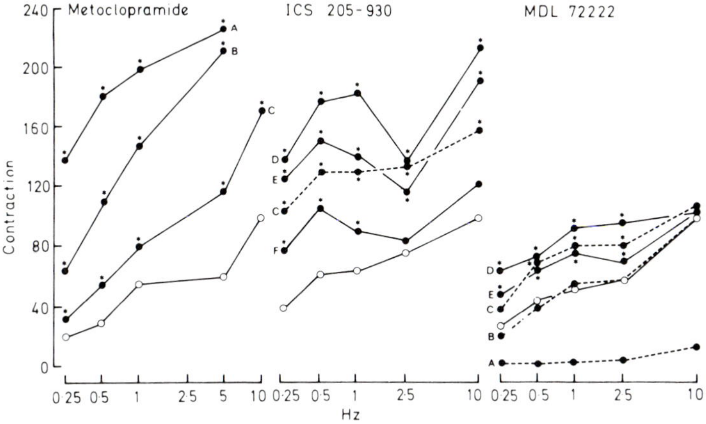

Fig. 13 Modification by metoclopramide, ICS 205-930 and MDL 72222 of contractions induced by electrical field stimulation (0.25–10Hz) of circular muscle taken from the body of the guinea pig stomach. Open circles, control values; solid circles joined by continuous line, dose-related enhancements of contractions by drug treatments; solid circles joined by dashed line, enhancements reduced as concentrations are increased. All values are expressed as a percentage of the control contractions occurring at 10 Hz which were designated as 100%. $n = 6$. SEMs on original data $< 12\%$. Enhancement of the contractions significant to *$p < 0.05$–$p < 0.001$ as analysed by the Mann–Whitney U-test. Concentrations (M): A 10^{-5}, B 10^{-6}, C 10^{-7}, D 10^{-8}, E 10^{-9}, F 10^{-10}. (From Buchheit et al., 1985a; reproduced by permission)

obtained with concentrations of the 5-HT$_3$ receptor antagonists which are far too low for it to be mediated by the type of receptor found in the guinea pig ileum. Thus, 5-HT$_3$ receptors located in the ENS are not a homogeneous population.

What kind of responses are mediated by 5-HT$_3$ receptors in the GIT? In the isolated guinea pig ileum, high concentrations of serotonin elicit a contraction which is produced by three different mechanisms (Fig. 14; Buchheit et al., 1985b). The major part is produced by the activation of neuronally located 5-HT$_3$ receptors. This response is resistant to methysergide and atropine, and can be selectively and competitively blocked by ICS 205-930 (Buchheit et al., 1985b), but not by MDL 72222 (Fozard, 1984a). Only a minor part of the contraction results from the interaction of serotonin with the 5 HT$_2$ receptors (i.e. the D-receptors of Gaddum and Picarelli's classification) located on smooth muscle cells (Engel et al., 1984, 1985). The remainder of the contraction results from the activation of another, as yet uncharacterized, neuronal 5-HT receptor. This occurs at very low serotonin concentrations

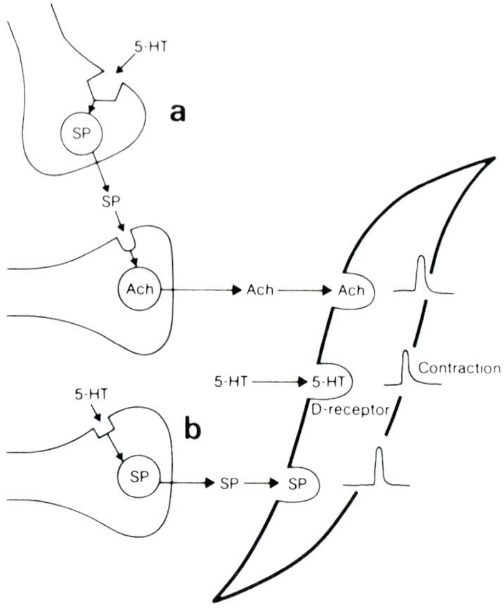

Fig. 14 Hypothetical mechanism for the contractile effect of 5-HT in the isolated guinea pig ileum. (a) = neuronal high-affinity receptor for 5-HT; (b) = 5-HT$_3$ receptor. For explanation see text. (From Buchheit *et al.*, 1985b; reproduced by permission)

and is insensitive to blockade by either methysergide or ICS 205-930, and thus cannot be mediated by either a 5-HT$_2$ or a 5-HT$_3$ receptor mechanism.

There is evidence that serotonin produces most of its contractile effect in the guinea pig ileum by releasing substance P from nerve endings within this tissue (Fig. 14). However, this substance P release is mediated by two different serotonin receptors. At very low concentrations (10^{-8}–10^{-7}M) serotonin activates a neuronally located receptor, resulting in substance P release. Once liberated, this substance P activates receptors located on cholinergic nerve endings, which in turn leads to the release of acetylcholine and contraction (Chahl, 1983). This is the mechanism which is not blocked by methysergide or ICS 205–930, referred to above. At higher concentrations (10^{-7}–10^{-5} mol/l), serotonin activates the ileal 5-HT$_3$ receptor which produces the release of sufficient quantities of substance P to activate substance P receptors on smooth muscle cells and thus cause contraction. This latter mechanism is the one which can be blocked by ICS 205–930 with a pA$_2$ value of 7.9. With regard to the remaining non-neuronal mechanism, supramaximal serotonin concentrations ($>5\times10^{-6}$ M) are required to activate the 5-HT$_2$ receptors located on smooth muscle cells, so that this is unlikely to be of great physiological relevance.

So how does the mechanism whereby serotonin induces substance P release via activation of 5-HT₃ receptors fit Gaddum and Picarelli's original concept of an acetylcholine-releasing 5-HT receptor? Gaddum and Picarelli ascribed the morphine-sensitive, phenoxybenzamine-insensitive contraction produced by serotonin to activation of a single receptor, the M-receptor. However, there are clearly two different receptor systems involved; when activated, both lead to substance P release and both are sensitive to morphine but, as indicated above, only one is blocked by 5-HT₃ antagonists (see Fig. 14).

So far, all effects that occur subsequent to activation of 5-HT₃ receptors and which have been reviewed in this chapter can be explained by the release of excitatory neurotransmitters. The basis for the enhancement of electrically induced contractions in guinea pig stomach strips produced by 5-HT₃ receptor antagonists, however, is more complicated (see above; and Buchheit *et al.*, 1985a; Gunning *et al.*, 1986). It appears as if contraction of gastric circular muscle is regulated by both an inhibitory and a facilitatory system and that 5-HT₃ receptors are involved in the inhibitory one. Thus activation of these receptors by endogenous serotonin released on electrical stimulation results in the liberation of an inhibitory transmitter which reduces twitch amplitude. Blockade of this system by a 5-HT₃ receptor antagonist causes disinhibition and consequently the potentiation of electrically stimulated contractile responses (see Fig. 13).

In conscious animals ICS 205-930 and BRL 24924 produce an increase in gastric motility and emptying (Fig. 15; Buchheit *et al.*, 1985a; Cooper *et al.*, 1986). There is good evidence that this can be at least partly explained by an effect of these compounds on 5-HT₃ receptors in the ENS. An additional central site of action, however, may be involved since low, systemically ineffective, doses of ICS 205-930 facilitate gastric emptying when injected into the hypothalamus. In contrast, serotonin and the specific 5-HT₃ receptor agonist 2-methyl-5-HT when applied in the same way, inhibit gastric emptying in guinea pigs (Costall *et al.*, 1986b).

Sensory neurones

Serotonin has been shown by many investigators to stimulate afferent nerve fibres in a wide variety of locations, although a systematic analysis of the receptor type involved has rarely been performed (reviewed by Fozard, 1984b). As shown in Table 3, however, there are a few exceptions.

In the nodose ganglion of the rabbit, serotonin induces rapid depolarization of a majority of C-type neurones (Higashi, 1977; Higashi and Nishi, 1982; Wallis *et al.*, 1982; Stansfeld and Wallis, 1982). Studies with antagonist drugs indicate that the 5-HT₁ and 5-HT₂ receptor antagonist methysergide does

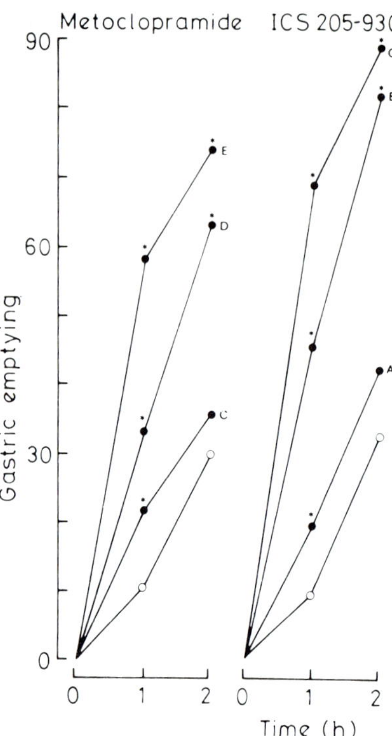

Fig. 15 Modification by metoclopramide and ICS 205–930 of gastric
emptying in the guinea pig. Gastric
emptying is expressed as the
percentage of spheroids leaving the
stomach over a 2 h period. Open
circles, control values; solid circles
enhancement of gastric emptying by
drug treatments, significant to $*p <$
$0.005–p < 0.001$ as analysed by the
Mann–Whitney U-test. $n = 5$–6.
SEMs on original data <13%. Dose
(mg kg^{-1} i.p): A 0.01, B 0.1, C 1.0,
D 5.0, E 10.0 . (From Buchheit *et al.*,
1985a, with permission)

not inhibit this action of serotonin (Higashi and Nishi, 1982). Recently, studies using MDL 72222 and ICS 205-930 have clearly shown that the depolarizing action of serotonin in this preparation is mediated by stimulation of 5-HT$_3$ receptors (Azami *et al.*, 1985; Round and Wallis, 1985, 1986). The pA$_2$ values obtained indicate that the type of 5-HT$_3$ receptor found on the cell bodies of these nerves is identical with that found on their axonal portions i.e. in the rabbit cervical vagus nerve (Round and Wallis 1985, 1986; Richardson and Engel, 1986; see also the section on bioassay systems). In the latter preparation all criteria for defining the depolarization response caused by serotonin as being mediated by 5-HT₃ receptors have been met. Similar studies have also been performed using the rat vagus nerve (Ireland *et al.*, 1982).

Several studies have been made of the action of serotonin on sensory nerve endings *in vivo*. Thus the intravenous injection of serotonin activates vagal afferents in the heart and lung, and this evokes reflex cardiac and respiratory effects (Comroe *et al.*, 1953; Douglas and Toh, 1953; Krayer, 1961; Jacob and Comroe, 1971; Paintal, 1973; Fozard, 1984b). One of these reflex actions of serotonin, the von Bezold–Jarisch reflex, has already been discussed in detail (see section on bioassay systems). All the criteria listed above for defining this effect as being mediated by a 5-HT₃ receptor have been fulfilled (Fozard and Host, 1982; Fozard, 1984a; Richardson *et al.*, 1985; Richardson and Engel, 1986).

Recent studies in anaesthetized rabbits have shown that the ability of serotonin, when injected intravenously, to induce pulmonary depressor and respiratory chemoreflexes is blocked by very low doses of MDL 72222 (Armstrong and Kay, 1985). These observations have been extended to situations where miliary pulmonary embolism has been experimentally induced. Under such circumstances MDL 72222 can also inhibit the reflex tachypnoea which presumably results as a consequence of serotonin release from pulmonary platelet aggregates (Armstrong *et al.*, 1986). Together these observations suggest that the receptors involved in the activation of vagal afferent nerve endings in the lung by serotonin are of the 5-HT₃ subtype.

The ability of serotonin to induce pain when applied to the base of a skin blister in humans and to potentiate the pain produced by local application of bradykinin have already been discussed in detail (see section on bioassay systems). These actions of serotonin are clearly mediated by 5-HT₃ receptors. In addition to its ability to induce pain by activation of cutaneous nociceptive afferent nerve endings, serotonin also induces a wheal and flare response when applied to the skin (Donatsch *et al.*, 1984c; Richardson *et al.*, 1985; Orwin and Fozard, 1986; Richardson and Engel, 1986). Again 5-HT₃ receptors are involved since this effect can be blocked by low doses of ICS 205-930 or MDL 72222. It is believed that the flare response which occurs when serotonin is applied to a blister base or when it is injected intradermally into humans results from activation of an axonal reflex mechanism (Greaves and Shuster, 1967). Like the pain response, 5-HT₃ receptors are involved in mediating this effect. Since pain and flare responses occur concomittantly (Donatsch *et al.*, 1984c; Richardson *et al.*, 1985), it is probable that activation of the same population of nociceptive nerve endings results in both phenomena.

Chemoreceptors in the carotid body of the cat are also sensitive to serotonin which, when injected directly into the carotid artery, causes a complex triphasic effect on the firing rate of the carotid sinus nerve (Kirby and McQueen, 1984). Initially there is a transient burst of electrical activity which is followed by neuronal depression. Finally there is a delayed, longer-lasting increase in firing rate. The 5-HT$_3$ receptor antagonist MDL 72222 virtually abolishes the first response and reduces the subsequent chemodepression, whereas the 5-HT$_2$ receptor antagonist ketanserin is without effect on the first response. Conversely, the delayed and sustained increase in firing rate induced by serotonin is not affected by MDL 72222, but can be blocked by low doses of ketanserin. Used in combination, the two antagonists completely abolish the triphasic response to serotonin. This shows that 5-HT$_3$ receptors are probably involved in the initial excitatory response. The subsequent reduction in neuronal firing may be merely a consequence of the initial depolarization, i.e. a depolarizing blockade (Nishi, 1975). MDL 72222 also blocked this response, which is consistent with such an interpretation.

5-HT$_3$ RECEPTOR–RESPONSE COUPLING MECHANISMS

As already indicated, activation of 5-HT$_3$ receptors invariably results in neuronal depolarization (Fig. 1), although the ionic mechanisms by which this is brought about may vary with the type of neurone studied.

Intracellular recordings from C-type neurones in the rabbit nodose ganglion show that the depolarization caused by serotonin in these cells is primarily due to an increase in membrane Na$^+$ and K$^+$ conductance (Higashi and Nishi, 1982). Similarly, serotonin-induced depolarization of sympathetic neurones in the rabbit superior cervical ganglion and of vagal C-fibres also results from increased Na$^+$ permeability (Wallis and Woodward, 1975; Neto, 1978). The molecular events involved between the occupation of the 5-HT$_3$ receptor by the agonist molecule and opening of ion channels in the nerve membrane have not been studied.

In contrast to the situation just described for sympathetic postganglionic and afferent neurones, depolarization of enteric neurones in the guinea pig myenteric plexus by serotonin is associated with an increased rather than a decreased membrane resistance. This depolarization seems to result from a decrease in resting potassium conductance (Johnson *et al.*, 1980, 1981). Again, the events occurring between activation of 5-HT$_3$ receptors and the closing of these K$^+$ channels are not understood.

FUTURE TRENDS

Much of the work on 5-HT$_3$ receptors reviewed in this chapter is very recent. This is because specific, potent agonist and antagonist drugs for these receptors have only just become available. These new tools will undoubtedly help define the involvement of 5-HT$_3$ receptors in various physiological and pathophysiological processes. They arrive at a time when considerable progress has been made regarding the selective activation and blockade of the other types of 5-HT receptor, namely 5-HT$_1$ and 5-HT$_2$ receptors (Bradley *et al.*, 1986; Fig. 1), and this should stimulate serotonin research considerably in the future.

It is already clear that at least three different subtypes of 5-HT$_3$ receptors can be identified using competitive antagonists. However, none of the currently available compounds possesses absolute specificity for one or other of them. Obviously the development of such compounds would greatly facilitate the definitive characterization of 5-HT$_3$ receptor subtypes. Another deficiency in the pharmacologist's armamentarium is the absence of the corresponding agonist drugs. Although 2-methyl-5-HT is a compound which shows considerable selectivity for 5-HT$_3$ receptors *vis-à-vis* 5-HT$_1$ or 5-HT$_2$ receptors, it does not distinguish between any particular 5-HT$_3$ receptor subtype. Moreover, the biostability of this compound is obviously inadequate for *in vivo* use and its lack of brain penetration severely limits its potential for studying the possible existence and function of central 5-HT$_3$ receptors in whole animal studies. However, it is to be expected that compounds lacking such deficiencies will eventually become available.

It can also be anticipated that radioligands suitable for receptor binding and autoradiographic studies will be produced. Neuroblastoma cell lines, such as N1E-115, may provide a convenient source of membranes for the former (Neijt *et al.*, 1986).

Seven different subtypes of serotonin receptors can now be identified pharmacologically and one wonders just how many more will appear over the next few years. Whether they are all different gene products, the results of varying post-translational modification to a common precursor protein, or derive their pharmacological specificities from certain aspects of the local membrane environment in which they are situated will only become clear when each of these receptors has been isolated as a pure protein and its amino acid sequence determined. Ultimately the cloning and expression of genes coding for these receptor proteins will provide sufficient amounts for studying drug–receptor interactions with nuclear magnetic resonance and other physicochemical techniques and provide a solid basis for truly rational drug design.

Very little is yet known of the therapeutic potential of 5-HT$_3$ receptor antagonists. Nevertheless, it would indeed be a precedent if potent selective antagonists of a major neurotransmitter were not to prove useful in clinical medicine. In this chapter we have indicated that the pain, and wheal and flare responses induced by serotonin are mediated by 5-HT$_3$ receptors. Since these phenomena occur in the cranial vasculature during a migraine attack and serotonin has been implicated in their genesis (Lance, 1982; Fozard,

1982; Moskowitz, 1984; Fozard, 1985b), the use of 5-HT$_3$ receptor antagonists in the symptomatic relief of migraine attacks would seem worth evaluating. In fact preliminary reports with MDL 72222 in this indication look encouraging (Fozard *et al.*, 1985; Loisy *et al.*, 1985). Recent reports that ICS 205–930, MDL 72222 and BRL 24924 can increase gastric emptying in animals (Buchheit *et al.*, 1985a; Cooper *et al.*, 1986) implicate 5-HT$_3$ receptors in the physiological control of gastric motility, and suggest the utility of such compounds in disturbances of gastric function. One of these is the nausea and vomiting associated with chemotherapy (Andrews *et al.*, 1986; Costall *et al.*, 1986a; Miner and Sanger, 1986; Miner *et al.*, 1986). The potential of the drugs in other indications, such as cardiovascular disease is, as yet, unclear.

Finally, the question of whether 5-HT$_3$ receptors occur only on peripheral neurones, or whether they are present in the central nervous system and on non-neuronal tissue, will become apparent very soon. Such studies will also establish whether 5-HT$_3$ receptors mediate exclusively excitatory actions of serotonin in these locations.

ACKNOWLEDGEMENTS

We thank Prof. E. Flückiger and Dr D. Hoyer for reading the manuscript, Messrs P. Gugger and F. Luh for preparing the figures and Mrs R. Heiniger for secretarial assistance.

REFERENCES

Aas, P. (1983) Serotonin induced release of acetylcholine from neurons in the bronchial smooth muscle of the rat, *Acta. Physiol. Scand.*, **117**, 477–480.

Aghajanian, G. K. (1981) The modulatory role of serotonin at multiple receptors in the brain, in *Serotonin Neurotransmission and Behaviour* (Eds B. L. Jacobs and A. Gelperin), pp. 156–186, MIT Press, Cambridge, Mass.

Andrews, P. L. R., Hawthorn, J., and Sanger, G. J. (1987) The effect of abdominal visceral nerve lesions and a novel 5-HT-M receptor antagonist on cytotoxic and radiation induced emesis in the ferret, *J. Physiol. (Lond.)*, **382**, 47P.

Apperley, E., Humphrey, P. P. A., and Levy, G. P. (1976) Receptors for 5-hydroxytryptamine and noradrenaline in rabbit isolated ear artery and aorta, *Brit. J. Pharmacol.*, **58**, 211–221.

Armstrong, D. J., and Kay, I. S. (1985) MDL 72222 (a 5-HT antagonist) antagonizes the pulmonary depressor and respiratory chemoreflexes evoked by phenylbiguanide in anaesthetized rabbits, *J. Physiol. (Lond.)*, **365**, 104P.

Armstrong, D. J., Kay, I. S., and Russell, N. J. W. (1986) MDL 72222 antagonizes the reflex tachypnoeic response to miliary pulmonary embolism in anaesthetized rabbits, *J. Physiol. (Lond.)*, **381**, 13P.

Arunlakshana, O., and Schild, H. O. (1959) Some quantitative uses of drug antagonists, *Brit. J. Pharmacol. Chemother.*, **14**, 48–58.

Azami, J., Fozard, J. R., Round, A. A., and Wallis, D. I. (1985) The depolarizing action of 5-hydroxytryptamine on rabbit vagal primary afferent and sympathetic neurones and its selective blockade by MDL 72222, *Naunyn-Schmiedeberg's Arch. Pharmacol.*, **328**, 423–429.

Beck, S. G., Clarke, W. P., and Goldfarb, J. (1985) Spiperone differentiates multiple 5-hydroxytryptamine responses in rat hippocampal slices in vitro, *Eur. J. Pharmacol.*, **116**, 195–197.

Bradley, P. B., Engel, G., Feniuk, W., Fozard, J. R., Humphrey, P. P. A., Middlemiss, D. N., Mylecharane, E. J., Richardson, B. P., and Saxena, P. R. (1986) Proposals for the classification and nomenclature of functional receptors for 5-hydroxytryptamine, *Neuropharmacology*, **25**, 563–576.

Branchek, T., Kates, M. and Gershon, M. D. (1984) Enteric receptors for 5-hydroxytryptamine, *Brain Res.*, **324**, 107–118.

Buchheit, K. H., Costall, B., Engel, G., Gunning, S. J., Naylor, R. J., and Richardson, B. P. (1985a) 5-Hydroxytryptamine receptor antagonism by metoclopramide and ICS 205-930 in the guinea pig leads to enhancement of contractions of stomach muscle strips induced by electrical stimulation and facilitation of gastric emptying *in vivo, J. Pharm. Pharmacol.*, **37**, 664–667.

Buchheit, K. H., Engel, G., Mutschler, E., and Richardson, B. (1985b) Study of the contractile effect of 5-hydroxytryptamine (5-HT) in the isolated longitudinal muscle strip from guinea-pig ileum, *Naunyn-Schmiedeberg's Arch. Pharmacol.*, **329**, 36–41.

Buchheit, K. H., Engel, G., Hagenbach, A., Hoyer, D., Kalkman, H. O., and Seiler, M. P. (1986) The rat isolated stomach fundus strip, a model for 5-HT₁C receptors, *Brit. J. Pharmacol.*, **88**, 367P.

Chahl, L. A. (1983) Substance P mediates atropine-sensitive response of guinea-pig ileum to serotonin, *Eur. J. Pharmacol.*, **87**, 485–489.

Charlton, K. G., Bond, R. A., and Clarke, D. E. (1985). An inhibitory prejunctional '5-HT-like' receptor in the isolated perfused rat kidney: apparent distinction from the 5-HT₁A, 5-HT₁B and 5-HT₁C subtypes, *Naunyn-Schmiedeberg's Arch. Pharmacol.*, **328**, 154–159.

Comroe, J. H., Jr., Van Lingen, B., Stroud, R. C., and Roncoroni, A. (1953) Reflex and direct cardiopulmonary effects of 5-OH-tryptamine (serotonin), *Am. J. Physiol.*, **173**, 379–389.

Cooper, S. M., McClelland, C. M., McRitchie, B., and Turner, D. H. (1986) BRL 24924: a new and potent gastric motility stimulant, *Brit. J. Pharmacol.*, **88**, 383P.

Costall, B., Domeney, A. M., Naylor, R. J., and Tattersall, F. D. (1986a) 5-Hydroxytryptamine M-receptor antagonism to prevent cisplatin-induced emesis, *Neuropharmacology*, **25**, 959–961.

Costall, B., Kelly, M. E., Naylor, R. J., Tan, C. C. W., and Tattersall, F. D. (1986b) 5-Hydroxytryptamine M-receptor antagonism in the hypothalamus facilitates gastric emptying in the guinea-pig. *Neuropharmacology*, **25**, 1293–1296.

Cubeddu, L. X., Barnes, E., Langer, S. Z., and Weiner, N. (1974) Release of noradrenaline and dopamine beta-hydroxylase by nerve stimulation. I. Role of neuronal and extraneuronal uptake and of alpha presynaptic receptors, *J. Pharmacol. Exp. Ther.*, **190**, 431–450.

Day, M., and Vane, J. R. (1963). An analysis of the direct and indirect actions of drugs on the isolated guinea-pig ileum, *Brit. J. Pharmacol.*, **20**, 150–170.

Donatsch, P., Engel, G., Richardson, B. P., and Stadler, P. A. (1984a) Subtypes of neuronal 5-hydroxytryptamine (5-HT) receptors as identified by competitive antagonists, *Brit. J. Pharmacol.*, **81**, 33P.

Donatsch, P., Engel, G., Richardson, B. P., and Stadler, P. (1984b) ICS 205–930: A highly selective and potent antagonist at peripheral neuronal 5-hydroxytryptamine (5-HT) receptors, *Brit. J. Pharmacol.*, **81**, 34P.

Donatsch, P., Engel, G., Richardson, B. P., and Stadler, P. A. (1984c) The inhibitory effects of neuronal 5-hydroxytryptamine (5-HT) receptor antagonists on experimental pain in humans, *Brit. J. Pharmacol.*, **81**, 35P.

Douglas, W. W., and Toh, C. C. (1953) The respiratory stimulant action of 5-hydroxytryptamine (serotonin) in the dog, *J. Physiol. (Lond.)*, **120**, 311–318.

Drakontides, A. B., and Gershon, M. D. (1968) 5-Hydroxytryptamine receptors in the mouse duodenum, *Brit. J. Pharmacol.*, **33**, 480–492.

Dunbar, A. W., McClelland, C. M., and Sanger, G. J. (1986) BRL 24924: A stimulant of gut motility which is also a potent antagonist of the Bezold–Jarisch reflex in anaesthetised rats, *Brit. J. Pharmacol.*, **88**, 319P.

Engel, G., Göthert, M., Hoyer, D., Schlicker, E., and Hillenbrand, K. (1986) Identity of inhibitory presynaptic 5-hydroxytryptamine (5-HT) autoreceptors in the rat brain cortex with 5-HT$_{1B}$ binding sites. *Naunyn-Schmiedeberg's Arch. Pharmacol.*, **332**, 1–7.

Engel, G., Göthert, M., Müller-Schweinitzer, E., Schlicker, E., Sistonen, L., and Stadler, P. A. (1983) Evidence for common pharmacological properties of ^{3}H-5-hydroxytryptamine binding sites, presynaptic 5-hydroxytryptamine autoreceptors in CNS and inhibitory presynaptic 5-hydroxytryptamine receptors on sympathetic nerves, *Naunyn-Schmiedeberg's Arch. Pharmacol.*, **324**, 116–124.

Engel, G., Hoyer, D., Kalkman, H. O., and Wick, M. B. (1984) Identification of 5-HT$_2$ receptors on longitudinal muscle of the guinea-pig ileum, *J. Recep. Res.*, **4**, 113–126.

Engel, G., Hoyer, D., Kalkman, H. O., and Wick, M. B. (1985) Pharmacological similarity between the 5-HT D-receptor on the guinea-pig ileum and the 5-HT$_2$ binding site, *Brit. J. Pharmacol.*, **84**, 106P.

Erspamer, V. (1966) Occurrence of indole alkylamines in nature, in *5-Hydroxytryptamine and Related Indole Alkylamines* (Ed. V. Erspamer), *Handbook of Experimental Pharmacology*, Vol. 19, pp. 132–182, Springer Verlag, New York.

Fallon, S. L., Kim, H. S., and Welch, J. J. (1983) Electrophysiological evidence for modulation of dopamine neurones by 8-hydroxy-2(di-*n*-propylamino)tetralin, *Neurosci. Abst.*, **9**, 716.

Feniuk, W., Humphrey, P. P. A., and Watts, A. D. (1979) Presynaptic inhibitory action of 5-hydroxytryptamine in dog isolated saphenous vein, *Brit. J. Pharmacol.*, **67**, 247–254.

Feniuk, W., Humphrey, P. A. A., and Watts, A. D. (1983) 5-Hydroxytryptamine-induced relaxation of isolated mammalian smooth muscle, *Eur. J. Pharmacol.*, **96**, 71–78.

Fozard, J. R. (1982) Basic mechanisms of antimigraine drugs, in *Headache: Physiopathological and Clinical Concepts* (Eds M. Critchley, A. Friedman, S. Gorini and F. Sicuteri), *Advances in Neurology*, Vol. 33, pp. 295–307, Raven Press, New York.

Fozard, J. R. (1984a) MDL 72222: a potent and highly selective antagonist at neuronal 5-hydroxytryptamine receptors, *Naunyn-Schmiedeberg's Arch. Pharmacol.*, **326**, 36–44.

Fozard, J. R. (1984b) Neuronal 5-HT receptors in the periphery, *Neuropharmacology*, **23**, 1473–1486.

Fozard, J. R. (1984c) Characteristics of the excitatory 5-HT receptor on the cholinergic nerves of the rabbit heart, in *Proceedings of the 9th International Congress of Pharmacology*, 1189P, Macmillan Press Ltd., London.

Fozard, J. R. (1985a) Peripheral neuronal 5-hydroxytryptamine receptors and their biological significance, in *Proceedings of the Third SCI/RSC Medicinal Chemistry*

Symposium (Ed. R. W. Lambert), pp. 37–56, The Royal Society of Chemistry, London.

Fozard, J. R. (1985b) 5-Hydroxytryptamine in the pathophysiology of migraine, in *Proceedings of 5th International Symposium on Vascular Neuroeffector Mechanisms, Paris 1984*, pp. 321–328, Elsevier, Amsterdam.

Fozard, J. R., and Gittos, M. W. (1983) Selective blockade of 5-hydroxytryptamine neuronal receptors by benzoic acid esters of tropine, *Brit. J. Pharmacol.*, **80**, 511P.

Fozard, J. R., and Host, M. (1982) Selective inhibition of the Bezold–Jarisch effect of 5-HT in the rat by antagonists at neuronal 5-HT receptors, *Brit. J. Pharmacol.*, **77**, 520P.

Fozard, J. R., and Mobarok Ali, A. T. M. (1978a) Blockade of neuronal tryptamine receptors by metoclopramide, *Eur. J. Pharmacol.*, **49**, 109–112.

Fozard, J. R., and Mobarok Ali, A. T. M. (1978b) Receptors for 5-hydroxytryptamine on the sympathetic nerves of the rabbit heart, *Naunyn-Schmiedeberg's Arch. Pharmacol.*, **301**, 223–235.

Fozard, J. R., and Mwaluko, G. M. P. (1976) Mechanism of the indirect sympathomimetic effect of 5-hydroxytryptamine on the isolated heart of the rabbit, *Brit. J. Pharmacol.*, **57**, 115–125.

Fozard, J. R., Loisy, C., and Tell, G. P. (1985) Blockade of neuronal 5-hydroxytryptamine receptors with MDL 72222. A novel approach to the symptomatic treatment of migraine, in *Proceedings of the 5th Migraine Symposium, London, 1984*, pp. 264–272, Karger, Basle.

Fozard, J. R., Mobarok Ali, A. T. M., and Newgrosh, G. (1979) Blockade of serotonin receptors on autonomic neurones by (−)-cocaine and some related compounds, *Eur. J. Pharmacol.*, **59**, 195–210.

Furchgott, R. F. (1966) The use of beta-haloalkylamines in the differentiation of receptors and in the determination of dissociation constants of receptor agonist complexes, *Adv. Drug Res.*, **3**, 21–55.

Gaddum, J. H., and Hameed, K. A. (1954) Drugs which antagonize 5-hydroxytryptamine, *Brit. J. Pharmacol. Chemother.*, **9**, 240–248.

Gaddum, J. H., and Picarelli, Z. P. (1957) Two kinds of tryptamine receptor, *Brit. J. Pharmacol. Chemother.*, **12**, 323–328.

Gershon, M. D. (1982) Enteric serotonergic neurons, in *Biology of Serotonergic Transmission* (Ed. N. N. Osborne), pp. 363–401, Wiley, Chichester.

Gershon, M. D., Takaki, M., Tamir, H., and Branchek, T. (1985) The enteric neural receptor for 5-hydroxytryptamine, *Experientia*, **41**, 863–868.

Giger, R. K. A., Donatsch, P., Engel, G., Richardson, B. P., and Stadler, P. A. (1985) Indole-3-carboxylic acid derivatives: A new class of highly selective and potent antagonists at 5-HT M receptors. *Proceedings of the VIIIth International Symposium on Medicinal Chemistry, Uppsala 1984* (Eds R. Dahlbom and J. L. G. Nilsson), Vol. 2, pp. 133–135.

Göthert, M., and Duehrsen, V. (1979) Effects of 5-hydroxytryptamine and related compounds on the sympathetic nerves of the rabbit heart, *Naunyn-Schmiedeberg's Arch. Pharmacol.*, **308**, 9–18.

Greaves, M., and Shuster, S. (1967) Responses of skin blood vessels to bradykinin, histamine and 5-hydroxytryptamine, *J. Physiol. (Lond.)*, **193**, 255–267.

Gunning, S. J., Bradbury, A. J., Costall, B., and Naylor, R. J. (1986) Evidence that 5-hydroxytryptamine may exert both facilitatory and inhibitory control of electrical field stimulation-evoked contractions in longitudinal muscle taken from the body of guinea-pig stomach, *J. Pharm. Pharmacol.*, **38**, 182–187.

Higashi, H. (1977) 5-Hydroxytryptamine receptors on visceral primary afferent neurones in the nodose ganglion of the rabbit, *Nature*, **267**, 448–450.

Higashi, H., and Nishi, S. (1982) 5-Hydroxytryptamine receptors on visceral primary afferent neurones on rabbit nodose ganglion, *J. Physiol. (Lond.)*, **323**, 543–567.

Hoyer, D., Engel, G., and Kalkman, H. O. (1985) Molecular pharmacology of 5-HT$_1$ and 5-HT$_2$ recognition sites in rat and pig brain membranes: Radioligand binding studies with [^{3}H]5-HT, [^{3}H]8-OH-DPAT, (−)[^{125}I] iodocyano-pindolol, [^{3}H]mesulergine and [^{3}H]ketanserin, *Eur. J. Pharmacol.*, **118**, 13–23.

Humphrey, P. P. A. (1983) Pharmacological characterization of cardiovascular 5-hydroxytryptamine receptors, in *Proceedings of IVth Vascular Neuroeffector Mechanisms Symposium* (Eds J. Bevan *et al.*), pp. 237–242, Raven Press, New York.

Humphrey, P. A. A. (1984) Peripheral 5-hydroxytryptamine receptors and their classification, *Neuropharmacology*, **23**, 1503–1510.

Humphrey, P. A. A., Feniuk, W., and Watts, A. D. (1982) Ketanserin – a novel anti-hypertensive drug?, *J. Pharm. Pharmacol.*, **34**, 541.

Ireland, S. J., Straughan, D. W., and Tyers, M. B. (1982) Antagonism by metoclopramide and quipazine of 5-hydroxytryptamine-induced depolarizations of rat isolated vagus nerve, *Brit. J. Pharmacol.*, **75**, 16P.

Jacob, L., and Comroe, J. H., Jr. (1971) Reflex apnoea, bradycardia and hypotension produced by serotonin and phenyldiguanide acting on the nodose ganglion of the cat, *Circulation Res.*, **24**, 145–155.

Jacob, J., and Poite-Bevierre, M. (1960) Actions de la serotonine et de la benzyl-1-dimethyl-2,5-serotonine sur le coeur isolé de lapin, *Arch. Int. Pharmacodyn. Ther.*, **127**, 11–26.

Johnson, S. M., Katayama, Y., Morita, K., and North, R. A. (1981) Mediators of slow synaptic potentials in the myenteric plexus of the guinea-pig ileum, *J. Physiol. (Lond.)*, **320**, 175–186.

Johnson, S. M., Katayama, Y., and North, R. A. (1980) Multiple actions of 5-hydroxytryptamine on myenteric neurons of the guinea-pig ileum, *J. Physiol. (Lond.)*, **304**, 459–470.

Keele, C. A., and Armstrong, D. (1964) Experimental methods, in *Substances Producing Pain and Itch*, pp. 30–66, Arnold, London.

Kirby, G. C., and McQueen, D. S. (1984) Effects of the antagonists MDL 72222 and ketanserin on responses of cat carotid body chemoreceptors to 5-hydroxytryptamine, *Brit. J. Pharmacol.*, **83**, 259–269.

Kosterlitz, H. W., and Robinson, J. A. (1958) The inhibitory action of morphine on the contraction of the longitudinal muscle coat of the isolated guinea-pig ileum, *Brit. J. Pharmacol.*, **13**, 296–303.

Krayer, O. (1961) The history of the Bezold–Jarisch effect, *Naunyn-Schmiedeberg's Arch. Pharmacol.*, **240**, 361–368.

Lance, J. W. (1982) *Mechanism and Management of Headache*, 4th edn, pp. 162–164, Butterworth, London.

Leysen, J. E., De Chaffoy De Courcelles, D., De Clerk, F., Niemegeers, C. J. E., and Van Nueten, J. M. (1984) Serotonin S$_2$ receptor binding sites and functional correlates, *Neuropharmacology*, **23**, 1493–1501.

Loisy, C., Beorchia, S., Centonze, V., Fozard, J. R., Scheeter, P. J., and Tell, G. P. (1985) Effects on migraine headache of MDL 72222 an antagonist at neuronal 5-HT receptors. Double-blind, placebo controlled study, *Cephalalgia*, **5**, 79–82.

Maayani, S., Wilkinson, C. W., and Stollak, J. S. (1984) 5-Hydroxytryptamine receptor in rabbit aorta: characterisation by butyrophenone analogs, *J. Pharmacol. Exp. Ther.*, **229**, 346–350.

Markstein, R., Hoyer, D., and Engel, G. (1986) 5-HT$_{1A}$ receptors mediate stimulation of adenylate cyclase in rat hippocampus, *Naunyn-Schmiedeberg's Arch. Pharmacol.*, **333**, 335–341.

Middlemiss, D. N. (1986) Functional correlates of the subtypes of the 5-HT$_1$ recognition site, *Trends Pharmacol. Sci.*, **7**, 52–53.

Miner, W. D., and Sanger, G. J. (1986) Inhibition of cisplatin-induced vomiting by selective 5-hydroxytryptamine M-receptor antagonism, *Brit. J. Pharmacol.*, **88**, 497–499.

Miner, W. D., Sanger, G. J., and Turner, D. H. (1986) Comparison of the effect of BRL 24924, metoclopramide and domperidone on cisplatin-induced emesis in the ferret, *Brit. J. Pharmacol.*, **88**, 374P.

Moskowitz, M. A. (1984) The neurobiology of vascular head pain, *Ann. Neurol.*, **16**, 157–168.

Neijt, H. C., Vijverberg, H. P. M., and Van Den Becken, J. (1986) The dopamine response in mouse neuroblastoma cells is mediated by serotonin 5-HT$_3$ receptors, *Eur. J. Pharmacol.*, **127**, 271–274.

Neto, F. R. (1978) The depolarizing action of 5-HT on mammalian non-myelinated nerve fibres, *Eur. J. Pharmacol.*, **49**, 351–356.

Nishi, K. (1975) The action of 5-hydroxytryptamine on chemoreceptor discharges of the cat's carotid body, *Brit. J. Pharmacol.*, **55**, 27–40.

Oakley, B., and Schafer, R. (1978) Compound action potential, in *Experimental Neurobiology. A laboratory manual*, pp. 85–96, University of Michigan Press, Ann Arbor.

Offermeier, J., and Ariëns, E. J. (1966) Serotonin I. Receptors involved in its action, *Arch. Int. Pharmacodyn. Ther.*, **164**, 192–215.

Orwin, J. M., and Fozard, J. R. (1986) Blockade of the flare response to intradermal 5-hydroxytryptamine in man by MDL 72222, a selective antagonist at neuronal 5-hydroxytryptamine receptors, *Eur. J. Clin. Pharmacol.*, **30**, 209–212.

Paintal, A. S. (1973) Vagal sensory receptors and their reflex effects, *Physiol. Rev.*, **53**, 159–227.

Pazos, A., Hoyer, D., and Palacios, J. M. (1984) The binding of serotonin ligands to the porcine choroid plexus: Characterisation of a new type of serotonin recognition site, *Eur. J. Pharmacol.*, **106**, 539–546.

Pedigo, N. W., Yamamura, H. I. and Nelson, D. L. (1981) Discrimination of multiple ³H-5-hydroxytryptamine binding sites by the neuroleptic spiperone in rat brain, *J. Neurochem.*, **36**, 220–226.

Peroutka, S. J., and Snyder, S. H. (1979) Multiple serotonin receptors: differential binding of ³H-5-hydroxytryptamine, ³H-lysergic acid diethylamide and ³H-spiroperidol, *Mol. Pharmacol.*, **16**, 687–699.

Peroutka, S. J., Lebovitz, R. M., and Snyder, S. H. (1981) Two distinct central serotonin receptors with different physiological functions, *Science*, **212**, 827–829.

Richardson, B. P., and Engel, G. (1986) The pharmacology and function of 5-HT₃ receptors, *Trends Neurosci.*, **9**, 424–428.

Richardson, B. P., Engel, G., Donatsch, P., and Stadler, P. A. (1985) Identification of 5-hydroxytryptamine M-receptor subtypes and their specific blockade by a new class of drugs, *Nature*, **316**, 126–131.

Rocha e Silva, M., Valle, J. R., and Picarelli, Z. P. (1953) A pharmacological analysis of the mode of action of serotonin (5-hydroxytryptamine) upon the guinea-pig ileum, *Brit. J. Pharmacol. Chemother.*, **8**, 378–388.

Round, A. A., and Wallis, D. I. (1985) Selective blockade by ICS 205-930 of 5-

hydroxytryptamine (5-HT) depolarizations of rabbit vagal afferent and sympathetic ganglion cells, *Brit. J. Pharmacol.*, **86**, 734P.

Round, A. A., and Wallis, D. I. (1986) The depolarizing action of 5-hydroxytryptamine on rabbit vagal afferent and sympathetic neurones *in vitro* and its selective blockade by ICS 205-930, *Brit. J. Pharmacol.*, **88**, 485–494.

Saxena, P. R., Heiligers, J., Mylecharane, E. J., and Tio, R. (1985) Excitatory 5-hydroxytryptamine receptors in the cat urinary bladder are of the M- and 5-HT$_2$ type, *J. Autonom. Pharmacol.*, **5**, 101–107.

Schuurkes, J. A. J., van Nueten, J. M., van Daele, P. G. H., Reyntiens, A. J., and Janssen, P. A. J. (1985) Motor-stimulating properties of cisapride on isolated gastrointestinal preparations of the guinea-pig, *J. Pharmacol. Exp. Ther.*, **234**, 775–783.

Sicuteri, F., Fanciullacci, M., Franchi, G., and Del Biancho, P. L. (1965) Serotonin-bradykinin potentiation on the pain receptors in man, *Life Sci.*, **4**, 303–316.

Stansfeld, C. E., and Wallis, D. I. (1982) A- and C-type cells of the rabbit nodose ganglion respond differently to 5-HT, GABA and dimethylphenylpiperazinium (DMPP). *J. Physiol. (Lond.)*, **328**, 14P–15P.

Wallis, D. (1983) Neuronal 5-hydroxytryptamine receptors outside the central nervous system, *Life Sci.*, **29**, 2345–2355.

Wallis, D. I., and Nash, H. L. (1980) The action of methylated derivatives of 5-hydroxytryptamine at ganglionic receptors, *Neuropharmacology*, **19**, 465–472.

Wallis, D. I., and Woodward, B. (1975) Membrane potential changes induced by 5-hydroxytryptamine in the rabbit superior cervical ganglion, *Brit. J. Pharmacol.*, **55**, 199–212.

Wallis, D. I., Stansfeld, C. E., and Nash, H. L. (1982) Depolarizing responses recorded from nodose ganglion cells of the rabbit evoked by 5-hydroxytryptamine and other substances, *Neuropharmacology*, **21**, 31–40.

NOTE ADDED IN PROOFS

Since completion of our chapter, more than 50 new publications on 5-HT$_3$ receptors have appeared. Here we briefly review those results which contribute significant new knowledge to the 5-HT$_3$ receptor field.

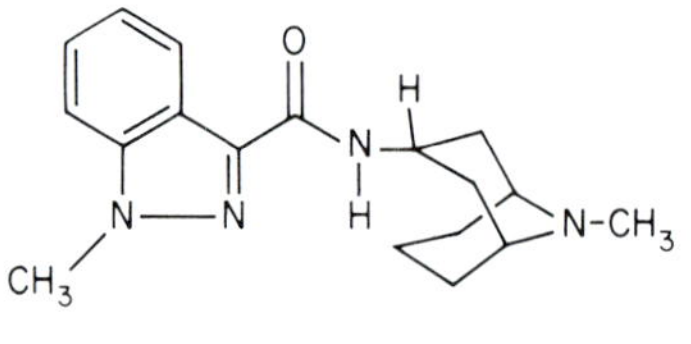

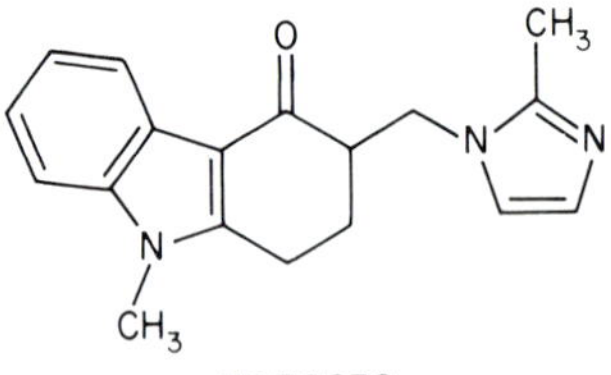

Fig. 16   New 5-HT$_3$ receptor antagonists

1. Most of the new results concern the development of two new 5-HT$_3$ receptor antagonists GR38032 (Glaxo) and BRL 43694 (Beecham: see Fig. 16). Neither compound differs significantly from ICS 205–930 in terms of potency, selectivity or affinity for 5-HT$_3$ receptors. Both have been characterized as 5-HT$_3$ receptor antagonists using the systems mentioned in the section on 'Bioassay systems'

(Brittain *et al.*, 1987; Fake *et al.*, 1987). Like ICS 205–930, both drugs are effective at inhibiting emesis induced by chemotherapeutic agents in ferrets.

2. Additional clinical applications for 5-HT₃ receptor antagonists are suggested by recent animal experiments. For instance, evidence for the involvement of 5-HT₃ receptors in intestinal secretion emerged from experiments in which ICS 205–930 inhibited antigen-induced chloride secretion from isolated intestinal epithelia (Baird and Cuthbert, 1987).

   Furthermore it has been shown that duodenal distension in rats elicits reflex hypotension, an effect which can be blocked by 5-HT₃ receptor antagonists like BRL 43694 and ICS 205–930 (Moss and Sanger, 1987). This reflex model might indicate that 5-HT₃ receptors are involved in visceral nociception.

   The most spectacular proposals for new clinical indications undoubtedly came from experiments suggesting that 5-HT₃ receptor antagonists might possess anxiolytic and antipsychotic properties. However, these reports have been confined to GR38032 and caution is necessary in extrapolating these results to all 5-HT₃ receptor antagonists. Thus GR38032 inhibits the increased central dopamine turnover caused by amphetamine or neurotensin in rats and marmosets and reduces the accompanying locomotor activity without producing dopamine receptor blockade (Costall *et al.*, 1987; Hagan *et al.*, 1987). In addition GR38032 induces behavioural effects similar to those produced by clinically active anxiolytics in rats and cynomolgus monkeys (Jones *et al.*, 1987). Evidence for potential central actions of 5-HT₃ receptor antagonists also comes from experiments where ICS 205–930 stimulated noradrenaline release from rabbit hippocampal slices *in vitro* (Feuerstein and Hertting, 1986).

3. A prerequisite for a central action of 5-HT₃ receptor antagonists is obviously the existence of 5-HT₃ receptors in the brain. Radioligand binding studies with [³H]-GR65630, a structural analogue of GR38032, provide evidence that a low density of 5-HT₃ binding sites does exist in certain regions of the rat brain (Kilpatrick *et al.*, 1987). Highest densities where found in the entorhinal cortex and in areas which are believed to be involved in eliciting behavioural disturbances in anxiety and schizophrenia.

   5-HT₃ binding sites have also been identified on neuronal tissue using [³H]-ICS 205–930 which labels a single high affinity site on neuroblastoma glioma cells (Hoyer and Neijt, 1987).

4. The therapeutic benefits of 5-HT₃ receptor antagonists have been demonstrated in two different pathophysiological conditions. Chemotherapy-induced emesis in cancer patients has been successfully prevented by pretreatment with ICS 205–930 (Leibundgut and Lancranjan, 1987) or GR38032 (Cunningham *et al.*, 1987). In addition, the excessive diarrhoea in patients suffering from carcinoid syndrome can be abolished by ICS 205–930 (Anderson *et al.*, 1987). As in the studies of blister base pain and cutaneous flare responses described in the main chapter (see section 'Bioassay systems'), these results show that 5-HT₃ receptors are found not only in animals but also in man.

Anderson, J. V., Coupe, M. O., Morris, J. A., Hodgson, H. J. F., and Bloom, S. R. (1987) Remission of symptoms in carcinoid syndrome with a new 5-hydroxytryptamine M receptor antagonist. *Brit. Med. J.*, **294**, 1129.

Baird, A. W., and Cuthbert, A. W. (1987) Neuronal involvement in type 1 hypersensitivity reactions in gut epithelia. *Brit. J. Pharmacol.*, **92**, 647–655.

Brittain, R. T., Butler, A., Coates, I. H., Fortune, D. H., Hagan, R., Hill, J. M., Humber, A., Humphrey, P. P. A., Hunter, D. C., Ireland, S. J., Jack, D., Jordan,

C. C., Oxford, A., and Tyers, M. B. (1987) GR38032F, A novel selective 5HT₃
receptor antagonist. *Brit. J. Pharmacol.*, **90**, 87P.

Costall, B., Domeney, A. M., Naylor, R. J., and Tyers, M. B. (1987) Effects of
the 5-HT₃ receptor antagonist, GR38032F, on raised dopaminergic activity in the
mesolimbic system of the rat and marmoset brain. *Brit. J. Pharmacol.*, **92**, 881–894.

Cunningham, D., Pople, A., Ford, H. T., Hawthorn, J., Gazet, J-C., and Challoner,
T. (1987) Prevention of emesis in patients receiving cytotoxic drugs by GR38032F,
a selective 5-HTc receptor antagonist. *Lancet*, **8548**, 1461–1463.

Fake, C. S., King, F. D., and Sanger, G. J. (1987) BRL 43694: A potent and novel
5-HT₃ receptor antagonist. *Brit. J. Pharmacol.*, **91**, 335P.

Feuerstein, T. J., and Hertting, G. (1986) Serotonin (5-HT) enhances hippocampal
noradrenaline (NA) release; evidence for facilitatory 5-HT receptors within the
CNS. *Naunyn Schmiedeberg's Arch. Pharmacol.*, **333**, 191–197.

Hagan, R. M., Butler, A., Hill, J. M., Jordan, C. C., Ireland, S. J., and Tyers, M.
B. (1987) Effect of the 5-HT₃ receptor antagonist, GR38032F, on responses to
injection of a neurokinin agonist into the ventral tegmental area of the rat brain.
Eur. J. Pharmacol., **138**, 303–305.

Hoyer, D., and Neijt, H. C. (1987) Identification of serotonin 5-HT₃ recognition
sites by radioligand binding in NG108–15 neuroblastoma-glioma cells. *Eur. J.
Pharmacol.*, **143**, 291–292.

Jones, B. J., Oakley, N. R., and Tyers, M. B. (1987) The anxiolytic activity of
GR38032F, a 5-HT₃ receptor antagonist in the rat and cynomolgus monkey. *Brit.
J. Pharmacol.*, **90**, 88P.

Kilpatrick, G. J., Jones, B. J., and Tyers, M. B. (1987) The identification and
distribution of 5-HT₃ receptors in rat brain using radioligand binding. *Nature*, **330**,
746–748.

Leibundgut, U., and Lancranjan, I. (1987) First results with ICS 205–930 (5-HT₃
receptor antagonist) in prevention of chemotherapy-induced emesis. *Lancet*, **8543**,
1198.

Moss, H. E., and Sanger, G. J. (1987) Antagonism by BRL 43694 of the pseudo-
affective reflex induced by duodenal distension. *Brit. J. Pharmacol.*, **92**, 531P.

Neuronal Serotonin
Edited by N. N. Osborne and M. Hamon
© 1988 John Wiley & Sons Ltd

CHAPTER 18

Autoradiography of Serotonin Receptors

A. Pazos
Dept. of Pharmacology and Therapeutics
School of Medicine
University of Cantabria
39071 Santander
Spain
and
D. Hoyer, M. M. Dietl and J. M. Palacios
Preclinical Research
Sandoz Ltd
CH-4002 Basle
Switzerland

INTRODUCTION

In the last twenty years, the application of a number of histochemical methods with a high anatomical resolution to the study of the different systems of neurotransmission has significantly improved our understanding of the chemical organization of the brain. In the case of serotonin (5-HT), cell bodies containing this transmitter as well as serotonergic pathways have been localized in detail by histofluorescence, autoradiography and immunohisto-chemistry (Dahlström and Fuxe, 1964; Parent *et al.*, 1981; Steinbush, 1981). In addition, the development and application of 'binding' techniques to the labeling of neurotransmitter receptors by using specific radioligands (Yama-mura *et al.*, 1985) has led to a much better knowledge of the pharmacological properties of 5-HT receptors, allowing the definition and characterization of several types and subtypes (Peroutka and Snyder, 1979; Pedigo *et al.*, 1981; Pazos *et al.*, 1984c). However, because of their limited resolution, binding procedures cannot be used to analyse the detailed anatomical distribution of drug and neurotransmitter receptors. In order to overcome this limitation, autoradiographic procedures to visualize these binding sites at the micro-scopic level were developed. Autoradiography is the process by which an amount of bound radioactivity can be localized in tissue sections. Receptor autoradiography adapts the methodology for the autoradiographic localiz-ation of diffusible substances (Stumpf and Roth, 1966) to the general prin-ciples of the radiometric labeling of receptors in membrane preparations (Kuhar, 1985). *In vitro* receptor autoradiography, originally developed by Young and Kuhar in 1979, involves incubation of slide-mounted microtome tissue sections with specific radioactive ligands, under similar conditions to those used in 'binding' studies. Autoradiograms are then generated by appo-sition to the labeled tissues of dry photographic emulsions, such as emulsion-coated flexible coverslips (Young and Kuhar, 1979) or, more recently, isotope sensitive sheetfilms (Penney *et al.*, 1981; Palacios *et al.*, 1981). The use of sensitive films, together with the availability of computer-assisted image-analysis systems, have simplified the process of quantitation of autoradio-graphic images (Palacios *et al.*, 1981; Unnerstall *et al.*, 1982). The contri-bution of receptor autoradiography studies to the knowledge of serotonergic neurotransmission is really impressive. In addition to providing a complete mapping of the distribution of these sites in the brain of laboratory animals and humans, these studies have played a crucial role in the clarification of the properties of the different 5-HT receptor subtypes (Pazos and Palacios, 1985; Pazos *et al.*, 1987a).

The goal of this chapter is to review some aspects of the autoradiographic labeling of 5-HT receptors. We will analyse the methodological procedures for serotonin receptor visualization, the contribution of these studies to the definition of 5-HT receptor subtypes and the anatomical distribution of these sites in the vertebrate brain. We will also illustrate the point of the species differences in distribution and pharmacology found for these receptors. Finally, the future trends in the field of serotonin receptor imaging, linked to the development of new methodological advances, will be commented upon.

METHODOLOGICAL PROCEDURES FOR THE AUTORADIOGRAPHIC LABELING OF SEROTONIN RECEPTORS IN THE BRAIN

On the basis of the results obtained in membrane binding studies in rat brain, central 5-HT receptors were first classified into 5-HT$_1$, those labeled with nanomolar affinity by [3H]5-HT and 5-HT$_2$, those labeled by [3H]spiperone. [3H]LSD recognized with high affinity both sites (Peroutka and Snyder, 1979). Initial autoradiographic studies of 5-HT receptors were carried out using these radioligands (Young and Kuhar, 1980; Meibach *et al.*, 1980). In the last few years, the increasing evidence for the existence of different 5-HT$_1$ receptor subtypes and the knowledge of their rank orders of drug affinities (Pedigo *et al.*, 1981; Middlemiss and Fozard, 1983; Pazos *et al.*, 1984c) led to the development of subtype selective radioligands, such as [3H]8-OH-DPAT (8-hydroxy-2-(*N,N*-di-n-propylamino)tetralin) (Gozlan *et al.*, 1983; Marcinkiewicz *et al.*, 1984), [3H]mesulergine (Closse, 1983), [125I]CYP (cyanopindolol) (Pazos *et al.*, 1985b) and [3H]TVX-Q 7821 (Glaser *et al.*, 1985). In addition, new 5-HT$_2$ radioactive compounds, such as [3H]ketanserin (Leysen *et al.*, 1982) and [125I]LSD (Engel *et al.*, 1984) were developed. All these radioligands are routinely used in the autoradiographic characterization and localization of 5-HT receptor sites. Table 1 summarizes the pharmacological selectivity for the different 5-HT recognition sites described up to date.

Table 2 shows the procedures used for *in vitro* autoradiography of sero-

Table 1 Pharmacological selectivity of the radioligands used in the autoradiographic labeling of serotonin receptors

Ligand	5-HT recognition site					
	5-HT$_{1A}$	5-HT$_{1B}$[a]	5-HT$_{1C}$	5-HT$_{1D}$[b]	5-HT$_2$	5-HT$_3$
[3H]5-HT	+	+	+	+		+
[3H]spiperone[c,d]					+	
[3H]LSD[c]	+		+	+	+	
[125I]LSD[c]			+		+	
[3H]ketanserin[d]					+	
[3H]8-OH-DPAT	+					
[3H]mesulergine			+		+[f]	
[125I]CYP[e]		+				
[3H]TVX-Q 7821	+					

[a] Absent in higher mammals
[b] Only present in higher mammals
[c] Labels dopamine D$_2$ receptors
[d] Labels a 'chemical recognition site'
[e] Labels β-adrenoceptors
[f] In the rodent brain

Table 2 Autoradiographic localization of serotonin receptors, ligands and incubation conditions

Ligand	(Ligand) (nM)	Preincubation protocol	Incubation			Washing protocol	Exposure time	Blank generation	Subtype labeled
			Buffer	Temperature	Time				
[^{3}H]5-HT	2	30 min at room temp. in incubation buffer	0.17 M Tris-HCl 4 mM CaCl$_2$ 0.01% ascorbic acid (pH 7.6)	room temp.	60 min	1 × 5 min (4°C)	60 days	10 nM 8-OH-DPAT 10 nM (−)21 009 100 nM 5-HT 100 nM 5-HT	5-HT$_{1A}$ 5-HT$_{1B}$ 5-HT$_{1C}$ 5-HT$_{1D}$
[^{3}H]5-HT	10	15 min at room temp. in incubation buffer	0.3 M Tris-HCl + 0.1 mM pargyline	25°C	30 min	3 × dipping	1–2 weeks	10^{-5} M 5-HT	5-HT$_3$
[^{3}H]8-OH-DPAT	2	30 min at room temp. in incubation buffer	0.17 M Tris-HCl 4 mM CaCl$_2$ 0.01% ascorbic acid (pH 7.6)	room temp.	60 min	2 × 5 min (4°C)	60 days	1 μM 5-HT	5-HT$_{1A}$
[^{3}H]TVX-Q 7821	9	30 min at 25°C in 0.17 M Tris-HCl (pH 7.7)	same as above (pH 7.7) plus 1 μM pargyline	25°C	60 min	4 × 15 sec (4°C)	35 days	10 μM 5-HT or 10 μM TVX-Q 7821	5-HT$_{1A}$
[^{3}H]LSD	5	5 min at room temp. in incubation buffer (-pargyline)	0.17 M Tris-HCl 4 mM CaCl$_2$ 0.01% ascorbic acid 0.15 M NaCl 10 μM pargyline	room temp.	60 min	2 × 10 min (4°C)	100 days	100 nM 5-HT 1 μM cinanserin	5-HT$_{1A}$ 5-HT$_2$

Radioligand	Concentration (nM)	Preincubation	Incubation buffer	Temperature	Time	Wash	Exposure time	Nonspecific binding	Receptor
		in incubation buffer	4 mM CaCl$_2$ 0.01% ascorbic acid (pH 7.6)						
[^{125}I]CYP	0.012	10 min at room temp. in Tris-HCl (pH 7.4) plus 150 mM NaCl	same as in preincubation	room temp.	120 min	2 × 15 min (4°C)	3 days	100 nM 5-HT	5-HT$_{1B}$
[^{125}I]LSD	0.05–1.5	30 min at room temp. in 0.17 mM Tris-HCl plus 1% BSA (pH 7.4)	50 mM Tris-HCl (pH 7.6) or same as in preincubation plus 150 mM NaCl	room temp.	120 or 60 min	2 × 15 min (4°C) or 4 × 20 min (4°C)	12–24 h	100 nM 5-HT 1 µM ketanserin	5-HT$_{1C}$ 5-HT$_2$
[^{3}H]mesulergine	5	15 min at room temp. in incubation buffer	0.17 M Tris-HCl (pH 7.7)	room temp.	2 h	2 × 10 min (4°C)	30 days	100 nM 5-HT 3 × 10^{-7} M cinanserin	5-HT$_{1C}$ 5-HT$_2$
[^{3}H]ketanserin	2	15 min at room temp. in incubation buffer	0.17 M Tris-HCl (pH 7.7)	room temp.	2 h	2 × 10 min (4°C)	30 days	10^{-6} M methysergide	5-HT$_2$
[^{3}H]spiperone	0.4	none	0.17 M Tris-HCl 1 mM MgCl$_2$ 2 mM CaCl$_2$ 5 mM KCl 120 mM NaCl (pH 7.7)	room temp.	30 min	2 × 5 min (4°C)	90 days	3 × 10^{-7} M ketanserin	5-HT$_2$

tonin receptors. The incubation conditions are initially tailored for those used in membrane assays and tested by performing biochemical measurements of the binding to tissue sections. The conditions providing the best total to non-specific binding ratio as well as the highest level of selectivity are finally selected.

PHARMACOLOGICAL PROPERTIES OF 5-HT$_1$ RECEPTOR SUBTYPES: THE CONTRIBUTION OF AUTORADIOGRAPHIC TECHNIQUES

The possibility of the existence of different subtypes of 5-HT$_1$ binding sites has received considerable support in the last few years. Pedigo *et al.* (1981), on the basis of the biphasic displacement of [^{3}H]5-HT binding by the neuroleptic spiperone, proposed the existence of two subtypes, 5-HT$_{1A}$ and 5-HT$_{1B}$, in the rat brain. A similar phenomenon was also described for other compounds, such as 8-OH-DPAT (Hjorth *et al.*, 1982; Middlemiss and Fozard, 1983; Hamon *et al.*, 1984), RU 24 969 (5-methoxy-3[1,2,3,6-tetrahydropyridin-4-yl]indole) (Euvrard and Boissier, 1980; Hunt and Oberlander, 1981) and some β-blockers (Middlemiss *et al.*, 1977; Nahorski and Willcocks, 1983). The information on the pharmacological properties and regional distribution of these two 5-HT$_1$ sites was very limited, mainly because most of the binding studies had been carried out in rat cerebral cortex, an area equally enriched in both 5-HT$_{1A}$ and 5-HT$_{1B}$ subtypes.

In addition, a new type of [^{3}H]5-HT binding sites, not recognized by the proposed 5-HT$_{1A}$ or 5-HT$_{1B}$ selective drugs, was first identified by autoradiography (Cortés *et al.*, 1984) and then characterized by both membrane binding and autoradiography (Pazos *et al.*, 1984c; Yagaloff and Hartig, 1985) in the mammalian choroid plexus. This site was named 5-HT$_{1C}$ (Pazos *et al.*, 1984c).

Taking advantage of the development of subtype-selective compounds, we used quantitative autoradiography to better characterize and localize these 5-HT$_1$ receptor subtypes in the rat brain (Pazos and Palacios, 1985). For that, the use of selective radioactive compounds was combined with the quantitative autoradiographic analysis of the pharmacology of [^{3}H]5-HT binding – competition profiles – in several microscopic brain areas. The analysis of the competition curves, some of them shown in Fig. 1, revealed the presence of three types of [^{3}H]5-HT binding sites, with different pharmacological properties (Table 3). The binding of [^{3}H]5-HT to the dentate gyrus of the hippocampus was inhibited with nanomolar affinity by 8-OH-DPAT, LSD, RU 24 969 and the β-blocker (−)21 009 (4-[3-terbutylamino-2-hydroxy-propoxy]-indol-2-carbonic acid isopropylester), while the other compounds used showed low to very low affinity for these binding sites. In contrast, [^{3}H]5-HT binding to the dorsal subiculum was displaced with nanomolar

Table 3 Affinities (K_i)[a] of several drugs for [³H]5-HT binding to rat brain areas measured by quantitative autoradiography

Drugs	Brain area						
	Dorsal subiculum	Globus pallidus	Substantia nigra (reticular)	Lateral septal nucleus (dorsal)	Dentate gyrus	CA$_1$ field hippocampus	Choroid plexus
LSD	29.9	25.2	44.9	—	3.6	7.2	4.1
RU 24 969	1.5	0.6	3.5	9.0	5.4	10.2	130.2
8-OH-DPAT	>3000	>3000	>5000	19.2	4.8	11.4	>4000
(−)21 009	0.6	1.7	0.9	6.0	7.2	48.1	>4000
Mesulergine	>5000	>3000	—	—	>1000	>3000	11.8
Spiperone	>3000	>3000	>3000	—	359.3	209.6	710.1
Ketanserin	1796.4	>1000	>1000	—	598.8	>1000	354.0
Mianserin	143.7	—	—	—	269.5	—	13.6
Cinanserin	>3000	>5000	—	>5000	>2000	—	2000
(±)Propranclol	149.7	173.7	—	—	257.5	509.0	>5000
Pirenperone	>5000	>5000	—	>2000	>5000	—	1000

[a] nM
— = not measured

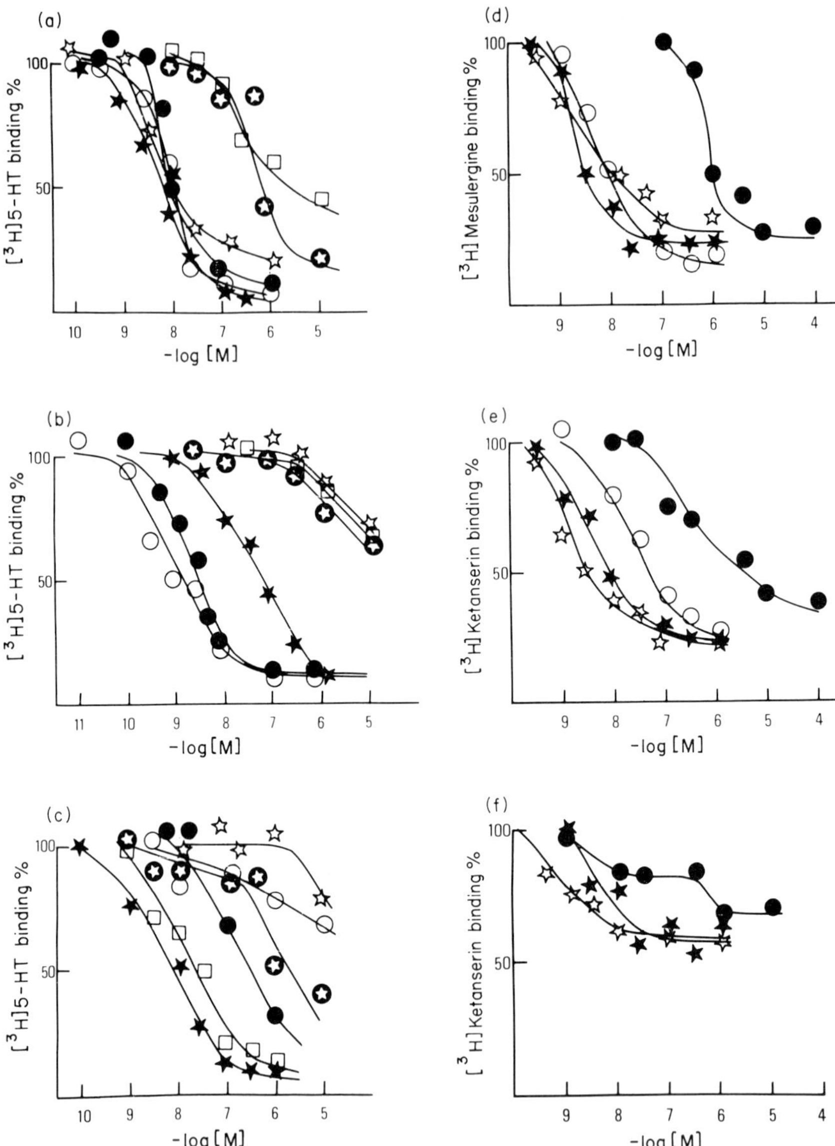

Fig. 1 Autoradiographic determination of the characteristics of the binding of 5-HT$_1$ and 5-HT$_2$ ligands to several regions of the rat brain. (a), (b), (c) Competition curves for [³H]5-HT binding to the dentate gyrus of the hippocampus, a 5-HT$_{1A}$-enriched area (a), dorsal subiculum, a 5-HT$_{1B}$ enriched area (b), and choroid plexus, a 5-HT$_{1C}$ enriched area (c). Symbols are (open circle) (−)21 009, (solid circle) RU-24 969, (open star) 8-OH-DPAT, (solid star) LSD, (open square) mesulergine, and (open star in solid circle) spiperone. Panels (d) and (e) illustrate the competition curves for [³H]mesulergine (d) and [³H]ketanserin (e) binding of

affinity by RU 24 969 and (−)21 009, while LSD showed an intermediate affinity for these sites, and the other drugs tested, including 8-OH-DPAT, inhibited [^{3}H]5-HT binding to this area with low or very low affinity (Table 3). Finally, in the choroid plexus, only LSD, mesulergine and mianserin recognized the binding of [^{3}H]5-HT with high affinity (Fig. 1 and Table 3). According to the definition of the different subtypes of 5-HT$_1$ recognition sites and to the affinities presented by the 'subtype selective' ligands, [^{3}H]5-HT sites in the dentate gyrus correspond to the 5-HT$_{1A}$ subtype, while those in the dorsal subiculum exhibit the characteristics of 5-HT$_{1B}$ sites. The pharmacology of [^{3}H]5-HT binding to the choroid plexus was similar to that of the 5-HT$_{1C}$ sites, as defined in membrane binding studies. In addition to these three anatomical areas, other rat brain nuclei presented a similar enrichment in one or the other subtype of 5-HT$_1$ sites (Table 3). A close correlation exists between K_i values for the different drugs in the dorsal subiculum, globus pallidus and substantia nigra, areas enriched in 5-HT$_{1B}$ sites. On the other hand, K_i values found in dentate gyrus, stratum oriens of the CA$_1$ field of the hippocampus and lateroseptal nucleus, 5-HT$_{1A}$ rich areas, are also very similar. This pharmacological characterization demonstrates the existence of the three subtypes of 5-HT$_1$ recognition sites in different nuclei of the rat brain. While these areas were enriched in one subtype of 5-HT sites, many other brain areas presented mixed populations of sites (Vagé et al., 1986).

The analysis of these competition experiments was further supported by the results obtained by the direct labeling of the different subtypes with selective radioligands. So, specific binding of [^{3}H]8-OH-DPAT was fully comparable, in terms of both anatomical localization and pharmacological profile, to that identified as the 5-HT$_{1A}$ component of [^{3}H]5-HT binding. Similarly, [^{3}H]mesulergine labeled in the choroid plexus a population of sites with the properties of 5-HT$_{1C}$ binding sites (Pazos and Palacios, 1985), although this radioligand also recognized, in other regions, 5-HT$_2$ receptors (see below).

The full understanding of the recognition of 5-HT$_{1B}$ receptors by β-adrenergic drugs in the rat brain is another issue where receptor autoradiography has been instrumental. As already mentioned several authors reported that the binding of [^{3}H]5-HT to rat brain membranes was inhibited by some β-blockers in a biphasic manner (Middlemiss et al., 1977; Nahorski and Willcocks, 1983). We demonstrated by using autoradiographic techniques that the β-blocker (−)21 009 showed the highest selectivity for the proposed

---

the 5-HT$_2$ rich lamina IV of the frontoparietal cortex. Panel (f) shows competition curves for ketanserin binding in the caudate-putamen, an area presenting high levels of non-displaceable binding. Symbols are (open circle) cinanserin, (solid circle) 5-HT, (open star) spiperone, and (solid star) pirenperone. (From Pazos et al. (1985a) and Pazos and Palacios (1985), with permission)

5-HT$_{1B}$ receptor subtype (see above). On the basis of these results, we labeled by autoradiography this population of sites directly with a radioactive β-blocker, [^{125}I]CYP, in rat brain sections (Pazos *et al.*, 1985b). It was observed that in some brain areas, such as the dorsal subiculum or the substantia nigra, binding of [^{125}I]CYP was blocked with high affinity by some serotonergic ligands. The anatomical distribution and pharmacological properties of these sites were identical to those previously defined for 5-HT$_{1B}$ binding as labeled by [^{3}H]5-HT.

From all these data, it is clear that the autoradiographic approach has been a critical tool in the clarification of the existence and properties of 5-HT receptor subtypes. This is due to the fact that receptor autoradiography allows the identification of microscopic areas enriched in only one receptor subtype, providing anatomical and pharmacological information which is very difficult to attain simply by using membrane binding assays (Pazos *et al.*, 1984a).

Receptor autoradiography has also played a relevant role in the studies dealing with the species differences observed in the characteristics of 5-HT receptors, specially in the identification of the 5-HT$_{1D}$ subtype, which is only present in certain higher mammalian species. We will analyse this issue in detail later.

ANATOMICAL DISTRIBUTION OF 5-HT RECEPTORS IN THE RAT BRAIN

Distribution of 5-HT$_1$ receptors

The presence of the different subtypes of 5-HT$_1$ receptors in each brain area was determined by comparing the direct labeling with selective radioligands with the inhibition of [^{3}H]5-HT binding induced by 10 nM (−)21 009 and 10 nM 8-OH-DPAT (Pazos and Palacios, 1985). The anatomical distribution of these sites is illustrated in Fig. 2. Maximal density values are shown in Table 4.

The external plexiform layer of the *olfactory bulb*, anterior olfactory nucleus, olfactory tubercle and piriform cortex presented intermediate concentrations of 5-HT$_1$ sites, while in the internal granular layer of the bulb, a somewhat lower density of sites was found. The piriform cortex, olfactory tubercle and anterior olfactory nucleus appeared to contain the three subtypes of 5-HT$_1$ receptors. In contrast, the external plexiform layer was enriched in 5-HT$_{1A}$ sites, and the internal granular layer contained predominantly 5-HT$_{1B}$ sites.

In the *neocortex*, intermediate concentrations of specific binding were found in the different layers of the frontal and frontoparietal cortex, while the striate cortex presented only low concentrations of sites. It can be said

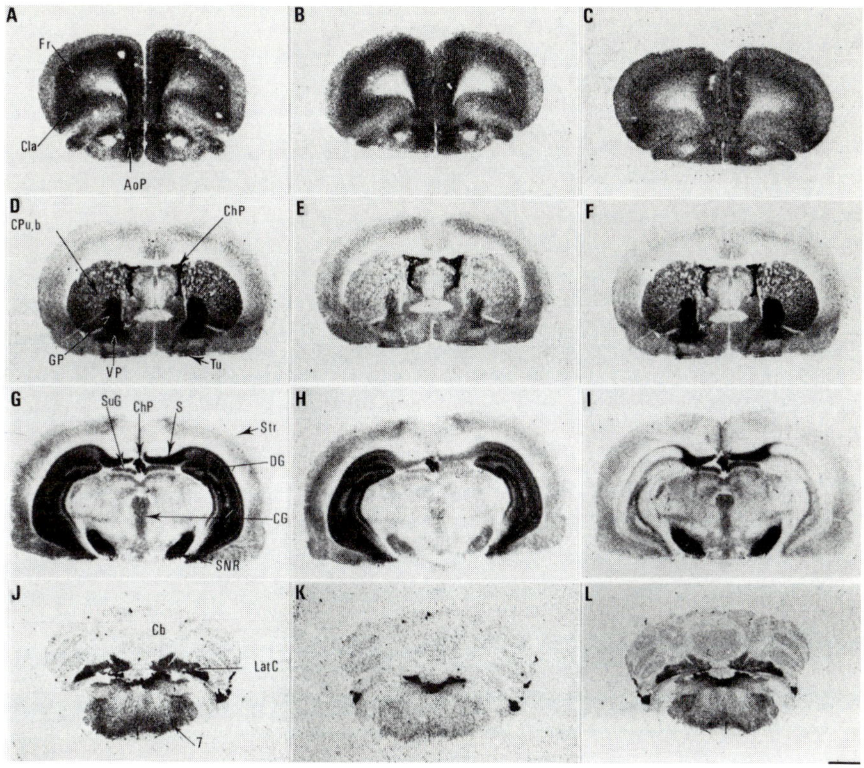

Fig. 2 [³H]5-HT binding in coronal sections of the rat brain through the frontal cortex (A–C), the globus pallidus (D–F), the upper midbrain (G–I) and the pons (J–L). Autoradiograms shown correspond to the total binding of [³H]5-HT (A,D,G,J), that in the presence of 10 nM (−)21 009 (B,E,H,K) and 100 nM 8-OH-DPAT (C,F,I,L). Structures where 5-HT₁ₐ sites are present include the internal laminae of the neocortex and the dentate gyrus of the hippocampus (DG). In contrast, 5-HT₁ᵦ sites are enriched in the globus pallidus (GP), the subiculum (S) and the substantia nigra reticulata (SNR). Finally, choroid plexuses (ChP) are very enriched in 5-HT₁c sites. Other abbreviations are: frontal cortex (Fr), claustrum (Cla), anterior olfactory nucleus, posterior part (AoP), caudate-putamen, body (CPu,b), ventral pallidum (VP), olfactory tubercle (Tu), superficial grey layer of the superior colliculus (SuG), striate cortex (Str), central grey (CG), lateral cerebellar nucleus (LatC), facial nucleus (7), cerebellum (Cb). Bar = 2 mm. (From Pazos and Palacios (1985), with permission)

that the internal laminae (IV–VI) were richer in 5-HT₁ binding than the external ones (I–III). In the frontal cortex, most of the binding to all the laminae was of the 5-HT₁ᵦ class, although 5-HT₁ₐ and 5-HT₁c sites were also present. In contrast, in the other regions studied, 5-HT₁ᵦ sites were predominant in the external laminae, but both 5-HT₁ₐ and 5-HT₁ᵦ sites

Table 4 Maximal densities of 5-HT$_1$ and 5-HT$_2$ receptors in different areas of the rat brain (fmol/mg protein)

Area	Specific 5-HT$_1$ binding			Specific 5-HT$_2$ binding
	[^{3}H]5-HT (mean ± SEM)	% binding[a] to 5-HT$_{1A}$	% binding[a] to 5-HT$_{1B}$	[^{3}H]ketanserin (mean ± SEM)
Olfactory system				
External plexiform layer of the olfactory bulb	1286.8 ± 175.8	85.7	57.6	107.2 ± 11.2
Anterior olfactory nucleus	1757.8 ± 180.2	65.9	49.1	281.1 ± 56.3
Olfactory tubercle	1787.3 ± 210.9	58.9	78.3	378.9 ± 73.8
Septal region				
Lateral septal nucleus, dorsal part	3047.3 ± 289.5	79.6	42.7	46.7 ± 3.7
Bed nucleus of the stria terminalis	1971.7 ± 118.7	52.1	74.8	32.2 ± 5.1
Amygdala				
Lateral amygdaloid nucleus	2075.6 ± 286.3	33.5	90.0	84.4 ± 15.0
Basomedial amygdaloid nucleus	1420.9 ± 139.0	41.9	100.0	64.9 ± 28.6
Medial amygdaloid nucleus	1627.3 ± 214.8	42.5	83.3	75.0 ± 14.8
Anterior cortical amygdaloid nucleus	2230.9 ± 362.2	45.8	67.0	101.0 ± 17.2
Posteromedial cortical amygdaloid nucleus	2439.0 ± 487.3	35.0	68.4	—
Basal ganglia				
Accumbens nucleus	1309.2 ± 9.0	52.7	97.2	139.8 ± 2.0
Caudate-putamen, head	1178.2 ± 114.6	42.4	96.2	0.0
Caudate-putamen, body	1230.9 ± 2.9	44.4	97.0	124.1 ± 38.3
Globus pallidus	3431.9 ± 180.4	36.3	100.0	10.6 ± 1.8
Entopeduncular nucleus	2266.0 ± 195.1	15.7	100.0	—
Hippocampus				
Dentate gyrus	3401.9 ± 168.7	71.1	16.3	102.8 ± 19.4
CA$_1$ field, oriens layer	2082.4 ± 80.4	71.9	32.3	19.4 ± 0.1
CA$_1$ field, stratum radiatum	1788.2 ± 386.3	64.7	41.1	28.6 ± 14.7
CA$_1$ field, lacunosum molecular layer	2501.2 ± 283.4	65.1	28.6	8.1 ± 2.5
CA$_2$ field, pyramidal layer	1341.9 ± 130.9	70.1	41.5	—

Area	Specific 5-HT$_1$ binding			Specific 5-HT$_2$ binding
	[³H]5-HT (mean ± SEM)	% binding[a] to 5-HT$_{1A}$	% binding[a] to 5-HT$_{1B}$	[³H]ketanserin (mean ± SEM)
CA$_3$ field, oriens layer	1860.7 ± 8.3	60.8	41.2	32.4 ± 20.6
CA$_3$ field, pyramidal layer	773.1 ± 55.6	84.4	59.8	25.0 ± 2.9
Dorsal subiculum	2841.7 ± 46.0	26.3	100.0	13.0 ± 9.6
Hypothalamus				
Ventromedial hypothalamic nucleus	2667.5 ± 234.1	43.2	69.3	68.5 ± 5.0
Lateral hypothalamic area	1363.6 ± 144.6	41.3	90.6	65.9 ± 17.2
Dorsomedial hypothalamic nucleus	1977.0 ± 247.3	47.3	86.3	—
Posterior hypothalamic nucleus	1355.6 ± 61.9	36.6	80.1	—
Medial mamillary nucleus	1792.9 ± 204.8	48.6	89.2	227.6 ± 63.4
Thalamus				
Lateral habenular nucleus	855.1 ± 111.9	56.7	63.8	58.4 ± 41.6
Olivary pretectal nucleus	2761.4 ± 331.2	32.7	93.2	—
Central medial thalamic nucleus	1707.3 ± 236.0	67.4	84.7	40.8 ± 21.4
Laterodorsal thalamic nucleus	1193.6 ± 87.0	48.6	95.2	56.3 ± 16.3
Ventrolateral thalamic nucleus	349.2 ± 32.9	42.0	87.9	44.7 ± 31.8
Reuniens thalamic nucleus	1529.5 ± 132.1	38.3	100.0	—
Reticular thalamic nucleus	181.2 ± 30.9	48.5	89.5	—
Dorsal lateral geniculate nucleus	1074.1 ± 146.8	24.9	100.0	11.5 ± 8.3
Claustrum, neo and cingulate cortex				
Claustrum	1491.4 ± 178.5	67.5	75.2	598.4 ± 80.5
Frontal cortex (Area 10)				
lamina I	1309.0 ± 69.5	69.6	92.3	299.2 ± 8.1
lamina II	1149.7 ± 38.3	59.3	91.9	132.2 ± 19.6
laminae III–IV	1681.9 ± 66.0	69.7	79.1	603.3 ± 30.4
lamina V	1715.3 ± 159.2	62.8	72.7	236.8 ± 38.1
lamina VI	1573.1 ± 48.7	57.3	73.2	98.5 ± 16.8

Area	Specific 5-HT$_1$ binding			Specific 5-HT$_2$ binding
	[³H]5-HT (mean ± SEM)	% binding[a] to 5-HT$_{1A}$	% binding[a] to 5-HT$_{1B}$	[³H]ketanserin (mean ± SEM)
Frontoparietal motor cortex (Area 2)				
lamina I	798.7 ± 115.8	65.4	90.5	117.1 ± 13.4
laminae II–III	1002.6 ± 166.8	60.8	93.0	165.7 ± 21.2
lamina IV	1554.6 ± 208.3	69.0	76.7	382.1 ± 18.9
lamina V	1744.6 ± 110.7	61.0	71.4	—
lamina Va	—	—	—	180.3 ± 60.9
lamina Vb	—	—	—	257.6 ± 32.4
lamina VI	1316.3 ± 90.0	53.3	61.1	61.5 ± 24.9
Striate cortex (Area 18)				
lamina I	600.4 ± 89.7	66.0	67.2	60.5 ± 5.5
laminae II–III	688.2 ± 195.8	56.2	72.1	0.7 ± 0.3
lamina IV	782.4 ± 172.9	65.3	50.4	77.4 ± 9.3
lamina V	715.8 ± 109.0	72.8	59.5	162.3 ± 12.9
lamina VI	686.8 ± 153.4	67.0	59.5	20.7 ± 2.4
Anterior cingulate cortex (Area 24)				
lamina I	1860.4 ± 31.9	28.9	72.2	137.2 ± 12.4
laminae II–III	1633.1 ± 203.4	32.1	59.9	104.6 ± 4.2
lamina IV	2333.4 ± 332.6	44.5	49.8	410.6 ± 26.2
lamina V	2109.2 ± 50.2	35.3	68.4	129.9 ± 30.1
lamina VI	—	—	—	20.0 ± 14.3
Entorhinal cortex				
Principalis external layer	2844.6 ± 140.2	69.8	49.5	—
Intermediate layer	1213.9 ± 40.7	28.2	63.9	—
Principalis internal layer	2499.0 ± 359.2	60.3	48.1	201.3 ± 28.7
Cerebellum				
Stratum moleculare	265.3 ± 48.7	32.7	72.8	43.6 ± 13.2
Stratum granulosum	374.6 ± 5.8	52.8	29.7	23.3 ± 8.5

Area	Specific 5-HT$_1$ binding			Specific 5-HT$_2$ binding
	[³H]5-HT (mean ± SEM)	% binding[a] to 5-HT$_{1A}$	% binding[a] to 5-HT$_{1B}$	[³H]ketanserin (mean ± SEM)
Midbrain				
Superficial grey layer of the superior colliculus	1604.3 ± 97.5	32.8	76.5	60.0 ± 3.9
Principal oculomotor nucleus	1266.5 ± 131.4	33.6	54.6	7.2 ± 6.1
Central grey (periaqueductal)	1456.0 ± 21.4	39.6	81.1	28.6 ± 8.1
Midbrain				
Dorsal raphe nucleus	2348.7 ± 317.3	60.1	43.6	23.1 ± 19.2
Median raphe nucleus	1150.0 ± 266.0	49.8	50.8	12.0± 6.1
Substantia nigra, pars reticulata	2742.4 ± 135.8	29.5	77.4	29.4 ± 11.7
Substantia nigra, pars compacta	819.7 ± 28.0	28.3	100.0	37.4 ± 25.2
Interpenduncular nucleus	1140.4 ± 297.5	39.4	51.8	7.6 ± 5.1
Pons				
Dorsal cochlear nucleus	—	—	—	11.1 ± 6.2
Nucleus of the spinal tract of the trigeminal nerve, oral part	707.0 ± 90.4	48.3	53.4	24.7 ± 11.2
Facial nucleus	773.9 ± 51.8	33.1	53.1	45.1 ± 10.6
Pontine nuclei	419.7 ± 54.3	21.8	71.0	5.7 ± 6.1
Nucleus of the solitary tract	1184.6 ± 106.8	50.4	59.3	69.3 ± 11.5
Medulla oblongata				
Ambiguus nucleus	745.8 ± 77.8	21.8	72.2	23.3 ± 17.4
Gracile nucleus	860.9 ± 79.5	18.4	57.2	87.8 ± 14.0
Inferior olive	793.7 ± 158.0	44.4	77.5	17.4 ± 11.2
Prepositus hypoglossal nucleus	—	—	—	65.9 ± 7.2
Hypoglossal nucleus	710.0 ± 3.6	26.4	95.4	27.3 ± 21.4
Nucleus of the spinal tract of the trigeminal nerve, gelatinosus subnucleus	1322.1 ± 193.9	59.3	43.1	—
Medial vestibular nucleus	791.7 ± 41.9	35.7	69.0	65.9 ± 7.2

Area	Specific 5-HT$_1$ binding			Specific 5-HT$_2$ binding
	[^{3}H]5-HT (mean ± SEM)	% binding[a] to 5-HT$_{1A}$	% binding[a] to 5-HT$_{1B}$	[^{3}H]ketanserin (mean ± SEM)
Spinal cord				
Layer 1 of Rexed	1011.2 ± 169.8	89.6	42.4	—
Layer 2 of Rexed	—	—	—	6.2 ± 5.1
Layer 7 of Rexed	—	—	—	6.4 ± 2.9
Layer 10 of Rexed	458.5 ± 63.2	54.3	78.4	—
Choroid plexus				
Lateral ventricle	3496.0 ± 530.2	90.3	88.5	42.3 ± 13.7
Fourth ventricle	4265.3 ± 279.5	76.1	94.7	—

[a] The concentrations of 8-OH-DPAT and (−)21 009 used in the study, although highly selective for 5-HT$_{1A}$ and 5-HT$_{1B}$ subtypes, respectively, still show a certain degree of affinity for the other subtype. This explains the fact that the sum in percent of the number of sites is higher than 100.

appeared to coexist in laminae IV–VI. As indicated above, the presence of low densities of 5-HT$_{1C}$ sites at these levels, cannot be excluded.

The *claustrum* presented an intermediate density of 5-HT$_1$ receptors. The three subtypes of 5-HT$_1$ sites appeared to be present in this structure.

Regarding the *cingulate cortex*, in the anterior cingular cortex intermediate to high concentrations of specific binding were found, the internal layers being the most enriched. A predominant but not exclusive presence of 5-HT$_{1B}$ sites was detected. Finally, the retrosplenial cortex presented low densities of 5-HT$_1$ receptors in all the layers.

The external layer of the *entorhinal cortex* presented a very high concentration of 5-HT$_1$ receptors. In the internal layer, a high density of autoradiographic grains was found, while only intermediate concentrations were present in the intermediate layer. Both 5-HT$_{1A}$ and 5-HT$_{1B}$ sites appeared to exist in these areas, although 5-HT$_{1A}$ sites were predominant in the internal and external layers.

The *lateral septum* presented very high concentrations of 5-HT$_1$ receptors, the majority being of the 5-HT$_{1A}$ class. The bed nucleus of the stria terminalis contained an intermediate density of specific binding.

The *hippocampal formation* was very enriched in 5-HT$_1$ receptors. The dentate gyrus and the dorsal subiculum contained very high concentrations of sites. In the different fields of the hippocampus (CA$_1$–CA$_4$), high to intermediate densities of specific binding were found, depending on the layer analysed. While binding to the dorsal subiculum showed the pharmacological characteristics of 5-HT$_{1B}$ sites, the other hippocampal areas mostly contained 5-HT$_{1A}$ sites.

The different nuclei of the *amygdaloid complex* presented an important concentration of autoradiographic grains. In the posteromedial and anterior cortical nuclei, high densities of 5-HT$_1$ receptors were found. The other nuclei showed intermediate concentration of sites. All these amygdaloid nuclei were mainly enriched in 5-HT$_{1B}$ recognition sites.

In the *basal ganglia*, very high concentrations of 5-HT$_1$ sites were present in the globus pallidus and ventral pallidum. While the entopeduncular nucleus presented a high density of autoradiographic grains, the nucleus accumbens and the caudate-putamen only contained intermediate concentrations of specific binding. Most of the binding found in the basal ganglia corresponded to 5-HT$_{1B}$ class.

In the *thalamus* marked differences were observed in the density of 5-HT$_1$ receptors. Some nuclei, such as the ventrolateral and the reticular nuclei, presented very low densities of specific binding. Low concentrations were found in the laterodorsal nucleus and the dorsal lateral geniculate nucleus. The central medial and the olivary pretectal nuclei were rich in 5-HT$_1$ sites. 5-HT$_1$ receptors in the thalamus presented the pharmacological profile of 5-

HT$_{1B}$ sites, although some 5-HT$_{1C}$ binding appeared to exist in the central medial nucleus.

The *hypothalamus* presented high concentrations of 5-HT$_{1B}$ receptors in the ventromedial hypothalamic nucleus; other hypothalamic nuclei presented intermediate densities of specific binding.

A heterogeneous distribution of 5-HT$_1$ receptors was found in the *midbrain*, the richest areas being the substantia nigra (reticular), the dorsal raphe and the superficial grey layer of the superior colliculus. In contrast, other structures, such as the optic layer of the superior colliculus or the ventral tegmental area, presented only low concentrations of specific binding. In most of the areas, 5-HT$_{1B}$ sites were predominant, although the other 5-HT$_1$ subtypes were also present. An exception was the dorsal raphe nucleus, which appeared to be richer in 5-HT$_{1A}$ sites. Only low concentrations of 5-HT$_1$ receptors were found in the *pons*, although the nucleus of the solitary tract contained a higher density of autoradiographic grains. As in midbrain, 5-HT$_{1B}$ sites appeared to be predominant.

The different nuclei of the *medulla oblongata* presented low concentrations of 5-HT$_1$ receptors, except the gelatinosus subnucleus of the nucleus of the spinal tract of the trigeminal nerve, where an intermediate density of specific binding was found. The presence of 5-HT$_{1A}$ and 5-HT$_{1B}$ sites was apparent in these areas, although in most of them, 5-HT$_{1B}$ was predominant. In the cervical part of the *spinal cord*, laminae of the dorsal horn presented intermediate densities of sites, which had the characteristics of 5-HT$_{1A}$ recognition sites. In contrast, in the ventral horn, the concentration of specific binding was low or very low, and of the 5-HT$_{1B}$ class. The presence of some 5-HT$_{1C}$ binding in the dorsal horn was also found.

In the *cerebellum* the density of receptors in the different layers was very low, the deep nuclei presented detectable amounts of specific binding ranging from intermediate (lateral nucleus) to low densities (infracerebellar and intermediate nuclei). Binding to these nuclei presented the pharmacological characteristics of 5-HT$_{1B}$ sites.

The *choroid plexuses* presented the highest concentrations of 5-HT$_1$ receptors, having the characteristics of 5-HT$_{1C}$ sites.

Distribution of 5-HT$_2$ receptors

The anatomical distribution of 5-HT$_2$ receptors in the rat brain was studied using different radioligands, such as [³H]ketanserin, [³H]LSD, [³H]spiperone and [³H]mesulergine (Pazos *et al.*, 1985a). Under the appropriate conditions, all these compounds label 5-HT$_2$ receptors, although most of them recognize, in addition, other populations of binding sites. [³H]Ketanserin appears to be the most selective ligand for these receptors. However, in some brain areas such as the caudate-putamen and the dorsal raphe, [³H]ketanserin presented a high density of non-specific binding, that could represent a 'chemical recognition site' (Pazos *et al.*, 1985a). Table 4 summarizes the quantitative values of specific [³H]ketanserin binding in the rat brain. Fig. 3 illustrates this distribution at different anatomical levels.

In the *olfactory system* the external plexiform layer presented a low density of 5-HT$_2$ sites, while very low concentrations of sites were found in the other layers of this structure, such as the internal granular layer. Very high levels of specific binding were present in the different layers of the olfactory tubercle. The anterior olfactory nucleus, specially in the posterior part, as well as the piriform cortex and the insulae Calleja, presented a high to intermediate density of autoradiographic grains.

A very high concentration of 5-HT$_2$ receptors was found in the *neocortex*. These sites were present in all the areas and laminae of the neocortex. The highest density of binding sites was localized to a continuous band which included lamina IV and extended to part of lamina III, depending upon the cortical area. This band is referred to as lamina IV for simplicity. A decreasing rostrocaudal gradient of 5-HT$_2$ sites was observed in this band. While very high densities were found in the frontal and in the frontoparietal motor cortex, only high concentrations of sites were present in the frontoparietal sensory and striate cortex. High to intermediate grain concentrations were present over the lamina I. Lamina V contained varying concentrations of sites in different cortical areas. The highest densities in this lamina were observed in frontoparietal motor cortex. Contrasting with that, the density of 5-HT$_2$ sites in this lamina in the auditory cortex was very low. Intermediate to low densities were found in lamina II, while low to very low were present in lamina VI.

In the *claustrum* very high concentrations of 5-HT$_2$ receptors were found.

In the *cingulate cortex*, the anterior cingular cortex presented a pattern of specific binding similar to that described in the neocortex, the highest concentrations of receptors being found in lamina IV and I. In contrast in the retrosplenial cortex, the binding was very low in all the laminae.

High to intermediate densities of sites were found in the *entorhinal cortex*, the highest concentration of specific binding being found in the principalis internal layer.

Over the *lateral septum*, as well as in the bed nucleus of the stria terminalis very low densities of autoradiographic grains were present.

In the *hippocampal formation*, the concentrations of 5-HT$_2$ receptors were low or very low in all the areas, except in the ventral dentate gyrus, which presented an intermediate density of binding sites.

Low levels of specific binding were present over the *amygdaloid complex*, although the lateral and the anterior cortical nuclei were somewhat more enriched in binding sites than the other nuclei.

In the *basal ganglia*, marked differences in the 5-HT$_2$ receptor densities

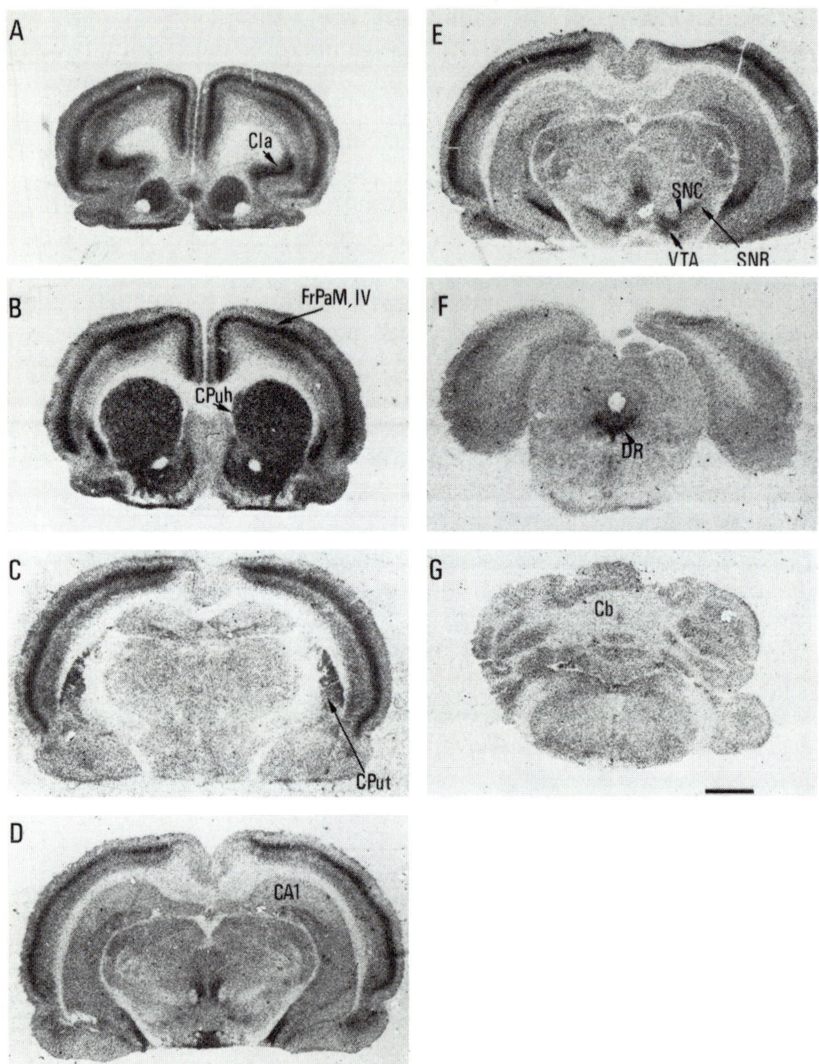

Fig. 3 Localization of [³H]ketanserin binding sites in the rat brain. The autoradiograms demonstrate the presence of high or intermediate concentrations of receptors in the same structures labeled by [³H]mesulergine. Note the very high concentrations of sites in areas such as the caudate-putamen, head (CPu,h), substantia nigra, pars compacta (SNC) and dorsal raphe nucleus (DR), where most of the binding is not displaced by serotonergic ligands. Other abbreviations are: frontoparietal motor cortex, lamina IV (FrPaM, IV), CA1 field of the hippocampus (CA1), substantia nigra, pars compacta (SNC), ventral tegmental area (VTA), substantia nigra, pars reticulata (SNR). Bar = 2 mm. (From Pazos *et al.* (1985a), with permission)

were observed. While intermediate levels of binding sites were found in the nucleus accumbens and in the body and tail of the caudate-putamen nucleus, low-to-very low concentrations of receptors were present in the head of the caudate-putamen, as well as in the globus pallidus.

Low densities of 5-HT$_2$ receptors were found in the *thalamus*, except in a small area, at the posterior level, probably corresponding to the parafascicular nucleus, where intermediate concentrations of binding sites were found. Areas such as the dorsal lateral geniculate nucleus, medial geniculate nucleus, ventrolateral and laterodorsal nuclei showed a very low concentration of autoradiographic grains. The central medial nucleus and the anterior pretectal area, as well as the lateral habenula presented low densities of specific binding. The binding of [^{3}H]mesulergine to the central medial nucleus displayed the pharmacological characteristics of a 5-HT$_1$ site.

In the *hypothalamus*, the medial and posterior part of the medial mamillary nucleus presented an intermediate to high concentration of autoradiographic grains, while the different nuclei of the hypothalamus showed low to very low levels of specific binding.

In the *midbrain*, *pons*, *medulla oblongata* and *spinal cord* the densities of 5-HT$_2$ ranged from low to very low. An exception was the inferior olive which showed low to intermediate densities of 5-HT$_2$ receptors.

In the *cerebellum* as well as in the deep cerebellar nuclei specific binding was very low.

AUTORADIOGRAPHIC STUDIES OF SEROTONIN RECEPTORS IN THE HUMAN BRAIN

The existence of both 5-HT$_1$ and 5-HT$_2$ receptors in the human brain has been studied by binding procedures. In general terms, the properties of these sites appeared to be similar to those previously described in the rat brain (Enna *et al.*, 1977; Cross, 1982; Schotte *et al.*, 1983). However, more detailed studies have shown the existence of important pharmacological differences between human and rat 5-HT recognition sites. Regarding 5-HT$_1$ receptors, we were unable to detect in membranes from human frontal cortex and hippocampus binding sites with the pharmacological profile correspondent to the 5-HT$_{1B}$ subtype (Hoyer *et al.*, 1986a). Secondly, [^{3}H]mesulergine, that labels with high affinity 5-HT$_2$ receptors in the rat brain, did not recognize these sites in human brain (Pazos *et al.*, 1984b). Therefore, the autoradiographic studies of 5-HT receptors in the human brain have not only provided a detailed mapping on their anatomical distribution, but also clarified the differences observed in the pharmacological properties (Palacios *et al.*, 1983; Hoyer *et al.*, 1986a, b; Köhler *et al.*, 1986; Biegon *et al.*, 1986; Pazos *et al.*, 1987a, b).

5-HT$_1$ receptors in human brain

The analysis of the binding of 5-HT$_1$ radioligands to human brain tissue sections revealed both similarities and differences when compared to rat brain. In many human brain areas, binding sites with the properties of 5-HT$_{1A}$ receptors were found. These areas, including the hippocampus, external layers of the cortex, raphe nuclei and thalamus, among others, were labeled by [^{3}H]5-HT and [^{3}H]8-OH-DPAT, but not by [^{3}H]mesulergine (Hoyer *et al.*, 1986a; Pazos *et al.*, 1987a). Areas with a pharmacological profile corresponding to the 5-HT$_{1C}$ receptor subtype were also identified with a very high density in the choroid plexus, but also present in the globus pallidus and the substantia nigra, among others (Hoyer *et al.*, 1986b; Pazos *et al.*, 1987a). In contrast to that, and confirming our previous findings in binding assays, we failed to identify in human brain sections a population of binding sites with the pharmacological properties of 5-HT$_{1B}$ receptors, as defined in the rat. The binding of [^{3}H]5-HT was not displaced with high affinity by the 5-HT$_{1B}$ selective compound (−)21 009 (Hoyer *et al.*, 1986a; Pazos *et al.*, 1987b). In addition, we were unable to label 5-HT$_{1B}$ sites using [^{125}I]CYP either in human brain membranes (Hoyer *et al.*, 1986a) or tissue sections (Pazos *et al.*, unpublished).

In some human brain areas, the binding of [^{3}H]5-HT was insensitive to nanomolar concentrations of 8-OH-DPAT (5-HT$_{1A}$ selective), (−)21 009 (5-HT$_{1B}$ selective) and mesulergine (5-HT$_{1C}$ selective). Very interestingly, these areas included structures such as the globus pallidus and the substantia nigra, very enriched in 5-HT$_{1B}$ receptors in the rat brain. These binding sites were called 'atypical' 5-HT$_{1C}$ receptors, in order to simplify the description of 5-HT receptors in the human brain. More recently it has been found that these sites correspond to a new type of 5-HT$_1$ receptor, the 5-HT$_{1D}$ site, whose pharmacological properties will be commented later (Pazos *et al.*, 1987a; Heuring and Peroutka, 1987; Hoyer *et al.*, 1987; Waeber *et al.*, unpublished observations.

In the following, we will describe the anatomical distribution of 5-HT$_1$ receptors in the human brain. Figure 4 illustrates this distribution at several levels. Table 5 summarizes the quantitative values of binding densities in each brain area.

In the *olfactory system* the olfactory bulb presented a low to very low density of 5-HT$_1$ receptors, the olfactory tubercle contained a low amount of 5-HT$_1$ binding, with the characteristics of 5-HT$_{1A}$ receptors.

High densities of 5-HT$_1$ binding were characteristic of the different *cortical areas*. There was some regional variation of the laminar organization of 5-HT$_1$ receptors in the entorhinal cortex. In the anterior part, intermediate to high densities of 5-HT$_1$ sites were found over the layers I–II and V, with low concentrations of receptors in the other layers. Most of these receptors

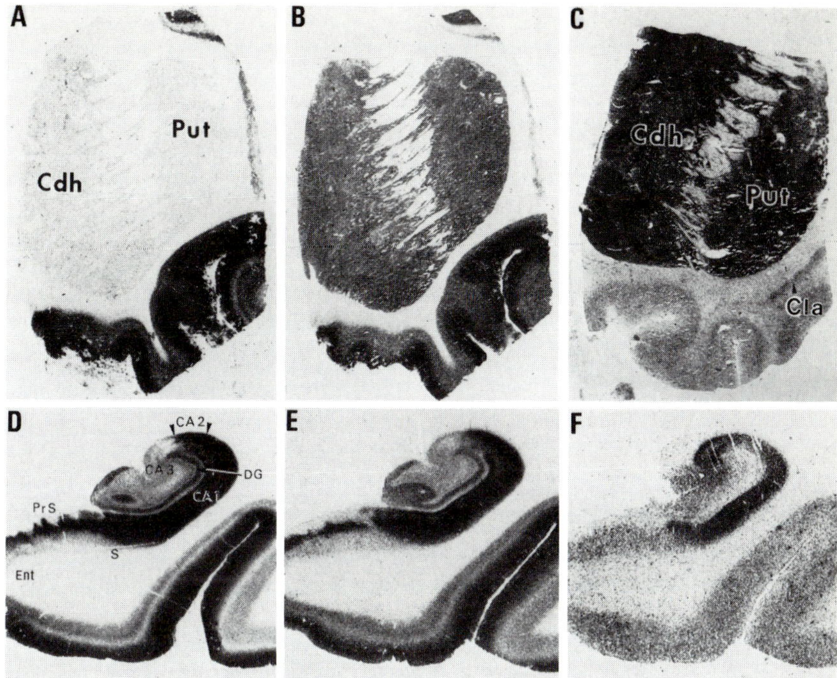

Fig. 4 Autoradiographic distribution of 5-HT$_1$ receptors in the human brain in consecutive coronal sections at the level of the basal ganglia (A–C) and the hippocampus (D–F). A and D show the total binding of [^{3}H]8-OH-DPAT. B and E represent the total binding of [^{3}H]5-HT, while C and F show the binding of [^{3}H]mesulergine. Areas labeled with [^{3}H]8 OH DPAT and [^{3}H]5-HT, such as the dentate gyrus (DG) and the CA$_1$ field of the hippocampus (CA1), are enriched in 5-HT$_{1A}$ sites. Areas labeled by [^{3}H]mesulergine and [^{3}H]5-HT, such as the caudatus (Cdh) and putamen (Put) contain binding sites belonging to the 5-HT$_{1C}$ class. Bar = 3 mm. Modified from Pazos *et al.* (1987a), with permission

showed pharmacological characteristics corresponding to 5-HT$_{1A}$ sites. In the medial and posterior entorhinal cortex, the high levels of 5-HT$_1$ sites were restricted to layer I–II, with a more moderate concentration of autoradiographic grains over layer III and low densities of binding sites in the internal layers. While in layer III, 5-HT$_{1C}$ receptors were predominant, the other layers mainly contained 5-HT$_{1A}$ receptors. In the proisocortex, we have studied the distribution of 5-HT$_1$ receptors in several areas of the cingulate region. In these areas, an intermediate level of binding was found in the external layers, specially over the layers I and II. In contrast, the internal layers presented only low concentrations of autoradiographic grains. Binding

Table 5   Densities of 5-HT$_1$ receptors in the human brain as labeled with [³H]-HT, [³H]8-OH-DPAT, [³H]mesulergine, and 5-HT$_2$ receptors as labeled by [³H]ketanserin. Specific binding is given in fmol/mg protein (mean ±SEM)

Area	[³H]5-HT[a] (5-HT$_{1A}$, 5-HT$_{1C}$, 5-HT$_{1D}$)	[³H]8-OH-DPAT[a] (5-HT$_{1A}$)	[³H]mesulergine[b] (5-HT$_{1C}$)	[³H]ketanserin[a]
Entorhinal cortex				
(Brodmann areas 28,34)				
layers I–II	531.8 ± 124.5	387.6 ± 30.0	—	317.8 ± 24.1
layer III	245.4 ± 33.7	233.9 ± 8.6	50.6 ± 6.8	375.6 ± 21.7
Frontal cortex				
(Brodmann areas 9,10,45)				
layer I	468.6 ± 58.3	209.5 ± 18.7	—	164.2 ± 20.6
layer II	594.1 ± 66.3	322.6 ± 12.2	—	147.0 ± 13.3
layer III	340.1 ± 46.3	37.9 ± 8.1	47.1 ± 4.7	268.8 ± 19.6
layer V	326.3 ± 44.2	42.5 ± 7.6	—	238.4 ± 21.0
layer VI	331.7 ± 37.3	87.3 ± 9.8	—	136.2 ± 14.4
Striate cortex				
(Brodmann area 17)				
layers I–II	370.7 ± 37.8	200.4 ± 19.5	—	207.8 ± 23.0
layers III–IVc	324.1 ± 39.6	15.5 ± 2.5	30.2 ± 18.0	283.0 ± 17.1
layer V	332.0 ± 49.6	27.0 ± 1.8	—	153.8 ± 8.0
Claustrum	316.7 ± 43.7	170.0 ± 18.7	37.0 ± 14.0	203.1 ± 21.0
Hippocampus				
dentate gyrus	359.7 ± 47.6	280.4 ± 35.2	69.5 ± 10.9	138.5 ± 15.5
CA$_1$ field, stratum pyramidalis	1033.5 ± 109.8	556.8 ± 55.4	141.8 ± 19.3	185.0 ± 10.2
CA$_1$ field, stratum moleculare	826.5 ± 94.8	426.8 ± 50.8	—	179.0 ± 16.1
CA$_3$ field, stratum pyramidalis	227.2 ± 33.0	102.0 ± 27.2	131.9 ± 25.1	148.3 ± 15.0
subiculum	669.5 ± 73.6	443.1 ± 24.7	55.2 ± 12.0	118.0 ± 18.5

Amygdala				
nucleus basalis	219.3 ± 34.7	128.2 ± 10.4	36.0 ± 16.9	141.8 ± 14.3
nucleus lateralis	146.7 ± 21.5	75.1 ± 17.3	68.1 ± 11.3	204.2 ± 15.4
nucleus corticalis	365.1 ± 38.2	248.6 ± 26.2	67.0 ± 13.3	132.1 ± 13.0
Basal ganglia				
nucleus caudatus	226.4 ± 14.1	12.8 ± 4.2	115.4 ± 13.6	154.0 ± 14.5
putamen	230.4 ± 24.7	16.6 ± 9.3	156.6 ± 31.2	102.8 ± 8.4
globus pallidus	492.9 ± 42.5	20.4 ± 1.5	187.0 ± 14.7	84.3 ± 10.2
nucleus accumbens	260.3 ± 21.6	13.2 ± 3.0	134.2 ± 13.7	139.5 ± 16.1
substantia innominata	227.8 ± 27.9	27.1 ± 11.6	73.4 ± 9.4	102.7 ± 11.1
Thalamus				
nucleus anterior	122.8 ± 9.4	13.3 ± 1.8	71.9 ± 6.9	100.0 ± 13.6
nucleus intralamellaris, pars centralis lateralis	386.0 ± 25.6	179.5 ± 11.4	67.0 ± 30.3	—
nucleus medialis	91.5 ± 12.3	9.9 ± 1.9	86.0 ± 19.6	87.2 ± 8.6
nucleus paramedianus	383.8 ± 45.2	210.6 ± 28.6	99.2 ± 13.9	—
nucleus lateralis posterior	69.7 ± 35.0	3.2 ± 1.1	17.8	66.5 ± 6.6
Hypothalamus				
nucleus ventromedialis	290.5 ± 118.2	69.9 ± 22.7	190.0 ± 62.1	86.1
nucleus posterior	240.1 ± 24.3	72.3 ± 19.1	49.7 ± 13.3	84.5 ± 19.8
area lateralis	275.7 ± 20.6	39.3 ± 10.7	85.1 ± 15.6	117.4 ± 28.9
corpus mamillare	211.5 ± 36.5	1.5	110.7 ± 29.5	282.6 ± 46.6
Midbrain				
griseum centrale mesencephali	246.7 ± 33.5	32.6 ± 6.6	65.2 ± 10.7	51.4 ± 9.8
nucleus raphe linearis	537.8 ± 150.2	220.4 ± 34.1	22.6 ± 7.5	30.7 ± 13.9
nucleus raphe dorsalis, pars supratrochlearis	853.0 ± 128.7	362.7 ± 77.7	56.8 ± 16.9	75.8 ± 8.6
substantia nigra	724.9 ± 50.8	3.3 ± 1.4	191.9 ± 15.0	96.3 ± 9.1
Pons				
nucleus raphe linearis	661.7 ± 56.2	53.8 ± 5.4	41.4	—
locus coeruleus	253.0 ± 29.5	182.7	89.6 ± 41.7	19.8 ± 19.8
griseum pontis	39.5 ± 13.3	17.0 ± 3.3	21.5 ± 7.5	26.6 ± 8.1

Area	[³H]5-HT[a] (5-HT$_{1A}$, 5-HT$_{1C}$, 5-HT$_{1D}$)	[³H]8-OH-DPAT[a] (5-HT$_{1A}$)	[³H]mesulergine[b] (5-HT$_{1C}$)	[³H]ketanserin[a]
Medulla oblongata				
nucleus solitarius	192.2 ± 43.9	82.3 ± 14.5	40.0 ± 0.7	124.1 ± 23.7
nucleus ambiguus	91.5 ± 36.8	72.8 ± 20.5	47.9	—
nucleus spinalis nerve trigemini, gelatinosa	151.1 ± 30.3	62.2 ± 5.5	39.1 ± 14.5	29.4 ± 8.8
nucleus raphe obscurus	581.3	147.7 ± 29.5	—	—
Spinal cord				
substantia gelatinosa	208.1 ± 43.2	—	36.7 ± 5.1	0.5 ± 0.5
lamina IX	81.5 ± 7.9	—	42.6 ± 23.8	118.7 ± 15.2
Cerebellum				
lamina molecularis	23.6 ± 12.4	—	31.8 ± 2.8	27.8 ± 4.0
nucleus dentatus	8.9 ± 9.0	—	0.0	60.1 ± 16.4
Choroid plexus				
ventriculus lateralis	454.2 ± 64.2	19.7	1696.3 ± 300.2	—

[a] Densities obtained at a concentration of 2 nM; [b] Densities obtained at a concentration of 5 nM

over the layers I and II was of the 5-HT$_{1A}$ class, while that found in layers III–VI appeared to belong to the 5-HT$_{1C}$ subtype. The analysis of the localization of 5-HT$_1$ receptors in several areas of the neocortex revealed a general pattern of anatomical distribution. It included high or very high densities of sites in layer II, intermediate densities in layer I and intermediate to low in layers III–VI. A different pattern was found in the striate area where the highest densities of 5-HT$_1$ receptors were found over the layer IVcβ, with more moderate levels of binding in the other layers. Although the amount of 5-HT$_1$ sites over the neocortex was uniform, slight regional differences were found, the frontal lobe being the most enriched.

The pharmacological properties of 5-HT$_1$ receptors in the neocortex were also similar in the different areas studied. Receptors in layers I and II belonged to the 5-HT$_{1A}$ subtype, while layers III–V showed a predominant presence of 5-HT$_{1D}$ sites and a small amount of 5-HT$_{1C}$ receptors. Binding sites over the layer IVcβ of the striate cortex belonged to the 5-HT$_{1D}$ class.

The *hippocampus* presented the highest densities of 5-HT$_{1A}$ receptors throughout the human brain, specially over the strata pyramidalis and lacunosum moleculare of the CA1 field. The stratum radiatum, the subiculum and the dentate gyrus contained high to intermediate levels of specific binding. Over the stratum pyramidalis, a significant level of 5-HT$_{1C}$ binding was observed. The *septal nuclei* contained only low amounts of 5-HT$_1$ specific binding.

The distribution of 5-HT$_1$ receptors in the *basal ganglia* was heterogeneous. While very high densities of specific binding were found in the globus pallidus, other areas such as caudatus, putamen, accumbens and substantia innominata, presented intermediate to low densities of autoradiographic grains. These nuclei contained a majority of 5-HT$_{1D}$ binding sites, as well as a relevant amount of sites with the properties of 5-HT$_{1C}$ receptors.

In the *amygdala*, most of the 5-HT$_1$ binding belonged to the 5-HT$_{1A}$ class. The nucleus corticalis, nucleus granularis and the reticular zone corresponding to the most rostral and lateral aspect of the lateral nucleus were the areas most enriched, with binding densities in the intermediate range.

In the *thalamus*, the density of 5-HT$_1$ receptors was low, with the exception of the formatio intralamellaris (centralis lateralis, parafascicularis) and paraventricularis (paramedianus, reuniens) which contained intermediate concentrations of 5-HT$_{1A}$ binding.

The nucleus ventromedialis, area lateralis and corpus mammillare of the *hypothalamus* presented both 5-HT$_{1C}$ and 5-HT$_{1D}$ receptors. In contrast, the nucleus infundibularis and the nucleus posterior contained binding sites of the 5-HT$_{1A}$ class.

In the *midbrain* the nucleus raphe dorsalis, pars supratrochlearis, was very rich in 5-HT$_{1A}$ receptors, its lateral aspect being more labeled than the central one. The nucleus raphe linearis and nucleus raphe centralis, pars annularis, also presented intermediate to high concentrations of autoradiographic grains. On the other hand, the substantia nigra showed a very high level of labeling with [³H]5-HT, presenting the characteristics of 5-HT$_{1D}$ receptors. Other mesencephalic structures were scarcely labeled by serotonergic ligands. In the *pons* the nucleus raphe dorsalis was again the most enriched area in 5-HT$_1$ receptors, followed by the nucleus raphe centralis and linearis. Other structures, such as the locus coeruleus or the griseum centrale pontis, presented low densities of sites.

In the *medulla oblongata*, the nucleus solitarius and the substantia gelatinosa of the nucleus spinalis nervi trigemini contained low densities of 5-HT$_{1A}$ receptors. A similar amount of specific binding was observed over the nucleus ambiguus which extended through the upper part of the nucleus medullae oblongatae centralis. The level of binding in other nuclei was very low. In the *spinal cord* the substantia gelatinosa showed an intermediate density of 5-HT$_{1A}$ receptors.

In the *cerebellum* the different layers presented very low densities of 5-HT$_1$ receptors.

Using [³H]mesulergine the *choroid plexus* showed a very high density of 5-HT$_{1C}$ receptors.

5-HT$_2$ receptors in human brain

The pharmacological characteristics of [³H]ketanserin binding to the human brain sections were similar to those seen in homogenates (Hoyer *et al.*, 1986b; Pazos *et al.*, 1987b). Table 5 summarizes the densities of 5-HT$_2$ receptors in the different areas of the human brain. This distribution is illustrated in Fig. 5.

In the *olfactory system* the olfactory bulb contained only a very low density of [³H]ketanserin binding sites. In contrast, intermediate densities of sites were found over the olfactory tubercle.

The *cerebral cortex* presented the highest densities of 5-HT$_2$ receptors, with only slight regional differences. In the *entorhinal cortex*, at the level of the gyrus parahippocampalis, very high densities of [³H]ketanserin specific binding were found over layers I, II, III and V, with the layer III being the most enriched.

In the *cingulate region*, several areas were examined. The anterogenual cortex presented a very high density of sites in layers III and V, with intermediate levels of specific binding over the other layers. In the medial cingulate area, where the ganglionic layer V is not clearly identifiable, very high concentrations of autoradiographic grains were restricted to the pyramidal layer III. A similar distribution of [³H]ketanserin binding was found in the area subcallosa. The distribution of 5-HT$_2$ receptors throughout the isocortex was studied in detail, analysing several areas from the frontal, temporal, parietal and occipital lobes. A general pattern of distribution of the specific binding was found. This pattern included the presence of very high densities of binding sites over the pyramidal (III) and ganglionic (V) layers, while the densities found in the other layers ranged from high to intermediate, depending on the field examined. In layer V, the outer part (Va) was the one presenting the highest level of specific binding. This general pattern of distribution was somewhat different in the striate area where the layer IVcβ, which contains the projections from the lateral geniculate body, was very enriched in 5-HT$_2$ receptors. Slight quantitative differences were found between the different areas studied, the gyrus rectus of the frontal pole being the most labeled, while the parietal cortex and the peristriate area presented more modest levels of binding.

A high level of [^{3}H]ketanserin binding was found along the claustrum, the caudal aspects of this nucleus being more labeled than the rostral ones.

A heterogeneous distribution of 5-HT$_2$ receptors was found in the *hippocampal formation*. While the pyramidal and molecular layers of the CA$_1$ subfield presented a high to intermediate density of binding sites, the dentate gyrus and the CA$_3$ subfield contained intermediate levels of specific binding. Low concentrations of autoradiographic grains were found over the subiculum. The *septal nuclei* presented a very low density of specific binding.

A high concentration of autoradiographic grains was observed over the nucleus lateralis of the *amygdala*. Other amygdaloid nuclei, such as the basalis, basalis accessorious and corticalis, presented an intermediate density of 5-HT$_2$ receptors. The nucleus granularis contained a low level of [^{3}H]ketanserin binding.

In the *basal ganglia* intermediate densities of specific binding were present over the nucleus caudatus and the putamen, as well as in the nucleus accumbens. In the putamen, an increasing rostrocaudal gradient could be observed in the distribution of these receptors. The globus pallidus contained a low level of specific binding. It must be underlined that the caudatus and putamen presented a very high level of non-specific binding, as previously mentioned.

In the basal forebrain intermediate to low densities of 5-HT$_2$ receptors were found in the *substantia innominata* (nucleus basalis of Meynert).

In the *thalamus*, the nucleus reuniens (nuclei mediani thalami) and the nucleus anterior presented an intermediate density of 5-HT$_2$ sites. While the levels of specific binding over the nucleus medialis and the corpus geniculatum mediale ranged from intermediate to low, only low amounts of binding sites were present in other thalamic nuclei, such as the nuclei of the formatio lateralis, the pulvinar and the corpus geniculatum laterale. In the hypothalamus, the corpus mamillare, both the nucleus medialis and lateralis, presented a very high density of [^{3}H]ketanserin binding sites, this structure being, together with the cortex, the most enriched area in 5-HT$_2$ receptors throughout the human brain. In other hypothalamic nuclei, the levels of specific binding ranged from intermediate to low.

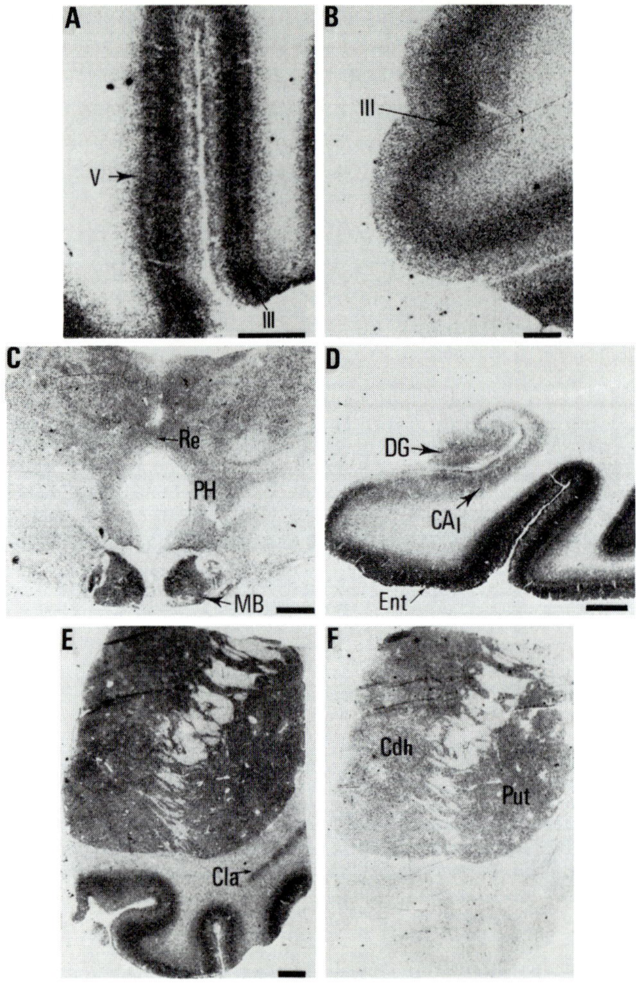

Fig. 5 Autoradiographic distribution of 5-HT$_2$ receptors, labeled by [^{3}H]ketanserin in the human brain. The figure shows the total binding at the level of the frontal cortex and medial cingulate cortex (A,B), the posterior hypothalamus (C), the hippocampus (D) and the caudate-putamen (E). F represents the binding of [^{3}H]ketanserin to the caudate-putamen in the presence of 10^{-6} M mianserin. Note the intense labeling over layers III and V of the frontal cortex (A) and over the layer III in the cingulate cortex (B). The high densities found over the mamillary bodies (MB) and the entorhinal cortex (Ent) are in contrast to the intermediate levels of the thalamus and the CA1 field of the hippocampus (CA1). Note the high level of non-specific binding over the caudate-putamen (F). Other abbreviations are: lamina V (V), lamina III (III), nucleus

The amount of specific [³H]ketanserin binding in the *midbrain* structures ranged from low to very low. However, there was a very high level of labeling over the substantia nigra, the nucleus raphe dorsalis (pars supratrochlearis) and the nucleus raphe centralis (pars annularis), but this binding was not displaceable by pirenperone, mianserin or any other serotonin-2 compound, showing its non-specific nature. Very low amounts of specific binding were found at the different levels of the *pons*. Structures, such as the raphe nuclei (centralis, dorsalis, linearis), griseum centrale metencephali and griseum pontis were very poor in 5-HT$_2$ receptors, although high levels of non-specific binding were seen in some nuclei. This was especially the case in the locus coeruleus. In the upper and middle part of the *medulla oblongata*, low densities were found in the nucleus nervi hypoglossi, the nucleus olivaris inferior and nucleus olivaris accessorius dorsalis. The nucleus solitarius, spinalis nervi trigemini (pars interpolaris) and nucleus raphe obscurus were very poor in 5-HT$_2$ receptors. At the level of the pyramidal decussation, the densities of [³H]ketanserin binding sites were low, although the substantia gelatinosa of the trigeminal nerve contained a remarkable level of non-displaceable binding. At this level the most caudal part of the nucleus solitarius was notable in presenting an intermediate to low concentration of specific binding. In the spinal cord the substantia gelatinosa was very poor in specific binding, while the anterior horn (motor neurons) contained an intermediate to low level of 5-HT$_2$ receptors.

The *cerebellum* showed only low or very low levels of specific binding both over the different layers of the cerebellar cortex and the nucleus dentatus, although a significant level of [³H]ketanserin non-displaceable binding was evident in the latter structure.

SPECIES DIFFERENCES IN BRAIN 5-HT RECEPTORS: AUTORADIOGRAPHIC STUDIES

In the recent years, evidence has accumulated from biochemical and functional studies supporting the existence of species differences in the characteristics of serotonin receptors. From the autoradiographic studies in rat and human brain (see above) differences in both distribution and pharmacology were also evident. In order to clarify this question, we have carried out a detailed analysis of the properties and localization of both 5-HT$_1$ and 5-HT$_2$ receptors in the brain of several laboratory animals (Dietl and Palacios, 1987).

---

reuniens of the thalamus (Re), nucleus posterior of the hypothalamus (PH), mamillary bodies (MB), dentate gyrus (DG), CA1 field of the hippocampus (CA1), entorhinal cortex (Ent), claustrum (Cla), nucleus caudatus, head (Cdh), putamen (Put). Bar = 3 mm. Modified from Pazos *et al.* (1987b), with permission

Regarding 5-HT$_2$ receptors, we have shown by binding assays that the properties of these sites in pig and human cortical membranes were somehow different from those seen in rat cortex (Pazos *et al.*, 1984b). The most striking difference observed was that [³H]mesulergine, a ligand showing high affinity for these receptors in the rat, does not label them in pig and human cortical tissue. Binding and functional data comparing rat and cat 5-HT$_2$ receptors also demonstrated the lack of high affinity of mesulergine for these sites in cat brain (Palacios *et al.*, unpublished). Our autoradiographic studies have confirmed and extended these findings showing that, while in the rodent brain [³H]mesulergine labels 5-HT$_{1C}$ and 5-HT$_2$ sites, in the avian and higher mammalian brain this radioactive compound only recognizes with high affinity 5-HT$_{1C}$ receptors (Pazos *et al.*, 1987b; Dietl and Palacios, 1987). In addition, some differences in anatomical distribution of 5-HT$_2$ sites were also evident from the autoradiographic data. Although in both rat and human brain [³H]ketanserin binding was predominant in cortical areas, the laminar distribution was different: while in the rat the lamina IV was the most intensely labeled, in the human cortex, the highest density of specific [³H]ketanserin binding was over laminae III and V (Pazos *et al.*, 1985a; Pazos *et al.*, 1987b).

With respect to 5-HT$_1$ receptors, the autoradiographic analysis of their phylogenetic evolution revealed that 5-HT$_{1A}$ sites, labeled by both [³H]5-HT and [³H]8-OH-DPAT, are the only subtype clearly present in the lower vertebrates (fish, frog, snake); moreover, pharmacological properties and anatomical distribution of these 5-HT$_{1A}$ receptors are well preserved in evolution from fish to human (Dietl and Palacios, 1987) (Fig. 6). In contrast serotonin sites with the characteristics of 5-HT$_{1B}$ and 5-HT$_{1C}$ sites were seen first in avian brain. While 5-HT$_{1C}$ binding appears to be conserved in higher mammals (Dietl and Palacios, 1987; Pazos *et al.*, 1987a), the sites with the characteristics of 5-HT$_{1B}$ binding are not present in higher mammals, as already commented (see above). Very interestingly, in our autoradiographic mapping of 5-HT$_1$ receptors in the human brain, we were able to identify in these anatomical areas which in the rat were rich in 5-HT$_{1B}$ sites, a new subtype of 5-HT$_1$ binding (see above). This subtype is recognized with high affinity by 5-HT, LSD and 5-carboxyamidotryptamine, but not by the selective 5-HT$_{1A}$, 5-HT$_{1B}$ or 5-HT$_{1C}$ compounds (Pazos *et al.*, 1987b; Hoyer *et al.*, 1987). This site has recently been called 5-HT$_{1D}$ by Heuring and Peroutka (1987) in bovine striatal membranes, and we have characterized this site in detail in human, pig and bovine brain tissues by both membrane binding and autoradiography (Hoyer *et al.*, 1987; Waeber *et al.*, unpublished observations). Taking in consideration the localization of these 5-HT$_{1D}$ sites (striatum, substantia nigra, neocortex, hypothalamus) the functional significance of this subtype is of potential relevance.

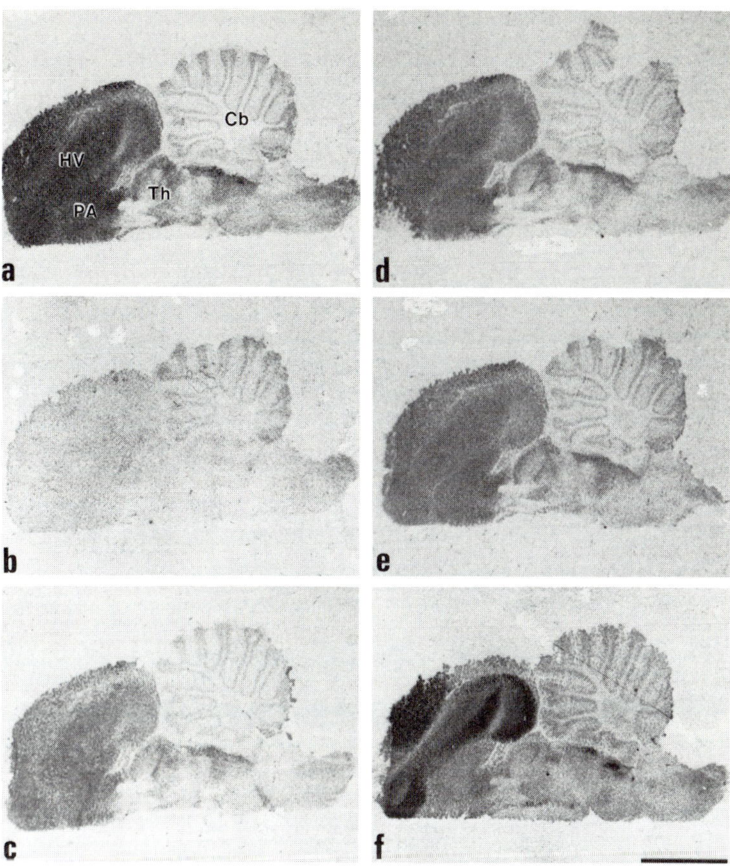

Fig. 6 Photomicrographs of autoradiograms generated by incubating consecutive sagittal sections of the pigeon brain in a 2 nM [^{3}H]5-HT, b [^{3}H]5-HT plus 1 μM unlabeled 5-HT, and c [^{3}H]5-HT plus 100 nM 8-OH-DPAT, d [^{3}H]5-HT plus 100 nM mesulergine, e [^{3}H]5-HT plus 100 nM (−)21 009 and f [^{3}H]8-OH-DPAT. Cb = cerebellum, HV = hyperstriatum ventrale, PA = paleostriatum augmentatum, Th = thalamus. Bar = 5 mm

CONCLUSIONS AND FUTURE TRENDS

Receptor autoradiographic techniques have enhanced our knowledge of the distribution, properties and regulation of serotonin receptors. They have been a crucial tool in the clarification of the existence and properties of 5-HT$_1$ receptor subtypes. All this information has greatly contributed to improve our understanding of brain serotonergic neurotransmission. These results provide basis for the development of new drugs which should specifically stimulate or block the different 5-HT receptor subtypes. The highly heterogeneous distribution of these 5-HT sites in the brain and their association with well established anatomical and functional pathways suggest that subtype selective drugs have the potential to affect differentially brain functions such as motor behaviour, limbic system function, neuroendocrine regulation, or blood pressure.

Our knowledge of brain 5-HT mechanisms is far from being completed. Receptor imaging procedures will be used in the future in order to increase our understanding of the neurobiology of serotonin. Relevant tasks for that future will be the development of electron microscopic autoradiography, the *in vivo* visualization of serotonin receptors in human brain using positron emission tomography procedures and the cloning of 5-HT receptor's genes (see Lübbert *et al.*, 1987a, b), which will provide probes to examine their cellular and subcellular localization, and the cellular expression of these genes using *in situ* hybridization.

REFERENCES

Biegon, A., Kargman, S., Snyder, L., and McEwen, B. S. (1986) Characterization and localization of serotonin receptors in human brain postmortem, *Brain Res.*, **363**, 91–98.

Closse, A. (1983) [³H]Mesulergine, a selective ligand for serotonin-2 receptors, *Life Sci.*, **32**, 2485–2495.

Cortés, R., Palacios, J. M., and Pazos, A. (1984) Visualization of multiple serotonin receptors in the rat brain by autoradiography, *Br. J. Pharmacol.*, **82**, 902P.

Cross, A. J. (1982) Interaction of [³H]LSD with serotonin receptors in human brain, *Eur. J. Pharmacol.*, **82**, 77–80.

Dahlström, A., and Fuxe, K. (1963) Evidence of the existence of monoamine containing neurons in the central nervous system. I. Demonstration of monoamines in the cell bodies of brainstem neurons, *Acta Physiol. Scand.*, **62**, Suppl. 232, 1–55.

Dietl, M., and Palacios, J. M. (1987) Receptor autoradiography as a tool for the study of the phylogeny of the basal ganglia, *J. Recept. Res.*, in press.

Engel, G., Müller-Schweinitzer, E., and Palacios, J. M. (1984) 2-[¹²⁵Iodo]LSD, a new ligand for the characterization and localization of 5-HT₂ receptors, *Naunyn-Schmiedeberg's Arch. Pharmacol.*, **325**, 328–336.

Enna, S. J., Bennett, J. P., Bylund, D. B., Creese, I., Burt, D. R., Charness, M. E., Yamamura, H. I., Simantov, R., and Snyder, S. H. (1977) Neurotransmitter receptor binding: regional distribution in human brain, *J. Neurochem.*, **28**, 233–236.

Euvrard, C., and Boissier, J. R. (1980) Biochemical assessment of the central 5-HT agonist activity of RU-24 969 (a piperidinylindole), *Eur. J. Pharmacol.*, **63**, 65–72.

Glaser, T., Rath, M., Traber, J., Zilles, K., and Schleicher, A. (1985) Autoradiographic identification and topographical analyses of high affinity serotonin receptor subtypes as a target for the novel putative anxiolytic TVX-Q 7821, *Brain Res.*, **358**, 129–136.

Gozlan, H., El Mestikawy, S., Pichat, L., Glowinski, J., and Hamon, M. (1983). Identification of presynaptic serotonin autoreceptors using a new ligand: [³H]PAT, *Nature*, **305**, 140–142.

Heuring, R. E., and Peroutka, S. J. (1987) Characterization of a novel [³H]5-hydroxytryptamine binding site subtype in bovine brain membranes, *J. Neurosci.*, **7**, 894–903.

Hjorth, S., Carlsson, A., Lindberg, P., Sanchez, D., Wikström, H., Arvidson, L. E., Hacksell, U., and Nilsson, J. L. G. (1982) 8-Hydroxy-2-(di-n-propylamino)-tetralin, 8-OH-DPAT, a potent and selective simplified ergot congener with central 5-HT receptor stimulating activity, *J. Neural Transm.*, **55**, 169–188.

Hoyer, D., Pazos, A., Probst, A., and Palacios, J. M. (1986a) Serotonin receptors in the human brain. I. Characterization and autoradiographic localization of 5-HT$_{1A}$ recognition sites. Apparent absence of 5-HT$_{1B}$ recognition sites, *Brain Res.*, **376**, 85–96.

Hoyer, D., Pazos, A., Probst, A., and Palacios, J. M. (1986b) Serotonin receptors in the human brain. II. Characterization and autoradiographic localization of 5-HT$_{1C}$ and 5-HT$_2$ recognition sites, *Brain Res.*, **376**, 97–107.

Hoyer, D., Waeber, C., Pazos, A. Probst, A., and Palacios, J. M. (1987) Identification of a 5-HT$_1$ recognition site in human brain membranes different from 5-HT$_{1A}$, 5-HT$_{1B}$ and 5-HT$_{1C}$ sites. *Neurosci. Lett.* submitted for publication.

Hunt, P., and Oberlander, C. (1981) The interaction of indole derivatives with the serotonin receptor and non-dopaminergic circling behavior, in *Serotonin-Current Aspects of Neurochemistry and Function* (Eds B. Haber, S. Gabay, M. R. Issidorides and S. G. A. Alivisatos), pp. 547–562, Plenum Press, New York.

Köhler, C., Radesäter, A.-C., Lang, W., and Chan-Palay, V. (1986) Distribution of serotonin-1A receptors in the monkey and the postmortem human hippocampal region. A quantitative autoradiographic study using the selective agonist [^{3}H]8-OH-DPAT, *Neurosci. Lett.*, **72**, 43–48.

Kuhar, M. J. (1985) Receptor localization with the microscope, in *Neurotransmitter Receptor Binding*, 2nd ed. (Ed. H. I. Yamamura), pp. 153–180, Raven Press, New York.

Leysen, J. E., Niemegeers, C. J. E., Van Nueten, J. M., and Laduron, P. M. (1982) [^{3}H]Ketanserin (R 41 468), a selective [^{3}H]-ligand for serotonin$_2$ receptor binding sites, *Mol. Pharmacol.*, **21**, 301–314.

Lübbert, H., Snutch, T. P., Dascal, N., Lester, H. A., and Davidson, N. (1987a) Rat brain 5-HT$_{1C}$ receptors are encoded by a 5–6 kbase mRNA size class and are functionally expressed in injected *Xenopus* oocytes, *J. Neurosci.*, **7**, 1159–1165.

Lübbert, H., Hoffman, B. J., Snutch, T. P., van Dyke, T., Levine, A. J., Hartig, P. R., Lester, H. A., and Davidson, N. (1987b) cDNA cloning of a serotonin 5-HT$_{1C}$ receptor by electrophysiological assays of mRNA-injected *Xenopus* oocytes, *Proc. Natl. Acad. Sci. USA*, **84**, 4332–4336.

Marcinkiewicz, M., Vergé, D., Gozlan, H., Pichat, L., and Hamon, M. (1984) Autoradiographic evidence for the heterogeneity of 5-HT$_1$ sites in the rat brain, *Brain Res.*, **291**, 159–163.

Meibach, R. C., Maayani, S., and Green, J. P. (1980) Characterization and radioautography of [^{3}H]LSD binding by rat brain slices *in vitro*: the effect of 5-hydroxytryptamine, *Eur. J. Pharmacol.*, **67**, 371–382.

Middlemiss, D. N., Blakeborough, L., and Leather, S. R. (1977) Direct evidence for an interaction of β-adrenergic blockers with the 5-HT receptor, *Nature*, **267**, 289–290.

Middlemiss, D. N., and Fozard, J. (1983) 8-Hydroxy-2-(di-n-propylamino)tetralin discriminates between subtypes of the 5-HT$_1$ recognition site, *Eur. J. Pharmacol.*, **90**, 151–153.

Nahorski, S. R., and Willcocks, A. L. (1983) Interactions of β-adrenoceptor antagonists with 5-hydroxytryptamine receptor subtypes in rat cerebral cortex, *Br. J. Pharmacol.*, Suppl. 78, 107P.

Palacios, J. M., Niehoff, D. L., and Kuhar, M. J. (1981) Receptor autoradiography with tritium-sensitive film: potential for computerized densitometry, *Neurosci. Lett.*, **24**, 111–116.

Palacios, J. M., Probst, A., and Cortés, R. (1983) The distribution of serotonin receptors in the human brain: high density of [³H]LSD binding sites in the raphe nuclei of the brainstem, *Brain Res.*, **274**, 150–155.

Parent, A., Descarries, L., and Beaudet, A. (1981) Organization of ascending serotonin systems in the adult rat brain. A radioautographic study after intraventricular administration of [³H]5-hydroxytryptamine, *Neuroscience.*, **6**, 115–138.

Pazos, A., Cortés, R., and Palacios, J. M. (1984a) Quantitative receptor autoradiography: application to the characterization of multiple receptor subtypes, *J. Recept. Res.*, **4**, 645–656.

Pazos, A., Hoyer, D., and Palacios, J. M. (1984b) Mesulergine, a selective serotonin-2 ligand in the rat cortex, does not label these receptors in porcine and human cortex: evidences for species differences on brain serotonin-2 receptors, *Eur. J. Pharmacol.*, **106**, 531–538.

Pazos, A., Hoyer, D., and Palacios, J. M. (1984c) The binding of serotoninergic ligands to the porcine choroid plexus: characterization of a new type of serotonin recognition site, *Eur. J. Pharmacol.*, **106**, 539–546.

Pazos, A., and Palacios, J. M. (1985) Quantitative autoradiographic mapping of serotonin receptors in the rat brain. I. Serotonin-1 receptors, *Brain Res.*, **346**, 205–230.

Pazos, A., Cortés, R., and Palacios, J. M. (1985a) Quantitative autoradiographic mapping of serotonin receptors in the rat brain. II. Serotonin-2 receptors, *Brain Res.*, **346**, 231–249.

Pazos, A., Engel, G., and Palacios, J. M. (1985b) β-Adrenoceptor blocking agents recognize a subpopulation of serotonin receptors in brain, *Brain Res.*, **343**, 403–408.

Pazos, A., Probst, A., and Palacios, J. M. (1987a) Serotonin receptors in the human brain. III. Autoradiographic mapping of serotonin-1 receptors, *Neuroscience.*, **21**, 97–122.

Pazos, A., Probst, A., and Palacios, J. M. (1987b) Serotonin receptors in the human brain. IV. Autoradiographic mapping of serotonin-2 receptors, *Neuroscience.*, **21**, 123–139.

Pedigo, N. W., Yamamura, H. I., and Nelson, D. L. (1981) Discrimination of multiple [³H]5-hydroxytryptamine binding sites by the neuroleptic spiperone in rat brain, *J. Neurochem.*, **36**, 220–226.

Penney, J. B., Pan, H. S., Young, A. B., Frey, K. A., and Dauth, G. W. (1981). Quantitative autoradiography of [³H]muscimol binding in rat brain, *Science*, **214**, 1036–1038.

Peroutka, S. J., and Snyder, S. H. (1979) Multiple serotonin receptors: differential binding of [³H]5-hydroxytryptamine, [³H]lysergic acid diethylamide and [³H]spiroperidol, *Mol. Pharmacol.*, **16**, 687–699.

Schotte, A., Maloteaux, J. M., and Laduron, P. M. (1983) Characterization and regional distribution of serotonin-S₂ receptors in human brain, *Brain Res.*, **276**, 231–235.

Steinbush, H. W. M. (1981) Distribution of serotonin-immunoreactivity in the central nervous system of the rat – cell bodies and terminals, *Neuroscience.*, **6**, 557–618.

Stumpf, W. E., and Roth, L. G. (1966) High resolution autoradiography with dry-mounted, freeze-dried frozen sections: comparative study of six methods using two diffusible compounds, [³H]estradiol and [³H]mesobilirubinogen, *J. Histochem. Cytochem.*, **14**, 274–287.

Unnerstall, J. R., Niehoff, D. L., Kuhar, M. J., and Palacios, J. M. (1982) Quantitative receptor autoradiography using [³H]ultrofilm: application to multiple benzodiazepine receptors, *J. Neurosci. Methods*, **6**, 59–73.

Yagaloff, K. A., and Hartig, P. R. (1985) [¹²⁵I]Lysergic acid diethylamide binds to a novel serotonergic site on rat choroid plexus epithelial cells, *J. Neurosci.*, **12**, 3178–3183.

Yamamura, H. I., Enna, S. J., and Kuhar, M. J. (1985) *Neurotransmitter Receptor Binding*, Raven Press, New York.

Young, W. S., III, and Kuhar, M. J. (1979) A new method for receptor autoradiography: [³H]opioid receptor labeling in mounted tissue sections, *Brain Res.*, **179**, 255–270.

Young, W. S., and Kuhar, M. J. (1980) Serotonin receptor localization in rat brain by light microscopic autoradiography, *Eur. J. Pharmacol.*, **62**, 237–239.

Index